Thin-Film Optical Filters
Fifth Edition

SERIES IN OPTICS AND OPTOELECTRONICS

Series Editors: **Robert G W Brown**, University of California, Irvine, USA
E Roy Pike, Kings College, London, UK

Thin-Film Optical Filters
Fifth Edition

H. Angus Macleod

Thin Film Center Inc., Tucson, Arizona, USA
Professor Emeritus of Optical Sciences,
University of Arizona, Tucson, USA

CRC Press
Taylor & Francis Group
Boca Raton London New York

CRC Press is an imprint of the
Taylor & Francis Group, an **informa** business

CRC Press
Taylor & Francis Group
6000 Broken Sound Parkway NW, Suite 300
Boca Raton, FL 33487-2742

First issued in paperback 2020

© 2018 by Taylor & Francis Group, LLC
CRC Press is an imprint of Taylor & Francis Group, an Informa business

No claim to original U.S. Government works

ISBN-13: 978-1-138-19824-1 (hbk)
ISBN-13: 978-0-367-78160-6 (pbk)

Library of Congress Cataloging-in-Publication Data

Names: Macleod, H. A. (Hugh Angus), author.
Title: Thin-film optical filters / H. Angus Macleod.
Other titles: Series in optics and optoelectronics (CRC Press)
Description: Fifth edition. | Boca Raton, FL : CRC Press, Taylor & Francis
Group, [2018] | Series: Series in optics and optoelectronics | Includes
bibliographical references and index.
Identifiers: LCCN 2017025765| ISBN 9781138198241 (hardback ; alk. paper) |
ISBN 1138198242 (hardback ; alk. paper)
Subjects: LCSH: Light filters. | Thin films–Optical properties.
Classification: LCC QC373.L5 M34 2018 | DDC 681/.42–dc23
LC record available at https://lccn.loc.gov/2017025765

Visit the Taylor & Francis Web site at
http://www.taylorandfrancis.com

and the CRC Press Web site at
http://www.crcpress.com

In memory of my Mother and Father

Agnes Donaldson Macleod
John Macleod

Brief Contents

Detailed Contents

Series Preface

This international series covers all aspects of theoretical and applied optics and optoelectronics. Active since 1986, eminent authors have long been choosing to publish with this series, and it is now established as a premier forum for high-impact monographs and textbooks. The editors are proud of the breadth and depth showcased by published works, with levels ranging from advanced undergraduate and graduate student texts to professional references. Topics addressed are both cutting edge and fundamental, basic science and applications-oriented, on subject matter that includes lasers, photonic devices, nonlinear optics, interferometry, waves, crystals, optical materials, biomedical optics, optical tweezers, optical metrology, solid-state lighting, nanophotonics, and silicon photonics. Readers of the series are students, scientists, and engineers working in optics, optoelectronics, and related fields in the industry.

Proposals for new volumes in the series may be directed to Lu Han, senior publishing editor at CRC Press, Taylor & Francis Group (lu.han@taylorandfrancis.com).

Preface to the Fifth Edition

Although I did eventually write this fifth edition, I never would have done it without the encouragement and support of many friends and colleagues. It was Lu Han of Taylor & Francis who eventually managed to push me over the edge—in the nicest possible way. And then there are four people, whose identities are hidden from me, but, since they are all clearly expert in thin-film optics, must certainly be people I know well. Their constructive comments based on my proposal and initial efforts were enormously helpful. I owe them and Lu a special debt of grateful thanks.

The field of thin-film optical coatings is still expanding and will continue to do so as long as there is optics. But the term *expansion* is really an oversimplification because there are substantial and significant changes also taking place. We are getting better at what we do is probably the best way to summarize them. There is a great deal of automation in our design, processes, and testing. Computers are everywhere. But automation does not decrease the need for understanding, particularly when there is a problem, and I have still tried to make understanding the primary thrust of this book.

The fundamentals have not changed, and so these are where we start. Then we have our well-known antireflection, high reflectance, bandpass, and related coatings. Although the subjects of these early chapters are still the same, their contents have been reordered and revised. The later chapters are almost entirely reorganized. Of course, the subject has not changed to the extent that everything needs to be completely rewritten, but I hope you will feel that the new arrangement is more logical. It has allowed me to expand on a good number of the important topics. There are also some older coatings that, although perhaps not often used today, are nevertheless instructive in their designs. Then more and more, we are seeing what can be considered as optical thin-film developments in other areas. I am thinking of metamaterials, coherent perfect absorbers, and the like. They tend to appear in journals other than the core ones for our community, and the technical language is often unfamiliar. Some are simply translations of what we already know well, but others could have important implications. I have tried to explain some of these in more familiar optical coating terms. These descriptions should not be thought of as in any way complete. They are simply intended to show how certain aspects might be understood in the context of optical coatings where they could eventually play some role. That there are more than 150 new figures may give some idea of the extent of the changes.

All the additional material has to be accommodated somehow and that implies that some material has to go. That has proved to be the most difficult task of all. Which old friends do I remove? I am helped a little in this by the fact that old books do not completely die but migrate to electronic versions, perhaps in the clouds. Previous editions are not lost forever. The topics I have eliminated will still exist, although they may be a little less readily accessible.

I continue to marvel at the incredibly good fortune that so long ago led me into this wonderfully welcoming, open, generous international optical thin-film community that has supported me all these years. Once again, I thank you, all of you, my readers, publishers, friends, colleagues, family, and especially my wife, Ann.

H. Angus Macleod
Tucson, Arizona

Preface to the Fourth Edition

In some ways first editions are easier, or perhaps I should say less difficult, to prepare than subsequent editions. By the time a fourth edition is required, there is a strong expectation among readers of the character and content of the book. Thus, the author must somehow try to maintain the style at the same time as bringing the book up to date. What to omit and what to include are very difficult questions. Modern optical coating design is virtually entirely performed by computer, frequently using automatic techniques. However, computers do not remove the need for understanding, and I think it is understanding that readers look for in the book. Also, I am conscious that a reader, having perhaps rejected an earlier edition in favor of a later and remembering something important in the earlier, might well expect to find it in the later. I made the decision, therefore, to retain most of the descriptions of the earlier design techniques because of their importance in understanding how designs work. Then, although some of the applications that I describe are rather old, nevertheless they do illustrate how optical coatings are incorporated into a system, and so I retained them. I have tried to incorporate a reasonable amount of new material throughout the book. I added a chapter on color because it is increasing in importance in optical coatings, and, although it is of largely academic interest, I could not resist a section on the effects of gain in optical coatings, because I find it a fascinating topic. Then I struggled with coatings for the soft x-ray region and, with some regret, decided not to include them at this time. It is the old design synthesis problem: one has to stop somewhere.

I am fortunate in my friends and colleagues who have helped me immeasurably with suggestions, advice, and, I have to admit it, corrections. The field of optical thin films has been very good to me. I cannot imagine a more friendly, supportive, and open group of people than the international optical thin-film community. It sets an example the rest of the world would do well to follow.

Thank you, all of you, my readers, publishers, friends, colleagues, family, and especially my wife, Ann.

H. Angus Macleod

Preface to the Third Edition

The foreword to the second edition of this book identified increasing computer power and availability as especially significant influences in optical coating design. This has continued to the point where any description I might give of current computing speed and capacity would be completely out of date by the time this work is in print. Software for coating design (and for other tasks) is now so advanced that commercial packages have almost completely replaced individually written programs. I have often heard it suggested that this removes all need for skill or even knowledge from the act of coating design. I firmly believe that the need for skill and understanding is actually increased by the availability of such powerful tools. The designer who knows very well what he or she is doing is always able to achieve better results than the individual who does not. Coating design still contains compromises. Some aspects of performance are impossible to attain. The results offered by an automatic process that is attempting to reach impossible goals are usually substantially poorer than those when the goals are realistic. The aim of the book, therefore, is still to improve understanding.

During the years since publication of the second edition, the energetic processes, and particularly ion-assisted deposition, have been widely adopted. There are several consequences. The improved stability of optical constants of the materials has enabled the reliable production of coatings of continuously increasing complexity. We even see coatings produced now purely for their aesthetic appeal. Then the enormous improvement in environmental stability has opened up new applications, especially in communications. Unprecedented temperature stability of optical coatings can now be achieved. Specially designed coatings have simplified the construction of ultrafast lasers. Banknotes of many countries inhibit counterfeiting by carrying patches exhibiting the typical iridescence of optical coatings. Coatings to inhibit the effects of glare are now integral parts of visual display units.

I mentioned in my previous foreword the difficulty I experienced in bringing the earlier edition up to date. This time the task has been even more difficult. The volume of literature has expanded to the extent that it is almost impossible to keep up with all of it. The pressure on workers to publish has in many cases reached almost intolerable levels. I regret I do not remember exactly who introduced the idea of the half-life of a publication after which it sinks into obscurity but it is clear that the half-life has become quite short. Comprehensively to review this vast volume of material that has appeared and continues to appear would have changed completely the style of the book. The continuing demand for the now out-of-print second edition of the book suggests that it is used much more as a learning tool than a research reference and so my aim has been to try to keep it so. There have been few fundamental changes that affect our basic understanding of optical coatings and so this third edition reflects that.

I appreciate very much the help of various organizations and individuals who provided material. Many are named in the foreword to the second edition and in the apologia to the first. Additional names include Shincron Company Ltd, Ion-Tech Inc, Applied Vision Ltd, Professor Frank Placido of the University of Paisley, and Roger Hunneman of the Department of Cybernetics, University of Reading.

Again I am grateful for all the helpful comments and suggestions from all my friends and colleagues. The enormous list of names is beyond what can be reproduced here but I must mention my debt to my old friend Professor Lee Cheng-Chung, who took the trouble to work completely through the book and provided me with what has to be the most detailed list of misprints and mistakes, and Professor Shigetaro Ogura, who was instrumental in the translation of the second edition into Japanese. The people at Adam Hilger must be the most patient people on earth. I think finally it was my shame at so trying the endurance of Kathryn Cantley who simply responded with encouragement and understanding that drove me to complete the work.

My eternal and grateful thanks to my wife. She did not write the book but she made sure that I did.

H. Angus Macleod
Tucson, Arizona

Preface to the Second Edition

A great deal has happened in the subject of optical coatings since the first edition of this book. This is especially true of facilities for thin-film calculations. In 1969 my thin-film computing was performed on an IBM 1130 computer that had a random access memory of 10 kbytes. Time had to be booked in advance, sometimes days in advance. Calculations remote from this computer were performed either by slide rule, log tables or electromechanical calculator. Nowadays my students scarcely know what a slide rule is, my pocket calculator accommodates programs that can calculate the properties of thin-film multilayers, and I have on my desk a microcomputer with a random access memory of 0.5 Mbytes, which I can use as and when I like. The earlier parts of this revision were written on a mechanical typewriter. The final parts were completed on my own word processor. These advances in data processing and computing are without precedent and, of course, have had a profound and irreversible effect on many aspects of everyday life as well as on the whole field of science and technology.

There have been major developments, too, in the deposition of thin-film coatings, and although these lack the spectacular, almost explosive, character of computing programs, nevertheless important and significant advances have been made. Electron-beam sources have become the norm rather than the exception, with performance and reliability beyond anything available in 1969. Pumping systems are enormously improved, and the box-coater is now standard rather than unusual. Microprocessors control the entire operation of the pumping system and, frequently, even the deposition process. We have come to understand that many of our problems are inherent in the properties of our thin films rather than in the complexity of our designs. Microstructure and its influence on material properties is especially important. Ultimate coating performance is determined by the losses and instabilities of our films rather than the accuracy and precision of our monitoring systems.

My own circumstances have changed too. I wrote the first edition in industry. I finish the second as a university professor in a different country.

All this change has presented me with difficult problems in the revision of this book. I want to bring it up to date but do not want to lose what was useful in the first edition. I believe that in spite of the great advances in computers, there is still an important place for the appreciation of the fundamentals of thin-film coating design. Powerful synthesis and refinement techniques are available and are enormously useful, but an understanding of thin-film coating performance and the important design parameters is still an essential ingredient of success. The computer frees us from much of the previous drudgery and puts in our hands more powerful tools for improving our understanding. The availability of programmable calculators and of microcomputers implies easy handling of more complex expressions and formulae in design and performance calculations. The book, therefore, contains many more of these than did the first edition. I hope they are found useful. I have included a great deal of detail on the admittance diagram and admittance loci. I use them in my teaching and research and have taken this opportunity to write them up. SI units, rather than Gaussian, have been adopted, and I think Chapter 2 is much the better for the change. There is more on coatings for oblique incidence including the admittance diagram beyond the critical angle, which explains and predicts many of the resonant effects that are observed in connection with surface plasmons, effects used by Greenland and Billington (Chapter 8, reference 12) in the late 1940s and early 1950s for monitoring thin-film deposition.

Inevitably, the first edition contained a number of mistakes and misprints and I apologise for them. Many were picked up by friends and colleagues who kindly pointed them out to me. Perhaps the worse mistake was in Figure 9.4 on uniformity. The results were quoted as for a flat plate but, in fact, referred to a spherical work holder. These errors have been corrected in this edition and I hope that I have avoided making too many fresh ones. I am immensely grateful to all the people who helped in this correction process. I hope they will forgive me for not including the huge list of their names here. My thanks are also due to J. J. Apfel, G. DeBell, E. Pelletier, and W. T. Welford, who read and commented on various parts of the manuscript.

To the list in the foreword of the first edition of organisations kindly providing material should be added the names Leybold-Heraeus GmbH, and Optical Coating Laboratory Inc. Airco-Temescal is now known as Temescal, a Division of the BOC Group Inc., and the British Scientific Instrument Research Association as Sira Institute.

My publisher is still the same Adam Hilger, but now part of the Institute of Physics. I owe a very great debt to Neville Goodman, who was responsible for the first edition and who also persuaded and encouraged me into the second. He retired while it was still in preparation, and the task of extracting the final manuscript from me became Jim Revill's. Ian Kingston and Brian McMahon did a tremendous job on the manuscript at a distance of 3000 miles. Their patience with me in the delays I have caused them has been amazing.

My wife and family have once again been a great source of support and encouragement.

H. Angus Macleod
Newcastle upon Tyne and Tucson, Arizona

Apologia to the First Edition

When I first became involved with the manufacture of thin-film optical filters, I was particularly fortunate to be closely associated with Oliver Heavens, who gave me invaluable help and guidance. Although I had not at that time met him, Dr. L. Holland also helped me through his book, *The Vacuum Deposition of Thin Films*. Lacking, however, was a book devoted to the design and production of multilayer thin-film optical filters, a lack which I have since felt especially when introducing others to the field. Like many others in similar situations I produced from time to time notes on the subject purely for my own use. Then in 1967, I met Neville Goodman of Adam Hilger, who had apparently long been hoping for a book on optical filters in general. I was certainly not competent to write a book on this wide subject, but, in the course of conversation, the possibility of a book solely on thin-film optical filters arose. Neville Goodman's enthusiasm was infectious, and with his considerable encouragement, I dug out my notes and began writing. This, some two years and much labour later, is the result. I have tried to make it the book that I would like to have had myself when I first started in the field, and I hope it may help to satisfy also the needs of others. It is not in any way intended to compete with the existing works on optical thin films, but rather to supplement them, by dealing with one aspect of the subject which seems to be only lightly covered elsewhere.

It will be immediately obvious to even the most casual of readers that a very large proportion of the book is a review of the work of others. I have tried to acknowledge this fully throughout the text. Many of the results have been recast to fit in with the unified approach which I have attempted to adopt throughout the book. Some of the work is, I fondly imagine, completely my own, but at least a proportion of it may, unknown to me, have been anticipated elsewhere. To any authors concerned I humbly apologise, my only excuse being that I also thought of it. I promise, as far as I can, to correct the situation if ever there is a second edition. I can, however, say with complete confidence that any shortcomings of the book are entirely my own work.

Even the mere writing of the book would have been impossible without the willing help, so freely given, of a large number of friends and colleagues. Neville Goodman started the whole thing off and has always been ready with just the right sort of encouragement. David Tomlinson, also of Adam Hilger, edited the work and adjusted it where necessary so that all sounded just as I had meant it to, but had not quite managed to achieve. The drawings were the work of Mrs Jacobi. At Grubb Parsons, Jim Mills performed all the calculations, using an IBM 1130 (he appears in the frontispiece for which I am also grateful), Fred Ritchie kindly gave me permission to quote many of his results and helped considerably by reading the manuscript, and Helen Davis transformed my almost illegible first manuscript into one which could be read without considerable strain. Stimulating discussions with John Little and other colleagues over the years have also been invaluable. Desmond Smith of Reading University kindly gave me much material especially connected with the section on atmospheric temperature sounding which he was good enough to read and correct. John Seeley and Alan Thetford, both of Reading University, helped me by amplifying and explaining their methods of design. Jim Ring, of Imperial College, read and commented on the section on astronomical applications and Dr. J. Meaburn kindly provided the photographs for it. Dr. A. F. Turner gave me much information on the early history of multiple halfwave filters. It is impossible to mention by name all those others who have helped but they include: M. J. Shadbolt, S. W. Warren, A. J. N. Hope, H. Bucher, and all the authors who led the way and whose work I have used and quoted.

Journals, publishers, and organisations which provided and gave permission for the reproduction of material were

- *Journal of the Optical Society of America* (The Optical Society of America)
- *Applied Optics* (The Optical Society of America)
- *Optica Acta* (Taylor and Francis Limited)
- *Proceedings of the Physical Society* (The Institute of Physics and the Physical Society)
- *IEEE Transactions on Aerospace* (The Institute of Electrical and Electronics Engineers, Inc.)
- *Zeitshrift für Physik* (Springer-Verlag)
- *Bell System Technical Journal* (The American Telephone and Telegraph Co.)
- *Philips Engineering Technical Journal* (Philips Research Laboratories)
- Methuen & Co. Ltd
- OCLI Optical Coatings Limited
- Standard Telephones and Cables Limited
- Balzers Aktiengesellschaft für Hochvacuumtechnik und dünne Schichten
- Edwards High Vacuum Limited
- Airco Temescal (A Division of Air Reduction Company Inc.)
- Hawker Siddeley Dynamics Limited
- System Computers Limited
- Ferranti Limited
- British Scientific Instrument Research Association
- And lastly, but far from least, the management of Sir Howard Grubb, Parsons & Co. Ltd, particularly Mr G M Sisson and Mr G E Manville, for much material, for facilities, and for permission to write this book.

To all these and to all the others, who are too numerous to name and who I hope will excuse me for not attempting to name them, I am truly grateful.

I should add that my wife and children have been particularly patient with me during the long writing process, which has taken up so much of the time that would normally have been theirs. Indeed, my children eventually began to worry if ever I appeared to be slacking and, by their comments, prodded me into redoubled efforts.

H. Angus Macleod
Newcastle upon Tyne

Author

H. Angus Macleod is president of Thin Film Center Inc., in Tucson, Arizona, and professor emeritus of Optical Sciences at the University of Arizona, Tucson, Arizona. Dr. Macleod is a graduate of Glasgow University. He holds the degree of Doctor of Technology from the Council for National Academic Awards (London) and an honorary doctorate from the University of Aix-Marseille in France. He is the author of over 200 academic publications and has taught courses on optical topics all over the world, with class audiences ranging from 1 person to over 200 people. He specializes in teaching techniques for understanding and logical thinking that avoid complicated theory without oversimplification.

Dr. Macleod is the recipient of numerous professional honors, including the Gold Medal (1987) from the International Society for Optics and Photonics, the Esther Hoffman Beller Medal (1997) from the Optical Society of America, the Nathaniel H. Sugerman Memorial Award (2002) from the Society of Vacuum Coaters, and the Senator Award (2008) from the European Vacuum Coaters.

Symbols and Abbreviations

The following list gives those more important symbols used in at least several places in the text. We have tried as far as possible to create a consistent set of symbols, but there are several well-known and accepted symbols that are universally used in the field for certain quantities, and changing them would probably lead to even greater confusion than would retaining them. This has meant that in some cases, the same symbol is used in different places for different quantities. We hope the table will make it clear. Less important symbols, defined and used only in very short sections, have been omitted.

A	Absorptance. The ratio of the power absorbed in the structure to the power incident on it.
$\mathscr{A}$	Potential absorptance. A quantity used in the calculation of the absorptance of coatings. It is equivalent to $(1 - \psi)$, where ψ is the potential transmittance.
B	The normalized total tangential electric field at an interface, usually the front interface of an assembly of layers. It is also very briefly used at the beginning of Chapter 2 as the magnetic induction.
C	The normalized total tangential magnetic field at an interface, usually the front interface of an assembly of layers.
d_q	The physical thickness of the qth layer in a thin film coating.
E	The electric vector in the electromagnetic field.
$\mathcal{E}$	The total tangential electric field amplitude, that is, the field parallel to the thin film boundaries.
$\mathscr{E}$	The electric field amplitude.
$\mathcal{E}$	The equivalent admittance of a symmetrical arrangement of layers.
F	A function used in the theory of the Fabry–Perot interferometer.
$\mathscr{F}$	Finesse. The ratio of the separation of adjacent fringes to the halfwidth of the fringe in the Fabry–Perot interferometer.
g	$g = \lambda_0/\lambda = \nu/\nu_0$, sometimes called the relative wavelength, or the relative wavenumber, or the wavelength ratio. λ_0 and ν_0 are the reference wavelength and reference wavenumber, respectively. The optical thicknesses of the layers in a coating are defined with respect to these quantities that are usually chosen to make the more important layers in the coating as close to quarter waves as possible.
H	The magnetic vector in the electromagnetic field.
$\mathcal{H}$	The total tangential magnetic field amplitude, that is, the field parallel to the thin film boundaries.
$\mathscr{H}$	The magnetic field amplitude.
H	Represents a quarter wave of high index in shorthand notation.
I	The irradiance of the wave, that is, power per unit area. Unfortunately, the standard International System of Units (SI) symbol for irradiance is E, but to use E would cause great confusion between irradiance and electric field. It is even more unfortunate that I is the SI symbol for intensity that is the power per unit solid angle from a point source. Doubly unfortunate is that the older definition of intensity is identical to the current definition of irradiance.
k	The extinction coefficient. The extinction coefficient denotes the presence of absorption. The complex refractive index N is given by $N = n - ik$.
L	Represents a quarter wave of low index in shorthand notation.

M	Represents a quarter wave of intermediate index in shorthand notation. Also used for a matrix element or to indicate an array of matrix elements.
N	Denotes the complex refractive index $n - ik$.
n	The refractive index or, sometimes, the real part of refractive index.
n^*	The effective index of a narrowband filter, that is, the index of an equivalent layer that yields a shift of its fringes in wavelength, by the same amount as the peak of the narrowband filter, when tilted with respect to the direction of incidence.
p	Packing density, that is, the ratio of the solid volume of a film to its total volume.
p	p-Polarization, that is, the polarization where the electric field direction is in the plane of incidence. It is sometimes known as TM for transverse magnetic.
R	The reflectance. The ratio at a boundary of the normal components of reflected and incident irradiance or, alternatively, the ratio of the total reflected beam power to the total incident beam power.
s	s-Polarization, that is, the polarization where the electric field direction is normal to the plane of incidence. It is sometimes known as TE for transverse electric.
T	The transmittance. The ratio of the normal components of transmitted and incident irradiance or, alternatively, the ratio of the total transmitted beam power to the total incident beam power.
TE	See s for s-polarization.
TM	See p for p-polarization.
x, y, z	The coordinate axes. In the case of a thin film or surface, the z-axis is usually taken positive into the surface in the direction of incidence. The x-axis is usually arranged in the plane of incidence, and the x-, y-, and z-axes, in that order, make a right-handed set.
$\bar{x}, \bar{y}, \bar{z}$	The three color matching functions that define the CIE 1931 Standard Colorimetric Observer.
X, Y, Z	The tristimulus values. They are the three basic responses defining a color.
x, y, z	The chromaticity coordinates, $X/(X+Y+Z)$, $Y/(X+Y+Z)$, and $Z/(X+Y+Z)$. Usually, z is omitted because they are normalized to add to unity.
$X + iZ$	The complex surface admittance.
y	The characteristic admittance of a material given in SI units (siemens) by $N\mathcal{Y}$, that is, $(n - ik)\mathcal{Y}$ and in units of the admittance of free space $\mathcal{Y}$ by N or $n - ik$.
Y	The surface admittance, that is, the ratio of the total tangential components of magnetic and electric field at any surface parallel to the film boundaries. $Y = C/B$.
$\mathcal{Y}$	The admittance of free space (2.6544×10^{-3} S).
y_0	The characteristic admittance of the incident medium.
y_m or y_{sub}	The characteristic admittance of the emergent medium or substrate.
α	The absorption coefficient, given by $4\pi k/\lambda$, usually in units of cm^{-1}.
α, β, γ	The three direction cosines, that is, the cosines of the angle the direction makes with the three coordinate axes.
β	Symbol for $2\pi kd/\lambda$, usually with reference to a metal.
γ	The equivalent phase thickness of a symmetrical arrangement of layers.
Δ	The relative retardation. It is given by $\varphi_p - \varphi_s \pm 180°$ in reflection and $\varphi_p - \varphi_s$ in transmission, where the normal thin-film sign convention for φ_p is used.
Δ	η_p/η_s, where η is the modified tilted admittance. The quantity is used in the design of polarization-free coatings.
δ	The phase thickness of a coating, given by $2\pi(n - ik)d/\lambda$.
ε	Indicates a small error in the discussion of tolerances etc.
ε	The permittivity of a medium.
η	The tilted optical admittance.
ϑ	The angle of incidence.

κ Sometimes called the wavenumber, κ is given by $2\pi(n - ik)/\lambda$, where λ is the free space wavelength. Note the confusing use of the term *wavenumber*. It is also applied to ν.

λ The wavelength of light. In the book, except at the very beginning of Chapter 2, it always indicates the wavelength in free space.

λ_0 The reference wavelength. The optical thicknesses of the layers in a coating are defined with respect to the reference wavelength that is usually chosen to make the more important layers in the coating as close to quarter waves as possible.

ν The wavenumber. $\nu = 1/\lambda$ and is frequently expressed in units of inverse centimetes (also sometimes known as kayser. The SI unit is strictly inverse meters or m^{-1}).

ν_0 The reference wavenumber, $1/\lambda_0$.

μ Permeability. Used in the early part of Chapter 2.

ρ The amplitude reflection coefficient; also used as electric charge density in the early part of Chapter 2.

τ The amplitude transmission coefficient.

φ Phase difference, often in reflection or transmission.

ψ The potential transmittance $T/(1 - R)$ or the ratio of the quantities $\mathrm{Re}(BC^*)$, evaluated at two different interfaces. It represents the net power emerging from a system divided by the net power entering and is unity if there is no loss.

1

Introduction

When this book was first written, the question of the title arose. *Optical Thin Films* was the obvious choice, but the publisher feared that it might then be confused with some other existing titles, and so, eventually to avoid confusion, *Thin-Film Optical Filters* was chosen and has remained the name through four editions and, now, into a fifth. It was never intended that the subject should be limited to the narrow designation of filters but should encompass as much of the field of optical thin films as possible. This is still the intention, but in a work of this size, it is not possible to cover the entire field of thin-film optical devices in the detail that some of them may deserve. The selection of topics is due, at least in part, to the author's own preferences and knowledge.

The intention of the book has always been to form an introduction to thin-film optical coatings for both the manufacturer and the user. The topics covered are a mixture of design, manufacture, performance, and features important in applications. It begins with enough of the basic mathematics of optical thin films for the reader to carry out thin-film calculations. The aim has been to present, as far as possible, a unified treatment, and there are some alternative methods of analysis that are not discussed.

When the book first appeared, there were just a few books available that covered aspects of the field. Now the situation has changed somewhat, and there is an array of relevant books. Some of these are listed in the bibliography at the end of this chapter. However, the half-life of a work these days is so short that knowledge can actually disappear. It is well worthwhile to take the time to go back to some of the earlier books. Heavens [1], Holland [2], Anders [3], and Knittl [4] are just some of those that will repay study, and they are listed in the bibliography along with some more recent volumes.

1.1 Early History

History is impossibly complicated, and we can have only an imperfect view of it, told generally through the medium of an historian who will, because of culture changes, attempt to interpret it in a way the intended audience can understand. The history of technology is no exception. To simplify the telling, we will usually pick certain events and individuals and connect them as a kind of series, one depending on the other. Technology, however, develops over a very broad front, rather like the advance of the tide. It depends on a network of effort. If one individual does not make a required advance, another certainly will. Technology adopts the advances it needs at the time and ignores those that it does not. Over and over again, we find that discoveries credited to a particular individual were actually anticipated by others, but the time was not right, and so little or no notice was taken of them. The abbreviated account of the history of the subject that follows is no exception. A true account is beyond us, and so we pick a few events and a few individuals and connect them, but there are many other routes through history. This one is based on the preferences and limited knowledge of your author.

Thin metal layers were known from very early times, but if we consider interference as the hallmark of modern thin-film optics, then the earliest of what might be called modern thin-film optics was the work of Robert Boyle and Robert Hooke on colors exhibited by materials in thin-film form. Sir Isaac Newton [5] related the colors to exact measurements of film thickness and placed the subject on a firm quantitative base with his brilliant technique now known as Newton's rings. The explanation of the colors is nowadays thought to be a very simple matter, being due to interference in a single thin film of varying thickness. However, at that time, the theory of the nature of light was not sufficiently far advanced. Newton struggled with his concept of the interval of fits that as we now understand is a half wavelength, but it was a further 100 years before the idea of light as a wave would be accepted. On November 12, 1801, in a Bakerian Lecture to the Royal Society, Thomas Young [6] enunciated the principle of the interference of light and produced a satisfactory explanation of the effect. As Henry Crew [7] has put it, "This simple but tremendously important fact that two rays of light incident upon a single point can be added together to produce darkness at that point is, as I see it, the one outstanding discovery which the world owes to Thomas Young."

Young's theory was far from achieving universal acceptance. Indeed Young became the victim of a bitter personal attack, against which he had the greatest difficulty defending himself. Recognition came slowly and depended much on the work of Augustin Jean Fresnel [8], who, quite independently, also arrived at a wave theory of light. Fresnel's discovery, in 1816, that two beams of light that are polarized at right angles could never interfere established the transverse nature of light waves. Then Fresnel combined Young's interference principle and Huygens's ideas of light propagation into an elegant theory of diffraction. It was Fresnel who put the wave theory of light on such a firm foundation that it has never been shaken. For the thin-film worker, Fresnel's laws, governing the amplitude and phase of light reflected and transmitted at a single boundary, are of major importance. Knittl [9] reminded us that Fresnel already knew that the sum of an infinite series of rays is necessary to determine the transmittance of a thick sheet of glass and that it was Simeon Denis Poisson, in correspondence with Fresnel, who included interference effects in the summation to arrive at the important results that a half-wave-thick film does not change the reflectance of a surface and that a quarter-wave-thick film of index $(n_0 n_1)^{1/2}$ will reduce to zero the reflectance of a surface between two media of indices n_1 and n_0, results extended to oblique incidence by Fresnel. Fresnel died in 1827, at the early age of 39.

In 1873, the great work of James Clerk Maxwell, *A Treatise on Electricity and Magnetism* [10], was published, and in his system of equations, we have all the basic theory for the analysis of thin-film optical problems.

Meanwhile, in 1817, Joseph Fraunhofer [11] made what are probably the first ever antireflection coatings. It is worth quoting his observations at some length because they show the considerable insight that he had, even at that early date, into the physical causes of the effects that were produced. The following is a translation of part of the paper as it appears in the collected works:

> Before I quote the experiments which I have made on this I will give the method which I have made use of to tell in a short time whether the glass will withstand the influence of the atmosphere. If one grinds and then polishes, as finely as possible, one surface of glass which has become etched through long exposure to the atmosphere, then wets one part of the surface, for example half, with concentrated sulfuric or nitric acid and lets it work on the surface for twenty-four hours, one finds after cleaning away the acid that that part of the surface on which the acid was, reflects much less light than the other half, that is it shines less although it is not in the least etched and still transmits as much light as the other half, so that one can detect no difference on looking through. The difference in the amount of reflected light will be most easily detected if one lets the light strike approximately vertically. It is the greater the more the glass is liable to tarnish and become etched. If the polish on the glass is not very good, this difference will be less noticeable. On glass which is not liable to tarnish, the sulfuric and nitric acid does not work. . . . Through this treatment with sulfuric or nitric acid some types of glasses get on their surfaces beautiful vivid colors which alter like soap bubbles if one lets the light strike at different angles.

Then, in an appendix to the paper added in 1819:

> Colors on reflection always occur with all transparent media if they are very thin. If for example, one spreads polished glass thinly with alcohol and lets it gradually evaporate, towards the end of the evaporation, colors appear as with tarnished glass. If one spreads a solution of gum-lac in a comparatively large quantity of alcohol very thinly over polished warmed metal the alcohol will very quickly evaporate, and the gum-lac remains behind as a transparent hard varnish which shows colors if it is thinly enough laid on. Since the colors, in glasses which have been colored through tarnishing, alter themselves if the inclination of the incident light becomes greater or smaller, there is no doubt that these colors are quite of the same nature as those of soap bubbles, and those which occur through the contact of two polished flat glass surfaces, or generally as thin transparent flakes of material. Thus, there must be on the surface of tarnished glass that shows colors, a thin layer of glass that is different in refractive power from the underlying. Such a situation must occur if a component is partly removed from the surface of the glass or if a component of the glass combines at the surface with a related material into a new transparent product.

It seems that Fraunhofer did not follow up this particular line into the development of an antireflection coating for glass, perhaps because optical components were not, at that time, sufficiently complicated for the need for antireflection coatings to be obvious. Possibly the important point that not only was the reflectance less but the transmittance also greater had escaped him.

In 1886, Lord Rayleigh [12] reported to the Royal Society an experimental verification of Fresnel's reflection law at off-normal incidence. In order to attain a sufficiently satisfactory agreement between measurement and prediction, he had found it necessary to use freshly polished glass because the reflectance of older material, even without any visible signs of tarnish, was too low. One possible explanation, which he suggested, was the formation, on the surface, of a thin layer of different refractive indices from the underlying material. He was apparently unaware of the earlier work of Fraunhofer and of the identical difficulties experienced by Malus and by Brewster in determining the polarizing angle, or Brewster angle, of glass.

Then, in 1891, Dennis Taylor [13,14] published the first edition of his famous book *On the Adjustment and Testing of Telescopic Objectives* and mentioned that, "As regards the tarnish which we have above alluded to as being noticeable upon the flint lens of an ordinary objective after a few years of use, we are very glad to be able to reassure the owner of such a flint that this film of tarnish, generally looked upon with suspicion, is really a very good friend to the observer, inasmuch as it increases the transparency of his objective."

In fact, Taylor [15] went on to develop a method of artificially producing the tarnish by chemical etching. This work was followed up by Kollmorgen [16], who developed the chemical process still further for different types of glasses.

At the same time, in the nineteenth century, a great deal of progress was being made in the field of interferometry. The most significant development, from the thin-film point of view, was the Fabry–Perot [17] interferometer described in 1899, which has become one of the basic structures for thin-film filters.

Developments became much more rapid in the 1930s, and indeed, it is in this period that we can recognize the beginnings of modern thin-film optical coating. In 1932, Rouard [18] observed that a very thin metallic film reduced the internal reflectance of a glass plate, although the external reflectance was increased. In 1934, Bauer [19], in the course of fundamental investigations of the optical properties of halides, produced reflection-reducing coatings, and Pfund [20] evaporated zinc sulfide layers to make low-loss beam splitters for Michelson interferometers, noting, incidentally, that titanium dioxide could be a better material. In 1936, John Strong [21] produced antireflection coatings by the evaporation of fluorite to give inhomogeneous films, which reduced the reflectance of glass to visible light by as much as 89%, a most impressive figure. At the same time, Alexander Smakula at the Carl Zeiss company, in Jena, developed antireflection coatings that were kept secret because of their military implications. Then, in 1939, Geffcken [22] constructed the first thin-film metal–dielectric interference filters. A fascinating account of Geffcken's work is given by Thelen [23], who described Geffcken's search for improved antireflection coatings and his creation of the famous quarter–half-quarter design.

Several factors were probably responsible for this sudden expansion of the field. Optical systems, particularly photographic objectives were becoming more complex, bringing a need for antireflection coatings. Telescopes and binoculars, especially for military applications, were also much improved by antireflection coatings. Then the manufacturing process was also becoming more reliable. Although sputtering was discovered about the middle of the nineteenth century, and vacuum evaporation around the beginning of the twentieth century, they had not yet been adopted as useful manufacturing processes. One difficulty was the lack of really suitable pumps, and it was not until the early 1930s that the work of C. R. Burch on diffusion pump oils introduced the oil diffusion pump. This enormously helped, although, particularly in Germany, mercury diffusion pumps were still used very effectively for some time. World War II saw a great expansion in the production of antireflection coatings. This certainly accelerated developments, but the expansion would have taken place without any war, because optics had now reached the stage where coatings were necessary. Since then, tremendous strides have been made. Modern optics without coating is unthinkable. It is almost impossible to imagine an optical instrument that would not rely on optical coatings to assure its performance. Filters with greater than 100 layers are not uncommon, and uses have been found for them in almost every branch of science and technology.

1.2 Thin-Film Filters

First of all, we assume for the purposes of this section that the materials in thin-film form are free from absorption or other loss. Then to understand in a qualitative way the performance of thin-film optical devices, it is necessary to accept several simple statements. The first is that the amplitude reflectance of light at any boundary between two media is given by $(1 - \rho)/(1 + \rho)$, where ρ is the ratio of the optical admittances at the boundary, which, in the optical region, is also the ratio of the refractive indices. The reflectance (the ratio of irradiances) is the square of this quantity. The second is that there is a phase shift of 180° when the reflectance takes place in a medium of lower refractive index than the adjoining medium and zero if the medium has a higher index than the one adjoining it. The third is that if light is split into two components by reflection at the top and bottom surfaces of a thin film, then the beams will recombine in such a way that the resultant amplitude will be the difference of the amplitudes of the two components, if the relative phase shift is 180°, or the sum of the amplitudes, if the relative phase shift is either zero or a multiple of 360°. In the former case, we say that the beams interfere destructively, and in the latter, constructively. Other cases where the phase shift is different will be intermediate between these two possibilities.

The antireflection coating depends on the more or less complete cancellation of the light reflected at the upper and lower of the two surfaces of the thin film for its operation (Figure 1.1). Let the index of the substrate be n_m; that of the film, n_1; and that of the incident medium, which will in almost all cases be air, n_0. For a completely accurate calculation, we should consider multiple beams, as in the subsequent chapters of this book, but for the moment, we adopt an approximation. We assume that although the reflection at the front surface has diminished the transmitted light a little, we shall completely neglect that loss. Then for complete cancellation of the two beams of light, they should be 180° out of phase and their amplitudes should be equal, which implies that the ratios of the refractive indices at each boundary should be equal, i.e., $n_0/n_1 = n_1/n_m$, or $n_1 = (n_0 n_m)^{1/2}$. This shows that the index of the thin film should be intermediate between the indices of air, which may be taken as unity, and of the substrate, which may be taken as at least 1.52. At both the upper and lower boundaries of the antireflection film, the reflection takes place in a medium of lower refractive index than the adjoining medium. Thus, to ensure that

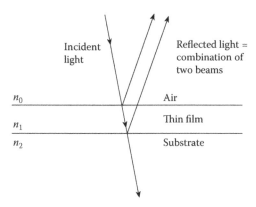

FIGURE 1.1
Single thin film.

the relative phase shift is 180° so that the beams cancel, the optical thickness of the film should be made one-quarter wavelength.

A simple antireflection coating should, therefore, consist of a single film of refractive index equal to the square root of that of the substrate and of optical thickness one-quarter of a wavelength. As will be explained in the chapter on antireflection coatings (Chapter 4), there are other improved coatings covering wider wavelength ranges involving greater numbers of layers.

Another basic type of thin-film structure is a stack of alternate high- and low-index films, all one-quarter wavelength thick (see Figure 1.2). Light reflected within the high-index layers will not suffer any phase shift on reflection, while that reflected within the low-index layers will suffer a change of 180°. It is fairly easy to see that the various components of the incident light produced by reflection at successive boundaries throughout the assembly will reappear at the front surface all in phase so that they will constructively recombine. This implies that the effective reflectance of the assembly can be made very high indeed, as high as may be desired, merely by increasing the number of layers. This is the basic form of the high-reflectance coating. When such a coating is constructed, it is found that the reflectance remains high over only a limited range of wavelengths, depending on the ratio of high and low refractive indices. Outside this zone, the reflectance abruptly changes to a low value. Because of this behavior, the quarter-wave stack, as it is called, is used as a basic building block for many types of thin-film filters. It can be used as a longwave-pass

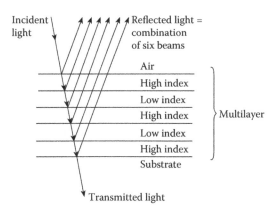

FIGURE 1.2
Multilayer.

filter; a shortwave-pass filter; a bandstop filter; a straightforward high-reflectance coating, for example, in laser mirrors; and a reflector in a thin-film Fabry–Perot interferometer (Figure 1.3), which is another basic filter type described in some detail in Chapters 6 and 8. Here, it is sufficient to say that it consists of a cavity layer, sometimes called a spacer layer, that is usually half a wavelength thick, bounded by two high-reflectance coatings. Multiple-beam interference in the cavity layer causes the transmission of the filter to be extremely high over a narrow band of wavelengths around that for which the cavity is a multiple of one-half wavelength thick. It is possible, as with lumped electric circuits, to couple two or more Fabry–Perot filters in series to give a more rectangular shape to the pass band.

Our assumption of vanishingly small absorption and other losses so that the films are completely transparent is true in the great majority of cases. Since no energy is lost, the filter characteristic in reflection is the complement of that in transmission. This fact is used in the construction of such devices as dichroic beam splitters for color separation in, for example, color projection engines.

This brief description has neglected the effect of multiple reflections in most of the layers, and for an accurate evaluation of the performance of a filter, these extra reflections must be taken into account. This involves extremely complex calculations, and an alternative, and more effective, approach has been found in the development of entirely new forms of solution of Maxwell's equations in stratified media. This is, in fact, the principal method used in Chapter 2 where basic theory is considered. The solution appears as a very elegant product of 2 × 2 matrices, each matrix representing a single film. Unfortunately, in spite of the apparent simplicity of the matrices, the calculation by hand of the properties of a given multilayer, particularly if there are absorbing layers present and a wide spectral region is involved, is an extremely tedious and time-consuming task. The preferred method of calculation is to use a computer. This makes calculation so rapid and straightforward that it makes little sense to use anything else. Even pocket calculators, especially the programmable kind, can be used to great effect. However, despite the enormous power of the modern computer, it is still true that skill and experience play a major part in successful coating design. The computer brings little in the way of understanding. Understanding is the emphasis in the bulk of this book. There are many techniques that date back to times when computers were expensive, cumbersome, and scarce, and alternatives, usually approximate, were required. These

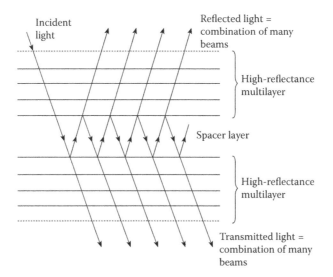

FIGURE 1.3
Fabry–Perot filter showing multiple reflections in the spacer or cavity layer.

would not be used for calculation today, but they bring an insight that straightforward calculation cannot deliver, even if it is very fast. Thus, we include many such techniques, and it is convenient to introduce them often in an historical context. The matrix method itself brings many advantages. For example, it has made possible the development of exceedingly powerful design techniques based on the algebraic manipulation of the matrices. These are also included. Graphical techniques are of considerable usefulness in the visualization of the properties of coatings. There are many such techniques, but in this book, we pay particular attention to one such method known as the admittance diagram. This is one that your author has found of considerable assistance over the years. It is an accurate technique in the sense that it contains no approximations other than those involved perhaps in sketching it, but it is normally used as an aid to understanding rather than as a calculation tool.

In the design of a thin-film multilayer, we are required to find an arrangement of layers to give a performance specified in advance, and this is much more difficult than the straightforward calculation of the properties of a given multilayer. There is no precise analytical solution to the general problem. The normal method of design is to arrive at a possible structure for a filter, using techniques to be described, which consist of a mixture of analysis, experience, and use of well-known building blocks. The evaluation is then completed by calculating the performance on a computer. Depending on the results of the computations, adjustments to the proposed design may be made and then recomputed, until a satisfactory solution is found. This adjustment process can itself be undertaken by a computer and is usually known by the term *refinement*. A related term is *synthesis*, which implies an element of construction as well as adjustment. The ultimate in synthesis is the complete construction of a design with no starting information beyond the performance specification, but it is normal to provide some starting information, such as materials to be used and, possibly, total thickness of coating or a very rough starting design.

The successful application of refinement techniques largely depends on a starting solution that has a performance close to that required. Under these conditions, it has been made to work exceedingly well. The operation of a refinement process involves the adjustment of the parameters of the system to minimize a merit coefficient (in some less common versions, a measure of merit may be maximized) representing the gap between the performance achieved by the design at any stage and the desired performance. The main difference between the various techniques is in the details of the rules used in adjusting the design. A major problem is the enormous number of parameters that can potentially be involved. Refinement is usually kept within bounds by limiting the search to small changes in an almost acceptable starting design. In synthesis with no starting design, the possibilities are virtually infinite, and so the rules governing the search procedure have to be very carefully organized. The most effective techniques incorporate two elements, an effective refinement technique that operates until it reaches a limit and a procedure for complicating the design that is then applied. These two elements alternate as the design is gradually constructed. Automatic design synthesis is undoubtedly increasing in importance in step with developments in computers, but it is still true that in the hands of a skilled practitioner, the achievements of both refinement and synthesis are much more impressive than when no skill is involved. Someone who knows well what he or she is doing will always succeed much better than someone who does not. This branch of the subject is much more a matter of computing techniques rather than fundamental to the understanding of thin-film filters, and so it is largely outside the scope of this book. The book by Liddell [24] and the more recent text by Furman and Tikhonravov [25] give good introductory accounts of various methods. The real limitation to what is, at the present time, possible in optical thin-film filters and coatings is the capability of the manufacturing process to produce layers of precisely the correct optical constants and thickness, rather than any deficiency in design techniques.

The common techniques for the construction of thin-film optical coatings can be classified as physical vapor deposition. They are vacuum processes where a solid film condenses from the vapor phase. The most straightforward, and the traditional method, is known as thermal evaporation, and this is still much used. Because of the defects of solidity possessed by thermally evaporated films, there has, in recent years, been a shift, now accelerating, toward what are described as the energetic processes. Here, mechanical momentum is transferred to the growing film, either by deliberate bombardment or by an increase in the momentum of the arriving film material, and this added momentum drives the outermost material deeper into the film, increasing its solidity. These processes are briefly described in the later chapters of the book, but much more information will be found in the books listed in the bibliography at the end of this chapter.

Then some words of explanation might be useful. Except for some deliberately simplified, and therefore approximate, techniques that will be positively identified, the theory that will be presented of our thin-film interference effects is exact and of perfect precision so that any numerical results can reflect the precision of the numerical data that are entered. However, in almost the entire field of thin-film activity, we do not have perfect precision. For example, characteristic values of material parameters fluctuate with deposition conditions and are difficult to measure with extreme accuracy. Reflectance and transmittance measurements are similarly limited. Thus, although the theory can accommodate any degree of precision that we might wish, in practice, we are rather limited by our imperfect knowledge and the behavior of our real samples. We reflect this in our calculations in this book where we will frequently suggest numerical values for our demonstrations that will rarely be of greater precision than two places of decimal for refractive indices. In particular, we will assume a value of the refractive index of air as unity, although in reality, air has a very slightly higher index, at around 1.00029, that varies with factors such as humidity, composition and temperature, and wavelength. Thus, our air will be indistinguishable from vacuum in our numerical results. This is common practice in the thin-film field. The theory will certainly support whatever precision the user should require.

In Chapter 2, which deals with theory, it will become clear that an optical material is characterized by two different physical parameters, its refractive index and its characteristic admittance, both of which may be complex. In the optical region, from soft X-rays to the far infrared, in other words, the region that interests us throughout this book, the characteristic admittance is proportional to the refractive index, the constant of proportionality being the characteristic admittance of free space. The refractive index is unitless, while the SI unit of admittance is the siemens. By changing the units of the characteristic admittance to units of the free space admittance, the numerical value of characteristic admittance becomes equal to the refractive index. Therefore, a specification of refractive index is also a specification of characteristic admittance. When discussing the properties of a thin film, therefore, there is no need slavishly to state both quantities. The theoretical expressions will indicate the correct choice. Unless it is important that we distinguish between the two parameters, we shall usually follow normal practice in tending to use refractive index when referring to a material.

References

1. O. S. Heavens. 1955. *Optical Properties of Thin Solid Films*. London: Butterworth Scientific Publications.
2. L. Holland. 1956. *Vacuum Deposition of Thin Films*. London: Chapman & Hall.
3. H. Anders. 1965. *Dünne Schichten für die Optik* (English translation: *Thin Films in Optics*. Waltham, MA: Focal Press, 1967). Stuttgart: Wissenschaftliche Verlagsgesellschaft mbH.
4. Z. Knittl. 1976. *Optics of Thin Films*. Hoboken, NJ: John Wiley & Sons; Berlin: SNTL.
5. I. Newton. 1704. *Opticks or a Treatise of the Reflections, Refractions, Inflections and Colours of Light*. London: The Royal Society.
6. T. Young. 1802. On the theory of light and colours (The 1801 Bakerian Lecture). *Philosophical Transactions of the Royal Society of London* 92:12–48.

7. H. Crew. 1930. Thomas Young's place in the history of the wave theory of light. *Journal of the Optical Society of America* 20:3–10.

8. H. de Sénarmont, E. Verdet, and L. Fresnel, eds. 1866–1870. *Oeuvres complètes d'Augustin Fresnel.* Paris: Imprimerie Impériale.

9. Z. Knittl. 1978. Fresnel historique et actuel. *Optica Acta* 25:167–173.

10. J. C. Maxwell. 1873. *A Treatise on Electricity and Magnetism.* Wotton-under-Edge: Clarendon Press.

11. J. von Fraunhofer. 1817. Versuche über die Ursachen des Anlaufens und Mattwerdens des Glases und die Mittel, denselben zuvorzukommen. In J. von Fraunhofer. 1888. *Gesammelte Schriften*, pp. 33–49. Munich: Verlag der Königlich Bayerischen Akademie der Wissenschaften.

12. Lord Rayleigh. 1886. On the intensity of light reflected from certain surfaces at nearly perpendicular incidence. *Proceedings of the Royal Society of London* 41:275–294.

13. H. D. Taylor. 1891. *On the Adjustment and Testing of Telescopic Objectives.* York: T. Cooke & Sons.

14. H. D. Taylor. 1983. *On the Adjustment and Testing of Telescopic Objectives.* Fifth ed. Bristol: Adam Hilger.

15. H. D. Taylor. 1904. *Lenses.* UK Patent, 29561.

16. F. Kollmorgen. 1916. Light transmission through telescopes. *Transactions of the American Illumination Engineering Society* 11:220–228.

17. C. Fabry and A. Perot. 1899. Théorie et applications d'une nouvelle méthode de spectroscopie interférentielle. *Annales de Chimie et de Physique, Paris, 7th series* 16:115–144.

18. P. Rouard. 1932. Sur le pouvoir réflecteur des métaux en lames très minces. *Comptes Rendus de l'Academie de Science* 195:869–872.

19. G. Bauer. 1934. Absolutwerte der optischen Absorptionskonstanten von Alkalihalogenidkristallen im Gebiet ihrer ultravioletten Eigenfrequenzen. *Annalen der Physik (Leipzig), 5th series* 19:434–464.

20. A. H. Pfund. 1934. Highly reflecting films of zinc sulphide. *Journal of the Optical Society of America* 24:99–102.

21. J. Strong. 1936. On a method of decreasing the reflection from non-metallic substances. *Journal of the Optical Society of America* 26:73–74.

22. W. Geffcken. 1939. Interferenzlichtfilter. *German Patent, DE 716 153.*

23. A. Thelen. 1997. The pioneering contributions of W, Geffcken. In *Thin Films on Glass,* H. Bach and D. Krause (eds), pp. 227–239. Berlin: Springer-Verlag.

24. H. M. Liddell. 1981. *Computer-Aided Techniques for the Design of Multilayer Filters.* Bristol: Adam Hilger.

25. S. A. Furman and A. V. Tikhonravov. 1992. *Basics of Optics of Multilayer Systems.* First ed. Gif-sur-Yvette: Editions Frontières.

Bibliography

A complete bibliography of primary references would stretch to an enormous length. This list is, therefore, primarily one of secondary references.

1. H. Anders. 1965. *Dünne Schichten für die Optik* (English translation: *Thin Films in Optics.* Waltham, MA: Focal Press, 1967). Stuttgart: Wissenschaftliche Verlagsgesellschaft mbH.

2. H. Bach and D. Krause, eds. 1997. *Thin Films on Glass.* Berlin, Heidelberg: Springer-Verlag.

3. P. W. Baumeister. 2004. *Optical Coating Technology.* Bellingham, WA: SPIE Press.

4. D. H. Cushing. 2011. *Enhanced Optical Filter Design.* Bellingham, WA: SPIE Press.

5. J. A. Dobrowolski. 1995. Optical properties of films and coatings. In *Handbook of Optics,* M. Bass, E. W. V. Stryland, D. R. Williams et al. (eds), pp. 42.1–42.130. New York: McGraw-Hill.

6. F. R. Flory, ed. 1995. *Thin Films for Optical Systems.* New York: Marcel Dekker.

7. H. Frey and G. Kienel, eds. 1987. *Dünnschicht Technologie.* Düsseldorf: VDI-Verlag GmbH.

8. S. A. Furman and A. V. Tikhonravov. 1992. *Basics of Optics of Multilayer Systems.* First ed. Gif-sur-Yvette: Editions Frontières.

9. H. L. Hartnagel, A. L. Dawar, A. K. Jain et al. 1995. *Semiconducting Transparent Thin Films.* Bristol: Institute of Physics.

10. O. S. Heavens. 1955. *Optical Properties of Thin Solid Films.* London: Butterworth Scientific Publications.

11. I. J. Hodgkinson and Q. H. Wu. 1997. *Birefringent Thin Films and Polarizing Elements*. First ed. Singapore: World Scientific.
12. L. Holland. 1956. *Vacuum Deposition of Thin Films*. London: Chapman & Hall.
13. R. E. Hummel and K. H. Guenther, eds. 1995. *Thin Films for Optical Coatings*. Boca Raton, FL: CRC Press.
14. M. R. Jacobson, ed. 1988. *Modeling of Optical Thin Films*, vol. 821. Bellingham: SPIE Press.
15. M. R. Jacobson, ed. 1989. *Deposition of Optical Coatings*. Bellingham: SPIE Press.
16. M. R. Jacobson, ed. 1990. *Design of Optical Coatings*. Bellingham: SPIE Press.
17. M. R. Jacobson, ed. 1992. *Characterization of Optical Coatings*. Bellingham: SPIE Press.
18. N. Kaiser and H. K. Pulker, eds. 2003. *Optical Interference Coatings*. Berlin: Springer-Verlag.
19. Z. Knittl. 1976. *Optics of Thin Films*. Hoboken, NJ: John Wiley & Sons; Berlin: SNTL.
20. H. M. Liddell. 1981. *Computer-Aided Techniques for the Design of Multilayer Filters*. Bristol: Adam Hilger.
21. P. H. Lissberger. 1970. Optical applications of dielectric thin films. *Reports of Progress in Physics* 33:197–268.
22. B. E. Perilloux. 2002. *Thin Film Design: Modulated Thickness and Other Stopband Design Methods*, vol. TT57. Bellingham: SPIE Optical Engineering Press.
23. A. Piegari and F. Flory, eds. 2013. *Optical Thin Films and Coatings. From Materials to Applications*. Oxford: Woodhead.
24. H. K. Pulker. 1999. *Coatings on Glass*. Second ed. Amsterdam: Elsevier.
25. J. D. Rancourt. 1987. *Optical Thin Films: Users' Handbook*. New York: Macmillan.
26. D. Ristau, ed. 2015. *Laser-Induced Damage in Optical Materials*. Boca Raton, FL: CRC Press.
27. O. Stenzel. 2014. *Optical Coatings: Material Aspects in Theory and Practice*. Berlin: Springer-Verlag.
28. A. Thelen. 1988. *Design of Optical Interference Coatings*. First ed. New York: McGraw-Hill.
29. A. Vasicek. 1960. *Optics of Thin Films*. Amsterdam: North Holland.
30. R. R. Willey. 2002. *Practical Design and Production of Optical Thin Films*. Second ed. New York: Marcel Dekker.

2

Basic Theory

The next part of the book is a long and rather tedious account of some basic theory that is necessary in order to make calculations of the properties of multilayer thin-film coatings. It is perhaps worth reading just once or when some deeper insight into thin-film calculations is required. In order to make it easier for those who have read it to find the basic results or for those who do not wish to read it at all, to proceed with the remainder of the book, the principal results are summarized in Section 2.9.

2.1 Maxwell's Equations and Plane Electromagnetic Waves

For those readers who are still with us, we begin our attack on thin-film problems by solving Maxwell's equations together with the appropriate material equations. In isotropic media, these are

$$\operatorname{curl}H = \nabla \times H = j + \partial D/\partial t, \tag{2.1}$$

$$\operatorname{curl}E = \nabla \times E = -\partial B/\partial t, \tag{2.2}$$

$$\operatorname{div}D = \nabla \cdot D = \rho, \tag{2.3}$$

$$\operatorname{div}B = \nabla \cdot B = 0, \tag{2.4}$$

$$j = \sigma E, \tag{2.5}$$

$$D = \varepsilon E, \tag{2.6}$$

$$B = \mu H, \tag{2.7}$$

where the symbols in bold are vector quantities. In anisotropic media, Equations 2.1 through 2.7 become much more complicated with σ, ε, and μ being tensor rather than scalar quantities.

Anisotropic media are covered by Yeh [1] and Hodgkinson and Wu [2]. They are discussed quite briefly in Chapter 15.

The International System of Units (SI) is used as far as possible throughout this book. Table 2.1 shows the definitions of the quantities in the equations together with the appropriate SI units.

To the equations, we can add

$$\varepsilon = \varepsilon_r \varepsilon_0, \tag{2.8}$$

$$\mu = \mu_r \mu_0, \tag{2.9}$$

$$\varepsilon_0 = 1/(\mu_0 c^2), \tag{2.10}$$

where ε_0 and μ_0 are the permittivity and permeability of free space, respectively. ε_r and μ_r are the relative permittivity and permeability, respectively, and c is a constant that can be identified as the velocity of light in free space. ε_0, μ_0, and c are important constants, the values of which are given in Table 2.2.

In the normal way the parameters in Equations 2.8 through 2.10 do not depend on either E or H, and so the phenomena are linear. Note that there is a branch of optics dealing with non-linear effects, but the electromagnetic power density necessary for the production of such effects is usually enormous, and the materials, rather special. Such effects are outside the scope of this book.

Linear implies that the response to the sum of a set of stimuli is the sum of the responses to each stimulus separately. Thus, we can divide any arbitrary electromagnetic wave into components that can be separately considered. The usefulness of this approach lies in the fact that what we call a harmonic wave, that is a wave with a sine or cosine profile, propagates through any dispersive

TABLE 2.1

Electromagnetic Parameters

Symbol	Physical Quantity	SI Unit	Symbol for SI Unit
E	Electric field strength	Volts per meter	V/m
D	Electric displacement	Coulombs per square meter	C/m^2
H	Magnetic field strength	Amperes per meter	A/m
j	Electric current density	Amperes per square meter	A/m^2
B	Magnetic flux density or magnetic induction	Tesla	T
ρ	Electric charge density	Coulombs per cubic meter	C/m^3
σ	Electric conductivity	Siemens per meter	S/m
μ	Permeability	Henrys per meter	H/m
ε	Permittivity	Farads per meter	F/m

TABLE 2.2

Physical Constants

Symbol	Physical Quantity	Value
c	Velocity of light in free space	2.99792458×10^8 m/s
μ_0	Permeability of free space	$4\pi \times 10^{-7}$ H/m
ε_0	Permittivity of free space $= 1/(\mu_0 c^2)$	$8.854187817 \times 10^{-12}$ F/m

medium with no change of frequency and, therefore, retains its shape and has a precise velocity. There is a complete body of theory, known as Fourier, that permits a profile to be broken down into a set of sine and/or cosine functions. Thus, we use the harmonic wave as our basic component, and the collection of harmonic components that makes up our primary wave is known as its spectrum. We are quite used to breaking any light input into its spectrum and following the spectral components through the system separately, and this is the way we will normally operate in this book. In our theoretical analysis, therefore, we will concentrate on a single, general, spectral component, that is a harmonic wave, and we will usually derive what is known as the spectral performance of our coatings. Also we will tend to use the simplest type of harmonic wave, the linearly polarized, plane, harmonic wave.

The following analysis is brief and incomplete. For a full, rigorous treatment of the electromagnetic field equations, the reader is referred to Born and Wolf [3].

First, we assume an absence of space charge so that ρ is zero. This implies

$$\mathrm{div}\boldsymbol{D} = \varepsilon(\nabla \cdot \boldsymbol{E}) = 0, \tag{2.11}$$

and solving for $\boldsymbol{E}$,

$$\nabla \times (\nabla \times \boldsymbol{E}) = \nabla(\nabla \cdot \boldsymbol{E}) - \nabla^2 \boldsymbol{E} = -\mu \frac{\partial}{\partial t}(\nabla \times \boldsymbol{H}) = -\mu\sigma \frac{\partial \boldsymbol{E}}{\partial t} - \mu\varepsilon \frac{\partial^2 \boldsymbol{E}}{\partial t^2}, \tag{2.12}$$

i.e.,

$$\nabla^2 \boldsymbol{E} = \varepsilon\mu \frac{\partial^2 \boldsymbol{E}}{\partial t^2} + \mu\sigma \frac{\partial \boldsymbol{E}}{\partial t}. \tag{2.13}$$

A similar expression holds for $\boldsymbol{H}$.

First of all, we look for a solution of Equation 2.13 in the form of a linearly polarized plane harmonic wave (or *plane polarized*, a term meaning the same as linearly polarized), and we choose the complex form of this wave, the physical meaning being associated with either the real or the imaginary part of the expression:

$$\boldsymbol{E} = \boldsymbol{\mathcal{E}} \exp[i\omega(t - z/v)] \tag{2.14}$$

represents such a wave propagating along the z-axis with velocity v. $\boldsymbol{\mathcal{E}}$ is the vector amplitude and ω is the angular frequency of this wave. Note that since we are dealing with linear phenomena, ω is invariant as the wave propagates through media with differing properties. The advantage of the complex form of the wave is that phase changes can be dealt with very readily by including them in a complex amplitude. If we include a relative phase φ in Equation 2.14, then it becomes

$$\boldsymbol{E} = \boldsymbol{\mathcal{E}} \exp[i\{\omega(t - z/v) + \varphi\}] = \boldsymbol{\mathcal{E}} \exp(i\varphi) \exp[i\omega(t - z/v)], \tag{2.15}$$

where $\boldsymbol{\mathcal{E}} \exp(i\varphi)$ is the complex vector amplitude. The complex scalar amplitude is given by $\mathcal{E} \exp(i\varphi)$, where $\mathcal{E} = |\boldsymbol{\mathcal{E}}|$. Equation 2.15, which has phase φ relative to Equation 2.14, is simply Equation 2.14 with the amplitude replaced by the complex amplitude.

In Equation 2.14, we chose to place the time variable first and the spatial variable second in the argument of the exponential. This is a convention, because we could have chosen the alternative of the spatial variable first. However, to reverse the direction of the wave in this convention, we simply change the minus sign to a plus, reversing the spatial direction. In the alternative convention, it is tempting to reverse the wave once again by changing the sign from minus to plus, but that would reverse the time axis, not the spatial direction. We shall stick to the convention in Equation 2.14 throughout this book.

For Equation 2.14 to be a solution of Equation 2.13, it is necessary that

$$\omega^2/v^2 = \omega^2 \varepsilon\mu - i\omega\mu\sigma. \tag{2.16}$$

In a vacuum, we have $\sigma = 0$ and $v = c$, so that from Equation 2.16,

$$c^2 = 1/\varepsilon_0\mu_0, \tag{2.17}$$

which is identical to Equation 2.10. By multiplying Equation 2.15 by Equation 2.17 and dividing through by ω^2, we obtain

$$\frac{c^2}{v^2} = \frac{\varepsilon\mu}{\varepsilon_0\mu_0} - i\frac{\mu\sigma}{\omega\varepsilon_0\mu_0},$$

where c/v is clearly a dimensionless parameter of the medium, which we denote by N:

$$N^2 = \varepsilon_r\mu_r - i\frac{\mu_r\sigma}{\omega\varepsilon_0}. \tag{2.18}$$

This implies that N is of the form

$$N = c/v = n - ik. \tag{2.19}$$

There are two possible values of N from Equation 2.18, but for physical reasons, we choose that which gives a positive value of n. N is known as the complex refractive index; n, as the real part of the refractive index (or, often simply, as the refractive index, because N is real in an ideal dielectric material), and k is known as the extinction coefficient.

If the various parameters are real (not always the case), then from Equations 2.18 and 2.19,

$$n^2 - k^2 = \varepsilon_r\mu_r, \tag{2.20}$$

$$2nk = \frac{\mu_r\sigma}{\omega\varepsilon_0}. \tag{2.21}$$

Equation 2.14 can now be written as

$$E = \mathcal{E}\exp\{i[\omega t - (2\pi N/\lambda)z]\}, \tag{2.22}$$

where we have introduced the wavelength in free space $\lambda \,(= 2\pi c/\omega)$.

Substituting $n - ik$ for N in Equation 2.22 gives

$$E = \mathcal{E}\exp[-(2\pi k/\lambda)z]\exp\{i[\omega t - (2\pi n/\lambda)z]\}, \tag{2.23}$$

and the significance of k emerges as being a measure of loss in the medium. The distance $\lambda/(2\pi k)$ is that in which the amplitude of the wave falls to $1/e$ of its original value. The way in which the power carried by the wave falls off will be considered shortly.

In passing, we also note that, provided we continue dealing with a harmonic wave, Equation 2.18 can also be written as

$$N^2 = \mu_r\left(\varepsilon_r - i\frac{\sigma}{\omega\varepsilon_0}\right) = \hat{\varepsilon}_r\mu_r, \tag{2.24}$$

where $\hat{\varepsilon}_r$ is not the same as ε_r but contains the entire contents of the bracketed quantity and, of course, is complex. This complex permittivity is a function of frequency. It is much used because it avoids the complication in theoretical studies of the conductivity, which it simply contains. Unfortunately, it is normally consistent with the convention, opposite to our current one, which treats the complex index as $n + ik$.

The change in phase produced by a traversal of distance z in the medium is the same as that produced by a distance nz in a vacuum. Because of this, nz is known as the optical distance, as distinct from the physical or geometrical distance. Generally, in thin-film optics, one is more interested in optical distances and optical thicknesses than in physical ones.

Since $\mathcal{E}$ is constant, Equation 2.22 represents a linearly polarized plane wave propagating along the z-axis. For a similar wave propagating in a direction given by direction cosines (α, β, γ), the expression becomes

$$E = \mathcal{E}\exp\{i[\omega t - (2\pi N/\lambda)(\alpha x + \beta y + \gamma z)]\}. \tag{2.25}$$

This is the simplest type of wave in an absorbing medium. In an assembly of absorbing thin films, we shall see that we are occasionally forced to adopt a slightly more complicated expression for the wave.

There are some important relationships for this type of wave which can be derived from Maxwell's equations. Let the direction of propagation of the wave be given by unit vector $\hat{s}$ where

$$\hat{s} = \alpha i + \beta j + \gamma k$$

and where i, j, and k are unit vectors along the x, y, and z axes, respectively. From Equation 2.25, we have

$$\partial E/\partial t = i\omega E,$$

and from Equations 2.1, 2.5, and 2.6,

$$\mathrm{curl}H = \sigma E + \varepsilon\,\partial E/\partial t$$

$$= (\sigma + i\omega\varepsilon)E$$

$$= i\frac{\omega N^2}{c^2\mu}E.$$

Now

$$\mathrm{curl} = \left(\frac{\partial}{\partial x}i + \frac{\partial}{\partial y}j + \frac{\partial}{\partial z}k\right)\times,$$

where $\times$ denotes the vector product. But

$$\frac{\partial}{\partial x} = -i\frac{2\pi N}{\lambda}\alpha = -i\frac{\omega N}{c}\alpha,$$

$$\frac{\partial}{\partial y} = -i\frac{\omega N}{c}\beta, \quad \frac{\partial}{\partial z} = -i\frac{\omega N}{c}\gamma,$$

so that

$$\mathrm{curl}H = -i\frac{\omega N}{c}(\hat{s}\times H).$$

Then,

$$-i\frac{\omega N}{c}(\hat{s} \times H) = i\frac{\omega N^2}{c^2\mu}E,$$

i.e.,

$$(\hat{s} \times H) = -\frac{N}{c\mu}E, \tag{2.26}$$

and similarly

$$\frac{N}{c\mu}(\hat{s} \times E) = H. \tag{2.27}$$

For this type of wave, therefore, E, H, and $\hat{s}$ are mutually perpendicular and form a right-handed set. The quantity $N/c\mu$ has the dimensions of an admittance and is known as the characteristic optical admittance of the medium, written as y. In free space, it can be readily shown that the optical admittance is given by

$$\mathscr{Y} = (\varepsilon_0/\mu_0)^{1/2} = 2.6544 \times 10^{-3}\,\text{S}. \tag{2.28}$$

Now

$$\mu = \mu_r\mu_0. \tag{2.29}$$

Direct magnetic interactions at optical frequencies are vanishingly small so that μ_r is effectively unity. Thus we can write

$$y = N\mathscr{Y}, \tag{2.30}$$

and

$$H = y(\hat{s} \times E) = N\mathscr{Y}(\hat{s} \times E). \tag{2.31}$$

We recall that since μ_r is unity, we can write Equation 2.8 as

$$N^2 = \varepsilon_r - i\frac{\sigma}{\omega\varepsilon_0} = \hat{\varepsilon}_r, \tag{2.32}$$

where $\hat{\varepsilon}_r$ is a complex relative permittivity. We have already mentioned this in Equation 2.24 and noted that it can usefully simplify related analytical expressions and so is frequently employed. We emphasize once again that, in the literature, the sign convention commonly used puts a plus sign in Equation 2.32 rather than the minus. The complex permittivity is also called the dielectric function, and the complex relative permittivity, the relative dielectric function.

2.1.1 Poynting Vector

An important feature of electromagnetic radiation is that it is a form of energy transport, and it is the energy associated with the wave that is normally observed. The instantaneous rate of flow of energy across a unit area is given by the Poynting vector:

$$S = E \times H. \tag{2.33}$$

The direction of the vector is the direction of energy flow.

When we add or subtract complex numbers, or multiply them by a real number, the real parts and imaginary parts remain independent. Such operations are known as linear. Interference calculations involve adding the waves, and so we can happily use the complex wave with all its advantages in such calculations. The multiplication of two complex numbers, however, mixes the real and imaginary parts in the result. Such operations are known as nonlinear, and we are unable to directly use the complex form of the wave in them. The Poynting expression is a nonlinear one (E is *multiplied* by H), and so we have a problem with the complex form of the wave. Either the real or the imaginary part of the wave expression should be used. The real sine or cosine form of the wave implies its square in the result, and so the instantaneous value of the Poynting vector must oscillate at twice the frequency of the wave. We turn our attention to the mean value because it is the mean that is significant in our measurements. This is defined as the irradiance or, in the older systems of units, intensity. (Beware. Intensity is defined differently in the SI system as the power per unit solid angle from a point source.) In the SI system of units, irradiance is measured in watts per square meter. An unfortunate feature of the SI system, for our purposes, is that the symbol for irradiance is E. The use of this symbol would make it very difficult for us to distinguish between irradiance and electric field. Since both are extremely important in almost everything we do, we must be able to differentiate between them, and so we adopt a nonstandard symbol I for irradiance (which, unfortunately, is the SI symbol for intensity). The mean of the Poynting vector involves integrating the real expression over a cycle, but the complex form of the wave actually comes to our rescue. For a harmonic wave, we find that we can derive a very attractive and simple expression for the irradiance using the complex form of the wave and thus avoiding the integration. This is

$$I = \frac{1}{2} \, \mathrm{Re}(\boldsymbol{E} \times \boldsymbol{H}^*), \qquad (2.34)$$

where * denotes complex conjugate. It should be emphasized that the complex form *must* be used in Equation 2.34. The irradiance $\boldsymbol{I}$ is written in Equation 2.34 as a vector quantity, when it has the same direction as the flow of energy of the wave. The more usual scalar irradiance I is simply the magnitude of $\boldsymbol{I}$. Since $\boldsymbol{E}$ and $\boldsymbol{H}$ are perpendicular, Equation 2.34 can be written as

$$I = \frac{1}{2} \, \mathrm{Re}(EH^*), \qquad (2.35)$$

where E and H are the scalar magnitudes.

It is important to note that for the net irradiance, the electric and magnetic vectors in Equation 2.34 should be the total resultant fields due to all the waves involved. This is implicit in the derivation of the Poynting vector expression. We will return to this point when calculating reflectance and transmittance.

For a single, homogeneous, harmonic wave of the form shown in Equation 2.25,

$$\boldsymbol{H} = y(\hat{s} \times \boldsymbol{E}),$$

so that

$$I = \mathrm{Re}\left(\frac{1}{2} y EE^* \hat{s}\right) = \frac{1}{2} n \mathcal{Y} EE^* \hat{s}. \qquad (2.36)$$

Now, from Equation 2.25, the magnitude of E is given by

$$E = \mathcal{E} \exp\{i \, [\omega t - (2\pi[n - ik]/\lambda)(\alpha x + \beta y + \gamma z)]\}$$
$$= \mathcal{E} \exp[-(2\pi k/\lambda)(\alpha x + \beta y + \gamma z)] \exp\{i[\omega t - (2\pi n/\lambda)(\alpha x + \beta y + \gamma z)]\},$$

implying that

$$EE^* = \mathcal{E}\mathcal{E}^* \exp[-(4\pi k/\lambda)(\alpha x + \beta y + \gamma z)]$$

and

$$I = \frac{1}{2}n\mathcal{Y}|\mathcal{E}|^2 \exp[-(4\pi k/\lambda)(\alpha x + \beta y + \gamma z)].$$

The expression $(\alpha x + \beta y + \gamma z)$ is simply the distance along the direction of propagation, and thus, the irradiance drops to $1/e$ of its initial value in a distance given by $\lambda/4\pi k$. The inverse of this distance is defined as the absorption coefficient α; that is,

$$\alpha = 4\pi k/\lambda. \tag{2.37}$$

The absorption coefficient α should not be confused with the direction cosine. However,

$$|\mathcal{E}| \exp[-(2\pi k/\lambda)(\alpha x + \beta y + \gamma z)]$$

is really the amplitude of the wave at the point (x, y, z) so that a much simpler way of writing the expression for irradiance is

$$I = \frac{1}{2}n\mathcal{Y}(\text{amplitude})^2 \tag{2.38}$$

or

$$I \propto n \times (\text{amplitude})^2. \tag{2.39}$$

This expression is a better form than the more usual

$$I \propto (\text{amplitude})^2. \tag{2.40}$$

The expression will frequently be used for comparing irradiances, in calculating reflectance or transmittance, for example, and if the media in which the two waves are propagating are of different index, then errors will occur unless n is included as mentioned earlier.

2.2 Notation

Throughout the book, we will be dealing with assemblies of elements involving different materials and where their order is important. In many cases, these will consist of an incident medium and a substrate separated by a number of thin films. To identify these various entities and to make their order unambiguously clear, we shall endeavor to use a consistent notation involving suffices. The incident medium will have the subscript 0 as in y_0. The substrate, or emergent medium, will usually have the subscript m, as in y_m, although occasionally the subscript sub, as in y_{sub}. Layers will be numbered sequentially from the incident medium to the substrate or emergent medium so that the layer next to the incident medium has the subscript 1; that next to it, 2; and so on, as in y_1, y_2, and so on. Layer $q - 1$ will be next to layer q and will be situated on the side toward the incident medium (Figure 2.1).

There will be occasions where we will want to identify interfaces rather than layers. To differentiate between an interface and a layer, we will often use letters, usually in the order of the alphabet, to refer to interfaces, as a, b, c, etc., but numbers will be used to refer to layers.

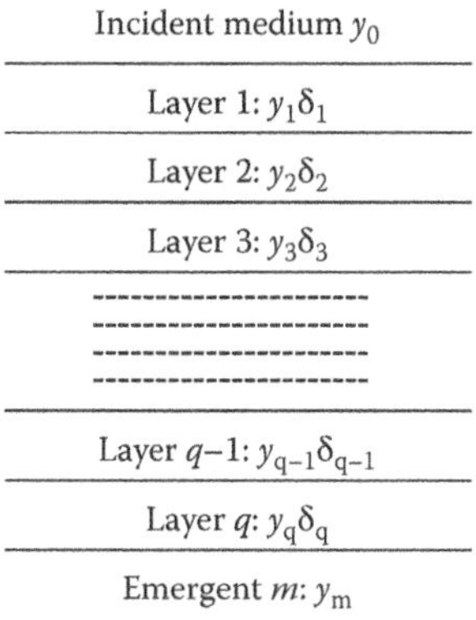

Incident medium y_0

Layer 1: $y_1\delta_1$

Layer 2: $y_2\delta_2$

Layer 3: $y_3\delta_3$

Layer $q-1$: $y_{q-1}\delta_{q-1}$

Layer q: $y_q\delta_q$

Emergent m: y_m

FIGURE 2.1
Numbering system for designs will normally follow the arrangement shown. A lower number will normally mean that the element is closer to the incident medium. The quantity δ is the phase thickness of the appropriate layer and will be defined shortly.

2.3 Simple Boundary

Thin-film filters usually consist of a number of boundaries between various homogeneous media, and it is the effect of these boundaries on an incident wave that we will wish to calculate. A single boundary is the simplest case. First of all, we consider absorption-free media, i.e., $k = 0$. The arrangement is sketched in Figure 2.2. A plane harmonic wave is incident on a plane surface, separating the incident medium from a second, or emergent, medium. The plane containing the normal to the surface and the direction of propagation of the incident wave is known as the plane of incidence, and the plane of the sketch corresponds to this plane. We take the z-axis as the normal into the surface in the sense of the incident wave and the x-axis as normal to it and on the plane of incidence. At a boundary, the tangential components of E and H, that is, the components along the

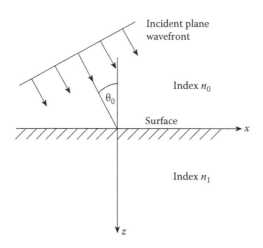

FIGURE 2.2
Plane wavefront incident on a single surface.

boundary, are continuous across it because there is no mechanism that will change them. This boundary condition is fundamental in our thin-film theory.

The first problem we have is that the boundary conditions are incompatible with a simple traversal of the boundary by the incident wave. The discontinuity in the characteristic admittance implies a power discontinuity impossible if the wave simply crosses the boundary with no other consequence. This difficulty is immediately solved by introducing a reflected wave in the incident medium, and this, of course, is directly in line with our experience. Our objective then becomes the calculation of the relative parameters of the three waves, incident, reflected, and transmitted. However, this introduces a further complication. We will use the boundary conditions to construct a set of equations from which we will extract the required relations. The complication is that the reflected wave will certainly be traveling in a different sense from the others so that there will be differences in the phase factors that will considerably complicate the calculations. We can enormously help ourselves by defining the boundary by $z = 0$, eliminating the z term from the phase factors at the boundary. Then the tangential components must be continuous for all values of x, y, and t.

We therefore have three harmonic waves, an incident, a reflected, and a transmitted wave. The incident wave, in the plane of incidence, has direction cosines $(\cos \vartheta_0, 0, \sin \vartheta_0)$. Let the direction cosines of the $\hat{s}$ vectors of the transmitted and reflected waves be given by $(\alpha_t, \beta_t, \gamma_t)$ and $(\alpha_r, \beta_r, \gamma_r)$, respectively. We can therefore write the phase factors in the following forms:

- Incident wave: $\exp\{i[\omega t - (2\pi n_0/\lambda_i)(x \sin \vartheta_0 + z \cos \vartheta_0)]\}$
- Reflected wave: $\exp\{i[\omega t - (2\pi n_0/\lambda_r)(\alpha_r x + \beta_r y + \gamma_r z)]\}$
- Transmitted wave: $\exp\{i[\omega t - (2\pi n_1/\lambda_t)(\alpha_t x + \beta_t y + \gamma_t z)]\}$

The relative phases of these waves are included in the complex amplitudes. For waves with these phase factors to satisfy the boundary conditions for all x, y, and t at $z = 0$ implies that the coefficients of these variables must be separately identically equal. Had we not already known that there would be no change in frequency, this would have confirmed it. Since the frequencies are constant, so too will be the free space wavelengths. Next,

$$0 \equiv n_0 \beta_r \equiv n_1 \beta_t; \tag{2.41}$$

that is, the directions of the reflected and transmitted or refracted beams are confined to the plane of incidence. This, in turn, means that the direction cosines of the reflected and transmitted waves are of the forms

$$\alpha = \sin \vartheta, \quad \gamma = \cos \vartheta. \tag{2.42}$$

Also,

$$n_0 \sin \vartheta_0 \equiv n_0 \alpha_r \equiv n_1 \alpha_t,$$

so that if the angles of reflection and refraction are ϑ_r and ϑ_t, respectively, then

$$\vartheta_0 = \vartheta_r, \tag{2.43}$$

that is, the angle of reflection equals the angle of incidence, and

$$n_0 \sin \vartheta_0 = n_1 \sin \vartheta_t.$$

The result appears more symmetrical if we replace ϑ_t with ϑ_1, giving

$$n_0 \sin \vartheta_0 = n_1 \sin \vartheta_1. \tag{2.44}$$

This is the familiar relationship known as Snell's law. γ_r and γ_t are then given either by Equation 2.42 or by

$$\alpha_r^2 + \gamma_r^2 = 1 \quad \text{and} \quad \alpha_t^2 + \gamma_t^2 = 1. \tag{2.45}$$

Note that for the reflected beam, we must choose the *negative* root of Equation 2.45 so that the beam will propagate in the correct direction.

2.3.1 Normal Incidence in Absorption-Free Media

Let us limit our initial discussion to normal incidence, and let the incident wave be a linearly polarized plane harmonic wave. The coordinate axes are shown in Figure 2.3. The xy plane is the plane of the boundary. We can take the incident as propagating along the z-axis with the positive direction of the E vector along the x-axis. Then the positive direction of the H vector will be the y-axis. It is clear that the only waves that satisfy the boundary conditions are linearly polarized in the same sense as the incident wave.

A quoted phase difference between two waves travelling in the same direction is immediately meaningful. A phase difference between two waves travelling in opposite directions is absolutely meaningless, unless a reference plane at which the phase difference is measured is first defined. This is simply because the phase difference between oppositely propagating waves of the same frequency has a term ($\pm 4\pi n s / \lambda$) in it, where s is a distance measured along the direction of propagation. Before proceeding further, therefore, we need to define the reference point for measurements of relative phase between the oppositely propagating beams. Since we have already used the device of defining the boundary as $z = 0$, we can continue this idea and define the boundary as that plane where the reflected phase shift should be defined.

Then there is another problem. The waves have electric and magnetic fields that with the direction of propagation form right-handed sets. Since the direction of propagation is reversed in the reflected beam, the orientation of electric and magnetic fields cannot remain the same as that in the incident beam; otherwise, we would no longer have a right-handed set. We need to decide on

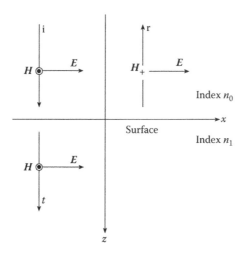

FIGURE 2.3
Convention defining positive directions of the electric and magnetic vectors for reflection and transmission at an interface at normal incidence.

how we are going to handle this. Since the electric field is the one that is most important from the point of view of interaction with matter, we will define our directions with respect to it.

The matter of phase references and electric field directions are what we call conventions because we do have complete freedom of choice, and any self-consistent arrangement is possible. We must simply ensure that once we have made our choice, we adhere to it. A good rule, however, is to never make things difficult when we can make them easy, and so we will normally choose the rule that is most convenient and least complicated. We define the positive direction of E along the x-axis for all the beams that are involved. Because of this choice, the positive direction of the magnetic vector will be along the y-axis for the incident and transmitted waves, but along the negative direction of the y-axis for the reflected wave.

We now consider the boundary conditions. Since we have already made sure that the phase factors are satisfactory, we can cancel out the $i\omega t$ terms and have only to consider the amplitudes, and we will be including any phase changes in these:

1. Electric vector continuous across the boundary

$$\mathcal{E}_i + \mathcal{E}_r = \mathcal{E}_t. \tag{2.46}$$

2. Magnetic vector continuous across the boundary

$$\mathcal{H}_i - \mathcal{H}_r = \mathcal{H}_t,$$

where we must use a minus sign because of our convention for positive directions. The relationship between magnetic and electric fields through the characteristic admittance, gives

$$y_0\mathcal{E}_i - y_0\mathcal{E}_r = y_1\mathcal{E}_t. \tag{2.47}$$

This can also be derived using the vector relationship in Equations 2.31 and 2.46. We can eliminate $\mathcal{E}_t$ to give

$$y_1(\mathcal{E}_i + \mathcal{E}_r) = y_0(\mathcal{E}_i - \mathcal{E}_r),$$

i.e.,

$$\frac{\mathcal{E}_r}{\mathcal{E}_i} = \frac{y_0 - y_1}{y_0 + y_1} = \frac{n_0 - n_1}{n_0 + n_1}, \tag{2.48}$$

the second part of the relationship being correct only because at optical frequencies, we can write

$$y = n\mathcal{Y}.$$

Similarly, eliminating $\mathcal{E}_r$,

$$\frac{\mathcal{E}_t}{\mathcal{E}_i} = \frac{2y_0}{y_0 + y_1} = \frac{2n_0}{n_0 + n_1}. \tag{2.49}$$

These quantities are called the amplitude reflection and transmission coefficients and are denoted by ρ and τ, respectively. Thus

$$\rho = \frac{y_0 - y_1}{y_0 + y_1} = \frac{n_0 - n_1}{n_0 + n_1}, \tag{2.50}$$

$$\tau = \frac{2y_0}{y_0 + y_1} = \frac{2n_0}{n_0 + n_1}. \tag{2.51}$$

We are still assuming zero for k, and so in this particular case, all y are real, and these two derived quantities are therefore real. τ is always a positive real number, indicating that according to our phase convention, there is no phase shift between the incident and transmitted beams at the interface. The behavior of ρ indicates that there will be no phase shift between the incident and reflected beams at the interface provided $n_0 > n_1$, but that if $n_0 < n_1$, there will be a phase change of π because the value of ρ becomes negative.

We now examine the energy balance at the boundary. The total tangential components of electric and magnetic field are not only continuous across the boundary, but also, since the boundary is of zero thickness, it can neither supply energy to nor extract energy from the various waves. On both counts, the Poynting vector, that is the net irradiance, will be continuous across the boundary, so that we can write

$$\text{Net irradiance} = \text{Re}\left[\frac{1}{2}(\mathcal{E}_i + \mathcal{E}_r)(y_0\mathcal{E}_i - y_0\mathcal{E}_r)^*\right]$$

$$= \text{Re}\left[\frac{1}{2}\mathcal{E}_i(y_1\mathcal{E}_t)^*\right]$$

(using $\text{Re}\left(\frac{1}{2}E \times H^*\right)$ and Equations 2.46 and 2.47). Now

$$\mathcal{E}_r = \rho\mathcal{E}_i \quad \text{and} \quad \mathcal{E}_t = \tau\mathcal{E}_i,$$

i.e.,

$$\text{Net irradiance} = \frac{1}{2}y_0\mathcal{E}_i\mathcal{E}_i^*(1 - \rho^2) = \frac{1}{2}y_0\mathcal{E}_i\mathcal{E}_i^*(y_1/y_0)\tau^2. \tag{2.52}$$

We recognize $(1/2)y_0\mathcal{E}_i\mathcal{E}_i^*$ as the irradiance of the incident beam I_i. We can identify $\rho^2(1/2)y_0 \mathcal{E}_i\mathcal{E}_i^* = \rho^2 I_i$ as the irradiance of the reflected beam I_r and $(y_1/y_0) \times \tau^2(1/2)y_0\mathcal{E}_i\mathcal{E}_i^* = (y_1/y_0)\tau^2 I_i$ as the irradiance of the transmitted beam I_t. We define the reflectance R as the ratio of the reflected and incident irradiances and the transmittance T as the ratio of the transmitted and incident irradiances. Then,

$$T = \frac{I_t}{I_i} = \frac{y_1}{y_0}\tau^2 = \frac{4y_0y_1}{(y_0 + y_1)^2} = \frac{4n_0n_1}{(n_0 + n_1)^2},$$

$$R = \frac{I_r}{I_i} = \rho^2 = \left(\frac{y_0 - y_1}{y_0 + y_1}\right)^2 = \left(\frac{n_0 - n_1}{n_0 + n_1}\right)^2. \tag{2.53}$$

From Equation 2.52, we have, using Equation 2.53,

$$(1 - R) = T. \tag{2.54}$$

Equations 2.52 through 2.54 are therefore consistent with our ideas of splitting the irradiances into incident, reflected, and transmitted irradiances which can be treated as separate waves, the energy flow into the second medium being simply the difference of the incident and reflected irradiances. Remember that all this, so far, assumes that there is no absorption. We shall shortly see that the situation changes when absorption is present.

2.3.2 Oblique Incidence in Absorption-Free Media

Now let us consider oblique incidence, still retaining our absorption-free media. For any general direction of the vector amplitude of the incident wave, we quickly find that the application of the boundary conditions leads us into complicated and difficult expressions for the vector amplitudes of the reflected and transmitted waves. Fortunately, there are two orientations of the incident

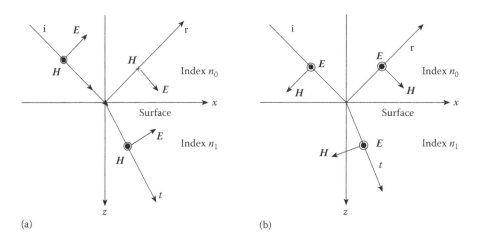

FIGURE 2.4
(a) Convention defining the positive directions of the electric and magnetic vectors for p-polarized light (TM waves).
(b) Convention defining the positive directions of the electric and magnetic vectors for s-polarized light (TE waves).

wave, which lead to reasonably straightforward calculations, the vector electrical amplitudes aligned in the plane of incidence (i.e., the xy plane in Figure 2.2) and the vector electrical amplitudes aligned normal to the plane of incidence (i.e., parallel to the y-axis in Figure 2.2). In each of these cases, the orientations of the transmitted and reflected vector amplitudes are the same as for the incident wave. Any incident wave of arbitrary polarization can therefore be split into two components having these simple orientations. The transmitted and reflected components can be separately calculated for each orientation and then combined to yield the resultant. Since, therefore, it is necessary to consider two orientations only, they have been given special names. A wave with the electric vector in the plane of incidence is known as p-polarized or, sometimes, as TM (for transverse magnetic), and a wave with the electric vector normal to the plane of incidence, as s-polarized or, sometimes, TE (for transverse electric). p and s are derived from the German *parallel* and *senkrecht* (perpendicular). Before we can actually proceed to the calculation of the reflected and transmitted amplitudes, we must choose the various reference directions of the vectors from which any phase differences will be calculated. We have, once again, complete freedom of choice, but once we have established the convention, we must adhere to it, just as in the normal incidence case. The conventions which we will use in this book are illustrated in Figure 2.4. They have been chosen to be compatible with those for normal incidence already established. In some works, an opposite convention for the p-polarized reflected beam has been adopted, but this leads to an incompatibility with results derived for normal incidence, and we prefer to avoid this situation. Note that for reasons connected with consistency of reference directions for elliptically polarized light, the convention normal in ellipsometric calculations is opposite to that of Figure 2.4 for reflected p-polarized light. When ellipsometric parameters (in reflection) are compared with the results of the expressions we shall use, it will usually be necessary to introduce a shift of 180° in the p-polarized reflected results.

We can now apply the boundary conditions. Since we have already ensured that the phase factors will be correct, we need only consider the vector amplitudes.

2.3.2.1 p-Polarized Light

1. Electric component parallel to the boundary; continuous across it:

$$\mathcal{E}_i \cos \vartheta_0 + \mathcal{E}_r \cos \vartheta_0 = \mathcal{E}_t \cos \vartheta_1. \tag{2.55}$$

2. Magnetic component parallel to the boundary; continuous across it: Here we need to calculate the magnetic vector amplitudes, and we can do this by either using Equation 2.31 to operate on Equation 2.55 directly, or, since the magnetic vectors are already parallel to the boundary, using Figure 2.4 and then converting, since $\mathcal{H} = y\mathcal{E}$:

$$y_0 \mathcal{E}_i - y_0 \mathcal{E}_r = y_1 \mathcal{E}_t. \tag{2.56}$$

At first sight, it seems logical just to eliminate first $\mathcal{E}_t$ and then $\mathcal{E}_r$ from these two equations to obtain $\mathcal{E}_r / \mathcal{E}_i$ and $\mathcal{E}_t / \mathcal{E}_i$:

$$\frac{\mathcal{E}_r}{\mathcal{E}_i} = \frac{y_0 \cos \vartheta_1 - y_1 \cos \vartheta_0}{y_0 \cos \vartheta_1 + y_1 \cos \vartheta_0},$$

$$\frac{\mathcal{E}_t}{\mathcal{E}_i} = \frac{2y_0 \cos \vartheta_0}{y_0 \cos \vartheta_1 + y_1 \cos \vartheta_0}, \tag{2.57}$$

and then simply to set

$$R = \left(\frac{\mathcal{E}_r}{\mathcal{E}_i}\right)^2 \quad \text{and} \quad T = \frac{y_1}{y_0}\left(\frac{\mathcal{E}_t}{\mathcal{E}_i}\right)^2,$$

but when we calculate the expressions which result, we find that $R + T \neq 1$. In fact, there is no mistake in the calculations. We have computed the irradiances measured along the direction of propagation of the waves, and the transmitted wave is inclined at an angle which differs from that of the incident wave. This leaves us with the problem that adopting these definitions will involve the rejection of the $(R + T = 1)$ rule.

We could correct this situation by modifying the definition of T to include this angular dependence, but an alternative, preferable, and generally adopted approach is to use the components of the energy flows that are normal to the boundary. The E and H vectors that are involved in these calculations are then parallel to the boundary. Since these are those that directly enter into the boundary, it seems appropriate to concentrate on them when we are dealing with the amplitudes of the waves. Note that reflectance and transmittance defined for infinite plane waves in terms of normal flows of irradiance are absolutely consistent with reflectance and transmittance defined in terms of the ratios of total beam power when using confined beams such as the output from a laser.

The thin-film approach to all this, then, is to use the components of E and H parallel to the boundary, what are called the tangential components, in the expressions ρ and τ that involve amplitudes. Note that the normal approach in other areas of optics is to use the full components of E and H in amplitude expressions but to use the components of irradiance in reflectance and transmittance. The amplitude coefficients are then known as the Fresnel coefficients. The thin-film coefficients are *not* the Fresnel coefficients except at normal incidence, although the only coefficient that actually has a different value is the amplitude transmission coefficient for p-polarization.

The tangential components of E and H, that is, the components parallel to the boundary, have already been calculated for use in Equations 2.55 and 2.56. However, it is convenient to introduce special symbols for them, E and $\mathcal{H}$.

Then, we can write

$$E_i = \mathcal{E}_i \cos \vartheta_0, \quad \mathcal{H}_i = \mathcal{H}_i = y_0 \mathcal{E}_i = \frac{y_0}{\cos \vartheta_0} E_i, \tag{2.58}$$

$$E_r = \mathcal{E}_r \cos \vartheta_0, \quad \mathcal{H}_r = \frac{y_0}{\cos \vartheta_0} E_r, \tag{2.59}$$

$$E_t = \mathcal{E}_t \cos \vartheta_1, \quad \mathcal{H}_t = \frac{y_1}{\cos \vartheta_1} E_t. \tag{2.60}$$

The orientations of these vectors are exactly the same as for normally incident light.

Equations 2.55 and 2.56 can then be written as follows:

1. Electric field parallel to the boundary

$$E_i + E_r = E_t.$$

2. Magnetic field parallel to the boundary

$$\frac{y_0}{\cos \vartheta_0} \mathcal{H}_i - \frac{y_0}{\cos \vartheta_0} \mathcal{H}_r = \frac{y_1}{\cos \vartheta_1} \mathcal{H}_t,$$

giving us, by a process exactly similar to that we have already used for normal incidence,

$$\rho_p = \frac{E_r}{E_i} = \left(\frac{y_0}{\cos \vartheta_0} - \frac{y_1}{\cos \vartheta_1} \right) \Big/ \left(\frac{y_0}{\cos \vartheta_0} + \frac{y_1}{\cos \vartheta_1} \right), \tag{2.61}$$

$$\tau_p = \frac{E_t}{E_i} = \left(\frac{2y_0}{\cos \vartheta_0} \right) \Big/ \left(\frac{y_0}{\cos \vartheta_0} + \frac{y_1}{\cos \vartheta_1} \right), \tag{2.62}$$

$$R_p = \left[\left(\frac{y_0}{\cos \vartheta_0} - \frac{y_1}{\cos \vartheta_1} \right) \Big/ \left(\frac{y_0}{\cos \vartheta_0} + \frac{y_1}{\cos \vartheta_1} \right) \right]^2, \tag{2.63}$$

$$T_p = \left(\frac{4y_0 y_1}{\cos \vartheta_0 \cos \vartheta_1} \right) \Big/ \left(\frac{y_0}{\cos \vartheta_0} + \frac{y_1}{\cos \vartheta_1} \right)^2, \tag{2.64}$$

where $y_0 = n_0 \mathcal{Y}$ and $y_1 = n_1 \mathcal{Y}$ and the ($R + T = 1$) rule is retained. The subscript p has been used in the expressions mentioned earlier to denote p-polarization.

It should be noted that the expression for τ_p is now different from that in Equation 2.57, the form of the Fresnel amplitude transmission coefficient. Fortunately, the reflection coefficients in Equations 2.57 and 2.63 are identical, and since much more use is made of reflection coefficients, confusion is rare.

2.3.2.2 s-Polarized Light

In the case of s-polarization, the amplitudes of the components of the waves parallel to the boundary are

$$E_i = \mathcal{E}_i, \quad \mathcal{H}_i = \mathcal{H}_i \cos \vartheta_0 = (y_0 \cos \vartheta_0) E_i,$$

$$E_r = \mathcal{E}_r, \quad \mathcal{H}_r = \mathcal{H}_r \cos \vartheta_0 = (y_0 \cos \vartheta_0) E_r,$$

$$E_t = \mathcal{E}_t, \quad \mathcal{H}_t = (y_1 \cos \vartheta_1) E_t,$$

and here we have again an orientation of the tangential components exactly as for normally incident light, and so a similar analysis leads to

$$\rho_s = \frac{E_r}{E_i} = (y_0 \cos \vartheta_0 - y_1 \cos \vartheta_1)/(y_0 \cos \vartheta_0 + y_1 \cos \vartheta_1), \tag{2.65}$$

$$\tau_s = \frac{E_t}{E_i} = (2y_0 \cos\vartheta_0)/(y_0 \cos\vartheta_0 + y_1 \cos\vartheta_1), \tag{2.66}$$

$$R_s = [(y_0 \cos\vartheta_0 - y_1 \cos\vartheta_1)/(y_0 \cos\vartheta_0 + y_1 \cos\vartheta_1)]^2, \tag{2.67}$$

$$T_s = (4y_0 \cos\vartheta_0 y_1 \cos\vartheta_1)/(y_0 \cos\vartheta_0 + y_1 \cos\vartheta_1)^2, \tag{2.68}$$

where once again $y_0 = n_0\mathcal{Y}$ and $y_1 = n_1\mathcal{Y}$ and the ($R + T = 1$) rule is retained. The subscript s is used in the preceding expressions to denote s-polarization.

2.3.3 Optical Admittance for Oblique Incidence

The expressions which we have derived so far have been in their traditional form (except for the use of the tangential components rather than the full vector amplitudes), and they involve the characteristic admittances of the various media or their refractive indices together with the admittance of free space $\mathcal{Y}$. However, the notation is becoming increasingly cumbersome and will appear even more so when we consider the behavior of thin films.

Equation 2.31 gives $H = y(\hat{s} \times E)$, where $y = N\mathcal{Y}$ is the optical admittance. We have found it convenient to deal with $\mathcal{E}$ and $\mathcal{H}$, the components of $\mathcal{E}$ and $\mathcal{H}$ parallel to the boundary, and so we introduce a tilted optical admittance η which connects E and $\mathcal{H}$ as

$$\eta = \frac{\mathcal{H}}{E}. \tag{2.69}$$

At normal incidence, $\eta = y = n\mathcal{Y}$ while at oblique incidence,

$$\eta_p = \frac{y}{\cos\vartheta} = \frac{n\mathcal{Y}}{\cos\vartheta}, \tag{2.70}$$

$$\eta_s = y\cos\vartheta = n\mathcal{Y}\cos\vartheta, \tag{2.71}$$

where the ϑ and the y in Equations 2.70 and 2.71 are those appropriate to the particular medium. In particular, Snell's law (Equation 2.44) must be used to calculate $\tilde{\vartheta}$.

Then, in all cases, we can write

$$\rho = \left(\frac{\eta_0 - \eta_1}{\eta_0 + \eta_1}\right), \quad \tau = \left(\frac{2\eta_0}{\eta_0 + \eta_1}\right), \tag{2.72}$$

$$R = \left(\frac{\eta_0 - \eta_1}{\eta_0 + \eta_1}\right)^2, \quad T = \frac{4\eta_0\eta_1}{(\eta_0 + \eta_1)^2}. \tag{2.73}$$

These expressions can be used to compute the variation of reflectance of simple boundaries between extended media. Examples are shown in Figure 2.5 of the variation of reflectance with the angle of incidence. In this case, there is no absorption in the material, and it can be seen that the reflectance for p-polarized light (TM) falls to zero at a definite angle. This particular angle is known as the Brewster angle and is of some importance. There are many applications where the windows of a cell must have close to zero reflection loss. When it can be arranged that the light will be linearly polarized, a plate tilted at the Brewster angle will be a good solution. The light that is reflected at the Brewster angle is also linearly polarized with electric vector normal to the plane of incidence. This affords a way of identifying the absolute direction of polarizers and analyzers—very difficult in any other way.

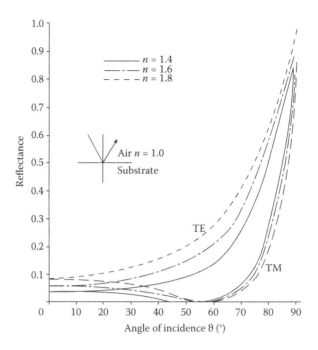

FIGURE 2.5
Variation of reflectance with angle of incidence for various values of refractive index. *TE* is *s*-polarization and *TM* is *p*-polarization.

The expression for the Brewster angle can be derived as follows. For the *p*-reflectance to be zero, from Equation 2.63,

$$\frac{y_0}{\cos\vartheta_0} = \frac{n_0\mathscr{Y}}{\cos\vartheta_0} = \frac{y_1}{\cos\vartheta_1} = \frac{n_1\mathscr{Y}}{\cos\vartheta_1}.$$

Snell's law gives another relationship between ϑ_0 and ϑ_1:

$$n_0\sin\vartheta_0 = n_1\sin\vartheta_1.$$

Eliminating ϑ_1 from these two equations gives an expression for ϑ_0:

$$\tan\vartheta_0 = n_1/n_0. \tag{2.74}$$

Note that this derivation depends on the relationship $y = n\mathscr{Y}$, valid at optical frequencies.

Figure 2.6 shows the variation of tilted admittance of a number of dielectric materials as a function of the angle of incidence in air. Note that the divergence of the two tilted admittances, the polarization splitting, becomes less as the index of refraction increases.

2.3.4 Normal Incidence in Absorbing Media

We must now examine the modifications necessary in our results in the presence of absorption. First, we consider the case of normal incidence and write

$$N_0 = n_0 - ik_0,$$

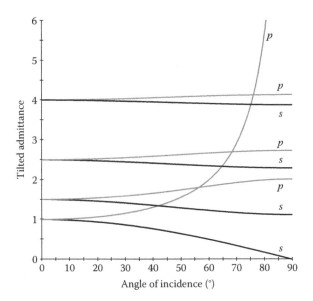

FIGURE 2.6
Tilted admittances of several dielectric (absorption-free) materials as a function of the angle of incidence in air.

$$N_1 = n_1 - ik_1,$$

$$y_0 = N_0\mathscr{Y} = (n_0 - ik_0)\mathscr{Y},$$

$$y_1 = N_1\mathscr{Y} = (n_1 - ik_1)\mathscr{Y}.$$

The analysis follows that for absorption-free media. The boundaries are, as before,

1. Electric vector continuous across the boundary

$$\mathcal{E}_i + \mathcal{E}_r = \mathcal{E}_t.$$

2. Magnetic vector continuous across the boundary

$$y_0\mathcal{E}_i - y_0\mathcal{E}_r = y_1\mathcal{E}_t,$$

and by eliminating first $\mathcal{E}_t$ and then $\mathcal{E}_r$, we obtain the expressions for the amplitude coefficients

$$\rho = \frac{\mathcal{E}_r}{\mathcal{E}_i} = \frac{y_0 - y_1}{y_0 + y_1} = \frac{(n_0 - ik_0)\mathscr{Y} - (n_1 - ik_1)\mathscr{Y}}{(n_0 - ik_0)\mathscr{Y} + (n_1 - ik_1)\mathscr{Y}} = \frac{(n_0 - n_1) - i(k_0 - k_1)}{(n_0 + n_1) - i(k_0 + k_1)}, \tag{2.75}$$

$$\tau = \frac{\mathcal{E}_t}{\mathcal{E}_i} = \frac{2y_0}{y_0 - y_1} = \frac{2(n_0 - ik_0)\mathscr{Y}}{(n_0 - ik_0)\mathscr{Y} + (n_1 - ik_1)\mathscr{Y}} = \frac{2(n_0 - ik_0)}{(n_0 + n_1) - i(k_0 + k_1)}. \tag{2.76}$$

Our troubles begin when we try to extend this to reflectance and transmittance. We remain at normal incidence. Following the method for the absorption-free case, we compute the Poynting vector at the boundary in each medium and equate the two values obtained. In the incident medium, the resultant electric and magnetic fields are

$$\mathcal{E}_i + \mathcal{E}_r = \mathcal{E}_i(1 + \rho)$$

and

$$\mathcal{H}_i - \mathcal{H}_r = y_0(1 - \rho)\mathcal{E}_i,$$

respectively, where we have used the notation for tangential components, and in the second medium, the fields are

$$\tau\mathcal{E}_i \quad \text{and} \quad y_1\tau\mathcal{E}_i,$$

respectively. Then the net irradiances on either side of the boundary are as follows:

- Medium 0: $I = \text{Re}\left\{\frac{1}{2}\left[\mathcal{E}_i(1 + \rho)\right]\left[y_0^*(1 - \rho^*)\mathcal{E}_i^*\right]\right\}$
- Medium 1: $I = \text{Re}\left\{\frac{1}{2}\left[\tau\mathcal{E}_i\right]\left[y_1^*\tau^*\mathcal{E}_i^*\right]\right\}$

We then equate these two values which gives, at the boundary,

$$\text{Re}\left[\frac{1}{2}y_0^*\mathcal{E}_i\mathcal{E}_i^*(1 + \rho - \rho^* - \rho\rho^*)\right] = \frac{1}{2}\text{Re}(y_1)\tau\tau^*\mathcal{E}_i\mathcal{E}_i^*,$$

$$\frac{1}{2}\text{Re}(y_0^*)\mathcal{E}_i\mathcal{E}_i^* - \frac{1}{2}\text{Re}(y_0^*)\rho\rho^*\mathcal{E}_i\mathcal{E}_i^* + \frac{1}{2}\text{Re}[y_0^*(\rho - \rho^*)]\mathcal{E}_i\mathcal{E}_i^* = \frac{1}{2}\text{Re}(y_1)\tau\tau^*\mathcal{E}_i\mathcal{E}_i^*. \qquad (2.77)$$

We can replace the different parts of Equation 2.77 with their normal interpretations to give

$$I_i - RI_i + \frac{1}{2}\text{Re}[y_0^*(\rho - \rho^*)]\mathcal{E}_i\mathcal{E}_i^* = TI_i. \qquad (2.78)$$

$(\rho - \rho^*)$ is imaginary. This implies that if y_0 is real, the third term in Equation 2.78 is zero. The other terms then make up the incident, the reflected, and the transmitted irradiances, and these balance. If y_0 is complex, then its imaginary part will combine with the imaginary $(\rho - \rho^*)$ to produce a real result that will imply that $T + R \neq 1$. The irradiances involved in the analysis are those actually at the boundary, which is of zero thickness, and it is impossible that it should either remove or donate energy to the waves. Our assumption that the irradiances can be divided into separate incident, reflected, and transmitted irradiances is therefore incorrect. The source of the difficulty is a coupling between the incident and reflected fields which occurs only in an absorbing medium and which must be taken into account when computing energy transport. The expressions for the amplitude coefficients are perfectly correct. The phenomenon is well understood and has been described in a number of contributions, for example, that by Berning [4]. We shall return to the problem in more general terms in Section 2.13 once we have had a brief look at coherence in Section 2.12. It is, however, convenient to look now at some consequences.

The extra term is on the order of (k^2/n^2). For any reasonable experiment to be carried out, the incident medium must be sufficiently free of absorption for the necessary comparative measurements to be performed with acceptably small errors. In such cases, the error is vanishingly small. Although we will certainly be dealing with absorbing media in thin-film assemblies, our incident media will never be heavily absorbing, and it will not be a serious lack of generality if we assume that our incident media are absorption free. Since our expressions for the amplitude coefficients are valid, then any calculations of amplitudes in absorbing media will be correct. We simply have to ensure that the calculations of reflectances are carried out in a transparent medium. With this restriction, then, we have

$$R = \left(\frac{y_0 - y_1}{y_0 + y_1}\right)\left(\frac{y_0 - y_1}{y_0 + y_1}\right)^*, \tag{2.79}$$

$$T = \frac{4y_0 \operatorname{Re}(y_1)}{(y_0 + y_1)(y_0 + y_1)^*}, \tag{2.80}$$

where y_0 is real.

2.3.4.1 Rear Surface of Absorbing Substrate

We have avoided the problem connected with the definition of reflectance in a medium with complex y_0 simply by not defining it unless the incident medium is sufficiently free of either gain or absorption. Without a definition of reflectance, however, we have trouble with the meaning of antireflection, and there are cases such as the rear surface of an absorbing substrate where an antireflection coating would be relevant. We do need to deal with this problem, and although we have not yet discussed antireflection coatings, it is most convenient to include the discussion here where we have already the basis for the theory. The discussion was originally published in a paper of Macleod [5].

The usual purpose of an antireflection coating is the reduction of reflectance. But frequently the objective of the reflectance reduction is the corresponding increase in transmittance. Although an absorbing or amplifying medium will rarely present us with a problem in terms of a reflectance measurement, we must occasionally treat a slab of such material on both sides to increase overall transmittance. In this context, therefore, we define an antireflection coating as one that increases transmittance and, in the ideal case, maximizes it. But to accomplish that, we need to define what we mean by *transmittance*.

We have no problem with the measurement of irradiance at the emergent side of our system, even if the emergent medium is absorbing. The incident irradiance is more difficult. We can define this as the irradiance we would measure at the position of the surface if the transmitting structure were removed and replaced by an infinite extent of incident medium material. Then the transmittance will simply be the ratio of these two values.

That is,

$$I_{\text{inc}} = \frac{1}{2}\operatorname{Re}(y_0)\mathcal{E}_i\mathcal{E}_i^*,$$

and then

$$T = \frac{\dfrac{1}{2}\operatorname{Re}(y_1)\mathcal{E}_t\mathcal{E}_t^*}{\dfrac{1}{2}\operatorname{Re}(y_0)\mathcal{E}_i\mathcal{E}_i^*}.$$

This is completely consistent with Equation 2.78, that is, with a slight manipulation,

$$T = 1 - \rho\rho^* + \frac{\operatorname{Re}[y_0^*(\rho - \rho^*)]}{\operatorname{Re}(y_0)}. \tag{2.81}$$

An alternative form uses

$$\mathcal{E}_t = \frac{2y_0}{(y_0 + y_1)} \mathcal{E}_i,$$

so that

$$T = \frac{4y_0 y_0^* \operatorname{Re}(y_1)}{\operatorname{Re}(y_0) \cdot [(y_0 + y_1)(y_0 + y_1)^*]}. \tag{2.82}$$

Now let the surface be coated with a dielectric system so that it presents the surface admittance Y. We have not, so far, introduced the idea of surface admittance, and we shall deal with it in more detail in Section 2.5. The ratio of the total tangential magnetic amplitude to the total tangential electric amplitude at a surface is an admittance that we can consider to be a property of the surface and call it the surface admittance. In the case of a simple boundary, the surface admittance is simply the characteristic admittance (tilted if necessary) of the emergent medium. In the case of an optical coating, it plays the same role, but its value is now a function of the interference effects in the coating. For the moment, let us accept that the surface admittance of the rear surface of our absorbing substrate, because of the coating, now presents a surface admittance of Y that will be interpreted by the incident wave as if it were a simple surface before a medium of characteristic admittance Y.

Then, since, in the absence of absorption, the net irradiance entering the thin-film system must also be the emergent irradiance,

$$T = \frac{4y_0 y_0^* \operatorname{Re}(Y)}{\operatorname{Re}(y_0) \cdot [(y_0 + Y)(y_0 + Y)^*]}. \tag{2.83}$$

Let $Y = \alpha + i\beta$; then

$$T = \frac{4\alpha(n_0^2 + k_0^2)}{n_0 [(n_0 + \alpha)^2 + (k_0 - \beta)^2]},$$

and T can readily be shown to be a maximum when

$$Y = \alpha + i\beta = n_0 + ik_0 = (n_0 - ik_0)^*. \tag{2.84}$$

The matching admittance should therefore be the *complex conjugate* of the incident admittance. For this perfect matching, the transmittance becomes

$$T = \left(1 + \frac{k_0^2}{n_0^2}\right),$$

and this is greater than unity. This is not a mistake but rather a consequence of the definition of transmittance. Irradiance falls by a factor of roughly $4\pi k_0$ in a distance of one wavelength, rather larger than any normal value of k_0^2/n_0^2, so that the effect is quite small. It originates in a curious pattern in the otherwise exponentially falling irradiance. It is caused by the presence of the interface and is a cyclic fluctuation in the rate of irradiance reduction. Note that the transmittance is unity if the coating is designed to match $n_0 - ik_0$ rather than its complex conjugate.

A dielectric coating that transforms an admittance of y_1 to an admittance of y_0^* will also, when reversed, exactly transform an admittance of y_0 to y_1^*. This is dealt with in more detail later (Section 8.6) when induced transmission filters are discussed. Thus, the optimum coating to give highest transmittance will be the same in both directions. This implies that an absorbing substrate in identical dielectric incident and emergent media should have exactly similar antireflection coatings on both front and rear surfaces.

Although also a little premature, it is convenient to mention here that the calculation of the properties of a coated slice of material involves multiple beams that are combined either coherently or incoherently. The coherent case considers the slice as an ordinary absorbing thin film and is simply the usual interference calculation, and we will return to it in considerable detail when we deal with induced transmission filters. We will see then that as the absorbing film becomes thicker, the matching rules for an induced transmission filter tend to approach Equation 2.84. The incoherent case is at first sight less obvious. An estimate of the reflected beam is necessary for a multiple beam calculation. Such calculations imply that the absorption is not sufficiently high to completely eliminate a beam that suffers two traversals of the system. This implies, in turn, a negligible absorption in the space of one wavelength, in other words, $4\pi k_0$ is very small. The upper limit on the size of the effect under discussion is k_0^2/n_0^2, and this will still be less significant. For an incoherent calculation to be appropriate, there must be a jumbling of phase that washes out its effect. We can suppose for this discussion that the jumbling comes from a variation in the position of the reflecting surface over the aperture, although in the normal way, there will also be some variation of the incident angle. The variation of the extra term in Equation 2.84 is locked for its phase to the reflecting surface, and so at any exactly plane surface that may be chosen as a reference, an average of the extra term is appropriate, and this will be zero because ρ will have a phase that varies throughout the four quadrants. For multiple beam calculations, therefore, the reflectance can be taken simply as $\rho\rho^*$. Where k_0^2/n_0^2 is significant, the absorption will be very high and certainly enough for the influence of the multiple beams to be automatically negligible.

2.3.5 Oblique Incidence in Absorbing Media

Remembering what we said in the Section 2.3.4, we limit ourselves to a transparent incident medium and an absorbing second, or emergent, medium. Our first aim must be to ensure that the phase factors are consistent. Taking advantage of some of the earlier results, we can write the phase factors as follows:

- Incident:

$$\exp\{i[\omega t - (2\pi n_0/\lambda)(x\sin\vartheta_0 + z\cos\vartheta_0)]\}$$

- Reflected:

$$\exp\{i[\omega t - (2\pi n_0/\lambda)(x\sin\vartheta_0 - z\cos\vartheta_0)]\} \qquad (2.85)$$

- Transmitted:

$$\exp\{i[\omega t - (2\pi\{n_1 - ik_1\}/\lambda)(\alpha x + \gamma z)]\},$$

where α and γ in the transmitted phase factors are the only unknowns. The phase factors must be identically equal for all x and t with $z = 0$. This implies

$$\alpha = \frac{n_0 \sin\vartheta_0}{(n_1 - ik_1)},$$

and since $\alpha^2 + \gamma^2 = 1$,

$$\gamma = \left(1 - \alpha^2\right)^{1/2}.$$

There are two solutions to this equation, and we must decide which is to be adopted. We note that it is strictly $(n_1 - ik_1)\alpha$ and $(n_1 - ik_1)\gamma$ that are required:

$$(n_1 - ik_1)\gamma = \left[(n_1 - ik_1)^2 - n_0^2 \sin^2\vartheta_0\right]^{1/2}$$
$$= \left[n_1^2 - k_1^2 - n_0^2 \sin^2\vartheta_0 - i2n_1k_1\right]^{1/2}. \tag{2.86}$$

The quantity within the square root is in either the third or fourth quadrant, and so the square roots are in the second quadrant (of the form $-a + ib$) and in the fourth quadrant (of the form $a - ib$). If we consider what happens when these values are substituted into the phase factors, we see that the fourth quadrant solution must be correct because this leads to an exponential falloff with z of amplitude together with a change in the phase of the correct sense. The second quadrant solution would lead to an increase with z and a change in phase of the incorrect sense, which would imply a wave travelling in the opposite direction. The fourth quadrant solution is also consistent with the solution for the absorption-free case. The transmitted phase factor is therefore of the form

$$\exp\{i[\omega t - (2\pi n_0 \sin\vartheta_0 x/\lambda) - (2\pi/\lambda)(a - ib)z]\}$$
$$= \exp(-2\pi bz/\lambda)\exp\{i[\omega t - (2\pi n_0 \sin\vartheta_0 x/\lambda) - (2\pi az/\lambda)]\},$$

where

$$(a - ib) = \left[n_1^2 - k_1^2 - n_0^2 \sin^2\vartheta_0 - i2n_1k_1\right]^{1/2}.$$

A wave which possesses such a phase factor is known as inhomogeneous. The exponential falloff in amplitude is along the z-axis, while the propagation direction in terms of phase is determined by the direction cosines, which can be extracted from

$$(2\pi n_0 \sin\vartheta_0 x/\lambda) + (2\pi az/\lambda).$$

The existence of such waves is another good reason for our choosing to consider the components of the fields parallel to the boundary and the flow of energy normal to the boundary.

We should note at this stage that provided we include the possibility of complex angles, the formulation of the absorption-free case applies equally well to absorbing media, and we can write

$$(n_1 - ik_1)\sin\vartheta_1 = n_0\sin\vartheta_0,$$
$$\alpha = \sin\vartheta_1,$$
$$\gamma = \cos\vartheta_1,$$
$$(a - ib) = (n_1 - ik_1)\cos\vartheta_1.$$

The calculation of amplitudes follows the same pattern as before. However, we have not previously examined the implications of an inhomogeneous wave. Our main concern is the calculation of the tilted admittance connected with such a wave. Since the x, y, and t variations of the wave are contained in the phase factor, we can write

$$\text{curl} \equiv \left(\frac{\partial}{\partial x}\boldsymbol{i} + \frac{\partial}{\partial x}\boldsymbol{j} + \frac{\partial}{\partial x}\boldsymbol{k}\right) \times$$
$$\equiv \left(-i\frac{2\pi N}{\lambda}\alpha\boldsymbol{i} - i\frac{2\pi N}{\lambda}\gamma\boldsymbol{k}\right) \times$$

and

$$\frac{\partial}{\partial t} \equiv i\omega,$$

where the k is a unit vector in the z-direction and should not be confused with the extinction coefficient k.

For p-waves, the H vector is parallel to the boundary in the y-direction, so $H = \mathcal{H}_y j$. The component of E parallel to the boundary will then be in the x-direction $\mathcal{E}_x i$. We follow the analysis leading up to Equation 2.26 and as before

$$\text{curl}H = \sigma E + \varepsilon \frac{\partial E}{\partial t}$$

$$= (\sigma + i\omega\varepsilon)E$$

$$= \frac{i\omega N^2}{c^2\mu} E.$$

Now the tangential component of curl H is in the x-direction so that

$$-i\frac{2\pi N}{\lambda}\gamma(k \times j)\mathcal{H}_y = i\frac{\omega N^2}{c^2\mu}\mathcal{E}_x i.$$

But

$$-(k \times j) = i,$$

so that

$$\eta_p = \frac{\mathcal{H}_y}{\mathcal{E}_x} = \frac{\omega N\lambda}{2\pi c^2\mu\gamma} = \frac{N}{c\mu\gamma}$$

$$= \frac{N\mathcal{Y}}{\gamma} = \frac{y}{\gamma}.$$

For the s-waves, we use

$$\text{curl}E = -\frac{\partial B}{\partial t} = -\mu\frac{\partial H}{\partial t}.$$

E is now along the y-axis, and a similar analysis to that for p-waves yields

$$\eta_s = \frac{\mathcal{H}_x}{\mathcal{E}_y} = N\mathcal{Y}\gamma = y\gamma. \tag{2.87}$$

Now γ can be identified as $\cos\vartheta$, provided that ϑ is permitted to be complex, and so

$$\eta_p = y/\cos\vartheta,$$
$$\eta_s = y\cos\vartheta. \tag{2.88}$$

Alternatively, we can use the expressions in Equations 2.86 and 2.87, together with the fact that $y = (n - ik)\mathcal{Y}$, to give

$$\eta_s = \mathcal{Y}\left[n_1^2 - k_1^2 - n_0^2\sin^2\vartheta_0 - i2n_1k_1\right]^{1/2}. \tag{2.89}$$

The fourth quadrant being the correct solution, and then

$$\eta_p = \frac{y^2}{\eta_s}.$$ (2.90)

This second form is completely consistent with Equation 2.28 but avoids any problems with the quadrant. Then the amplitude and irradiance coefficients become as before:

$$\rho = \frac{\eta_0 - \eta_1}{\eta_0 + \eta_1},$$ (2.91)

$$\tau = \frac{2\eta_0}{\eta_0 + \eta_1},$$ (2.92)

$$R = \left(\frac{\eta_0 - \eta_1}{\eta_0 + \eta_1}\right)\left(\frac{\eta_0 - \eta_1}{\eta_0 + \eta_1}\right)^*,$$ (2.93)

$$T = \frac{4\eta_0 \operatorname{Re}(\eta_1)}{(\eta_0 + \eta_1)(\eta_0 + \eta_1)^*},$$ (2.94)

And, of course, these expressions are valid for absorption-free media as well.

2.4 Critical Angle and Beyond

Let us consider the case of a simple boundary between two dielectric materials with refractive indices n_0 and n_1 and corresponding characteristic admittances y_0 and y_1. Given ϑ_0, Snell's law (Equation 2.44) allows us to calculate the corresponding propagation angle ϑ_1 in the second medium:

$$n_0 \sin \vartheta_0 = n_1 \sin \vartheta_1.$$

Suppose, however, that n_0 is greater than n_1. Since $\sin \vartheta_1 = (n_0/n_1) \sin \vartheta_0$, as the angle ϑ_0 increases from $0°$ to $90°$, there comes a point where $\sin \vartheta_1$ becomes greater than unity. The value of ϑ_0 where this first occurs is known as the critical angle. Beyond the critical angle, ϑ_1 is imaginary. The critical angle is given by

$$\vartheta_0 = \arcsin\left(\frac{n_1}{n_0}\right).$$ (2.95)

It is difficult to visualize an imaginary angle. Fortunately, that is not necessary. What we actually need is the cosine of the angle, and if we already know the sine, a simple relationship gives us the cosine:

$$\cos^2 \vartheta_1 = 1 - \sin^2 \vartheta_1 = 1 - \frac{n_0^2}{n_1^2} \sin^2 \vartheta_0.$$ (2.96)

It is not only simply $\cos \vartheta_1$ that is required, but also $n_1 \cos \vartheta_1$. Then,

$$n_1 \cos \vartheta_1 = \sqrt{n_1^2 - n_0^2 \sin^2 \vartheta_0} = i\sqrt{n_0^2 \sin^2 \vartheta_0 - n_1^2}, \tag{2.97}$$

where the quantity under the root sign is positive and where, as yet, we have not assigned a positive or negative character to the root. Equation 2.86 comes to our rescue. We are actually solving for the direction cosine γ times the refractive index, a combination that appears in the spatial part of the phase factor. We have already established that even in the case of miniscule absorption, we should take the fourth quadrant solution rather than the second in Equation 2.86 and that applies to Equation 2.97 so that

$$n_1 \cos \vartheta_1 = -i\sqrt{n_0^2 \sin^2 \vartheta_0 - n_1^2} = \gamma n_1. \tag{2.98}$$

The phase factor of the transmitted wave is then

$$\begin{aligned}
&\exp\left\{i\left[\omega t - \tfrac{2\pi}{\lambda}(n_0 \sin \vartheta_0)x - \tfrac{2\pi}{\lambda}\left(-i\sqrt{n_0^2 \sin^2 \vartheta_0 - n_1^2}\right)z\right]\right\} \\
&= \exp\left\{-\tfrac{2\pi}{\lambda}\left(\sqrt{n_0^2 \sin^2 \vartheta_0 - n_1^2}\right)z\right\}\exp\left\{i\left[\omega t - \tfrac{2\pi}{\lambda}(n_0 \sin \vartheta_0)x\right]\right\}.
\end{aligned} \tag{2.99}$$

The wave is known as evanescent. It is pinned to the surface along which it propagates with a wavelength smaller than could be supported by a progressive wave, and it decays exponentially away from the surface in the z-direction. The tilted admittance is imaginary, and so the component of the magnetic field that also propagates in the z-direction is 90° out of phase with the electric field so that no energy actually propagates in the z-direction. It is not surprising, therefore, that since there is no loss in the emergent medium, the reflectance is total. It is usually termed *total internal reflectance* and is sometimes abbreviated to TIR.

The tilted admittances, in free space units, are given by

$$\eta_s = -i\left[n_0^2 \sin^2 \vartheta_0 - n_1^2\right]^{1/2} \tag{2.100}$$

and

$$\eta_p = \frac{y^2}{\eta_s} = +i\frac{n_0^2}{\left[n_0^2 \sin^2 \vartheta_0 - n_1^2\right]^{1/2}}. \tag{2.101}$$

It is easy to see that since both are imaginary, the reflectance for both p- and s-polarizations will be 100% (or unity in absolute terms). However, the phase change on reflection will vary with the angle of incidence. At the critical angle where η_s is zero and η_p is infinite, it is zero for s-polarization and 180° for p-polarization. The s-polarization phase shift then moves through the first into the second quadrant with increasing incidence while the p-polarization phase shift moves through the third into the fourth quadrant. The relative retardation, or delta, we recall, has to be corrected by 180° to take account of the sign convention, and it rises from zero to a maximum value that depends on the incident index and then drops back to zero at grazing incidence.

Figure 2.7 shows calculated properties for a glass incident medium (1.52 index) and air emergent medium as a function of angle of incidence.

Strictly, the critical angle is a phenomenon that is related to completely dielectric media only. However, if the k value of the emergent medium is quite small, then a reference angle, usually simply referred to as the critical angle, is sometimes defined by dropping k and using Equation 2.95. This is an especially frequent practice when gain rather than absorption is involved (gain is covered in Chapter 11), but it should be used with caution because it may be thought to imply some kind of abrupt transition when there is none. Otherwise, the behavior of absorbing materials even at high angles of incidence in a high-index incident medium is already covered in Section 2.3.5.

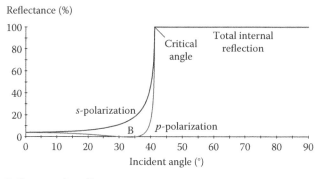

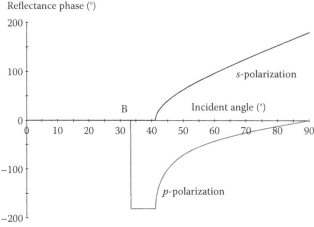

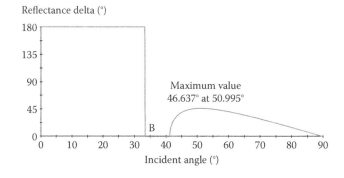

FIGURE 2.7
Internal reflectance (*top*), phase shifts (*middle*), and reflectance delta or relative retardation (*bottom*) of the internal surface of glass of index 1.52 against an emergent medium of air. *B* marks the Brewster angle where the *p*-phase shift flips by 180°. Remember that the relative retardation in reflection is given by $\varphi_p - \varphi_s \pm 180°$.

2.5 Reflectance of a Thin Film

A simple extension of the analysis mentioned earlier occurs in the case of a thin, plane parallel film of material covering the surface of a substrate. The presence of two (or more) interfaces means that a number of beams will be produced by successive reflections, and the properties of the film will be determined by the summation of these beams. We say that the film is thin when full interference effects can be detected in the reflected or transmitted light. We describe such a case as coherent. When no interference effects can be detected, the film is described as thick, and the case described as incoherent. The coherent and incoherent cases depend on the presence or absence of a constant phase relationship between the various beams, and this will depend on the nature of the light and the receiver and on the quality of the film. The same film can appear thin or thick under differing illumination conditions. Normally, we will find that the films on the substrates can be treated as thin, while the substrates supporting the films can be considered thick. Thick films and substrates will be considered toward the end of this chapter. Here we concentrate on the thin case.

The arrangement is illustrated in Figure 2.8. At this stage, it is convenient to introduce a new notation already referred to in Section 2.2. We denote waves in the direction of incidence by the symbol + (that is, positive-going) and waves in the opposite direction by − (that is, negative-going).

The interfaces between the film and the incident medium and the substrate, denoted by the symbols a and b, can be treated in exactly the same way as the simple boundary already discussed. We consider the tangential components of the fields. We shall consider the substrate, or emergent, medium, to be semiinfinite, and so there is no negative-going wave in the substrate. The waves in the film can be summed into one resultant positive-going wave and one resultant negative-going wave. At interface b, then the tangential components of E and H are

$$\mathcal{E}_b = \mathcal{E}_{1b}^+ + \mathcal{E}_{1b}^-,$$

$$\mathcal{H}_b = \eta_1 \mathcal{E}_{1b}^+ - \eta_1 \mathcal{E}_{1b}^-,$$

where we are neglecting the common phase factors and where $\mathcal{E}_b$ and $\mathcal{H}_b$ represent the resultants. Hence,

$$\mathcal{E}_{1b}^+ = \frac{1}{2}(\mathcal{H}_b/\eta_1 + \mathcal{E}_b), \tag{2.102}$$

$$\mathcal{E}_{1b}^- = \frac{1}{2}(-\mathcal{H}_b/\eta_1 + \mathcal{E}_b), \tag{2.103}$$

$$\mathcal{H}_{1b}^+ = \eta_1 \mathcal{E}_{1b}^+ = \frac{1}{2}(\mathcal{H}_b + \eta_1 \mathcal{E}_b), \tag{2.104}$$

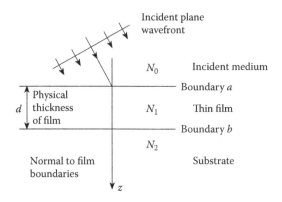

FIGURE 2.8
Plane wave incident on a thin film.

$$\mathcal{H}_{1b}^{-} = -\eta_1 \mathcal{E}_{1b}^{-} = \frac{1}{2}(\mathcal{H}_b - \eta_1 \mathcal{E}_b). \tag{2.105}$$

The fields at the other interface a at the same instant and at a point with identical x and y coordinates can be determined by altering the phase factors of the waves to allow for a shift in the z coordinate from 0 to $-d$. The phase factor of the positive-going wave will be multiplied by $\exp(i\delta)$, where

$$\delta = 2\pi N_1 d \cos \vartheta_1 / \lambda,$$

and ϑ_1 may be complex, while the negative-going phase factor will be multiplied by $\exp(-i\delta)$. We imply that this is a valid procedure when we say that the film is thin.

δ is a particularly important parameter and is known as the phase thickness of the film. We shall shortly see that it and the appropriate admittance define the properties of the film.

The values of $\mathcal{E}$ and $\mathcal{H}$ at interface a are now, using Equations 2.102 through 2.105,

$$\mathcal{E}_{1a}^{+} = \mathcal{E}_{1b}^{+}e^{i\delta} = \frac{1}{2}(\mathcal{H}_b/\eta_1 + \mathcal{E}_b)e^{i\delta},$$

$$\mathcal{E}_{1a}^{-} = \mathcal{E}_{1b}^{-}e^{-i\delta} = \frac{1}{2}(-\mathcal{H}_b/\eta_1 + \mathcal{E}_b)e^{-i\delta},$$

$$\mathcal{H}_{1a}^{+} = \mathcal{H}_{1b}^{+}e^{i\delta} = \frac{1}{2}(\mathcal{H}_b + \eta_1 \mathcal{E}_b)e^{i\delta},$$

$$\mathcal{H}_{1a}^{-} = \mathcal{H}_{1b}^{-}e^{-i\delta} = \frac{1}{2}(\mathcal{H}_b - \eta_1 \mathcal{E}_b)e^{-i\delta},$$

so that

$$\mathcal{E}_a = \mathcal{E}_{1a}^{+} + \mathcal{E}_{1a}^{-}$$

$$= \mathcal{E}_b\left(\frac{e^{i\delta} + e^{-i\delta}}{2}\right) + \mathcal{H}_b\left(\frac{e^{i\delta} - e^{-i\delta}}{2\eta_1}\right)$$

$$= \mathcal{E}_b \cos \delta + \mathcal{H}_b \frac{i \sin \delta}{\eta_1},$$

$$\mathcal{H}_a = \mathcal{H}_{1a}^{+} + \mathcal{H}_{1a}^{-}$$

$$= \mathcal{E}_b\eta_1\left(\frac{e^{i\delta} - e^{-i\delta}}{2}\right) + \mathcal{H}_b\left(\frac{e^{i\delta} + e^{-i\delta}}{2}\right)$$

$$= \mathcal{E}_b i\eta_1 \sin \delta + \mathcal{H}_b \cos \delta.$$

These two simultaneous equations can be written in matrix notation as

$$\begin{bmatrix} \mathcal{E}_a \\ \mathcal{H}_a \end{bmatrix} = \begin{bmatrix} \cos \delta & (i \sin \delta)/\eta_1 \\ i\eta_1 \sin \delta & \cos \delta \end{bmatrix} \begin{bmatrix} \mathcal{E}_b \\ \mathcal{H}_b \end{bmatrix}. \tag{2.106}$$

Since the tangential components of E and H are continuous across a boundary, and since there is only a positive-going wave in the substrate, this relationship connects the tangential components of E and H at the incident interface with the tangential components of E and H transmitted through the final interface. The 2×2 matrix on the right-hand side of Equation 2.106 is known as the characteristic matrix of the thin film.

When we dealt with a simple interface, we saw that the total tangential electric and magnetic fields were indistinguishable from the fields of the emergent wave at the interface. Their ratio was, therefore, that of the admittance y_1 or η_1 of the emergent medium and, as such, appeared in the expressions for the amplitude reflection coefficient and the reflectance. Here we have interface a with its total tangential fields. If we think purely of the amplitude reflection coefficient and the reflectance, we can see that the ratio of the total tangential components at a now performs the role that was previously performed by the admittance y_1 or η_1. We call this ratio the surface admittance, because it is both an admittance and a property of the surface, and we define it by analogy with Equation 2.69 as

$$Y = \mathcal{H}_a / \mathcal{E}_a, \tag{2.107}$$

when the problem becomes merely that of finding the reflectance of a simple interface between an incident medium of admittance η_0 and a medium of admittance Y, i.e.,

$$\rho = \frac{\eta_0 - Y}{\eta_0 + Y},$$

$$R = \left(\frac{\eta_0 - Y}{\eta_0 + Y} \right) \left(\frac{\eta_0 - Y}{\eta_0 + Y} \right)^*. \tag{2.108}$$

Because we now have a structure beyond the front interface, transmittance is a little more complicated, and we leave it until later. We can normalize Equation 2.106 by dividing through by $\mathcal{E}_b$ to give

$$\begin{bmatrix} \mathcal{E}_a / \mathcal{E}_b \\ \mathcal{H}_a / \mathcal{E}_b \end{bmatrix} = \begin{bmatrix} B \\ C \end{bmatrix} = \begin{bmatrix} \cos \delta & (i \sin \delta)/\eta_1 \\ i\eta_1 \sin \delta & \cos \delta \end{bmatrix} \begin{bmatrix} 1 \\ \eta_2 \end{bmatrix}, \tag{2.109}$$

and B and C, the normalized electric and magnetic fields at the front interface, are the quantities from which we will be extracting the properties of the thin-film system. Clearly, from Equations 2.107 and 2.109, we can write

$$Y = \frac{\mathcal{H}_a}{\mathcal{E}_a} = \frac{C}{B} = \frac{\eta_2 \cos \delta + i\eta_1 \sin \delta}{\cos \delta + i(\eta_2/\eta_1) \sin \delta}, \tag{2.110}$$

and from Equations 2.108 and 2.110, we can calculate the reflectance.

$$\begin{bmatrix} B \\ C \end{bmatrix}$$

is known as the characteristic matrix of the assembly.

In this section, the surface admittance Y is also the input admittance of the thin-film system, but the concept of surface admittance can be applied to any interface, real or notional, in an assembly of this films. Then, we will write it with an appropriate subscript indicating the particular interface. We will make much use of the concept in subsequent chapters.

2.6 Reflectance of an Assembly of Thin Films

Let another film be added to the single film of the previous section so that the final interface is now denoted by c, as shown in Figure 2.9. The characteristic matrix of the film nearest the substrate is

$$\begin{bmatrix} \cos \delta_2 & (i \sin \delta_2)/\eta_2 \\ i\eta_2 \sin \delta_2 & \cos \delta_2 \end{bmatrix}, \tag{2.111}$$

and from Equation 2.106,

$$\begin{bmatrix} \mathcal{E}_b \\ \mathcal{H}_b \end{bmatrix} = \begin{bmatrix} \cos \delta_2 & (i \sin \delta_2)/\eta_2 \\ i\eta_2 \sin \delta_2 & \cos \delta_2 \end{bmatrix} \begin{bmatrix} \mathcal{E}_c \\ \mathcal{H}_c \end{bmatrix}.$$

We can apply Equation 2.106 again to give the parameters at interface a, i.e.,

$$\begin{bmatrix} \mathcal{E}_a \\ \mathcal{H}_a \end{bmatrix} = \begin{bmatrix} \cos \delta_1 & (i \sin \delta_1)/\eta_1 \\ i\eta_1 \sin \delta_1 & \cos \delta_1 \end{bmatrix} \begin{bmatrix} \cos \delta_2 & (i \sin \delta_2)/\eta_2 \\ i\eta_2 \sin \delta_2 & \cos \delta_2 \end{bmatrix} \begin{bmatrix} \mathcal{E}_c \\ \mathcal{H}_c \end{bmatrix},$$

and the characteristic matrix of the assembly, by analogy with Equation 2.109, is

$$\begin{bmatrix} B \\ C \end{bmatrix} = \begin{bmatrix} \cos \delta_1 & (i \sin \delta_1)/\eta_1 \\ i\eta_1 \sin \delta_1 & \cos \delta_1 \end{bmatrix} \begin{bmatrix} \cos \delta_2 & (i \sin \delta_2)/\eta_2 \\ i\eta_2 \sin \delta_2 & \cos \delta_2 \end{bmatrix} \begin{bmatrix} 1 \\ \eta_3 \end{bmatrix}.$$

Y is, as before, C/B, and the amplitude reflection coefficient and the reflectance are, from Equation 2.108.,

$$\rho = \frac{\eta_0 - Y}{\eta_0 + Y},$$

$$R = \left(\frac{\eta_0 - Y}{\eta_0 + Y}\right) \left(\frac{\eta_0 - Y}{\eta_0 + Y}\right)^{*}. \tag{2.112}$$

This result can be immediately extended to the general case of an assembly of q layers over an emergent medium of admittance η_m, when the characteristic matrix is simply the product of the individual matrices taken in the correct order, i.e.,

$$\begin{bmatrix} B \\ C \end{bmatrix} = \left\{ \prod_{r=1}^{q} \begin{bmatrix} \cos \delta_r & (i \sin \delta_r)/\eta_r \\ i\eta_r \sin \delta_r & \cos \delta_r \end{bmatrix} \right\} \begin{bmatrix} 1 \\ \eta_m \end{bmatrix}, \tag{2.113}$$

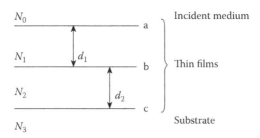

FIGURE 2.9
Notation for two films on a surface.

where

$$\delta_r = \frac{2\pi N_r d_r \cos \vartheta_r}{\lambda},$$

$$\eta_r = \mathscr{Y} N_r \cos \vartheta_r \quad \text{for } s\text{-polarisation (TE),}$$

$$\eta_r = \mathscr{Y} N_r / \cos \vartheta_r \quad \text{for } p\text{-polarisation (TM),}$$

and where we have now used the subscript m to denote the substrate or emergent medium.

$$\eta_m = \mathscr{Y} N_m \cos \vartheta_m \quad \text{for } s\text{-polarisation (TE),}$$

$$\eta_m = \mathscr{Y} N_m / \cos \vartheta_m \quad \text{for } p\text{-polarisation (TM).}$$

If the angle of incidence ϑ_0 is given, the values of ϑ_r can be found from Snell's law, i.e.,

$$N_0 \sin \vartheta_0 = N_r \sin \vartheta_r = N_m \sin \vartheta_m. \tag{2.114}$$

Equation 2.113 is of prime importance in optical thin-film work and forms the basis of almost all calculations.

A useful property of the characteristic matrix of a thin film is that the determinant is unity. This means that the determinant of the product of any number of these matrices is also unity.

It avoids difficulties over signs and quadrants if, in the case of absorbing media, the arrangement used for computing phase thicknesses and admittances is

$$\delta_r = (2\pi/\lambda) d_r \left(n_r^2 - k_r^2 - n_0^2 \sin^2 \vartheta_0 - 2i n_r k_r \right)^{1/2}, \tag{2.115}$$

the correct solution being in the fourth quadrant. Then,

$$\eta_{rs} = \mathscr{Y} \left(n_r^2 - k_r^2 - n_0^2 \sin^2 \vartheta_0 - 2i n_r k_r \right)^{1/2}, \tag{2.116}$$

also in the fourth quadrant, and

$$\eta_{rp} = \frac{y_r^2}{\eta_{rs}} = \frac{\mathscr{Y}^2 (n_r - ik_r)^2}{\eta_{rs}}. \tag{2.117}$$

It is useful to examine the phase shift associated with the reflected beam. Let $Y = a + ib$. Then, with η_0 real,

$$\rho = \frac{\eta_0 - a - ib}{\eta_0 + a + ib}$$

$$= \frac{\left(\eta_0^2 - a^2 - b^2 \right) - i(2b\eta_0)}{\left(\eta_0 + a \right)^2 + b^2},$$

i.e.,

$$\tan \varphi = \frac{(-2b\eta_0)}{\left(\eta_0^2 - a^2 - b^2 \right)}, \tag{2.118}$$

where φ is the phase shift. This must be interpreted, of course, on the basis of the sign convention we have already established in Figure 2.4. It is important to preserve the signs of the numerator and the denominator separately as shown; otherwise, the quadrant cannot be uniquely specified. The rule is simple. It is the quadrant in which the vector associated with ρ lies, and the following scheme can be derived by treating the denominator as the x coordinate and the numerator as the y coordinate.

Numerator	+	+	−	−
Denominator	+	−	+	−
Quadrant	1st	2nd	4th	3rd

Note particularly that the reference surface for the calculation of phase shift on reflection is the front surface of the multilayer.

2.7 Reflectance, Transmittance, and Absorptance

Sufficient information is included in Equation 2.113 to allow the transmittance and absorptance of a thin-film assembly to be calculated. For this to have a physical meaning, as we have already seen, the incident medium should be transparent; that is, η_0 must be real. The substrate need not be transparent, but the transmittance calculated will be the transmittance into, rather than through, the substrate.

First of all, we calculate the net irradiance at the exit side of the assembly, which we take as the kth interface. This is given by

$$I_k = \frac{1}{2} \operatorname{Re}(\mathcal{E}_k \mathcal{H}_k^*),$$

where, once again, we are dealing with the component of irradiance normal to the interfaces and where the admittance of the exit, or emergent, material is η_m.

$$
\begin{aligned}
I_k &= \tfrac{1}{2} \operatorname{Re}(\mathcal{E}_k \eta_m^* \mathcal{E}_k^*) \\
&= \tfrac{1}{2} \operatorname{Re}(\eta_m^*) \mathcal{E}_k \mathcal{E}_k^*.
\end{aligned}
\tag{2.119}
$$

If the characteristic matrix of the assembly is

$$\begin{bmatrix} B \\ C \end{bmatrix},$$

then the net irradiance at the entrance to the assembly is

$$I_a = \frac{1}{2} \operatorname{Re}(BC^*) \mathcal{E}_k \mathcal{E}_k^*, \tag{2.120}$$

where we recall that B and C are calculated with respect to unity emergent tangential electric field amplitude. Let the incident irradiance be denoted by I_i, then Equation 2.120 represents the irradiance actually entering the assembly, which is $(1 - R)I_i$:

$$(1 - R)I_i = \frac{1}{2} \operatorname{Re}(BC^*) \mathcal{E}_k \mathcal{E}_k^*,$$

i.e.,

$$I_i = \frac{\text{Re}(BC^*)\mathcal{E}_k\mathcal{E}_k^*}{2(1-R)}.$$

Equation 2.119 represents the irradiance leaving the assembly and entering the substrate, and so the transmittance T is

$$T = \frac{I_k}{I_i} = \frac{\text{Re}(\eta_m)(1-R)}{\text{Re}(BC^*)}. \tag{2.121}$$

The absorptance A in the multilayer is connected with R and T by the relationship

$$1 = R + T + A,$$

so that

$$A = 1 - R - T = (1 - R)\left(1 - \frac{\text{Re}(\eta_m)}{\text{Re}(BC^*)}\right). \tag{2.122}$$

In the absence of absorption in any of the layers, it can readily be shown that the expressions mentioned earlier are consistent with $A = 0$ and $T + R = 1$, for the individual film matrices will have determinants of unity and the product of any number of these matrices will also have a determinant of unity. The product of the matrices can be expressed as

$$\begin{bmatrix} \alpha & i\beta \\ i\gamma & \delta \end{bmatrix},$$

where $\alpha\delta + \gamma\beta = 1$ and, because there is no absorption, α, β, γ, and δ are all real;

$$\begin{bmatrix} B \\ C \end{bmatrix} = \begin{bmatrix} \alpha & i\beta \\ i\gamma & \delta \end{bmatrix}\begin{bmatrix} 1 \\ \eta_m \end{bmatrix} = \begin{bmatrix} \alpha + i\beta\eta_m \\ \delta\eta_m + i\gamma \end{bmatrix},$$

$$\text{Re}(BC^*) = \text{Re}[(\alpha + i\beta\eta_m)(\delta\eta_m - i\gamma)] = (\alpha\delta + \gamma\beta)\,\text{Re}(\eta_m),$$

$$= \text{Re}(\eta_m)$$

and the result follows.

We can manipulate Equations 2.121 and 2.122 into slightly better forms. From Equation 2.108,

$$R = \left(\frac{\eta_0 B - C}{\eta_0 B + C}\right)\left(\frac{\eta_0 B - C}{\eta_0 B + C}\right)^*, \tag{2.123}$$

so that

$$(1 - R) = \frac{2\eta_0(BC^* + B^*C)}{(\eta_0 B + C)(\eta_0 B + C)^*}. \tag{2.124}$$

Inserting this result in Equation 2.12, we obtain

$$T = \frac{4\eta_0\,\text{Re}(\eta_m)}{(\eta_0 B + C)(\eta_0 B + C)^*}, \tag{2.125}$$

and in Equation 2.122,

$$A = \frac{4\eta_0\,\text{Re}(BC^* - \eta_m)}{(\eta_0 B + C)(\eta_0 B + C)^*}. \tag{2.126}$$

Equations 2.123, 2.125, and 2.126 are the most useful forms of the expressions for R, T, and A.

An important quantity which we shall discuss in a later section of this chapter is $T/(1 - R)$, known as the potential transmittance ψ. From Equation 2.121,

$$\psi = \frac{T}{(1 - R)} = \frac{\text{Re}(\eta_m)}{\text{Re}(BC^*)}. \tag{2.127}$$

The phase change on reflection (Equation 2.118) can also be put in a form compatible with Equations 2.123 through 2.126:

$$\varphi = \arctan\left(\frac{\text{Im}[\eta_m(BC^* - CB^*)]}{(\eta_m^2 BB^* - CC^*)}\right). \tag{2.128}$$

The quadrant of φ is given by the same arrangement of signs of numerator and denominator as Equation 2.118. The phase change on reflection is measured at the front surface of the multilayer.

Phase shift on transmission is sometimes important. This can be obtained in a way similar to the phase shift on reflection. We denote the phase shift by ζ, and we define it as the difference in phase between the resultant transmitted wave as it enters the emergent medium and the incident wave exactly at the front surface, that is, as it enters the multilayer. The electric field amplitude at the emergent surface has been normalized to unity, and so the phase may be taken as zero. Then we simply have to find the expression, which will involve B and C, for the incident amplitude. These are the normalized total tangential electric and magnetic fields. So we can write

$$\mathcal{E}_i + \mathcal{E}_r = B,$$

$$\eta_0 \mathcal{E}_i - \eta_0 \mathcal{E}_r = C.$$

Then, we eliminate $\mathcal{E}_r$ to give

$$\mathcal{E}_i = \frac{1}{2}\left(B + \frac{C}{\eta_0}\right),$$

and the amplitude transmission coefficient as

$$\tau = \frac{2\eta_0}{(\eta_0 B + C)} = \frac{2\eta_0(\eta_0 B + C)^*}{(\eta_0 B + C)(\eta_0 B + C)^*},$$

so that

$$\zeta = \arctan\left[\frac{-\text{Im}(\eta_0 B + C)}{\text{Re}(\eta_0 B + C)}\right]. \tag{2.129}$$

Again, it is important to keep the signs of the numerator and the denominator separately. Then the quadrant is given by the same arrangement of signs as Equation 2.118.

2.8 Units

We have been using the International System of Units (SI) in the work so far. In this system y, η, and Y are measured in siemens. Much thin-film literature, especially the early literature, has been written in Gaussian units. In Gaussian units, the admittance of free space $\mathcal{Y}$ is unity, and so, since at optical frequencies $y = N\mathcal{Y}$, y (the optical admittance) and N (the refractive index) are numerically equal at normal incidence, although N is a number without units. The position is different in SI units, where $\mathscr{Y}$ is 2.6544×10^{-3} S. We could, if we choose, measure y and η in units of $\mathscr{Y}$ siemens, which we can call free space units, and in this case, y becomes numerically equal to N, just as in the Gaussian system. This is a perfectly valid procedure, and all the expressions for ratioed quantities, notably reflectance, transmittance, absorptance, and potential transmittance, are unchanged. This applies particularly to Equations 2.113 and 2.123 through 2.127. We must simply take due care when calculating absolute rather than relative irradiance and when deriving the magnetic field. In particular, Equation 2.106 becomes

$$\begin{bmatrix} \mathcal{E}_a \\ \mathcal{H}_a/\mathscr{Y} \end{bmatrix} = \begin{bmatrix} \cos\delta & (i\sin\delta)/\eta_1 \\ i\eta_1\sin\delta & \cos\delta \end{bmatrix} \begin{bmatrix} \mathcal{E}_b \\ \mathcal{H}_b/\mathscr{Y} \end{bmatrix}, \tag{2.130}$$

where η is now measured in free space units. In most cases in this book, either arrangement can be used. In many cases, particularly where we are using graphical techniques, we shall use free space units, because otherwise, the scales become quite cumbersome.

2.9 Summary of Important Results

We have now covered the basic theory necessary for the understanding of the remainder of the book. It has been a somewhat long and involved discussion, and so we now summarize the principal results. The statement numbers refer to those in the text where the particular quantities were originally introduced.

The refractive index is defined as the ratio of the velocity of light in free space c to the velocity of light in the medium v. When refractive index is real, it is denoted by n, but it is frequently complex and then is denoted by N:

$$N = c/v = n - ik. \tag{2.19}$$

N is often called the complex refractive index; n, the real refractive index (or often simply refractive index); and k, the extinction coefficient. N is always a function of λ.

k is related to the absorption coefficient α by $\alpha = 4\pi k/\lambda$ (Equation 2.37). Light waves are electromagnetic, and a homogeneous, plane, linearly polarized harmonic (or monochromatic) wave may be represented by expressions of the form

$$E = \mathcal{E}\exp\{i[\omega t - (2\pi N/\lambda)z]\}, \tag{2.22}$$

where z is the distance along the direction of propagation, E is the electric field, $\mathcal{E}$ is the electric amplitude, and φ is an arbitrary phase. A similar expression holds for H, the magnetic field:

$$H = \mathcal{H}\exp\{i[\omega t - (2\pi N/\lambda)z + \varphi\prime]\}, \tag{2.131}$$

where φ, φ', and N are not independent. The physical significance is attached to the real parts of the preceding expressions (or the imaginary parts).

The phase change suffered by the wave on traversing a distance d of the medium is, therefore,

$$-\frac{2\pi N d}{\lambda} = -\frac{2\pi n d}{\lambda} + i\frac{2\pi k d}{\lambda}, \tag{2.132}$$

and the imaginary part can be interpreted as a reduction in amplitude (by substituting in Equation 2.22). Note that in our convention, a wave suffers a phase lag on the traversal of the distance d.

The optical admittance is given by the ratio of the magnetic and electric fields, $y = H/E$ (Equations 2.26 through 2.31),and y is usually complex. In free space, y is real and is denoted by $\mathscr{Y}$.

$$\mathscr{Y} = 2.6544 \times 10^{-3}\,\text{S}. \tag{2.133}$$

The optical admittance of a medium is connected with the refractive index by

$$y = N\mathscr{Y}. \tag{2.134}$$

(In Gaussian units, $\mathscr{Y}$ is unity and y and N are numerically the same. In SI units, we can make y and N numerically equal by expressing y in units of $\mathscr{Y}$, i.e., free space units. All expressions for reflectance, transmittance, etc., involving ratios will remain valid, but care must be taken when computing absolute irradiances, although these are not often needed in thin-film optics, except where damage studies are involved.)

The irradiance of the light, defined as the mean rate of flow of energy per unit area carried by the wave, is given by

$$I = \frac{1}{2}\,\text{Re}(EH^*). \tag{2.35}$$

This can also be written as

$$I = \frac{1}{2}n\mathscr{Y}EE^*, \tag{2.135}$$

where * denotes complex conjugate.

At a boundary between two media, denoted by the subscript 0 for the incident medium and by the subscript 1 for the exit medium, the incident beam is split into a reflected beam and a transmitted beam. For *normal incidence*, we have,

$$\rho = \frac{\mathcal{E}_r}{\mathcal{E}_i} = \frac{y_0 - y_1}{y_0 + y_1} = \frac{(n_0 - ik_0)\mathscr{Y} - (n_1 - ik_1)\mathscr{Y}}{(n_0 - ik_0)\mathscr{Y} + (n_1 - ik_1)\mathscr{Y}} = \frac{(n_0 - n_1) - i(k_0 - k_1)}{(n_0 + n_1) - i(k_0 + k_1)}, \tag{2.75}$$

$$\tau = \frac{\mathcal{E}_t}{\mathcal{E}_i} = \frac{2y_0}{y_0 - y_1} = \frac{2(n_0 - ik_0)\mathscr{Y}}{(n_0 - ik_0)\mathscr{Y} + (n_1 - ik_1)\mathscr{Y}} = \frac{2(n_0 - ik_0)}{(n_0 + n_1) - i(k_0 + k_1)}, \tag{2.76}$$

where ρ is the amplitude reflection coefficient and τ is the amplitude transmission coefficient.

There are fundamental difficulties associated with the definitions of reflectance and transmittance unless the incident medium is absorption free; i.e., N_0 and y_0 are real. For that case,

$$R = \rho\rho^* = \left(\frac{y_0 - y_1}{y_0 + y_1}\right)\left(\frac{y_0 - y_1}{y_0 + y_1}\right)^*, \tag{2.79}$$

$$T = \frac{4y_0\,\text{Re}(y_1)}{(y_0 + y_1)(y_0 + y_1)^*}. \tag{2.80}$$

Oblique incidence calculations are simpler if the wave is split into two linearly polarized components, one with the electric vector in the plane of incidence, known as p-polarized (or TM, for transverse magnetic field), and one with the electric vector normal to the plane of incidence, known as s-polarized (or TE, for transverse electric field). The propagation of each of these two waves can be treated quite independently of the other. Calculations are further simplified if only energy flows normal to the boundaries and electric and magnetic fields parallel to the boundaries are considered, because then we have a formulation which is equivalent to a homogeneous wave.

We must introduce the idea of a tilted optical admittance η, which is given by

$$\eta_p = \frac{N\mathscr{Y}}{\cos\vartheta} \quad \text{(for p-waves)}$$

$$\eta_s = N\mathscr{Y}\cos\vartheta \quad \text{(for s-waves)},$$

(2.88)

where N and ϑ denote either N_0 and ϑ_0 or N_1 and ϑ_1 as appropriate. ϑ_1 is given by Snell's law, in which complex angles may be included:

$$N_0 \sin\vartheta_0 = N_1 \sin\vartheta_1.$$

(2.136)

Denoting η_p or η_s by η, we have, for either direction of polarization,

$$\rho = \frac{\eta_0 - \eta_1}{\eta_0 + \eta_1},$$

(2.91)

$$\tau = \frac{2\eta_0}{\eta_0 + \eta_1},$$

(2.92)

If η_0 is real, we can write

$$R = \left(\frac{\eta_0 - \eta_1}{\eta_0 + \eta_1}\right)\left(\frac{\eta_0 - \eta_1}{\eta_0 + \eta_1}\right)^*,$$

(2.93)

$$T = \frac{4\eta_0 \,\mathrm{Re}(\eta_1)}{(\eta_0 + \eta_1)(\eta_0 + \eta_1)^*}.$$

(2.94)

The phase shift experienced by the wave as it traverses a distance d normal to the boundary is then given by $-2\pi Nd\cos\vartheta/\lambda$.

The reflectance of an assembly of thin films is calculated through the concept of optical admittance. We replace the multilayer by a single surface which presents an admittance Y, which is the ratio of the total tangential magnetic and electric fields and is given by

$$Y = C/B,$$

(2.137)

where

$$\begin{bmatrix} B \\ C \end{bmatrix} = \left\{ \prod_{r=1}^{q} \begin{bmatrix} \cos\delta_r & (i\sin\delta_r)/\eta_r \\ i\eta_r\sin\delta_r & \cos\delta_r \end{bmatrix} \right\} \begin{bmatrix} 1 \\ \eta_m \end{bmatrix}.$$

(2.113)

$\delta_r = 2\pi Nd\cos\vartheta/\lambda$ and η_m = substrate admittance. δ is the phase thickness of the film. In this case, it is the tilted phase thickness containing the $\cos\vartheta$ factor.

The order of multiplication is important. If q is the layer next to the substrate, then the order is

$$\begin{bmatrix} B \\ C \end{bmatrix} = [M_1][M_2]...[M_q]\begin{bmatrix} 1 \\ \eta_m \end{bmatrix}. \tag{2.138}$$

M_1 indicates the matrix associated with layer 1 and so on. Y and η are in the same units. If η is in siemens, then so is Y, or if η is in free space units (i.e., units of $\mathscr{Y}$), then Y will be in free space units. As in the case of a single surface, η_0 must be real for reflectance and transmittance to have a valid meaning. With that proviso, then

$$R = \left(\frac{\eta_0 B - C}{\eta_0 B + C}\right)\left(\frac{\eta_0 B - C}{\eta_0 B + C}\right)^*, \tag{2.123}$$

$$T = \frac{4\eta_0 \operatorname{Re}(\eta_m)}{(\eta_0 B + C)(\eta_0 B + C)^*}, \tag{2.125}$$

$$A = \frac{4\eta_0 \operatorname{Re}(BC^* - \eta_m)}{(\eta_0 B + C)(\eta_0 B + C)^*}, \tag{2.126}$$

$$\psi = \text{potential transmittance} = \frac{T}{(1 - R)} = \frac{\operatorname{Re}(\eta_m)}{\operatorname{Re}(BC^*)}. \tag{2.127}$$

Phase shift on reflection, measured at the front surface of the multilayer, is given by

$$\varphi = \arctan\left(\frac{\operatorname{Im}[\eta_m(BC^* - CB^*)]}{(\eta_m^2 BB^* - CC^*)}\right), \tag{2.128}$$

and that on transmission is measured between the emergent wave as it leaves the multilayer and the incident wave as it enters, by

$$\zeta = \arctan\left[\frac{-\operatorname{Im}(\eta_0 B + C)}{\operatorname{Re}(\eta_0 B + C)}\right]. \tag{2.129}$$

The signs of the numerator and the denominator in these expressions must be preserved separately. Then, the quadrants are given by the arrangement in the following:

Numerator	+	+	−	−
Denominator	+	−	+	−
Quadrant	1st	2nd	4th	3rd

Despite the apparent simplicity of Equation 2.113, numerical calculations without some automatic aid are tedious in the extreme. Even with the help of a calculator, the labor involved in determining the performance of an assembly of more than a very few transparent layers at one or two wavelengths is completely discouraging. At the very least, a programmable calculator of reasonable capacity is required. Extended calculations are usually carried out on a computer.

However, insight into the properties of thin-film assemblies cannot easily be gained simply by feeding the calculations into a computer, and insight is necessary if filters are to be designed and if their limitations in use are to be fully understood. Studies of the properties of the characteristic matrices have been carried out, and some results which are particularly helpful in this context have been obtained. Approximate methods, especially graphical ones, have also been found useful.

2.10 Potential Transmittance

We find transmittance a rather less accessible parameter from the point of view of the theory than reflectance. Reflectance is immediately a function of the admittance of the front surface of the multilayer. To calculate transmittance, we need further information. A useful concept related to transmittance, but rather more accessible and susceptible to theoretical manipulation, is potential transmittance.

The potential transmittance of a layer, or an assembly of layers, is the ratio of the irradiance leaving by the rear, or exit, interface to that entering by the front interface. The concept was introduced by Berning and Turner [6], and we will make considerable use of it in designing metal–dielectric filters and in calculating losses in all-dielectric multilayers. Potential transmittance is denoted by ψ and is given by

$$\psi = \frac{I_{\text{exit}}}{I_{\text{enter}}} = \frac{T}{(1-R)}, \tag{2.139}$$

that is the ratio between the irradiance leaving the assembly and the net irradiance actually entering. For the entire system, the net irradiance actually entering is the difference between the incident and reflected irradiances. Note that in accordance with the definition of reflectance and transmittance, the irradiances concerned are the normal components.

The potential transmittance of a series of subassemblies of layers is simply the product of the individual potential transmittances. Figure 2.10 shows a series of film subunits making up a complete system. Clearly,

$$\psi = \frac{I_e}{I_i} = \frac{I_d}{I_a} = \frac{I_b}{I_a}\frac{I_c}{I_b}\frac{I_b}{I_c} = \psi_1\psi_2\psi_3. \tag{2.140}$$

The potential transmittance is fixed by the parameters of the layer, or combination of layers, involved, and by the characteristics of the structure at the exit interface, and it represents the transmittance, which this particular combination would give if there were no reflection losses.

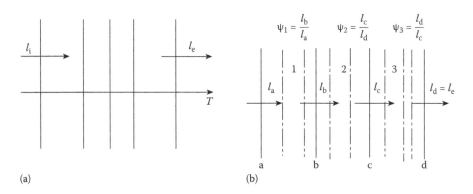

FIGURE 2.10
(a) An assembly of thin films. (b) The potential transmittance of an assembly of thin film consisting of a number of subunits.

Thus, it is a measure of the maximum transmittance which could be expected from the arrangement. By definition, the potential transmittance is unaffected by any transparent structure deposited over the front surface—which can affect the transmittance as distinct from the potential transmittance—and to ensure that the transmittance is equal to the potential transmittance, the layers added to the front surface must maximize the irradiance actually entering the assembly. This implies reducing the reflectance of the complete assembly to zero or, in other words, adding an antireflection coating. The potential transmittance is, however, affected by any changes in the structure at the exit interface, and it is possible to maximize the potential transmittance of a subassembly in this way.

There is, however, a problem with Equation 2.139. Reflectance, as we have seen, is defined only in media that are free of absorption. Equation 2.140 is correct as long as we interpret the irradiances as the net irradiance passing through the appropriate interface. In our matrix expression for the properties of any combination of thin films, we use B and C to denote the appropriate normalized electric and magnetic fields at any particular interface. Then, the net normalized irradiance is given by $(1/2)\,\mathrm{Re}\,(BC^*)$. Since we are dealing with ratios, we can drop the factor $(1/2)$ and replace the irradiances in Equations 2.139 and 2.140 by $\mathrm{Re}\,(BC^*)$.

We now show that the parameters of the layer, or subassembly of layers, together with the optical admittance at the rear surface, are sufficient to define the potential transmittance. Let the complete multilayer performance be given by

$$\begin{bmatrix} B \\ C \end{bmatrix} = [M_1][M_2]\dots[M_a][M_b][M_c]\dots[M_p]\begin{bmatrix} M_q \end{bmatrix}\begin{bmatrix} 1 \\ \eta_m \end{bmatrix},$$

where we want to calculate the potential transmittance of the subassembly $[M_a][M_b][M_c]$. Let the product of the matrices to the right of the subassembly be given by

$$\begin{bmatrix} B_e \\ C_e \end{bmatrix}.$$

Now, if

$$\begin{bmatrix} B_i \\ C_i \end{bmatrix} = [M_a][M_b][M_c]\begin{bmatrix} B_e \\ C_e \end{bmatrix}, \tag{2.141}$$

then

$$\psi = \frac{\mathrm{Re}(B_e C_e^*)}{\mathrm{Re}(B_i C_i^*)}. \tag{2.142}$$

By dividing Equation 2.141 by B_e, we have

$$\begin{bmatrix} B'_i \\ C'_i \end{bmatrix} = [M_a][M_b][M_c]\begin{bmatrix} 1 \\ Y_e \end{bmatrix},$$

where $Y_e = C_e/B_e$, $B'_1 = B_1/B_e$, $C'_1 = C_1/C_e$, and the potential transmittance is

$$\psi = \frac{\mathrm{Re}(Y_e)}{\mathrm{Re}(B'_i C'^*_i)}$$

$$= \frac{\mathrm{Re}(C_e/B_e)}{\mathrm{Re}[(B_i/B_e)(C_i^*/B_e^*)]} = \frac{B_e B_e^* \,\mathrm{Re}(C_e/B_e)}{\mathrm{Re}(B_i C_i^*)}$$

$$= \frac{\mathrm{Re}(B_e^* C_e)}{\mathrm{Re}(B_i C_i^*)} = \frac{\mathrm{Re}(B_e C_e^*)}{\mathrm{Re}(B_i C_i^*)},$$

which is identical with Equation 2.142. Thus, the potential transmittance of any subassembly is determined solely by the characteristics of the layer or layers of the subassembly, together with the optical admittance of the structure at the exit interface.

Further expressions involving potential transmittance will be derived as they are required.

2.11 Theorem on the Transmittance of a Thin-Film Assembly

The transmittance of a thin-film assembly is independent of the direction of propagation of the light. This applies regardless of whether or not the layers are absorbing.

A proof of this result, due to Abelès [7,8], who was responsible for the development of the matrix approach to the analysis of thin films, quickly follows from the properties of the matrices.

Let the matrices of the various layers in the assembly be denoted by

$$[M_1], [M_2], ..., \left[M_q \right],$$

and let the two massive media on either side be transparent. The two products of the matrices corresponding to the two possible directions of propagation can be written as

$$[M] = [M_1][M_2][M_3], ..., \left[M_q \right]$$

and

$$[M'] = \left[M_q \right] \left[M_{q-1} \right] ... [M_2][M_1].$$

Now, because the form of the matrices is such that the diagonal terms are equal, regardless of whether there is absorption or not, we can show that if we write

$$[M] = \left[a_{ij} \right] \quad \text{and} \quad [M'] = \left[a'_{ij} \right],$$

then

$$a_{ij} = a'_{ij} \ (i \neq j), \quad a_{11} = a'_{22}, \quad \text{and} \quad a_{22} = a'_{11}.$$

This can be proved simply by induction.

We denote the medium on one side of the assembly by η_0 and on the other by η_m, where η_0 is next to layer 1. In the case of the first direction the characteristic matrix is given by (Equation 2.113)

$$\begin{bmatrix} B \\ C \end{bmatrix} = [M] \begin{bmatrix} 1 \\ \eta_m \end{bmatrix}$$

and

$$B = a_{11} + a_{12}\eta_m, \quad C = a_{21} + a_{22}\eta_m.$$

In the second case,

$$B = a'_{11} + a'_{12}\eta_0 = a_{22} + a_{12}\eta_0,$$

$$C = a'_{21} + a'_{22}\eta_0 = a_{21} + a_{11}\eta_0.$$

The two expressions for the transmittance of the assembly are then, from Equation 2.125,

$$T = \frac{4\eta_0\eta_m}{\left|\eta_0(a_{11} + a_{12}\eta_m) + a_{21} + a_{22}\eta_m\right|^2},$$

$$T' = \frac{4\eta_m\eta_0}{\left|\eta_m(a_{22} + a_{12}\eta_0) + a_{21} + a_{11}\eta_0\right|^2},$$

which are identical.

This rule does not, of course, apply to the *reflectance* of an assembly, which will necessarily be the same on both sides of the assembly only if there is no absorption in any of the layers.

Among other things, this expression shows that the one-way mirror, which allows light to travel through it in one direction only, cannot be constructed from simple optical thin films. The common so-called one-way mirror has a high reflectance with some transmittance and relies on an appreciable difference in the illumination conditions existing on either side for its operation.

2.12 Coherence

Coherence is a concept that quantifies the ability to produce detectable interference effects. Although often used to describe the properties of a beam of light, it is much better considered as a property of the complete optical system. We usually use the adjective *coherent* to describe the presence of maximum interference effects, and *incoherent*, their complete absence. Cases in between these limits are described as partially coherent. Here we take a brief look at some of the aspects of coherence that are important in optical coating applications. There are few preconditions. We can imagine that we are carrying out an experiment. We would find it difficult if there were serious fluctuations in the measurements. The same considerations apply to the ideas of coherence. We assume that in all the cases we discuss, we can describe the phenomena as stationary. This implies that although the waves may exhibit variations, their statistical properties do not vary with time. Thus, in what follows, we will simply average them over a sufficiently long time, or sufficiently large number of cases, just as though we were conducting an experiment.

As always, we take advantage of the linear nature of the interactions and represent an arbitrary wave by a corresponding spectrum of harmonic waves. We first consider two beams derived from one single spectral element. Let two beams be linearly polarized plane waves with identical polarization, propagating together along the z-direction, and with complex amplitudes $\mathcal{E}_1$ and $\mathcal{E}_2$. Let the phases of these two waves at $z = 0$ and $t = 0$ be φ_1 and φ_2, with φ_1 and φ_2 contained in the complex amplitudes. Now let us combine the waves. Since the polarizations are exactly equal, the electric and magnetic fields of the resultants will be the simple sums of the components. The resultant irradiance will therefore be

$$I = \tfrac{1}{2}\,\mathrm{Re}[(E_1 + E_2)(H_1 + H_2)^*] = \tfrac{1}{2}\,\mathrm{Re}[(E_1 + E_2)y^*(E_1 + E_2)^*]$$

$$= \tfrac{1}{2}\,\mathrm{Re}(y)E_1 E_1^* + \tfrac{1}{2}\,\mathrm{Re}(y)E_2 E_2^* + \mathrm{Re}(y)\,\mathrm{Re}(E_1 E_2^*). \tag{2.143}$$

The common phase factor cancels to give

$$I = I_1 + I_2 + \mathrm{Re}(y)\,\mathrm{Re}(\mathcal{E}_1\mathcal{E}_2^*)$$

$$= I_1 + I_2 + 2\sqrt{I_1 I_2}\cos(\varphi_1 - \varphi_2). \tag{2.144}$$

This is the sum of the two independent irradiances plus a third term representing the interference effect and known as the interference term. This interference term can be positive or negative, depending on the argument of the cosine factor.

Now imagine that many more light waves are involved. These may have different frequencies and/or different phases, and/or different amplitudes. They will also contribute to the resultant irradiance that will then be the sum of the individual irradiances together with a sum of interference terms. Everything depends on the spread of parameters across the various light waves. The interference term may remain just as strong or, since the terms may be positive or negative, may cancel out altogether, or may take on some intermediate value. We describe the first case as coherent, the second as incoherent, and the intermediate case as partially coherent.

The jumbling of the interference term may result from variations in the source of illumination but equally well from variations in any other feature of the system. For example, if a thin film is involved, there could be variations in thickness across the aperture of the film. Substrates are usually thick enough and sufficiently variable in their thicknesses that no interference effects are observed; that is, substrates normally exhibit incoherent behavior.

The light in any real system is never a single infinite, plane, linearly polarized harmonic wave, and the components of the system are never perfect. There are some simple parameters that help in our assessment of coherence. For convenience, we will first think of them as a function of the light yet bear in mind that the system is ultimately what is important. Let us extract two rays from exactly the same point in a beam of light and let them produce some kind of interference effect involving fringes. Now let us gradually move apart the points where we extract the two rays. As the distance between the two points increases, the interference fringes become gradually less pronounced. When the fringes just disappear, we can take the distance between the two points as a measure of coherence. If the two points are separated along the direction of propagation of the primary light, then we call that distance the coherence length. If the light, instead of being extracted from two points, is taken from a complete area of the primary light, then that area where the fringes just disappear is called the coherence area, and if from a volume, the coherence volume. If, instead of moving the points, we delay one sample by a variable time, then the delay at which the fringes just disappear is the coherence time. Clearly, coherence time is just the time it takes for the light to travel the coherence length. Coherence length and coherence time are particularly useful concepts. They are quite simple and yet can explain many otherwise puzzling phenomena.

Let us express our primary light in terms of its spectral components and let the spectrum be a continuous band of wavelengths centered on λ and with a bandwidth of $\Delta\lambda$. We produce interference fringes using this light with a variable path difference. Each elemental wavelength will produce its own set of fringes, and as we increase the path difference, the fringes will be smeared out and the contrast will fall. The path difference at that point where they just disappear will be a measure of the coherence length. We can define the point as that where the fringes will become smeared out over the interval between fringes. In other words, a fringe of order m at the smallest wavelength will just coincide with one of order $m - 1$ at the longest wavelength.

$$\text{Path difference} = m\left(\lambda - \frac{\Delta\lambda}{2}\right) = (m - 1)\left(\lambda + \frac{\Delta\lambda}{2}\right), \tag{2.145}$$

and some mild analysis arrives at the result

$$\text{Coherence length} = \left(m - \frac{1}{2}\right) \cdot \lambda = \frac{\lambda^2}{\Delta\lambda}. \tag{2.146}$$

This is an exceptionally important and useful result. The coherence time will be this result divided by the velocity of light. Since these are all vacuum wavelengths, the coherence length is automatically referred to vacuum and the coherence time will be given by

$$\text{Coherence time} = \frac{\lambda^2}{c\Delta\lambda}, \tag{2.147}$$

where c is 299.792458 nm/fs.

Cones of illumination also have coherence effects, and we can calculate an effective coherence length that can be useful. If ϑ is the cone semiangle and the cone axis is at normal incidence, then, using an approximate two-dimensional model, we can rewrite Equation 2.145 as

$$m\lambda = (m+1)\lambda\cos\vartheta, \tag{2.148}$$

and a similar extraction process yields

$$\text{Coherence length} = m\lambda = \frac{\lambda\cos\vartheta}{1-\cos\vartheta} \simeq \frac{2\lambda}{\vartheta^2}, \tag{2.149}$$

or, if ϑ is given in degrees,

$$\text{Coherence length} = \frac{7\times10^3\lambda}{\vartheta^2}. \tag{2.150}$$

This ϑ is the angle within the particular thin-film structure. If the incident medium is of a different index, then we can make an approximate correction to Equation 2.150 to give

$$\text{Coherence length} = \frac{7\times10^3\lambda n_1^2}{n_0^2\vartheta^2}, \tag{2.151}$$

where n_0 is the incident medium index and n_1 is either the index of the film concerned or the effective index of the coating. This particular coherence length is a function of the properties of the cone and the properties of the film.

When several independent effects are present, such as both cone and bandwidth, then a reasonable rule for combining coherence lengths is

$$\frac{1}{L} = \sqrt{\frac{1}{L_a^2} + \frac{1}{L_b^2} + \cdots}. \tag{2.152}$$

Figure 2.11 shows the appearance of fringes calculated for a 1 mm thick glass substrate where the uncoated surfaces are perfectly parallel. The fringes are scanned with light of zero bandwidth, that is infinite coherence length, where the fringes have their maximum theoretical amplitude. This is the completely coherent case. Scanning with 0.2 nm bandwidth corresponds to a coherence length of 5 mm, and the fringes are roughly halved in amplitude. This is partial coherence. Finally, with a bandwidth of 2 nm, the coherence length becomes 0.5 mm and the fringes virtually disappear to give the incoherent case.

In the bulk of this book, we shall assume complete coherence when discussing the properties of the thin-film coatings and complete incoherence when dealing with the substrates, and we return to this point in Chapter 16. However, it is perhaps worthwhile to look just a little further at the concept of partial coherence. The key lies in the interference term of Equation 2.143, $\text{Re}(y)\,\text{Re}(E_1 E_2^*)$. Although the discussion of this expression assumed that E_1 and E_2 represented harmonic waves, the derivation is quite general. E_1 and E_2 can represent any arbitrary waves, the only condition

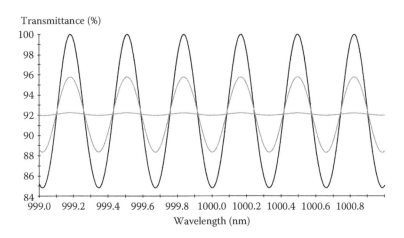

Transmittance (%)

Wavelength (nm)

FIGURE 2.11
Appearance of fringes in a perfectly parallel 1 mm thick glass substrate illuminated with light of zero bandwidth, 0.2 nm bandwidth (5 mm coherence length), and 2 nm bandwidth (0.5 mm coherence length). The fringes show reducing amplitude with decreasing coherence length.

being that the characteristic admittance of the medium y should be constant so that the magnetic field H can be replaced by yE.

We began by assuming that the essential difference between E_2 and E_1 was the phase angle φ, and the interference phenomenon was, therefore, a function of φ. If we are now permitting E_1 and E_2 to represent quite general waves, φ has no meaning and makes no sense, because it applies only to a harmonic wave. However, E_1 and E_2 are both functions of time, and so we can introduce a variable time delay τ to take the place of φ. To keep matters simple, we take just the $E_1 E_2{}^*$ part. Now that we have made the waves much more general, we should take an average over a sufficiently long time or over a sufficiently large number of measurements. We then obtain what is called the mutual coherence function:

$$\Gamma_{12}(\tau) = \langle E_1(t+\tau)E_2^*(t)\rangle = \lim_{T\to\infty}\frac{1}{2T}\int_{-T}^{T}E_1(t+\tau)E_2^*(t)\mathrm{d}t. \tag{2.153}$$

Should we multiply E_1 and E_2 by an identical factor, then we multiply the mutual coherence function by the square of that same factor, but there is no corresponding change in the nature of the interference phenomenon. To make it more representative of the interference phenomenon, and, hence, the coherence, we normalize it to give

$$\gamma_{12}(\tau) = \frac{\Gamma_{12}(\tau)}{\sqrt{\Gamma_{11}(0)}\sqrt{\Gamma_{22}(0)}} = \frac{\langle E_1(t+\tau)E_2^*(t)\rangle}{\sqrt{\langle E_1(t)E_1^*(t)\rangle}\sqrt{\langle E_2(t)E_2^*(t)\rangle}}, \tag{2.154}$$

where $\gamma_{12}(\tau)$ is known as the complex degree of coherence.

We can now rewrite Equation 2.143 as

$$\begin{aligned}
I &= \tfrac{1}{2}\operatorname{Re}(y)\langle E_1(t)E_1^*(t)\rangle + \tfrac{1}{2}\operatorname{Re}(y)\langle E_2(t)E_2^*(t)\rangle + \operatorname{Re}(y)\operatorname{Re}(\langle E_1(t+\tau)E_2^*(t)\rangle) \\
&= \tfrac{1}{2}\operatorname{Re}(y)\langle E_1(t)E_1^*(t)\rangle + \tfrac{1}{2}\operatorname{Re}(y)\langle E_2(t)E_2^*(t)\rangle \\
&\quad + \operatorname{Re}(y)\sqrt{\langle E_1(t)E_1^*(t)\rangle\langle E_2(t)E_2^*(t)\rangle}\operatorname{Re}[\gamma_{12}(\tau)] \\
&= I_1 + I_2 + 2\sqrt{I_1 I_2}\operatorname{Re}[\gamma_{12}(\tau)],
\end{aligned} \tag{2.155}$$

when we can see that the real part of the complex degree of coherence is related to the fringe shape. When we have fringes that are not too far from a cosine or a sine profile, we can define a fringe visibility that is given by

$$V = \frac{I_{max} - I_{min}}{I_{max} + I_{min}}. \tag{2.156}$$

The fringe is a cyclic function. How do we find the maximum and minimum extrema? This is where the complex degree of coherence turns out to be particularly useful. Although the real part is a fluctuating function, the full complex form is essentially a rotating vector. The magnitude, therefore represents the amplitude of the fringe function, and the visibility is given by

$$V = \frac{2\sqrt{I_1 I_2}|\gamma_{12}(\tau)|}{I_1 + I_2}. \tag{2.157}$$

When I_1 and I_2 are equal, the fringe visibility is at a maximum with value $|\gamma|$.

Direct evaluation of the complex degree of coherence from Equation 2.154 is not easy, but Equation 2.157 connects it simply to the visibility of the fringes.

We have concentrated on time as the variable in generating the coherence functions, but clearly, this could also be in terms of distance. The coherence time and coherence length are simply that value of time or distance at which the magnitude of the complex degree of coherence falls to a sufficiently low figure. This level tends to be chosen as whatever is most convenient in the particular calculation concerned.

There is much more to coherence, but this abbreviated account has covered the major aspects of importance in thin-film optical coatings.

2.13 Mixed Poynting Vector

We have already alluded to the mixed Poynting vector when discussing absorbing media. Now that we have briefly looked at the phenomenon of coherence, it is useful to briefly return to the vector. We extensively rely on the paper by Macleod [9].

Let there be two waves of identical frequency defined only as possessing such polarization and direction that the resultant fields are simply the sum of the individual fields. Then, denoting the fields of the waves by the subscripts 1 and 2, we find the complex Poynting expression:

$$\begin{aligned}
I &= \tfrac{1}{2}\,\text{Re}[(E_1 + E_2)(H_1 + H_2)^*] \\
&= \tfrac{1}{2}\,\text{Re}(E_1 H_1^*) + \tfrac{1}{2}\,\text{Re}(E_2 H_2^*) + \tfrac{1}{2}\,\text{Re}(E_2 H_1^* + E_1 H_2^*) \\
&= I_1 + I_2 + \tfrac{1}{2}\,\text{Re}(E_2 H_1^* + E_1 H_2^*),
\end{aligned} \tag{2.158}$$

where we have treated the individual irradiances I_1 and I_2 as we would were they propagating completely independently. The third term, involving the mixed products, is the mixed Poynting expression, or mixed Poynting vector when written as a vector quantity. The mixed Poynting vector is usually invoked when reflection at an interface is involved, and hence, the waves are counterpropagating. However, there is no good reason to limit the concept to one particular configuration. It can be readily recognized as equivalent to the interference term of the previous section.

A k of zero exactly leads to the interference term when the waves are monochromatic and propagating in the same direction or to the mutual coherence function when there is a mixture of waves of differing frequency and relative phase. The particular interest of the mixed Poynting vector is when the propagation is in opposite directions. We retain zero k but reverse wave 2 so that it propagates along the negative z-direction. The phase factors become

$$\text{Wave 1: } \exp[i(\omega t - \kappa z)]; \quad \text{wave 2: } \exp[i(\omega t + \kappa z + \varphi)], \tag{2.159}$$

where κ is real. In addition to the change in sign in the phase factor, we must also make sure that the Poynting vector for wave 2 points along the negative z-direction, and we do that by flipping the direction of the magnetic field so that

$$I = \tfrac{1}{2} \, \text{Re}[(E_1 + E_2)(H_1 - H_2)^*]$$

$$= I_1 - I_2 + \tfrac{1}{2} \, \text{Re}[\mathcal{E}_2 \mathcal{H}_1 \exp\{i(2\kappa z + \varphi)\} - \mathcal{E}_1 \mathcal{H}_2 \exp\{-i(2\kappa z + \varphi)\}] \tag{2.160}$$

$$= I_1 - I_2 + \tfrac{1}{2} y \mathcal{E}_2 \mathcal{E}_1 \, \text{Re}[\exp\{i(2\kappa z + \varphi)\} - \exp\{-i(2\kappa z + \varphi)\}],$$

where we are retaining our definition of what we understand as I_1 and I_2. In this case, because of the minus sign, the bracketed term in the mixed Poynting vector is totally imaginary, and so its real part is zero, and therefore,

$$I = I_1 - I_2. \tag{2.161}$$

Now let us introduce absorption into our supporting medium. This implies that both refractive index and characteristic admittance will be complex, and, in our sign convention, of the forms $(n - ik)$ and $y = (n - ik)\mathcal{Y}, \mathcal{Y}$ indicating the admittance of free space. There is little effect in the result for waves propagating in parallel except for an exponential decay, but for counterpropagating waves, there is a major change.

$$\text{Wave 1: } E_1 = \mathcal{E}_1 \exp\left[i\left\{\omega t - \frac{2\pi(n - ik)}{\lambda} z\right\}\right]$$

$$= \mathcal{E}_1 \exp\left[-\frac{\alpha z}{2}\right] \exp[i\{\omega t - \kappa z\}],$$

$$\text{Wave 2: } E_2 = \mathcal{E}_2 \exp\left[i\left\{\omega t + \frac{2\pi(n - ik)}{\lambda} z + \varphi\right\}\right] \tag{2.162}$$

$$= \mathcal{E}_2 \exp\left[\frac{\alpha z}{2}\right] \exp[i\{\omega t + \kappa z + \varphi\}].$$

By expressing H immediately in terms of E, we avoid the complication of the phase difference between them:

$$I = \tfrac{1}{2} \, \text{Re}(y)E_1 E_1^* - \tfrac{1}{2} \, \text{Re}(y)E_2 E_2^* + \tfrac{1}{2} \, \text{Re}[y^*(E_2 E_1^* - E_1 E_2^*)]$$

$$= I_1 e^{-\alpha z} - I_2 e^{\alpha z} + \tfrac{1}{2} \, \text{Re}[y^*(E_2 E_1^* - E_1 E_2^*)], \tag{2.163}$$

giving, for the third term, the mixed Poynting vector:

$$\tfrac{1}{2} \mathcal{Y} \mathcal{E}_1 \mathcal{E}_2 \, \text{Re}[(n + ik)(\exp[i\{2\kappa z + \varphi\}] - \exp[-i\{2\kappa z + \varphi\}])]$$

$$= -k \mathcal{Y} \mathcal{E}_1 \mathcal{E}_2 \sin(2\kappa z + \varphi) \tag{2.164}$$

$$= -2\frac{k}{n} \sqrt{I_1 I_2} \sin(2\kappa z + \varphi).$$

Thus, the mixed Poynting vector is essentially a general interference term demonstrating that counterpropagating waves can actually interfere provided there is absorption present. It prevents us from separating the irradiances of the two waves as we can in a medium free from absorption. Curiously, unlike the interference term for parallel waves in an absorbing medium, its amplitude of fluctuation remains constant with z.

The strange behavior is confined to the irradiances, and there is no coupling of the amplitudes. The effect is well understood within our normal electromagnetic theory, although its history did not always demonstrate a correct understanding. We emphasize that our thin-film theory includes without modification all the effects of the mixed Poynting expression without any additional effort on our part. It is included here not only for completeness, but also for interest. More detail can be found in the paper by Macleod [9].

2.14 Incoherent Reflection at Two or More Surfaces

So far, we have treated substrates as being one-sided slabs of material of infinite depth. In almost all practical cases, the substrate will have finite depth with rear surfaces that reflect some of the energy and affect the performance of the assembly.

The depth of the substrate will usually be much greater than the wavelength of the light and variations in the flatness and parallelism of the two surfaces will be appreciable fractions of a wavelength. Generally, the incident light will not be particularly well collimated. Under these conditions, it will not be possible with a finite aperture to observe interference effects between light reflected at the front and rear surfaces of the substrate, and because of this, the substrate is known as thick. The coherence length is small compared with the double traversal of the substrate that is the basic path difference in any interference effects. We describe the addition of the various waves as incoherent rather than coherent. In Equation 2.144, the interference term vanishes, and we are left with the sum of the irradiances instead of the vector sum of the amplitudes.

The symbols used are illustrated in Figure 2.12. Waves are successively reflected at the front and rear surfaces. The sums of the irradiances are given by

$$R = R_a^+ + T_a^+ R_b^+ T_a^- \left[1 + R_a^- R_b^+ + (R_a^- R_b^+)^2 + \ldots \right]$$
$$= R_a^+ + [T_a^+ R_b^+ T_a^- / (1 - R_a^- R_b^+)];$$

i.e., since T^+ and T^- are always identical:

$$T_a^+ = T_a^- = T_a,$$

and so

$$R = \frac{R_a^+ + R_b^+ (T_a^2 - R_a^- R_a^+)}{1 - R_a^- R_b^+}.$$

If there is no absorption in the layers,

$$R_a^+ = R_a^- = R_a \quad \text{and} \quad 1 = R_a + T_a,$$

so that

$$R = \frac{R_a + R_b - 2 R_a R_b}{1 - R_a R_b}.$$

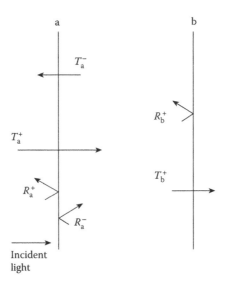

FIGURE 2.12
Symbols used in calculation of incoherent reflection at two or more surfaces.

Similarly,

$$T = T_a^+ T_b^+ \left[1 + R_a^- R_b^+ + (R_a^- R_b^+)^2 + \ldots\right]$$

$$= \frac{T_a T_b}{1 - R_a^- R_b^+},$$

and again, if there is no absorption,

$$T = \frac{T_a T_b}{1 - R_a R_b} \tag{2.165}$$

or

$$T = \left(\frac{1}{T_a} + \frac{1}{T_b} - 1\right)^{-1} \tag{2.166}$$

since

$$R_a = 1 - T_a, \quad R_b = 1 - T_b.$$

A nomogram for solving Equation 2.166 can be easily constructed. Two axes at right angles are laid out on a sheet of graph paper, and taking the point of intersection as the zero, two linear equal scales of transmittance are marked out on the axes. One of these is labeled T_a, and the other, T_b. The angle between T_a and T_b is bisected by a third axis which is to have the T scale marked out on it. To do this, a straight edge is placed so that it passes through the 100% transmittance value on, say, the T_a axis and any chosen transmittance on the T_b-axis. The value of T to be associated with the point where the straight edge crosses the T-axis is then that of the intercept with the T_b-axis. The entire scale can be marked out in this way. A completed nomogram of this type is shown in Figure 2.13.

In the absence of absorption, the analysis can be very simply extended to further surfaces. Consider the case of two substrates, i.e., four surfaces. We can label these T_a, T_b, T_c, and T_d. Then, from Equation 2.166, we have for the first substrate

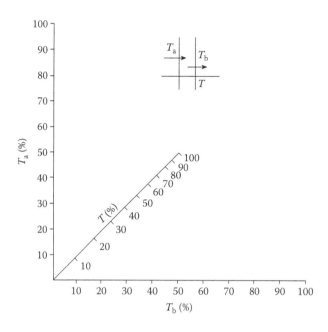

FIGURE 2.13
A nomogram for calculating the overall transmittance of a thick transparent plate given the transmittance of each individual surface.

$$T_1 = \left(\frac{1}{T_a} + \frac{1}{T_b} - 1 \right)^{-1},$$

i.e.,

$$\frac{1}{T_1} = \frac{1}{T_a} + \frac{1}{T_b} - 1,$$

and similarly for the second,

$$\frac{1}{T_2} = \frac{1}{T_c} + \frac{1}{T_d} - 1.$$

The transmittance through all four surfaces is then obtained by applying Equation 2.166 once again:

$$\frac{1}{T} = \frac{1}{T_1} + \frac{1}{T_2} - 1,$$

i.e.,

$$T = \left(\frac{1}{T_a} + \frac{1}{T_b} + \frac{1}{T_c} + \frac{1}{T_d} - 3 \right)^{-1}. \tag{2.167}$$

The iterative nature of these calculations can be clumsy when dealing with a succession of surfaces. A technique based on a study by Baumeister et al. [10] yields a rather more useful matrix form of the calculation. The emphasis is placed on the flows of irradiance. Absorption in the media between the coated surfaces is supposed to be sufficiently small so that the coupling problem mentioned earlier is negligible. The symbols are defined in Figure 2.14.

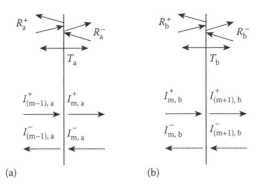

FIGURE 2.14
Symbols defining two successive coatings with intervening medium in a stack. The source of light is on the left. Panels (a) and (b) represent two coated surfaces belonging to an intervening medium that is denoted by subscript m. The medium on the left is denoted by $m-1$ and on the right by $m+1$. R and T represent the reflectance and transmittance of the coated surfaces and are identified by subscripts a and b. I indicates irradiance at a surface, indicated by subscript a or b and measured just inside or outside the surface indicated by the additional subscript $m-1$, m, or $m+1$. The plus superscript represents a property of light in the incident direction and a minus superscript in the opposite direction. The added arrows indicating directions are intended to help in identification of the symbols.

The direction of the light is denoted by the usual plus and minus signs. a and b are two coatings separated by a medium m with internal transmittance T_{mint}. The final medium will be the emergent medium, and there, the negative-going irradiance will be zero. The procedure to be outlined will derive the values of I^+_{ma} and I^-_{ma} from $I^+_{(m+1)b}$ and $I^-_{(m+1)b}$. The rest is straightforward.

The irradiances on either side of the coating with label b are related through the equations

$$I^+_{(m+1)b} = T_b I^+_{mb} + R^-_b I^-_{(m+1)b},$$

$$I^-_{mb} = R^+_b I^+_{mb} + T_b I^-_{(m+1)b}.$$

These can be manipulated into the forms

$$I^-_{mb} = \frac{1}{T_b}\left\{ R^+_b I^+_{(m+1)b} + \left(T^2_b - R^-_b R^+_b\right) I^-_{(m+1)b} \right\},$$

$$I^+_{mb} = \frac{1}{T_b}\left\{ I^+_{(m+1)b} - R^-_b I^-_{(m+1)b} \right\},$$

and in matrix form this is

$$\begin{bmatrix} I^-_{mb} \\ I^+_{mb} \end{bmatrix} = \begin{bmatrix} \dfrac{\left(T^2_b - R^-_b R^+_b\right)}{T_b} & \dfrac{R^+_b}{T_b} \\ \dfrac{-R^-_b}{T_b} & \dfrac{1}{T_b} \end{bmatrix} \begin{bmatrix} I^-_{(m+1)b} \\ I^+_{(m+1)b} \end{bmatrix}. \tag{2.168}$$

The conversion through the medium is given by

$$\begin{bmatrix} I^-_{ma} \\ I^+_{ma} \end{bmatrix} = \begin{bmatrix} T_{\mathrm{mint}} & 0 \\ 0 & \dfrac{1}{T_{\mathrm{mint}}} \end{bmatrix} \begin{bmatrix} I^-_{mb} \\ I^+_{mb} \end{bmatrix}. \tag{2.169}$$

Equations 2.168 and 2.169 can be applied to the various coatings and intervening media in succession.

References

1. P. Yeh. 1988. *Optical Waves in Layered Media*. Hoboken, NJ: John Wiley & Sons.
2. I. J. Hodgkinson and Q. H. Wu. 1997. *Birefringent Thin Films and Polarizing Elements*. First ed. Singapore: World Scientific.
3. M. Born and E. Wolf. 2002. *Principles of Optics: Electromagnetic Theory of Propagation, Interference and Diffraction of Light*. Seventh ed. New York: Cambridge University Press.
4. P. H. Berning. 1963. Theory and calculations of optical thin films. In *Physics of Thin Films*, G. Hass (ed.), pp. 69–121. New York: Academic Press.
5. H. A. Macleod. 1995. Antireflection coatings on absorbing substrates. In *38th Annual Technical Conference*, pp. 172–175. Chicago, IL: Society of Vacuum Coaters.
6. P. H. Berning and A. F. Turner. 1957. Induced transmission in absorbing films applied to band pass filter design. *Journal of the Optical Society of America* 47:230–239.
7. F. Abelès. 1950. Recherches sur la propagation des ondes électromagnétiques sinusoïdales dans les milieus stratifiés: Applications aux couches minces: I. *Annales de Physique, 12ième Serie* 5:596–640.
8. F. Abelès. 1950. Recherches sur la propagation des ondes électromagnétiques sinusoïdales dans les milieus stratifiés: Applications aux couches minces: II. *Annales de Physique, 12ième Serie* 5:706–784.
9. A. Macleod. 2014. The mixed Poynting vector. In *Society of Vacuum Coaters Bulletin* 14(Summer):30–34.
10. P. Baumeister, R. Hahn, and D. Harrison. 1972. The radiant transmittance of tandem arrays of filters. *Optica Acta* 19:853–864.

3

Theoretical Techniques

The previous chapter deals with the fundamental theory culminating in exact expressions for the basic properties of thin-film systems. These are the expressions that we normally use in computing coating performance. There is more to designing, manufacturing, and using coatings than simple calculation. There are many additional techniques that can help us. Some of these are based on older approximate methods of calculation that predate digital computers and are used more nowadays for rapid understanding than for performance calculation. Others are accurate techniques that we will tend to use in an approximate way. Some are based on properties of the characteristic matrices themselves.

3.1 Quarter- and Half-Wave Optical Thicknesses

The characteristic matrix of a dielectric thin film takes on a very simple form if the optical thickness is an integral number of quarter or half waves. That is, if

$$\delta = m(\pi/4), \quad m = 0, 1, 2, 3, \ldots .$$

For m even, $\cos \delta = \pm 1$ and $\sin \delta = 0$, so that the layer is an integral number of half wavelengths thick, and the matrix becomes

$$\pm \begin{bmatrix} 1 & 0 \\ 0 & 1 \end{bmatrix}.$$

This is the unity matrix, and it can have no effect on the reflectance or transmittance of an assembly. It is as if the layer were completely absent. This is a particularly useful result, and because of it, half-wave layers are sometimes referred to as *absentee* layers. In the computation of the properties of any assembly, layers that are an integral number of half wavelengths thick can be completely omitted without altering the result. Of course, this is true only at the wavelength for which the layers are half waves.

For m odd, $\sin \delta = \pm 1$ and $\cos \delta = 0$, so that the layer is an odd number of quarter wavelengths thick, and the matrix becomes

$$\pm \begin{bmatrix} 0 & i/\eta \\ i\eta & 0 \end{bmatrix}.$$

This is not quite as simple as the half-wave case, but such a matrix is still easy to handle in calculations. In particular, if a substrate, or a combination of thin films, has an admittance of Y, then the addition of an odd number of quarter waves of admittance η alters the admittance of the assembly to η^2/Y. This is known as the quarter wave rule and makes the properties of a succession of quarter-wave layers very easy to calculate. The admittance of, say, a stack of five quarter-wave layers is

$$Y = \frac{\eta_1^2 \, \eta_3^2 \, \eta_5^2}{\eta_2^2 \, \eta_4^2 \, \eta_m},$$

where the symbols have their usual meanings.

Because of the simplicity of assemblies involving quarter- and half-wave optical thicknesses, designs are often specified in terms of fractions of quarter waves at a reference wavelength. Usually only two, or perhaps three, different materials are involved in designs, and a convenient shorthand notation for quarter-wave optical thicknesses is H, M, or L, where H refers to the highest of the three indices; M, the intermediate; and L, the lowest. Half waves are denoted by HH, MM, LL or $2H$, $2M$, and so on.

3.2 Admittance Loci

The admittance diagram, in common with the Smith chart and the reflection circle diagram, described later, is a graphical technique based on an exact solution of the appropriate equations. We imagine that the multilayer is gradually built up on the substrate layer by layer, immersed all the time in the final incident medium. As each layer in turn increases from zero thickness to its final value, the admittance of the multilayer at that stage of its construction, that is, the surface admittance of the final interface, is calculated, and the locus is plotted. Alternatively, we may imagine the multilayer as already constructed and then a reference plane is continuously slid through the layers. The locus of surface admittance at that plane, or we can think of it as the admittance of the structure up to that point, is plotted. Either of these views is equally valid, and the results are identical. (Note that only the first possibility applies to the reflection circle diagram and only the second to the Smith chart.) The loci for dielectric layers take the form of a series of circular arcs or even complete circles, each corresponding to a single layer, which are connected at points corresponding to the interfaces between the different layers. Perfect metals are also represented by arcs of circles. Absorbing materials give spiral loci. Although the technique can be used for quantitative calculation, it cannot compete even with a small programmable calculator, and its great value is in the visualization of the characteristics of a particular multilayer.

As the reference plane moves from the surface of the substrate to the front surface of the multilayer, let us calculate and plot the variation of the surface admittance at the reference plane. The matrix expression is

$$\begin{bmatrix} B \\ C \end{bmatrix} = \left\{ \prod_{r=1}^{q} \begin{bmatrix} \cos\delta_r & (i\sin\delta_r)/\eta_r \\ i\eta_r \sin\delta_r & \cos\delta_r \end{bmatrix} \right\} \begin{bmatrix} 1 \\ \eta_m \end{bmatrix},$$

where $Y = C/B$ is the surface admittance of the front interface. For the rth layer, we can write

$$\begin{bmatrix} B \\ C \end{bmatrix} = \begin{bmatrix} \cos\delta_r & (i\sin\delta_r)/\eta_r \\ i\eta_r \sin\delta_r & \cos\delta_r \end{bmatrix} \begin{bmatrix} B' \\ C' \end{bmatrix},$$

and since it is optical admittance that we are interested in, we can divide throughout by B' to give

$$\begin{bmatrix} B/B' \\ C/B' \end{bmatrix} = \begin{bmatrix} \cos\delta_r & (i\sin\delta_r)/\eta_r \\ i\eta_r \sin\delta_r & \cos\delta_r \end{bmatrix} \begin{bmatrix} 1 \\ Y' \end{bmatrix},$$

where $Y' = C/B'$ represents the surface admittance of the rear surface, that is, the front surface of the structure at the exit side of the layer. We now find the locus of the surface admittance:

$$Y = \frac{C}{B} = \frac{C/B'}{B/B'}.$$

Let

$$Y = x + iy$$

and

$$Y' = \alpha + i\beta,$$

and let the layer in question be dielectric so that η_r and δ_r are both real. Then,

$$Y = x + iy = \frac{(\alpha + i\beta)\cos\delta_r + i\eta_r\sin\delta_r}{\cos\delta_r + (\alpha + i\beta)(i\sin\delta_r)/\eta_r}$$

$$= \frac{\alpha\cos\delta_r + i(\beta\cos\delta_r + \eta_r\sin\delta_r)}{[\cos\delta_r - (\beta/\eta_r)\sin\delta_r] + i(\alpha/\eta_r)\sin\delta_r}.$$

Equating real and imaginary parts:

$$x[\cos\delta_r - (\beta/\eta_r)\sin\delta_r] - (y\alpha/\eta_r)\sin\delta_r = \alpha\cos\delta_r \tag{3.1}$$

$$y[\cos\delta_r - (\beta/\eta_r)\sin\delta_r] + (x\alpha/\eta_r)\sin\delta_r = \beta\cos\delta_r + \eta_r\sin\delta_r. \tag{3.2}$$

Eliminating δ_r yields

$$x^2 + y^2 - x[(\alpha^2 + \beta^2 + \eta_r^2)/\alpha] + \eta_r^2 = 0, \tag{3.3}$$

that is, the equation of a circle with center $((\alpha^2 + \beta^2 + \eta_r^2)/2\alpha, 0)$, i.e., on the real axis and with radius such that it passes through point (α, β), i.e., its starting point. The circle is traced out in a clockwise direction, which can be shown by setting $\beta = 0$ in Equation 3.2.

The scale of δ_r can also be plotted on the diagram. Let $\beta = 0$, and then, from Equations 3.1 and 3.2,

$$x - (y\alpha/\eta_r)\tan\delta_r = \alpha,$$

$$y + (x\alpha/\eta_r)\tan\delta_r = \eta_r\tan\delta_r.$$

Eliminating α, we have

$$x^2 + y^2 - y\eta_r(\tan\delta_r - 1/\tan\delta_r) - \eta_r^2 = 0. \tag{3.4}$$

This is a circle with center

$$(0, (\eta_r/2)(\tan\delta_r - 1/\tan\delta_r)),$$

i.e., on the imaginary axis and passing through point $(\eta_r, 0)$. The simplest contours of equal δ_r are $\delta_r = 0, \pi/2, \pi, 3\pi/2, \ldots$, which coincide with the real axis, and $\delta_r = \pi/4, 3\pi/4, 5\pi/4, \ldots$, which is the circle with center at the origin passing through point $(\eta_r, 0)$. For layers starting at a point not on the real axis, the same set of contours of equal δ_r will still apply, with a correction to the value of δ_r that each represents.

Figure 3.1a shows the locus of a film deposited on a transparent substrate of admittance α. The starting point is $(\alpha, 0)$, and as the thickness is increased to a quarter wave, a semicircle is traced out clockwise, which finally intersects the real axis again at point $(\eta_r^2/\alpha, 0)$. A second quarter wave completes the circle. We could have had any point on the locus as starting point without changing its form. The only difference would have been an offset in the scale of δ_r.

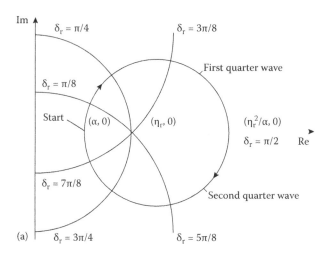

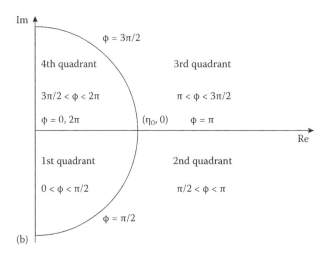

FIGURE 3.1
(a) Admittance locus of a single dielectric film. The locus is a circle centered on the real axis and described clockwise. The film of characteristic admittance y is assumed to be deposited over a substrate or structure with real admittance α. Note that the product of the admittance of the two points of intersection of the locus with the real axis is always y^2, the square of the characteristic admittance of the film. Equiphase-thickness contours have also been added to the diagram. (b) Contours of constant phase shift on reflection ϕ can be added to the admittance diagram. These contours are all circles with centers on the imaginary axis and passing through the point on the real axis corresponding to the admittance η_0 of the incident medium. The four most important contours correspond to 0, $\pi/2$, π, and $3\pi/2$, and these are represented by portions of the real axis and the circle centered on the origin and passing through point η_0. These are indicated on the diagram, and the regions corresponding to the various quadrants of ϕ are marked.

We could add isoreflectance contours to the diagram if we wished. These are circles with centers on the real axis, centers and radii being given by

$$(\eta_0(1+R)/(1-R),0) \quad \text{and} \quad 2\eta_0(R)^{1/2}/(1-R), \tag{3.5}$$

respectively, where η_0 is the admittance of the incident medium. Since the addition of incident medium material to a surface has no effect on its reflectance, the isoreflectance circles must also be admittance circles of material with admittance η_0.

The phase of the reflectance can also be important, and isophase contours are not unlike the contours of constant δ_r. We can carry through a similar procedure to determine the contours, and the most important ones are 0, $\pi/2$, π, and $3\pi/2$, that is, the boundaries between the quadrants. The boundary between the first and fourth and between the second and third is simply the real axis, while that between the first and second and between the third and fourth is a circle with center at the origin which passes through point $(\eta_0, 0)$. These contours are shown in Figure 3.1b where the various quadrants are labeled.

For the purpose of drawing an admittance diagram, it is most convenient to set η in units of $\mathscr{Y}$, the admittance of free space. Then the optical admittances will have the same numerical value as the refractive indices (at normal incidence only, of course).

The method can be illustrated by a simple example that is also used later in Section 3.4.5 on amplitude reflection coefficient loci:

$$\text{Air}|LH|\text{Glass},$$

where we assign an index of 1.52 to glass so that its characteristic admittance is also 1.52 but in free space units; 1.0 to air; 2.35 to H, which we imagine might be zinc sulfide; and 1.35 to L, which, in turn, we might think of as cryolite.

In free space units, the starting admittance is simply 1.52, the admittance of glass. The termination of the first layer, since it is a quarter wave, will be at an admittance of $2.35^2/1.52 = 3.633$ on the real axis, and that of the second, which is also a quarter wave, at $1.35^2/3.633 = 0.5016$ on the real axis. The circles are traced out clockwise, and each is a semicircle with center on the real axis. Figure 3.2 shows the complete locus.

Metal and other absorbing layers can also be included, although we find the calculations sufficiently involved to require the assistance of a computer. Figure 3.3 shows two loci applying to metal layers, one starting from an admittance of 1.0 and the other from 1.52 (free space units). The higher the ratio k/n for the metal, the nearer the locus is to a circle with center at the origin. In the case of Figure 3.3, the locus is somewhat distorted from the ideal case, with a loop bowing out along the direction of the real axis. If we were to add isoreflectance contours to the diagram, corresponding to an admittance of 1.52 for the starting admittance of 1.0 and an admittance of 1.0 for the starting admittance of 1.52, so that the loci correspond to internal and external reflections from such a metal layer on glass in air, we would see that the observed reduction in internal reflectance when the metal is very thin is predicted by the diagram as well as the constantly

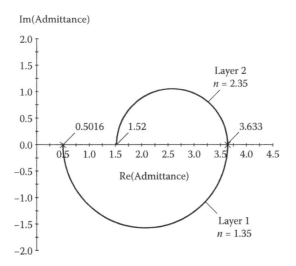

FIGURE 3.2
Admittance of the coating, Air | LH | Glass, with L a quarter wave of index 1.35 and H of 2.35. The indices of air and glass are 1.00 and 1.52, respectively. This is the same coating as in Figure 3.12. Note the similarity in shape to that figure.

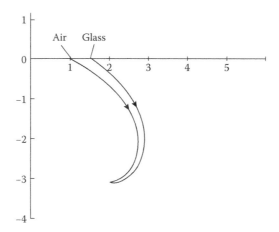

FIGURE 3.3
Admittance loci corresponding to a metal such as chromium with $n - ik = 2 - i3$. Loci are shown for starting points 1.00 and 1.52, corresponding to air and glass, respectively. Note that the initial direction toward the lower right of the diagram implies that in the case of the internal reflectance of the film deposited on glass (i.e., air as substrate and glass as incident medium and the left of the two loci), the reflectance initially falls and then rises, whereas the external reflectance (glass as substrate and air as incident medium and the right of the two admittance loci) always increases, even for very thin layers. When the layers are very thick, they terminate at point $2 - i3$, so that the film is optically indistinguishable from the bulk material.

increasing external reflectance for the same range of thicknesses (we can see such an effect in Figure 5.8). Metals with still lower ratios of k/n still depart further from the ideal circle, and in fact, those starting at 1.0 can initially loop into the first quadrant so that they actually cut the real axis again, even sometimes at point 1.52 to give zero internal reflectance.

We have gained much in simplicity by choosing to deal in terms of optical admittance throughout the assembly. It has not affected in any way our ability to calculate either the amplitude reflection coefficient or reflectance. Transmittance is another matter. Strictly, we need to separately preserve the values of B and C in the matrix calculation. The optical admittance is not sufficient. For dielectric assemblies, we know that the transmittance is given by $(1 - R)$, but for assemblies containing absorbing layers, subsidiary calculations are necessary. For many purposes, reflectance is sufficient, and since the graphical technique is used for visualization rather than calculation, a lack of transmission information is not normally a serious defect. Nevertheless, there are concepts that do yield useful information about transmittance and about losses in layers, directly from the admittance diagram. These are dealt with in the following section.

3.2.1 Electric Field and Losses in the Admittance Diagram

The optical properties of any material are largely determined by the electrons and their interaction with electromagnetic disturbances. Any optical material is made up of atoms or molecules consisting of heavy positively charged masses surrounded by negatively charged electrons. The outer electrons are light and mobile compared with the heavy positively charged nuclei with their tightly bound inner electrons. An electric field can exert a force on a charged particle even while it is stationary, but a magnetic field can interact only when the charged particle moves, and for any significant interaction, the particle must be moving at an appreciable fraction of the speed of light. At the very high frequencies of optical waves, the magnetic interaction with the electrons is virtually zero. We have already used the fact that the relative permeability is unity in setting up the basic theory. The interaction between light and a material is, therefore, entirely through the electric field. Where the electric field amplitude is high, the potential for interaction is high. When thin-film optical coatings are illuminated by light, standing wave patterns form that can exhibit considerable variations in electric field amplitude both in terms of wavelength and in terms of position within the coating. The admittance diagram permits a simple technique for assessing these amplitude variations, and from them, deductions about losses can be made, sometimes with surprising results.

In this discussion, we limit ourselves to normal incidence. Oblique incidence represents only a very slight extension.

The basic matrix technique for the calculation of the properties of an optical coating already actually contains the electric field, and so only a slight modification is required to extract it. The matrix expression, with the usual meaning for the symbols, is

$$\begin{bmatrix} B \\ C \end{bmatrix} = \begin{bmatrix} \cos\delta & \dfrac{i\sin\delta}{y} \\ iy\sin\delta & \cos\delta \end{bmatrix} \begin{bmatrix} 1 \\ y_{\text{exit}} \end{bmatrix}.$$

In this expression, B and C and the corresponding terms in the other column matrix are normalized total tangential electric and magnetic fields. The admittances are normalized too so that they are in free space units rather than in SI units. The first thing we do, therefore, is to restore the expressions to their fundamental form:

$$\begin{bmatrix} E' \\ H' \end{bmatrix} = \begin{bmatrix} \cos\delta & \dfrac{i\sin\delta}{y} \\ iy\sin\delta & \cos\delta \end{bmatrix} \begin{bmatrix} E \\ H \end{bmatrix}.$$

Here y is in free space units, and so to change it to SI units, we must write

$$y = (n - ik)\mathcal{Y},$$

where $\mathcal{Y}$ is the admittance of free space. E and H indicate the complex tangential amplitudes that include the relative phase.

To have absolute values for the total tangential electric field amplitude through the multilayer, it remains simply to give an absolute value to one of the Es. This can be done in a number of ways. The easiest is to put a value on the final tangential component at the emergent interface, that is, the interface with the substrate. This is related to the incident irradiance through the transmittance. If the incident irradiance is I_{inc}, then

$$\frac{1}{2}\operatorname{Re}(E_{\text{exit}} \cdot H^*_{\text{exit}}) = T \cdot I_{\text{inc}},$$

but

$$H_{\text{exit}} = y_{\text{exit}} E_{\text{exit}},$$

and so

$$\frac{1}{2}\operatorname{Re}(E_{\text{exit}} \cdot y^*_{\text{exit}} E^*_{\text{exit}}) = T \cdot I_{\text{inc}}.$$

Now

$$E \cdot E^* = \mathcal{E}^2,$$

giving, with a little manipulation,

$$E_{\text{exit}} = \mathcal{E}_{\text{exit}} = \sqrt{\frac{2T \cdot I_{\text{inc}}}{y_{\text{exit}}}},$$

where y_{exit} must be in SI units, that is, siemens.

If the multilayer system is completely free of absorption, then there is a simple connection between the variation of admittance through the multilayer, which is the quantity we plot in the admittance diagram, and the electric field amplitude.

The admittance at any point in the multilayer is simply the ratio of the total tangential magnetic and electric fields. These total tangential fields also yield the total net irradiance transmitted by the multilayer. Since this multilayer is free of losses, the transmitted irradiance is constant through the multilayer. Putting all this together gives

$$I_{out} = \frac{1}{2} \mathrm{Re}(E \cdot H^*)$$

$$= \frac{1}{2} \mathrm{Re}(E \cdot Y^* E^*)$$

$$= \frac{1}{2} \mathcal{E}^2 \cdot \mathrm{Re}(Y),$$

i.e.,

$$\mathcal{E} = \sqrt{\frac{2I_{out}}{\mathrm{Re}(Y)}} = \sqrt{\frac{2T \cdot I_{inc}}{\mathrm{Re}(Y)}} \propto \frac{1}{\sqrt{\mathrm{Re}(Y)}}. \tag{3.6}$$

Contours of constant electric field are therefore lines, normal to the real axis in the admittance diagram, as in Figure 3.4. If we put Y in free space units, then Equation 3.6 becomes

$$\mathcal{E} = 27.46 \sqrt{\frac{T \cdot I_{inc}}{\mathrm{Re}(Y)}} \text{ (in volts/meter)}. \tag{3.7}$$

Now let us consider a very thin slice of absorbing material embedded in a multilayer. What can we say about the absorption of this slice?

The result is contained in the expression

$$\begin{bmatrix} E' \\ H' \end{bmatrix} = \begin{bmatrix} \cos \delta & \dfrac{i \sin \delta}{y} \\ iy \sin \delta & \cos \delta \end{bmatrix} \begin{bmatrix} E \\ H \end{bmatrix},$$

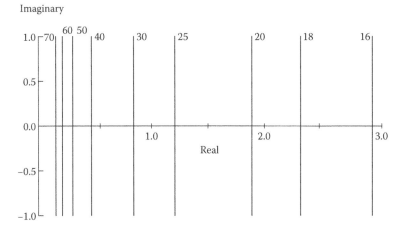

FIGURE 3.4
Lines of constant electric field amplitude for dielectric materials in the admittance diagram. The figures are in volts per meter if the transmitted irradiance is 1 W/m^2.

where the input and exit irradiances are given by

$$I_{in} = \frac{1}{2} \operatorname{Re}(E' \cdot H'^*) \quad \text{and} \quad I_{exit} = \frac{1}{2} \operatorname{Re}(E \cdot H^*).$$

The irradiance lost by absorption in the layer is the difference between these two quantities. Now let the layer be extremely thin. Since the layer is absorbing, δ is given by

$$\delta = \frac{2\pi(n - ik)d}{\lambda} = \alpha - i\beta. \tag{3.8}$$

Equation 3.8 defines the quantities α and β. By extremely thin, we mean that d/λ should be sufficiently small to make both α and β vanishingly small, whatever the size of either n or k. Then,

$$\begin{bmatrix} E' \\ H' \end{bmatrix} = \begin{bmatrix} \cos(\alpha - i\beta) & \dfrac{i\sin(\alpha - i\beta)}{y} \\ iy\sin(\alpha - i\beta) & \cos(\alpha - i\beta) \end{bmatrix} \begin{bmatrix} E \\ H \end{bmatrix}$$

$$= \begin{bmatrix} 1 & \dfrac{i(\alpha - i\beta)}{(n - ik)y} \\ i(\alpha - i\beta)(n - ik)y & 1 \end{bmatrix} \begin{bmatrix} E \\ H \end{bmatrix}$$

$$= \begin{bmatrix} E + \dfrac{i(\alpha - i\beta)H}{(n - ik)y} \\ i(\alpha - i\beta)(n - ik)yE + H \end{bmatrix},$$

where we are including terms up to the first order only in α and β.

The irradiance at the entrance to this thin layer will then be given by

$$I_{in} = \frac{1}{2} \operatorname{Re}\left[\left\{E + \frac{i(\alpha - i\beta)H}{(n - ik)y}\right\} \cdot \{i(\alpha - i\beta)(n - ik)yE + H\}^*\right]$$

$$= \frac{1}{2} \operatorname{Re}[E \cdot H^* + E \cdot \{-i(\alpha + i\beta)(n + ik)yE^*\}] \tag{3.9}$$

$$+ \frac{1}{2} \operatorname{Re}\left[\frac{i(\alpha - i\beta)H \cdot H^*}{(n - ik)y}\right].$$

The second of the two terms in Equation 3.9 simplifies to

$$\frac{1}{2} \operatorname{Re}\left[\frac{i(\alpha - i\beta)H \cdot H^*}{(n - ik)y}\right] = \frac{1}{2} \operatorname{Re}\left[\frac{i(\alpha - i\beta)(n + ik)H \cdot H^*}{(n^2 + k^2)y}\right]$$

$$= \frac{1}{2} \operatorname{Re}\left[\frac{\{\beta n - \alpha k + i(\alpha n + \beta k)\}H \cdot H^*}{(n^2 + k^2)y}\right]$$

$$= \frac{1}{2}\left[\frac{(\beta n - \alpha k)H \cdot H^*}{(n^2 + k^2)y}\right].$$

But

$$\beta n - \alpha k = \frac{2\pi k d}{\lambda}n - \frac{2\pi n d}{\lambda}k = 0.$$

The first term gives

$$I_{in} = \frac{1}{2} \, \text{Re}[E \cdot H^* + E \cdot \{-i(\alpha + i\beta)(n + ik)\mathcal{Y}E^*\}]$$

$$= \frac{1}{2} \, \text{Re}[E \cdot H^*] + \frac{1}{2}[(\alpha k + \beta n)\mathcal{Y}E \cdot E^*],$$

where

$$\alpha k + \beta n = \frac{4\pi nkd}{\lambda} \quad \text{and} \quad E \cdot E^* = \mathcal{E}^2.$$

The irradiance that has been absorbed is therefore given by the difference between the irradiance incident on the thickness element I_{in} and that emerging on the exit side I_{exit}. And this is

$$I_{absorbed} = \frac{2\pi nkd}{\lambda} \cdot \mathcal{Y} \cdot \mathcal{E}^2. \tag{3.10}$$

The magnitude of the absorbed energy is directly proportional to the product of n and k. Both must be nonzero for absorption to occur. The absorption will be small for both a metal with vanishingly small n and a dielectric with vanishingly small k. The factor involving n and k may be thought of as a phase thickness multiplied by k or as a quantity β multiplied by n. The quantity nk is therefore a useful indicator of the potential for loss in any given material.

Now we need to consider the contribution to the absorptance A of the multilayer. This is a little more difficult, and we need to introduce a further concept that will be used in subsequent chapters.

The potential transmittance ψ of any element of a coating system is defined as the ratio of the output to the input irradiances, the input being the net irradiance rather than the incident. Potential transmittance has several advantages over transmittance when dealing with absorbing systems because it completely avoids any problems associated with the mixed Poynting vector in absorbing media. The potential transmittance of a complete system is simply the product of the individual potential transmittances:

$$\psi = \frac{I_{exit}}{I_{in}},$$

$$\psi_{system} = \psi_1 \cdot \psi_2 \cdot \psi_3 \cdot \psi_4 \cdot \psi_5 \cdot \ldots \cdot \psi_q,$$

with the eventual overall transmittance given by

$$T = (1 - R) \cdot \psi_{system}.$$

The potential transmittance of the thin elemental film is given by

$$\psi = \frac{I_{exit}}{I_{in}} = 1 - \frac{I_{absorbed}}{I_{in}} = 1 - \mathcal{A},$$

where A is the potential absorptance. But

$$I_{in} = \frac{1}{2}\mathcal{Y} \cdot \text{Re}(Y) \cdot \mathcal{E}^2,$$

where Y is given in free space units. Then,

$$\psi = 1 - \mathcal{A} = 1 - \frac{2\pi nkd}{\lambda} \cdot \frac{2}{\text{Re}(Y)}. \tag{3.11}$$

This result allows the interpretation of an admittance locus in terms of potential absorptance.

To move from potential absorptance to absorptance is straightforward when the absorption is confined to a very thin layer, the rest of the multilayer being essentially transparent. Then, the absorptance A is given by

$$A = (1 - R)\mathcal{A}.$$

If, however, the absorption is distributed through the layer, then the calculation is rather more involved. Normally, the absorptance would be calculated by the normal matrix expression for the entire film and then would be completely accurate. We, however, are looking for a way of estimating the absorptance and its variation through a layer given the locus in the admittance diagram or the electric field distribution. Let us assume that any absorption is rather small. The layer may be considered as a succession of slices of equal optical thickness and extinction coefficient, and so the first factor in the expression for A is a constant. Each slice has a potential absorptance that depends on the real part of the optical admittance, following Equation 3.11. Then, the potential transmittance is given by the product of the individual potential transmittances:

$$\psi = \psi_1 \cdot \psi_2 \cdot \psi_3 \cdot \psi_4 \cdot \ldots$$
$$= (1 - \mathcal{A}_1) \cdot (1 - \mathcal{A}_2) \cdot (1 - \mathcal{A}_3) \cdot \ldots$$
$$= 1 - (\mathcal{A}_1 + \mathcal{A}_2 + \mathcal{A}_3 + \mathcal{A}_4 + \ldots) + \mathcal{A}_1\mathcal{A}_2 + \ldots.$$

Provided the potential absorptances are small enough, the product terms can be neglected, and then the total potential absorptance is given by the sum of the individual absorptances:

$$\mathcal{A} = \mathcal{A}_1 + \mathcal{A}_2 + \mathcal{A}_3 + \mathcal{A}_4 + \ldots. \tag{3.12}$$

In terms of an integral, this can be written as

$$\mathcal{A} = \sum_j \mathcal{A}_j = \int_\delta \frac{2k}{\mathrm{Re}(Y)}\, d\delta = \int_\beta \frac{2n}{\mathrm{Re}(Y)}\, d\beta. \tag{3.13}$$

If an accurate answer is required, we will always turn to the computer and a very simple rapid calculation. For understanding the result, usually, we would like to know what to do either to increase or decrease the absorptance or to find sensitive regions where contamination or scattering roughness is especially to be avoided. To answer such questions, usually, a rough answer that shows trends is all that is needed.

3.3 Vector Method

The vector method is a valuable technique, especially in design work associated with antireflection coatings. Two assumptions are involved: first, that there is no absorption in the layers, and second, that the behavior of a multilayer can be determined by considering one reflection of the incident wave at each interface only. The errors involved in using this method can be, in some cases, significant, especially where high overall reflectance from the multilayer exists, but they are small in most types of antireflection coating.

Consider the assembly sketched in Figure 3.5. If there is no absorption in the layers, then $N_r = n_r$ and $k_r = 0$. The amplitude reflection coefficient at each interface is given by

$$\rho = \frac{n_{r-1} - n_r}{n_{r-1} + n_r},$$

which may be positive or negative depending on the relative magnitudes of n_{r-1} and n_r.

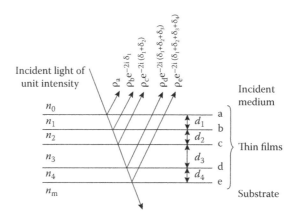

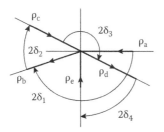

Polar diagram showing
vector directions

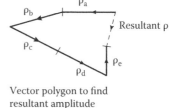

Vector polygon to find
resultant amplitude

FIGURE 3.5
Vector method. The lengths of the vectors and the phase angles are given by

$$\rho_a = (n_0 - n_1)/(n_0 + n_1), \quad \delta_1 = 2\pi n_1 d_1/\lambda,$$
$$\rho_b = (n_1 - n_2)/(n_1 + n_2), \quad \delta_2 = 2\pi n_2 d_2/\lambda,$$

etc. Note that the sign of the expression for the vector lengths is important and must be included. In the diagram, ρ_a, ρ_c, and ρ_e are shown as having a negative sign. Note also that the angles between successive vectors are phase lags so that the sense of all the angles in the polar diagram is also negative.

The phase thicknesses of the layers are given by $\delta_1, \delta_2, \ldots,$ where

$$\delta_r = 2\pi n_r d_r/\lambda.$$

A quarter-wave optical thickness is represented by 90° and a half wave by 180°.

The assumption that only one reflection at each interface is sufficient is certainly an approximation. Let us now adopt another. We will further assume that each reflected beam suffers no loss at any of the other interfaces it traverses. Then, as the diagram shows, the resultant amplitude reflection coefficient is given by the vector sum of the coefficients for each interface, where each is associated with the appropriate phase lag corresponding to the passage of the wave from the front surface to the interface and back to the front surface again:

$$\rho = \rho_a + \rho_b \exp(-2i\delta_1) + \rho_c \exp[-2i(\delta_1 + \delta_2)]$$
$$+ \rho_d \exp[-2i(\delta_1 + \delta_2 + \delta_3)] + \ldots.$$

The sum can be found analytically or, as is more usual, graphically. The graphical case is easier because the angles between successive vectors are merely $2\delta_1$, $2\delta_2$, $2\delta_3$, and so on.

The keen observer will immediately notice that there is a further complication. That we are neglecting any reflection loss at all interfaces does not imply equal amplitudes on either side of each interface but equal irradiances. Even in the presence of zero loss, the amplitudes will be changed in the ratio of η_p/η_q. Does this not invalidate the entire procedure? Each ray that passes through an interface actually traverses it twice, in the direction of incidence and in the reverse direction. The ratio that is to be applied in the forward direction is the inverse of that in the negative direction, and the correction cancels out in the double passage. We can therefore ignore it. Within the limitations of our approximations, our sum is correct.

The calculation of the angles for any wavelength is simplified if, as is usual, the optical thicknesses of the layers are given in terms of quarter-wave optical thicknesses at a reference wavelength λ_0. If the optical thickness of the rth layer is t_r quarter waves at λ_0, then the value of δ_r at λ is just $\delta_r = (90° t_r \lambda_0/\lambda)$ degrees of arc.

In practice, it will be found extremely easy to confuse angles and directions, particularly where negative reflection coefficients are involved. The task of drawing the vector diagram is greatly eased by first plotting the vectors with directions on a polar diagram and then transferring the vectors to a vector polygon, rather than attempting to draw the vector polygon straight away. An important point to remember is that the resultant vector represents the amplitude reflection coefficient and its length must be squared in order to give the reflectance.

A typical arrangement is shown in Figure 3.5. The vector method is used to a considerable extent in Chapter 4, which deals with antireflection coatings.

3.4 Other Techniques

Great progress was made in the subject of thin-film optics well before computers became both exceedingly powerful and generally available. Many techniques for assisting in the creation and assessment of designs were developed at a time when accurate extended calculations were so time consuming as to be out of the question. Their usefulness has not ceased with the advent of the personal computer because they bring an insight that is completely lacking in pure numerical calculations. We will use some of these techniques from time to time in the remainder of the book. Others are commonly encountered in the literature of the subject. The fact that we collect a number of them together under the appellation of *other* should not be taken as an indication of a reduced usefulness or ranking, but rather as an admission that there is a limit to the size of this book. There are many others that we have simply been unable to include.

3.4.1 Herpin Index

An extremely important result for filter design is derived in Chapter 7, which deals with edge filters. Briefly, this is the fact that any symmetrical product of three thin-film matrices can be replaced by a single matrix which has the same form as that of a single film and which therefore possesses an equivalent thickness and an equivalent optical admittance. Of course, this is a mathematical device rather than a case of true physical equivalence, but the result is of considerable use in giving an insight into the properties of a great number of filter designs, which can be split into a series of symmetrical combinations. The method also allows the replacement, under certain conditions, of a layer of intermediate index by a symmetrical combination of high- and low-index materials. This is especially useful in the design of antireflection coatings, which frequently require quarter-wave thicknesses of unobtainable intermediate indices. These difficult layers can be replaced by symmetrical combinations of existing materials with the additional advantage of limiting the total number of materials required for the structure.

The equivalent admittance is frequently known as the Herpin index, after the originator, and the symmetrical combination as an Epstein period, after the author of two of the most important early papers dealing with the application of the result to the design of filters.

The detailed derivation of the relevant formulas is left until Chapter 7, which will make considerable use of the concept.

3.4.2 Alternative Method of Calculation

The success of the vector method prompts one to ask whether it can be made more accurate by considering second and subsequent reflections at the various boundaries instead of just one. In fact, an alternative solution of the thin-film problem can be obtained in this way, and this was the earlier way of formulating film properties dating back to Poisson (Chapter 1). It is simpler to consider normal incidence only. The expressions can be adapted for nonnormal incidence quite simply when the materials are transparent and with some difficulty when they are absorbing. We first consider the case of a single film. Figure 3.6 defines the various parameters.

The resultant amplitude reflection coefficient is given by

$$\rho^+ = \rho_a^+ + \tau_a^+ \rho_b^+ \tau_a^- e^{-2i\delta} + \tau_a^+ \rho_b^+ \rho_a^- \rho_b^+ \tau_a^- e^{-4i\delta}$$

$$+ \tau_a^+ \rho_b^+ \rho_a^- \rho_b^+ \rho_a^- \rho_b^+ \tau_a^- e^{-6i\delta}$$

$$= \rho_a^+ + \frac{\rho_b^+ \tau_a^+ \tau_a^- e^{-2i\delta}}{1 - \rho_b^+ \rho_a^- e^{-2i\delta}}.$$

However,

$$\tau_a^+ \tau_a^- = \frac{4N_0 N_1}{(N_0 + N_1)^2} = 1 - \rho,$$

and $\rho_a^- = -\rho_a^+$, so that

$$\rho^+ = \frac{\rho_a^+ + \rho_b^+ e^{-2i\delta}}{1 + \rho_b^+ \rho_a^+ e^{-2i\delta}}. \tag{3.14}$$

Similarly,

$$\tau^+ = \tau_a^+ \tau_b^+ \rho_b^+ e^{-i\delta} + \tau_a^+ \rho_b^+ \rho_a^- \tau_b^+ e^{-3i\delta} + \tau_a^+ \rho_b^+ \rho_a^- \rho_b^+ \rho_a^- \tau_b^+ e^{-5i\delta},$$

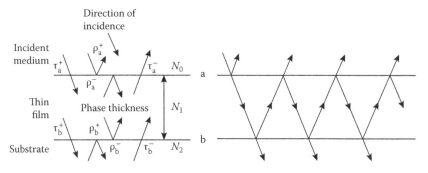

FIGURE 3.6
Parameters in the multiple-beam summation.

which reduces to

$$\tau^+ = \frac{\tau_a^+ \tau_b^+ e^{-i\delta}}{1 - \rho_a^- \rho_b^+ e^{-2i\delta}}$$

$$= \frac{\tau_a^+ \tau_b^+ e^{-i\delta}}{1 + \rho_a^+ \rho_b^+ e^{-2i\delta}}.$$

(3.15)

These expressions can be used in calculations of assemblies of more than one film by applying them successively, first to the final two interfaces, which can then be replaced by a single interface with the resultant coefficients; and then to this equivalent interface and the third last interface, and so on.

The resultant amplitude transmission and reflection coefficients τ^+ and ρ^+ can be converted into transmittance and reflectance by using the expressions

$$R = (\rho^+)(\rho^+)^*,$$

$$T = \frac{n_2}{n_0} (\tau^+)(\tau^+)^*.$$

n_2 and n_0 are the refractive indices of the substrate, or exit medium, and the incident medium, respectively. For these expressions to be meaningful, we must, as before, restrict the incident medium to be transparent so that $N_0 = n_0$. No such restriction applies to the exit medium, which can have complex $N_2 = n_1 - ik_2$, the real part being used in the preceding expression for T.

It is also possible to develop a matrix approach along these lines. The electric field vectors $\mathcal{E}_0^+$ and $\mathcal{E}_0^-$ in medium 0 at interface a can be expressed in terms of $\mathcal{E}_1^+$ and $\mathcal{E}_1^-$ in film 1 at interface b (see Figure 3.7).

$$\begin{bmatrix} \mathcal{E}_0^+ \\ \mathcal{E}_0^- \end{bmatrix} = \frac{1}{\tau_a^+} \begin{bmatrix} e^{i\delta_1} & \rho_a^+ e^{-i\delta_1} \\ \rho_a^+ e^{i\delta_1} & e^{-i\delta_1} \end{bmatrix} \begin{bmatrix} \mathcal{E}_1^+ \\ \mathcal{E}_1^- \end{bmatrix}.$$

(3.16)

If $\mathcal{E}_2^+$ is the tangential component of amplitude in medium 2, then, since there is only a positive-going wave in that medium,

$$\begin{bmatrix} \mathcal{E}_1^+ \\ \mathcal{E}_1^- \end{bmatrix} = \frac{1}{\tau_b^+} \begin{bmatrix} 1 \\ \rho_b^+ \end{bmatrix} \mathcal{E}_2^+.$$

(3.17)

Equations 3.16 and 3.17 can be extended in the normal way to cover the case of many layers. The only point to watch is that ρ_a^+ and τ_a^+ must refer to the coefficients of the boundary in the correct medium. That is, all the reflection coefficients ρ and transmission coefficients τ must be calculated for the boundaries as they exist in the multilayer. Thus, if we take an existing multilayer and add an extra layer, not only do we add an extra interface, but we also alter the amplitude reflection and

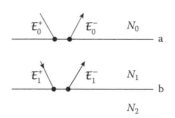

FIGURE 3.7
Positive- and negative-going waves at the two interfaces.

transmission coefficients of what now becomes the second last interface. Thus, two layers must be recomputed and not just one.

If absorption is included, the formulas remain the same, but the parameters ρ, τ, and δ become complex.

3.4.3 Smith's Method of Multilayer Design

In 1958, Smith [1], then of the University of Reading, published a useful design method based on Equation 3.15. The technique is also known as the *method of effective interfaces*. It consists of choosing any layer in the multilayer and then considering multiple reflections within it, the reflection and transmission coefficients at its boundaries being the resultant coefficients of the complete structures on either side. The method of summing multiple beams is, of course, quite old, and the novel feature of the present technique is the way in which it is applied. Although the technique described by Smith was principally concerned with dielectric multilayers, it can be extended to deal with absorbing layers. As before, we limit ourselves, in the derivation, to normal incidence. When the layers are transparent, the expressions can be extended to oblique incidence without major difficulty. The notation is illustrated in Figure 3.8.

From Equation 3.15,

$$\tau^+ = \frac{\tau_a^+ \tau_b^+ e^{-i\delta}}{1 - \rho_a^- \rho_b^+ e^{-2i\delta}},$$

where

$$\delta = 2\pi N d / \lambda.$$

Now $N = n - ik$, and we can write δ as

$$\delta = 2\pi(n - ik)d/\lambda = \alpha + i\beta,$$

and

$$e^{-i\delta} = e^{-\beta} e^{-i\alpha},$$

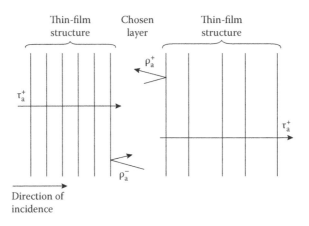

FIGURE 3.8
Quantities associated with the effective interfaces in Smith's technique.

where $\alpha = 2\pi nd/\lambda$, the phase thickness of the layer, and $\beta = 2\pi kd/\lambda$. Now

$$T = \frac{n_m}{n_0}(\tau^+)(\tau^+)^*,$$

where n_m is the real part of the exit medium index and n_0 is the refractive index of the incident medium:

$$T = \frac{n_m}{n_0}\frac{(\tau_a^+)(\tau_a^+)^*(\tau_b^+)(\tau_b^+)^*e^{-2\beta}}{(1-\rho_a^-\rho_b^+e^{-2\beta}e^{-2i\alpha})(1-\rho_a^-\rho_b^+e^{-2\beta}e^{-2i\alpha})^*}.$$

Now, let

$$\tau_a^+ = |\tau_a^+|e^{i\varphi'_a}, \quad \rho_a^- = |\rho_a^-|e^{i\varphi_a},$$
$$\tau_b^+ = |\tau_b^+|e^{i\varphi'_b}, \quad \rho_b^+ = |\rho_b^+|e^{i\varphi_b}.$$

Then,

$$T = \frac{n_m}{n_0}\frac{|\tau_a^+|^2|\tau_b^+|^2e^{-2\beta}}{\left(1-|\rho_a^-|^2|\rho_b^+|^2e^{i(\varphi_a+\varphi_b)}e^{-2\beta}e^{-2i\alpha}\right)\left(1-|\rho_a^-|^2|\rho_b^+|^2e^{-i(\varphi_a+\varphi_b)}e^{-2\beta}e^{2i\alpha}\right)},$$

i.e.,

$$T = \frac{n_m}{n_0}\frac{|\tau_a^+|^2|\tau_b^+|^2e^{-2\beta}}{\left[1-|\rho_a^-|^2|\rho_b^+|^2e^{-4\beta}-2|\rho_a^-||\rho_b^+|e^{-2\beta}\cos(\varphi_a+\varphi_b-2\alpha)\right]}. \tag{3.18}$$

A marginally more convenient form of the expression can be obtained by substituting $1-2\sin^2[(\varphi_a+\varphi_b)/2-\alpha]$ for $\cos(\varphi_a+\varphi_b-2\alpha)$, and with some rearrangement,

$$T = \frac{n_m}{n_0}\frac{|\tau_a^+|^2|\tau_b^+|^2e^{-2\beta}}{\left(1-|\rho_a^-||\rho_b^+|e^{-2\beta}\right)^2}\cdot\left[1+\frac{4|\rho_a^-||\rho_b^+|e^{-2\beta}}{\left(1-|\rho_a^-||\rho_b^+|e^{-2\beta}\right)^2}\times\sin^2\left(\frac{\varphi_a+\varphi_b}{2}-\frac{2\pi nd}{\lambda}\right)\right]^{-1}. \tag{3.19}$$

If there is no absorption in the chosen layer, i.e., $\beta = 0$, then the restrictions on reflectances in absorbing media no longer apply, and we can write

$$T_a = \frac{n}{n_0}|\tau_a^+|^2, \quad R_a^- = |\rho_a^-|^2,$$
$$T_b = \frac{n_m}{n}|\tau_b^+|^2, \quad R_a^- = |\rho_b^+|^2,$$

$$T = \frac{T_aT_b}{\left[1-\left(R_a^-R_b^+\right)^{1/2}\right]^2}\cdot\left[1+\frac{4R_a^-R_b^+}{\left[1-\left(R_a^-R_b^+\right)^{1/2}\right]^2}\times\sin^2\left(\frac{\varphi_a+\varphi_b}{2}-\frac{2\pi nd}{\lambda}\right)\right]^{-1}, \tag{3.20}$$

which is the more usually quoted version.

The usefulness of this method is mainly in providing an insight into the properties of a particular type of filter, and its major usefulness is in the creation of a design. It is certainly not the easiest method of determining the performance of a given multilayer—this is best tackled by a straightforward application of the matrix method. What Equation 3.19 or 3.20 does is to make it possible to isolate a layer, or a combination of several layers, and to examine the influence that these layers and any changes in them have on the performance of the filter as a whole. Smith's original paper includes a large number of examples of this approach and repays close study.

3.4.4 Smith Chart

The Smith chart [2,3] is one of a number of different devices of the same broad type that were originally intended to simplify calculation. The Smith chart is the one that most frequently appears in the literature, and so it is included here, although little use is made of it in the remainder of the book. The method depends on three properties of a thin-film structure:

1. Since the tangential components of E and H are continuous across a boundary, so also is the equivalent admittance. This has been implied in the section dealing with the matrix method, but has not, perhaps, been explicitly stated there.

2. In any thin film, for example, layer q in Figure 3.9, the amplitude reflectance ρ at any plane within the layer is related to that at the edge of the layer remote from the incident wave ρ_m by

$$\rho = \rho_m e^{-2i\delta}, \tag{3.21}$$

   where δ is the phase thickness of that part of the layer between the far boundary m and the plane in question.

   This second point is almost self-evident, but may be shown by putting $\rho_a^+ = 0$ in Equation 3.15, since the boundary under consideration is an imaginary one between two media of identical admittance.

3. The amplitude reflection coefficient of any thin-film assembly, with optical admittance at the front surface of Y, is given by Equation 3.14, i.e.,

$$\rho = \frac{\eta_0 - Y}{\eta_0 + Y} = \frac{1 - Y/\eta_0}{1 + Y/\eta_0}, \tag{3.22}$$

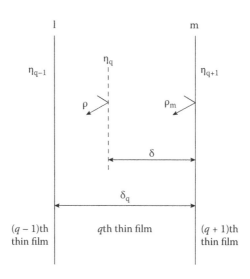

FIGURE 3.9
Parameters used in the Smith chart description.

where η_0 is the admittance of the incident medium. Y/η_0 is sometimes known as the reduced admittance.

The procedure for calculating the effect of any layer in a thin-film assembly by using these properties is as follows:

1. ρ_m, the amplitude reflection coefficient at the boundary of the layer remote from the side of incidence, is given.

2. The amplitude reflection coefficient within the layer just inside the boundary l is then given by Equation 3.21:

$$\rho = \rho_m e^{-2i\delta_q}. \tag{3.23}$$

3. The optical admittance just inside the boundary l is given by Equation 3.22:

$$\rho = \frac{1 - Y/\eta_q}{1 + Y/\eta_q}, \tag{3.24}$$

i.e.,

$$\frac{Y}{\eta_q} = \frac{1 - \rho}{1 + \rho}. \tag{3.25}$$

4. The optical admittance on the incident side of the boundary l is still Y because of condition 1. The reduced admittance is Y/η_{q-1}, where

$$\frac{Y}{\eta_{q-1}} = \frac{\eta_q}{\eta_{q-1}} \cdot \frac{Y}{\eta_q}. \tag{3.26}$$

5. The amplitude reflection coefficient ρ_l on the incident side of the boundary l is given by

$$\rho_l = \frac{1 - Y/\eta_{q-1}}{1 + Y/\eta_{q-1}}. \tag{3.27}$$

The calculation of the amplitude reflection coefficient of any thin-film assembly is merely the successive application of Equations 3.23 through 3.27 to each layer in the system, starting with that at the end of the assembly, remote from the incident wave.

The calculation can be carried out in any convenient way and can even be used as the basis for a computer program. The problem is similar to one found in the study of high-frequency transmission lines, and a simple graphical approach has been devised. The most awkward parts of the calculation are in Equations 3.25 and 3.27. A chart connecting values of X and Z, where

$$X = \frac{1 - Z}{1 + Z}, \tag{3.28}$$

is shown in Figure 3.10 [4] and is known as a Smith chart after the originator P. H. Smith [2] (not to be confused with the S. D. Smith of the previous section). Z is plotted in polar coordinates on the diagram, and the corresponding real and imaginary parts of X are read off from the sets of orthogonal circles. A slide rule is capable of the other part of the calculation, the multiplication by η_q/η_{q-1}.

A scale is provided around the outside of the chart to enable the calculation involved in Equation 3.23 to be very simply carried out by rotating the point corresponding to ρ_m around the center of the chart through the appropriate angle $2\delta_q$. The scale is calibrated in terms of the optical thickness measured in fractions of a wavelength, taking into account that the angle is actually $2 \times \delta_q$.

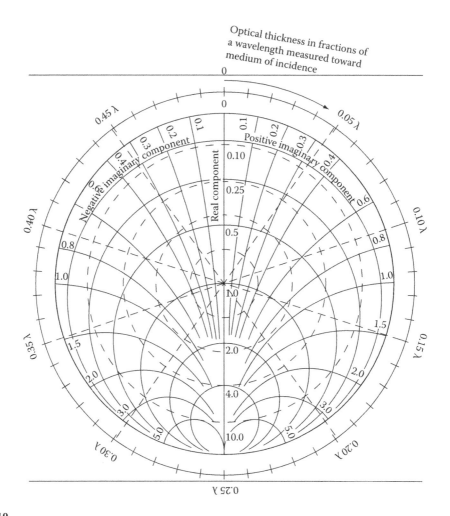

FIGURE 3.10
Smith chart. Broken circles are circles of constant amplitude reflection coefficient ρ. From the smallest to the largest, they correspond to ρ = 0.2, 0.4, 0.6, 0.8, and 1.0, the outer solid circle. Solid circles are circles of constant real part and constant imaginary part of the reduced optical admittance. Note: An optical thickness of 0.25λ corresponds to a phase thickness of 90°. (Constructed using the details given by W. Jackson. 1951. *High Frequency Transmission Lines*, pp. 129, 146. Third ed. London: Methuen.)

3.4.5 Reflection Circle Diagrams

This technique, sometimes referred to simply as a circle diagram, was described by Berning [5], and its use in coating design was considerably developed and described in much detail by Apfel [6]. According to Apfel, Frank Rock originated this technique in the mid-1950s. The technique results in diagrams that have an appearance similar to that of the admittance diagram.

The scale and shape of the diagram is similar to that of the Smith chart, and indeed, the identical set of coordinates and prepared graph paper may be used for both. This leads to a confusion of the two techniques, with the name Smith chart being applied to the circle diagram. They are really quite different. The Smith chart slides a reference plane through an already existing multilayer and plots the net amplitude reflection coefficient at the plane. There are discontinuities in the locus, therefore, when an interface is crossed. Dielectric loci are circles centered at the origin. The circle diagram assumes that the multilayer is under construction so that the incident medium for the amplitude reflection coefficient is the incident medium for the entire multilayer. This also results in circles, but there are no discontinuities in the resulting locus, and the individual dielectric circles are no longer centered at the origin.

Equation 3.14 gives an expression for calculating the change in amplitude reflection coefficient resulting from the addition of a single layer:

$$\rho^+ = \frac{\rho_a^+ + \rho_b^+ e^{-2i\delta}}{1 + \rho_b^+ \rho_a^+ e^{-2i\delta}}.$$

We can calculate the properties of a multilayer by successive applications of this formula, as has already been indicated. Let us imagine that we have arrived at the pth layer in the calculation. The quantities involved are indicated in Figure 3.11. ρ_f^+ is the amplitude reflection coefficient of the $(p - 1)$th layer at the outer interface, which we have labeled f:

$$\rho_f^+ = \frac{\eta_{p-1} - \eta_p}{\eta_{p-1} + \eta_p}.$$

ρ' in Figure 3.11 is the resultant amplitude reflection coefficient at the inner interface of the pth layer due to the entire structure on that side and is not to be confused with ρ_q, the amplitude reflection coefficient of the qth interface. The resultant amplitude reflection coefficient ρ at the fth interface is given by

$$\rho = \frac{\rho_f^+ + \rho' e^{-2i\delta}}{1 + \rho_f^+ \rho' e^{-2i\delta}}. \tag{3.29}$$

Provided we are dealing with dielectric materials, ρ_f^+ will be real. ρ' may be complex, but we can include any phase angle due to ρ' in the factor $e^{-2i\delta}$. Let us plot the locus of ρ on the complex plane as δ varies. To simplify the analysis, we can replace ρ by $x + iy$ and $\rho' e^{-2i\delta}$ by $\alpha + i\beta$, where

$$(\alpha^2 + \beta^2)^{1/2} = |\rho'|.$$

Then,

$$x + iy = \frac{\rho_f^+ + \alpha + i\beta}{1 + \rho_f^+ (\alpha + i\beta)}.$$

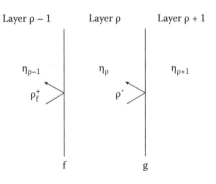

FIGURE 3.11
Quantities in the method of reflection circles.

Multiplying both sides by the denominator of the right-hand side and then equating real and imaginary parts of the resulting expressions yields

$$x(1 + \rho_f^+ \alpha) - y\rho_f^+ \beta = \rho_f^+ + \alpha,$$
$$y(1 + \rho_f^+ \alpha) + x\rho_f^+ \beta = \beta,$$

i.e.,

$$(x - \rho_f^+) = \alpha(1 - x\rho_f^+) + \beta y\rho_f^+,$$
$$y = -\alpha y\rho_f^+ + \beta(1 - x\rho_f^+).$$

To find the locus, we square and add these equations to give

$$(x - \rho_f^+)^2 + y^2 = (\alpha^2 + \beta^2)\left[(1 - x\rho_f^+)^2 + (\rho_f^+ y)^2\right]$$
$$= |\rho'|^2\left[(1 - x\rho_f^+)^2 + (\rho_f^+ y)^2\right],$$

which can be manipulated to

$$x^2\left(1 - |\rho'|^2\rho_f^{+2}\right) + y^2\left(1 - |\rho'|^2\rho_f^{+2}\right) - 2x\rho_f^+\left(1 - |\rho'|^2\right) + \rho_f^{+2} - |\rho'|^2 = 0. \quad (3.30)$$

This is the equation of a circle with center

$$\left(\frac{\rho_f^+\left(1 - |\rho'|^2\right)}{\left(1 - |\rho'|^2\rho_f^{+2}\right)}, 0\right),$$

i.e., on the real axis, and radius

$$\frac{|\rho'|\left(1 - \rho_f^{+2}\right)}{\left(1 - |\rho'|^2\rho_f^{+2}\right)}.$$

The locus of the reflection coefficient as the layer thickness is allowed to steadily increase from zero is therefore a circle. A half-wave layer traces out a complete circle, while a quarter-wave layer, if it starts on the real axis, will trace out a semicircle; otherwise, it will be slightly more or less than a semicircle, depending on the exact starting point. In all cases, the circle is traced clockwise.

The locus corresponding to a single layer is straightforward. The plotting of the locus corresponding to two or more layers is slightly more complicated. The form of the locus of each layer is an arc of a circle traced from the terminal point of the previous layer. The complication arises from the subsidiary calculation that must be performed each time to calculate the current value of ρ' from the terminal value of the previous layer. An example will serve to illustrate the point.

Let us consider a glass substrate of index 1.52, on which a layer of zinc sulfide of index 2.35 and thickness of one quarter wave is first deposited, followed by a layer of cryolite of index 1.35 and thickness of also one quarter wave. Air, of index 1.0, is the incident medium.

The calculation of the circles is most easily performed by using Equation 3.29 to calculate the terminal points. The starting point is known, and that, together with the fact that the center is on the real axis, completes the specification of the circles.

The values of ρ_f^+ and ρ' for the first layer are

$$\rho_f^+ = \frac{1.0 - 2.35}{1.0 + 2.35} = -0.4030,$$

$$\rho' = \frac{2.35 - 1.52}{2.35 + 1.52} = 0.2144.$$

The starting point for the layer is

$$\rho = \frac{\rho_f^+ + \rho'}{1 + \rho_f^+ \rho'} = -0.2063,$$

which corresponds to the amplitude reflection coefficient of bare glass in air.

For a quarter-wave layer, $e^{-2i\delta} = -1$, and so the terminal value of ρ is given by

$$\rho = \frac{\rho_f^+ - \rho'}{1 - \rho_f^+ \rho'} = -0.5683,$$

and the locus up to this point is a semicircle. This value of ρ corresponds to the amplitude reflection coefficient of a quarter wave of zinc sulfide on glass in air. To continue the locus into the next layer, we need new values of ρ_f^+ and ρ'.

$(\rho_f^+)_{new}$ is straightforward, being the external reflection coefficient at an air–cryolite boundary:

$$(\rho_f^+)_{new} = \frac{1.0 - 1.52}{1.0 + 1.52} = -0.1489.$$

$(\rho')_{new}$ is more difficult. This is the amplitude reflection coefficient that the substrate plus a quarter wave of zinc sulfide will have, no longer in a medium of air, but in one of cryolite. It can be calculated either by using the normal matrix method or simply by inverting the following equation:

$$\rho = (\rho)_{old} = \frac{(\rho_f^+)_{new} + (\rho')_{new}}{1 + (\rho_f^+)_{new}(\rho')_{new}},$$

which must be satisfied if the start of the new layer is to coincide with $(\rho)_{old}$, the termination of the old:

$$(\rho')_{new} = \frac{(\rho)_{old} - (\rho_f^+)_{new}}{1 - (\rho)_{old}(\rho_f^+)_{new}},$$

and in this case, $(\rho)_{old}$ is -0.5683, so that

$$(\rho')_{new} = \frac{-0.5683 - (-0.1489)}{1 - (-0.5683)(-0.1489)} = -0.4582.$$

The new locus, which is another semicircle, then starts at point -0.5683 on the real axis and terminates at

$$\rho = \frac{(\rho_f^+)_{new} - (\rho')_{new}}{1 - (\rho_f^+)_{new}(\rho')_{new}} = 0.3319.$$

The loci are shown in Figure 3.12.

The advantage of the technique over the Smith chart is especially that the locus is a continuous one, since the termination of each layer is the starting point for the next. All possible loci corresponding to a particular refractive index form a set of nested circles centered on the real axis of the diagram. Enough of these circles can be drawn to form a separate template or overlay for each of the materials involved in a design, and these can considerably ease the task of drawing the diagram.

Since the method of the Smith chart is based on the real and imaginary axes of the amplitude reflection coefficient, the loci can actually be drawn on the same diagram as a Smith chart. Strictly, in that case, the chart should not be referred to as a Smith chart because it is not being used in that way.

Many examples of the use of this technique in design are given by Apfel [6], who also extended it to include absorbing layers such as metals.

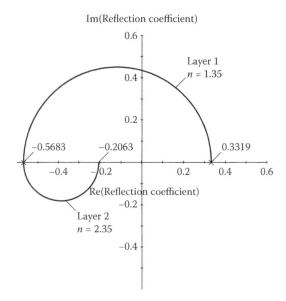

FIGURE 3.12
Reflection circles, or amplitude reflection locus, for the coating Air | *LH* | Glass, where *L* indicates a quarter wave of index 1.35 and *H* of 2.35, and the indices of air and glass are 1.00 and 1.52, respectively.

3.4.6 Automatic Design

Given a possible solution to a thin-film design problem, can we devise an objective method to change the parameters so that it becomes a better design? Can we continue the process to make the design as good as possible? And, of course, can we finally devise a way of achieving all this by using an automatic computer? The answer to all these questions is a conditional affirmative. The first description of such a technique in optical coating design is that of Baumeister in 1958 [7].

An automatic process that makes adjustments to an already existing design without making major changes is known as *refinement*. An automatic process that involves an element of design construction is usually known as *synthesis*. The term *synthesis* may denote anything from a mild complication of an almost acceptable design to a process that builds an acceptable design from nothing more than a list of materials and a performance specification. The term *optimization* simply means improving performance and includes both refinement and synthesis. These are not by any means universal definitions, and there is no universal agreement on the meanings of the terms.

Before we can make a coating better, we must define what we mean by *better*, and our definition must be one that can be applied to automatic methods. All that a computer can understand is whether or not one number is larger or smaller than another, and so, at the current stage of development of the subject, the concept *better* must invariably be expressed in terms of changes in a single number, the *figure of merit*. The usual arrangement is for a smaller figure of merit to be better than a larger one and a figure of merit to be zero if the coating has exactly the desired performance. The figure of merit is then, essentially, a measure of the current error in performance. However, automatic processes can work just as well with a figure of merit that increases as the merit improves. The figure of merit is derived by what is sometimes called a *function of merit*, or an *objective function*, consisting of a comparison of the actual calculated performance of a design and a specification of a desired performance. The derivation involves the application of a set of rules, and it is important that the rules should yield a completely unambiguous figure of merit.

Performance may include any attributes of the coating that can be quantified, but it is frequently taken as the reflectance, or transmittance, or phase shift, or color coordinates, or some such normal expressions of performance, at specified points over a prescribed range of parameters. Each individual expression of performance is known as a *target*. Usually, the form of the rules for calculating the figure of merit will be similar to the following expression:

$$F = \frac{\sum\limits_{j}\left[W_j\left|\frac{\left(T_j - P_j\right)}{Tol_j}\right|^q\right]}{\sum\limits_{j}W_j},$$ (3.31)

where F is the figure of merit, T_j is the jth target, P_j is the corresponding calculated value of performance, Tol_j is a tolerance, and W_j is a weight. The modulus sign around the error ensures that negative errors do not cancel out positive ones. The tolerance parameter serves two purposes. It cancels out the units that may be mixed, with degrees in one target, percent in another, and color parameters in a third, and it normalizes the magnitude of the error associated with each target so that they are all of magnitude unity when the error is equal to the tolerance, simplifying the termination criteria. The weight attached to the target then indicates the relative importance of that particular target, and it can be adjusted to influence the direction of the refinement process based on progress. Sometimes, the weight is arranged to include the tolerance, and in that case, the units are stripped off the error. It is usual to normalize the expression so that the refinement or synthesis process has always approximately the same working range, and this is indicated in Equation 3.31 by dividing by the sum of the weights. The quantity q, the power to which the performance gap is raised, may be completely free for the user to choose or may, in some procedures, be constrained to a particular value. Experience shows that a value of q of 2 works well in most cases. An increase in the value of q makes the process more responsive to larger performance gaps at the expense of smaller.

The function of merit is ultimately a function of the particular set of design parameters that we can consider as the set of variables. For efficient and reliable optimization, the function of merit should be a continuous, single-valued function of these variables. Abrupt changes in the function of merit as design parameters vary inhibit efficient refinement and should, if possible, be avoided. Hard constraints on the process can have the same effect as abrupt changes, and so it is often more efficient to soften the constraints by expressing their effect in terms of penalty functions attached to the function of merit rather than rigid boundaries. The design parameters are perhaps most often the thicknesses of the various layers, but virtually all the techniques can also accommodate optical constants, particularly the index of refraction.

If we have the same number of targets in the definition of the merit function as we have parameters in the design, then in principle, provided that the targets are attainable and not mutually exclusive, the problem should be completely soluble, although it may require impossible optical constants or thicknesses. In most cases, however, the desired performance cannot be completely attained whatever our adjustments are, and then the objective of the optimization process becomes making the figure of merit as small as possible. We can visualize the function of merit as represented by a surface in multidimensional space, one dimension for each adjustable parameter and one for the figure of merit. Making the figure of merit as small as possible, then, is translated into finding a minimum of the merit function and thence into finding the lowest possible minimum or, as it is known, the *global minimum*. If there are constraints on the parameters, such as permissible ranges, then the lowest possible minimum within the constraints is known as the *constrained global minimum*. Since there always are constraints (we cannot permit infinite thicknesses, for instance), the minimum that concerns us will be the constrained global minimum. Unfortunately, although it is relatively easy to find a minimum of the merit function, it is not nearly as easy to find, or even to be sure that one has found, the constrained global minimum. Unless the function of merit is analytically friendly, the only way to be absolutely sure is to carry out an exhaustive search of the given parameter region. We can illustrate the problems involved in this, by assuming a 20-layer design with 20 possible values of thickness for each layer, where refractive indices are already prescribed. Assume that one complete figure of merit can be generated in 1 ns. Then an exhaustive search of all possible designs will occupy a time of 20^{20} ns, that is, around 2×10^9 years. This problem is considerably constrained, but it already gives some idea of what is involved in an exhaustive search. All optimization techniques, therefore, carry out a more limited procedure that arrives at a local minimum that may be as good a minimum as is economically possible. The adjective *global* is sometimes applied to processes that essentially search in constrained parameter space for more than one merit function minimum so that they have an improved chance of finding the constrained global minimum.

We may have major gaps in our ideas of a starting design. Perhaps we do not have any idea of the indices for the layers beyond the range of possibilities that are available, or we may not know the number of layers beyond perhaps a prescribed maximum. In that case, we have the synthesis problem. If we have a reasonably good design which simply needs minor adjustment, then we have refinement. Synthesis clearly has rather greater dimensions than refinement. To begin, we will concentrate on refinement and assume that we have a starting design of a certain number of layers that the process will alter only in some limited way such as in terms of layer thicknesses or refractive indices or possibly both.

In optical thin-film design, we do have many techniques capable of establishing good designs that can be already almost satisfactory. In other words, they are already in the region of an acceptable minimum of the merit function, and all that is required is to reach the actual minimum as quickly as possible. This is the objective of many of the refinement techniques that are used in optical coating work. Such is the complicated nature of the function of merit that all do not necessarily find the same minimum from the same starting design. Then there are techniques especially designed so that they do not necessarily choose a neighboring minimum. Instead, they range over a region of the parameter space, in a gradually more and more constrained manner. This permits them the opportunity of discovering any other merit function minimum that might offer improved performance over that nearest to the point of departure.

There are many ways of classifying the various refinement techniques. They can be divided into those that use a single design that is gradually altered in prescribed ways until a minimum is reached and those that use a family of designs, rejecting members of the family and replacing them by other designs and reaching the minimum in this way. They may also be classified as those that continuously attempt to move toward a minimum of the merit function and those that may take some time before they finally choose the particular merit function minimum and, therefore, have a greater chance of finding a more satisfactory minimum.

Only an analytical technique can involve continuous alteration of parameters. In computer optimization, the parameters are altered in finite steps that are usually adjusted in size as the process continues. It consists, essentially, of probing the merit function surface. The results of previous probings are used to guide the choice of future ones. Optimization is normally divided into repeated units called iterations. Each iteration will usually involve a single or multiple adjustments of the design or designs according to a set prescription and a reassessment of a new figure of merit. The process is continued until either a satisfactory outcome is attained or fresh iterations are unable to achieve any further improvement. The nature of the adjustment of the design and the way in which it is predicted is what principally distinguishes the various techniques [8].

It is tempting to find the best slope of the merit function as a function of the adjustable design parameters and to simply move down this slope as quickly as possible by changing the design parameters depending on the steepness of the slope. However, it is easy for the technique to become violently unstable with one overcorrection following another if precautions are not taken. The *steepest descent* method picks the maximum slope and follows it, but the parameter changes are usually restrained according to the derivative of the slope. If this is high, indicating that the slope appears to be rapidly changing, then the parameter changes are kept small. The steepest slope may not directly lead to the desired minimum. A zigzag path is a frequent feature of the convergence. The method of *damped least squares* is more efficient. Here the path is chosen to minimize a figure of merit based on the squares of the deviation between actual and target performance values. The calculation of the direction requires details of the local merit surface slope. Ideal convergence is achieved if the merit surface has a quadratic form. Stability is assured by restraining the movement by an adjustable damping parameter. Then there are several *univariate search techniques* in which only one parameter is altered at each iteration. The commonest is probably the *golden section* technique. Here a minimum of the merit function is achieved for each parameter in turn. The parameters may be chosen in the order of some prescribed scheme or at random. The search for the minimum in each case involves the process of bracketing, where three values of the parameter are maintained, with the figure of merit of the central one less than either of the two outer values. This means that a minimum exists between the two outer parameters. By always dividing the appropriate region in the ratio of $1:(3 - \sqrt{5})/2$, that is, $1:0.382$, the golden section, the most efficient search can be performed. *Linear search techniques* are like univariate search techniques, but they may freely choose the directions along which they search in parameter space. The most effective techniques change the directions from time to time based on previous progress. They are usually called *direction set methods*. The most efficient ones try to find a set of conjugate directions; that is, a set of directions that are decoupled from each other with respect to the minimization process—minimizing along a second direction after a first—should not alter the minimum of the first direction. Just one pass through the directions is then sufficient to reach the minimum. This works perfectly for simple quadratic functions. Unfortunately, the thin-film functions are very complicated, and they usually have to be searched over quite large regions so they rarely reach the final minimum in just one pass, but the search can be made more efficient if a continuous attempt is made to achieve conjugate directions.

Flip-flop optimization [9] is a digital technique, in a sense. A design is set up consisting of a large number of very thin layers of equal geometrical or optical thickness. These thin layers may have either of only two possible indices, or admittances, usually a high value and a low value. A merit function is set up and the figure of merit calculated. Now the layers of the design, from one end to the other, are scanned. At each iteration step, the figure of merit of the coating is assessed, with the index of the appropriate layer set to both the permitted values in turn. The better arrangement, in the sense of a lower figure of merit, is chosen, and the index of the layer is set to that value. The process then passes to the adjacent layer, and so on. Several complete passes of the design may be employed, and the order in which the layers are examined may be changed. Usually, the design stabilizes at a minimum of the merit function after only a few passes. The designs often consist of quite long blocks of one or the other index, corresponding to normal discrete layers, separated by blocks that clearly correspond to discrete layers of intermediate index, and occasionally, a structure that represents a thicker inhomogeneous layer is obtained. The process appears very stable. It is relatively easy to take a normal discrete layer design and turn it into a suitable starting design for this process, although it appears to work quite well with all layers initially set to one or the other of the two indices.

A process that does not immediately necessarily choose the minimum toward which it shall move is *simulated annealing* [8]. This uses a Boltzmann probability distribution:

$$\text{Prob}\,(E) = \exp\,(-E/kT), \tag{3.32}$$

where E is replaced by a merit function, and kT, by an annealing parameter T. Then, if the existing figure of merit is E_1 and a suggested new design has E_2, the probability that the new design is accepted in place of the old is

$$p = \text{probability} = \exp[-(E_2 - E_1)/T], \tag{3.33}$$

except that for $E2 < E1$, the probability is unity. The process involves calculating a new figure of merit based on a random choice of parameters within an assigned domain. If the merit function is less than the old, the new design replaces the old. If the merit function is greater than the old, it will be accepted with probability p based on the drawing of a random number. An *annealing schedule* is required that decides on the way in which T is allowed to fall until no further improvement is achieved.

One of the better techniques that uses a family of designs rather than a single one is the *simplex* technique, sometimes called *nonlinear simplex* to distinguish it from a similarly named technique in linear programming. The family of designs is known as the simplex, and the number of designs is one more than the number of design parameters involved. It will usually consist of the original design plus a set where each of the variables is offset in turn. At each iteration, the worst design, that is, the design with the greatest figure of merit, is rejected in favor of a new better design. The alternative new designs are generated in three possible ways. First, the worst design is reflected at the center of gravity of the simplex and the figure of merit is calculated. If this yields a better design, then a further equal move is made in the same direction and, again, the corresponding figure of merit is calculated. The better of these two designs replaces the existing worst design. If the first move fails to yield a better performance, then the worst design is moved halfway toward the center of gravity, which will then normally be an improvement. In the rare cases where none of the alternatives yields a better design, a completely new simplex is generated by moving all the designs halfway toward the existing best design [8].

The *statistical testing* method of Tang and Zheng [10] also involves a family of designs. Like simulated annealing, it does not move immediately down a particular slope but takes rather longer and so has a better chance of finding a more acceptable minimum. A starting region of parameter space is chosen, and then this region gradually shrinks around, it is hoped, a good, and perhaps even a global, minimum. Designs are chosen at random within the starting domain until a prescribed number have been found with merit function less than a starting target. The region then shrinks around those designs, and a new target that is now the mean of the merit functions is chosen. The process is repeated until a final minimum is reached.

The *genetic algorithm* is a refinement process based on ideas inherent in natural selection [11,12]. A design is a sequence of layers and, for the purposes of the genetic algorithm, is considered to be representing a strand of deoxyribonucleic acid. The techniques differ in detail and are very sensitive to the choices of parameters, but they operate more or less as follows. Rather like real species, there are three operations that mimic cloning, parenting, and mutation. A set of different designs, known as a generation, or population, is created, usually randomly and involving a reasonably large number of designs, and the merit, or fitness as it is sometimes called, of every design is calculated. Fitness may be arranged to be larger when better. The next generation, of equal size, is then created, first by a process of cloning, in which the better designs are permitted to produce a greater number of clones; then by a process of recombination, where the sequences from a certain number of pairs of parents, chosen at random from the clones, are each used to generate, from a linear combination of their designs, two replacement members of the next generation; and finally by a process of mutation, where some parameters of some of the members are randomly changed. The next generation then becomes the current generation, and the process repeats. The relative importance of the various parts of the process is the key to the success, or not, of the technique. The cloning, for instance, increases the number of better designs in each generation. To improve the chances of finding a good solution, the method is usually set to be fairly slow. Gradually, the individuals in successive generations exhibit improvements in merit, and the operation continues until either a set number of generations is reached or a satisfactory performance is achieved.

Differential evolution is a related process, also involving generations and selection, but is rather more complex [13]. First of all, a generation of designs is set up, usually randomly in a given domain of parameter space, and the merit of each is established. Next, a new generation is created by picking each member of the existing generation in turn along with three other members of the same generation selected randomly. A new intermediate design is then constructed. On the basis of the value of a random number, each layer of the intermediate design is assigned the corresponding layer parameters from the old design or a combination of the corresponding layer parameters of the three other designs involving the sum of the corresponding parameter from the first of the three together with a given factor times the difference of the corresponding parameters in the other two. Finally, the old design and the new design are crossed by a process involving random selection of the appropriate layer from the old or from the new design to produce a candidate member of the new generation. Then, the candidate or the old design, depending on which has the better merit, forms part of the new generation. The genetic algorithm and differential evolution have an advantage that the starting design is relatively unimportant, but they take rather longer than many of the other techniques.

Reduced dependence on the starting design can also be achieved simply by varying it in a random fashion followed by a complete refinement of each variation. The technique relies on the constantly increasing speed of computers because with n starting designs, the process takes n times as long as with a single design. It permits different paths over the merit surface to be followed and increases the chance of finding a good, if not the best, minimum. Some of the global techniques, already mentioned, can be of this nature. Others might involve a perturbation of each finishing design to form the next start.

There is a great deal of debate about which technique is better than another, and it is clear that there are differences in performance for different starting designs and coating types. A few comparative studies of optical coating design have been performed [14,15], but they have not unambiguously identified any technique always superior to all others. The secret of success in refinement is a good starting design that still offers scope for improvement. In that context, there is little difference between the various methods.

Synthesis is similar to refinement but involves some construction of the design beyond the adjustment of the existing layers. There are, of course, some analytical techniques of synthesis, but here, in this section, we limit ourselves to automatic techniques that extend what is possible with refinement. The number of possible designs is infinite, and so the synthesis problem can be solved only by introducing some constraints. Imagine that we have a very efficient refinement technique that is capable of dealing with starting designs that are rather far from ideal. Let us now set up targets and merit function in the normal way. Next, we create a starting design that uses a very small number of layers, perhaps only one. We refine this design until it is optimum. Then, we add layers according to some prescribed rules. Perhaps the figure of merit will now be rather larger than before, but we refine again and eventually achieve an optimum figure of merit that is lower. Again, we add layers according to our prescription and refine as before. We continue this process until we reach a stage where no improvement is taking place, and at that stage, we accept the best design. This is a viable synthesis technique and represents fairly well the few techniques that are sometimes used in practice. The way in which layers are added is the major difference between them. Dobrowolski [16] was the major pioneer in this field. He recognized that the addition of one single layer is often ineffective and the addition of more layers is indicated. His technique of evolution, now usually called *gradual evolution*, is still much used. In its basic form, it adds layers to either end of an existing layer sequence. This is often combined with other synthesis techniques.

Some spectacular results have been obtained by the *needle variation* method devised by Furman and Tikhonravov [17] and Tikhonravov et al. [18]. This searches the design for the best place to add a thin slice of material, initially of zero thickness. The definition of best is a maximum negative derivative of the merit function with respect to the added layer thickness. The addition of this vanishingly thin slice, known as the needle, effectively adds two layers because it cuts the existing layer in two. Sometimes, it is arranged to add varying numbers of layers depending on the stage of the synthesis and on the constraints. All this depends on a powerful and efficient refinement technique. The statistical refinement techniques tend to be less suitable because they already use considerable computer time, and it is more usual to use either the gradient, damped least squares, or linear search techniques in synthesis.

It may sometimes be said in support of a particular technique that it opens up new possibilities in design and arrives at performance levels that cannot be achieved in any other way. However, any design, however achieved, lies in the constrained parameter space. We may think of it as already existing. All that the various techniques can do is to search the constrained parameter space to find a suitable merit function minimum. They cannot find a minimum that does not exist. Although it may seem that synthesis is an ideal technique, the difficulties in finding the constrained global, or even a very good, minimum, which are compounded by the rapid increase in complexity as layers are added, mean that the final design may not be as good as one arrived at by a process of establishing a very good starting design and then carrying out a minimum of refinement [19]. In some techniques, quite thin layers that are difficult to manufacture may form part of the final design that must then be processed to remove them. The needle method, for example, introduces such thin layers as a necessary part of the process, and they may remain at termination. Synthesis is therefore best used when the designer is hard pressed with little idea of how to proceed, and it most effectively works when the total number of layers is not large.

Refinement and synthesis work best when the targets call for high transmittance. The performance of an optical coating is essentially a set of interference fringes. High reflectance presents certain problems because the fringes can be very much narrower than those in high transmittance. Refinement targets should be set so that they are closer together than the fringe spacing; otherwise, the performance between the targets may be seriously in error, and the number required for this can present problems in reflectance. The problem is sometimes called aliasing. For sine or cosine fringe profiles, avoidance of aliasing roughly implies that if the film is m quarter waves thick, then the spacing for wavelength target points should be λ/m or less. We often tend to work in constant increments of wavelength rather than wavenumber, and so the target for a film m quarter waves thick at λ should have $m + 1$ points to cover the octave λ to 2λ. A film that is 25 wavelengths thick should then have a target function with 100 wavelength points per octave.

This modest requirement is adequate for coatings with low reflectance but, unfortunately, completely inadequate for coatings where reflectance must be high [20]. The reason is that fringe profiles are not always approximately sine or cosine functions. In an antireflection coating, the reflectance is small and multiple-beam interference is weak. The fringes are then virtually sinusoidal, and so the simple calculation applies. In high-reflectance coatings, the fringes are invariably the result of multiple-beam interference and, therefore, are very narrow. This enormously increases the required number of targets necessary to ensure that a fringe cannot creep in between them. Additionally, there is a definite tendency for narrow fringes of lower reflectance to appear in coatings where high reflectance is required. We can readily understand the reason. Figure 3.13 shows the reflectance curves of two similar coatings. One is a quarter-wave stack with high reflectance. The other is derived from it by increasing the thickness of one of the central quarter waves to one half wave. Although this converts the coating into a single-cavity narrow band filter, the width of the high reflectance zone is considerably increased. The price is a very narrow central fringe. A density curve, Figure 3.14, of the same filter, shows that there is really no fundamental

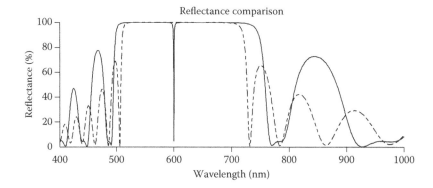

FIGURE 3.13
The insertion of a narrow fringe into the center of a high-reflectance coating can actually cause an apparent increase in the width of the high-reflectance zone. The basic quarter-wave stack high reflector is the broken line.

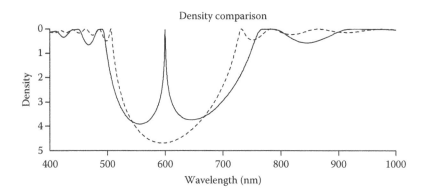

FIGURE 3.14
A look at the density variation shows that the performance is not better, but most merit functions are based on transmittance or reflectance, not density, and would prefer the broader zone in Figure 3.13.

gain, but most merit functions are based on reflectance or transmittance and would assign a lower figure of merit to the broader curve. Small changes in the thickness of the nominal half-wave layer can then adjust the lateral position of the fringe with virtually no other changes. Thus, the appearance of such features, sitting in between the target points in broadband reflectors, is not surprising. They are persistent and exceedingly difficult to eliminate, particularly by automatic means. Adding extra target points at the fringe is not very successful because a simple adjustment of the cavity layer thickness can move the fringe to where the target points are wider. It is therefore a very simple process for the refinement to slightly alter the thickness of one layer and move the sharp fringe exactly midway between two target points, with resulting substantial decrease of the figure of merit. This is a much easier operation for the process than the removal of a fringe, and sharp deep fringes are therefore persistent features that naturally position themselves between the target points, because a small change in the thickness of virtually any layer, but especially the cavity layer, will simply translate the fringe with almost no change in shape.

The fringe peaks are at their narrowest when the coating takes the form of a single cavity at the center of the coating surrounded by maximum reflectors. Let us assume a total thickness for the coating of x full waves and arrange it as a series of quarter waves of alternate high and low indices and with a central half-wave cavity layer. The half width of such an assembly is approximately given by

$$\frac{\Delta\lambda}{\lambda} = \frac{4y_L^{2x-1}y_{sub}}{\pi y_H^{2x}}, \tag{3.34}$$

where we neglect any dispersion of phase shift. The spacing of the wavelength points should be perhaps half this value:

$$\frac{\Delta\lambda}{\lambda} = \frac{2}{\pi} \cdot \frac{y_L^{2x}}{y_H^{2x}}, \tag{3.35}$$

where we have assumed the substrate admittance to be equal to y_L. We can take the wavelength interval as, say, λ to 2λ and the ratio of admittances as $\sqrt{2}$, so that the total number of points in the specification becomes

$$N = \pi 2^{x-1} \approx 2^x. \tag{3.36}$$

Every time another full wave is added, the number of points in the specification for the merit function should double.

It can be argued that the calculations are too pessimistic, but it is certainly clear that there is an inexorable increase in computing requirements with coating thickness. The increased burden of calculation becomes rapidly severe if not impossible. Many of the newer processes are capable of very large numbers of layers, and especially in the case of polymeric films, coatings with thousands of layers are achievable. A technique that your author has found helpful in this situation, although it is not a complete cure, is to include some targets of zero derivative in the high-reflectance zone. Some trial and error involving weights and tolerances and number of targets is required.

Automatic methods have revolutionized the design of coatings. They have not eliminated the older techniques but have rather changed their role. The drudgery of hand calculation has been completely removed. However, as the complexity of optical coatings increases, the completely automatic methods approach a barrier to further progress in the form of suitable measures of merit, and further developments in design techniques are required. The advent of the computer has certainly not reduced the need for the skill, experience, and innovation that has characterized the field until now.

References

1. S. D. Smith. 1958. Design of multilayer filters by considering two effective interfaces. *Journal of the Optical Society of America* 48:43–50.
2. P. H. Smith. 1939. Transmission line calculator. *Electronics* 12:29–31.
3. P. H. Smith. 1969. *Electronic Applications of the Smith Chart.* New York: McGraw-Hill.
4. W. Jackson. 1951. *High Frequency Transmission Lines*, pp. 129, 146. Third ed. London: Methuen.
5. P. H. Berning. 1963. Theory and calculations of optical thin films. In *Physics of Thin Films*, G. Hass (ed.), pp. 69–121. New York: Academic Press.
6. J. H. Apfel. 1972. Graphics in optical coating design. *Applied Optics* 11:1303–1312.
7. P. W. Baumeister. 1958. Design of multilayer filters by successive approximations. *Journal of the Optical Society of America* 48:955–958.
8. W. H. Press, B. P. Flannery, S. A. Teukolsky et al. *1986.* Numerical Recipes: The Art of Scientific Computing. *First ed. Cambridge, UK: Cambridge University Press.*
9. W. H. Southwell. 1985. Coating design using very thin high- and low-index layers. *Applied Optics* 24:457–460.
10. J. F. Tang and Q. Zheng. 1982. Automatic design of optical thin-film systems—Merit function and numerical optimization method. *Journal of the Optical Society of America* 72:1522–1528.
11. H. Greiner. 1996. Robust optical coating design with evolutionary strategies. *Applied Optics* 35:5477–5483.
12. S. Martin, J. Rivory, and M. Schoenauer. 1995. Synthesis of optical multilayer systems using genetic algorithms. *Applied Optics* 34:2247–2254.
13. R. Storn and K. Price. 1997. Differential evolution—A simple and efficient heuristic for global optimization over continuous spaces. *Journal of Global Optimization* 11:341–359.
14. J. A. Aguilera, J. Aguilera, P. Baumeister et al. *1988. Antireflection coatings for germanium IR optics: A comparison of numerical design methods.* Applied Optics 27:2832–2840.
15. J. A. Dobrowolski and R. A. Kemp. 1990. Refinement of optical multilayer systems with different optimization procedures. *Applied Optics* 29:2876–2893.
16. J. A. Dobrowolski. 1965. Completely automatic synthesis of optical thin film systems. *Applied Optics* 4:937–946.
17. S. A. Furman and A. V. Tikhonravov. 1992. *Basics of Optics of Multilayer Systems.* First ed. Gif-sur-Yvette: Editions Frontières.
18. A. V. Tikhonravov, M. K. Trubetskov, and G. W. DeBell. 1996. Application of the needle optimization technique to the design of optical coatings. *Applied Optics* 35:5493–5508.
19. A. Thelen. 1998. Computer aided design. In *Optical Interference Coatings*, pp. 268–270. Washington, DC: Optical Society of America.
20. H. A. Macleod. 1996. Recent trends in optical thin films. *Review of Laser Engineering* 24:3–10.

4

Antireflection Coatings

As has already been mentioned in Chapter 1, antireflection coatings were the principal objective of much of the early work in thin-film optics. Of all the possible applications, antireflection coatings have had the greatest impact on technical optics, and even today, in sheer volume of production, they still exceed all other types of coating. In some applications, antireflection coatings are simply required for the reduction of surface reflection. In others, not only must the surface reflection be reduced, but the transmittance must also be increased. The crown glass elements in a compound lens have a transmittance of only 96% per untreated surface, while the flint components can have a surface transmittance of as low as 90%. The net transmittance of even a modest number of untreated elements in series can therefore be quite low. Additionally, part of the light reflected at the various surfaces eventually reaches the focal plane, where it appears as ghosts or as a veiling glare, thus reducing the contrast of the images. This is especially true of the zoom lenses used in television or photography, where 20 or more elements may be included and which would be completely unusable without antireflection coatings.

Antireflection coatings can range from a simple single layer, having virtually zero reflectance at just one wavelength, to a multilayer system of more than a dozen layers, having reduced reflectance over a range of several octaves. The type used in any particular application will depend on a variety of factors, including the substrate material, the wavelength region, the required performance, and the cost. In the visible region, crown glass, which has a refractive index of around 1.52, is most commonly used. As we shall see, this presents a very different problem from infrared materials, which can have very much higher refractive indices.

There is no systematic method for the design of antireflection coatings. Trial and error, assisted by approximate techniques (frequently, one or other of the graphical methods mentioned in Chapter 3) and backed up by accurate computer calculation, is frequently employed. Very promising designs can be further improved by computer refinement. Complete synthesis where the design is arrived at with virtually no input from the designer other than a specification accompanied by possible materials is also common. Here we concentrate on understanding. Several different approaches are used in this chapter, partly to illustrate their use and partly because they are complementary. All the performance curves have been computed by the application of the matrix method. In most cases, the materials are considered completely transparent.

The vast majority of antireflection coatings are required for matching an optical element into air. Air has an index of around 1.0003 at standard temperature and pressure, which, for practical purposes, can be considered as unity.

The earliest antireflection coatings were on glass for use in the visible region of the spectrum. A single-layer antireflection coating on glass, for the center of the visible region, has a distinct magenta tinge when visually examined in reflection. This gives an appearance not unlike tarnish; indeed, in Chapter 1, we mentioned the beneficial effects of the tarnish layer on aged flint objectives, and so the term *bloom*, in the sense of tarnish, has been used in this connection. Particularly in the older literature, the action of applying the coating is sometimes referred to as *blooming*, and the element is said to be *bloomed*.

There is an enormously wide range of refractive indices in optical materials, and so it is not surprising that antireflection coatings exhibit a similarly wide range of designs. We can roughly divide them into those for high-index substrates and those for low-index substrates. The term *high-index* in this context cannot be precisely defined in the sense of a range with a definite lower bound. It implies that the substrate has an index that is sufficiently high within the range of available thin-film materials to enable the design of high-performance antireflection coatings consisting entirely, or almost entirely, of layers with indices lower than that of the substrate. These high-index substrates are principally of use in the infrared. Semiconductors, such as germanium, with an index of around 4.0, giving a reflection loss of around 36% per surface, and silicon, with an index around 3.5 and a reflection loss of 31%, are common, and it would be completely impossible to use them in the vast majority of applications without some form of antireflection coating. For many purposes, the reduction of a 30% reflection loss to one of a few percentage would be considered adequate. It is only in a limited number of applications where the reflection loss must be reduced to less than 1%. Low-index substrates, such as glass and silica, have much lower uncoated reflectance but present the quite difficult problem of the very limited range of available coating materials with still lower refractive index. The necessary employment of indices higher than those of the substrate complicates the designs and limits the possible level of performance.

We will give examples of the various techniques that we will describe. In much of the discussion of antireflection coatings for low-index materials, we will use glass as the example substrate and the visible region as the target. We shall assume 400–700 nm as the extent of the visible region. The techniques should not be thought of as limited to the visible region, however, but the use of a consistent example will allow easier comparison of the various levels of achieved performance.

All materials, substrates included, exhibit dispersion that is a variation in optical properties with wavelength or frequency. Normal dispersion is characterized by an increasing index of refraction (and usually extinction coefficient) as frequency increases and wavelength reduces. The higher the index of the material, the more pronounced the effect. To include the effects of dispersion in analytical techniques of design would complicate them to an almost impossible degree, completely obscuring the regular details of any involved model. Therefore, in most of the following techniques, we have assumed zero dispersion, but that does not imply that it should be completely neglected. Its primary effect is to increase the contrast between high- and low-index materials at shorter wavelengths so that performance falls off slightly more rapidly than that in the dispersion-free case. In practice, we deal with dispersion by introducing it at some suitable stage after which we will use refinement to complete the process. Since we are concentrating on understanding, in most cases, in this chapter where we examine theoretical performance, we will content ourselves with the dispersion-free case.

4.1 Single Layer

Let us start by examining a single-layer coating. This is so simple that it hardly needs analysis, but it will help to introduce various methods that we will find valuable when discussing more complex coatings.

First, let us use our knowledge of a quarter-wave film. We recall, from Chapter 3, the quarter-wave rule that a surface admittance of Y is transformed to one of η^2/Y when a quarter wave of characteristic admittance η is added. In this case, the surface to be coated has an admittance of η_m, and so for perfect antireflection, we need a quarter-wave film of admittance η, where η^2/η_m is equal to η_0, the admittance of the incident medium. At normal incidence, that is

$$\frac{y^2}{y_m} = y_0$$

or

$$y = \sqrt{y_0 y_m}.$$ (4.1)

This rule for the perfect reflection-reducing coating dates back to the early nineteenth century.

Let us now carry out a similar analysis using the vector method. Consider Figure 4.1. Here we have a vector diagram that, since two interfaces are involved, contains two vectors, each representing the amplitude reflection coefficient at an interface.

If the incident medium is air, then provided the index of the film is lower than the index of the substrate, the reflection coefficient at each interface will be negative, denoting a phase change of 180°. The resultant locus is a circle with a minimum at the wavelength for which the phase thickness of the layer is 90°, that is, a quarter-wave optical thickness, when the two vectors are completely opposed. Complete cancellation at this wavelength, that is, zero reflectance, will occur if the vectors are of equal length. This condition, in the notation of Figure 4.1, is

$$\frac{y_0 - y_1}{y_0 + y_1} = \frac{y_1 - y_m}{y_1 + y_m},$$

which requires

$$\frac{y_1}{y_0} = \frac{y_m}{y_1}$$

or

$$y_1 = (y_0 y_m)^{1/2}.$$ (4.2)

We immediately recognize this as identical to Equation 4.1. This is one of those unusual cases where an approximate technique gives an exactly correct answer.

The result at optical frequencies can also be written

$$n_1 = (n_0 n_m)^{1/2}.$$

At oblique incidence, the admittances y in Equation 4.2 should be replaced by the appropriate tilted values η.

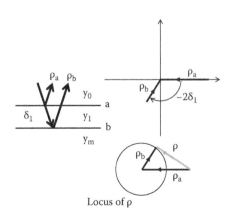

Locus of ρ

FIGURE 4.1
Vector diagram of a single-layer antireflection coating.

The condition for a perfect single-layer antireflection coating is, therefore, a quarter-wave optical thickness of material with optical admittance equal to the square root of the product of the admittances of substrate and medium. It is seldom possible to find a material of exactly the optical admittance that is required. If there is a small error ε in y_1 such that

$$y_1 = (1 + \varepsilon) \cdot (y_0 y_m)^{1/2},$$

then at that wavelength for which the antireflecting layer is a quarter wave,

$$R = \left[\frac{y_0 - y_0 y_m (1 + \varepsilon)^2 / y_m}{y_0 + y_0 y_m (1 + \varepsilon)^2 / y_m} \right]^2 = \left[\frac{-2\varepsilon - \varepsilon^2}{2 + 2\varepsilon + \varepsilon^2} \right]^2 \approx \varepsilon^2, \tag{4.3}$$

provided that ε is small. Thus, provided the layer remains a quarter wave, a 10% error in y_1 will lead to a residual reflectance of around 1%.

It is fairly straightforward to derive an analytical expression for the variation in reflectance of a single lossless dielectric layer on a dielectric substrate. We use the matrix method and a general approach that includes any effect of polarization and angle of incidence. The characteristic matrix of a single film on a substrate is given by

$$\begin{bmatrix} B \\ C \end{bmatrix} = \begin{bmatrix} \cos \delta_1 & \dfrac{i \sin \delta_1}{\eta_1} \\ i \eta_1 \sin \delta_1 & \cos \delta_1 \end{bmatrix} \begin{bmatrix} 1 \\ \eta_m \end{bmatrix},$$

i.e.,

$$\begin{bmatrix} B \\ C \end{bmatrix} = \begin{bmatrix} \cos \delta_1 + i(\eta_m/\eta_1) \sin \delta_1 \\ \eta_m \cos \delta_1 + i \eta_1 \sin \delta_1 \end{bmatrix},$$

where the symbols have the following meanings, defined in Chapter 2:

$$\left. \begin{aligned} \eta_p &= y/\cos \vartheta \\ \eta_s &= y \cos \vartheta \end{aligned} \right\} \quad \text{for each material}$$

$$\delta_1 = (2\pi n_1 d_1 \cos \vartheta_1)/\lambda$$

and where

$$n_0 \sin \vartheta_0 = n_1 \sin \vartheta_1 = n_m \sin \vartheta_m.$$

If λ_0 is the wavelength for which the layer is a quarter-wave optical thickness at normal incidence, then $n_1 d_1 = \lambda_0/4$ and

$$\delta_1 = \frac{\pi}{2} \left(\frac{\lambda_0}{\lambda} \right) \cos \vartheta_1$$

so that the new optimum wavelength is shifted to $\lambda_0 \cos \vartheta_1$.

The amplitude reflection coefficient is

$$\rho = \frac{\eta - Y}{\eta + Y} = \frac{\eta_0 - C/B}{\eta_0 + C/B}$$

$$= \frac{(\eta_0 - \eta_m) \cos \delta_1 + i[(\eta_0 \eta_m/\eta_1) - \eta_1] \sin \delta_1}{(\eta + \eta_m) \cos \delta_1 + i[(\eta_0 \eta_m/\eta_1) + \eta_1] \sin \delta_1},$$
(4.4)

and the reflectance is

$$R = \frac{(\eta_0 - \eta_m)^2 \cos^2 \delta_1 + [(\eta_0 \eta_m/\eta_1) - \eta_1]^2 \sin^2 \delta_1}{(\eta_0 + \eta_m)^2 \cos^2 \delta_1 + [(\eta_0 \eta_m/\eta_1) + \eta_1]^2 \sin^2 \delta_1}.$$
(4.5)

This expression is deceptively simple. An increase in the number of layers or a move to an absorbing system immediately increases the complexity of an analytical approach to a degree that is completely discouraging.

Summarizing the results so far, we can state that the simplest antireflection coating based on interference is a quarter wave of index given by the square root of the product of substrate and incident medium indices. Such a coating is perfect at the wavelength for which the layer is a quarter wave. Let us consider the visible region of the spectrum for a moment. The visible region can be taken as stretching from 400 to 700 nm. Our interference fringes in our single layer are symmetrical with respect to frequency but not with wavelength. In order for the reflectance at 400 nm to not be too different from that at 700 nm, we should make our reference wavelength at which the layer is a quarter wave correspond to the center of the frequency scale, implying that it should be 510 nm rather than 550 nm.

Immediately, we see that there is a problem with low-index substrates that are common in the visible region. For crown glass, say, of index 1.52, and a medium of air, with index 1.00, the required material index is 1.233. With a silica substrate of index 1.45, it is even lower at 1.204. We do not possess solid thin-film materials with such low indices of refraction. There are some applications where porous materials can be arranged to present lower indices, and we will deal with some of these later, but for normal applications, where the coating is exposed to the outside world, we need tough environmentally resistant layers. Cryolite (Na_3AlF_6) has one of the lowest indices at 1.35 but is not a very rugged material, and magnesium fluoride (MgF_2) is preferred with its index of 1.38. If we assume that there is no dispersion, then we can expect a variation of reflectance with wavelength for one single antireflected glass surface as shown in Figure 4.2.

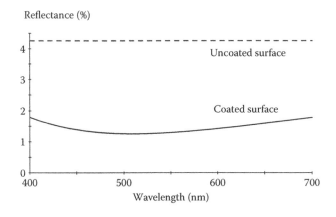

FIGURE 4.2
Calculated reflectance of one surface of glass ($n = 1.52$) coated with a quarter wave of MgF_2 ($n = 1.38$). The incident medium is air ($n = 1.00$) and the reference wavelength λ_0 is 510 nm.

Although the single-layer coating does not give us zero reflectance, nevertheless, the improvement in performance with such a simple coating is impressive, and single-layer coatings of this type are produced in massive quantities.

It is instructive to prepare an admittance diagram (Figure 4.3) for the single-layer coating. Admittance loci were discussed in Chapter 3. We consider normal incidence only and use free space units for the admittances so that they are numerically equal to the refractive indices. The locus for a single layer is a circle that cuts the real axis in two points with the product of the corresponding admittances being the square of the characteristic admittance of the layer, that is, the quarter-wave rule. For this admittance diagram, let us choose values that are more characteristic of the infrared than the visible region. We can imagine a substrate of germanium with an index of 4.0 and a film of zinc sulfide of index 2.2 in the infrared at a wavelength of perhaps around 4.0 μm. The starting point for the locus is the substrate surface, and so in this case, it begins at point 4.0 on the real axis. The center of the circle is on the real axis, and the circle cuts the real axis again at point $2.2^2/4.0 = 1.21$, corresponding to a quarter-wave optical thickness. Note especially that since the two points of intersection with the real axis are defined, we do not need to calculate the position of the center.

We can mark a scale of δ_1 along the locus. Contours of constant phase thickness are all circles centered on the imaginary axis and passing through the point on the real axis corresponding to the characteristic admittance, in this case, point 2.2. Usually, the circle centered on the origin that cuts the locus into eighth waves is sufficient. Since $\delta_1 = 2\pi n_1 d_1/\lambda$, we can either assume λ to be constant and replace the scale with one of optical thickness or, provided that we assume that the refractive index remains constant with wavelength, for a given layer optical thickness, we can mark the scale in terms of g $(=\lambda_0/\lambda)$. These various scales have been added. The scale of g assumes that λ_0 is the wavelength for which the layer has an optical thickness of one quarter of a wave.

This is a particularly simple admittance locus, and it is principally included to illustrate the method. We can add isoreflectance contours to the diagram (Figure 4.4). We will make some use of admittance diagrams in this chapter. Normally, these will be drawn for one value of wavelength and for one value of optical thickness for each layer. In the normal way, because they add somewhat confusing detail to an admittance diagram of any reasonable degree of complexity, we

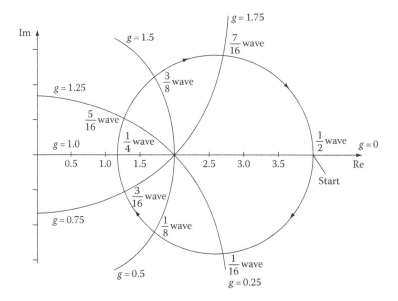

FIGURE 4.3
Admittance diagram for a single-layer zinc sulfide ($n = 2.2$) coating on germanium ($n = 4.0$).

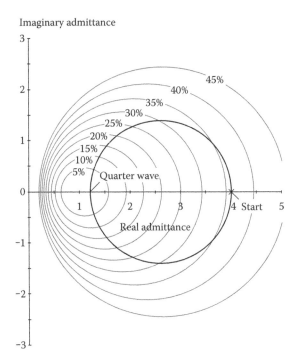

Imaginary admittance

FIGURE 4.4

Isoreflectance contours for an incident medium of $n = 1.00$ have been added to the admittance locus of Figure 4.3. To keep the diagram simple, the phase thickness contours have been removed.

will normally omit the isoreflectance contours. They coincide with a set of nested admittance circles for incident medium material.

Zinc sulfide has an index of around 2.2 at 2 μm and 2.15 at 15 μm. It has sufficient transparency for use as a quarter-wave antireflection coating over the range of 0.4–25 μm. Germanium, silicon, gallium arsenide, indium arsenide, and indium antimonide can all be treated satisfactorily by a single layer of zinc sulfide. The procedure to be followed for hard, rugged zinc sulfide films is described in a paper by Cox and Hass [1]. The substrate should be maintained at around 150°C during coating and cleaned by a glow discharge immediately before coating. The transmittance of a germanium plate with a single-layer zinc sulfide antireflection coating is shown in Figure 4.5.

Zinc sulfide, even if deposited under the best conditions, can deteriorate after prolonged exposure to humid atmospheres. Somewhat harder and more robust coatings are produced with cerium oxide or silicon monoxide. Cerium oxide, when deposited at a substrate temperature of 200°C or more, forms very hard and durable films of refractive index 2.2 at 2 μm. Unfortunately, in common with many other materials, it displays a slight absorption band at 3 μm, owing to adsorbed water vapor. Silicon monoxide does not show this water vapor band to the same degree, and so Cox and Hass have recommended this material as the most satisfactory for coating germanium and silicon in the near infrared. The index of silicon monoxide evaporated in a good vacuum at a high rate is around 1.9. The transmittance of a silicon plate coated on both sides with silicon monoxide is shown in Figure 4.6.

So far, we have considered only normal incidence in our numerical calculations. At angles of incidence other than normal, the behavior is similar, but the effective phase thickness of the layer is reduced as the incidence increases due to the cosine term in the phase thickness. Also, as the angle of incidence increases, the contrast between the admittances of materials increases for s-polarization but decreases for p-polarization until the angles are quite large. In our examples so far of a single-layer antireflection coating, the layer has already an index that is a little too high.

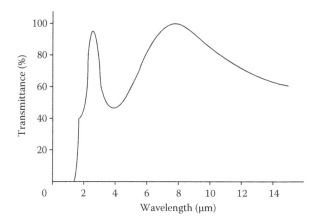

FIGURE 4.5
Transmittance of a germanium plate antireflected on both sides with zinc sulfide for 8 μm. (Courtesy of Sir Howard Grubb, Parsons and Co. Ltd., Newcastle upon Tyne.)

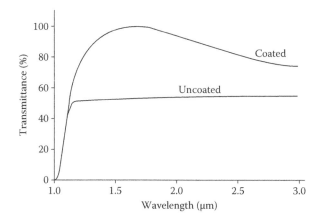

FIGURE 4.6
Transmittance of a 1.5 mm thick silicon plate with and without antireflection coatings of silicon monoxide, a quarter wavelength thick at 1.7 μm. (After J. T. Cox and G. Hass, *Journal of the Optical Society of America*, 48, 677–680, 1958. With permission of Optical Society of America.)

As the incidence increases, this situation will worsen for *s*-polarization and, up to a certain angle, will improve for *p*-polarization. We take the visible region antireflection coating of Figure 4.2 as our example and plot contours of *p*- and *s*-reflectance in Figure 4.7 that precisely show this behavior. This is a quite normal trend for most antireflection coatings.

4.2 Two-Layer Antireflection Coatings

The performance of our antireflection coatings for low-index substrates suffers because we cannot obtain solid hard films of a low enough refractive index. That of our coatings for high-index substrates suffers because even though we have a better chance of finding a suitable film material, the performance is optimum only at one wavelength with the reflectance rising on either side.

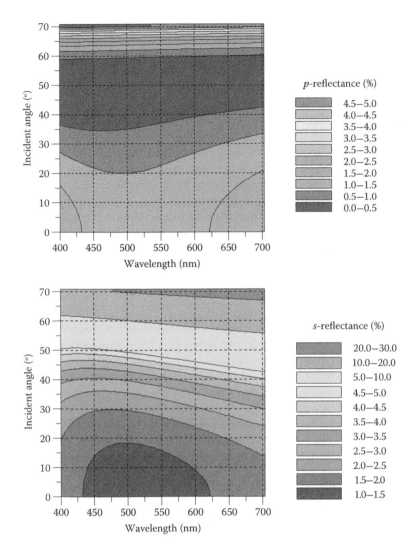

FIGURE 4.7
Contour plots showing the variation in reflectance for *p*-polarization (*upper*) and *s*-polarization (*lower*) for angles of incidence up to 70°. The design is the MgF₂ quarter wave over glass in an air incident medium from Figure 4.2. Note the rapid rise in reflectance for *s*-polarization for angles greater than 60°.

To do better, more adjustable parameters are required for the design, and that implies more layers. Let us examine what we can achieve with two layers. We will start with our visible region problem where our lowest suitable index is still too large, and in the first instance, we will limit ourselves to normal incidence.

Let us begin with a particularly simple approach using two quarter-wave layers. Two dielectric quarter-wave layers y_1 and y_2 over a substrate y_m yield, by a double application of the quarter-wave rule, an admittance for the front surface of $(y_1^2 y_m / y_2^2)$. For a perfect antireflection coating, this should be equal to the incident medium admittance y_0 so that we must have

$$y_1/y_2 = (y_0/y_m)^{1/2}. \tag{4.6}$$

Materials available in the visible region have indices of refraction in thin-film form ranging from around 1.35 for cryolite (Na₃AlF₆) to around 2.4 for titania (TiO₂). The range is a little uncertain because the index of a thin film depends on its microstructure that, in turn, depends on the method and conditions of deposition. However, it is clear that if the substrate has an index of 1.52, corresponding to crown glass, and if the incident medium is air, then there is an infinite number of possible combinations of materials that will satisfy Equation 4.6. Which should we choose? There could be all kinds of reasons for choosing one combination rather than another. However, from the optical point of view, we probably want the broadest characteristic. We are using two quarter waves, twice the thickness of the single-layer coating. The thicker the coating, the narrower the fringes, and so the characteristic of the coating is already rather narrower than that of our earlier single layer. We can perhaps instinctively see that the ideal coating would likely have interfaces with roughly equal contribution to reflectance, but when we examine our available range of indices, we see that the outermost interface, next to the incident medium, has the largest contribution, and the best we can do is to minimize it. This implies cryolite as our outer layer, but magnesium fluoride is rather more robust, and so, although it has slightly higher index, we follow normal practice and choose it at its value of 1.38. This requires 1.70 for the index of the second layer. The performance is shown in Figure 4.8 along with a similar coating with indices of 1.50 and 1.85, illustrating the consequent narrowing of the fringe. Antireflection coatings of this type where there is only one minimum are often referred to as V-coats because the characteristic shape might be perceived as similar to the letter *V*.

This design approach is quite straightforward, but presents us with a common problem in optical coating design: when a particular value of refractive index is prescribed, what material can we find to deliver it? Here we have a required index of 1.70. Aluminum oxide (Al_2O_3, the composition of sapphire) can be produced with an index as high as 1.65, but not easily at 1.70, although its bulk index is still higher. It is a very tough material, and so it is often used in this application. An index of 1.65 gives a surface admittance of 1.063 and, therefore, a reflectance at the minimum of 0.094%, low enough for a wide range of applications.

However, we would much prefer to choose what our preferred materials are and then design a suitable coating using them. Here the admittance diagram comes to our aid. Let us continue our examination of a single zero antireflection characteristic and start by choosing our MgF_2 with an index of 1.38 as the outermost material. A good trick with a design problem like this is to imagine that the problem is solved and then examine the consequences. If we have zero reflectance, then the admittance locus must terminate at the point corresponding to the incident admittance, in this case, 1.00. Since this point is on axis and since the two intersections with the axis obey the quarter-wave rule, the other point of intersection must have admittance of $1.38^2/1.00$, that is, 1.90. We can now draw the corresponding circular locus (Figure 4.9). Of course, the locus does not go near the substrate admittance of 1.52.

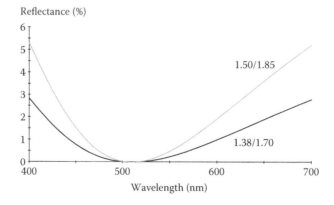

FIGURE 4.8
Performance of the quarter–quarter antireflection coating on glass ($n = 1.52$) in air ($n = 1.00$). The lower and broader curve is the performance of the combination of indices 1.38 and 1.70. The higher and narrower curve has indices 1.50 and 1.85.

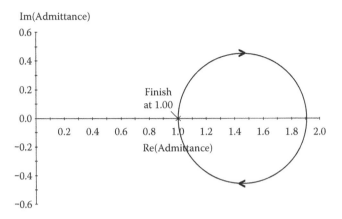

FIGURE 4.9
Circular locus corresponding to the outer magnesium fluoride layer in the antireflection coating (and described clockwise). The termination for zero reflectance must be as shown, but the starting point of the locus has yet to be determined.

The locus thus far illustrates a small complication that we will sometimes experience. The part that matters is over to the right of the diagram. The left-hand part is empty. In the future, we shall sometimes displace the imaginary axis along the real axis so that the important details are nearer the center of the diagram.

We now move to the starting point for the complete locus, the surface admittance of the substrate, 1.52. Somehow, we have to link that point to the already existing MgF$_2$ locus. The existing locus cuts the real axis at 1.90. Clearly, whatever we use for the layer next to the substrate must have a locus that cuts the real axis at 1.90 or greater; otherwise, the substrate admittance will not be joined to the MgF$_2$, and the locus will not be valid. Our already identified 1.70 index quarter wave cuts the real axis at 1.90 and is the limiting case. Anything of lower index does not close the gap, but anything of 1.70 and higher will be suitable. We choose Ta$_2$O$_5$ with an index of 2.15 as our preferred material. This presents a circular locus, starting at 1.52 and cutting the real axis again in $2.15^2/1.52$, that is, 3.04. Figure 4.10 shows the loci.

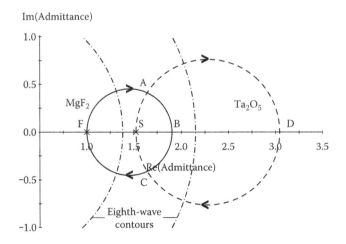

FIGURE 4.10
We add the 2.15 locus of Ta$_2$O$_5$ to the admittance diagram in Figure 4.9. There are now two major unbroken paths from substrate admittance S to the incident medium admittance at F, $SABCF$ and $SADCF$. We choose the thinner solution $SABCF$. Note that the origin of the plot corresponds to point 0.5 on the real axis.

Now, neglecting any paths that go round the circles multiple times, we have two possibilities shown in Figure 4.10 as $SABCF$ and $SADCF$. Usually we will want the broader response, and so we will choose the thinner solution, in this case, clearly, $SABCF$. Added to the diagram are the two contours of constant eighth-wave thicknesses. We recall that such a contour is a circle, centered on the origin and passing through the characteristic admittance of the layer on the real axis. From these, we can judge that the thickness of the high-index Ta_2O_5 layer is perhaps about one half of an eighth wave, and that of the low-index MgF_2 layer, about a quarter wave plus half an eighth wave. For a starting design for refinement, these thicknesses will be near enough.

Although automatic design will yield the correct thicknesses very quickly and easily, it is still possible to find an analytical solution. This was performed in 1962 by Catalan [2], and we will repeat the analysis here as a useful exercise. We will assume normal incidence to avoid the problems of polarization. The matrix expression is our starting point and the notation is our usual. Also, as usual, we assume a complete absence of any loss.

$$
\begin{bmatrix} B \\ C \end{bmatrix} = \begin{bmatrix} \cos\delta_1 & \dfrac{i\sin\delta_1}{y_1} \\ iy_1\sin\delta_1 & \cos\delta_1 \end{bmatrix} \begin{bmatrix} \cos\delta_2 & \dfrac{i\sin\delta_2}{y_2} \\ iy_2\sin\delta_2 & \cos\delta_2 \end{bmatrix} \begin{bmatrix} 1 \\ y_m \end{bmatrix}
$$

$$
= \begin{bmatrix} \cos\delta_1[\cos\delta_2 + i(y_m/y_2)\sin\delta_2] + i\sin\delta_1(y_m\cos\delta_2 + iy_2\sin\delta_2)/y_1 \\ iy_1\sin\delta_1[\cos\delta_2 + i(y_m/y_2)\sin\delta_2] + \cos\delta_1(y_m\cos\delta_2 + iy_2\sin\delta_2) \end{bmatrix}.
$$

The reflectance will be zero if the optical admittance Y is equal to y_0, i.e.,

$$
iy_1\sin\delta_1[\cos\delta_2 + i(y_m/y_2)\sin\delta_2] + \cos\delta_1(y_m\cos\delta_2 + iy_2\sin\delta_2)
$$
$$
= y_0\{\cos\delta_1[\cos\delta_2 + i(y_m/y_2)\sin\delta_2] + i\sin\delta_1(y_m\cos\delta_2 + iy_2\sin\delta_2)/y_1\}.
$$

The real and imaginary parts of these expressions must be equated separately, giving

$$
-(y_1y_m/y_2)\sin\delta_1\sin\delta_2 + y_m\cos\delta_1\cos\delta_2
$$
$$
= y_0\cos\delta_1\cos\delta_2 - (y_0y_2/y_1)\sin\delta_1\sin\delta_2
$$

and

$$
y_1\sin\delta_1\cos\delta_2 + y_2\cos\delta_1\sin\delta_2
$$
$$
= (y_0y_m/y_2)\cos\delta_1\sin\delta_2 + (y_0y_m/y_1)\sin\delta_1\cos\delta_2,
$$

i.e.,

$$
\tan\delta_1\tan\delta_2 = (y_m - y_0)/[(y_1y_m/y_2) - (y_0y_2/y_1)]
$$
$$
= y_1y_2(y_m - y_0)/(y_1^2 y_m - y_0 y_2^2)
\tag{4.7}
$$

and

$$
\tan\delta_2/\tan\delta_1 = y_2(y_0y_m - y_1^2)/[y_1(y_2^2 - y_0y_m)],
\tag{4.8}
$$

giving

$$
\tan^2\delta_1 = \frac{(y_m - y_0)(y_2^2 - y_0y_m)y_1^2}{(y_1^2 y_m - y_0 y_2^2)(y_0 y_m - y_1^2)},
$$
$$
\tan^2\delta_2 = \frac{(y_m - y_0)(y_0 y_m - y_1^2)y_2^2}{(y_1^2 y_m - y_0 y_2^2)(y_2^2 - y_0 y_m)}.
\tag{4.9}
$$

The values of δ_1 and δ_2 found from these equations must be correctly paired, and this is most easily done either by ensuring that they also satisfy the two preceding equations or by sketching a rough admittance diagram as in Figure 4.11.

For solutions to exist, or, putting it in another way, for the circles in the admittance diagram to intersect, the right-hand sides of Equation 4.9 must be positive. δ_1 and δ_2 are then real. If we assume that y_m is greater than y_0, then this requires that, of the following expressions,

$$\left(y_2^2 - y_0 y_m\right),\tag{4.10}$$

$$\left(y_1^2 y_m - y_0 y_2^2\right),\tag{4.11}$$

$$\left(y_0 y_m - y_1^2\right),\tag{4.12}$$

either all three must be positive or any two are negative and the third is positive. Should y_0 be greater than y_m, then the requirement becomes all three negative or two positive and one negative. The former case can be summarized in a useful diagram (Figure 4.12) known as a Schuster

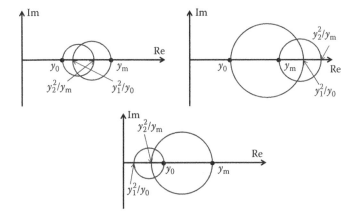

FIGURE 4.11
Admittance diagram for the double-layer antireflection coating. Valid solutions are possible only when the two circles touch or intersect. This diagram shows the three possible valid configurations when $y_m > y_0$. There are three others for $y_m < y_0$.

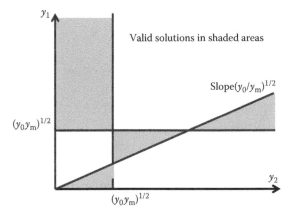

FIGURE 4.12
Construction of a Schuster diagram, where y_m is greater than y_0. Equations 4.10 through 4.12 are combined in one diagram, and the shaded areas are those in which real solutions exist.

diagram after one of the originators [3]. The diagram is constructed on the basis of the conditions involving Equations 4.10 through 4.12. The diagram does not include the actual thicknesses that are required. It simply shows that valid solutions exist for certain combinations of y_1 and y_2.

We can demonstrate the coating with our combination of MgF_2 and Ta_2O_5. The thinner solution assuming glass with index 1.52 and air as media has optical thicknesses in units of wavelength of 0.319 and 0.063 respectively. The reflectance of such a coating at a reference wavelength, λ_0, of 510 nm, and as a function of wavelength is shown in Figure 4.13. This is also known as a V-coat.

We can follow Catalan [2] and plot curves showing how the values of δ_1 and δ_2 vary with the index of the layer next to the substrate. Such curves are shown in Figure 4.14, and from them, several points of interest emerge. First, as already predicted by the Schuster plot, there is a region in which no solution is possible. Second, and more importantly, the curves flatten out as the index of the layer increases and changes in refractive index are accompanied by only small changes in optical thickness. One of the problems in manufacturing coatings is the control of the refractive index of the layers, particularly of the high-index layers, and the curves indicate good stability of the performance of the coating in this respect.

Let us return to Figure 4.10. We have so far chosen the thinnest solution because the interference fringes are broadest for the thinnest solution. There are many more solutions. The next thickest is *SADCF*, where the high-index layer is slightly less than a half wave and the low index is slightly less than a quarter wave. The next thickest again is *SADCSABCF*, where the high-index layer is now slightly greater than a half wave and the low index is slightly greater than a quarter wave. This progression in optical thicknesses, both layers increasing, is exactly what we obtain with gradually reducing wavelength. Can we combine these two solutions into one coating? An exact solution demands very particular values, but, as with most optical coatings, we are not looking for an exact solution but rather a desirable improvement in performance. This leads us to the idea of a coating consisting of a high-index half-wave layer next to the substrate followed by a low-index quarter-wave layer next to the incident medium. The reflectance at the reference wavelength λ_0 is that of the quarter wave since the half wave is an absentee, but the reflectance drops to a minimum on either side. It is thus a simple broadening improvement over the single quarter wave, which retains its reflectance at λ_0, and it is very easy to manufacture (Figure 4.15). Because of the shape of the response, it is usually called a W-coat. In practice, its main use is in the antireflection of glasses of higher index where the peak at λ_0 is lower.

Two-layer coatings are particularly useful for high-index substrates. Here, once again, the admittance diagram helps us with existence conditions. We can assume that we have air as incident medium so that $y_m \gg y_0$. The two possible solutions are shown in Figure 4.16. The existence of

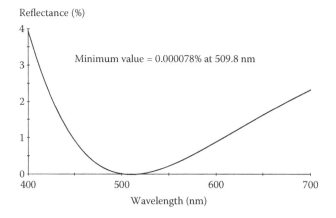

FIGURE 4.13
Performance of the V-coat with materials and thickness as specified in the text. The offset in the minimum wavelength and the residual value of reflectance at the minimum is a consequence of the rounding of the thickness values to three places.

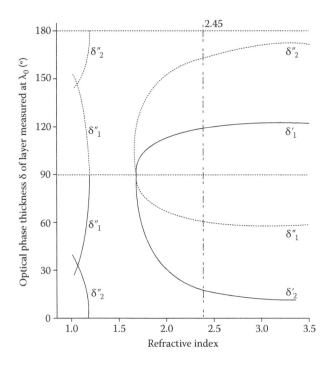

FIGURE 4.14
Optimum thicknesses of the layers in a double-layer antireflection coating at normal incidence. δ_1 and δ_2, the optical phase thicknesses, given by Equation 4.9, are plotted against n_2, the refractive index of the high-index layer. The low-index layer is assumed to be magnesium fluoride of index 1.38, and the coating is deposited on glass of index 1.50. Two pairs of solutions are possible for each set of refractive indices and are denoted by δ'_1 and δ'_2 and δ''_1 and δ''_2. The value of refractive index, 2.45, shown by the broken line, corresponds to bismuth oxide and was used by Catalan in his calculations. (After L. A. Catalan, *Journal of the Optical Society of America* 52, 437–440, 1962. With permission of Optical Society of America.)

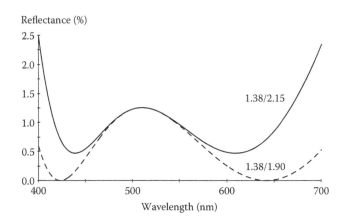

FIGURE 4.15
Performance of the W-coat on a glass substrate ($n = 1.52$) in an incident medium of air. An index of 2.15 (Ta_2O_5) for the half wave is a little high, and 1.90 (Y_2O_3 depending on deposition conditions) is rather better.

a solution depends on the intersection of the two admittance circles. Then there are two solutions, *SAF* and *SABCDAF*. Either can be arranged to give zero reflectance in the manner of a V-coat, and we can make a sufficiently accurate estimate of thicknesses for a good starting design for refinement, but the symmetry again suggests a coating of two quarter waves would have a characteristic similar to the W-coat already introduced.

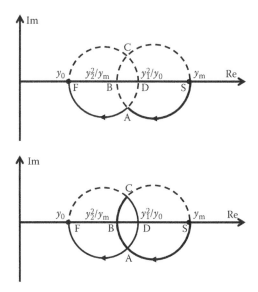

FIGURE 4.16
Existence of a solution depends on the intersection of the two admittance circles. Then there are two solutions, *SAF* and *SABCDAF*.

Let us illustrate the use of two quarter waves with a vector diagram as in Figure 4.17. We choose a quite simple arrangement of three vectors of the same length. Provided the angle between them is either 60°, corresponding to a wavelength of $3\lambda_0/2$, or 120°, corresponding to a wavelength of $3\lambda_0/4$, the condition for the characteristic admittances is

$$\frac{y_0}{y_1} = \frac{y_1}{y_2} = \frac{y_2}{y_m}. \tag{4.13}$$

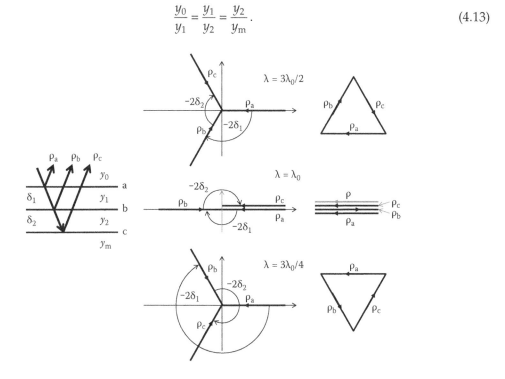

FIGURE 4.17
Vector diagram for the quarter–quarter coating. The vectors are arranged to be of the same length, and with angles between them of 60° or 120°, zero reflectance will be attained.

This can be solved for y_1 and y_2 as

$$y_1 = \sqrt[3]{y_0^2 y_m},$$

$$y_2 = \sqrt[3]{y_0 y_m^2} .$$

(4.14)

The reflectance at λ_0 is then

$$R = \left(\frac{y_0 - (y_1^2/y_2^2)y_m}{y_0 + (y_1^2/y_2^2)y_m} \right)^2$$

$$= \left(\frac{1 - (y_m/y_0)^{1/3}}{1 + (y_m/y_0)^{1/3}} \right)^2 .$$

(4.15)

We can illustrate the coating with a substrate of germanium ($n = 4.0$) in an incident medium of air. Equation 4.14 gives values for the layer admittances of 1.59 for y_1 and 2.52 for y_2. Since λ_0 is 3.0 µm, the zeros should be at 2.25 µm and at 4.5 µm with the peak at λ_0 rising to 5.15%. Figure 4.18 confirms these values. Unfortunately, these values of characteristic admittance, or index, are exactly prescribed, and we are unlikely to find materials that match them exactly. Figure 4.17 gives us some idea of how changing material parameters might affect the performance. Imagine that we move the admittances of the two layers closer together. This will decrease ρ_b and increase ρ_a and ρ_c. The imbalance at λ_0 is increased resulting in an increased reflectance, and the two minima will move away from λ_0. Moving the admittances of the layers further apart has the opposite effect. Figure 4.19 shows some examples of coatings based on this design approach.

This suggests a more general coating where the layers are of equal thickness. To compute the general conditions, it is easiest to return to the analysis leading up to Equation 4.9.

Let δ_1 be set equal to δ_2 and denoted by δ, where we recall that if λ_0 is the wavelength for which the layers are quarter waves, then

$$\delta = \frac{\pi}{2} \left(\frac{\lambda_0}{\lambda} \right).$$

From Equation 4.8,

$$y_2 \left(y_0 y_m - y_1^2 \right) = y_1 \left(y_2^2 - y_0 y_m \right),$$

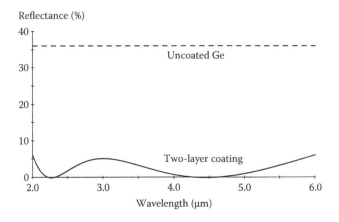

FIGURE 4.18
Calculated reflectance (single side) of a quarter–quarter antireflection coating on germanium ($n = 4.0$) in air. The reference wavelength λ_0 is 3.0 µm, and the layer admittances are $y_1 = 1.59$ and $y_2 = 2.52$.

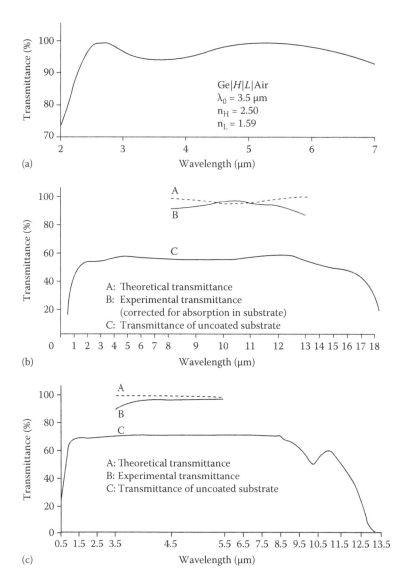

FIGURE 4.19
Double-layer antireflection coatings for high-index substrates. (a) Theoretical transmittance of a quarter–quarter coating on germanium (single surface). (b) Theoretical and measured transmittance of a similar coating on arsenic trisulfide glass (double surface). (c) Theoretical and measured transmittance of a similar coating on arsenic triselenide glass (double surface). (Courtesy of Barr and Stroud Ltd., Suffollk, UK.)

i.e.,

$$y_0 y_m = y_1 y_2 . \tag{4.16}$$

This is a necessary condition for zero reflectance.
From Equation 4.7, we find the wavelengths λ corresponding to zero reflectance:

$$\tan^2 \delta = \frac{y_1 y_2 (y_m - y_0)}{y_1^2 y_m - y_0 y_2^2} = \frac{y_0 y_m (y_m - y_0)}{y_1^2 y_m - y_0 y_2^2} .$$

If δ is the solution in the first quadrant, then there are two solutions:

$$\delta = \delta' \quad \text{or} \quad \delta = \pi - \delta'.$$

And the two values of λ are

$$\lambda = \left(\frac{\pi/2}{\delta}\right)\lambda_0.$$

In all practical cases, y_m will be greater than y_0, and the equation mentioned earlier for $\tan^2 \delta$ will have a real solution, provided

$$y_1^2 y_m - y_0 y_2^2 \geq 0. \tag{4.17}$$

The left-hand side of this inequality is identical to Equation 4.11.

Figure 4.20 gives the allowed values of y_1 and y_2 for germanium in air plotted on a Schuster diagram assuming normal incidence. The form of the characteristic curve of the coating is similar to that of Figure 4.19. The reflectance rises to a maximum value at the reference wavelength λ_0 situated between the two zeros. The reflectance at λ_0 can be found quite simply. At this wavelength, $\delta = \pi/2$ and the layers are quarter waves. The optical admittance is given, therefore, by

$$\frac{y_1^2}{y_2^2} y_m;$$

and the reflectance, by

$$R = \left(\frac{y_0 - (y_1^2/y_2^2)y_m}{y_0 + (y_1^2/y_2^2)y_m}\right)^2. \tag{4.18}$$

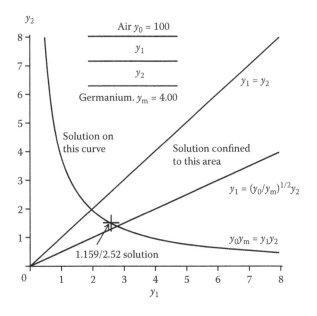

FIGURE 4.20
Schuster diagram showing possible values of film indices for a quarter–quarter coating on germanium ($n = 4.0$). The curve corresponds to Equation 4.16, and the lower line, to the lower limit of Equation 4.17. The upper line is the condition for the reflectance in the center to be not greater than the uncoated reflectance.

We are considering cases where y_m is large. For $y_1 = y_2$, the reflectance at λ_0 is that of the bare substrate. If $y_1 > y_2$, the reflectance is even higher. Thus, for the solution to be useful at all, y_1 should be less than y_2, and the region where this condition holds is indicated on the diagram.

Oblique incidence brings more problems. Performance calculation at oblique incidence is straightforward and we have already covered that in detail, and computers have made it all very easy. It was not always so, and many of the early papers reproduced calculations of oblique performance that were immediately of great usefulness. Typical of these is the paper of Catalan [2], who assumed bismuth oxide with an index of 2.45 for the high index and magnesium fluoride for the low index in the design of a V-coat for glass. Two of the plots for glass of index 1.50 are shown as Figures 4.21 and 4.22. The performance is very good up to an angle of incidence of 20°, but beyond that, it begins to fall off.

As far as design for oblique incidence is concerned, the design techniques that we have followed so far are immediately applicable provided the phase thickness and characteristic admittances are replaced by the values that apply to the particular angle of incidence. However, there is the additional complication of polarization. As long as we deal with one angle of incidence with one

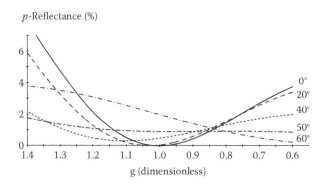

FIGURE 4.21
Theoretical p-reflectance as a function of wavelength ratio g (= λ_0/λ) of a double-layer antireflection coating. $n_0 = 1.00$, $n_1 = 1.38$, $n_2 = 2.45$, and $n_m = 1.50$. (After L. A. Catalan, *Journal of the Optical Society of America*, 52, 437–440, 1962. With permission of Optical Society of America.)

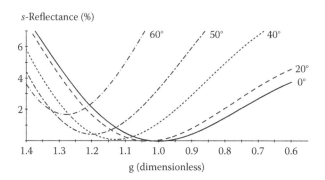

FIGURE 4.22
Theoretical s-reflectance as a function of wavelength ratio g (= λ_0/λ) of a double-layer antireflection coating. $n_0 = 1.00$, $n_1 = 1.38$, $n_2 = 2.45$, and $n_m = 1.50$. (After L. A. Catalan, *Journal of the Optical Society of America*, 52, 437–440, 1962. With permission of Optical Society of America.)

TABLE 4.1

Turbadar's Four Solutions

		Bismuth Oxide	Magnesium Fluoride
s-Polarization	s'	$0.065\lambda_0$	$0.376\lambda_0$
	s''	$0.457\lambda_0$	$0.206\lambda_0$
p-Polarization	p'	$0.021\lambda_0$	$0.382\lambda_0$
	p''	$0.501\lambda_0$	$0.201\lambda_0$

Source: Turbadar, T., *Optica Acta*, 11, 159–170, 1964.

polarization mode, the design follows the normal incidence process with the tilted parameter values. Serious difficulties emerge when we extend the requirements either to a range of angles of incidence, or to both polarization modes, or, still more difficult, to a combination of the two. Oblique incidence is treated in greater detail in a subsequent chapter, but here we mention some design calculations made by Turbadar [4], who examined designs for the angle of incidence 45°. The materials were once again bismuth oxide and magnesium fluoride, of indices 2.45 and 1.38, respectively, on glass of index 1.5. Four possible solutions were given, which are reproduced as Table 4.1, where the bismuth oxide is next to the glass. Many performance curves of the various designs under different conditions, including the effect of errors, were produced. Today, this is something we can do at great speed on a desktop computer. At the time, this was not possible and the plots that were included of equireflectance contours over a grid of angle of incidence against wavelength were particularly valuable. The fact that they can be more readily created does not reduce their usefulness. and so they are given in Figure 4.23.

4.3 Multilayer Antireflection Coatings

There is little further improvement in performance that can be achieved with two-layer coatings, given the limitations existing in usable film indices. For higher performance, further layers are required. With high-index substrates, we have seen that the availability, or otherwise, of a range of materials with indices in between those of the incident medium and substrate, dominates the design process. The problems associated with lower-index materials such as glass differ from those associated with high-index materials such as germanium and silicon. The differences become even more pronounced with increasing numbers of layers. More parameters implies more flexibility in design, and antireflection coatings are actually good vehicles for refinement and synthesis. However, an understanding of the structure of designs is still very important, especially when unexpected problems intrude. We therefore continue with our discussion of analytical methods.

We start with a high-index substrate such as germanium. Our success with the vector method in designing a two-layer coating consisting of quarter waves suggests a similar approach with three layers. A vector diagram is shown in Figure 4.24.

The coating is similar to the quarter–quarter coating in Figure 4.17, but where the two zeros of the two-layer coating are situated at $(3/4)\lambda_0$ and $(3/2)\lambda_0$, those of this three-layer coating stretch from $(2/3)\lambda_0$ to $2\lambda_0$, a much broader region.

The condition for the vectors to be of equal length is

$$\frac{y_1}{y_0} = \frac{y_2}{y_1} = \frac{y_3}{y_2} = \frac{y_m}{y_3},$$

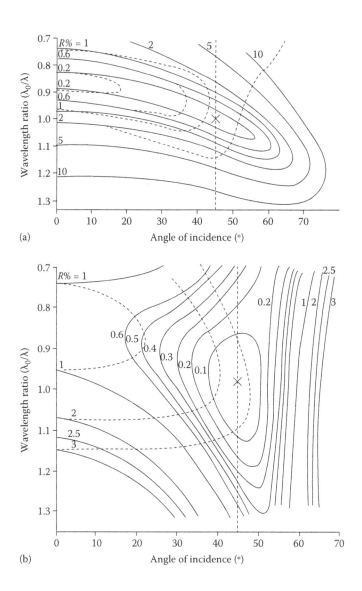

FIGURE 4.23

(a) Equireflectance contours for double-layer antireflection coatings on glass. $n_0 = 1.00$, $n_1 = 1.38$, $n_2 = 2.45$, and $n_m = 1.50$, with layer thicknesses optimized for s-polarization at a 45° angle of incidence, given by s' in Table 4.1. *Solid curves*, s-reflectance; *broken curves*, p-reflectance. (b) Equireflectance contours for double-layer antireflection coatings on glass. $n_0 = 1.00$, $n_1 = 1.38$, $n_2 = 2.45$, $n_m = 1.50$, with layer thicknesses optimized for p-polarization at a 45° angle of incidence, given by p' in Table 4.1. *Solid curves*, p-reflectance; *broken curves*, s-reflectance. (After T. Turbadar, *Optica Acta*, 11, 159–170, 1964.)

which, with some manipulation, becomes

$$y_1^4 = y_0^3 y_m,$$
$$y_2^4 = y_0^2 y_m^2, \tag{4.19}$$
$$y_3^4 = y_0 y_m^3.$$

For germanium in air at normal incidence,

$$n_0 = 1.00, \quad n_m = 4.00,$$

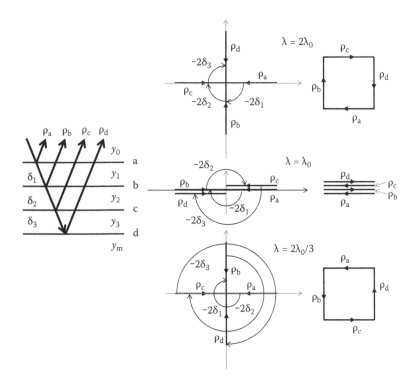

FIGURE 4.24
Vector diagram for a three-layer coating on a high-index substrate such as germanium. Each layer is a quarter-wave thick at λ_0. If $y_m > y_3 > y_2 > y_1 > y_0$, then the vectors will oppose each other, as shown, at $(2/3)\lambda_0$, λ_0, and $2\lambda_0$ and, provided the vectors are all of equal length, will completely cancel at these wavelengths, giving zero reflectance.

and the refractive indices required for the layers are

$$n_1 = 1.41,$$
$$n_2 = 2.00, \tag{4.20}$$
$$n_3 = 2.83.$$

The theoretical reflectance of such a coating with exactly these parameters is shown in Figure 4.25 with a reference wavelength of 3 µm.

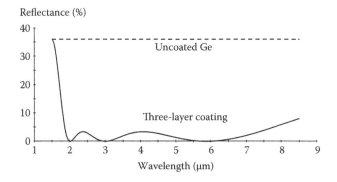

FIGURE 4.25
Theoretical performance of a three-quarter wave coating on germanium ($n = 4.0$) in air using the ideal index values from Equation 4.20. At 1.5 µm, the layers are all half waves and, therefore, absentees, causing the rise in reflectance to that of the uncoated substrate.

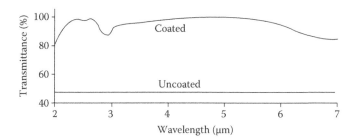

FIGURE 4.26
Measured transmittance of a germanium plate with coatings consisting of MgF_2 + CeO_2 + Si ($n_1d_1 = n_2d_2 = n_3d_3 = \lambda/4$ at 3.5 µm). (After J. T. Cox, G. Hass, and G. F. Jacobus, *Journal of the Optical Society of America*, 51, 714–718, 1961. With permission of Optical Society of America.)

A coating, which is not far removed from these theoretical figures, is silicon, next to the substrate of index 3.3, followed by cerium oxide of index 2.2, followed by magnesium fluoride of index 1.35. The performance of such a coating with $\lambda_0 = 3.5$ µm is shown in Figure 4.26. This coating, along with other one- and two-layer coatings for the infrared, was described by Cox et al. [5]. The exact theory of this coating may be developed in the same way as that of the two-layer coating, but the calculations are very much more involved and not very rewarding.

It is relatively easy to extend the vector method to deal with four layers, where the zeros of reflectance are found at $(5/8)\lambda_0$, $(5/6)\lambda_0$, $(5/4)\lambda_0$, and $(5/2)\lambda_0$, an even broader region than the three-layer coating. Five layers are equally straightforward. Whether or not such coatings are of practical value depends very much on the application. For many purposes, the two-layer coating is quite adequate.

The addition of an extra layer makes the exact theory of an arbitrary three-layer coating very much more involved than that of the two-layer. The number of possible groups of designs is enormous. It therefore becomes profitable to employ techniques that, rather than calculate performance in detail, indicate arrangements that are likely to be capable of acceptable performance and eliminate those that are not. Performance can then be accurately established by automatic means.

We continue, for the moment, with substrates of rather high index. A particularly useful technique was developed by Musset and Thelen [6]. It is based on Smith's method, that is, the method of effective interfaces. We recall from Chapter 3 that this involves the breaking down of the assembly into two subsystems on either side of a given layer. We can label these a and b, and then if δ denotes the phase thickness of the given layer, the overall transmittance of the multilayer is given by

$$T = \left(\frac{T_a T_b}{\left(1 - R_a^{1/2} R_b^{1/2}\right)^2}\right) \times \left[1 + \frac{4 R_a^{1/2} R_b^{1/2}}{\left(1 - R_a^{1/2} R_b^{1/2}\right)^2} \sin^2\left(\frac{\varphi_a + \varphi_b - 2\delta}{2}\right)\right]^{-1}. \tag{4.21}$$

We assume that there is no absorption, so that $T_a = 1 - R_a$ and $T_b = 1 - R_b$.

Both of the expressions multiplied together on the right-hand side of Equation 4.21 have maximum possible values of unity, and for maximum transmittance, therefore, both must be separately maximized. The first expression,

$$\frac{T_a T_b}{\left(1 - R_a^{1/2} R_b^{1/2}\right)^2}, \tag{4.22}$$

will be unity if and only if $R_a = R_b$, while the second,

$$\left[1 + \frac{4R_a^{1/2}R_b^{1/2}}{\left(1 - R_a^{1/2}R_b^{1/2}\right)^2} \sin^2\left(\frac{\varphi_a + \varphi_b - 2\delta}{2}\right) \right]^{-1},$$

will be unity if and only if

$$\sin^2\left(\frac{\varphi_a + \varphi_b - 2\delta}{2}\right) = 0. \tag{4.23}$$

The conditions for a perfect antireflection coating are then

$$R_a = R_b, \tag{4.24}$$

called the *amplitude condition* by Musset and Thelen, and

$$\frac{\varphi_a + \varphi_b - 2\delta}{2} = m\pi, \tag{4.25}$$

called the *phase condition*. The amplitude condition is a function of the two subsystems. The phase condition can be satisfied by adjusting the thickness of the given layer, or phase-matching layer, that separates the two subsystems. The amplitude condition can, using a method devised by Musset and Thelen, be satisfied for all wavelengths, but it is difficult to satisfy the phase condition except at a limited number of discrete wavelengths. At other wavelengths, the performance departs from ideal to a varying degree.

The transmittance and reflectance of a multilayer remain constant when the optical admittances are all multiplied by a constant factor or when they are all replaced by their reciprocals, in both cases keeping the optical thicknesses constant. These properties can be readily demonstrated from the structure of the characteristic matrices [7]. They enable the design of pairs of substructures having identical reflectance so that only the phase condition need be satisfied for perfect anti-reflection. We can, following Musset and Thelen, imagine a multilayer consisting of two subsections a and b, as shown in Figure 4.27, with the phase-matching layer of admittance y_i in between. At this stage, we put no restrictions on this layer in terms of either refractive index or thickness, but, as we shall see, they will become defined at a later stage. Subsection a is bounded by y_0 on one side and y_i on the other, while b is bounded in the same way by y_i and y_m. We can now apply the appropriate rules for ensuring that the amplitude condition is satisfied. We set up any subsystem a and then convert it into subsystem b by retaining the optical thicknesses and either multiplying the admittances by a constant multiplier or taking the reciprocals of the admittances and multiplying them by a constant multiplier. Systems derived by the former procedure are classified by Musset and Thelen as of type I, and those by the latter, as type II.

FIGURE 4.27
Multilayer antireflection coating consisting of two subsystems a and b separated by a phase-matching layer.

For type I systems, we must have the same relationship between y_m and y_i as between y_i and y_0, and between the layers of b and those of a. That is,

$$y_m f = y_i,$$
$$y_i f = y_0,$$
$$y_b f = y_a,$$

so that

$$y_i = (y_0 y_m)^{1/2} \quad \text{and} \quad f = (y_0/y_m)^{1/2}. \tag{4.26}$$

In this way, any y_b gives a corresponding y_a of $y_b(y_0/y_m)^{1/2}$.

Type II systems, on the other hand, obey the reciprocal relationship, so that their relationships are

$$f/y_m = y_0,$$
$$f/y_i = y_i,$$
$$f/y_a = y_b,$$

where we have had to reverse the order of sequence a when comparing it with the sequence b because, otherwise, the relationships collapse. Then

$$y_i = (y_0 y_m)^{1/2} \quad \text{and} \quad f = y_0 y_m, \tag{4.27}$$

so that any y_a gives a corresponding y_b of $y_0 y_m / y_a$.

There are no restrictions on layer thickness or on the number of layers in each subsystem except, of course, that they must be equal in number, and it is simpler if quarter-wave layers are used. Once the individual subsystems a and b are established, the amplitude condition is automatically satisfied at all wavelengths, and it remains to satisfy the phase condition. This involves the coupling arrangement through the phase-matching layer. It is impossible to meet the phase condition at all wavelengths, and the problem is so complex that it is best to take the easy way out and adopt a layer of admittance y_i with a thickness of zero, in which case, the layer is omitted or we can make it a quarter wave like the remaining layers of the assembly.

The method can be illustrated by application to the antireflection of germanium at normal incidence. In this case, $n_0 = 1.00$ and $n_m = 4.00$. Hence, $n_i = (n_0 n_m)^{1/2} = 2.0$ in both types I and II systems. First of all, we take, for subsystem b, a straightforward single quarter wave matching the substrate to the coupling medium:

n_1	n_b	n_m
	$(n_i n_m)^{1/2}$	
2.0	2.826	4.0 .

Subsystem a is then, for both types I and II systems,

n_0	n_a	n_i
	$(n_0 n_i)^{1/2}$	
1.0	1.414	2.0 .

Putting the two subsystems together, we have either a two-layer coating if we permit the thickness of the coupling layer to shrink to zero or a three-layer coating if the coupling layer is a quarter wave. In the former case, we have the design

$$\begin{array}{c|c|c|c} \text{Air} & 1.414 & 2.282 & \text{Ge} \\ \hline 1.0 & 0.25\lambda_0 & 0.25\lambda_0 & 4.0, \end{array}$$

and, in the latter, we have

$$\begin{array}{c|c|c|c|c} \text{Air} & 1.414 & 2.0 & 2.282 & \text{Ge} \\ \hline 1.0 & 0.25\lambda_0 & 0.25\lambda_0 & 0.25\lambda_0 & 4.0. \end{array}$$

The first design gives a single minimum. The second, which is similar to the three-layer design, already obtained by the vector method, has a broad three-minimum characteristic (Figure 4.28).

The subsystems need not be perfect matching systems for n_m to n_i and n_i to n_0. We could, for instance, use

$$n_0 = 1.0,$$

$$n_a = (1.0 \times 4.0)^{1/3} = 1.587,$$

$$n_m = 2.0$$

from the two-layer coating derived by the vector method. This gives complete two- and three-layer coatings, as follows:

- Type I

$$\begin{array}{c|c|c|c} \text{Air} & 1.587 & 3.174 & \text{Ge} \\ \hline 1.0 & 0.25\lambda_0 & 0.25\lambda_0 & 4.0, \end{array}$$

$$\begin{array}{c|c|c|c|c} \text{Air} & 1.587 & 2.0 & 3.174 & \text{Ge} \\ \hline 1.0 & 0.25\lambda_0 & 0.25\lambda_0 & 0.25\lambda_0 & 4.0. \end{array}$$

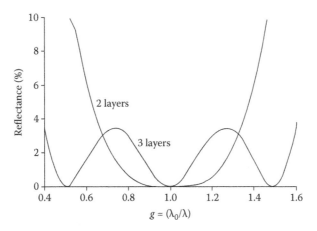

FIGURE 4.28
Theoretical performance of antireflection coatings on germanium designed by the method of Mussett and Thelen [6]. (From A. Musset and Thelen. 1966. Multilayer antireflection coatings. In *Progress in Optics*, E. Wolf (ed.), pp. 201–237. Amsterdam: North Holland.)

$$\begin{array}{c|c|c|c|c} \text{Two} & \text{Air} & 1.414 & 2.828 & \text{Ge} \\ \hline \text{layers:} & 1.00 & 0.25\lambda_0 & 0.25\lambda_0 & 4.00 \end{array}$$

$$\begin{array}{c|c|c|c|c|c} \text{Three} & \text{Air} & 1.414 & 2.00 & 2.828 & \text{Ge} \\ \hline \text{layers:} & 1.00 & 0.25\lambda_0 & 0.25\lambda_0 & 0.25\lambda_0 & 4.00 \end{array}$$

- Type II

$$\text{Air} \mid 1.587 \mid 2.520 \mid \text{Ge}$$
$$1.0 \mid 0.25\lambda_0 \mid 0.25\lambda_0 \mid 4.0,$$

$$\text{Air} \mid 1.587 \mid 2.0 \mid 2.520 \mid \text{Ge}$$
$$1.0 \mid 0.25\lambda_0 \mid 0.25\lambda_0 \mid 0.25\lambda_0 \mid 4.0$$

The first of the type II designs is identical with the vector method coating. Performance curves are given in Figure 4.29.

Analytical expressions for calculating the positions of the zeros and the residual reflectance maxima of two- and three-layer coatings of the types mentioned earlier are given by Musset and Thelen. The method can be readily extended to four and more layers.

Young [8] developed alternative techniques for coatings consisting of quarter-wave optical thicknesses based on the correspondence between the theory of thin-film multilayers and that of microwave transmission lines. He gave a useful set of tables for the design of multilayer coatings where all thicknesses are quarter waves. Given the bandwidth and the maximum permissible reflectance, it is possible to quickly derive the coating that meets the specification with the least number of layers. The method, of course, takes no account of the possibility of achieving the given indices in practice, as with many of the other methods we have been discussing, but the optimum solution is a very useful point of departure in the design of coatings using real indices.

Now let us turn to substrates of lower index. Due to the simplicity of calculation and under-standing when thicknesses are quarter waves, or integral numbers of quarter waves, much attention has been directed toward coatings that involve layers of either quarter-wave or half-wave optical thicknesses. We start with some that can be looked upon as modifications of some of the two-layer designs already considered.

First, we take the two-layer W-coat consisting of a half-wave layer next to the substrate followed by a quarter-wave layer (Figure 4.15). This has a peak reflectance in the center of the low-reflectance region. This peak corresponds to the minimum reflectance of a single-layer coating because the inner layer, being a half wave at that wavelength, is an absentee. We can reduce the peak but retain to some extent the flattening effect of the half-wave layer by splitting it into two quarter waves, only slightly different in index. We can retain the first layer as 1.9, although it is in no way critical, and then if we make the second quarter wave of slightly higher index, say, 2.0, the design now becoming

$$\text{Air} \mid 1.38 \mid 2.0 \mid 1.9 \mid \text{Glass}$$
$$1.0 \mid 0.25\lambda_0 \mid 0.25\lambda_0 \mid 0.25\lambda_0 \mid 1.52$$

We find a reduction in the reflectance at λ_0 from 1.26% to 0.38%. The characteristic remains fairly broad. Increasing the index of the central layer still further, to 2.13, i.e., the following design, reduces the reflectance at λ_0 to virtually zero, but the width of the coating becomes much more significantly reduced:

$$\text{Air} \mid 1.38 \mid 2.13 \mid 1.9 \mid \text{Glass}$$
$$1.0 \mid 0.25\lambda_0 \mid 0.25\lambda_0 \mid 0.25\lambda_0 \mid 1.52$$

The characteristic curves of these coatings are shown in Figure 4.30.

It is useful to employ the admittance locus in the examination of the flattening action of the half-wave layer in the W-coat. Figure 4.31 shows the locus at the reference wavelength 510 nm and at 640 nm in the vicinity of the longer wavelength minimum. The opening of the half-wave locus

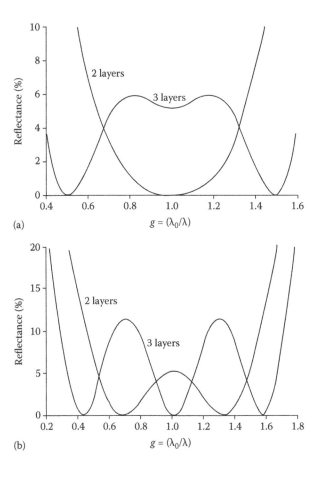

(a)

(b)

FIGURE 4.29
(a) Theoretical performance of type I antireflection coatings on germanium designed by the method of Mussett and Thelen. (From A. Musset and Thelen. 1966. Multilayer antireflection coatings. In *Progress in Optics*, E. Wolf (ed.), pp. 201–237. Amsterdam: North Holland.)

Two	Air	1.587	3.174	Ge
layers:	1.00	$0.25\lambda_0$	$0.25\lambda_0$	4.00

Three	Air	1.587	2.00	2.174	Ge
layers:	1.00	$0.25\lambda_0$	$0.25\lambda_0$	$0.25\lambda_0$	4.00

(b) Theoretical performance of type II antireflection coatings on germanium designed by the method of Mussett and Thelen [6]. (From A. Musset and Thelen. 1966. Multilayer antireflection coatings. In *Progress in Optics*, E. Wolf (ed.), pp. 201–237. Amsterdam: North Holland.)

Two	Air	1.587	2.520	Ge
layers:	1.00	$0.25\lambda_0$	$0.25\lambda_0$	4.00

Three	Air	1.587	2.00	2.5204	Ge
layers:	1.00	$0.25\lambda_0$	$0.25\lambda_0$	$0.25\lambda_0$	4.00

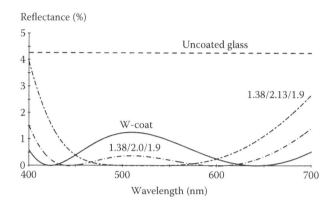

FIGURE 4.30
Progressive changes in an antireflection coating consisting of three-quarter wave layers. The original coating is the W-coat with following design, where the two 1.90 index layers combine to form a single half-wave layer:

Air	1.38	1.90	1.90	Glass
1.00	$0.25\lambda_0$	$0.25\lambda_0$	$0.25\lambda_0$	1.52

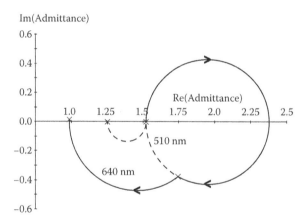

FIGURE 4.31
Admittance diagram for the W-coat:

Air	1.38	1.90	Glass
1.00	$0.25\lambda_0$	$0.5\lambda_0$	1.52

matches the shortening of the quarter-wave 1.38 locus so that its termination remains close to the real axis and moves toward the incident medium admittance at 1.0. The compensation, therefore, is assured by the termination of one layer and the start of the next on the same side of the real axis. Should they be on opposite sides, the effect is narrowing rather than broadening. This allows us to determine where the insertion of a half-wave might be useful in broadening performance.

Yet a further increase in the width of the coating can be achieved by adding a half-wave layer of low index next to the substrate. We take the intermediate design of Figure 4.30 and insert a half wave of index 1.38 next to the substrate. A half-wave layer in the same position with index higher than the substrate would be ineffective. The design becomes

Air	1.38	2.00	1.90	1.38	Glass
1.0	$0.25\lambda_0$	$0.25\lambda_0$	$0.25\lambda_0$	$0.5\lambda_0$	1.52

The admittance plot is shown in Figure 4.32, and we see the characteristic shape where the final part of the locus of the half-wave layer and the start of the following layer are on the same side of the real axis. The performance is shown in Figure 4.33.

A certain amount of additional trial and error leads to the designs shown in Figure 4.34. These are

Design	Air	1.38	1.905	1.76	1.38	Glass
A	1.0	$0.25\lambda_0$	$0.25\lambda_0$	$0.5\lambda_0$	$0.5\lambda_0$	1.52

and

Design	Air	1.38	2.13	1.9	1.38	Glass
B	1.0	$0.25\lambda_0$	$0.25\lambda_0$	$0.25\lambda_0$	$0.5\lambda_0$	1.52

An alternative approach is to broaden the wider quarter–quarter design of Figure 4.8 by inserting a half-wave layer between the two quarter waves. In order to achieve the broadening effect, it must, of course, be of high index, so that the admittance plot, Figure 4.35, will exhibit the

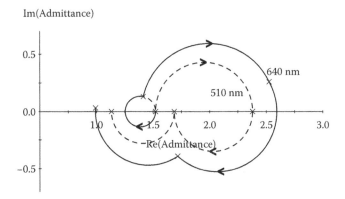

FIGURE 4.32
Admittance diagram at reference wavelength 510 nm and at 640 nm of the coating with following design:

Air	1.38	2.00	1.90	1.38	Glass
1.00	$0.25\lambda_0$	$0.25\lambda_0$	$0.25\lambda_0$	$0.5\lambda_0$	1.52

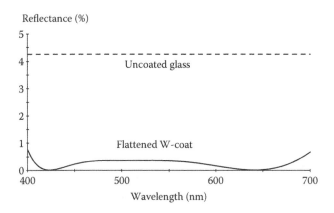

FIGURE 4.33
Reflectance of the coating in Figure 4.32.

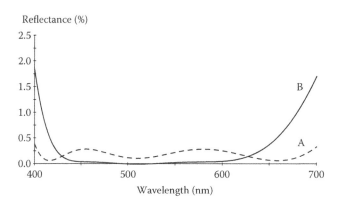

FIGURE 4.34
Performance of the two four-layer designs labeled A and B in the text.

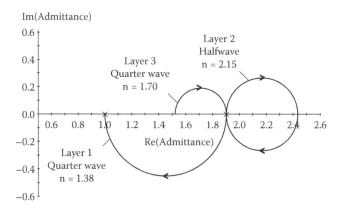

FIGURE 4.35
Admittance locus of the quarter–half–quarter coating:

Air	1.38	1.90	1.70	Glass
1.0	$0.25\lambda_0$	$0.5\lambda_0$	$0.25\lambda_0$	1.52

broadening features as in Figure 4.32. The coating is frequently referred to as the quarter–half–quarter coating with typical performance in Figure 4.36. It has a long history and has greatly influenced the development of antireflection coatings for substrates such as glass. Coatings that fit into this general type date back to the 1940s and were described by Lockhart and King [9]. A design technique explaining the functions of the various layers, however, had not been available until a study by Cox et al. [10]. The authors also investigated the effect of varying the indices of the various layers to achieve best results on crown glass. Good results were obtained with values of the index of the half-wave layer in the range of 2.0–2.4, while the outermost layer index should be between 1.35 and 1.45, and the innermost layer index, between 1.65 and 1.70. The outermost layer is the critical one in the design.

Figure 4.37, in Cox et al. [10], shows the measured reflectance of an experimental coating consisting of magnesium fluoride, index of 1.38; zirconium oxide, index of 2.1; and cerium fluoride, which was evaporated rather too slowly and had an index of 1.63, accounting for the slight rise in the middle of the range. The coating is a good practical confirmation of the theory.

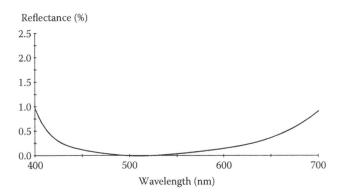

FIGURE 4.36
Calculated reflectance of the quarter–half-quarter coating shown in Figure 4.35.

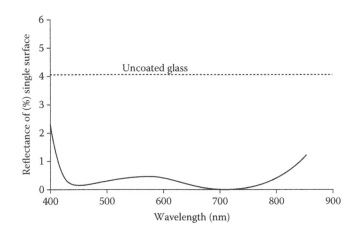

FIGURE 4.37
Measured reflectance of a quarter–half-quarter antireflection coating of MgF_2 + ZrO_2 + CeF_3 on crown glass. $\lambda_0 = 550$ nm. (After J. T. Cox, G. Hass, and A. Thelen, *Journal of the Optical Society of America*, 52, 965–969, 1962. With permission of Optical Society of America.)

The effect of variations in the angle of incidence has also been examined. The results of the study by Cox et al. [10] for tilts of up to 50° of a coating designed for normal incidence are shown in Figure 4.38. The performance of the coating is excellent up to 20° but begins to fall off beyond 30°. The coatings can, of course, be designed for use at angles of incidence other than normal, and Turbadar [11] published a full account of a design for use at 45°. The particular design depends on whether light is *s*- or *p*-polarized, and Figure 4.39 shows sets of equireflectance contours for both designs.

The quarter–half-quarter coating is certainly the most significant of the early multilayer coatings for low-index glass, and it has had considerable influence on the development of the field.

The success of the broadening effect of the half-wave layer on the quarter–quarter coating prompts us to consider inserting a similar half wave in the two-layer coating of Figure 4.13. In this case, there is an advantage in using a layer of the same index as that next to the substrate. Here we cannot split the coating at the interface between the high- and the low-index layers, because the admittance plot would not show the correct broadening configuration. Instead, we must split the

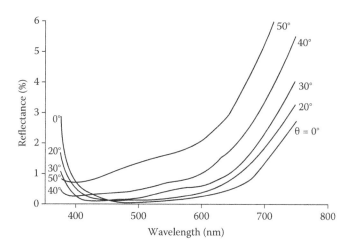

FIGURE 4.38
Calculated reflectance as a function of wavelength for quarter–half-quarter antireflection coatings on glass at various angles of incidence. $n_0 = 1.00$, $n_1 = 1.38$, $n_2 = 2.2$, $n_3 = 1.70$, $n_m = 1.51$. (After J. T. Cox, G. Hass, and A. Thelen, *Journal of the Optical Society of America*, 52, 965–969, 1962. With permission of Optical Society of America.)

coating at the point where the low-index locus cuts the real axis so that the plot appears as in Figure 4.40. The design of the coating is then

Air	1.38	2.15	1.38	2.15	Glass
1.0	$0.25\lambda_0$	$0.5\lambda_0$	$0.069\lambda_0$	$0.063\lambda_0$	1.52

The performance is shown in Figure 4.41. There is a considerable resemblance between this admittance plot and that of the quarter–half-quarter design. This design approach can be originally attributed to Frank Rock, who used the properties of reflection circles in deriving it rather than admittance loci.

Vermeulen [12] independently arrived at an ultimately similar design in a completely different way. There is a difficulty in achieving the correct value for the intermediate index in the quarter–half-quarter design in practice, and Vermeulen realized that the deposition of a low-index layer over a high-index layer of less than a quarter wave would lead to a maximum turning value in reflectance rather lower than would have been achieved with a quarter wave of high index on its own. He therefore designed a two-layer high–low combination to give an identical turning value to that which should be obtained with the 1.70 index layer of the quarter–half-quarter coating, and he discovered that good performance was maintained. The turning value in reflectance must, of course, correspond to the intersection of the locus with the real axis, and the rest follows. We shall return to this coating later.

The quarter–half-quarter coating can be further improved by replacing the layer of intermediate index by two quarter-wave layers. The layer next to the substrate should have an index lower than that of the substrate. A practical coating of this general type is shown in Figure 4.42. Trial and error leads to the following design, the theoretical performance of which is shown in Figure 4.43:

Air	1.38	2.05	1.60	1.45	Glass
1.0	$0.25\lambda_0$	$0.5\lambda_0$	$0.25\lambda_0$	$0.25\lambda_0$	1.52

Similar designs with slightly different index values were given by Cox and Hass [13] and by Musset and Thelen [6]. Ward [14] published a useful version of this coating with indices chosen to

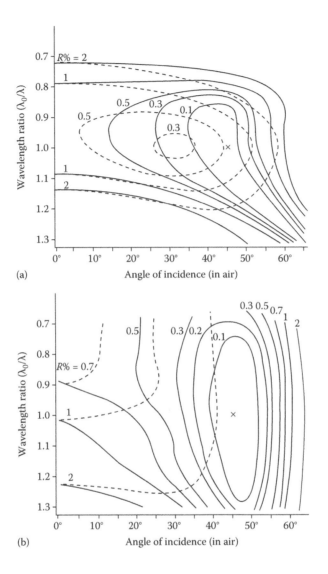

FIGURE 4.39
(a) Equireflectance contours for a quarter–half–quarter antireflection coating designed for use at 45° on crown glass. The indices are chosen for best performance with s-polarization. $n_0 = 1.00$, $n_1 = 1.35$, $n_2 = 2.45$, $n_3 = 1.70$, and $n_m = 1.50$. *Solid curves*, s-polarization; *broken curves*, p-polarization. (b) Equireflectance contours for a quarter–half–quarter antireflection coating designed for use at 45° on crown glass. The indices are chosen for best performance with polarization. $n_0 = 1.00$, $n_1 = 1.40$, $n_2 = 1.75$, $n_3 = 1.58$, and $n_m = 1.50$. *Solid curves*, s-polarization; *broken curves*, p-polarization. (After T. Turbadar, *Optica Acta*, 11, 195–205, 1964.)

match those of available materials rather than to achieve optimum performance. Examples of four-layer coatings for substrates of indices other than 1.52 are also given by Ward and by Musset and Thelen [6].

Yet a further four-layer design can be obtained by splitting the half-wave layer of the quarter–half–quarter coating into two quarter waves and adjusting the indices to improve the performance. A five-layer design derived in a similar way from the design in Figure 4.43 is shown in Figure 4.44.

Air	1.38	1.86	1.94	1.65	1.47	Glass
1.0	$0.25\lambda_0$	$0.25\lambda_0$	$0.25\lambda_0$	$0.25\lambda_0$	$0.25\lambda_0$	1.52

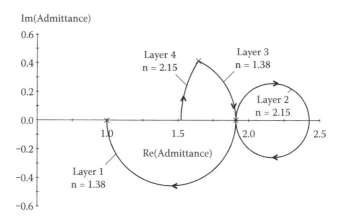

FIGURE 4.40
Two-layer coating of Figure 4.13 with the low-index layer split where it intersects the real axis and a high-index flattening layer is inserted. The design is as follows:

Air	1.38	2.15	1.38	2.15	Glass
1.0	$0.25\lambda_0$	$0.5\lambda_0$	$0.069\lambda_0$	$0.063\lambda_0$	1.52

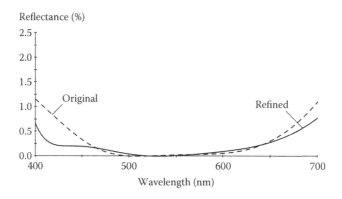

FIGURE 4.41
Performance of the four-layer coating. The broken curve is from the coating in Figure 4.40. The full line is the product of slight refinement. Although essentially arrived at by way of the admittance plot in Figure 4.40, the design is virtually identical to one published by Vermeulen whose design technique was quite different (see text).

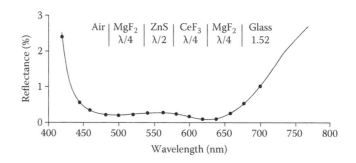

FIGURE 4.42
Measured reflectance of a four-layer antireflection coating on crown glass. The results are for a single surface. (Courtesy of M. J. Shadbolt, Sira Institute, Chislehurst, Kent, UK, 1967.)

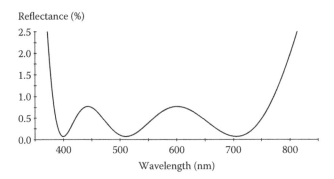

FIGURE 4.43
Performance of the four-layer coating of the following design, where $\lambda_0 = 510$ nm:

Air	1.38	2.05	1.60	1.45	Glass
1.00	$0.25\lambda_0$	$0.5\lambda_0$	$0.25\lambda_0$	$0.25\lambda_0$	1.52

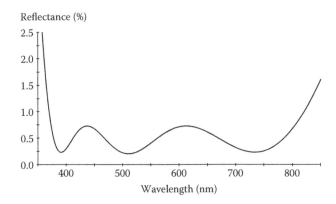

FIGURE 4.44
Five-layer design derived from Figure 4.43 by replacing the half-wave layer by two quarter-wave layers and adjusting the values of the indices. The performance is broadened slightly. The design is as follows, where $\lambda_0 = 510$ nm:

Air	1.38	1.86	1.94	1.65	1.47	Glass
1.00	$0.25\lambda_0$	$0.25\lambda_0$	$0.25\lambda_0$	$0.25\lambda_0$	$0.25\lambda_0$	1.52

These two latter designs have performance extending over a little more than the visible region. They also have the advantage that the layer thicknesses are either half waves or quarter waves, and the possibilities are clearly enormous. However, it is clear that any problems will be found much more in the construction of the coatings than in their design, because not all the required indices are readily available. This is addressed in a following section.

A rather interesting design based on four layers of alternate high and low indices has been published in a patent by the company C. Reichert Optische Werke AG [15]. Full details of the design method are, unfortunately, not given. The thicknesses and materials are given in Table 4.2. Note that the thicknesses are quoted as optical. The reflectance of this coating (Figure 4.45) is similar to the unrefined performance of Figure 4.41 but a little inferior to the refined curve.

TABLE 4.2

Reichert Design

Material	Index	Optical Thickness (nm)
Air	1.00	Massive
MgF_2	1.37	161
TiO_2	2.28	78.5
MgF_2	1.37	56.5
TiO_2	2.28	54
Glass	1.52	Massive

Source: Reichert, C., Optishche, A.G., *Improvements in or relating to optical components having reflection-reducing coatings*, UK Patent 991,635,1962.

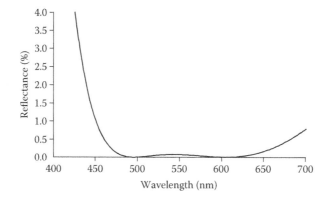

FIGURE 4.45
Reichert four-layer two-material antireflection coating.

Although the Reichert design technique is not described, nevertheless, it is a good exercise to attempt to understand how the coating functions. For this, it is easiest if we simply draw an admittance diagram. Since the coating is clearly centered on 550 nm, we draw the diagram for that wavelength.

The admittance diagram (Figure 4.46) shows that the Reichert design can be considered to be derived by applying two Vermeulen equivalents to the W-coat and its three-layer variations in

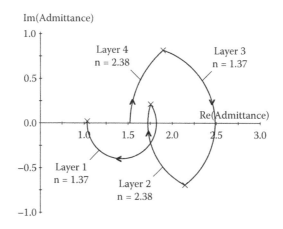

FIGURE 4.46
Admittance locus of the Reichert design at 550 nm.

Figure 4.30. A particularly interesting feature of the Reichert coating is that it is quite thin compared with the W-coat from which it is derived. This three-layer double Vermeulen equivalent is a powerful replacement for a flattening half wave in a design. We shall return to this structure later when we consider buffer layers.

4.4 Equivalent Layers

There are great advantages in using a series of quarter waves or multiples of quarter waves in the first stages of the design of antireflection coatings because the characteristic curves of such coatings are symmetrical about $g = 1.0$. However, problems are presented in construction because the indices specified in this way do not often correspond to indices that are readily available. Using mixtures of materials of higher and lower indices to produce a layer of intermediate index is a technique that has been successfully used (see Chapter 11), but a more straightforward method is to replace the layers by equivalent combinations involving only two materials, one of high index and one of low. These two materials can be well tried and stable, with characteristics established over many production runs in the machine that will be used for, and under the conditions that will apply to, the production of the coatings. To illustrate the method, we assume two materials of index 2.30 and 1.38, approximately corresponding to titanium dioxide and magnesium fluoride, respectively, but it will be obvious that the magnesium fluoride could readily be replaced by the tougher silica (SiO_2), though of slightly higher index.

The first technique to mention is the already referenced one by Vermeulen [12]. It involves the replacement of a quarter wave by a two-layer equivalent. The analysis is exactly that already given for the two-layer antireflection coating, and it is assumed that the quarter wave to be replaced has a locus that starts and terminates at predetermined points on the real axis. The replacement is, therefore, valid for the particular starting and terminating points used in its derivation only and for that single wavelength for which the original layer is a quarter wave. Under conditions that are increasingly remote from these ideal ones, the two-layer replacement becomes increasingly less satisfactory. It is advisable, when calculating the parameters of the layers, to sketch a rough admittance plot because, otherwise, there is a real danger of picking incorrect values of layer thickness. In the particular case we are considering, the starting admittance is 1.52 on the real axis and the terminating admittance is 1.9044, which will ensure that the outermost 1.38 index quarter-wave layer will terminate at point 1.00 on the real axis. Clearly, the high-index layer should be next to the substrate. The thicknesses are then, using Equation 4.9 and selecting the appropriate pair of solutions, 0.05217 and 0.07339 full waves for the high- and low-index layers, respectively. We complete the design by adding a half-wave of index 2.30 and a quarter wave of index 1.38. The admittance locus of this coating, although using slightly different indices, is essentially that of Figure 4.40 with its characteristic curve in Figure 4.41, all of which, we recall, was arrived at in a slightly different way.

As already mentioned, the four-layer Reichert coating (Table 4.2) can be thought of as a Vermeulen equivalent of the coatings of Figure 4.30.

To obtain a replacement for a quarter wave that does not depend on the starting point, we turn to a technique originated by Epstein [16] involving the symmetrical periods and the Herpin admittance briefly mentioned in Chapter 3. We recall that any symmetrical combination of layers acts as a single layer with an equivalent phase thickness and equivalent optical admittance. In this particular application, we consider combinations of the form ABA only. We choose those of the two materials from which the coating is to be constructed for the indices of A and B. Then for each quarter-wave layer of the coating, we construct a three-layer symmetrical period that has an equivalent thickness of one quarter wave and an equivalent admittance equal to that required by the original design.

To proceed further, we need expressions for the equivalent thickness and admittance of a symmetrical period. These are derived later in Chapter 7. Since the symmetrical period is of the form ABA, then

$$y_E = y_A$$

$$\times \left(\frac{\sin 2\delta_A \cos \delta_B + \frac{1}{2}[(y_B/y_A) + (y_A/y_B)] \cos 2\delta_A \sin \delta_B + \frac{1}{2}[(y_B/y_A) - (y_A/y_B)] \sin \delta_B}{\sin 2\delta_A \cos \delta_B + \frac{1}{2}[(y_B/y_A) + (y_A/y_B)] \cos 2\delta_A \sin \delta_B - \frac{1}{2}[(y_B/y_A) - (y_A/y_B)] \sin \delta_B} \right)^{1/2} , \quad (4.28)$$

$$\cos \gamma = \cos 2\delta_A \cos \delta_B - \frac{1}{2}[(y_B/y_A) + (y_A/y_B)] \sin 2\delta_A \sin \delta_B, \quad (4.29)$$

where y_E is the equivalent optical admittance and γ is the equivalent phase thickness. The important feature of the symmetrical combination is that it behaves as a single layer of phase thickness γ and admittance y_E regardless of the starting point for the admittance locus.

In our particular case here, we have a partial simplification in that the equivalent thickness of the combination should be a quarter wave. That is,

$$\cos \gamma = \cos(\pi/2) = 0$$

$$= \cos 2\delta_A \cos \delta_B - \frac{1}{2}[(y_B/y_A) + (y_A/y_B)] \sin 2\delta_A \sin \delta_B,$$

which gives

$$\tan 2\delta_A \tan \delta_B = \frac{2y_A y_B}{y_A^2 + y_B^2} . \quad (4.30)$$

By substituting in Equation 428 and manipulating the expression, then

$$y_E = y_A \left(\frac{1 + [(y_B^2 - y_A^2)/(y_B^2 + y_A^2)] \cos 2\delta_A}{1 - [(y_B^2 - y_A^2)/(y_B^2 + y_A^2)] \cos 2\delta_A} \right)^{1/2} , \quad (4.31)$$

yielding

$$\cos 2\delta_A = \frac{(y_B^2 + y_A^2)(y_E^2 - y_A^2)}{(y_B^2 - y_A^2)(y_E^2 + y_A^2)} , \quad (4.32)$$

and, by Equation 4.30, δ_B,

$$\tan \delta_B = \frac{2y_A y_B}{y_A^2 + y_B^2} \cdot \frac{1}{\tan 2\delta_A} . \quad (4.33)$$

The optical thicknesses are then

$$\frac{n_A d_A}{\lambda_0} = \frac{\delta_A}{2\pi} \text{ full waves at } \lambda_0,$$

$$\frac{n_B d_B}{\lambda_0} = \frac{\delta_B}{2\pi} \text{ full waves at } \lambda_0. \quad (4.34)$$

If an equivalent combination for a half-wave layer is required, then it is considered as two quarter waves in series.

As an example of the application of this technique, we take the four-layer coating B of Figure 4.34.

Air	1.38	2.13	1.9	1.38	Glass
1.0	$0.25\lambda_0$	$0.25\lambda_0$	$0.25\lambda_0$	$0.5\lambda_0$	1.52

The layers that must be replaced are the quarter waves with indices 2.13 and 1.90. There are two possible combinations, *HLH* or *LHL*, for each of these layers.

$$
\begin{matrix} 2.13 \\ 0.25\lambda_0 \end{matrix} \rightarrow \left\{ \begin{matrix} 1.38 & 2.30 & 1.38 \\ 0.04128\lambda_0 & 0.15861\lambda_0 & 0.04128\lambda_0 \\ 2.30 & 1.38 & 2.30 \\ 0.11198\lambda_0 & 0.02302\lambda_0 & 0.11198\lambda_0, \end{matrix} \right.
$$

$$
\begin{matrix} 1.90 \\ 0.25\lambda_0 \end{matrix} \rightarrow \left\{ \begin{matrix} 1.38 & 2.30 & 1.38 \\ 0.06793\lambda_0 & 0.10438\lambda_0 & 0.06793\lambda_0 \\ 2.30 & 1.38 & 2.30 \\ 0.09216\lambda_0 & 0.05868\lambda_0 & 0.09216\lambda_0. \end{matrix} \right.
$$

As an indication of the closeness of fit between the symmetrical periods and the layers they replace, the variation, with *g*, of equivalent admittance and equivalent optical thickness is plotted in Figure 4.47.

We can now replace the layers in the actual design of antireflection coating. There are two possible replacements for each of the relevant layers, but where *HLH* and *LHL* combinations are mixed, there is a tendency toward an excessive number of layers in the final design, and so we consider two possibilities only, one based on *HLH* periods and one based on *LHL*. These are shown in Table 4.3.

The spectral characteristics of these coatings along with the original design are shown in Figure 4.48. The replacements have a slightly inferior performance due to the asymmetric dispersion that can be seen in Figure 4.47. It is interesting to note that the five-layer design with the *LHL* replacement structures has a similar structure to the Reichert design of Figure 4.46 but with the additional half-wave (slightly greater than a half wave) low-index layer next to the substrate (Figure 4.49).

The process of design need not stop at this point, however, because the designs are excellent starting points for refinement. Figure 4.50 shows the performance of a refined version of one of the coatings. In practice, the refinement will include an allowance for the dispersion of the indices of the materials, and there will be a certain amount of adjustment of the coating during the production trials.

If performance over a much wider region is required, then the apparent dispersion of the equivalent periods may become a problem. This dispersion can be reduced by using equivalent periods of eighth-wave thickness instead of a quarter wave. Each quarter wave in the original design is then replaced by two periods in series. This considerably adds to the number of layers, and the solution of the appropriate equations is no longer quite as simple.

We have been using a low index of 1.38 in most of our examples so far. Frequently silica (SiO_2) is preferred as the low-index material. It is tougher than magnesium fluoride and has the advantage that it can be readily sputtered. Its index of around 1.45 is a little greater than the 1.38 of magnesium fluoride, and so the attainable performance is slightly inferior. It is a straightforward matter to replace the magnesium fluoride in the designs—simple replacement in the case of those depending on quarter-wave thicknesses and with some additional refinement in the case of the

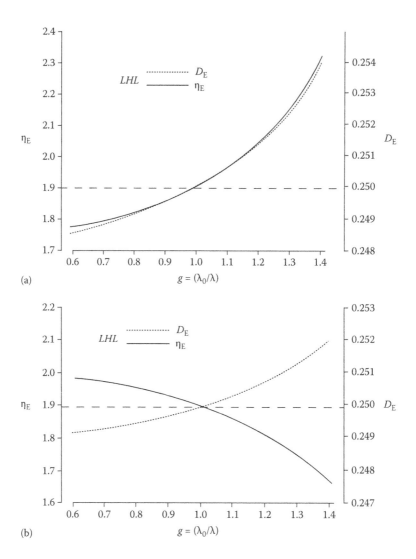

FIGURE 4.47
Equivalent admittances and optical thickness as a function of g ($= \lambda_0/\lambda$) of symmetrical period replacements for a single quarter wave of index 1.90. The indices used in the symmetrical replacement are 2.30 for the high index and 1.38 for the low index. (a) *LHL* combination. (b) *HLH* combination. D_E is the equivalent optical thickness expressed in full waves. For a perfect match, D_E and y_E should both be constant at $0.25\lambda_0$ and 1.9, respectively, whatever the value of g.

others. Figure 4.51 shows an example. The performance of the replacement design is, as expected, slightly affected mainly because the half wave-broadening low-index layer next to the substrate is less effective in its correction.

An interesting design approach involving equivalent layers was devised by Schulz et al. [17]. We start with a substrate that has an index higher than that of the incident medium. Then, provided we have the materials, we can construct very efficient antireflection coatings consisting of a series of quarter waves of index gradually descending from that of the substrate to that of the incident medium. The earlier part of this chapter described several coatings of this general type. The most important aspect of the index variation is that it should be reasonably smooth and should especially not show a sudden reversal in its sense. There are also many systematic techniques for arriving at optimum values depending on the details of the desired performance. Further information on these can be found in Thelen's book [18]. Here we shall use a quite simple distribution based on some of the ideas already expressed in this chapter.

TABLE 4.3

Designs with Symmetrical Period Replacements

Layer Number	Design Based on *LHL* Periods		Design Based on *HLH* Periods	
	Index	Thickness	Index	Thickness
0	1.0	Incident medium	1.0	Incident medium
1	1.38	$0.29128\lambda_0$	1.38	$0.25\lambda_0$
2	2.30	$0.15861\lambda_0$	2.30	$0.11198\lambda_0$
3	1.38	$0.10921\lambda_0$	1.38	$0.02302\lambda_0$
4	2.30	$0.10438\lambda_0$	2.30	$0.20414\lambda_0$
5	1.38	$0.56793\lambda_0$	1.38	$0.05868\lambda_0$
6	1.52	Substrate	2.30	$0.09216\lambda_0$
7			1.38	$0.5\lambda_0$
8			1.52	Substrate

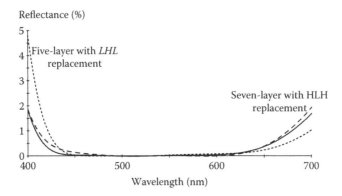

FIGURE 4.48
Performance of the designs of Table 4.3: *Dotted line*, five-layer design based on *LHL* periods; *dashed line*, seven-layer design based on *HLH* periods; *full line*, original four-layer design from which the other two were derived.

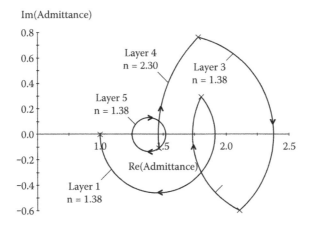

FIGURE 4.49
Admittance locus of the five-layer equivalent design using the *LHL* symmetrical replacement from Figure 4.48. Clearly, it is of a structure quite similar to the Reichert design in Figure 4.46 with an additional half wave of low index next to the substrate.

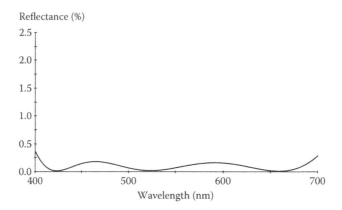

FIGURE 4.50
Refined version of the seven-layer design in Figure 4.48. Design:

Air	1.38	2.30	1.38	2.30	1.38	2.30	1.38	Glass
1.0	$0.25\lambda_0$	$0.151\lambda_0$	$0.0361\lambda_0$	$0.230\lambda_0$	$0.080\lambda_0$	$0.064\lambda_0$	$0.485\lambda_0$	1.52

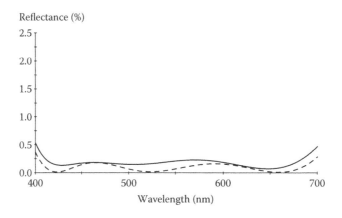

FIGURE 4.51
Dashed line, design in Figure 4.50; *full line*, refined version with index 1.38 (magnesium fluoride) replaced by 1.45 (silica). Design:

Air	1.45	2.30	1.45	2.30	1.45	2.30	1.45	Glass
1.0	$0.257\lambda_0$	$0.206\lambda_0$	$0.025\lambda_0$	$0.220\lambda_0$	$0.094\lambda_0$	$0.049\lambda_0$	$0.447\lambda_0$	1.52

Let the quarter-wave indices be based on an equal amplitude reflection coefficient at all interfaces. Then,

$$\frac{y_0}{y_1} = \frac{y_1}{y_2} = \frac{y_2}{y_3} = \cdots = \frac{y_q}{y_{\text{sub}}} = f \tag{4.35}$$

and

$$f = \sqrt[q+1]{\frac{y_0}{y_{\text{sub}}}}, \tag{4.36}$$

allowing the calculation of the various necessary layer indices.

The performance of a six-layer coating designed in this way is shown in Figures 4.52 and 4.53. The performance as a function of g is excellent except at those values of g where the individual layers have optical thicknesses of an integral number of half waves.

Unfortunately, we cannot construct such a coating because we lack the necessary low-index materials. However, as shown in Figure 4.52, the performance repeats itself at every odd value of g. The region between values of g of 2 and 4, where the layers are three quarter waves thick, is wide enough just to cover the visible region (Figure 4.53). Although we are unable to replace the low-index quarter waves, we have no such difficulty where three-quarter wave layers are concerned. There are several possible solutions, but a particularly useful one is where the bulk of the combination is of low index, and there is a central quite thin high-index layer.

Replacing the layers by a three-layer symmetrical arrangement and then gently refining the resulting design yields the profile shown in Figure 4.54 with performance in Figure 4.55.

Although this present design was based on symmetrical periods, the principle of thin high-index layers in a matrix of low index can be extended to systems that are not so based and with broader regions of low reflectance. The designs are normally fairly thick, 1.6 µm physical thickness in the case of the design of Figure 4.54, but this is of particular advantage in the case of antireflection coatings for plastic materials, because the coating itself can also play the role of what is called a hard coat. The material of normal antireflection coatings is brittle and tends to easily crack when point loads are applied because of the yielding nature of the underlying plastic. It is normal to

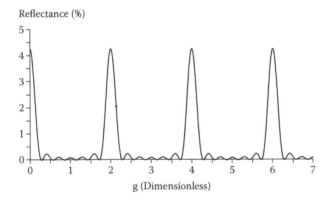

FIGURE 4.52
The antireflection region between g of 2 and 4 in Figure 4.53 just covers the visible region.

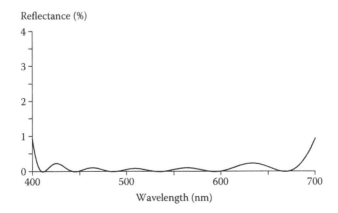

FIGURE 4.53
Replacement of the three-quarter wave layers by their three-layer equivalents followed by slight refinement gives this design with performance shown in Figure 4.56.

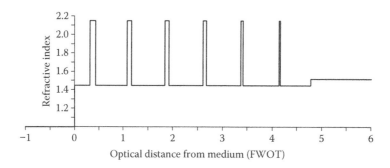

FIGURE 4.54
Performance of the design in Figure 4.55. The width of the coating can be increased at the expense of a slight increase in residual reflectance.

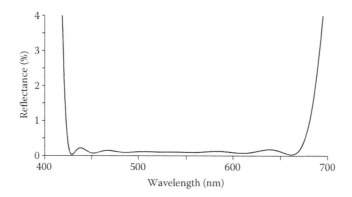

FIGURE 4.55
Index profile of a six-layer stepped antireflection coating to match index of 1.52 (glass) to 1.00 (air).

strengthen the plastic surface by applying a hard coat, which is a layer of hard, less yielding material, often acrylic. The thick silicon dioxide material of the present coating is tough and hard and can support point loads much better than normal antireflection coatings, and so the hard coat can be omitted.

So far, most of our equivalents have been based on symmetrical structures. A different kind of equivalence was proposed by Schallenberg et al. [19] and Schallenberg [20,21]. We can reduce the calculation of the reflectance of any arbitrary multilayer to the following matrix expression, where the individual characteristic matrices have been multiplied to give just one product matrix representing the entire multilayer. As long as the layers are perfect dielectrics, then the quantities M_{xy} will all be real.

$$\begin{bmatrix} B \\ C \end{bmatrix} = \begin{bmatrix} M_{11} & iM_{12} \\ iM_{21} & M_{22} \end{bmatrix} \begin{bmatrix} 1 \\ y_{sub} \end{bmatrix} = \begin{bmatrix} M_{11} + iy_{sub}M_{12} \\ y_{sub}M_{22} + iM_{21} \end{bmatrix}. \tag{4.37}$$

The reflectance is given by

$$R = \left| \frac{y_0 B - C}{y_0 B + C} \right|^2 = \frac{M_{11}^2 \left(y_0 - y_{sub} \dfrac{M_{22}}{M_{11}} \right)^2 + M_{12}^2 \left(y_0 y_{sub} - \dfrac{M_{21}}{M_{12}} \right)^2}{M_{11}^2 \left(y_0 + y_{sub} \dfrac{M_{22}}{M_{11}} \right)^2 + M_{12}^2 \left(y_0 y_{sub} + \dfrac{M_{21}}{M_{12}} \right)^2}. \tag{4.38}$$

We can define two new quantities

$$E = \sqrt{\frac{M_{21}}{M_{12}}} \quad \text{and} \quad S = y_{\text{sub}} \frac{M_{22}}{M_{11}}, \tag{4.39}$$

where E is known as the equivalent stack admittance, and S, as the equivalent substrate admittance. Substituting these values into Equation 4.38 gives

$$R = \frac{M_{11}^2 (y_0 - S)^2 + M_{12}^2 (y_0 y_{\text{sub}} - E^2)^2}{M_{11}^2 (y_0 + S)^2 + M_{12}^2 (y_0 y_{\text{sub}} + E^2)^2}. \tag{4.40}$$

Clearly, the reflectance will be zero if

$$S = y_0 \quad \text{and} \quad E = \sqrt{y_0 y_{\text{sub}}}. \tag{4.41}$$

However, if the conditions in Equation 4.41 are not met, then if the coating consists of a series of quarter waves, M_{11} and M_{12} will oscillate such that each shows a maximum value while the other is zero. Then the reflectance will vary between two extrema R_E and R_S given by

$$R_s = \frac{(y_0 - S)^2}{(y_0 + S)^2} \quad \text{and} \quad R_E = \frac{(y_0 y_{\text{sub}} - E^2)^2}{(y_0 y_{\text{sub}} + E^2)^2}. \tag{4.42}$$

It is possible, without too much pain, to develop analytical solutions for coatings with a few layers, and some examples are given by Schallenberg.

We digress for a moment to consider designs that consist of a descending series of quarter waves, on the lines of the design of Figure 4.56. The series mimics an inhomogeneous layer (see later in this chapter), and a true inhomogeneous layer antireflects all wavelengths shorter than that for which the total thickness is one-half wavelength. Thus, if all the layers are quarter waves at $g = 1$, then a system of q quarter waves would exhibit a lower limit at roughly $g = 1/q$. In this case, though, the antireflection breaks down when the individual layers become half waves, that is, as we reach $g = 2$. Since the interference condition around $g = 2$ is identical to that around $g = 0$, the antireflection region will stretch from $g = 1/q$ to $g = 2 - 1/q$, that is, a ratio of $2q - 1$. Clearly, the greater the q, the greater the region of low reflectance. This rough value applies reasonably well to

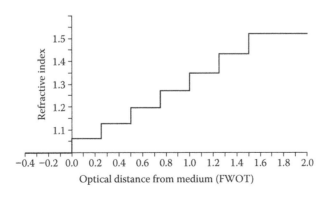

FIGURE 4.56
Performance of the six-layer stepped antireflection coating in Figure 4.52 as a function of g. The peaks of reflectance occur where the layers are an integral number of half waves thick.

TABLE 4.4

Schallenberg's Arrangement for Maximally Flat Antireflection Coatings

One Layer	Two Layers	Three Layers	Four Layers	Five Layers	Six Layers
					y_0
				y_0	
			y_0		$y_0^{63/64}y_m^{1/64}$
		y_0		$y_0^{31/32}y_m^{1/32}$	
	y_0		$y_0^{15/16}y_m^{1/16}$		$y_0^{57/64}y_m^{7/64}$
y_0		$y_0^{7/8}y_m^{1/8}$		$y_0^{26/32}y_m^{6/32}$	
	$y_0^{3/4}y_m^{1/4}$		$y_0^{11/16}y_m^{5/16}$		$y_0^{42/64}y_m^{22/64}$
$y_0^{1/2}y_m^{1/2}$		$y_0^{4/8}y_m^{4/8}$		$y_0^{16/32}y_m^{16/32}$	
	$y_0^{1/4}y_m^{3/4}$		$y_0^{5/16}y_m^{11/16}$		$y_0^{22/64}y_m^{42/64}$
y_m		$y_0^{1/8}y_m^{7/8}$		$y_0^{6/32}y_m^{26/32}$	
	y_m		$y_0^{1/16}y_m^{15/16}$		$y_0^{7/64}y_m^{57/64}$
		y_m		$y_0^{1/32}y_m^{31/32}$	
			y_m		$y_0^{1/64}y_m^{63/64}$
				y_m	
					y_m

TABLE 4.5

Six-Layer Design Using Schallenberg's Method

Layer Number	Refractive Index
Incident	1.00
1	1.0219
2	1.1637
3	1.6105
4	2.4837
5	3.4372
6	3.9143
Substrate	4.00

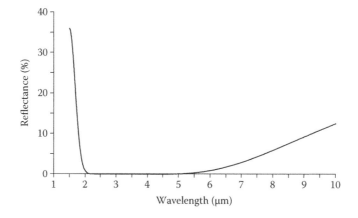

FIGURE 4.57

Performance of the coating in Table 4.5 assuming a reference wavelength of 3 μm. Unfortunately, the coating as it stands, requires unrealistic material indices and so could not be directly constructed.

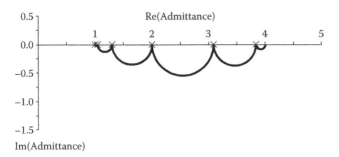

FIGURE 4.58
Admittance diagram of the maximally flat coating in Figure 4.57 at the reference wavelength. The locus of layer 1 at the reference wavelength is so small to be hardly visible.

sequences of quarter waves such as that in Figure 4.56 but gives a rather too optimistic estimate when a class of antireflection coatings known as maximally flat is concerned. Maximally flat antireflection coatings also consist of a descending series of quarter waves, but the particular sequence is chosen not to exhibit ripple in the antireflected region. Although they have wide bands of low reflectance, the removal of the ripple implies somewhat smaller bandwidths than that given earlier in this paragraph.

Schallenberg [21] noted that the maximally flat designs also obey the expressions in Equation 4.42, and in particular, the equivalent stack admittance must be $(y_0 y_m)^{1/2}$. The expression for R_E in Equation 4.42 is just the quarter-wave rule. Adding a quarter wave to the substrate transforms the admittance to E^2/y_m. This new admittance can be antireflected in the normal way using another quarter wave with admittance given by the square root of $y_0 \times E^2/y_m$. We can apply this in a quite regular way to a series of quarter waves. A single quarter wave will bridge from y_m to y_0 and so must have an admittance of $y_1 = (y_0 y_m)^{1/2}$. A two-quarter-wave design can bridge with the first quarter wave from y_m to y_1 and then with the second quarter wave from y_1 to y_0. The admittances of the two quarter waves will be $y_m^{3/4} y_0^{1/4}$ and $y_m^{1/4} y_0^{3/4}$. For a three-layer coating, we bridge the gap from y_m to $y_m^{3/4} y_0^{1/4}$, from $y_m^{3/4} y_0^{1/4}$ to $y_m^{1/4} y_0^{3/4}$, and from $y_m^{1/4} y_0^{3/4}$ to y_0. This gives layers of admittance $y_m^{7/8} y_0^{1/8}$, $y_m^{4/8} y_0^{4/8}$, and $y_m^{1/8} y_0^{7/8}$. Schallenberg's arrangement is shown in Table 4.4 and can be very easily extended to greater numbers of layers. It avoids the complicated calculations that are usually required for these coatings.

The required indices for a six-layer coating of this type on germanium, assuming an index of 4.00 and air as the incident medium, are given in Table 4.5. Because of the unobtainable indices, the design, as it stands, could not be constructed, but it will serve to illustrate the method, and it could be used as the basis for a replacement technique, such as has just been described. The calculated performance with a reference wavelength of 3 μm is shown in Figure 4.57, and the corresponding admittance locus, in Figure 4.58.

4.5 Antireflection Coatings for Two Zeros

There are occasional applications where antireflection coatings are required, which have zeros at certain well-defined wavelengths rather than over a wide spectral region. One of the most frequent of these applications is frequency doubling, where antireflection is required at two wavelengths, one of which is twice the other.

The simplest coating that will satisfy this requirement is the quarter–quarter coating that has already been considered (Figure 4.17). We recall that the coating has two zeros at $\lambda = 3\lambda_0/4$ and $\lambda = 3\lambda_0/2$, just what is required. The conditions are

$$y_1 = \left(y_0^2 y_m\right)^{1/3},$$
$$y_2 = \left(y_0 y_m^2\right)^{1/3}. \tag{4.43}$$

The principal problem with this coating is once again the low-index substrate. With an index of 1.38 as the lowest value for film index, the lowest value of substrate index that can be accommodated, from Equation 4.43, is $1.38^3 = 2.63$. Thus, the coating is suitable only for high-index substrates.

A common material that requires antireflection coatings at λ and 2λ is lithium niobate, which has an index of around 2.25. The quarter–quarter coating should have indices of 1.310 and 1.717. Indices of 1.38 and 1.717 give a reflection loss of 0.2%, which will probably be adequate for many applications, and indeed similar performance is obtained with any index between 1.7 and 1.8 for the high-index layer.

Should this performance be inadequate, then an additional layer can be added. Provided we keep to quarter waves and multiples of quarter waves, we retain the symmetry of about $g = 1$ and have to consider the performance at $g = 2/3$ only, since that at $g = 3/4$ will be automatically equivalent. From the point of view of the vector diagram, the problem with the quarter–quarter coating is ρ_a, the amplitude reflection coefficient from the first interface, which is too large. The vectors are inclined at 120° to each other, and for zero reflectance, they should be of equal length so that they form an equilateral triangle. If an extra quarter wave y_3 is added, there will be four vectors, and the fourth ρ_d will be along the same direction as ρ_a. If ρ_d is made to be of opposite sense to ρ_a, that is, if $y_3 > y_m$, then it is possible to reduce the resultant of the two vectors to the same length as the other two. This can be achieved by the design

Air	1.38	1.808	2.368	Lithium niobate
1.0	$0.25\lambda_0$	$0.25\lambda_0$	$0.25\lambda_0$	2.25

We can take 2.35, the index of zinc sulfide, as the value of y_3, and then any value in the range 1.75–1.85 for y_2, to keep the minimum reflectance at $g = 2/3$ to below 0.1%.

There are many other possible arrangements. A coating with the first layer that is a half wave, instead of a quarter wave, can give a similar improvement, this time through a combination with ρ_c, which means that $y_2 > y_3$. Here the ideal design is

Air	1.38	1.81	1.72	Lithium niobate
1.0	$0.5\lambda_0$	$0.25\lambda_0$	$0.25\lambda_0$	2.25

And once again, there is reasonable flexibility in the values of y_2 and y_3 if the aim is simply a reflectance of less than 0.1%. It is interesting to note the similarity between this coating and the quarter–quarter coating. The quarter–quarter coating has another zero at $g = 8/3$. If the inner quarter waves in the design mentioned earlier were merged into a single half wave of index around 1.75, then the coating would be identical with the quarter–quarter coating used at $g = 4/3$ and $g = 8/3$. Figure 4.59 shows the performance of these coatings.

This idea of using the fourth vector to trim the length of one of the other three so that a low reflectance is obtained can be extended to low-index substrates. The coating now, of course,

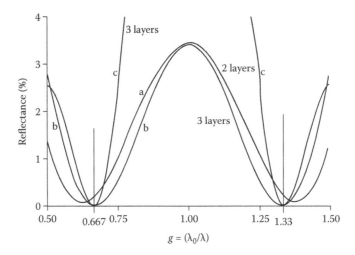

FIGURE 4.59
Performance of various two-zero 2:1 antireflection coatings on a high-index substrate such as lithium niobate with $n = 2.25$. The ideal positions for the two zeros are $g = 0.667$ and $g = 1.333$.

(a)

Air	1.38	1.72	Lithium niobate
1.0	$0.25\lambda_0$	$0.25\lambda_0$	2.25

(b)

Air	1.38	1.808	2.368	Lithium niobate
1.0	$0.25\lambda_0$	$0.25\lambda_0$	$0.25\lambda_0$	2.25

(c)

Air	1.38	1.81	1.72	Lithium niobate
1.0	$0.5\lambda_0$	$0.25\lambda_0$	$0.25\lambda_0$	2.25

considerably departs from the original quarter–quarter coating. A quarter–quarter–quarter design based on this approach is as follows:

Air	1.38	1.585	1.82	Glass
1.0	$0.25\lambda_0$	$0.25\lambda_0$	$0.25\lambda_0$	1.52

And its performance is shown in Figure 4.60, where with monitoring wavelength at 707 nm, the two zeros would be situated at 530 nm and 1.06 μm.

The method can be extended to four and even more quarter waves, although the derivation of the final designs is very much more of a trial and error process because of the rather cumbersome expressions that cannot be reduced to explicit formulas for the various indices. Indeed, there are now too many parameters for there to be just one solution, and the surplus can be used in an optimizing process for broadening the reflectance minima. A number of interesting designs were given by Baumeister et al. [22].

Mouchart [23] too considered the derivation of antireflection coatings intended to eliminate reflection at two wavelengths. In coatings where all layers have thicknesses that are specified in advance to be multiples of a quarter wave at $g = 1$, it is arbitrarily possible to choose the indices of all the layers except the final two, which can then be calculated from the values given to the others. The calculation involves the solution of an eighth-order equation that can be set up using expressions derived by Mouchart. The values of $\partial^2 R / \partial \lambda^2$ at the antireflection wavelength, which is inversely related to the bandwidth of the coating, can be used to assist in choosing the more promising designs from the enormous number that can be produced. Mouchart considers three-layer coatings of this type in some detail.

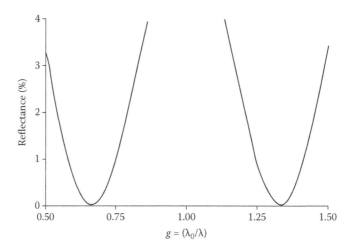

FIGURE 4.60
Three-layer two-zero 2:1 antireflection coating for a low-index substrate. Design:

Air	1.38	1.585	1.82	1.82	Glass
1.0	$0.25\lambda_0$	$0.25\lambda_0$	$0.25\lambda_0$	$0.25\lambda_0$	1.52

4.6 Antireflection Coatings for the Visible and the Infrared Regions

There are frequent requirements for coatings that span the visible region and reduce the reflectance at an infrared wavelength corresponding to a laser line. Such coatings are required in instruments where visual information and laser light share common elements, such as surgical instruments, surveying devices and the like. There are very many designs for such coatings, and manufacturers seldom publish them. Design is often a process of trial and error, with refinement and synthesis playing important parts. In this section, we consider the fundamental design process only, neglecting dispersion and, in most cases, retaining the ideal values of index. We assume that the substrate is always glass of index 1.52 and that, as usual, the incident medium is air of index 1.0. Again, for the sake of continuity with the other sections of this chapter, we shall make much use of magnesium fluoride with its 1.38 index, although silica at 1.45 is becoming a much more common low-index material.

The simplest type of coating that has low reflectance in the visible region and at a wavelength in the near infrared is a single layer of low-index material that is three quarter waves thick. This has low reflectance at both λ_0 and $3\lambda_0$. Unfortunately, the lowest index, of 1.38, corresponding to magnesium fluoride, gives a residual reflectance of 1.25% at the minima, and the performance in the visible region is rather narrower than that for the single quarter-wave coating since the layer is three times thicker. The magnesium fluoride layer could be considered as an outer quarter wave over an inner half wave, and a high-index half-wave flattening layer, of index 1.8, could be introduced between them giving the design

Air|LHHLL|Glass.

Unfortunately, the half-wave layer, while it flattens the performance in the visible region, destroys the performance in the infrared at $3\lambda_0$, where it is two thirds of a quarter wave thick. The solution is to make the layer three half waves thick in the visible, so that it is still a half wave, and therefore an absentee, at $3\lambda_0$. The design then becomes

Air|L6H2L|Glass,

and the performance is shown in Figure 4.61, where the reference wavelength is 510 nm. The performance in the visible region is indeed flattened in the normal way, although, because the flattening layer is three times thicker than normal, the characteristic sharply rises in the blue and red regions. The minimum in the infrared around 1.53 μm is still present, although slightly skewed because of the half-wave layer. However, perhaps the most surprising feature is the appearance of a third and very deep minimum at 840 nm. We use the admittance diagram to help in understanding the origin of this dip.

Figure 4.62 shows the admittance diagram for the coating at the wavelength 840 nm. Layer 2, the 1.8 index layer, is almost two half waves thick at this wavelength and so describes almost two complete revolutions, linking the ends of the loci of the two 1.38 index layers in such a way that almost zero reflectance is obtained. The loci of the two low-index layers are not very sensitive to changes in wavelength, and therefore, the position of the dip is fixed almost entirely by the high-index layer. Changes in its thickness will change the position of the dip. Making it thinner, 1.0 full

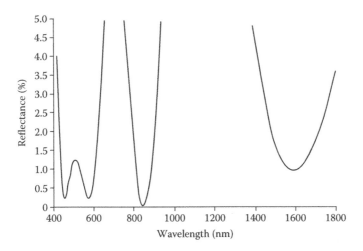

FIGURE 4.61
Performance of the coating: Air (1.0) | L6H2L | Glass (1.52) with L a quarter wave of index 1.38 and H of 1.8. λ_0 is 510 nm.

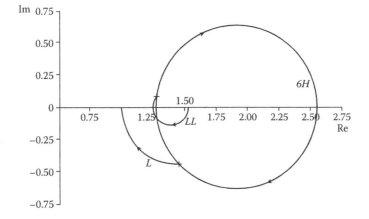

FIGURE 4.62
Admittance diagram for the coating in Figure 4.61 at 840 nm, corresponding to the unexpected sharp zero, explains the occurrence of the dip.

waves instead of 1.5, for example, will move the dip to a longer wavelength. The performance characteristic of the following coating of design is shown in Figure 4.63:

$$\text{Air}|L4H2L|\text{Glass}.$$

The dip is now fairly near the desired wavelength of 1.06 µm.

A coating that gives good performance over the visible region but has high reflectance at 1.06 µm is the quarter–half–quarter coating. The admittance diagram at λ_0 for such a coating is shown in Figure 4.35. The locus intersects or crosses the real axis at points 1.9 and 2.45. It is possible to insert layers of index 1.9 or 2.45, respectively, at these points in the design without any effect on the performance at λ_0 at all. The loci of these layers, whatever their thicknesses are, would simply be points. Such layers are known as *buffer layers* and were devised by Mouchart [24]. For a layer to be a buffer layer, the reflectance of at least one of its surfaces must be zero. The admittance locus assures that the surface toward the medium treated as emergent has zero reflectance, although in a perfect antireflection coating, both surfaces will automatically be antireflected. At the reference wavelength, buffer layers exert no influence on coating reflectance, but at other wavelengths, where the starting points of their loci move away from their reference wavelength positions, the loci appear in the normal way and can have important effects on performance. They are similar in some respects to half-wave layers that, by virtue of their precise thickness, are absentees at λ_0 but that have considerable influence at other wavelengths. The index can be chosen to sharpen or flatten a characteristic. The buffer layer has a precise value of index but can have any thickness, which can be chosen to adjust performance at wavelengths other than λ_0. Here we attempt to use buffer layers to alter the performance at 1.06 µm. One buffer layer is not sufficient, and we need to insert the two possible 1.9 index layers so that the design becomes

$$\text{Air}|LB'HHB''N|\text{Glass},$$

where $y_L = 1.38$, $y_H = 2.15$, and $y_N = 1.70$. B' and B'' are buffer layers of admittance 1.9. Trial and error establishes thicknesses for B' of $0.342\lambda_0$, and for B'', of $0.084\lambda_0$. However, although the reflectance at 1.06 µm is considerably reduced, the buffer layers do distort the performance characteristic somewhat in the visible region (Figure 4.64), and only by refining the design is a completely satisfactory performance obtained. The final design, also illustrated in Figure 4.64, is

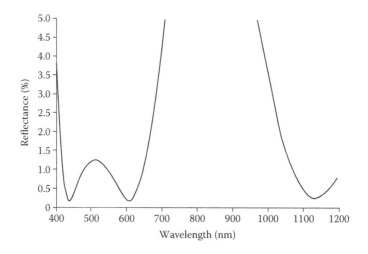

FIGURE 4.63
Performance of the coating: Air (1.0) | $L4H2L$ | Glass (1.52) with L a quarter wave of index 1.38, H of 1.8, and reference wavelength λ_0 of 510 nm. Note that the dip has moved to a longer wavelength than in Figure 4.61.

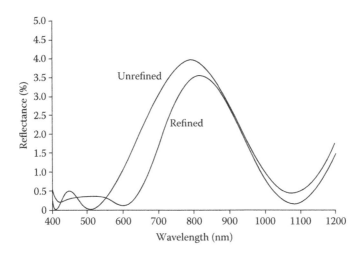

FIGURE 4.64
Performance of the design Air (1.0) | $LB'HHB''M$ | Glass (1.52) with L, H, and M quarter waves of indices 1.38, 2.15, and 1.70 respectively. B' and B'' are buffer layers of index 1.9 (see text) and thicknesses $0.342\lambda_0$ and $0.084\lambda_0$, respectively. λ_0 is 510 nm. The design has also been refined to yield the second performance curve. The refined design is given in the text.

Air	1.38	1.90	2.15	1.90	1.70	Glass
1.00	$0.2667\lambda_0$	$0.3085\lambda_0$	$0.5395\lambda_0$	$0.1316\lambda_0$	$0.1796\lambda_0$	

Many of the designs currently used for the visible and 1.06 μm involve just two materials of high and low index. Designs of this type can be arrived at in a number of ways. The arrangements mentioned earlier that use ideal layers can be replaced by symmetrical periods in the way already discussed. This type of design is seldom immediately acceptable because the very wide wavelength range makes it difficult to exactly match the layers with symmetrical periods, and they are therefore usually refined by computer.

Figure 4.65 shows the performance of a six-layer design arrived at by computer synthesis:

Air	1.38	2.25	1.38	2.25	1.38	2.25	Glass
1.00	$0.03003\lambda_0$	$0.1281\lambda_0$	$0.0657\lambda_0$	$0.6789\lambda_0$	$0.0718\lambda_0$	$0.0840\lambda_0$	

Buffer layers are very useful in such coatings. Half-wave absentee layers rapidly correct performance as the wavelength moves from that for which they are half waves. Buffer layers react more slowly and are therefore very helpful when reflectance must remain low over a wide spectral region. The difficulty with buffer layers is that their refractive index is fixed by the axis crossings of the admittance locus of the coating in which they are to be inserted. We normally have a limited set of indices corresponding to the particular materials we are using, and in order to employ such layers as buffers, we must engineer an axis crossing at the appropriate value of admittance. The double Vermeulen structure makes this possible. In Figure 4.46, the axis crossing on the extreme right can be moved by simply adjusting the thicknesses of the layers making up the structure. It is straightforward to arrange that the axis crossing should actually coincide with the index of the high-index layer already used in the design. This has been achieved with the first of the designs in Table 4.6. Note that the thicknesses are optical so that they can be directly compared with those in Table 4.2.

Figure 4.66 shows the admittance locus of the adjusted coating. The axis crossing has been arranged and the final three layers of the design have been adjusted to give good performance over the visible region. The performance of the coating is shown in Figure 4.67. The buffer design is given in the first design column of Table 4.6. Then the buffer layer of TiO_2 is added and the

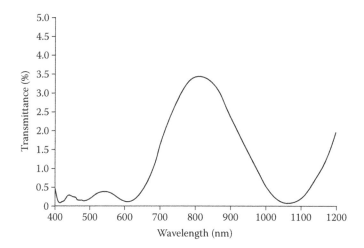

FIGURE 4.65
Performance of a six-layer design of antireflection coating for the visible region and 1.06 μm, arrived at purely by computer synthesis. T he reference wavelength is 510 nm, and the design is given in the text.

TABLE 4.6

Introduction of Buffer Layers

Material	Index	Optical Thickness (nm)		
		Starting Design	**With Buffer**	**With Buffer and Absentee**
Air	1.00	Massive	Massive	Massive
MgF$_2$	1.37	154.47	154.47	140.80
TiO$_2$	2.28	57.96	57.96	50.70
MgF$_2$	1.37	22.66	22.66	17.46
TiO$_2$	2.28	–	247.50	240.84
MgF$_2$	1.37	35.06	35.06	44.31
TiO$_2$	2.28	49.23	49.23	39.99
MgF$_2$	1.37	–	–	294.54
Glass	1.52	Massive	Massive	Massive

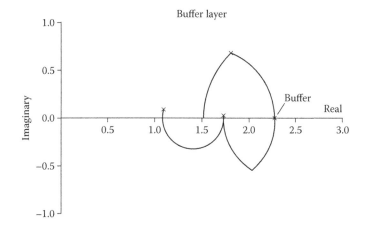

FIGURE 4.66
Admittance locus of the adjusted coating showing the axis crossing at 2.28. A buffer layer has been inserted there.

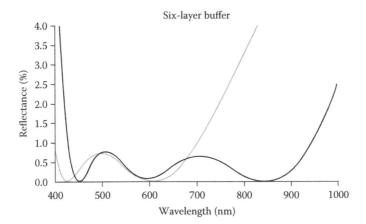

FIGURE 4.67
Starting four-layer coating performance is shown in gray. The addition of the buffer layer makes the coating into a six-layer system. Adjustment of the buffer layer thickness until just less than a half wave gives the performance shown by the black line. The designs are given in Table 4.6.

appearance of the admittance locus does not change with buffer layer thickness. The adjustment of the buffer layer by trial and error gives the improvement shown in Figure 4.67.

The addition of a half-wave layer of low index between the coating and the glass substrate followed by the refinement of all layers yields the performance shown in Figure 4.68. This is as good a performance as we are likely to get with seven layers of the given indices. Significant improvement in performance demands more layers.

The major determinant of antireflection-coating performance for low-index substrates is the lowest index of refraction of the design materials. Magnesium fluoride, mentioned before in this chapter, is a frequent choice, but, unfortunately, it is not ideal. It suffers from high tensile stress and, for reasonable durability, must be deposited on a heated substrate. Silicon dioxide, with its slightly higher index, is much tougher and more stable and can be sputtered as well as evaporated. The critical interface in antireflection coatings is the outermost one. In multilayer applications where performance is critical, it is possible to use silicon dioxide as the low-index material inside the coating and to continue with magnesium fluoride as the outermost layer.

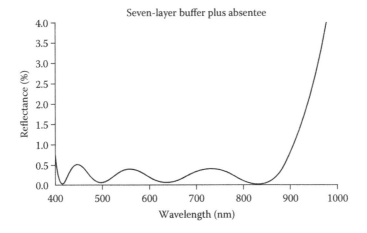

FIGURE 4.68
Performance of the seven-layer coating found in Table 4.6.

4.7 Inhomogeneous Layers

Inhomogeneous layers are ones in which the refractive index varies through the thickness of the layer. As we shall see in Chapter 9, many of the thin-film materials that are commonly used give films that are inhomogeneous. This inhomogeneity is often quite small and the layers can safely be treated as if they were homogeneous in all but the most precise and exacting coatings. There is, however, a number of films that show sufficient inhomogeneity to perceptibly affect the performance of an antireflection coating. If such a layer is used instead of a homogeneous one in a well-corrected antireflection coating, then a reduction in performance is the normal result. Provided the inhomogeneity is not large, an adjustment of the indices of the other layers is usually sufficient correction, and as Ogura [25] pointed out, an index that slightly decreases with thickness associated with the high-index layer in the quarter–half–quarter coating can actually broaden the characteristic. Zirconium oxide is a much used material that exhibits an index increasing with film thickness when deposited at room temperature, but decreasing with thickness when deposited at substrate temperatures above 200°C. Vermeulen [26] considered the effect of the inhomogeneity of zirconium oxide on the quarter–half–quarter coating and has shown how it is possible to correct for the inhomogeneity by varying the index of the intermediate-index layer, which, for virtually complete compensation, should be of the two-layer composite type [12] already referred to in this chapter. This type of inhomogeneity is one that is intrinsic and relatively small. By arranging for the evaporation of mixtures of composition varying with film thickness, it is possible to produce layers which show an enormous degree of inhomogeneity and which permit the construction of entirely new types of antireflection coating.

Accurate calculation techniques for such layers were reviewed by Jacobsson [27] and Knittl [28]. The simplest method involves the splitting of the inhomogeneous layer into a very large number of thin sublayers. Each sublayer is then replaced by a homogeneous layer of the same thickness and mean refractive index so that the smoothly varying index of the inhomogeneous layer is represented by a series of small steps. Computation can then be carried out as for a multilayer of homogeneous layers. There is no difficulty with modern computers in accommodating very large numbers of sublayers so that, although an approximation, the method can be made to yield results identical for all practical purposes with those which would have been obtained by exact calculation (in cases where exact calculation techniques exist). A good indicator that the layers are sufficiently thin is to recalculate the performance with still slightly thinner layers. No change in performance shows that the thickness is suitable.

For our purposes, we can approach the theory of such coatings from the starting point of the multilayer antireflection coating for high-index substrates. As more and more layers are added to the coating, the performance, both the bandwidth and the maximum reflectance in the low-reflectance region, steadily improves. In the limit, there will be an infinite number of layers with infinitesimal steps in optical admittance from one layer to the next. If, as layers are added, the total optical thickness of the multilayer is kept constant, the thickness of the individual layers will tend to zero and the multilayers will become indistinguishable from a single layer of identical optical thickness, but with optical admittance smoothly varying from that of the substrate to that of the incident medium.

If there are n layers in the multilayer, then let the total optical thickness of the coating be $n\lambda_0/4$, which may be denoted by T. There will be n zeros of reflectance extending from a shortwave limit

$$\lambda_S = \frac{(n+1)}{n} \cdot \frac{\lambda_0}{2}$$

to a longwave limit

$$\lambda_L = (n+1) \cdot \frac{\lambda_0}{2}.$$

In terms of T, the total optical thickness, these limits are

$$\lambda_S = \frac{2(n+1)}{n^2} \cdot T,$$

$$\lambda_L = \frac{2(n+1)}{n} \cdot T.$$

At wavelengths of $2\lambda_L$ or longer, the arrows in the vector diagram are confined to the third and fourth quadrant so that the antireflection coating is no longer effective.

If now n tends to infinity but T remains finite, the multilayer tends to a single inhomogeneous layer, λ_S tends to zero, and λ_L tends to $2T$. For all wavelengths between these limits, the reflectance of the assembly is zero. Thus, the inhomogeneous film with smoothly varying refractive index is a perfect antireflection coating for all wavelengths shorter than twice the optical thickness of the film. At wavelengths longer than this limit, the performance falls off, and at the wavelength given by four times the optical thickness of the film, the coating is no longer effective.

Of course, in practice, there is no useful thin-film material with refractive index as low as unity and any inhomogeneous thin film must terminate with an index of around, say, 1.35, which, in the infrared, is the index of magnesium fluoride. The reflectance of such a coated component will be equal to that of a plate of magnesium fluoride, 2.2% per surface.

Jacobsson and Martensson [29] actually produced an inhomogeneous antireflection coating of this type on a germanium plate. The films were manufactured by the simultaneous evaporation of germanium and magnesium fluoride, the relative proportions of which were varied throughout the deposition to give a smooth transition between the indices of the two materials. An example of the performance attained is shown in Figure 4.69. For this particular coating, the physical thickness is quoted as 1.2 μm. To find the optical thickness, we assume that the variation of refractive index with physical thickness is linear (mainly because any other assumed law of variation would lead to very difficult calculations, although possibly more accurate results). The optical thickness is then given by the physical thickness times the mean of the two terminal indices. For this present film, starting with an index of 4.0 and finishing with 1.35, the mean is 2.68 and the optical thickness

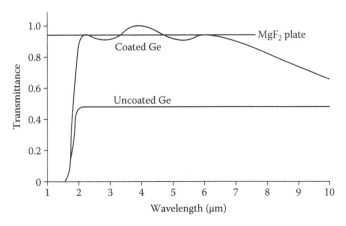

FIGURE 4.69
Measured transmittance of a germanium plate coated on both sides with an inhomogeneous Ge–MgF₂ film with geometrical thickness of 1.2 μm. (After R. Jacobsson and J. O. Martensson, *Applied Optics*, 5, 29–34, 1966. With permission of Optical Society of America.)

is therefore 2.68 × 1.2 μm, i.e., 3.2 μm. This implies that the coating should give excellent antireflection for wavelengths out to 6.4 μm; after which, it should show a gradually reducing transmission until a wavelength of 4 × 3.2 μm, i.e., 12.8 μm. The curve of the coated component in Figure 4.69 shows that this is indeed the case.

Berning [30] has suggested the use of the Herpin index concept for the design of antireflection coatings composed of homogeneous layers of two materials, one of high index and the other of low index, which are step approximations to the inhomogeneous layer and which, because they involve homogeneous layers of well understood and stable materials, might be easier to manufacture than the ideal inhomogeneous layers. He has suggested designs for the antireflection coating of germanium consisting of up to 39 alternate layers of germanium and magnesium fluoride equivalent to 20 quarter waves of gradually decreasing index.

As with coatings consisting of homogeneous layers, the most serious limitation is the lack of low-index materials. A single inhomogeneous layer to match a substrate to air must terminate at an index of around 1.38, which means that the best that can be done with such a layer is a residual reflectance of 2.5%. This limits their direct use to high-index substrates. For low-index substrates, it is likely that their role of improving the performance of designs incorporating homogeneous materials will remain.

However, there is another approach. Imagine a surface that has been treated so that it presents a tightly packed array of pyramids or cones, each gradually shrinking in cross section toward the outside. Provided the pyramids or cones are sufficiently small compared with the wavelength, they will present a composite index of refraction decreasing from that of the substrate to that of the surrounding medium, usually air. Provided the total optical thickness of this treatment is at least one half wave, then it will act as an antireflection coating. One advantage of inhomogeneous antireflection coatings of this type is that they can have excellent angular properties. Provided they remain at least one half-wave thick and that they do not exhibit regions of slope consistent with a still thinner layer, their performance will be maintained even at very oblique incidence. However, there are some dimensional requirements. Once the unit spacing becomes equal to, or greater than, a wavelength, they will exhibit significant scatter. Thus, the lateral dimension fixes the shortest wavelength of the antireflected range. The length of the features then fixes the longest wavelength. Around an octave is reasonable but greater than that is difficult. The technique dates back at least to Clapham and Hutley [31], who etched an array of pyramids in a layer of photoresist to create what they termed a *moth-eye coating*, named after a similar structure on the eye of moths to stop glinting in the dark. The adoption of the moth-eye coatings has been slow. A major problem is their mechanical weakness that makes them less suitable for applications where ruggedness is important. However, applications, particularly where the coating is protected from the environment in a sealed system, have gradually appeared.

The early approaches mostly used etching controlled by photolithographic masking to create the coatings. Embossing has also been tried [32]. More recently, plasma treatments without the need for photolithographic masking have been developed. There are also deposition techniques involving oblique incidence that reduce the packing density and, hence, index of the deposited film.

Schulz et al. [33] have developed a plasma etching process for a wide range of plastic materials using oxygen as the reactive gas. Originally, they discovered that the treatment of polymethylmethacrylate naturally produced an array of needlelike features over the surface that acted as a graded index antireflection coating. The process, however, was not, at first, effective on other types of optical plastics. The solution was found to be the initial application of a thin (0.5–3 nm thick) dielectric layer over the surface that appears to organize itself in the initial stages of etching into an effective mask. A wide range of plastics of many different compositions can be treated in this way. The surface treatment sometimes results in what are essentially bumps and sometimes holes, holes being more mechanically resistant. Durability can be improved by the addition of a thin (20–80 nm) film of silica without a too great degradation in the antireflecting properties.

The index profile of the surface treatments is not necessarily the completely graded optimum. Sometimes the treatment results in what is closer to a single quarter-wave antireflection coating. Thus, extremely broad antireflection properties are difficult to achieve. Then, also, not all optical surfaces can be etched in this way. An alternative technique by Schulz et al. [34], suitable even for glass and bringing additional advantages, applies an organic layer that itself can be etched. There are two main approaches. The fundamental optical idea is to achieve an index profile much closer to the ideal variation. The surface may be initially treated or initially coated to create that part of the coating where the index is closer to that of the surface. The organic etchable layer is then applied over this. Alternatively, the etchable organic layer may contain the complete profile in itself without the need for the initial treatments. The etched organic layer is, inevitably rather weak, and protection by silica considerably strengthens it without unacceptable degradation of the optical properties. Particular attention has been paid to two organic materials, melamine and 5,5′-bis(4-biphenylyl)-2,2′-bithiophene (BP2T). With a glass substrate, two initial layers of silica and magnesium fluoride are deposited to form the upper index part of the film. The melamine is deposited over these, together with a thin silica masking layer, and then plasma etched, with oxygen as the reactive gas, to produce a film with index as low as 1.1 that is then protected by a final thin silica film. The BP2T is used in a somewhat different way. A silica layer starts the coating followed by a BP2T layer that is then etched so that it almost disappears but leaving a pattern of small islands that then act as a template for the deposition of a final silica layer exhibiting a pattern of bumps to form the necessary inhomogeneous profile. Antireflection extending over a region as wide as 400–1200 nm is possible with both of these arrangements.

Glancing angle deposition, where the angle of incidence of depositing vapor is exceedingly oblique, results in a material with pronounced columnar structure and low packing density yielding, in turn, low index. This permits the deposition of coatings, some multilayer, that approach the desired inhomogeneous profile [35,36]. Again the porous film is rather weak mechanically, and so its application is limited to those where some protection from the environment can be assured. A limitation is the difficulty of uniform deposition over large areas.

4.8 Further Information

It has not been possible in a single chapter in this book to completely cover the field of antireflection coatings. Further information will be found in the studies by Cox and Hass [13] and Musset and Thelen [6]. There is also a very useful account of antireflection coatings in Knittl's book [28], which contains some alternative techniques.

References

1. J. T. Cox and G. Hass. 1958. Antireflection coatings for germanium and silicon in the infrared. *Journal of the Optical Society of America* 48:677–680.
2. L. A. Catalan. 1962. Some computed optical properties of antireflection coatings. *Journal of the Optical Society of America* 52:437–440.
3. K. Schuster. 1949. Anwendung der Vierpoltheorie auf die Probleme der optischen Reflexionsminderung, Reflexionsverstarkung, und der interferenzfilter. *Annalen der Physik, 6th Series* 4:352–356.
4. T. Turbadar. 1964. Equireflectance contours of double layer antireflection coatings. *Optica Acta* 11:159–170.

5. J. T. Cox, G. Hass, and G. F. Jacobus. 1961. Infrared filters of antireflected Si, Ge, InAs and InSb. *Journal of the Optical Society of America* 51:714–718.
6. A. Musset and Thelen. 1966. Multilayer antireflection coatings. In *Progress in Optics*, E. Wolf (ed.), pp. 201–237. Amsterdam: North Holland.
7. A. Thelen. 1969. Design of multilayer interference filters. In *Physics of Thin Films*, G. Hass and R. E. Thun (eds), pp. 47–86. New York: Academic Press.
8. L. Young. 1961. Synthesis of multiple antireflection films over a prescribed frequency band. *Journal of the Optical Society of America* 51:967–974.
9. L. B. Lockhart and P. King. 1947. Three-layered reflection-reducing coatings. *Journal of the Optical Society of America* 37:689–694.
10. J. T. Cox, G. Hass, and A. Thelen. 1962. Triple-layer antireflection coating on glass for the visible and near infrared. *Journal of the Optical Society of America* 52:965–969.
11. T. Turbadar. 1964. Equireflectance contours of triple-layer antireflection coatings. *Optica Acta* 11:195–205.
12. A. J. Vermeulen. 1971. Some phenomena connected with the optical monitoring of thin-film deposition and their application to optical coatings. *Optica Acta* 18:531–538.
13. J. T. Cox and G. Hass. 1964. Antireflection coatings. In *Physics of Thin Films*, G. Hass and R. E. Thun (eds), pp. 239–304. New York: Academic Press.
14. J. Ward. 1972. Towards invisible glass. *Vacuum* 22:369–375.
15. C. Reichert Optische Werke AG. 1962. *Improvements in or relating to optical components having reflection-reducing coatings*. UK Patent, 991,635.
16. L. I. Epstein. 1952. The design of optical filters. *Journal of the Optical Society of America* 42:806–810.
17. U. Schulz, U. B. Schallenberg, and N. Kaiser. 2003. Symmetrical periods in antireflective coatings for plastic optics. *Applied Optics* 42:1346–1351.
18. A. Thelen. 1988. *Design of Optical Interference Coatings*. First ed. New York: McGraw-Hill.
19. U. B. Schallenberg, U. Schulz, and N. Kaiser. 2004. Multicycle AR coatings: A theoretical approach. *Proceedings of SPIE* 5250:357–366.
20. U. B. Schallenberg. 2006. Antireflection design concepts with equivalent layers. *Applied Optics* 45:1507–1514.
21. U. Schallenberg. 2008. Design principles for broadband AR coatings. *Proceedings of SPIE* 7101:710103-1–710103-8.
22. P. W. Baumeister, R. Moore, and K. Walsh. 1977. Application of linear programming to antireflection coating design. *Journal of the Optical Society of America* 67:1039–1045.
23. J. Mouchart. 1978. Thin film optical coatings: 6: Design method for two given wavelength antireflection coatings. *Applied Optics* 17:1458–1465.
24. J. Mouchart. 1978. Thin film optical coatings: 5: Buffer layer theory. *Applied Optics* 17:72–75.
25. S. Ogura. 1975. Some features of the behaviour of optical thin films. PhD Thesis. Newcastle upon Tyne, UK: Northumbria University.
26. A. J. Vermeulen. 1976. Influence of inhomogeneous refractive indices in multilayer anti-reflection coatings. *Optica Acta* 23:71–79.
27. R. Jacobsson. 1975. Inhomogeneous and coevaporated homogeneous films for optical applications. *Physics of Thin Films* 8:51–98.
28. Z. Knittl. 1976. *Optics of Thin Films*. London: John Wiley & Sons and SNTL.
29. R. Jacobsson and J. O. Martensson. 1966. Evaporated inhomogeneous thin films. *Applied Optics* 5:29–34.
30. P. H. Berning. 1962. Use of equivalent films in the design of infrared multilayer antireflection coatings. *Journal of the Optical Society of America* 52:431–436.
31. P. B. Clapham and M. C. Hutley. 1973. Reduction of lens reflexion by the "moth eye" principle. *Nature* 244:281–282.
32. S. J. Wilson and M. C. Hutley. 1982. The optical properties of "moth eye" antireflection coatings. *Optica Acta* 29:993–1009.
33. U. Schulz, P. Munzert, R. Leitel et al. 2007. Antireflection of transparent polymers by advanced plasma etching procedures. *Optics Express* 15:13108–13113.
34. U. Schulz, F. Rickelt, H. Ludwig et al. 2015. Gradient index antireflection coatings on glass containing plasma-etched organic layers. *Optical Materials Express* 5:1259–1265.

35. S.-H. Woo, Y. J. Park, D.-H. Chang et al. 2007. Wideband antireflection coatings of porous MgF$_2$ films by using glancing angle deposition. *Journal of the Korean Physical Society* 51:1501–1506.
36. S. Chhajed, M. F. Schubert, J. K. Kim et al. 2008. Nanostructured multilayer graded-index antireflection coating for Si solar cells with broadband and omnidirectional characteristics. *Applied Physics Letters* 93:251108-1–251108-3.

5

Neutral Mirrors and Beam Splitters

5.1 High-Reflectance Mirror Coatings

Mirrors are probably the oldest optical instruments known to humans. Mostly, they consisted of a smooth reflecting surface that exhibited the natural reflectance of whatever material was involved. Then the fifteenth century with its development of outstandingly clear glass saw the introduction of the essentially modern mirror consisting of a glass plate with a tin amalgam coating. Initially, this was a product of the island of Murano, where the Venetian glass was manufactured, but despite the greatly enforced secrecy, the invention spread. However, it was not until the middle of the nineteenth century that the reflecting astronomical telescope exchanged its metallic speculum mirrors for glass chemically coated with silver. Astronomical telescopes require front surface mirrors where the reflecting coating is carried on the front surface rather than on the rear. The second surface mirror with its coating on the rear of the substrate yields a reflecting surface with essentially the quality of the substrate, while the front surface mirror possesses the properties of the front surface of the metal film rather than that of the substrate. With chemical deposition, this was not of the same high quality as the substrate, and so the silver coatings had to be hand polished, a difficult and somewhat uncertain process and even more uncertain when a beam splitter rather than a mirror was involved. The revolutionary development came in the early twentieth century with the discovery that vapor deposition under vacuum resulted in a metallic film exhibiting a surface of exactly the same quality as the supporting substrate.

Reflectors are still of immense importance. In the vast majority of cases, the sole requirement is that the specular reflectance should be as high as conveniently possible, although, as we shall see, there are specialized applications where not only should the reflectance be high, but also the absorptance should be low. For mirrors in optical instruments, simple metallic layers usually give adequate performance, and these will be examined first. For some applications where the reflectance must be higher than what can be achieved with simple metallic layers, their reflectance can be increased by the addition of extra dielectric layers. Multilayer all-dielectric reflectors, which combine maximum reflectance with minimum absorption and which transmit the energy that they do not reflect, are reserved for the next chapter.

5.1.1 Metallic Layers

The simplest thin-film reflector is a single metallic layer. Metallic layers of quite modest thickness completely block light and so are indistinguishable from bulk metal. Their reflectance is therefore that of the bulk metal, and since they contour the substrate surface more or less perfectly, they present a mirror surface of quality equal to that of the substrate. The resulting mirror possesses not only the high stability of the substrate, but also the high reflectance of the metal.

The optical properties of metals are dominated by the interaction of light with their free electrons. This interaction is vanishingly small at very high frequencies but becomes very significant with reducing frequency and increasing wavelength. Over much of the wavelength range where the interaction is significant, metals have rather high extinction coefficient that, in most metals, is roughly proportional to the wavelength, and since absorption coefficient is given by

$$\alpha = \frac{4\pi k}{\lambda},\tag{5.1}$$

it is essentially constant. This is why a metal just thick enough to strongly reflect at shorter wavelengths is still an effective reflector at much longer ones.

The performance of the commonest metals used as reflecting coatings is shown [1] in Figure 5.1.

Silver was once the most popular material of all. Until the 1930s, it was the principal reflecting material for precision optics, usually chemically deposited in a liquid process. Thermal evaporation under vacuum was then found to be a much more useful and satisfactory process for metallic film deposition, especially since the front surface replicated the quality of the substrate surface. The liquid chemical deposition of silver is still sometimes used for the production of second surface architectural mirrors. In that application, the bonding of the silver to the glass surface is assured by a very thin layer of tin, and the silver is protected by an outer layer of copper followed by an additional layer of paint. In front surface applications, silver reacts with atmospheric oxygen and tarnishes by the formation of silver sulfide, but the initial high reflectance and the extreme ease of evaporation still make it a common choice for components used only for a short period. Silver is also useful where it is necessary to temporarily coat a component, such as an interferometer plate, for a test of flatness. We shall deal more fully with protected silver in the next section.

John Strong [2], working with astronomical mirrors, pioneered in the 1930s, the replacement of chemically deposited silver by evaporated aluminum. Aluminum is easy to evaporate and has good ultraviolet, visible, and infrared reflectances, together with the additional advantage of strongly adhering to most substances, including plastics. As a result, it is the most frequently used

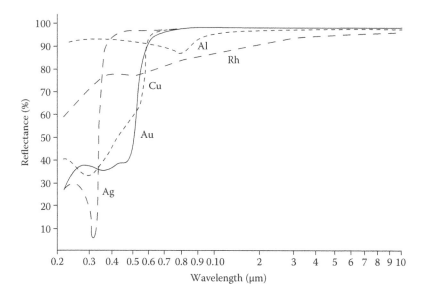

FIGURE 5.1
Reflectance of freshly deposited films of aluminum, copper, gold, rhodium, and silver as a function of wavelength from 0.2 to 10 μm. (After G. Hass, *Journal of the Optical Society of America*, 45, 945–952, 1955. With permission of Optical Society of America.)

film material for the production of reflecting coatings. The reflectance of an aluminum coating does gradually drop in use, although the thin oxide layer, which always very quickly forms on the surface after coating, helps to protect it from further corrosion. In use, especially if the mirror is exposed at all, dust and dirt invariably collect on the surface and cause a fall in reflectance. The performance of most instruments is not seriously affected by a slight drop in reflectance, but in some cases where it is important to collect the maximum amount of light, as it is difficult to clean the coatings without damaging them, the components are periodically recoated. This particularly applies to the mirrors of large astronomical reflecting telescopes. The primary mirrors of these are periodically lightly washed but then frequently recoated with aluminum around once a year in coating machines that are installed in the observatories for this purpose. Because the primaries are very large and heavy, it is not usual to rotate them during coating, and the uniformity of coating is achieved using multiple sources. Aluminum is still used for the vast majority of telescope mirrors, but those in some of the very latest telescopes are being coated with a more advanced coating consisting of protected silver.

Gold is probably the best material for infrared-reflecting coatings. Its reflectance rapidly drops off in the visible region, and it is really useful only beyond 700 nm. On glass, gold tends to form rather soft, easily damaged films, but it strongly adheres to a film of chromium or Nichrome (trade name of a resistance wire usually of 80% nickel and 20% chromium composition), and this is often used as an underlayer between the gold and the glass substrate.

The reflectance of rhodium and platinum is rather less than that of the other metals mentioned, and these metals are used only where stable films very resistant to corrosion are required. Both materials very strongly adhere to glass. Dental mirrors that are subjected to a quite difficult environment and have to be sterilized by heating are usually rhodium coated. Rhodium is also used for some automobile rearview mirrors. These are frequently front-surface mirrors and, on the outside of a vehicle, are subjected to the weather, washing and cleaning, and a general level of serious abuse. An early paper pointing out the advantages of rhodium with its great stability over films of aluminum is that of Auwärter [3].

Chromium is somewhat cheaper than rhodium and is much used for automobile rearview mirrors.

5.1.2 Protection of Metal Films

Most metal films are rather softer than hard dielectric films and can be easily scratched. Unprotected evaporated aluminum layers, for example, can be badly damaged if wiped with a cloth, while gold and silver films are even softer. This is a serious disadvantage, especially when periodic cleaning of the mirrors is necessary. One solution, as we have seen, is periodic recoating. An alternative, which improves the ruggedness of the coatings and protects them from atmospheric corrosion, is overcoating with an additional dielectric layer. The behavior of a single dielectric layer on a metal is a useful illustration of the calculation techniques in Chapter 2. We shall also require some related results later, and so it is useful to spend a little time on the problem.

First of all, the admittance diagram (Figure 5.2) gives us a qualitative picture of the behavior of the system as the dielectric layer is added. The metal layer will normally be thick enough for the optical admittance at its front surface to be simply that of the metal, the substrate optical constants having no effect. The optical admittance of the metal will always be in the fourth quadrant, and so, as a dielectric layer is added, the reflectance must fall until the locus of the admittance of the assembly crosses the real axis. (We recall that the reflectance associated with the locus of a dielectric layer in an incident medium of lower admittance always falls as the locus is traced out in the fourth quadrant [that is, below the real axis] and always rises in the first [that is, above the real

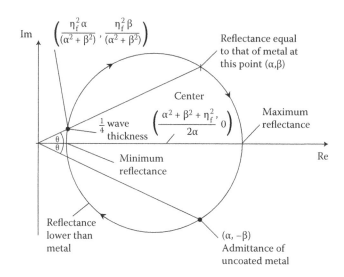

FIGURE 5.2
Admittance diagram of a dielectric layer deposited over a metal. The metal admittance would usually be much closer to the imaginary axis but has been moved away from it for greater clarity in the diagram. The dielectric locus starts at the admittance of the uncoated metal. The construction to find the quarter-wave point is explained in the text, as are the other parameters.

axis].) This minimum of reflectance will occur at a dielectric layer thickness of less than a quarter wave, but, for most metals, greater than an eighth wave. For layer thicknesses of up to twice this figure, therefore, the reflectance of the protected metal film will be reduced. The reduction in reflectance very much depends on the particular metal and the index of the dielectric film. For most high-performance metals, the first intersection of the dielectric overcoat with the real axis will be to the left of the incident admittance, and since loci of higher-index films are nested about points to the right of those of low-index films, the high-index locus will cut the real axis closer to the incident admittance and therefore exhibit a lower minimum reflectance. In general, therefore, low-index protecting films are to be preferred over those of high index.

We can mark the position of the quarter-wave dielectric layer thickness by a simple construction. We draw the line from the origin to the starting point of the dielectric locus, that is, the metal admittance $\alpha - i\beta$, which lies in the fourth quadrant. This line makes an angle ϑ with the real axis. Then, also through the origin, we draw a line in the first quadrant making the same angle ϑ with the real axis. This cuts the dielectric locus in two points. One is point $\alpha + i\beta$, the image of the starting point in the real axis, and at this point, the reflectance of the assembly is identical to that of the uncoated metal. The second point of intersection is

$$\left(\frac{\eta_f^2 \alpha}{(\alpha^2 + \beta^2)} , \frac{\eta_f^2 \beta}{(\alpha^2 + \beta^2)} \right), \quad \text{i.e.,} \quad \frac{\eta_f^2}{(\alpha - i\beta)},$$

and here the layer is clearly one-quarter-wave thick. In fact, this is a useful construction to find the terminal point of a quarter-wave layer.

We can derive straightforward analytical expressions for the various parameters and, in particular, the points of intersection of the locus with the real axis, which, as we know, correspond to the points of maximum and minimum reflectances.

The characteristic matrix is given by

$$
\begin{bmatrix} B \\ C \end{bmatrix} = \begin{bmatrix} \cos\delta_f & i(\sin\delta_f/\eta_f) \\ i\eta_f\sin\delta_f & \cos\delta_f \end{bmatrix} \begin{bmatrix} 1 \\ \alpha - i\beta \end{bmatrix}, \tag{5.2}
$$

where $\alpha - i\beta$ is the characteristic admittance of the metal, i.e., $(n_m - ik_m)$ in free space units and at normal incidence; $\delta_f = 2\pi n_f d_f \cos\vartheta_f/\lambda$; and η_f is the characteristic admittance of the film material. Then,

$$
\begin{bmatrix} B \\ C \end{bmatrix} = \begin{bmatrix} \cos\delta_f + (\beta\sin\delta_f/\eta_f) + i(\alpha\sin\delta_f/\eta_f) \\ \alpha\cos\delta_f + i(\eta_f\sin\delta_f - \beta\cos\delta_f) \end{bmatrix}.
$$

Now, at the points of intersection of the locus with the real axis, we must have that the admittance, which we can denote by μ, is real. But

$$
\mu = C/B.
$$

And, equating real and imaginary parts,

$$
\alpha\cos\delta_f = \mu[\cos\delta_f + (\beta\sin\delta_f/\eta_f)], \tag{5.3}
$$

$$
\eta_f\sin\delta_f - \beta\cos\delta_f = \mu(\alpha\sin\delta_f/\eta_f). \tag{5.4}
$$

Hence, first eliminating μ,

$$
(\alpha\cos\delta_f)(\alpha\sin\delta_f)/\eta_f = (\eta_f\sin\delta_f - \beta\cos\delta_f)[\cos\delta_f + (\beta\sin\delta_f/\eta_f)],
$$

i.e.,

$$
[(\alpha^2 + \beta^2 - \eta_f^2)/(2\eta_f)]\sin(2\delta_f) = -\beta\cos(2\delta_f).
$$

Thus,

$$
\tan(2\delta_f) = 2\beta\eta_f/(\eta_f^2 - \alpha^2 - \beta^2),
$$

so that

$$
\delta_f = \frac{1}{2}\arctan[2\beta\eta_f/(\eta_f^2 - \alpha^2 - \beta^2)] + \frac{m\pi}{2}, \quad m = 0, 1, 2, 3, \ldots. \tag{5.5}
$$

Or, in full waves,

$$
D_f/\lambda_0 = [1/(4\pi)]\arctan[2\beta\eta_f/(\eta_f^2 - \alpha^2 - \beta^2)] + m/4, \tag{5.6}
$$

where the arctangent is to be taken in either the first or second quadrant so that δ_f for $m = 0$ is positive and represents the first intersection with the real axis where the film is less than, or at the very most, equal to a quarter wave. A similar result was derived by Park [4] by using a slightly different technique.

The value of μ can be found by slightly rearranging Equations 5.3 and 5.4,

$$
(\mu - \alpha)\cos\delta_f + (\beta\mu/\eta_f)\sin\delta_f = 0,
$$

$$\beta \cos \delta_f + [(\mu\alpha/\eta_f) - \eta_f] \sin \delta_f = 0,$$

and, eliminating δ_f,

$$(\mu - \alpha)[(\mu\alpha/\eta_f) - \eta_f] - \beta(\mu\beta/\eta_f) = 0.$$

The two solutions are

$$\mu = [(\alpha^2 + \beta^2 + \eta_f^2)/(2\alpha)] \pm \left\{ [(\alpha^2 + \beta^2 + \eta_f^2)/(2\alpha)]^2 - \eta_f^2 \right\}^{1/2},$$

but this is not the best form for calculation. We know that the two solutions μ_1 and μ_2 are related by $\mu_1\mu_2 = \eta_f^2$, and so we write

$$\mu_1 = 2\alpha\eta_f^2 / \left\{ (\alpha^2 + \beta^2 + \eta_f^2) + \left[(\alpha^2 + \beta^2 + \eta_f^2)^2 - 4\alpha^2\eta_f^2 \right]^{1/2} \right\}, \tag{5.7}$$

$$\mu_2 = [(\alpha^2 + \beta^2 + \eta_f^2)/(2\alpha)] + \left\{ [(\alpha^2 + \beta^2 + \eta_f^2)/(2\alpha)]^2 - \eta_f^2 \right\}^{1/2}, \tag{5.8}$$

and the value that corresponds to the first intersection ($m = 0$ in Equation 5.5) is

$$\mu_1 = 2\alpha\eta_f^2 / \left\{ (\alpha^2 + \beta^2 + \eta_f^2) + \left[(\alpha^2 + \beta^2 + \eta_f^2)^2 - 4\alpha^2\eta_f^2 \right]^{1/2} \right\} \quad \text{(Equation 5.7)}.$$

Often,

$$(\alpha^2 + \beta^2 + \eta_f^2)^2 \gg 4\alpha^2\eta_f^2,$$

and in that case,

$$\mu_1 = \alpha\eta_f^2 / (\alpha^2 + \beta^2 + \eta_f^2), \tag{5.9}$$

$$\mu_2 = (\alpha^2 + \beta^2 + \eta_f^2)/\alpha. \tag{5.10}$$

The limits of reflectance are given by

$$R_{minimum} = [(\eta_0 - \mu_1)/(\eta_0 + \mu_1)]^2. \tag{5.11}$$

$$R_{minimum} = [(\eta_0 - \mu_2)/(\eta_0 + \mu_2)]^2. \tag{5.12}$$

As already remarked, the higher the index of the dielectric film, the greater the fall in reflectance at the minimum. The reflectance rises above that of the bare metal at the maximum, but for the metals commonly used as reflectors, the increase is not great, and so the lower-index films are to be preferred as protecting layers. As an example, we can consider aluminum, which has a refractive index of $0.82 - i5.99$ at 546 nm [5] with protecting layers of silica of index 1.45 or a high-index layer (2.3) such as cerium oxide. The results in Table 5.1 were calculated from Equations 5.6 through 5.8, 5.11, and 5.12. Clearly, if high-index films are used for protecting metal layers, then the monitoring of layer thickness must be accurate; otherwise, there is a risk of a sharp drop in reflectance. Also, the high reflectance region associated with the high-index protecting layer is rather narrow.

Aluminum is probably the commonest mirror coating material for the visible region, and in addition to the silicon dioxide and cerium oxide mentioned earlier, there is a large number of materials that can be used for protecting it. The best overall material is definitely silicon dioxide,

TABLE 5.1

Protected Aluminum

Aluminum (0.82 − i5.99)	$R_{uncoated}$ (%)	R_{min} (%)	D_{min} (Full Waves)	R_{max} (%)	D_{max} (Full Waves)
SiO$_2$ (1.45)	91.63	83.64	0.2128	91.86	0.4628
CeO$_2$ (2.30)	91.63	65.90	0.1925	92.44	0.4425

often referred to as silica or quartz (although strictly quartz refers to the natural mineral). If highest luminous reflectance is required, then a half wave of silicon dioxide is best. (Figure 5.3). Silicon monoxide (SiO), for example, is also a very effective protecting material, but it has strong absorption at the blue end of the spectrum, where it causes the reflectance of the composite coating to be rather low, and since it has a higher index, it presents a greater reduction of reflectance at the minimum than silicon dioxide does. Another useful coating is aluminum oxide (Al$_2$O$_3$), also often known as sapphire (although sapphire is strictly a crystalline form). This can be vacuum deposited, or the aluminum at the surface of the coating can be anodized by an electrolytic technique [1], forming a very hard layer of aluminum oxide. Gold and silver are more difficult to protect because of the difficulty of getting films to stick to them. Gold is sometimes protected with silicon monoxide, but the sticking is not always entirely satisfactory. Since gold resists tarnishing much better than silver does, if it is not going to be handled, it is often preferable not to overcoat it. It has been found that aluminum oxide sticks very well to silver [6,7]. Aluminum oxide does not appear to be a very effective barrier against moisture, and so it has been principally used as a bonding layer between the silver and a layer of silicon oxide which affords good moisture resistance and which, although it adheres only weakly to silver, strongly adheres to the aluminum oxide. Further details of the coating are given by Hass et al. [7]. The starting material can be silicon monoxide, but to reduce the absorption at the blue end of the spectrum, the silicon monoxide should be reactively deposited (see Chapter 11) when the actual oxide produced will lie between SiO and SiO$_2$. With such a coating, it is possible to achieve a reflectance greater than 95% over the visible and infrared regions from 0.45 to 20 μm.

Aluminum oxide and silicon oxide are absorbing at wavelengths longer than 8 μm, and it was discovered by Pellicori [8] and confirmed theoretically by Cox and Hass [9] that reflectors protected by these materials exhibit a sharp dip in reflectance at high angles of incidence, that is, 45°

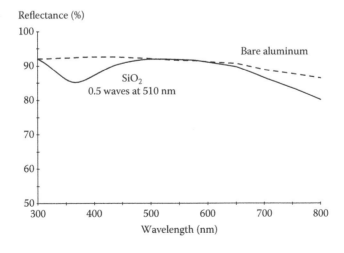

FIGURE 5.3
Calculated reflectance of a half wave of SiO$_2$ (λ_0 = 510 nm) over aluminum. Over the visible region (400–700 nm), the reflectance remains high but falls outside that region.

and above. The dip can be avoided by the use of a protecting material that does not absorb in this region. Magnesium fluoride is such a material, but it must be deposited on a hot substrate (temperatures in excess of 200°C) if it is to be robust. The metals have their best performance if deposited at room temperature, and thus, the substrates should be heated only after they have been coated with the metal.

In the 1930s, John Strong [2] introduced the evaporated aluminum coating for the mirrors in reflecting telescopes, replacing chemically deposited silver. This aluminum coating has remained essentially unchanged for 70 years. Astronomical telescopes are normally expected to operate over as wide a spectral region as possible, and aluminum covers the ultraviolet, visible, and infrared to a wider extent than any other material. However, there is a problem, particularly in the infrared. The faintness of an infrared source that can be observed is largely determined by the thermal radiation from the mirrors themselves, and this, of course, depends on the thermal emittance of the mirror coatings. The material with the highest reflectance and lowest emissivity in the infrared is silver. Silver has some disadvantages. It has low reflectance in the ultraviolet, and it is environmentally weak. Attempts have been made to develop a silver-based coating that would be more rugged and would more strongly reflect in the ultraviolet [10]. So far, no universal coating based on silver has been adopted. However, some of the new generation of telescopes are optimized for the infrared, and the coating being used for these instruments is based on protected silver [11–13]. The two Gemini telescopes sited in Chile and Hawaii with their 8 m diameter primaries have already been coated with this silver-based coating. The technique used is magnetron sputtering. The mirror is rotated underneath the magnetron targets that cover the entire mirror in three passes. A thin layer of $NiCrN_x$ over the glass acts to bond the silver layer. The silver is then protected by a very thin layer of $NiCrN_x$ followed by a thicker layer of silicon nitride. The standard coating [12,13] consists, from substrate to incident medium, of 6.5 nm of $NiCrN_x$, 110 nm of silver, 0.6 nm of $NiCrN_x$, and 8.5 nm of SiN_x. The uniformity of the coating is better than 1 nm. The success of the Gemini coatings is encouraging its use for other telescopes.

5.1.3 Overall System Performance, Enhanced Reflectance

In optical instruments of any degree of complexity, there will be a number of reflecting components in series, and the overall throughput of the system will be given by the product of the reflectances of the various elements. Figure 5.4 gives the overall transmittance of any system with a number of components in series, with identical values of reflectance. It is obvious from the diagram that even with the best metal coatings, the performance with, say, 10 elements is low. If the instrument is to be used over a wide range, little can be done to alleviate the situation. Most spectrometers, for instance, have 10 or more reflections with a consequent severe drop in throughput, but are required to work over a wide region—possibly as much as a 25:1 variation in wavelength. The spectrometer designer normally just accepts this loss and designs the rest of the instrument accordingly.

In cases where the wavelength range is rather more limited, say, to the visible region or to a single wavelength, it is possible to increase the reflectance of a simple metal layer by overcoating it with extra dielectric layers.

The characteristic admittance of a metal can be written as $n - ik$, and the reflectance in air at normal incidence is

$$R = \left| \frac{1 - (n - ik)}{1 + (n - ik)} \right|^2 = \frac{(1 - n)^2 + k^2}{(1 + n)^2 + k^2} = \frac{1 - \left[2n/\left(1 + n^2 + k^2\right)\right]}{1 + \left[2n/\left(1 + n^2 + k^2\right)\right]}. \tag{5.13}$$

The quarter-wave rule tells us that the optical admittance of an assembly Y becomes n^2/Y when a quarter-wave optical thickness of index n, that is, admittance in free space units, is added.

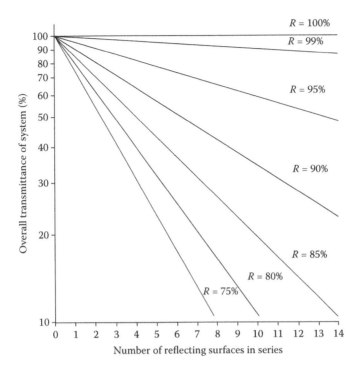

FIGURE 5.4
Overall transmittance of an optical system with a number of reflecting elements in series.

If the metal is overcoated with two quarter waves of material of indices n_1 and n_2, n_2 being next to the metal, then the optical admittance at normal incidence is

$$\left(\frac{n_1}{n_2}\right)^2 (n - ik),$$

and the reflectance in air, also at normal incidence, is

$$R = \left|\frac{1 - (n_1/n_2)^2(n - ik)}{1 + (n_1/n_2)^2(n - ik)}\right|^2 = \frac{\left[1 - (n_1/n_2)^2 n\right]^2 + (n_1/n_2)^4 k^2}{\left[1 + (n_1/n_2)^2 n\right]^2 + (n_1/n_2)^4 k^2},$$

i.e.,

$$R = \frac{1 - \left[2(n_1/n_2)^2 n\right] / \left[1 + (n_1/n_2)^4 (n^2 + k^2)\right]}{1 + \left[2(n_1/n_2)^2 n\right] / \left[1 + (n_1/n_2)^4 (n^2 + k^2)\right]}. \tag{5.14}$$

This will be greater than the reflectance of the bare metal given by Equation 5.13 if

$$\frac{2(n_1/n_2)^2 n}{1 + (n_1/n_2)^4 (n^2 + k^2)} < \frac{2n}{1 + n^2 + k^2}, \tag{5.15}$$

which is satisfied by either

$$\left(\frac{n_1}{n_2}\right)^2 > 1 \quad \text{or} \quad \left(\frac{n_1}{n_2}\right)^2 < \frac{1}{n^2 + k^2}, \tag{5.16}$$

assuming that $n^2 + k^2 \geq 1$.

The first solution is of greater practical value than the second, which can be ignored. This shows that the reflectance of any metal can be enhanced by a pair of quarter-wave layers for which $(n_1/n_2) > 1$, n_1 being on the outside, and n_2, next to the metal. The higher this ratio, the greater the increase in reflectance.

As an example, consider aluminum at 550 nm with $n - ik = 0.92 - i5.99$. From Equation 5.31, the untreated reflectance of this is approximately 91.6%. If the aluminum is covered by two quarter waves consisting of silicon oxide of index 1.45, next to the aluminum, followed by titanium oxide of index 2.40, then $(n_1/n_2) = 1.655$, and from Equation 5.14, the reflectance jumps to 96.4%.

An approximate result can be very quickly obtained using $A = (1 - R)$. When the two layers are added, A is roughly reduced to $A/(n_1/n_2)^2$. Inserting the figures mentioned earlier, for aluminum, A is 9.4% initially and, on addition of the layers, drops to 3.45%, corresponding to an enhanced reflectance of 96.5% (instead of the more accurate figure of 96.4%).

A second similar pair of dielectric layers will raise the reflectance even higher, to approximately 99%, and greater numbers of quarter-wave pairs may be used to give a still higher reflectance.

Unfortunately, the region over which the reflectance is enhanced is limited. In the visible region, the gain is essentially in luminous reflectance. Outside the enhanced zone, the reflectance is less than it would be for the bare metal. Jenkins [14] measured the reflectance of an aluminum layer overcoated with six quarter-wave layers of cryolite, of index 1.35, and zinc sulfide of index 2.35. With layers monitored at 550 nm, the reflectance of the boosted aluminum was greater than 95% over a region 280 nm wide and greater than 99% over the major part. Coatings that are more robust can be obtained using magnesium fluoride, silicon dioxide, or aluminum oxide as the low-index layers and cerium oxide or titanium oxide as the high-index layers. To attain maximum toughness, the dielectric layers should be deposited on a hot substrate. Aluminum, however, if deposited hot, tends to badly scatter, and so the substrates should be heated only after the deposition of the aluminum is complete. Figure 5.5 shows the reflectance of aluminum enhanced by four quarter-wave layers, increasing the reflectance over the visible region.

We have already considered more exactly the behavior of a single dielectric layer on a metal and have shown, as did Park [4], that the thickness of the dielectric layer for minimum reflectance should be (Equation 5.6)

$$D_f = \left\{ \arctan\left[2\beta\eta_f/\left(\eta_f^2 - \alpha^2 - \beta^2\right)\right] \right\} [\lambda_0/(4\pi)],$$

where $(\alpha - i\beta)$ is the admittance of the metal and the angle is in the first or second quadrant. This is the thickness that the low-index layer next to the metal should have if the maximum possible increase in reflectance is to be achieved. A moment's consideration of the admittance diagram will show that this is indeed the case. Layers other than that next to the metal will, of course, retain their quarter-wave thicknesses. Usually, however, the layer is simply made a quarter wave because the resulting difference in performance over the visible region is not great.

5.1.4 Reflecting Coatings for the Ultraviolet Region

The production of high-reflectance coatings for the ultraviolet is a much more exacting task than that for the visible and infrared. Space-borne ultraviolet astronomical instrumentation drove a great deal of research on this topic starting in the 1960s. A very full review is given by Madden

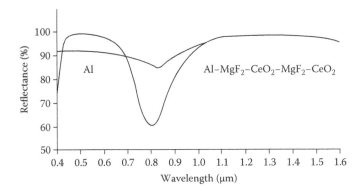

FIGURE 5.5
Reflectances of evaporated aluminum with (*solid curve*) and without (*broken curve*) two reflectance-increasing film pairs of MgF$_2$ and CeO$_2$ as a function of wavelength from 0.4 to 1.6μm. (After G. Hass, *Journal of the Optical Society of America*, 45, 945–952, 1955. With permission of Optical Society of America.)

[15], supplemented in great detail by a later account by Hass and Hunter [16]. The following is a very brief summary.

The most suitable material known for the production of reflecting coatings for the ultraviolet out to around 100 nm is aluminum. To achieve the best results, the aluminum should be evaporated at a very high rate, 40 nm/s or more, if possible, on to a cold substrate, the temperature of which should not be permitted to exceed 50°C, and at pressures of 1.33×10^{-4} Pa (1.33×10^{-6} mbar or 10^{-6} Torr as in the paper) or lower. The aluminum should be of the purest grade. Hass and Tousey [17] quoted results showing a significant improvement (as high as 10% at 150 nm) in the ultraviolet reflectance of aluminum films if 99.99% pure aluminum is used in preference to 99.5% pure. Aluminum should, in theory, have a much higher reflectance than is usually achieved in practice, particularly at the shortwave end of the range. This has been found to be due to the formation of a thin oxide layer on the surface, and as we have already shown, such a layer must, unless it is very thick, lead to a reduction in reflectance. This oxidation takes place even at partial pressures of oxygen below 1.33×10^{-4} Pa (1.33×10^{-6} mbar or 10^{-6} Torr as in the paper). Unprotected aluminum films, therefore, inevitably show a rapid fall in reflectance with time when exposed to the atmosphere. The reflectance stabilizes when the layer is of sufficient thickness to inhibit further oxidation, but this occurs only when the reflectance at short wavelengths has catastrophically fallen.

Attempts have been made to find a suitable protecting material for aluminum to prevent oxidation, and promising results have been obtained with magnesium fluoride (robust coatings) and lithium fluoride (less robust), which, in crystal form, are very useful window materials for the ultraviolet. Figures 5.6 and 5.7 show the effect of an extra protecting layer of magnesium fluoride [18] or lithium fluoride [19] on the reflectance of aluminum. The increase in reflectance is due not only to the lack of oxide layer, but also to interference effects.

It is necessary to evaporate the protecting layer immediately after the aluminum in order that the minimum amount of oxidation is allowed to take place. This is usually achieved by simultaneously running the two sources and arranging for the shutter that covers the aluminum source at the end of the deposition of the aluminum layer to uncover at the same time the magnesium fluoride or lithium fluoride source. The use of magnesium fluoride-overcoated aluminum as a reflecting coating for the ultraviolet is now becoming standard practice.

The aluminum and magnesium fluoride coating was examined in some detail by Canfield et al. [18]. Among other results, they showed that provided that the magnesium fluoride is thicker than

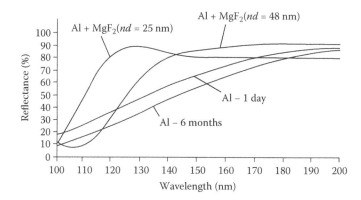

FIGURE 5.6
Reflectance of evaporated aluminum from 100 to 200 nm with and without protective layers of MgF_2 of two different thicknesses. (After L. R. Canfield, G. Hass, and J. E. Waylonis, *Applied Optics*, 5, 45–50, 1966. With permission of Optical Society of America.)

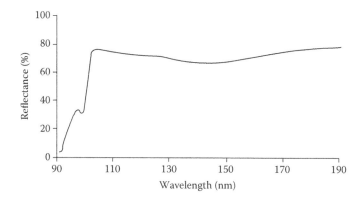

FIGURE 5.7
Reflectance of an evaporated aluminum film with a 14 nm thick LiF overcoating in the region of 90–190 nm. Measurements were begun 10 minutes after the evaporation was completed. (After T. Cox, G. Hass, and J. E. Waylonis, *Applied Optics*, 7, 1535–1539, 1968. With permission of Optical Society of America.)

10 nm, the coatings will withstand, without deterioration, exposure to ultraviolet radiation and to electrons (up to 10^{16} 1 MeV electrons/cm^2) and protons (up to 10^{12} 5 MeV protons/cm^2).

5.2 Neutral Beam Splitters

A device that divides a beam of light into two parts is known as a beam splitter. The functional part of a beam splitter generally consists of a plane surface coated to have a specified reflectance and transmittance over a certain wavelength range. The incident light is split into a transmitted and a reflected portion at the surface, which is usually tilted so that the incident and reflected beams are separated. The ideal values of reflectance and transmittance may vary from one application to another. The beam splitters considered in this section are known as neutral beam splitters, because the reflectance and transmittance should ideally be constant over the wavelength range concerned. There are different types of beam splitter, considered in later chapters, that separate the light into different wavelength regions and are usually known as dichroic or that separate the light according to polarization.

Neutral beam splitters are usually specified by the ideal values of transmittance and reflectance expressed as a percentage and written as T/R. Probably the most common are the 50/50 beam splitters.

5.2.1 Beam Splitters Using Metallic Layers

Apart from a single uncoated surface, which is sometimes used, the simplest type of beam splitter consists of a metal layer deposited on a glass plate. Silver, which has least absorption of all the common metals used in the visible region, is traditionally the most popular material for this. Although commercial beam splitters nowadays are usually constructed from metals, such as chromium, which are less prone to damage by abrasion and corrosion, 50/50 beam splitters are frequently referred to as being "half silvered."

All metallic beam splitters suffer from absorption. The transmittance of a metal film is the same, regardless of the direction in which it is measured. This is not so for reflectance, and that measured at the air side is slightly higher than that measured at the glass side. This effect does not appear with a transparent film. Since $T + A + R = 1$, the reduction in reflectance at the substrate side means that the absorption from that side must always be higher. Figure 5.8 shows curves for platinum demonstrating this behavior [20]. Because of this difference in reflectance, metallic beam splitters should always be used in the manner shown in Figure 5.9 if the highest efficiency is to be achieved.

It is possible to decrease the absorption in metallic beam splitters by adding an extra dielectric layer. The method was applied to chromium films by Pohlack [21], and Figure 5.10 gives some of the measurements made. The first pair of results is for a simple chromium film on glass of index 1.52 measured both from the air side and from the glass side. The second pair of results shows how the absorption in the chromium can be reduced by the presence of a quarter-wave layer of high-refractive index material (zinc sulfide of index approximately 2.4 in this case) between the metal and the glass. This layer forms an antireflection coating on the rear surface of the metal, and the effect can be seen particularly strongly in the results for reflectance and transmittance from the glass side. There, the transmittance remains exactly as before, but the

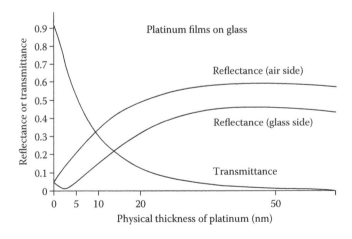

FIGURE 5.8
Reflectance and transmittance curves for a platinum film on glass, calculated from the optical constants on the bulk metal. (After O. S. Heavens, *Optical Properties of Thin Solid Films*, Butterworth Scientific Publications, London, 1955.)

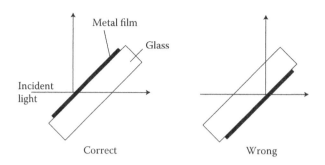

FIGURE 5.9
Correct use of a metallic beam splitter.

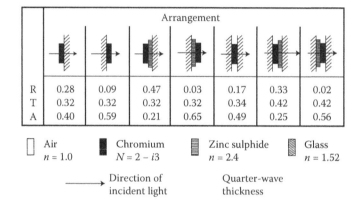

FIGURE 5.10
Values of reflectance, transmittance, and absorptance at 550 nm and normal incidence for semireflecting films of chromium on glass showing the effect of adding a quarter-wave layer of zinc sulfide. (After H. Pohlack, *Jenaer Jahrbuch*, Gustav Fischer Verlag, Jena, pp. 241–245, 1953.)

reflectance is considerably reduced. Results are also given for a chromium layer protected by a glass cover cemented on the front surface with and without the antireflecting layer. The metallic absorption again is very much less when the antireflection layer is on the side of the metal remote from the incident light. The dielectric layers partially act as matching layers, but primarily, they reduce the electric field amplitude in the metal and so reduce the losses.

Pohlack's results are for one wavelength, normal incidence and for a particular set of optical constants. Metal optical properties very much depend on interaction with free electrons. Their mobility is quite sensitive to impurities and microstructure so that metal optical constants are not as predictable as those of dielectric materials that depend on bound electrons. In any particular case, therefore, some small adjustment will certainly be required to take full advantage of Pohlack's techniques. Figures 5.11 and 5.12 show the performance of a beam splitter using Pohlack's design in the third column from the left in Figure 5.10 but adjusted for other optical constants. Dispersive chromium optical constants were taken from Palik's handbook [22] (3.12–i4.42 at 550 nm), and typical dispersive optical constants were used for a zinc sulfide film (2.356 at 550 nm). The rear surface of the glass substrate was assumed to be perfectly antireflected. The eventual chromium physical thickness was 6 nm, and the zinc sulfide optical thickness at 550 nm was 0.29 full waves, that is, a quarter wave at a slightly longer wavelength. At oblique incidence, polarizing splitting occurs. At 45° incidence, this is quite serious but manageable at 30°, as illustrated in Figure 5.12. The performance is not far from Pohlack's.

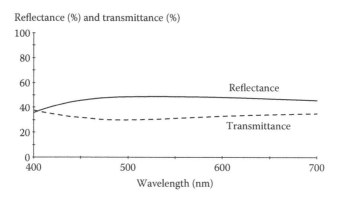

FIGURE 5.11
Pohlack beam splitter with a zinc sulfide layer behind the chromium with the design adjusted to take account of the optical constants for chromium extracted from Palik's handbook [22]. (From E. D. Palik, ed. 1985. *Handbook of Optical Constants of Solids I*. Orlando, FL: Academic Press.) The incident medium is air and the rear surface of the glass substrate is perfectly antireflected. Total transmittance and reflectance for unpolarized incident light at 30° is shown.

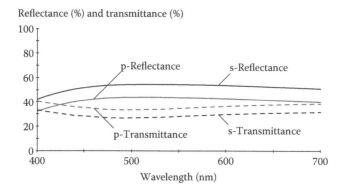

FIGURE 5.12
Pohlack beam splitter of Figure 5.11 calculated at 30° incidence showing the polarization sensitivity.

Shkliarevskii and Avdeenko [23] increased the transparency and decreased the absorption in metallic coatings using an antireflection coating in a similar manner. The antireflection coating in this case, instead of being dielectric, was a thin metallic layer. They found that a layer of silver deposited on a substrate heated to around 300°C increased the transparency of an aluminum coating, deposited on top of the silver at room temperature, by a factor as high as 3.5 at 1 μm and 2.5 at 700 nm without any decrease in reflectance at the aluminum–air interface. Since a simple layer of silver would not normally achieve this, there is presumably some microstructural effect involved or the silver at this high temperature is perhaps being converted toward a dielectric.

If the beam splitter is correctly used, the reduction in reflectance at the glass–film interface can help reduce the stray light derived from multiple reflections in the substrate.

As demonstrated in Figure 5.12, a complication found with beam splitters is a difference in the values of reflectance for the two planes of polarization when the beam splitter is tilted. The *s*-reflectance is usually higher than the *p*-reflectance. Anders [24] described a method for calculating efficiency and stray light performance.

It is not always possible to use the flat-plate beam splitter in some optical systems. Reflections from the rear surface can be a problem despite the antireflection layer behind the metal film, and in applications where the light passing through the plate is not collimated, aberrations are introduced. To overcome these difficulties, a beam-splitting cube, as shown in Figure 5.13, can be used, although

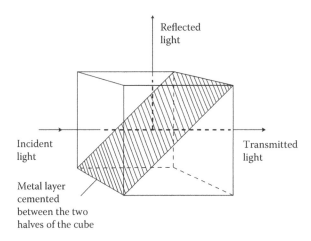

FIGURE 5.13
Cube beam splitter.

the absorption in the metal is greater in this configuration because both surfaces, instead of just one, are now in contact with a medium whose index is greater than unity. Since the cemented assembly protects the metal layers, the choice of materials is wide. Silver is probably most frequently used, although chromium, aluminum, and gold are also popular.

Chromium gives almost neutral beam splitting over the visible region, with an absorptance of approximately 55% for both planes of polarization, the s-reflectance being approximately 30%, and the p-reflectance, 15%. Silver varies more with wavelength, the reflectance falling toward the blue end of the spectrum, but the absorptance is rather less than for chromium, around 15% at 550 nm, with s-reflectance of 50% and p-reflectance of 30%. Curves of the performance of several different metallic beam splitters were given by Anders [24].

We can adopt Pohlack's ideas in the design of a cube beam splitter using silver. We assume a borosilicate glass cube with a 45° incidence on the buried coating in the manner of Figure 5.13. For the coating, we take a silver layer surrounded by layers of titania (2.349 at 510 nm). The silver optical constants are taken from Palik's handbook [22] (0.13 − i3.009 at 510 nm). We include dispersion in all the optical constants. Trial and error suggests a physical thickness for the silver of 60 nm. With that fixed, we refine the titania thicknesses to equalize as far as possible the polarization splitting, rather than assure an exact reflectance or transmittance value. The results are shown in Figure 5.14. With a simple coating, it is difficult to do better.

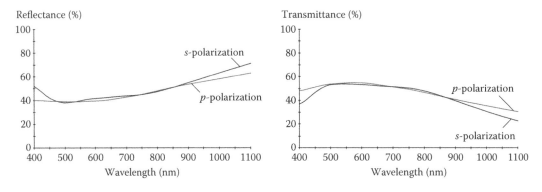

FIGURE 5.14
Reflectance and transmittance at 45° of the cube beam splitter based on silver. The coating design is Glass | 37.5 nm TiO$_2$ | 60.0 nm Ag | 80.6 nm TiO$_2$ | Glass.

5.2.2 Beam Splitters Using Dielectric Layers

There are many optical instruments where the light undergoes a transmission followed by a reflection, or vice versa, both at the same, or at the same type of, beam splitter. In two-beam interferometers, for example, the beams are first of all separated by one pass through a beam splitter and then combined again by a further pass either through the same beam splitter, as in the Michelson interferometer, or through a second beam splitter, as in the Mach–Zehnder interferometer. The effective transmittance of the instrument is given by the product of the transmittance and the reflectance of the beam splitter, taking into account the particular polarization involved. For a perfect beam splitter, the product TR would be 0.25; for most simple metallic beam splitters, it is around 0.08 or 0.10. The absorption in the film is the primary source of loss.

A beam splitter of improved performance, as far as the TR product is concerned, can be obtained by replacing the metallic layer with a transparent high-index quarter wave. At normal incidence, the reflectance of a quarter wave is given by

$$R = \left[\frac{y_0 - y_1^2/y_{\text{sub}}}{y_0 + y_1^2/y_{\text{sub}}}\right]^2 = \left[\frac{n_0 - n_1^2/n_{\text{sub}}}{n_0 + n_1^2/n_{\text{sub}}}\right]^2.$$

At a 45° angle of incidence, the position of the peak is shifted to a shorter wavelength, and the appropriate optical admittances must be used in calculating peak reflectance:

$$R = \left[\frac{\eta_0 - \eta_1^2/\eta_{\text{sub}}}{\eta_0 + \eta_1^2/\eta_{\text{sub}}}\right]^2,$$

and since η varies with the plane of polarization, R will have two values, R_s and R_p.

Figure 5.15 shows the peak reflectance of a quarter wave of index between 1.0 and 3.0 on glass of index 1.52 for both 45° incidence and normal incidence. At 45°, the peak reflectance for unpolarized light, $\frac{1}{2}(R_s + R_p)$, is within 1.5% of the peak value for normal incidence.

Zinc sulfide, with an index of 2.35, is a popular material for beam splitters. At 45°, we have

$$(TR)_s = (0.46 \times 0.54) = 0.248,$$

$$(TR)_p = (0.185 \times 0.815) = 0.151,$$

and

$$(TR)_{\text{unpolarized}} = \frac{1}{2}(0.248 + 0.151) = 0.200.$$

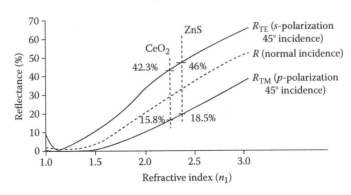

FIGURE 5.15
Peak reflectance in air of a quarter wave of index on glass of index n_1 on glass of index 1.52 at normal and 45° incidence.

$(TR)_{\text{unpolarized}}$ cannot be calculated using $T_{\text{mean}}R_{\text{mean}}$ because the light, after having undergone one reflection or transmission, is then partly polarized.

If a more robust film is required, titanium dioxide or cerium dioxide is a good choice. Cerium dioxide, with an index approximately 2.25, gives

$$(TR)_s = (0.423 \times 0.577) = 0.244,$$

$$(TR)_p = (0.158 \times 0.842) = 0.133,$$

$$(TR)_{\text{unpolarized}} = 0.189.$$

Clearly, the dielectric beam splitter, even if it does tend to have characteristics that more nearly correspond to 70/30 rather than 50/50, has a considerably better performance than the metallic beam splitter. The transmittance curve of a typical 70/30 beam splitter in Figure 5.16 shows how the transmittance varies on either side of the peak.

Beam splitters with 55/45 characteristics can be made by evaporating pure titanium in a good vacuum and subsequently oxidizing it to TiO_2 by heating at 420°C in air at atmospheric pressure. The titanium oxide thus formed has rutile structure and a refractive index of 2.8. Titanium films produced in a poor vacuum subsequently oxidize to the anatase form, having rather lower refractive index. The production of very large beam splitters of this type, 17 × 13 in., is described in a paper by Holland et al. [25].

The single-layer beam splitter suffers from a fall in reflectance on either side of the central wavelength. Just as single-layer antireflection coatings can be broadened by adding a half-wave layer, so the single quarter-wave beam splitter can be broadened. The same basic pattern of admittance circles can be achieved either by a low-index half-wave layer between the high-index quarter wave and the glass substrate or an even higher-index half wave deposited over the quarter wave. Since no suitable materials for the latter solution exist in practice, the low-index half wave is the only feasible approach. The admittance diagram is shown in Figure 5.17, and the performance, in Figure 5.18. The technique is also effective for multilayer systems to give a higher reflectance. Approximately 50% reflectance can be achieved by a four-layer coating, Air|*LHLH*|Glass, and this can be flattened by an additional low-index half wave at the glass end of the multilayer, that is, Air| *LHLHLL*|Glass. Figure 5.18 also shows the performance calculated for this design of beam splitter. A detailed discussion of the role of half-wave layers was given by Knittl [26]. Figure 5.19 shows the polarization splitting that occurs at incident angle 45°. Whether or not this is acceptable will very much depend on the application.

As mentioned earlier, beam-splitting cubes must be used in some applications where plate beam splitters are unsuitable. Unfortunately, the main problem connected with dielectric beam splitters, the low reflectance for *p*-polarization, becomes even worse with cube beam splitters. The reason for this is simply that 45° incidence in glass is effectively a much greater angle of incidence than 45°

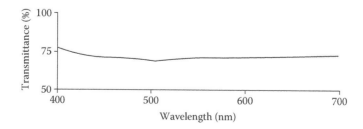

FIGURE 5.16
Measured transmittance curve of a dielectric 70/30 beam splitter at 45° angle of incidence. (Courtesy of Sir Howard Grubb, Parsons and Co. Ltd., Newcastle upon Tyne.)

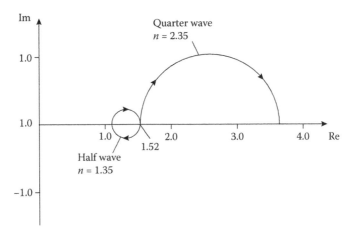

FIGURE 5.17
Admittance diagram at λ_0 of a two-layer beam splitter. The high-index quarter-wave layer gives the required high reflectance. The low-index half-wave layer flattens the performance over the visible region.

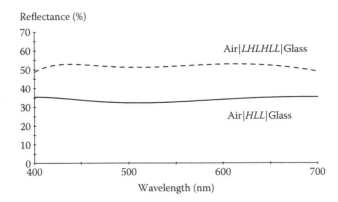

FIGURE 5.18
Performance at normal incidence of the beam splitter shown in Figure 5.17. The design is Air (1.00) | HLL | Glass (1.52) with L a quarter wave of index 1.35 and H of 2.35 together with the performance of a beam splitter of design Air (1.00) | LHLHLL | Glass (1.52) with the same indices.

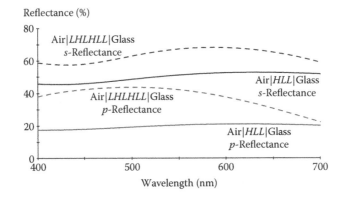

FIGURE 5.19
Performance of the beam splitters of Figure 5.18 at an angle of incidence 45°. Considerable polarization splitting is evident.

in air, and it is becoming closer to the Brewster angle. Consequently, for simple designs, the polarization splitting is even greater and the performance becomes so poor that the beam splitter is unusable in most applications. Metal layers are, therefore, required for straightforward cube beam splitters and combiners. This disadvantage of the dielectric layer can, however, be turned to an advantage by the construction of polarizers as we shall see in Chapter 9.

If the wavelength range is considerably restricted, then it is possible to achieve reasonable polarization performance with dielectric layers, although the designs are rather complex, and we will return to this in a later chapter on tilted performance. If we forget about practical issues connected with complex designs and think of theory, then we can see that with only two dielectric materials, there is little flexibility in the design of those coatings at oblique incidence, where a given performance is required for both polarizations. The p- and s-phase thicknesses are equal, and except for the outer surfaces of the coating where it contacts the two surrounding media, there is essentially only one kind of interface. The problem can be very slightly eased by adopting three, rather than two, materials, because now we have three different kinds of interfaces. This does not convert an impossible problem into an easy one. It simply converts virtually impossible into very difficult. Published designs achieved by powerful computer synthesis do exist but are, as we would expect, exceedingly complex. A useful paper principally dealing with a two-material plate beam splitter that not only is nonpolarizing over a very limited spectral region, but which also includes a useful commentary on the whole question of dielectric nonpolarizing beam splitters, is that by Ciosek et al. [27].

5.3 Neutral-Density Filters

A filter that is intended to reduce the power of an incident beam of light evenly over a wide spectral region is known as a neutral-density filter.

The performance of neutral-density filters is usually defined in terms of the optical density D:

$$D = \log_{10}(I_0/I_T) = -\log_{10}T,$$

where I_0 is the incident irradiance, I_T the transmitted irradiance, and T is given in a scale of 0–1.

Absorption and *absorptance* are terms not correctly used of neutral-density filters because they represent the fraction of energy actually absorbed in the film, and in neutral-density filters, a proportion of the incident energy is removed by reflection.

The advantage of using the logarithmic term if that the effect of placing two or more neutral-density filters in series is easily calculated. The overall density is simply the sum of the individual densities (provided that multiple reflections are not permitted to occur between the individual filters). Such reflections would affect the result in the way shown in Section 2.14.

Thin-film neutral-density filters consist of single metallic layers with thicknesses chosen to give the correct transmission values. Rhodium, palladium, tungsten, chromium, as well as other metals are all used to some extent, but very good performance is obtained by the evaporation of a nickel–chromium alloy, approximately 80% nickel and 20% chromium. Chromel A or Nichrome are trade names of standard resistance wires having suitable composition that can be readily obtained. The method was described by Banning [28]. Chromel or Nichrome should be evaporated at 1.33×10^{-2} Pa (1.33×10^{-4} mbar or 10^{-4} Torr) or better, from a thick tungsten spiral. Neutral films, having densities of up to around 1.5, corresponding to a transmittance of 3%, can be manufactured in this way. If the films are made thicker, they are not as neutral and tend to have higher transmittance in the red, owing to excess chromium. The films are very robust and do not need any protection, especially if they are heated to around 200°C after evaporation.

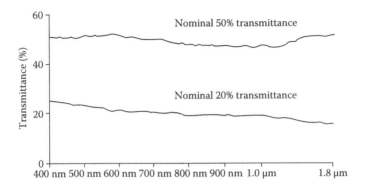

FIGURE 5.20

Measured transmittance curves of neutral density filters consisting of Nichrome films on glass substrates. (Courtesy of Sir Howard Grubb, Parsons and Co. Ltd., Newcastle upon Tyne.)

Figure 5.20 shows some response curves of neutral-density filters made from Nichrome on glass. The filters are reasonably neutral over the visible and near infrared out to 2 μm. In fact, if quartz substrates are used, the filters will be good over the range of 0.24–2 μm.

References

1. G. Hass. 1955. Filmed surfaces for reflecting optics. *Journal of the Optical Society of America* 45:945–952.
2. J. Strong. 1936. The evaporation process and its application to the aluminizing of large telescope mirrors. *Astrophysical Journal* 83:401–423.
3. M. Auwärter. 1939. Rhodium mirrors for scientific purposes. *Journal of Applied Physics* 10:705–710.
4. K. C. Park. 1964. The extreme values of reflectivity and the condition for zero reflection from thin dielectric films on metal. *Applied Optics* 3:877–881.
5. G. Hass and L. Hadley. 1972. Optical constants of metals. In *American Institute of Physics Handbook*, D. E. Gray (ed.), pp. 6.124–6.156. New York: McGraw-Hill.
6. J. T. Cox, G. Hass, and W. R. Hunter. 1975. Infrared reflectance of silicon oxide and magnesium fluoride protected aluminum mirrors at various angles of incidence from 8 μm to 12 μm. *Applied Optics* 14:1247–1250.
7. G. Hass, J. B. Heaney, H. Herzig et al. 1975. Reflectance and durability of Ag mirrors coated with thin layers of Al_2O_3 plus reactively deposited silicon oxide. *Applied Optics* 14:2639–2644.
8. S. F. Pellicori. 1974. Private communication (Santa Barbara Research Center, Goleta, CA).
9. J. T. Cox and G. Hass. 1978. Protected Al mirrors with high reflectance in the 8-12-μm region from normal to high angles of incidence. *Applied Optics* 17:2125–2126.
10. D. Y. Song, R. W. Sprague, H. A. Macleod et al. 1985. Progress in the development of a durable silver-based high-reflectance coating for astronomical telescopes. *Applied Optics* 24:1164–1170.
11. M. R. Jacobson, R. C. Kneale, F. C. Gillett et al. 1998. Development of silver coating options for the Gemini 8-m telescopes project. *Proceedings of SPIE* 3352:477–502.
12. M. Boccas, T. Vucina, C. Araya et al. 2006. Protected-silver coatings for the 8-m Gemini telescope mirrors. *Thin Solid Films* 502:275–280.
13. T. Schneider. 2016. Coating the Gemini telescopes with protected silver. *Society of Vacuum Coaters Bulletin* 16 (Issue: Fall):24–28.
14. F. A. Jenkins. 1958. Extension du domaine spectral de pouvoir réflecteur élevé des couches multiples diélectriques. *Journal de Physique et le Radium* 19:301–306.

15. R. P. Madden. 1963. Preparation and measurement of reflecting coatings for the vacuum ultraviolet. In *Physics of Thin films*, vol. 1, G. Hass (ed.), pp. 123–186. New York: Academic Press.
16. G. Hass and W. R. Hunter. 1978. The use of evaporated films for space applications—Extreme ultraviolet astronomy and temperature control of satellites. In *Physics of Thin Films*, G. Hass and M. H. Francombe (eds), pp. 71–166. New York: Academic Press.
17. G. Hass and R. Tousey. 1959. Reflecting coatings for the extreme ultraviolet. *Journal of the Optical Society of America* 49:593–602.
18. L. R. Canfield, G. Hass, and J. E. Waylonis. 1966. Further studies on MgF_2-overcoated aluminium mirrors with highest reflectance in the vacuum ultraviolet. *Applied Optics* 5:45–50.
19. J. T. Cox, G. Hass, and J. E. Waylonis. 1968. Further studies on LiF-overcoated aluminium mirrors with highest reflectance in the vacuum ultraviolet. *Applied Optics* 7:1535–1539.
20. O. S. Heavens. 1955. *Optical Properties of Thin Solid Films*. London: Butterworth Scientific Publications.
21. H. Pohlack. 1953. Beitrag zur Optik dünnster Metallschichten. In *Jenaer Jahrbuch*, pp. 241–245. Jena: Gustav Fischer Verlag.
22. E. D. Palik, ed. 1985. *Handbook of Optical Constants of Solids I*. Orlando, FL: Academic Press.
23. I. N. Shkliarevskii and A. A. Avdeenko. 1959. Increasing the transparency of metallic coatings. *Optics and Spectroscopy* 6:439–443.
24. H. Anders. 1965. *Dünne Schichten für die Optik* (English translation: *Thin Films in Optics*. Waltham, MA: Focal Press, 1967). Stuttgart: Wissenschaftliche Verlagsgesellschaft mbH.
25. L. Holland, K. Hacking, and T. Putner. 1953. The preparation of titanium dioxide beam-splitters of large surface area. *Vacuum* 3:159–161.
26. Z. Knittl. 1976. *Optics of Thin Films*. London: John Wiley & Sons and SNTL.
27. J. Ciosek, J. A. Dobrowolski, G. A. Clarke et al. 1999. Design and manufacture of all-dielectric nonpolarizing beam splitters. *Applied Optics* 38:1244–1250.
28. M. Banning. 1947. Neutral density filters of Chromel A. *Journal of the Optical Society of America* 37:686–687.

6

Multilayer High-Reflectance Coatings

There is no doubt that the broadest and simplest high-reflectance coatings are the front-surface metallic mirrors. A thin film of aluminum, for example, gives high reflectance from the extreme ultraviolet to the far infrared and beyond. Why, then, should we need anything else? Unfortunately, even metals such as silver, with highest possible performance, exhibit losses of at least several percent, and there are many applications where such levels of loss are completely unacceptable. There are also many applications where what is not reflected must be transmitted rather than absorbed. Interference structures using dielectric materials are the answer, but although interference can achieve incredible levels of performance over limited wavelength regions, the wide performance characteristic of metal layers is beyond it. We need both.

Our knowledge of the properties of quarter and half waves guides us to a qualitative understanding of the structures that lead to an interference-induced high reflectance. In an assembly of quarter-wave layers of alternate high and low indices, the beams reflected from the various interfaces alternately suffer zero and 180° phase shifts. The double traversal of each layer gives an additional half-wave path difference, so that all the reflected beams will be in phase, leading to strong constructive interference. Such an assembly can also be thought of as a repeated pair of high- and low-index quarter waves. Should we retain the total half-wave thickness of the basic pair but vary the relative thicknesses of the two layers, we will retain the equal phase change suffered by the beams reflected at the front and rear interfaces of the pair. Since these beams will suffer an additional path difference of a wavelength on their double traversal of each pair, they will constructively interfere. The intermediate interfaces will also show a repeat spacing of a half wave, and, therefore, constructive interference, which, although not quite in phase with the other resultant, will nevertheless be sufficiently close to contribute to a further increase in reflectance. We can see, therefore, that a repeated pair with a total thickness of a half wave will also lead to high reflectance, although a pair of quarter waves will yield the maximum effect. We can also readily visualize that this increased reflectance will also occur at those shorter wavelengths where the optical thickness is an integral number of half waves, unless the members of the structure happen to be half waves and, therefore, absentees. In fact, these ideas tend to apply to even more complex repeated components.

Many of the earlier and significant papers on high-reflectance coatings were focused on the use of the coatings in multiple-beam interferometry. The description that follows begins, therefore, with the most successful of the multiple-beam interferometers, the Fabry–Perot interferometer. As we shall see later, this interferometer is of considerable importance in the development of thin-film bandpass filters.

6.1 Fabry–Perot Interferometer

First described in 1899 by Fabry and Perot [1], the interferometer known by their names has profoundly influenced the development of thin-film optics. It belongs to the class of interferometers known as multiple-beam interferometers because a large number of beams are involved in the interference. The theory of each of the various types of multiple-beam interferometer is similar.

They mainly differ in physical form. Their common feature is that their fringes are much sharper than those in two-beam interferometers, thus improving both measuring accuracy and resolution. Multiple-beam interferometers are described in almost all textbooks on optics, for example, that by Born and Wolf [2].

The heart of the Fabry–Perot interferometer is the etalon consisting of two flat plates separated by a distance d_s and aligned so that they are parallel to a very high degree of accuracy. The separation is usually maintained by a spacer ring normally made of Invar or quartz. The inner surfaces of the two plates are generally coated to enhance their reflectance.

Figure 6.1 shows an etalon in diagrammatic form. The amplitude reflection and transmission coefficients are defined as shown. The basic theory has already been given in Chapter 3 (Equation 3.20), where it was shown that the transmittance for a plane wave is given by

$$T = \frac{T_a T_b}{\left[1 - (R_a^- R_b^+)^{1/2}\right]^2} \left[1 + \frac{4(R_a^- R_b^+)^{1/2}}{\left[1 - (R_a^- R_b^+)^{1/2}\right]^2} \sin^2\left(\frac{\varphi_a + \varphi_b}{2} - \delta\right)\right]^{-1}, \tag{6.1}$$

where $\delta = (2\pi n_s d_s \cos\vartheta_s)/\lambda$; d_s and n_s are the physical thickness and refractive index of the spacer layer, respectively; and T and R indicate the appropriate transmittance and reflectance, respectively. This is similar to Equation 3.20 except that δ has been modified to include oblique incidence ϑ_s. In order to simplify the discussion, let the reflectances and transmittances of the two surfaces be equal and let there be no phase change on reflection, i.e., let $\varphi_a = \varphi_b = 0$. Then,

$$T = \frac{T_s^2}{(1 - R_s)^2} \frac{1}{1 + \left[4R_s/(1 - R_s)^2\right]\sin^2\delta}, \tag{6.2}$$

and we can write

$$F = \frac{4R_s}{(1 - R_s)^2}. \tag{6.3}$$

Then,

$$T = \frac{T_s^2}{(1 - R_s)^2} \frac{1}{1 + F\sin^2\delta}. \tag{6.4}$$

If there is no absorption in the reflecting layers, then

$$1 - R_s = T_s$$

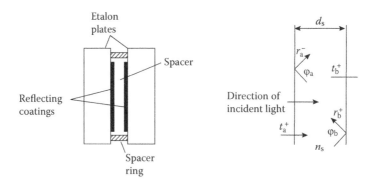

FIGURE 6.1
Fabry–Perot etalon. The amplitude coefficients in the diagram are converted to the coefficients of irradiance in Equation 6.1 as shown in Equation 3.20.

and

$$T = \frac{1}{1 + F\sin^2\delta}.$$ (6.5)

The form of this function is shown in Figure 6.2, where T is plotted against δ. T is a maximum for $\delta = m\pi$, where $m = 0, \pm 1, \pm 2, \ldots$, and a minimum halfway between these values. The successive peaks of T are multiple-beam fringes, and m is the order of the appropriate fringe. As F increases, the widths of the fringes become very much narrower. The ratio of the separation of adjacent fringes to the halfwidth (the fringe width measured at half the peak transmittance) is called the *finesse* of the interferometer and is written as $\mathcal{F}$. From Equation 6.5, the value of δ corresponding to a transmittance of half the peak value is given by

$$0.5 = \frac{1}{1 + F\sin^2\delta},$$

and if δ is sufficiently small so that we can replace $\sin^2\delta$ by δ^2, then

$$\delta = \frac{1}{F^{1/2}},$$

which is *half* the width of the fringe. The separation between values of δ representing successive fringes is π, so that

$$\mathcal{F} = \frac{\pi F^{1/2}}{2} = \frac{\pi R_s^{1/2}}{(1 - R_s)}.$$ (6.6)

The Fabry–Perot interferometer is principally used for the examination of the fine structure of spectral lines. The fringes are produced by passing light from the source in question through the interferometer. Measurement of the fringe pattern as a function of the physical parameters of the etalon can yield very precise values of the wavelengths of the various components of the line. The two most common arrangements are either to have the incident light highly collimated and incident normally, or at some constant angle, when the fringes can be scanned by varying the spacer thickness, or it is possible to keep the spacer thickness constant and scan the fringes by varying ϑ_s, the angle of incidence. Possible arrangements corresponding to these two methods are shown in Figure 6.3.

Practical considerations limit the achievable finesse to a maximum, normally of around 25, or perhaps 50 in exceptional cases. This is mainly due to imperfections in the plates themselves. It is extremely difficult to manufacture a plate with flatness better than $\lambda/100$ at, say, 546 nm. Variations in the flatness of the plates give rise to local variations in d_s and, hence, δ, causing the fringes to shift. These variations should not be greater than the fringe width; otherwise, the luminosity of the instrument will suffer. Chabbal [3] considered this problem in great detail, but for our present

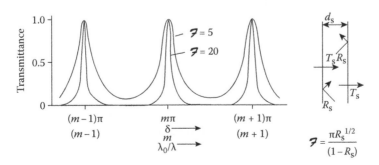

FIGURE 6.2
Fabry–Perot fringes.

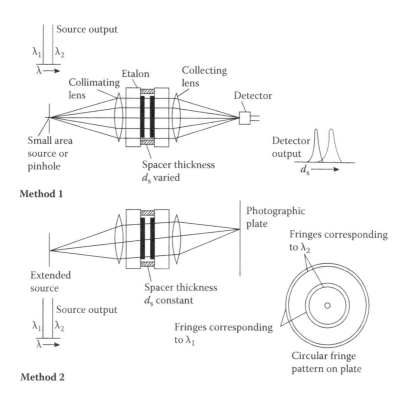

FIGURE 6.3
Two possible arrangements of a Fabry–Perot interferometer.

purpose, it is sufficient to assume that for a pair of $\lambda/100$ plates (i.e., having errors not greater than $\pm\lambda/200$ about the mean), the variation in the thickness of the spacer layer will be on the order of $\pm\lambda/100$ about the mean. This will occur when the defects in the plates are in the form of either spherical depressions in both plates or else protrusions. This, in turn, means a change in δ of $\pm 2\pi/100$ corresponding to a total excursion of $2\pi/50$. Any decrease in fringe width below this will not increase the resolution of the system but merely reduce the overall luminosity, so that $2\pi/50$ represents a lower limit on the fringe width. Since the interval between fringes is π, this condition is equivalent to an upper limit on finesse of $\pi/(2\pi/50)$, i.e., 25. In more general terms, if the plates are good enough to limit the total thickness variation in the space to λ/p (not quite the same as saying that each plate is good to λ/p), then the finesse should be not greater than $p/2$.

The resolving power of an optical instrument is normally determined by the Rayleigh criterion that is particularly concerned with power distributions of the form

$$I(\delta) = \left[\frac{\sin(\delta/2)}{\delta/2}\right]^2 I_{max},\qquad(6.7)$$

which are of a type produced by diffraction rather than interference effects. Two wavelengths are considered just resolved by the instrument if the power maximum of one component falls exactly over the first intensity zero of the other component. This implies that if the two components are of equal power, then, in the combined fringe pattern, the minimum that will exist between the two maxima will be of a power $8/\pi^2$ times that at either of them. In the Fabry–Perot interferometer, the fringes are of rather different form, and the pattern of zeros and successively weaker maxima associated with the Equation 6.7 function is missing. The Rayleigh criterion cannot, therefore, be applied directly. Born and Wolf [2] suggested that a suitable alternative form of the criterion that could be applied in this case might be that two equally intense lines are just resolved when the resultant response between the peaks in the combined fringe pattern is $8/\pi^2$ of either peak. On this basis, they showed that the resolving power of the Fabry–Perot interferometer is

$$\frac{\lambda}{\Delta\lambda} = 0.97m\mathscr{F},$$

which is virtually indistinguishable from

$$\frac{\lambda}{\Delta\lambda} = m\mathscr{F} \tag{6.8}$$

and which is the ratio of the peak wavelength of the appropriate order to the halfwidth of the fringe. Thus, the halfwidth of the fringe is a most useful parameter because it is directly related to the resolution of the instrument in a most simple manner. We shall make much use of the concept of halfwidth in Chapter 8.

Since resolving power is the product of finesse and order number, a low finesse does not necessarily mean low resolution, but it does mean that to achieve high resolution, the interferometer must be used in high order. This, in turn, means that the separation of neighboring orders in terms of wavelength is small—in high order, this is approximately given by λ/m. If steps are not taken to limit the range of wavelengths accepted by the interferometer, then the interpretation of the fringe patterns becomes impossible. This limiting of the range can be achieved by using a suitable filter in series with the etalon. This filter could be a thin-film filter of a type discussed in Chapter 8. Another method is to use, in series with the etalon, other etalons of lower order, and hence resolving power, arranged so that the fringes coincide only at the wavelength of interest and at wavelengths very far removed. The wide fringe interval, or, as it is also called, free spectral range, of the low-order, low-resolution instrument is thus combined with the high resolution and narrow free spectral range of the high-order instrument. A simpler and more convenient method, probably that most often employed, involves a spectrograph and is generally used in conjunction with the second method of scanning the interferometer: variation in ϑ_s keeping d_s constant. The resolving power of the spectrograph need not be high, and the entrance slit can be quite broad. It is usually placed where the photographic plate is in Figure 6.3, so that it accepts a broad strip down the center of the circular fringe pattern. The plate from the spectrograph then shows a low-resolution spectrum with a fringe pattern along each line corresponding to the fine-structure components within the line.

So far in our examination of the Fabry–Perot interferometer, we have neglected to consider absorption in the reflecting coatings. Equation 6.4 contains the information we need:

$$T = \frac{T_s^2}{(1 - R_s)^2} \frac{1}{1 + F\sin^2\delta}.$$

Let A_s be the absorptance of each coating. Then,

$$1 = R_s + T_s + A_s.$$

So that Equation 6.4 becomes

$$T = \frac{T_s^2}{(T_s + A_s)^2} \frac{1}{1 + F\sin^2\delta},$$

i.e.,

$$T = \frac{1}{(1 + A_s/T_s)^2} \frac{1}{1 + F\sin^2\delta}. \tag{6.9}$$

Clearly, the all-important parameter is A_s/T_s.

Curves are shown in Figure 6.4 connecting the transmittance of the etalon with finesse, given the absorptance of the coatings. It is possible on this diagram to plot the performance of any type of

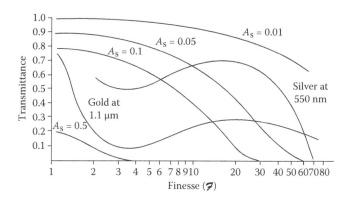

FIGURE 6.4
Etalon transmittance against finesse for various values of absorptance of the coatings (on a scale of 0–1).

coating if the way in which R_s, T_s, and A_s vary is known. This has been done for silver layers at 550 nm and gold at 1.1 μm. The figures from which these curves were plotted were taken from Mayer's book [4]. Other sources of information, particularly on silver films, are available [5–7], and results may differ from those plotted in some respects. However, the curves are adequate for their primary purpose, which is to show that the performance of silver, the best metal of all for the visible and near infrared, begins to rapidly fall off beyond a finesse of 20 and is inadequate for the very best interferometer plates. An enormous improvement is possible with all-dielectric multi-layer coatings.

6.2 Multilayer Dielectric Coatings

First of all, let us examine the simplest of the structures, a series of alternate high- and low-index quarter waves. An expression is given in Section 3.1 for the optical admittance of a series of quarter waves. If y_H and y_L are the admittances of the high- and low-index layers and if the stack is arranged so that the high-index layers are outermost at both sides, then

$$Y = \frac{y_H^{2p+2}}{y_L^{2p} y_m},$$
(6.10)

where y_m is the substrate admittance and $(2p + 1)$ is the number of layers in the stack.

The reflectance in a medium of admittance y_0 is then

$$R = \left[\frac{y_0 - y_H^{2p+2} / \left(y_L^{2p} y_m \right)}{y_0 + y_H^{2p+2} / \left(y_L^{2p} y_m \right)} \right]^2.$$
(6.11)

The greater the number of layers, the greater the reflectance. The incident medium admittance will usually be less than that of either material, and then the maximum reflectance for a given odd number of layers is obtained with the high-index layers outermost.

If the incident admittance is unity and there is no absorption,

$$\frac{y_H^{2p+2}}{y_L^{2p} y_m} \gg 1,$$

then

$$R \simeq 1 - 4\frac{y_L^{2p}y_m}{y_H^{2p+2}}$$

and

$$T \simeq 4\frac{y_L^{2p}y_m}{y_H^{2p+2}}, \tag{6.12}$$

which shows that when reflectance is high, then the addition of two extra layers reduces the transmittance by a factor of $(y_L/y_H)^2$.

Provided that the materials are transparent, the absorption in a multilayer stack can be made very small indeed. We shall return later to this topic, but we can note here that in the visible region of the spectrum, the absorptance can readily be considerably less than 0.01%.

Dielectric multilayers, however, do suffer from what might be considered as two defects. The first, more of a complication than a fault, is that there is a variable change in phase associated with the reflection. The second, more serious, is that the high reflectance is obtained over a limited range of wavelengths only.

We can see, qualitatively, how the phase shift varies, using the admittance diagram. If, as is usual, the multilayer consists of an odd number of layers with high-index layers on the outside, then at the outer surface of the final layer, the admittance will be on the real axis with a high positive value. This is diagrammatically shown in Figure 6.5. The quadrants are marked in the figure in the same way as done in Figure 3.1b. Clearly, the phase shift associated with the coating is π at the reference wavelength for which all the layers are quarter waves. For slightly longer wavelengths, the circles slightly shrink from the semicircles associated with the quarter waves, and so the terminal point of the locus moves upward into the region associated with the third quadrant. If the wavelength decreases, the terminal point moves into the second quadrant. The phase shift, therefore, increases with wavelength. If, on the other hand, the coating ends with a quarter wave of low-index material so that at the reference wavelength the admittance is real, but less than unity, then the phase shift on reflection will be zero, moving into the first quadrant as the wavelength increases or into the fourth as it decreases. In both cases, the phase shift on reflection increases with wavelength.

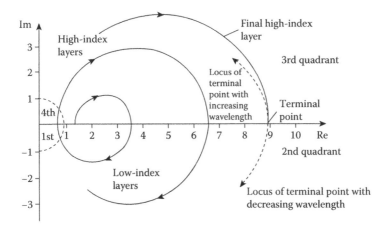

FIGURE 6.5
Admittance diagram for a quarter-wave stack ending with a high-index layer. The quadrants for the phase shift on reflection φ are marked on the diagram and correspond to those in Figure 3.1b. For decreasing wavelength, the terminal point moves into the region associated with values of φ in the second quadrant, while, for increasing wavelength, φ moves into the third quadrant.

The behavior of a typical quarter-wave stack is shown in Figure 6.6 [8]. The high-reflection zone can be seen to be limited in extent. On either side of a plateau, the reflectance abruptly falls to a low, oscillatory value. The addition of extra layers does not affect the width of the zone of high reflectance, but increases the reflectance within it and the number of oscillations outside. In fact, the number of extrema between 0° and 90°, including the reflectance peak, but not counting that at 0°, is equal to the number of quarter waves in the structure.

The width of the high-reflectance zone can be computed using the following method. If a multilayer consists of q repetitions of a fundamental period consisting of two, three, or indeed any number of layers, then the characteristic matrix of the multilayer is given by

$$[\mathcal{M}] = [M]^q,$$

where $[M]$ is the matrix of the fundamental period. Let $[M]$ be written as

$$\begin{bmatrix} M_{11} & M_{12} \\ M_{21} & M_{22} \end{bmatrix}.$$

Then it can be shown that for wavelengths which satisfy

$$\left| \frac{M_{11} + M_{22}}{2} \right| \geq 1, \tag{6.13}$$

the reflectance steadily increases with increasing number of periods. This is therefore the condition that a high-reflectance zone should exist, and the boundaries are given by

$$\left| \frac{M_{11} + M_{22}}{2} \right| = 1. \tag{6.14}$$

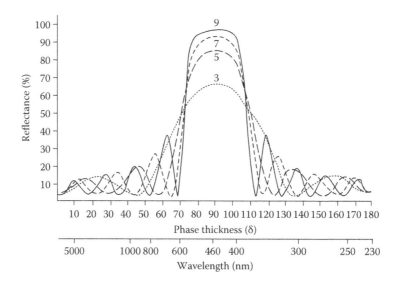

FIGURE 6.6
Reflectance R for normal incidence of alternating $\lambda/4$ layers of high-index (n_H = 2.3) and low-index (n_L = 1.38) dielectric materials on a transparent substrate (n_m = 1.52) as a function of the phase thickness $\delta = 2\pi nd/\lambda$ (*upper scale*) or the wavelength λ for λ_0 = 460 nm (*lower scale*). The number of layers is shown as a parameter on the curves. (With kind permission from Springer Science+Business Media: *Zeitschrift für Physik*, Fabry–Perot-Interferometerverspiegelungen aus dielektrischen Vielfachschichten, 142, 1955, 21–41, S. Penselin and A. Steudel.)

A rigorous proof of this result is somewhat involved. One version was given by Born and Wolf [2], and another, by Welford [9]. A justification of the result, rather than a proof, was given by Epstein [10], and it is his method that is followed here.

If the characteristic matrix of a thin-film assembly on a substrate of admittance y_m is given by

$$\begin{bmatrix} B \\ C \end{bmatrix},$$

then if y_m is real, Equation 2.125 shows that

$$T = \frac{4y_0 y_m}{(y_0 B + C)(y_0 B + C)^*} = \frac{4y_0 y_m}{|y_0 B + C|^2},$$

where y_0 is the admittance of the incident medium. Let the characteristic matrix of the assembly of thin films be, as mentioned earlier,

$$[\mathcal{M}] = \begin{bmatrix} \mathcal{M}_{11} & \mathcal{M}_{12} \\ \mathcal{M}_{21} & \mathcal{M}_{22} \end{bmatrix}.$$

Then,

$$\begin{bmatrix} B \\ C \end{bmatrix} = \begin{bmatrix} \mathcal{M}_{11} & \mathcal{M}_{12} \\ \mathcal{M}_{21} & \mathcal{M}_{22} \end{bmatrix} \begin{bmatrix} 1 \\ y_{sub} \end{bmatrix} = \begin{bmatrix} \mathcal{M}_{11} + y_{sub}\mathcal{M}_{12} \\ y_{sub}\mathcal{M}_{22} + \mathcal{M}_{21} \end{bmatrix},$$

giving

$$T = \frac{4y_0 y_m}{|y_0(\mathcal{M}_{11} + y_m\mathcal{M}_{12}) + y_m\mathcal{M}_{22} + \mathcal{M}_{21}|^2}.$$

If there is no absorption, $\mathcal{M}_{11}$ and $\mathcal{M}_{22}$ are real and $\mathcal{M}_{12}$ and $\mathcal{M}_{21}$ are imaginary. Then,

$$T = \frac{4y_0 y_m}{|y_0\mathcal{M}_{11} + y_m\mathcal{M}_{22}|^2 + |y_0 y_m\mathcal{M}_{12} + \mathcal{M}_{21}|^2}. \tag{6.15}$$

In the absence of the multilayer, the transmittance of the substrate will be

$$T_m = \frac{4y_0 y_m}{(y_0 + y_m)^2}. \tag{6.16}$$

To simplify the discussion, let $y_0 = y_m$. Then, from Equations 6.15 and 6.16, T will be less than T_m if

$$\left| \frac{\mathcal{M}_{11} + \mathcal{M}_{22}}{2} \right| > 1,$$

regardless of the values of $\mathcal{M}_{12}$ and $\mathcal{M}_{21}$. Now if

$$\left| \frac{M_{11} + M_{22}}{2} \right| > 1,$$

where $[M]$ is the matrix of the fundamental period in the multilayer, then, generally, as the number of periods increases, that is, as q tends to infinity,

$$\left| \frac{\mathcal{M}_{11} + \mathcal{M}_{22}}{2} \right| \rightarrow \infty.$$

That this is plausible may be seen by first of all squaring [M], whence, writing N_{ab} for the terms in $[M]^2$,

$$N_{11} + N_{22} = M_{11}^2 + 2M_{12}M_{21} + M_{22}^2.$$

Since det[M] = 1,

$$2M_{12}M_{21} = 2M_{11}M_{22} - 2,$$

so that

$$N_{11} + N_{22} = M_{11}^2 + 2M_{11}M_{22} + M_{22}^2 - 2 = (M_{11} + M_{22})^2 - 2.$$

Let

$$\left| \frac{M_{11} + M_{22}}{2} \right| = 1 + \varepsilon.$$

Then,

$$N_{11} + N_{22} = (2 + 2\varepsilon)^2 - 2 = 2 + 8\varepsilon + 4\varepsilon^2,$$

so that by squaring [M] and resquaring the result and so on, it can be seen that

$$\left| \frac{\mathcal{M}_{11} + \mathcal{M}_{22}}{2} \right| \rightarrow \infty \quad \text{as } q \rightarrow \infty.$$

The quarter-wave stack, which we have so far been considering, consists of a number of two-layer periods, together with one extra high-index layer. Each period has a characteristic matrix

$$[M] = \begin{bmatrix} \cos \delta & (i \sin \delta)/y_L \\ iy_L \sin \delta & \cos \delta \end{bmatrix} \begin{bmatrix} \cos \delta & (i \sin \delta)/y_H \\ iy_H \sin \delta & \cos \delta \end{bmatrix}.$$

Since the two layers are of equal optical thickness, δ without any subscript has been used for phase thickness

$$\frac{M_{11} + M_{22}}{2} = \cos^2 \delta - \frac{1}{2} \left(\frac{y_H}{y_L} + \frac{y_L}{y_H} \right) \sin^2 \delta.$$

The right-hand side of this expression cannot be greater than +1, and so to find the boundaries of the high-reflectance zone, we must set

$$-1 = \cos^2 \delta - \frac{1}{2} \left(\frac{y_H}{y_L} + \frac{y_L}{y_H} \right) \sin^2 \delta_e,$$

which, with some rearrangement, gives

$$\cos^2 \delta_e = \left(\frac{y_H - y_L}{y_H + y_L} \right)^2.$$

Now,

$$\delta = \frac{\pi}{2}\frac{\lambda_0}{\lambda} = \frac{\pi}{2}g,$$

where λ_0 is, as usual, the wavelength for which the layers have quarter-wave optical thickness and g is λ_0/λ. Let the edges of the high-reflectance zone be given by

$$\delta_e = \frac{\pi}{2}g_e = \frac{\pi}{2}(1 \pm \Delta g),$$

so that

$$\cos^2\delta_e = \sin^2\left(\pm\frac{\pi\Delta g}{2}\right)$$

and the width of the zone is $2\Delta g$. Then,

$$\Delta g = \frac{2}{\pi}\arcsin\left(\frac{y_H - y_L}{y_H + y_L}\right).$$

This shows that the width of the zone is a function of only the indices of the two materials used in the construction of the multilayer. The higher the ratio, the greater the width of the zone. Figure 6.7 shows Δg plotted against the ratio of refractive indices.

So far, we have considered only the fundamental reflectance zone for which all the layers are one-quarter of a wavelength thick. It is obvious that high-reflectance zones will exist at all wavelengths for which the layers are an odd number of quarter wavelengths thick. That is, if the center wavelength of the fundamental zone is λ_0, then there will also be high-reflectance zones with center wavelengths of $\lambda_0/3$, $\lambda_0/5$, $\lambda_0/7$, $\lambda_0/9$, and so on.

At wavelengths where the layers have optical thickness equivalent to an even number of quarter waves, which is the same as an integral number of half waves, the layers will all be absentee layers and the reflectance will be that of the uncoated substrate.

The analysis determining Δg for the fundamental zone is valid also for all higher-order zones so that the boundaries are given by

$$g_0 \pm \Delta g, \quad 3g_0 \pm \Delta g, \quad 5g_0 \pm \Delta g,$$

and so on. Higher-order reflectance curves are shown in Figure 6.8.

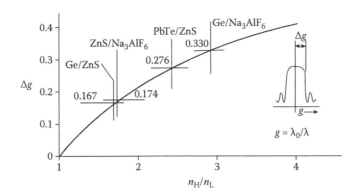

FIGURE 6.7
Width of the high-reflectance zone of a quarter-wave stack plotted against the ratio of the refractive indices n_H/n_L.

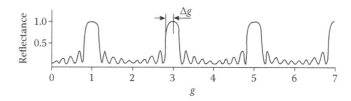

FIGURE 6.8

Reflectance of a nine-layer stack of zinc sulfide (n_H = 2.35) and cryolite (n_L = 1.35) on glass (n_m = 1.52) showing the high-reflectance bands.

The materials used in the visible region very much depend on the application. Zinc sulfide and cryolite is an old combination that is still often used. Although these materials present rather poorer environmental resistance than coatings based on oxides, they do possess some advantages. Both materials are easy to evaporate from simple thermal sources and give high optical performance even when evaporated onto a cold substrate. This means that the risk of distortion of very accurate interferometer plates through heating is eliminated. The layers are rather susceptible to attack by moisture, and care should be taken to avoid any condensation, such as what might happen when cold plates are exposed to a warmer atmosphere; otherwise, the coatings will be ruined. Touching by fingers is also to be avoided at all costs. The softness of the coatings can, however, be turned to an advantage. Etalon plates are extremely expensive, and if the coatings are easily removable, the plates can be recoated for use at other wavelengths. Prolonged soaking in warm water is often sufficient to bring zinc sulfide and cryolite coatings off. In cases where the coatings are not completely removed in this way, the addition of two or three drops of hydrochloric acid to the water will quickly complete the operation. This should obviously be done with great care, and the plates should be immediately rinsed in running water to avoid any risk of surface damage.

For more demanding applications, particularly where the coating may have to be exposed to a more aggressive environment, hard oxide layers would normally be chosen, silicon dioxide as low index and titanium dioxide, tantalum pentoxide, or niobium pentoxide as high-index materials. Hafnium oxide is frequently used as a high-index material when a high laser damage threshold is a requirement. Levels of absorption of less than 0.5% can be achieved with ease, 0.1% with some extra care and 0.001% with attention to detail. Still lower levels can be achieved and are required especially for the reflecting structures in more advanced narrowband filters, dealt with in a later chapter. In thermal evaporation, the oxide materials demand higher source temperatures, and the simple directly heated boat sources applicable to zinc sulfide and cryolite must be replaced by electron beam sources, described in more detail in a later chapter. Sputtering is an alternative process that has become very popular. Magnesium fluoride, the tough material much used in antireflection coatings for the visible region, has an attractive low index of refraction but does suffer from rather high intrinsic tensile stress and so can be a somewhat unreliable material in high-reflectance multilayers.

Zinc sulfide absorbs in the ultraviolet. In the 300–400 nm region, it can be replaced by antimony trioxide, which, with cryolite, can be evaporated onto a cold substrate from simple thermal sources. This combination should be handled at least as carefully as zinc sulfide and cryolite. Tougher materials are hafnium dioxide and silicon dioxide.

For the infrared, germanium for the region beyond 1.8 μm with an index around 4.0, or lead telluride for the region beyond 3.5 μm with an index around 5.7, are good high-index materials for filter applications. Both are useful beyond 12 μm. Zinc sulfide, with an index of 2.35, is a useful low-index material out to 20 μm. In the near infrared, silicon monoxide is frequently used as the low-index material accompanying germanium. Thorium fluoride has many desirable properties as a low-index material except that it is radioactive, and so its use is limited to those few applications where it is still necessary, notably high-power high-reflectance coatings for CO_2 laser applications, where it is frequently coupled with zinc selenide as a high-index material. Germanium, unfortunately, along with most high-index semiconductors, suffers from an extinction coefficient that significantly increases with temperature, and so it exhibits thermal runaway in high-power applications. Many fluorides, cerium fluoride and yttrium fluoride, for example, and mixtures of fluorides are also used as low-index materials out to around 12 μm. Materials are dealt with in much more detail later.

The losses experienced in the coatings are as much a function of the technique used as of the materials themselves. Great care in preparing the machine and substrates is needed. Everything should be scrupulously clean. Two papers, which will be found useful if the maximum performance is required, are by Perry [11] and Heitmann [12]. Both these authors are concerned with laser mirrors, where losses must be of an even lower order than in the case of the Fabry–Perot interferometer.

6.2.1 All-Dielectric Multilayers with Extended High-Reflectance Zones

The limited range over which high reflectance can be achieved with a quarter-wave stack is a difficulty in some applications, and there are ways of extending the range by complicating the design. Most of the techniques involve the staggering of the thicknesses of successive layers throughout the stack to form a regular progression, the aim being to ensure that at any wavelength in a fairly wide range, enough of the layers in the stack have optical thickness sufficiently near a quarter wave to give high reflectance. We begin by considering some historical attempts.

Penselin and Steudel [8] were probably the first workers to try this method. They produced a number of multilayers where the layer thicknesses were in a harmonic progression. Their best 13-layer results are shown as curve B in Figure 6.9. Baumeister and Stone [13] developed a simple computer-based technique to optimize their reflectors. Curve C in Figure 6.9 represents their best 15-layer design.

Heavens and Liddell [14] used a similar approach. By this time, computers had developed further, but the user still had to book time in advance and travel to the machine. They were able to compute a large number of reflection curves for assemblies of layers for which the thicknesses were in either arithmetic or geometric progression. With the same number of layers, the geometric progression gave very slightly broader reflection zones. In the computations, the high index was assumed to be 2.36 (zinc sulfide); the low index, 1.39 (magnesium fluoride); and the substrate

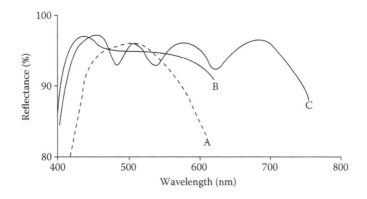

FIGURE 6.9
Broadband multilayer reflectors. A, computed curve for a seven-layer quarter-wave stack. B, measured reflectance of a 13-layer broadband design. (With kind permission from Springer Science+Business Media: *Zeitschrift für Physik*, Fabry–Perot-Interferometerverspiegelungen aus dielektrischen Vielfachschichten, 142, 1955, 21–41, S. Penselin and A. Steudel.) C, measured reflectance of a 15-layer alternative design. (From P. W. Baumeister and J. M. Stone, *Journal of the Optical Society of America*, 46, 228–229, 1956. With permission of Optical Society of America.)

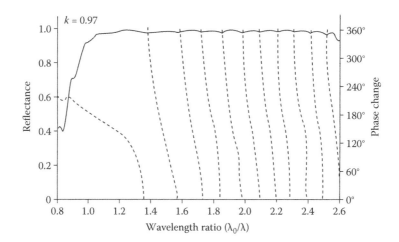

FIGURE 6.10
Reflectance of a 35-layer geometric stack on glass. Reflectance (*full curve*) and phase change on reflection (*broken curve*); $n_0 = 1.00$, $n_H = 2.36$, $n_L = 1.39$, $n_m = 1.53$, and common difference $k = 0.97$. (After O. S. Heavens and H. M. Liddell, *Applied Optics*, 5, 373–376, 1966. With permission of Optical Society of America.)

index, 1.53 (glass). Values of the common difference for the arithmetic progression ranged from −0.05 to +0.05 and, for the common ratio of the geometric progression, from 0.95 to 1.05. A 35-layer geometric curve is shown in Figure 6.10.

As in the case of antireflection coatings, computer refinement can be used to improve an initial, less satisfactory performance. Baumeister and Stone [13] and Baumeister [15] pioneered the use of this technique in optical thin films. By trial and error, they arrived at a preliminary 15-layer design with high reflectance over an extended range but with unacceptably large dips. The aim was to produce a reflectance of around 95% by using zinc sulfide ($n = 2.3$) and cryolite ($n = 1.35$) on glass (n not given), and the final result is shown as curve C in Figure 6.9 with design details listed in Table 6.1. Computer limitations forced the use of a very coarse net for the relaxation—only five points were involved—and arbitrary relationships between the various layers were used to reduce the number of independent variables to five. This was in 1956. The computer was described as an IBM card-programmed calculator. Since then, advances in the technique have kept pace with the increasing power of computers. By 1971, Pelletier et al. [16] had described the computer design and refinement of reflectors like that shown in Figure 6.11, where dispersion of the optical properties of the materials was included in the calculations.

A particularly simple method is to place a quarter-wave stack for one wavelength on top of another for a different wavelength. This process was considered in detail by Turner and Baumeister [17]. Unfortunately, if each stack consists of an odd number of layers with outermost layers of the same index, then a peak of transmission is found at the center of the high-reflectance zone. This peak arises because the two stacks act in much the same way as Fabry–Perot reflectors do. In a Fabry–Perot interferometer, as we have seen, provided that the reflectances and transmittances of the structures on either side of the spacer layer are equal in magnitude, the transmittance of the assembly will be unity for

$$\frac{\varphi_a + \varphi_b - 2\delta}{2} = q\pi,$$

where $q = 0, \pm1, \pm2, \ldots$.

The situation is sketched in Figure 6.12. The assembly of the two stacks is divided at the boundary between them and spaced apart leaving a layer of free space forming a spacer layer. The

TABLE 6.1

Fifteen-Layer Baumeister and Stone Design

Layer Number	Material	Index	Wavelength for Which Layer Is a Quarter Wave (nm)
	Air-incident medium	1.00	Massive
1	ZnS	2.30	414
2	Na_3AlF_6	1.35	414
3	ZnS	2.30	414
4	Na_3AlF_6	1.35	434.8
5	ZnS	2.30	463.7
6	Na_3AlF_6	1.35	463.7
7	ZnS	2.30	520.5
8	Na_3AlF_6	1.35	517
9	ZnS	2.30	626.2
10	Na_3AlF_6	1.35	701.3
11	ZnS	2.30	575.7
12	Na_3AlF_6	1.35	666.7
13	ZnS	2.30	690.8
14	Na_3AlF_6	1.35	690.8
15	ZnS	2.30	690.8
	Glass		Massive

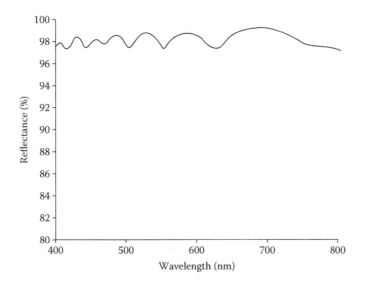

FIGURE 6.11

Calculated performance and the design of a 21-layer high-reflectance coating for the visible and near infrared. The dispersion of the indices of the materials has been taken into account in both design by refinement and in performance calculation. (After E. Pelletier, M. Klapisch, and P. Giacomo, *Nouvelle Revue d'Optique appliqué*, 2, 247–254, 1971, Institute of Physics.)

phase angle φ associated with each reflection coefficient is also shown. At one wavelength, given by the mean of the center wavelengths of the stacks, it can be seen that

$$\varphi_a + \varphi_b = 2\pi.$$

Also, by symmetry, at this wavelength, the reflectances of both stacks are equal, and therefore, the condition for unity transmittance will be completely satisfied if $2\delta = 0$, that is, if the intervening

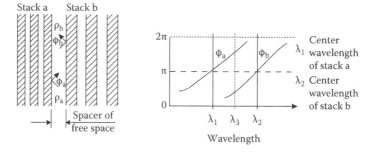

FIGURE 6.12
At λ_3, $(\varphi_a + \varphi_b) = \pi$. Also, by symmetry, at λ_3, $(\lambda_2/\lambda_3) - 1 = 1 - (\lambda_1/\lambda_3)$, i.e., $\lambda_3 = (\lambda_1 + \lambda_2)/2$.

layer is allowed to shrink until it completely vanishes. A peak of transmission will always therefore exist if two stacks are deposited so that they are overlapping at the mean of the two reference wavelengths. This is shown in Figure 6.13, which is reproduced from Turner and Baumeister's paper [17]. Curves A and B are measured reflectances of two high-reflectance quarter-wave stacks, each with the same odd number of layers, starting and finishing with a high-index layer. Curve C shows the measured reflectance of a coating made by combining the two stacks. The peak of transmission can be clearly seen as a dip in the reflectance curve. Experimental errors, in either monitoring or measurement, prevent it from reaching the theoretical minimum.

The dip can be removed by destroying the relationship

$$\frac{\varphi_a + \varphi_b - 2\delta}{2} = q\pi$$

in the region where both stacks have high reflectance. Turner and Baumeister achieved the result quite simply by adding a low-index layer, one-quarter-wave thick at the mean wavelength,

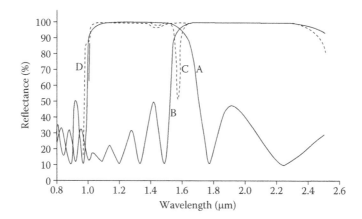

FIGURE 6.13
Measured reflectances of two quarter-wave stacks with slightly overlapping high-reflectance bands. Individual stacks (*full curves*): Curve A, A|0.8(*HLHLHLHLH*)|G. Curve B, A|1.2(*HLHLHLHLH*)|G. When these are combined in a single coating, there is a minimum in the overlap region resulting from the condition in Figure 6.12: Curve C (*broken curve*): A|0.8 (*HLHLHLHLH*) 1.2(*HLHLHLHLH*)|G. An inserted *L* layer eliminates the minimum by destroying the π phase shift. Curve D (*broken curve*): A|0.8(*HLHLHLHLHLH*) L 1.2(*HLHLHLHLH*)|G. G denotes the glass substrate (n = 1.52); A, the air incident medium (n = 1.00); H, the stibnite high-index films; and L, the chiolite low-index films. H and L are quarter-wave thicknesses at the reference wavelength λ_0 of 1.6 µm. (After A. F. Turner and P. W. Baumeister, *Applied Optics*, 5, 69–76, 1971. With permission of Optical Society of America.)

between the stacks. This gave a value for δ of $\pi/2$ and for $(\varphi_a + \varphi_b - 2\delta)/2$ of $\pi/2$, which corresponds to the minimum possible transmission and maximum reflectance. This is illustrated by curve D. The dip has completely disappeared, leaving a broad flat-topped reflectance curve.

Turner and Baumeister also considered the design of broadband reflectors from a slightly different point of view and achieved results similar to that mentioned earlier, although the reasoning is completely different. If a stack is made up of a number of symmetrical periods such as

$$\frac{H}{2}L\frac{H}{2} \quad \text{or} \quad \frac{L}{2}H\frac{L}{2},$$

it can be mathematically represented by a single layer of thickness similar to the actual thickness of the multilayer and with a real optical admittance. This relationship holds good for all regions except the zones of high reflectance, where the thickness and optical admittance are both imaginary. This result will be examined in much greater detail in the following two chapters. For our present purpose, it is sufficient to note that the relationship does exist. If a single layer of real refractive index is deposited on top of a 100% reflector, no interference maxima and minima can possibly exist. For reflectors falling short of the 100% condition, maxima and minima can exist, but are very weak. Thus, in the region where the overlapping stack has a real refractive index, the high reflectance of the lower stack remains virtually unchanged, provided that enough layers are used. The high-reflectance zones can either just touch without overlapping, in which case no reflectance minima will exist, or overlap, in which case the minima will be suppressed because the central layer, composed of an eighth-wave from each stack, is a quarter wavelength thick at the mean of the two monitoring wavelengths, and, as has been shown earlier, this effectively removes any reflectance minima. Figure 6.14 shows the measured reflectance of two stacks,

$$\left(\frac{L}{2}H\frac{L}{2}\right)^4,$$

on a barium fluoride substrate together with the measured reflectance of two similar stacks superimposed on the same substrate in such a way that the high-reflectance zones just touch.

So far, we have principally looked at analytical techniques of design. The more modern technique takes advantage of both the presence of powerful computer programs and the much larger number of layers possible with today's processes. Figure 6.15 shows the performance of a 96-layer extended zone reflector for the region of 400–1200 nm using materials TiO_2 and SiO_2. The reflectance is greater than 99% over the entire range. The design began as a tapered stack of layers with thickness in a geometric series from a quarter wave at 400 nm at the front to three quarters at the rear. This was then refined to give the performance in Figure 6.15. The TiO_2 has significant absorption at the shortwave end of the range, and its effect can be minimized by assuring that the shortwave light is reflected at the outermost layers that are therefore the thinnest. However, this situation is reversed when back reflectance is considered with the result shown in the lower plot in Figure 6.15.

6.2.2 Coating Uniformity Requirements

One feature of the broadband reflectors that we have been considering is that the change in phase on reflection varies very rapidly with wavelength, much more rapidly than in the case of the simple quarter-wave stack. The difficulty that this could cause if such coatings were used in the determination of wavelength in Fabry–Perot interferometer has been frequently mentioned. The method proposed by Stanley and Andrew [18] that uses two spacers does completely eliminate the effect of even the most rapid phase change with wavelength, but there is another effect, the subject of a dramatic report by Ramsay and Ciddor [19]. They used a 13-layer coating of a design similar to that of Baumeister and Stone, and it is listed in Table 6.2.

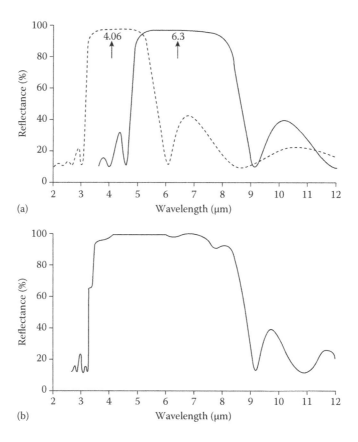

FIGURE 6.14
(a) Measured reflectances of two stacks A|(0.5L H 0.5L)^4|G on BaF$_2$ substrates. G denotes BaF$_2$, and A, air; H and L are films of stibnite and chiolite a quarter-wave thick at reference wavelength λ_0 = 4.06 μm (*broken curve*) or 6.3 μm (*full curve*). (b) Measured reflectance of the two stacks in (a) superimposed in a single coating for an extended high-reflectance region. (After A. F. Turner and P. W. Baumeister, *Applied Optics*, 5, 69–76, 1971. With permission of Optical Society of America.)

The coating was deposited with layer uniformity in the region of 1–2 nm from the center to the edge of the 75 mm diameter plates. When tested, however, after coating, the plates appeared to be λ/60 concave at 546 nm, very uniform at 588 nm, and λ/10 convex at 644 nm. This curvature is, of course, only apparent. Tests on the plates using silver layers showed that they were probably λ/60 concave. The apparent curvature results from changes both in the thickness of the coatings and in the phase change on reflection.

In fact, a theory sufficient to explain the effect was published, together with some estimates of required uniformity, by Giacomo [20] in 1958. He obtained the result that the apparent variation in spacer thickness (measured in units of phase) was equal to the error in the uniformity of the coating (measured as the variation in physical thickness) times a factor

$$\left(\frac{v}{e} \frac{\partial \varphi}{\partial v} + 4\pi v \right),$$

where e is the total thickness of the coating (physical thickness), $v = 1/\lambda$ is the wavenumber, and φ is the phase change on reflection at the surface of the coating. Another way of stating the result is to take $\Delta \rho_m$ as the maximum allowable error in spacer thickness (measured in units of phase) due to this cause, and then the uniformity in coating must be better than

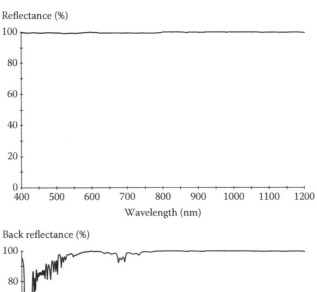

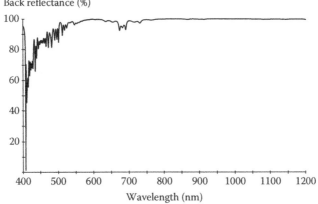

FIGURE 6.15
Upper: The performance in the incident medium of air of the 96-layer extended-zone high-reflectance coating of TiO_2 and SiO_2 and including dispersion and absorption. The reflectance is greater than 99% over the range of 400–1200 nm. *Lower*: The reflectance of the reverse side of the same coating in glass. The pronounced dip at the shortwave end is due to the absorption in the TiO_2 material. The explanation is in the text.

$$\frac{\Delta e}{e} = \frac{\Delta \rho_m}{[(\partial \varphi / \partial v) + 4\pi e]v}.$$

Giacomo [20] showed that the two terms in the expression $\partial \varphi / \partial v$ (which is generally negative) and $4\pi e$ could cancel, or partially cancel, so that some designs of coating would be more sensitive to uniformity errors than others. Ramsay and Ciddor [19] carried this further by pointing out that the two terms in the expression vary in magnitude throughout the high-reflectance zone of the coating, and, although the cancellation or partial cancellation does occur, in addition, the varying magnitudes mean that it is possible in some cases for the apparent curvature due to uniformity errors to vary from concave to convex, or vice versa, throughout the range. This is so for the particular coating they considered, and it is this change in apparent curvature that is particularly awkward, implying that the interferometer must be tested for flatness over the entire working range, not, as is normal, at one convenient wavelength.

For the conventional quarter-wave coating, the magnitude of $\partial \varphi / \partial v$ falls far short of $4\pi e$; for example, in the case of a seven-layer coating of zinc sulfide and cryolite, for the visible region, $\partial \varphi / \partial v$ is only −1.5 µm compared with $4\pi e$ of around +21.5 µm, and the uniformity that is required can be readily calculated from the finesse requirement and the physical thickness of the coating, neglecting the effect of the variations in phase angle altogether. In the case of the broadband

TABLE 6.2

Ramsay and Ciddor Design

Layer Number	Material	Wavelength for Which the Layer Is a Quarter Wave (nm)
	Air incident medium	Massive
1	ZnS	454
2	Na_3AlF_6	355
3	ZnS	385
4	Na_3AlF_6	392
5	ZnS	571
6	Na_3AlF_6	535
7	ZnS	539
8	Na_3AlF_6	573
9	ZnS	562
10	Na_3AlF_6	594
11	ZnS	720
12	Na_3AlF_6	671
13	ZnS	589
	Fused silica substrate	

Source: J.V. Ramsay and P.E. Ciddor. 1967. *Applied Optics* 6:2003–2004.

multilayer, however, the magnitude of $\partial \varphi / \partial v$ is very much greater and at some wavelengths will exceed the value of $4\pi e$. For example, Giacomo quoted a case where $\partial \varphi / \partial v$ reached -125 μm, completely swamping the thickness effect $4\pi e$. Heavens and Liddell [14] in their paper, quoted values of $\partial \varphi / \partial v$ varying from 10 to 26 μm for the staggered multilayers. The change in apparent curvature can therefore occur with these staggered systems, and it is dangerous to attempt to calculate the required uniformity simply from the coating thickness and the finesse requirement.

We can illustrate the problem of the broadband reflectors with the 96-layer coating in Figure 6.15. Refinement has destroyed any simple analytical description of the layer sequence, and so we simply calculate the error in the reflected wavefront. This involves the relative changes in reflected phase. We recall that the phase change on reflection is calculated at the front surface of the multilayer, and in a defect of uniformity, this surface will itself vary in depth and this must be taken into account in any calculations. The simplest way of achieving this is to add a suitable thickness of incident material to the front of the multilayer to bring the reference surface for phase calculations back to the unperturbed position. Figures 6.16 and 6.17 use this method.

A point raised in the paper by Ramsay and Ciddor [19] is the possibility of designing a coating where the two terms cancel almost completely throughout the entire working range. Ciddor [21] carried this a stage further and has now produced several possible designs. Particularly successful is a design for a reflector to give approximately 75% reflectance over the major part of the visible and which is approximately three times less sensitive to thickness variations than would be the case with a reflector exhibiting no phase change at all with change in thickness. The design is intended for film indices of 2.30 and 1.35 on a substrate of index 1.46, corresponding to zinc sulfide and cryolite on fused silica. The thicknesses are given in Table 6.3. The reflectance is constant within around ±2% over the region of 400–650 nm, and an interferometer plate with such a coating would behave as if it were much flatter than the purely geometrical lack of uniformity of the coating would suggest. Calculated reflectance and wavefront error are shown in Figures 6.18 and 6.19, respectively. At the date of the Ciddor paper, computer calculation was not as advanced as it is today. Nowadays, a set of targets involving both reflectance and limitations in permitted changes in phase would be the simplest way of approaching the problem.

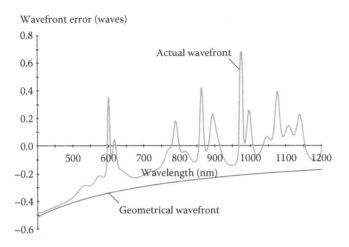

FIGURE 6.16
Uniformity error of −1% has been assumed in the 96-layer extended-zone reflector in Figure 6.15. The effect on the reflected wavefront, assuming no change in reflected phase so that the considerations are purely geometrical, is indicated along with that of including full phase calculations. Rapid changes in phase as a function of thickness result in the wavefront error oscillating between advancement (positive) and retardation (negative).

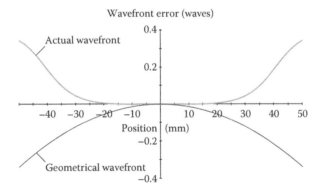

FIGURE 6.17
Here the coating in Figure 6.15 is assumed to be deposited over a flat, circular substrate of 100 mm diameter. There is a radially increasing uniformity error resulting in a spherical surface for the coating that falls from zero at the center to −1% at the periphery. We calculate the reflected figure shape at 600 nm wavelength by using simple geometry and a full phase calculation. At the edge of the substrate, the result is exactly that shown at 600 nm in Figure 6.16. But the relationship between wavefront and uniformity errors is not linear, and this is clear in the calculated wavefront shape.

6.3 Losses

If lossless materials are used, then the reflectance attainable by a quarter-wave stack solely depends on the number of layers. If the reflectance is high, then the addition of a further pair of layers reduces the transmittance by a factor $(n_L/n_H)^2$. In practice, the reflectance that can be ultimately achieved is limited by losses in the layers. There are many possible different sources of loss, but the primary ones are scattering and absorption.

TABLE 6.3

Ciddor's Improved Design

Layer Number	Index	Wavelength for Which Layer Is a Quarter Wave (nm)
	1.00	Massive
1	1.35	309
2	2.30	866
3	1.35	969
4	2.30	436
5	1.35	521
6	2.30	369
7	1.35	484
8	2.30	441
9	1.35	795
10	2.30	768
	1.46	Massive—substrate

Source: P.E. Ciddor. 1968. *Applied Optics* 7:2328–2329

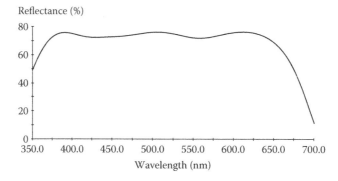

FIGURE 6.18
This is the reflectance coating in Table 6.3. The design range is 400–650 nm and the target reflectance is 75%.

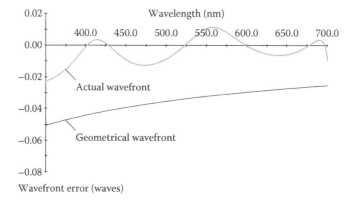

FIGURE 6.19
Here we show the calculated wavefront error corresponding to a uniformity error of −1% in the coating of Figure 6.18. Over the design range of 400–650 nm, the actual error is significantly less than that calculated by simple geometry.

Scattering losses are principally due to defects such as dust in the layers or surface roughness, and techniques for reducing them are essentially attention to detail and good housekeeping. Absorption losses are a property of the material and may be intrinsic or due to impurities or to composition or to structure. Absorption losses are related to the extinction coefficient of the material, and it is useful to consider the absorption losses of a quarter-wave stack composed of weakly absorbing layers having small but nonzero extinction coefficients. Expressions for this have been derived by several workers. The technique we use here is adapted from an approach devised by Hemingway and Lissberger [22]. Although scattering is a process different from absorption, they share the same dependence on the electric field of the wave and so, in simple measurements of reflectance and transmittance, are difficult to separate and are frequently lumped together in an extinction coefficient.

To estimate the losses in a quarter-wave stack structure, we use the extinction coefficient and the concept of potential transmittance introduced in Chapter 2. We split the multilayer into subassemblies of single layers, each with its own value of potential transmittance. The potential transmittance of the assembly is then the product of the individual transmittances.

For the entire multilayer, we can write

$$\psi = \frac{T}{1-R}.$$

Then, if A is the absorptance,

$$1 - \psi = \frac{1-R-T}{1-R} = \frac{A}{1-R}$$

and

$$A = (1-R)(1-\psi).$$

Now $0 \le \psi \le 1$, and so we can introduce a quantity $\mathscr{A}_f$ and write

$$\psi_f = 1 - \mathscr{A}_f$$

for each individual layer, and since we are considering only weak absorption, the potential transmittance will be very near unity, and so A_f will be very small. Then the potential transmittance of the entire assembly will be given by

$$\psi = \prod_{f=1}^{P} \psi_f = \prod_{f=1}^{P}(1 - \mathscr{A}_f)$$

$$= 1 - \sum_{f=1}^{P}\mathscr{A}_f + \cdots.$$

Now let us consider one single layer. The relevant parameters are contained in

$$\begin{bmatrix} B \\ C \end{bmatrix} = \begin{bmatrix} \cos\delta_f & i(\sin\delta_f)/y_f \\ iy_f \sin\delta_f & \cos\delta_f \end{bmatrix} \begin{bmatrix} 1 \\ y_e \end{bmatrix}$$

and

$$\psi = \frac{\mathrm{Re}(y_e)}{\mathrm{Re}(BC^*)}$$

from Equation 2.132. Also,

$$y_f = n_f - ik_f \quad \text{(free space units)},$$

$$\delta_f = 2\pi(n_f - ik_f)d_f/\lambda$$

$$= 2\pi n_f d_f/\lambda - i2\pi k_f d_f/\lambda$$

$$= \alpha - i\beta,$$

where k_f, and hence β, is small.

If we consider layers that are approximately quarter waves, we can set

$$\alpha = [(\pi/2) + \varepsilon],$$

where ε is small. Then,

$$\cos \delta_f \approx (-\varepsilon + i\beta),$$

$$\sin \delta_f \approx 1,$$

and the matrix expression becomes

$$\begin{bmatrix} B \\ C \end{bmatrix} = \begin{bmatrix} (-\varepsilon + i\beta) & i/(n - ik) \\ i(n - ik) & (-\varepsilon + i\beta) \end{bmatrix} \begin{bmatrix} 1 \\ y_e \end{bmatrix},$$

whence

$$\begin{bmatrix} B \\ C \end{bmatrix} = \begin{bmatrix} (-\varepsilon + i\beta) + iy_e/(n - ik) \\ i(n - ik) + y_e(-\varepsilon + i\beta) \end{bmatrix},$$

so that

$$BC^* = [(-\varepsilon + i\beta) + iy_e/(n - ik)][i(n - ik) + y_e(-\varepsilon + i\beta)]^*,$$

and, assuming that y_e is real, since we are dealing with a quarter-wave stack, and neglecting terms of the second order and above in k, β, and ε,

$$\text{Re}(BC^*) = (\beta n + y_e + y_e^2 \beta/n)$$

and

$$\psi_f = \frac{y_e}{(\beta n + y_e + y_e^2 \beta/n)} = \frac{1}{1 + \beta[(n/y_e) + (y_e/n)]}.$$

Then, since β is small,

$$\psi_f = 1 - \beta[(n/y_e) + (y_e/n)]$$

and

$$\mathscr{A}_f = 1 - \psi_f = \beta[(n/y_e) + (y_e/n)].$$

Next, we must find

$$(1 - R)\sum \mathscr{A}_f.$$

For this, we need the value of y_e at each interface. Let the stack of quarter-wave layers end with a high-index layer. Then the admittance of the whole assembly will be Y, where Y is large and real. If we denote the admittance of the incident medium by y_0 (n_0 in free space units and real), then

$$R = \left[\frac{y_0 - Y}{y_0 + Y}\right]^2.$$

If Y is sufficiently large,

$$1 - R = 1 - [1 - 4y_0/Y] = 4y_0/Y.$$

Further, since Y is the terminating admittance and the layers are all quarter waves, the admittances at each of the interfaces follow the pattern

$$
\begin{array}{ccccccc}
Y & \dfrac{y_H^2}{Y} & \dfrac{y_L^2 Y}{y_H^2} & \dfrac{y_H^4}{y_L^2 Y} & \dfrac{y_L^4 Y}{y_H^4} & \dfrac{y_H^6}{y_L^4 Y} & \dfrac{y_L^6 Y}{y_H^6}
\end{array}
$$

$$
\left|\; y_0 \;\right|\; n_H \;\left|\; n_L \;\right|\; n_H \;\left|\; n_L \;\right|\; n_H \;\left|\; n_L \;\right|\;\left|\;\right|
$$

Then,

$$A = (1 - R)\sum_{f=1}^{p} A_f$$

$$= \frac{4y_0}{Y}\left[\left(\frac{y_H}{y_H^2/Y} + \frac{y_H^2/Y}{y_H}\right)\beta_H + \left(\frac{y_L}{y_L^2 Y/y_H^2} + \frac{y_L^2 Y/y_H^2}{y_L}\right)\beta_L\right.$$

$$\left. + \left(\frac{y_H}{y_H^4/y_L^2 Y} + \frac{y_H^4/y_L^2 Y}{y_H}\right)\beta_H + \cdots\right],$$

i.e.,

$$A = 4y_0\left[\left(\frac{1}{y_H} + \frac{y_H}{Y^2}\right)\beta_H + \left(\frac{y_L}{y_H^2} + \frac{y_H^2}{y_L Y^2}\right)\beta_L + \left(\frac{y_L^2}{y_H^3} + \frac{y_H^3}{y_L^2 Y^2}\right)\beta_H + \cdots\right].$$

Since β_H and β_L are small and Y is large, we can neglect terms in β/Y^2, and the absorptance is then given by

$$A = 4y_0\left[\left(\frac{1}{y_H} + \frac{y_L^2}{y_H^3} + \frac{y_L^4}{y_H^5} + \cdots\right)\beta_H + \left(\frac{y_L}{y_H^2} + \frac{y_L^3}{y_H^4} + \frac{y_L^5}{y_H^6} + \cdots\right)\beta_L\right].$$

$(y_L/y_H)^2$ is less than unity, and although the series are not infinite, we can assume that they have a sufficiently large number of terms so that any error involved in assuming that they are in fact infinite is very small.

Thus,

$$A = 4y_0\left(\frac{\beta_H/y_H}{1 - (y_L/y_H)^2} + \frac{y_L\beta_L/y_H^2}{1 - (y_L/y_H)^2}\right) = \frac{4y_0(y_H\beta_H + y_L\beta_L)}{(y_H^2 - y_L^2)}.$$

Now

$$y\beta = y\left(\frac{2\pi kd}{\lambda}\right) = \left(\frac{2\pi nd}{\lambda}\right)k,$$

where, since we are working in free space units, we are replacing y by n. Since the layers are quarter waves,

$$\frac{2\pi n d}{\lambda} = \frac{\pi}{2},$$

so that

$$A = \frac{2\pi n_0 (k_H + k_L)}{\left(n_H^2 - n_L^2\right)} \quad \text{(final layer of high index)}.$$

The case of a multilayer terminating with a low-index layer can be dealt with in the same way. The final low-index layer acts to reduce the reflectance and so increases the absorptance that is given by

$$A = \frac{2\pi}{n_0} \left[\frac{\left(n_H^2 k_L + n_L^2 k_H\right)}{\left(n_H^2 - n_L^2\right)} \right] \quad \text{(final layer of low index)}.$$

As an example, we can consider a multilayer with $k_H = k_L = 0.0001$, $n_H = 2.35$, $n_L = 1.35$, and in incident medium air with $n_0 = 1.00$:

$$A = 0.03\% \text{ (high-index layer outermost)},$$

$$A = 0.12\% \text{ (low-index layer outermost)}.$$

In fact, in the red part of the spectrum, the losses in a zinc sulfide and cryolite stack can be less than 0.001%, indicating that the value of k must be less than 6×10^{-6}, assuming that the loss is entirely in one material. For tantalum pentoxide and silicon dioxide multilayer quarter-wave stacks, losses as low as 1 ppm at 1 μm, i.e., 0.0001%, have been reported. This is consistent with values of k an order of magnitude lower. At this level, small amounts of contamination on the reflector surfaces become important additional sources of loss.

In absolute terms, the absorption loss affects the reflectance more than the transmittance in any given quarter-wave stack. Giacomo [23,24] showed that $\Delta T/T$ and $\Delta R/R$ are of the same order, and therefore, since $R \gg T$, then $\Delta R \gg \Delta T$.

The analytical calculation of loss becomes much more difficult, if not impossible, in the case of more complex multilayers, especially when the layers have thicknesses that are not in a simple relationship. It is easier then to think about the losses qualitatively and to use the computer to carry out accurate calculations. This is the case with the extended zone reflector in Figure 6.15.

6.4 Reflectors with Multiple Peaks

Sometimes there is a requirement for a reflector with more than one high-reflectance peak. One way of achieving such a performance is through a structure known as a rugate. This is similar to a quarter-wave stack except that instead of a set of discrete quarter-wave layers, the rugate exhibits a sinusoidal variation in index through its thickness, the pitch of the sinusoid being equivalent to a half-wave thickness at the high-reflectance wavelength. We will deal with rugate structures later, but briefly, their advantages are that reflectance peaks of other orders are suppressed and that sinusoids of different pitches can be readily combined to give multiple peaks with no set numerical relationship among them. However, rugate structures present special difficulties in their realization. In this chapter, we concern ourselves with discrete layer solutions.

The simplest way of achieving such a performance is to place one quarter-wave stack over another, and this can work well. Unfortunately, the light at the peak corresponding to the inner reflector has to traverse the outer reflector twice. This results in a problem similar to that of the extended zone reflector already mentioned where absorption in the outer structure becomes important. Also, the electric field can be, and usually is, magnified in the double transmission, and this still further amplifies the losses and has serious implications for the handling of high beam power.

If there is a simple relationship between the wavelengths at which the high reflectance is required, it is sometimes possible to use a useful trick. We know from the discussion at the start of the chapter that any repeated structure, provided that there are enough repeats, will yield high reflectance provided that the basic structure is not itself an absentee. Figure 6.20 shows the performance of a repeated ($HH\,L\,0.5H\,0.5L$) structure. The reflectance peaks in the diagram occur wherever we have a thickness of an integral number of half waves except at $g = 4.0$ and 8.0, where all the individual layers are absentees. We note that the reflectance peaks are not all of equal reflectance. This is influenced by the exact thickness relationships in the basic structure, and a little trial and error can help improve the performance. As an example, we use a modest degree of refinement to equalize the peaks at values of g of 1.5 and 2.0 that, in Figure 6.20, are clearly slightly unbalanced. The modified performance together with the new design is shown in Figure 6.21.

A systematic technique of achieving a similar result involves what is known as thickness modulation, dealt with in more detail in a later chapter. There are scattered references to thickness modulation in the literature, but the major detailed work is a book by Perilloux [25], and this is the source of the method we now describe. Let us design a reflector for two wavelengths λ_0 and λ_1, where λ_1 is larger than λ_0. From our discussion so far, we know that a possible design would be a repeated period that is a whole number of half waves for both wavelengths, but an absentee for neither. However, we still have to obtain the correct reflectances, and although we can do this, as we have demonstrated, by trial and error, the great power of Perilloux's method is that it involves the systematic variation of just one single parameter. We will illustrate the technique with a simple example, similar to the one in the book of Perilloux [25], of high and equal reflectances at 1000 and at 1500 nm. Let us choose materials of low index 1.47 (with a quarter wave indicated by L) and

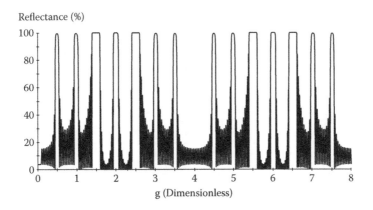

FIGURE 6.20
The design of this coating is Air | ($HH\,L\,0.5H\,0.5L$)^{12} | Glass with $n_H = 2.15$ and $n_L = 1.45$, roughly corresponding to Ta_2O_5 and SiO_2. The performance is shown in terms of $g = (\lambda_0/\lambda)$. High-reflectance peaks occur at those values where the thickness of the basic period is an integral number of half waves except for those values such as 4.0 and 8.0, where each layer in the basic period is itself an integral number of half waves.

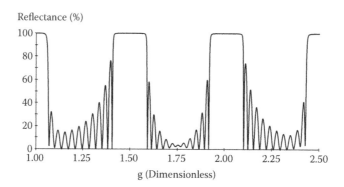

FIGURE 6.21
Design of Figure 6.20 has been modified to equalize the reflectance peaks at $g = 1.5$ and 2.0. The new design is Air | $(1.739H$ $1.152L\ 0.522H\ 0.575L)^{12}$ | Glass. The peak reflectances are both 99.999%.

high index 2.25 (with a quarter wave indicated by H) with air as incident medium and glass of index 1.52 as substrate. To keep it simple, we neglect any dispersion.

We start with two separate quarter-wave stacks, each tuned to one of the two wavelengths (Figure 6.22). First, we make sure that the optical thickness of a whole number of half waves in one stack is set equal to a whole number (that will be different) of half waves in the other. To achieve this, we may sometimes have to slightly adjust the wavelengths with the consequence that those specified may not be at the exact center of their reflecting bands. We can also include dispersion, if necessary. In this particular case, six quarter waves of the lesser wavelength are exactly equal to four quarter waves of the greater. Perilloux's method, then, is to gradually morph one quarter-wave stack into the other, using a parameter k that both retains the half-wave conditions, assuring a continuing high reflectance at each wavelength, and adjusts the relative reflectances.

The method is illustrated in Figure 6.22. The two stacks are aligned at either side each showing equally thick periods, A and B, each made up of quarter waves at the appropriate wavelengths and adding up to a whole number of half waves. The region between the periods shows the way in which the layers of the period will linearly vary with k, all the time retaining the same total optical thicknesses of the periods. Some layers expand and others contract and vanish. The way to

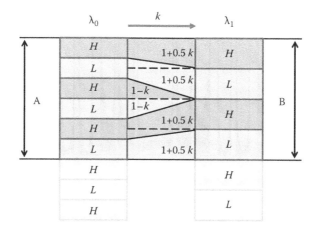

FIGURE 6.22
Arrangement of the two stacks and the transformation from one to the other depending on parameter k that ranges from zero to unity.

accomplish this will usually be obvious. An analytical approach is outlined in Perilloux's book [25]. As k varies from zero to unity, the reflectance at λ_0 falls and that at λ_1 rises. Finally, the adjustment of k and, usually, the total number of layers will achieve the desired performance (Figure 6.23).

Figure 6.24 shows the electric field distributions. At both wavelengths, the field at the front surface is low, and there is no field magnification through the thickness of the coating. This represents an enormous advantage of this type of approach.

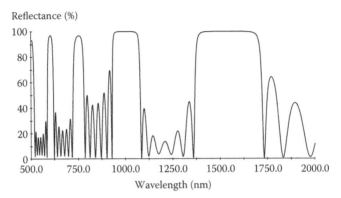

FIGURE 6.23
Final reflectance of the 42-layer dual-wavelength coating. The reflectance at 1000 and 1500 nm is 99.96%. The value of the factor k is 0.568.

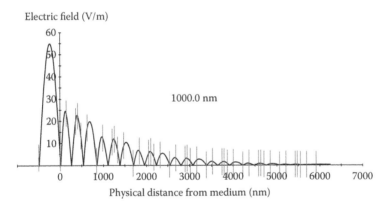

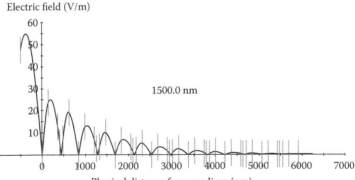

FIGURE 6.24
Distribution of electric field amplitude through the layers of the design in Figure 6.23 at 1000 nm (*upper*) and at 1500 nm (*lower*). In both cases, the distribution is typical of a single quarter-wave stack with no magnification. The vertical lines mark the layer boundaries and the incident surface is at the origin. The input power density is 1 W/m^2.

References

1. C. Fabry and A. Perot. 1899. Théorie et applications d'une nouvelle méthode de spectroscopie interférentielle. *Annales de Chimie et de Physique, Paris, 7th series* 16:115–144.
2. M. Born and E. Wolf. 2002. *Principles of Optics: Electromagnetic Theory of Propagation, Interference and Diffraction of Light.* Seventh ed. New York: Cambridge University Press.
3. R. Chabbal. 1958. V.—Qualité des lames et couches réfléchissantes pour le Fabry–Perot—Finesse limité d'un Fabry–Perot formé de lames imparfaites. *Journal de Physique et le Radium* 19:295–300.
4. H. Mayer. 1950. *Physik dünner Schichten.* Stuttgart: Wissenschaftliche Verlagsgesellschaft mbH.
5. H. Kuhn and B. A. Wilson. 1950. Reflectivity of thin silver films and their use in interferometry. *Proceedings of the Physical Society, B* 63:745–755.
6. U. Oppenheim. 1956. Semireflecting silver films for infrared interferometry. *Journal of the Optical Society of America* 46:628–633.
7. E. D. Palik, ed. 1985. *Handbook of Optical Constants of Solids I.* Orlando, FL: Academic Press.
8. S. Penselin and A. Steudel. 1955. Fabry–Perot-Interferometerverspiegelungen aus dielektrischen Vielfachschichten. *Zeitschrift für Physik* 142:21–41.
9. W. T. Welford (writing as W. Weinstein). 1954. Computations in thin film optics. *Vacuum* 4:3–19.
10. L. I. Epstein. 1955. Improvements in heat reflecting filters. *Journal of the Optical Society of America* 45:360–362.
11. D. L. Perry. 1965. Low loss multilayer dielectric mirrors. *Applied Optics* 4:987–991.
12. W. Heitmann. 1966. Extrem hochreflektierende dielektrische Spiegelschichten mit Zincselenid. *Zeitschrift für Angewandte Physik* 21:503–508.
13. P. W. Baumeister and J. M. Stone. 1956. Broad-band multilayer film for Fabry–Perot interferometers. *Journal of the Optical Society of America* 46:228–229.
14. O. S. Heavens and H. M. Liddell. 1966. Staggered broad-band reflecting multilayers. *Applied Optics* 5:373–376.
15. P. W. Baumeister. 1958. Design of multilayer filters by successive approximations. *Journal of the Optical Society of America* 48:955–958.
16. E. Pelletier, M. Klapisch, and P. Giacomo. 1971. Synthèse d'empilements de couches minces. *Nouvelle Revue d'Optique appliquée* 2:247–254.
17. A. F. Turner and P. W. Baumeister. 1966. Multilayer mirrors with high reflectance over an extended spectral region. *Applied Optics* 5:69–76.
18. R. W. Stanley and K. L. Andrew. 1964. Use of dielectric coatings in absolute wavelength measurements with a Fabry–Perot interferometer. *Journal of the Optical Society of America* 54:625–627.
19. J. V. Ramsay and P. E. Ciddor. 1967. Apparent shape of broad band multilayer reflecting surfaces. *Applied Optics* 6:2003–2004.
20. P. Giacomo. 1958. Proprietes chromatiques des couches reflechissantes multidielectriques. *Journal de Physique et le Radium* 19:307–311.
21. P. E. Ciddor. 1968. Minimization of the apparent curvature of multilayer reflecting surfaces. *Applied Optics* 7:2328–2329.
22. D. J. Hemingway and P. H. Lissberger. 1973. Properties of weakly absorbing multilayer systems in terms of the concept of potential transmittance. *Optica Acta* 20:85–96.
23. P. Giacomo. 1956. Les couches réfléchissantes multidiélectriques appliquées a l'interférometre de Fabry–Perot: Etude théorique et expérimentale des couches réelles: I. *Revue d'Optique* 35:317–354.
24. P. Giacomo. 1956. Les couches réfléchissantes multidiélectriques appliquées a l'interférometre de Fabry–Perot: Etude théorique et expérimentale des couches réelles: II. *Revue d'Optique* 35:442–467.
25. B. E. Perilloux. 2002. *Thin Film Design: Modulated Thickness and Other Stopband Design Methods,* vol. TT57. Bellingham, WA: SPIE Optical Engineering Press.

7

Edge Filters and Notch Filters

Filters in which the primary characteristic is an abrupt change between a region of rejection and a region of transmission are known as edge filters. Edge filters are divided into two principal groups, longwave pass and shortwave pass, the designations being descriptive. Note that sometimes the terms *high pass* and *low pass* may be employed. We advise against these alternative designations because it is not always clear whether they refer to wavelength or frequency and so can be ambiguous. To the edge filters we add the notch filter that consists of a rejection region surrounded by regions of high transmission on both sides. This is sometimes termed a *minus filter*, and sometimes a *band-stop filter*, but most frequently a notch filter.

The primary thrust in this book is, of course, thin-film interference, but interference depends on path difference that, in turn, depends on wavelength. A given interference condition is difficult to sustain over a wide wavelength region. This is particularly true of the rejection, or as it is often called, the blocking, region. Here the filter designer takes advantage of any useful absorbing materials, of any other filter types, of the sensitivity regions of receivers or the output of sources and anything else that can offer assistance. Excellent absorbing glass longwave pass filters for the near ultraviolet, visible, and near infrared that can be used to extend blocking regions are available, for example. However, we concentrate here on the thin-film aspects of the components. The principal problems that we must deal with are the ripple in the pass region and the extent of the blocking.

7.1 Thin-Film Absorption Filters

A thin-film absorption filter consists of a thin film of material with an absorption edge at the required wavelength and is usually a longwave pass in character. Semiconductors that exhibit a very rapid transition from opacity to transparency at the intrinsic edge are particularly useful in this respect, making reasonable longwave pass filters. The edges are not generally as steep as can be achieved with the interference filters to be described, but they are simple and easy to make. The primary complication is connected with the reflection loss in the pass region due to the high refractive index of the film. Germanium, for example, with an edge at 1.65 µm, has an index of 4.0, and as the thickness of germanium necessary to achieve useful rejection will be at least several quarter waves, there will be prominent interference fringes in the pass zone showing variations from substrate level, at the half-wave positions, to a reflectance of 68% (in the case of a glass substrate) at the quarter-wave position. The problem can be readily solved by placing antireflection coatings between the substrate and the germanium layer and between the germanium layer and the air. Single quarter-wave antireflection coatings are usually quite adequate. For optimum matching, the values required for the indices of the antireflecting layers are 2.46 between glass and germanium and 2.0 between germanium and air. The index of zinc sulfide (2.35) is

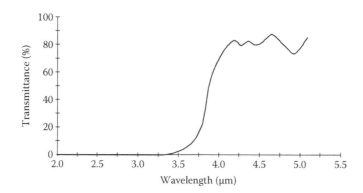

FIGURE 7.1
Measured characteristic of a lead telluride filter. Design:

$$\left| \begin{array}{c} \lambda/4 \\ ZnS \end{array} \right| \begin{array}{c} 30 \times \lambda/4 \\ PbTe \end{array} \left| \begin{array}{c} \lambda/4 \\ ZnS \end{array} \right| \begin{array}{c} CaF_2 \\ substrate \end{array} \right|$$

$\lambda_0 = 3$ µm. The small dip at 4.25 µm is probably due to a slight unbalance of the measuring spectrometer caused by atmospheric CO_2. (Courtesy of Sir Howard Grubb, Parsons and Co. Ltd., Newcastle upon Tyne.)

sufficiently near to both values. We can roughly estimate that the reflectance near the peak of the quarter-wave coatings will oscillate between

$$\left[\frac{1 - (2.35^4)/(4^2 \times 1.52)}{1 + (2.35^4)/(4^2 \times 1.52)} \right]^2 = 1.3\%$$

for wavelengths where the germanium layer is equal to an integral odd number of quarter waves, and 4%, that is, the reflectance of the bare glass substrate, where the germanium layer is an integral number of half waves thick (for at such a wavelength, the germanium layer acts as an absentee layer and the two zinc sulfide layers combine to form a half-wave and therefore an absentee layer).

Other materials used to form single-layer absorption filters in this way include cerium dioxide, giving an ultraviolet rejection-visible transmitting filter; silicon, giving a longwave pass filter with an edge at 1 µm; and lead telluride, giving a longwave pass filter at 3.4 µm.

A practical lead telluride filter characteristic is shown in Figure 7.1, along with its design. The two zinc sulfide layers were arranged to be quarter waves at 3.0 µm. Better results would probably have been obtained if the thicknesses had been increased to quarter waves at 4.5 µm.

7.2 Interference Edge Filters

The basic dielectric reflecting structure is the quarter-wave stack. It is characterized by a limited zone of high reflectance surrounded by regions of relatively high transmittance that exhibit rather pronounced ripple. The transition from reflecting to transmitting is a quite sharp one, the sharpness increasing with the number of layers in the structure. The performance is, therefore, already of the form of a shortwave or longwave pass filter. To convert this coating into a useful edge filter implies the reduction of the ripple in the pass region. This used to be a quite difficult task but is nowadays greatly simplified by the use of powerful computers in a process of refinement or even complete synthesis. We shall come to this process later in this chapter. Although older analytical techniques are being replaced by newer machine-aided design, nevertheless, their study is still useful in that it leads to an understanding that is difficult to obtain from computer-aided processes only. We therefore begin by looking at some of the fundamental theory.

7.2.1 Quarter-Wave Stack

The basic type of interference edge filter is the quarter-wave stack of the previous chapter. As was explained there, the principal characteristic of the optical transmission curve plotted as a function of wavelength is a series of high-reflection zones, i.e., low transmission, separated by regions of high transmission. The shape of the transmission curve of a quarter-wave stack is shown in Figure 7.2. The particular combination of materials shown is useful in the infrared beyond 2 µm, but the curve is typical of any pair of materials having a reasonably high ratio of refractive indices.

The system of Figure 7.2 can be used as either a longwave pass filter with an edge at 5.0 µm or a shortwave pass filter with an edge at 3.3 µm. These wavelengths can be altered at will by changing the reference wavelength.

It sometimes happens that the width of the rejection zone is adequate for the particular application, as, for example, where light of only a particularly narrow spectral region is to be eliminated or where the detector itself is insensitive to wavelength beyond the opposite edge of the rejection zone. In most cases, however, it is desirable to eliminate all wavelengths shorter than, or longer than, a particular value. The rejection zone, shown in Figure 7.2, must somehow be extended. This is usually done by adding additional filters.

Absorption filters usually have very high rejection in the stop region, but as they depend on the fundamental optical properties of the basic materials, they are inflexible in character, and the edge positions are fixed. Using interference and absorption filters together combines the best properties of both, the deep rejection of the absorption filter with the flexibility of the interference filter. The interference layers can be deposited over an absorption filter, which acts as the substrate, or the interference section can sometimes be made from material which itself has an absorption edge within the interference rejection zone. Within the absorption region, the filter behaves in much the same way as the single layers of the previous section.

Other methods of improving the width of the rejection zone will be dealt with shortly, but now, we must turn our attention to the more difficult problem created by the magnitude of the ripple in

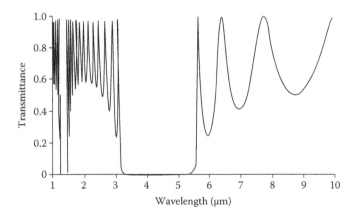

FIGURE 7.2
Computed characteristic of a 13-layer quarter-wave stack of germanium (index 4.0) and silicon monoxide (index 1.70) on a substrate of index 1.42. The reference wavelength λ_0 is 4.0 µm.

transmission in the pass region. As the curve of Figure 7.2 shows, the ripple is severe and the performance of the filter would be very much improved if somehow the ripple could be reduced.

Before we can reduce the ripple, we must first investigate the reason for its appearance, and this is not an easy task, because of the complexity of the mathematics. A paper published by Epstein [1] in 1952 is of immense importance, in that it lays the foundation of a method which gives the necessary insight into the problem to enable the performance to be not only predicted but also improved.

7.2.2 Symmetrical Multilayers and the Herpin Index

Epstein's 1952 paper [1] dealt with the mathematical equivalent of a symmetrical combination of films and a single layer and was the beginning of what has become one of the most powerful analytical design method to date for thin-film filters.

Any thin-film combination is known as symmetrical if each half is a mirror image of the other half. The simplest example of this is a three-layer combination in which a central layer is sandwiched between two identical outer layers. When such a symmetrical arrangement is used as a component of a thin-film design, it is usually called a symmetrical period. It can be shown that a symmetrical period can be treated as a single equivalent layer with a calculable phase thickness and equivalent characteristic admittance. If a multilayer consists of a number of repeats of an identical symmetrical period, then it becomes equivalent in performance to a single layer. Of course, analytical expressions for the equivalent optical properties are rather involved, but the basic form of the result can be relatively easily established and used as a qualitative guide while the accurate calculation can be left to the computer.

Consider first a symmetrical three-layer period pqp that we can think of as made up of dielectric materials free from absorption. The characteristic matrix of the combination is given by

$$
\begin{bmatrix} M_{11} & M_{12} \\ M_{21} & M_{22} \end{bmatrix} = \begin{bmatrix} \cos \delta_p & \left(i \sin \delta_p \right)/\eta_p \\ i\eta_p \sin \delta_p & \cos \delta_p \end{bmatrix} \begin{bmatrix} \cos \delta_q & \left(i \sin \delta_q \right)/\eta_q \\ i\eta_q \sin \delta_q & \cos \delta_q \end{bmatrix}
$$
$$
\times \begin{bmatrix} \cos \delta_p & \left(i \sin \delta_p \right)/\eta_p \\ i\eta_p \sin \delta_p & \cos \delta_p \end{bmatrix}
$$

(7.1)

(where we have used the more general tilted optical admittance η; note that the subscript p is an indication of layer p and not of polarization type). By performing the multiplication, we find

$$
M_{11} = \cos 2\delta_p \cos \delta_q - \frac{1}{2}\left(\frac{\eta_q}{\eta_p} + \frac{\eta_p}{\eta_q} \right) \sin 2\delta_p \sin \delta_q,
$$

(7.2)

$$
M_{12} = \frac{i}{\eta_p}\left[\sin 2\delta_p \cos \delta_q + \frac{1}{2}\left(\frac{\eta_q}{\eta_p} + \frac{\eta_p}{\eta_q} \right) \cos 2\delta_p \sin \delta_q \right.
$$
$$
\left. + \frac{1}{2}\left(\frac{\eta_p}{\eta_q} - \frac{\eta_q}{\eta_p} \right) \sin \delta_q \right],
$$

(7.3)

$$M_{12} = i\eta_p \left[\sin 2\delta_p \cos \delta_q + \frac{1}{2} \left(\frac{\eta_q}{\eta_p} + \frac{\eta_p}{\eta_q} \right) \cos 2\delta_p \sin \delta_q \right.$$
$$\left. - \frac{1}{2} \left(\frac{\eta_p}{\eta_q} - \frac{\eta_q}{\eta_p} \right) \sin \delta_q \right],$$

(7.4)

and

$$M_{22} = M_{11}.$$

(7.5)

There are four elements in this product matrix. However, Equation (7.5) implies that only three of the elements are independent. Then, it can be shown that since the determinant of each of the individual matrices is unity, the determinant of the product matrix is also unity. This adds one further relationship, and so the matrix can be represented by two independent variables. These two independent variables can be chosen in any way we wish. All that is necessary is that we should be able to use them to unambiguously reconstruct the product matrix.

Let the two independent variables be denoted by γ and E such that

$$M_{11} = \cos \gamma = M_{22},$$

(7.6)

$$M_{12} = \frac{i \sin \gamma}{E},$$

(7.7)

and

$$M_{21} = iE \sin \gamma.$$

(7.8)

The determinant of the matrix remains at unity, and the elements are single-valued functions of the two independent variables. These quantities have, intentionally, exactly the same form as a single layer of phase thickness γ and admittance E. We call these two quantities the equivalent phase thickness and the equivalent admittance, respectively. E is also sometimes known as the Herpin admittance after the name of an early worker in this area.

Equations 7.6 through 7.8 can be solved for γ and admittance E. Of course, the solutions are multivalued, but since we are going to use them to reconstruct the matrix, we can choose the values that are of greatest significance to us. We therefore choose γ to have a value that is nearest to the total phase thickness of the period. The solution for E is derived from a square root, and usually, we will take the positive root:

$$\gamma = \arccos M_{11},$$

(7.9)

$$E = \sqrt{\frac{M_{21}}{M_{12}}}.$$

(7.10)

M_{11} does not equal M_{22} in an unsymmetrical arrangement, and such a combination cannot therefore be replaced by a single layer.

It can easily be shown that this result can be extended to cover any symmetrical period consisting of any number of layers. First, the central three layers that, by definition, will form a symmetrical assembly on their own can be replaced by a single layer. This equivalent layer can then be taken along with the next layers on either side as a second symmetrical three-layer combination that can, in turn, be replaced by a further single layer. The process can be repeated until all the layers have been so replaced.

The importance of this result lies both in the ease of interpretation (the properties of a single layer can be much more readily visualized than those of a multilayer) and in the ease with which the result for a single period may be extended to that for a multilayer consisting of many periods.

If a multilayer is made up of, say, s identical symmetrical periods, each of which has an equivalent phase thickness γ and equivalent admittance E, then physical considerations show that the multilayer will be equivalent to a single layer of thickness $s\gamma$ and admittance E. This result also follows because of an easily derived analytical result:

$$\begin{bmatrix} \cos\gamma & i\sin\gamma/E \\ iE\sin\gamma & \cos\gamma \end{bmatrix}^s = \begin{bmatrix} \cos(s\gamma) & i\sin(s\gamma)/E \\ iE\sin(s\gamma) & \cos(s\gamma) \end{bmatrix}. \tag{7.11}$$

It must be emphasized that the equivalent single layer is not an exact replacement for the symmetrical combination in every respect physically. It is merely a mathematical expression for the product of a number of matrices and is valid only for those cases where such a product is involved. The effect of changes in angle of incidence, for instance, cannot be estimated by converting the multilayer to a single layer in this way.

So far, although we limited the discussion to dielectric layers at the start, what we have done actually applies to layers that are absorbing as well as those that are free from loss. Let us now strictly limit the discussion to lossless layers, that is, ones where k, the extinction coefficient, is zero. Now M_{11} and M_{22} (equal to M_{11}) will be real, and M_{12} and M_{21}, imaginary. In any practical case, when the matrix elements are computed, it will be found that there are regions where $M_{11} < -1$, i.e., $\cos\gamma < -1$. This expression cannot be solved for real γ, and in this region, γ must be imaginary. Since the determinant of the matrix must be unity, M_{12} and M_{21} must have opposite signs and E, therefore, must also be imaginary. We can think of this result in two different ways. The imaginary γ and E define a matrix that is not unlike the matrix of a perfect metallic layer. Increasing thickness of metal results in higher reflectance. Alternatively, since $|M_{11} + M_{22}|/2 > 1$, as the number of basic periods is increased, the reflectance must tend to unity. This condition of imaginary γ and E implies a high-reflectance zone or, in other words, a stop band. Outside the stop band, γ and E are real, and the multilayer appears as a dielectric slab of material exhibiting fringes, becoming more and more closely packed with increase in the number of periods. This dielectric region corresponds to a passband. The edges of the passbands and stop bands are given by $M_{11} = -1$.

We can write expressions for γ and E. The expression for the equivalent admittance in the passband is quite a complicated one. From Equations 7.3, 7.4, 7.7, and 7.8,

$$E = +\left(\frac{M_{21}}{M_{12}}\right)^{1/2}$$

$$= +\left[\frac{\eta_p^2\left\{\sin 2\delta_p \cos\delta_q + \dfrac{1}{2}\left(\dfrac{\eta_q}{\eta_p}+\dfrac{\eta_p}{\eta_q}\right)\cos 2\delta_p \sin\delta_q - \dfrac{1}{2}\left(\dfrac{\eta_p}{\eta_q}-\dfrac{\eta_q}{\eta_p}\right)\sin\delta_q\right\}}{\sin 2\delta_p \cos\delta_q + \dfrac{1}{2}\left(\dfrac{\eta_q}{\eta_p}+\dfrac{\eta_p}{\eta_q}\right)\cos 2\delta_p \sin\delta_q + \dfrac{1}{2}\left(\dfrac{\eta_p}{\eta_q}-\dfrac{\eta_q}{\eta_p}\right)\sin\delta_q}\right]^{1/2}. \tag{7.12}$$

For γ we have, from Equation 7.2,

$$\gamma = \arccos\left[\cos 2\delta_p \cos\delta_q - \frac{1}{2}\left(\frac{\eta_q}{\eta_p}+\frac{\eta_p}{\eta_q}\right)\sin 2\delta_p \sin\delta_q\right], \tag{7.13}$$

and we will take that value closest to the total phase thickness of the period $2\delta_p + \delta_q$.

Let us begin our discussion of these expressions with the particular and simpler case of the quarter-wave stack.

7.2.3 Application of the Herpin Index to the Quarter-Wave Stack

Returning for the moment to our quarter-wave stack, we see that it is possible to directly apply the results mentioned earlier if a simple alteration to the design is made. This is simply to add a pair of eighth-wave layers, to the stack, one at each end. Low-index layers are required if the basic stack begins and ends with quarter-wave high-index layers, and vice versa. The two possibilities are

$$\frac{H}{2}LHLHLH\ldots HL\frac{H}{2}$$

and

$$\frac{L}{2}HLHLHL\ldots LH\frac{L}{2}\ .$$

We can immediately replace these arrangements by

$$\frac{H}{2}L\frac{H}{2}\frac{H}{2}L\frac{H}{2}\frac{H}{2}L\frac{H}{2}\ \ldots\ \frac{H}{2}\frac{H}{2}L\frac{H}{2}$$

and

$$\frac{L}{2}H\frac{L}{2}\frac{L}{2}H\frac{L}{2}\frac{L}{2}H\frac{L}{2}\ \ldots\ \frac{L}{2}\frac{L}{2}H\frac{L}{2},$$

respectively, which can then be written as

$$\left[\frac{H}{2}L\frac{H}{2}\right]^{s}\quad\text{and}\quad\left[\frac{L}{2}H\frac{L}{2}\right]^{s},$$

$(H/2)L(H/2)$ and $(L/2)H(L/2)$ being the basic periods in each case. The results in Equations 7.1 through 7.11 can then be used to replace both the stacks mentioned earlier by single layers, making the performance in the passbands and the extent of the stop bands easily calculable. We shall first examine the width of the stop bands, given by $M_{11} = -1$. Equation 7.2 shows that this is equivalent to

$$\cos 2\delta_p \cos \delta_q - \frac{1}{2}\left(\frac{n_q}{n_p} + \frac{n_p}{n_q}\right)\sin 2\delta_p \sin \delta_q = -1, \tag{7.14}$$

and this yields exactly the same expression as was obtained in the previous chapter for the width of the unaltered quarter-wave stack. There, δ was replaced by $(\pi/2)g$, where $g = \lambda_0/\lambda$ (or v/v_0, where v is the wavenumber), and the edges of the stop band were defined by

$$\delta_e = \frac{\pi}{2}(1 \pm \Delta g).$$

The total width is therefore

$$2\Delta g = 2\Delta\left(\frac{\lambda_0}{\lambda}\right),$$

where

$$\Delta g = \frac{2}{\pi} \arcsin \left| \frac{\eta_q - \eta_p}{\eta_q + \eta_p} \right|. \tag{7.15}$$

This expression is already plotted in Figure 6.7. The width of the stop band is exactly the same regardless of whether the basic period is $(H/2)L(H/2)$ or $(L/2)H(L/2)$. Of course, it is possible to have other three-layer combinations where the width of the central layer is not equal to twice the thickness of the two outer layers, and some of the other possible arrangements will be examined, both in this chapter and the next, as they have some interesting properties, but as far as the width of the stop band is concerned, it has been shown by Vera [2] that the maximum width for a three-layer symmetrical period is obtained when the central layer is a quarter wave, and the outer layers, an eighth wave each.

The equivalent phase thickness of the period is given by Equation 7.13 as

$$\gamma = \arccos \left[\cos^2 \delta_q - \frac{1}{2} \left(\frac{\eta_p}{\eta_q} + \frac{\eta_q}{\eta_p} \right) \sin^2 \delta_q \right]. \tag{7.16}$$

This expression for γ is multivalued, and the value nearest to $2\delta_q$, the total phase thickness of the period, is the most easily interpreted value. Since γ will be used to reconstruct the original matrix product, we have the freedom to choose the value that makes most sense. It is clear from the expression for γ that the order of the admittances is unimportant.

Let us now turn our attention to the passband, first the equivalent admittance and then the equivalent optical thickness. The expression for the equivalent admittance in the passband is given in Equation 7.12, and for γ, in Equation 7.16. Figure 7.3 shows the equivalent admittance and optical thickness calculated for combinations of tantalum pentoxide and silica. The form of the curves is quite typical of such periods. The evaluation of the equivalent admittance and thickness helps in the understanding of the performance of the filter in the pass region and leads to its subsequent improvement.

From the point of view of modern coating design, the analytical manipulation of the expressions is less important than it used to be. Its primary usefulness now is in the understanding that the equivalence brings with it. For those interested in following the trail of analysis, Thelen [3] will be found most useful.

7.2.4 Application to Edge Filters

We can now make some performance calculations. The coating is equivalent to a thick layer of material with the equivalent admittance as its characteristic admittance. The thickness oscillates between a quarter wave and a half wave. Since these represent the maximum and minimum interference effects, they define the fringe envelopes. The reflectances are given by

$$R = \left(\frac{y_0 - y_{sub}}{y_0 + y_{sub}} \right)^2 \tag{7.17}$$

and

$$R = \left(\frac{y_0 - E^2/y_{sub}}{y_0 + E^2/y_{sub}} \right)^2 \tag{7.18}$$

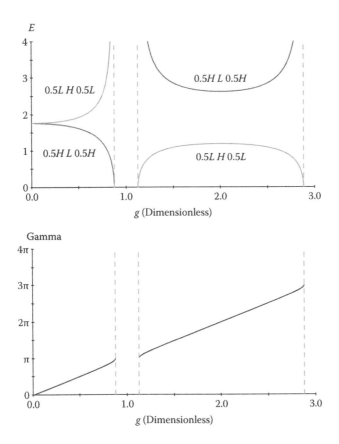

FIGURE 7.3
Equivalent optical admittance E and phase thickness γ of a symmetrical period of $n_H = 2.15$ and $n_L = 1.45$ at normal incidence. The materials are similar to Ta_2O_5 and SiO_2. The vertical broken lines mark the boundaries of the high-reflectance zones within which the values become imaginary and are not plotted. The values of equivalent phase thickness (gamma) are identical for both periods.

for half waves and quarter waves, respectively. We can readily see from Figure 7.3 that for a longwave pass filter with glass as substrate and air as incident medium, a design based on (0.5H L 0.5H) would present a better match than the alternate (0.5L H 0.5L) while the opposite is true for a shortwave pass design. For simple, undemanding edge filters, these structures, repeated a sufficient number of times, are quite adequate.

Figure 7.4 shows the performance of a longwave pass filter of design Air | (0.5H L 0.5H)^{10} | Glass, using the parameters from Figure 7.3. These are $n_H = 2.15$; $n_L = 1.45$, typical of Ta_2O_5 and SiO_2; and $n_{Glass} = 1.52$. The envelopes are superimposed over the fringes in the pass region. It will be noticed that the final half-wave thickness at the edge of the high-reflectance zone does not coincide with the appropriate envelope. Figure 7.3 shows that the equivalent admittance is zero at this point. At a half-wave condition, the sine of the phase thickness is also zero. Thus, we have a zero divided by zero in the corresponding characteristic matrix, implying that we must return to the original set of matrices for the performance calculation at this individual point. Then we find that in the product matrix, although the element M_{21} in the product matrix is zero, M_{12} is not, and this finite value of M_{12} implies that the absentee condition is not satisfied and the reflectance is high, as indicated. A similar condition exists with the use of the (0.5L H 0.5L) period in a shortwave pass configuration. Figure 7.3 also shows that the equivalent admittance can also become infinite at the high-reflectance zone edge. Again, one of the appropriate matrix elements is zero while the other is not,

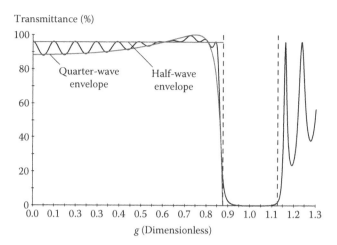

FIGURE 7.4
Performance of the simple longwave pass filter Air | $(0.5H\ L\ 0.5H)^{10}$ | Glass with the envelopes superimposed. The vertical broken lines represent the boundaries of the high-reflectance zone.

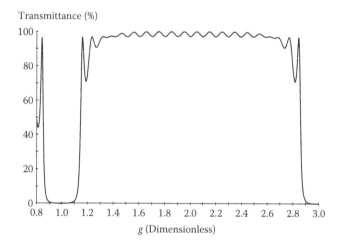

FIGURE 7.5
Calculated transmittance of a shortwave pass filter of design Air | $(0.5L\ H\ 0.5L)^{10}\ 0.5L$ | Glass with materials as in Figure 7.4. Note the extra $0.5L$ layer next to the substrate. This not only gives a weak antireflecting effect at $g = 2$, but it also makes the layer next to the substrate a quarter wave, rather than eighth wave, slightly easing the monitoring requirements.

and the reflectance does not coincide with the half-wave envelope. Although analytical expressions for the reflectance at this point exist and were fully detailed by Thelen [3], computer analysis is so simple and rapid that it is normally preferred.

Figure 7.5 shows the transmittance of a shortwave pass figure using the $(0.5L\ H\ 0.5L)$ period.

7.2.5 Application of the Herpin Index to Multilayers of Other than Quarter Waves

All the curves shown so far are for | eighth-wave | quarter-wave | eighth-wave | periods. If the relative thicknesses of the layers are varied from this arrangement, then the equivalent admittance is altered. It has already been mentioned that the reflectance zones for a combination other than those mentioned earlier must be narrower. Some idea of the way in which the equivalent admittance alters can be obtained from the value as $g \to 0$. Let $2\delta_p/\delta_q = \psi$. Then, from Equation 7.12,

$$E = \eta_p \left[\frac{\dfrac{\sin 2\delta_p}{\sin \delta_q} \cos \delta_q + \dfrac{1}{2}\left(\dfrac{\eta_q}{\eta_p} + \dfrac{\eta_p}{\eta_q}\right)\cos 2\delta_p - \dfrac{1}{2}\left(\dfrac{\eta_p}{\eta_q} - \dfrac{\eta_q}{\eta_p}\right)}{\dfrac{\sin 2\delta_p}{\sin \delta_q} \cos \delta_q + \dfrac{1}{2}\left(\dfrac{\eta_q}{\eta_p} + \dfrac{\eta_p}{\eta_q}\right)\cos 2\delta_p + \dfrac{1}{2}\left(\dfrac{\eta_p}{\eta_q} - \dfrac{\eta_q}{\eta_p}\right)} \right]^{1/2} . \tag{7.19}$$

Now $\sin 2\delta_p/\sin \delta_q \to \psi$ as $g \to 0$, i.e.,

$$E \to \eta_p \left[\frac{\psi + \dfrac{1}{2}\left(\dfrac{\eta_q}{\eta_p} + \dfrac{\eta_p}{\eta_q}\right) - \dfrac{1}{2}\left(\dfrac{\eta_p}{\eta_q} - \dfrac{\eta_q}{\eta_p}\right)}{\psi + \dfrac{1}{2}\left(\dfrac{\eta_q}{\eta_p} + \dfrac{\eta_p}{\eta_q}\right) + \dfrac{1}{2}\left(\dfrac{\eta_p}{\eta_q} - \dfrac{\eta_q}{\eta_p}\right)} \right]^{1/2} .$$

By rearranging this, we obtain

$$E = \eta_p \left[\frac{\psi + \left(\eta_q/\eta_p\right)}{\psi + \left(\eta_p/\eta_q\right)} \right]^{1/2} . \tag{7.20}$$

This result shows that for small g, it is possible, by varying the relative thicknesses of the two materials, to vary the equivalent admittance throughout the range of values between η_p and η_p, but not outside that range. This result has already been referred to in the chapter on antireflection coatings, where it was shown how to use the concept of equivalent admittance to create replacements for layers having indices difficult to reproduce.

Epstein [1] considered in more detail the variation of equivalent admittance by altering the thickness ratio and gave tables of results of zinc sulfide/cryolite multilayers. Ufford and Baumeister [4] gave sets of curves that assist in the use of equivalent admittance in a wide range of design problems.

Some results that are, at first sight, rather surprising are obtained when the value of the equivalent admittance around $g = 2$ is investigated. First of all, let the layers be the original quarter and eighth waves. As $g \to 2$, $2\delta_p \to \pi$, and $\delta_q \to \pi$ so that, from Equation 7.19,

$$E = \eta_p \left[\frac{-1 - \dfrac{1}{2}\left(\dfrac{\eta_q}{\eta_p} + \dfrac{\eta_p}{\eta_q}\right) - \dfrac{1}{2}\left(\dfrac{\eta_p}{\eta_q} - \dfrac{\eta_q}{\eta_p}\right)}{-1 - \dfrac{1}{2}\left(\dfrac{\eta_q}{\eta_p} + \dfrac{\eta_p}{\eta_q}\right) + \dfrac{1}{2}\left(\dfrac{\eta_p}{\eta_q} - \dfrac{\eta_q}{\eta_p}\right)} \right]^{1/2} = \left(\dfrac{\eta_p}{\eta_q}\right)^{1/2} . \tag{7.21}$$

This is quite a straightforward result. Now let $2\delta_p/\delta_q = \psi$, as in the case just considered where $g \to 0$. Let $g \to 2$ so that $2\delta_p + \delta_q \to 2\pi$. (In this case, we define $g = \lambda_0/\lambda$ by defining λ_0 as that wavelength making $2\delta_p + \delta_q = \pi$.)

We have, as $g \to 2$,

$$\cos 2\delta_p \to \cos\left(2\pi - \delta_q\right) = \cos \delta_q,$$

$$\sin 2\delta_p \to \sin\left(2\pi - \delta_q\right) = -\sin \delta_q,$$

and $\delta_q \to 2\pi/(1 + \psi)$ so that

$$E = \eta_p \left[\frac{-\sin\delta_q \cos\delta_q + \frac{1}{2}\left(\frac{\eta_q}{\eta_p} + \frac{\eta_p}{\eta_q}\right)\cos\delta_q \sin\delta_q - \frac{1}{2}\left(\frac{\eta_p}{\eta_q} - \frac{\eta_q}{\eta_p}\right)\sin\delta_q}{-\sin\delta_q \cos\delta_q + \frac{1}{2}\left(\frac{\eta_q}{\eta_p} + \frac{\eta_p}{\eta_q}\right)\cos\delta_q \sin\delta_q + \frac{1}{2}\left(\frac{\eta_p}{\eta_q} - \frac{\eta_q}{\eta_p}\right)\sin\delta_q} \right]^{1/2}$$

(7.22)

$$= \eta_p \left[\frac{-\cos\delta_q\left\{1 - \frac{1}{2}\left(\frac{\eta_q}{\eta_p} + \frac{\eta_p}{\eta_q}\right)\right\} - \frac{1}{2}\left(\frac{\eta_p}{\eta_q} - \frac{\eta_q}{\eta_p}\right)}{-\cos\delta_q\left\{1 - \frac{1}{2}\left(\frac{\eta_q}{\eta_p} + \frac{\eta_p}{\eta_q}\right)\right\} + \frac{1}{2}\left(\frac{\eta_p}{\eta_q} - \frac{\eta_q}{\eta_p}\right)} \right]^{1/2},$$

where $\cos\delta_q = \cos[2\pi/(1 + \psi)]$.

Whatever the value of ψ, the quantities within the square root brackets have opposite signs, which means that the equivalent admittance is imaginary. Even as $\psi \to 1$, where one would expect the limit to coincide with the result in Equation 7.21, the admittance is still imaginary. The explanation of this apparent paradox is as follows. An imaginary equivalent admittance, as we have seen, indicates a zone of high reflectance. Consider first the ideal eighth-wave | quarter-wave | eighth-wave stack of Equation 7.21. At the wavelength corresponding to $g = 2$, the straightforward theory predicts that the reflectance of the substrate shall not be altered by the presence of the multilayer, because each period of the multilayer is acting as a full wave of real admittance and is therefore an absentee layer. By looking more closely at the structure of the multilayer, we can see that this can also be explained by the fact the all the individual layers are a half-wavelength thick. If the ratio of the thicknesses is altered, the layers are no longer a half-wavelength thick and cannot act as absentees. In fact, the theory of the result mentioned earlier shows that a zone of high reflectance occurs.

The transmittance of a shortwave pass filter at the wavelength corresponding to $g = 2$ is therefore very sensitive to errors in the relative thicknesses of the layers. Even a small error leads to a peak of reflection. The width of this spurious high-reflectance zone is quite narrow if the error is small. Thus, the appearance of a pronounced narrow dip in the transmission curve of a shortwave pass filter is quite a common feature and is difficult to eliminate. The dip is sometimes referred to as a *half-wave hole*.

7.2.6 Reduction of Passband Ripple

Our discussion so far has shown that the ripple is a consequence of a mismatch between the thin-film structure and the surrounding media. The reduction of ripple therefore implies improved matching. We need what are essentially antireflection coatings at the front and at the rear of the structure. There is an enormous body of knowledge of such matching, most of it dating back to the time when analytical approaches were most important. We look first at a small selection.

The simplest approach is to start with the repeated symmetrical period and to insert the usual quarter-wave matching layers between it and the media. This requires materials of the correct index and somewhat limits the usefulness. Sometimes one of the materials in the symmetrical structure can be used for matching. A simple case is a shortwave pass filter consisting of zinc sulfide ($n_L = 2.35$) and germanium ($n_H = 4.0$) on a germanium substrate with air incident medium. We use a repeated (0.5L H 0.5L) period that matches air quite well close to the edge but exhibits a serious mismatch with the germanium substrate (Figure 7.6). However, the low-index zinc sulfide layer is of a suitable index for matching, and we choose a value of g of 1.25 for the quarter-wave condition. The resulting reflectance envelopes are indicated in Figure 7.6, and the transmittance of the filter is shown in Figure 7.7. The performance is reasonable, although not outstanding.

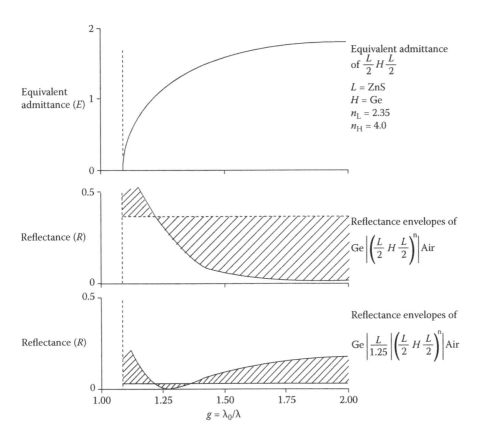

FIGURE 7.6
Steps in the design of a shortwave pass filter using zinc sulfide and germanium on a germanium substrate.

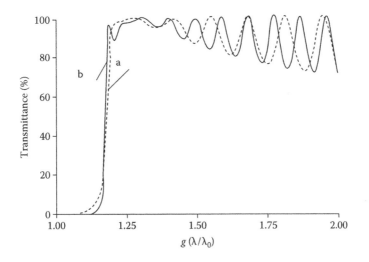

FIGURE 7.7
Calculated performance of filters designed according to Figure 6.9 with the design

$$\text{Air} | (0.5L\,H\,0.5L)^q\,L/1.25 | \text{Ge}$$

with $n_L = 2.35$, $n_H = 4.0$, $n_{Ge} = 4.0$, and $n_{Air} = 1.00$. $q = 7$ gives the 15-layer design (curve a), and $q = 10$, the 21-layer design (curve b).

A straightforward improved method was suggested by Welford [5], but it seems to have not been much used. This is simply to vary the thicknesses of the films in the basic period so that the equivalent admittance is altered to bring it nearer to the desired value for matching. For this method to be successful, the reflectance from the bare substrate must be kept low and the substrate should have a low index. Glass in the visible region is quite satisfactory, but the method is difficult to apply in the infrared to substrates such as silicon and germanium without modification.

Probably the ultimate antireflection coating is the inhomogeneous layer. Unfortunately, as we have seen, the major problem with inhomogeneous layers is the attainment of an index of refraction less than around 1.35, and we require this if we are to match to air. Jacobsson [6], however, briefly considered the matching of a multilayer longwave pass filter $[(H/2)L(H/2)]^6$, consisting of germanium with an index of 4.0 and silicon monoxide with an index of 1.80, to a germanium substrate by means of an inhomogeneous layer. His paper shows the three curves reproduced in Figure 7.8. Curve 1 is the multilayer on a glass substrate of index 1.52. Since, in the passband, the equivalent admittance of the multilayer falls gradually from $(1.8 \times 4.0)^{1/2} = 2.7$ to 0 as the wavelength approaches the edge, it will be a value not too different from the index of the substrate in the vicinity of the edge. The transmittance near the edge is, therefore, high, as we might expect. When, as in curve 2, the same multilayer is deposited on a germanium substrate of index 4.0; the severe mismatching causes a very large ripple to appear. With an inhomogeneous layer between the germanium substrate and the multilayer and with the index varying from that of germanium next to the substrate to 1.52 next to the multilayer, the performance achieved, curve 3, is almost exactly that of the original multilayer on the glass substrate.

Thelen [3] pointed out that the rapid variation of equivalent admittance near the edge of the filter is the major source of difficulty in edge filter design. It is a simple matter to match the multilayer to the substrate where the equivalent admittance curve is flat, some distance from the edge, but the variations near the edge usually give rise, with simple designs, to a pronounced dip in the transmission curve. Thelen devised an ingenious method of dealing with this dip, involving the equivalent of a single-layer antireflection coating. Between the substrate and/or the incident medium and the main or primary multilayer, which consists of a number of equal basic periods, Thelen placed a secondary multilayer, similar to the first but shifted in thickness so that at the

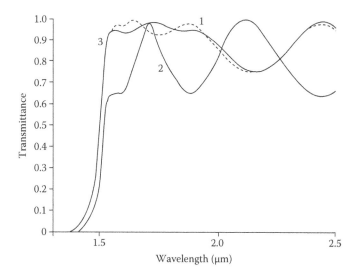

FIGURE 7.8
Reflectance versus wavelength of a multilayer on a substrate with indices $n_m = 1.52$ (curve 1) and $n_m = 4.00$ (curve 2) and on a substrate with $n_m = 4.00$ with an inhomogeneous layer between substrate and multilayer (curve 3). (After R. Jacobsson, *Journal of the Optical Society of America*, 54, 422–423, 1964. With permission of Optical Society of America.)

center of the steep portion of the admittance curve, the equivalent admittance of the secondary is made equal to the square root of the equivalent admittance of the primary times the admittance of the substrate. The number of secondary periods is chosen to make the thickness at this point an odd number of quarter waves and to completely satisfy the antireflection condition. Figure 7.9 shows the performance he achieved.

Other techniques are based on results derived in the microwave region. Young and Crystal [7] used symmetrical periods to mimic indices derived from microwave theory. Seeley et al. [8] and Seeley and Smith [9] adapted results from the synthesis of lumped electrical circuits. There are many studies of considerable interest, but for the practicing designer of today, computer-aided design, already discussed in Chapter 3, is the method of choice.

We have already mentioned computer refinement in Section 3.4.6. Computer refinement was introduced into optical coating design by Baumeister [10], who programmed a computer to estimate the effects of slight changes in the thicknesses of the individual layers on a merit function representing the deviation of the performance of the coating from the ideal. An initial design, not too far from ideal, was adopted and the thicknesses of the layers were gradually modified to improve the performance. The optimum thickness of any one layer is not independent of the thicknesses of the other layers so that the changes in thickness at each iteration could not be large without running the risk of instability. Computer speed and capacity has considerably increased since the early work of Baumeister, but the essentials of the method have not changed. The quality of the currently existing design is presented as a single figure of merit, and the design is changed so as to improve it. The way in which changes are made are assessed is the principal difference between the techniques in frequent use. An enormous number of different techniques exists. Comparisons of various methods have tended to show that there is no universal best technique [11]. Computer refinement is a very powerful design aid, but it can function only with an initial design. It then modifies the design to improve the performance. Less usual is a complete design

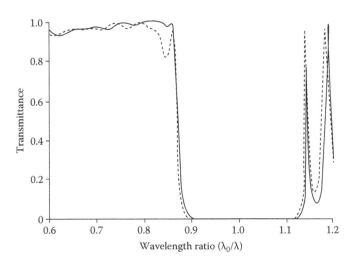

FIGURE 7.9
Comparison of the computed performance of the filters

$$1.00|(0.5\,H\,L\,0.5H)^{15}|1.52 \quad \textit{(dashed line)}$$

and

$$1.00|(0.5H\,L\,0.5H)^{12}\left[(1/1.05)(0.5H\,L\,0.5H)^{3}\right]|1.52 \quad \textit{(solid line)}$$

with $n_H = 2.3$ and $n_L = 1.56$. (After A. Thelen, *Journal of the Optical Society of America*, 56, 1533–1538, 1966. With permission of Optical Society of America.)

synthesis with no starting solution. This adds to refinement a process for complicating the design when the refinement reaches an impasse, that is, a minimum merit figure that is still unsatisfactory. Synthesis was introduced in the 1960s by Dobrowolski [12]. The use of both refinement and synthesis in design has grown in pace with improvements in and availability of computers.

Refinement processes, and synthesis, search parameter space for acceptable designs. They cannot find what is not present in parameter space. Computer techniques, no matter how powerful, are unable to find designs that do not, or cannot, exist. They have no appreciation of feasibility. Thus, there is still room for analytical techniques because they tell us about feasibility and existence of designs and save us from wasting time in unrealistic searches. They help us set up designs for computer refinement. Thus, in what follows, we will use what we have already learned about edge filters.

We know from the previous pages that the edge filter essentially consists of a structure based on the quarter-wave stack. The properties are derived from a mismatch between this structure and the surrounding media. In the rejection, or high-reflectance, zones the mismatch is so serious that high reflectance is achieved. The mismatch is in the nature of a real admittance and an imaginary admittance. Inserting dielectric layers in between these two media has little effect on the high reflectance. In the pass zones, or potential pass zones, the mismatch is essentially between two real admittances, and this leads to interference fringes that are usually known as ripple. Here dielectric layers in the form of matching assemblies can be used to reduce the mismatch and, hence, the ripple. Thus, a useful procedure is to establish a structure, usually based on the quarter-wave stack, that has the correct rejection and potential pass zones and to design suitable matching structures to reduce the ripple to acceptable levels. This is rather like the design of an antireflection coating except that the thin-film structure exhibits dispersion very much more complex than a simple dielectric material. However, the computer is just as successful at handling the complicated dispersion as it is at simple dispersion. The user simply has to set the problem up in the form of the correct instructions. We illustrate the method in the design of a straightforward edge filter of a longwave pass type. To simplify the demonstration, we will use dispersionless materials of indices 1.45 and 2.15 for the thin films and 1.52 for the substrate. We will arrange for the longwave edge of the rejection region to be at 700 nm.

Edge is a vague term. It means what the supplier of the filter intends it to mean. In some cases, the 5% transmittance point is taken as the edge wavelength. In other cases it can be the 50% wavelength. We use it here in the rather imprecise sense of the start of the pass region and, therefore, of rather high transmittance. If this is not satisfactory, then the reference wavelength can be readily adjusted to achieve whatever is the desired definition. We start with a quarter-wave stack with reference wavelength of 608 nm. How many layers to use is always a difficult question. Fortunately, with a relatively fast computer, a little trial and error is a completely feasible and simple method. The design process is going to construct matching assemblies on either side of the core of the filter that we will retain as a set of quarter waves. These matching assemblies will reduce a little, the steepness of edge. We start with a quarter-wave stack that has roughly the desired shape. This will be the core of the filter. Next, we add sufficient quarter-wave layers on either side to form the matching structures. We lock the quarter-wave stack core so that it does not take part in the refinement. We specify the transmittance in the pass region as the targets for refinement, and the result of this procedure with a 28-layer core and 10-layer matching structures on either side is shown in Figure 7.10.

This is normally a very fast procedure, and it has the advantage of avoiding the need for specifying the rejection region in the targets. It also avoids the appearance of spikes in between target points in the rejection region, which can be a problem when all layers take part in the refinement.

Some refinement processes, particularly very fast ones, can be rather eager in their travel over the merit surface, shutting down some layer thicknesses to zero. It can be difficult to recover from

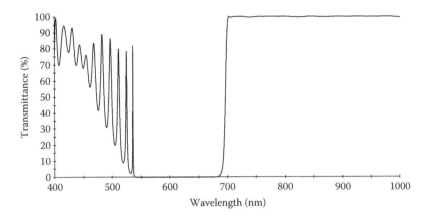

FIGURE 7.10
Transmittance of a 48-layer longwave pass filter. The outermost 10 layers on each side of the quarter-wave structure were refined to reduce the ripple in the pass region. The remaining layers form a quarter-wave core.

this. A good technique, especially in such cases, is to gradually open up the matching systems for refinement.

Once a design has been established, it is easy to check the rejection performance and make appropriate changes to the starting design by adding or subtracting layers.

Shortwave pass filters can be designed using essentially the same technique. It should be remembered, however, that the possible width of the pass region is limited by the appearance of the higher reflecting orders. They cannot be simply eliminated by the outer matching layers but need a rather different approach, which is dealt with a little later in this chapter.

7.2.7 Blocking

Because the stop band of the multilayer edge filter is limited in extent, it is usually necessary for practical filters to consist of a multilayer filter together with additional filters to give the broad rejection region that is almost always required. These additional blocking filters may be multilayer, and some methods of broadening the stop band in this way are mentioned in the following section. Usually, they are absorption filters having wide rejection regions but inflexible characteristics. These absorption filters may be combined with multilayer filters in a number of different ways. They may simply be placed in series with the substrates carrying the multilayers, the substrates themselves may be the absorption filters, or the multilayer materials may also act as thin-film absorption filters.

In the visible and near ultraviolet regions, there is a wide range of glass filters available which solve most of the problems connected with longwave pass filters. In Figure 7.11, the filter of Figure 7.10 has been deposited on a 3 mm thick substrate of RG645 filter glass (a product of Schott AG) with a four-layer antireflection coating on the opposite side. The lower plot with logarithmic scale shows the overlapping characteristics. Good shortwave pass characteristics are difficult to achieve with absorbing filters in the visible and near infrared, and the sidebands of shortwave pass filters are more difficult to block.

In the infrared, the position is rather more difficult, and often, the complete filter consists of several multilayers to connect the edge of the stop band to the nearest suitable absorption filter. Figure 7.12 shows a longwave pass filter for the infrared. Figure 7.13 gives some of the infrared absorption filters that have shortwave pass characteristics. Unfortunately, not all of these materials are currently easily available. For longwave pass characteristics, semiconductors such as silicon, with an edge at 1 μm, and germanium, with an edge at 1.65 μm, are the most suitable. Indium

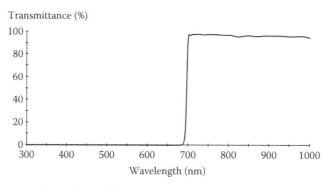

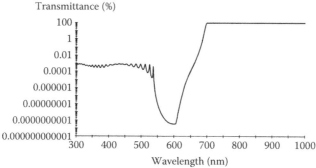

FIGURE 7.11
Transmittance of the filter of Figure 7.10 deposited over a substrate of RG645 filter glass (3 mm thick). A four-layer antireflection coating is deposited on the opposite side. The lower plot with logarithmic scale shows the overlapping characteristics.

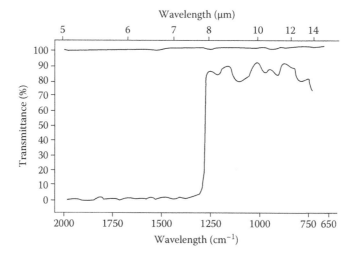

FIGURE 7.12
Measured transmittance of a practical longwave pass filter with edge at 1250 cm^{-1} (8 μm). (Courtesy of OCLI Optical Coatings Laboratory Ltd., Santa Rosa, CA.)

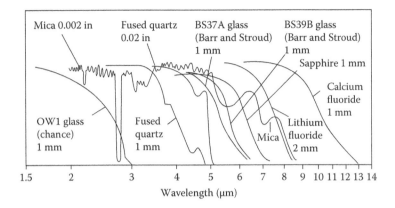

FIGURE 7.13
Selection of infrared materials that can be used as shortwave pass absorption filters. Note the Barr and Stroud materials are aluminate glasses. (Courtesy of Sir Howard Grubb, Parsons and Co. Ltd., Newcastle upon Tyne.)

arsenide, with an edge at 3.4 μm, and indium antimonide, with edge at 7.2 μm, are also useful, but because of the rather higher absorption, they can only be used in very thin slices, around 0.013 cm for indium antimonide and only a little thicker for indium arsenide. This means that they tend to be extremely fragile and can only be produced in a circular shape of rather limited diameter, not usually greater than 2.0 cm.

The measured transmittance for a longwave pass filter consisting of an edge filter together with an absorption filter is given in Figure 7.14. This filter was originally designed to block the shortwave sidebands of narrowband filters at 15 μm. It consists of two components, a multilayer filter made from a lead telluride and a zinc sulfide multilayer on a germanium substrate and placed in series with an indium antimonide filter. The very high rejection achieved can be seen from the logarithmic plot.

In the absence of suitable absorbing characteristics, the most convenient and straightforward way of extending the reflectance zone is to place a second quarter-wave stack in series with the

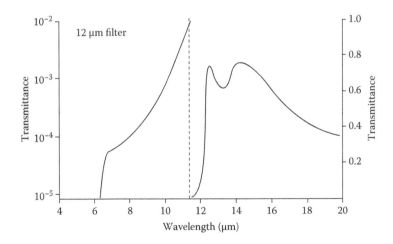

FIGURE 7.14
Measured transmittance of a multilayer blocking filter with edge at 12 μm. A subsidiary indium antimonide filter is included to ensure good blocking at wavelengths shorter than 7 μm. (After J. S. Seeley and S. D. Smith, *Applied Optics*, 5, 81–85, 1966. With permission of Optical Society of America.)

first, and to ensure that their rejection zones overlap. The second stack is most conveniently placed either on a second substrate or on the opposite side of the substrate from the first stack. This avoids some unfortunate interference consequences. However, there are also some multiple beam effects that can lead to disappointing performance.

We would like the transmittance of the combined components to be given by

$$T = T_a T_b,$$ (7.23)

where T_a and T_b are the individual transmittances. This will be the case if the transmission loss is purely by absorption, but if there are multiple beams reflected back and forth between the surfaces and these beams reach the receiver, then the transmittance will be higher.

The worst case occurs when all the multiple beams are collected and contribute to the signal. The full theory is given in Section 2.14. In the complete absence of absorption and with the collection of all beams, the transmittance is given by

$$T = \frac{1}{(1/T_a) + (1/T_b) - 1} .$$ (7.24)

If T_a and T_b are both small and equal, then instead of their product, the resultant transmittance is just one-half of their individual transmittance. Whatever the arrangement is, the net transmittance can never be larger than the smaller of T_a and T_b.

To avoid this, the substrate should be made reasonably thick and, if possible, slightly wedged. Anything that can be done to reduce the power of the reflected beams is also worthwhile. Absorbing material, even if only slightly absorbing, placed between the two surfaces can also help.

A more difficult situation occurs when it is impossible to place the stacks on separate surfaces, and one stack must be directly deposited on top of the other. In this case, it is necessary to take precautions to avoid the creation of transmission maxima. The problem has already been dealt with in Chapter 6, where the extension of the high-reflectance zone of a quarter-wave stack was discussed.

If we consider the assembly splitting into two separate multilayers, then a transmission maximum will occur at any wavelength for which $(\varphi_a + \varphi_b)/2 = m\pi$, where $m = 0, \pm1, \pm2, \ldots$. The height of this maximum is given by

$$T = \frac{|\tau_a^+|^2 |\tau_b^+|^2}{(1 - |\rho_a^-||\rho_b^+|)^2} = \frac{T_a T_b}{\left[1 - (R_a R_b)^{1/2}\right]^2} .$$ (7.25)

If there is no absorption, this expression implies that for low transmittance at the maxima, R_a and R_b should be as dissimilar as possible. This can be achieved by using many layers to keep the reflectance of one multilayer as high as possible in the pass region of the other.

In slightly more quantitative terms, from the reflectance envelope, which does not vary with the number of periods, we can find the highest reflectance in the pass region of either multilayers making up the composite filter. If we denote this reflectance by R_p, then we can be certain that the design will be acceptable if we choose a sufficiently high number of periods to make R_s, the lowest reflectance in the stop band of the other multilayer, sufficiently high to ensure that

$$\frac{\left(1 - R_p\right)(1 - R_s)}{\left[1 - \left(R_p R_s\right)^{1/2}\right]^2} \le T_c,$$ (7.26)

where T_c is some acceptable level for the transmission in the rejection zone of the complete filter. This formula will give a pessimistic result; the actual transmission achieved in practice will depend on the phase change as well as the reflectance.

The other danger area is the region where the two high-reflectance bands are overlapping. There, it must be arranged that on no account is $(\varphi_a + \varphi_b)/2 = m\pi$. The method for dealing with this was described in the previous chapter, where a layer of intermediate thickness was placed between the two quarter-wave stacks. The result is equivalent to placing two similar multilayers, both of the form $[(L/2)H(L/2)]^m$ or $[(H/2)L(H/2)]^m$, together.

More details, including the effects of phase, are given in Section 13.8 on performance envelopes.

Equation 7.26 also implies that some of the sections of the composite filter should have more periods than others. In the reduction of the ripple in the passband of the basic multilayer, the ripple on the other side of the stop band is almost invariably increased. Thus, in the combination of, say, two multilayers, the rejection zone of one stack will overlap a region of high ripple, while the rejection zone of the other stack will overlap a region of relatively low ripple. Since high ripple means that R_p is high, the former stack should have more periods than the latter if the same level of rejection is required throughout the combined rejection region. Figure 7.15 [13] shows two component edge filters which are combined in a single filter in Figure 7.16. The severe ripple that occurs in one of the multilayers can be seen reflected in the rejection zone of the composite filter. This ripple is limited to part of the rejection zone only, and in order to reduce the effect, more periods are necessary in the appropriate multilayer.

The same procedure is valid for shortwave pass filters, but there is the additional problem of the higher reflecting orders, which is dealt with in the next section.

A common current requirement is for a shortwave pass filter to block the near infrared and transmit the visible region. These filters are used in different applications, but a common one is the reduction of the infrared sensitivity of a silicon receiver that might be used in a digital camera. The width of the rejection region is beyond the capabilities of a single quarter-wave stack, and so we need to broaden the rejection zone. We can do this using one of the methods or others already discussed. Fortunately, the usual requirement can be satisfied without the need for the suppression of a third-order peak, and so a straightforward two-material structure is sufficient.

The filter in Figure 7.17 was created by the refinement of all layers, starting with a tapered design as shown and involving all layers in the refinement. A virtually indistinguishable final design can also be obtained by starting with air $|(LH)^{20}|$ Glass at $\lambda_0 = 800$ nm. The refinement process

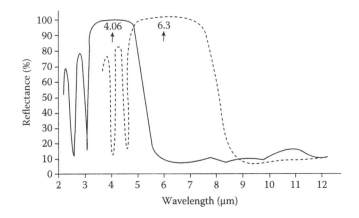

FIGURE 7.15
Measured reflectance of two longwave pass stacks:

$$A|(0.5H\,L\,0.5H)^4|BaF_2$$

H and L are films of stibnite and chiolite a quarter-wave thick at $\lambda_0 = 4.06$ or 6.3 μm. A is air and the substrate is barium fluoride. (After A. F. Turner and P. W. Baumeister, *Applied Optics*, 5, 69–76, 1966. With permission of Optical Society of America.)

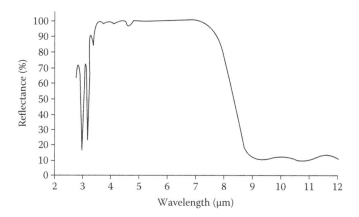

FIGURE 7.16
Measured reflectance of the two longwave pass stacks of Figure 7.15 superimposed in a single coating for an extended high-reflectance region. (After A. F. Turner and P. W. Baumeister, *Applied Optics*, 5, 69–76, 1966. With permission of Optical Society of America.)

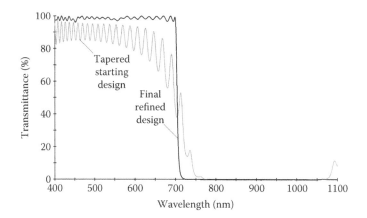

FIGURE 7.17
Performance of a shortwave pass filter with extended rejection zone. This is a 40-layer design with $n_L = 1.45$ and $n_H = 2.35$, roughly corresponding to SiO_2 and TiO_2. The substrate is glass with index 1.52 and the incident medium is air with $n = 1.00$. A linearly tapered 40-layer stack with low-index outermost was created as starting design, and then the entire system was refined to yield the final performance.

achieved the sharp edge by assembling the first half of the filter into what is essentially a quarter-wave stack with the required sharpness of shortwave edge. It adjusted the remaining layers to fill in the longwave part of the rejection region. By a slight tuning of all the layer thicknesses, it achieved the matching in the pass region. Different refinement techniques all appear to converge on much the same final design. Note that in this case, the targets must include the high-reflectance zone, unlike the edge filter using a single quarter-wave stack, and all the layers take part in the refinement.

This approach, using refinement, where a starting design, sufficiently close to the requirements can be created, and where there is reasonable certainty of the existence of a satisfactory end point, is typical of the more modern way of designing optical coatings. It does require knowledge of the structures likely to be successful, but it removes the tedious manual labor that used to be the primary feature of thin-film coating design.

7.2.8 Extending the Transmission Zone

The shortwave pass filter, as it has been described so far, possesses a limited passband because of the higher-order stop bands. These are not always particularly embarrassing, but occasionally, as, for example, with some types of heat-reflecting filters, a much wider passband is required. The problem was first considered by Epstein [14] and was more extensively studied by Thelen [15].

Epstein's analysis was as follows. Let the multilayer be represented by q periods each of the form

$$M = \begin{bmatrix} M_{11} & M_{12} \\ M_{21} & M_{22} \end{bmatrix}.$$

If a single period is considered as if it were immersed in a medium of admittance y, then the transmission coefficient of the period is given by

$$\tau = \frac{2y_0}{y_0 B + C} = \frac{2y}{y[(M_{11} + M_{22}) + \{yM_{12} + (M_{21}/y)\}]},$$

where y_0, B, and C have the usual meanings as explained in Chapter 2.

Let $\tau = |\tau|e^{i\varphi}$; then

$$\frac{1}{2}[(M_{11} + M_{22}) + \{yM_{12} + (M_{21}/y)\}] = \frac{\cos\varphi - i\sin\varphi}{|\tau|}.$$

If the period is transparent, equating real parts gives

$$\frac{1}{2}(M_{11} + M_{22}) = \frac{\cos\varphi}{|\tau|}.$$

Now, if light that has suffered two or more reflections at interfaces within the period is ignored, then

$$\varphi \simeq \sum \delta,$$

the total phase thickness of the period.

When $\Sigma\delta = m\pi$, $\cos\varphi = \pm 1$, and if $|\tau| < 1$, then

$$\frac{1}{2}|(M_{11} + M_{22})| > 1,$$

and a high-reflectance zone results. If, however, $|\tau| = 1$, then

$$\frac{1}{2}|(M_{11} + M_{22})| = 1,$$

and the high-reflectance zone is suppressed. In the simple form of stack, $[(L/2)H(L/2)]^q$ or $[(H/2)L(H/2)]^q$, $|\tau| = 1$ for $\tau = 2m\pi$ ($m = 1, 2, 3, 4, \ldots$), and the even-order high-reflectance zones are therefore suppressed. As noted earlier, only a slight change in the relative thicknesses of the layers is enough to reduce τ and turn the band into a high-reflectance zone.

Putting this result in another way, a zone of high reflectance potentially exists whenever the total optical thickness of an individual period of the multilayer is an integral number of half waves, and the high-reflectance zone is prevented from appearing if, and only if, $|\tau| = 1$. This result was then used by Epstein in his paper to design a multilayer in which the fourth- and fifth-order reflectance bands were suppressed. Thelen has extended Epstein's analysis to deal with cases where any two and any three successive orders are suppressed, and it is this method we shall follow.

Following Epstein, Thelen [15] assumed a five-layer form ABCBA involving three materials, as the basic period of the multilayer, and noted that if the period is thought of as immersed in medium M, combination AB becomes an antireflection coating for C in M at the wavelengths where suppression is required. In the construction of the final multilayer, medium M can be first considered to exist between successive periods and then to suffer a progressive decrease in thickness until it just vanishes. The shrinking procedure leaves the suppression of the various orders that has been arranged unchanged. M can therefore be chosen quite arbitrarily during the design procedure to be discarded later. The antireflection coating AB is of a type originally studied by Muchmore [16], and Thelen adapted his results as follows.

The various parameters of the layers are denoted by the usual symbols with the appropriate subscripts A, B, C, and M.

Let layers A and B be of equal optical thickness, i.e.,

$$\delta_A = \delta_B, \tag{7.27}$$

and let

$$y_A y_B = y_C y_M . \tag{7.28}$$

Then the wavelengths for which unity transmittance will be achieved will be given by

$$\tan^2 \delta'_A = \frac{y_A y_B - y_C^2}{y_B^2 - (y_A y_C^2 / y_B)} . \tag{7.29}$$

(This result can be derived from Equations 4.7 and 4.8. If we replace the subscripts 1, 2, m, and 0 in these equations by A, B, C, and M, respectively, then the condition for $\delta_A = \delta_B$ is, from Equation 4.8, $y_A y_B = y_C y_M$, and Equation 7.29 then immediately follows from Equation 4.7.)

Two solutions given by Equation 7.29, δ'_A and $(\pi - \delta'_A)$, are possible. We can specify that δ'_A corresponds to λ_1, and $(\pi - \delta'_A)$, to λ_2, where λ_1 and λ_2 are the two wavelengths where suppression is to be obtained. Solving these two equations for δ'_A gives

$$\delta'_A = \frac{\pi}{1 + (\lambda_1/\lambda_2)} . \tag{7.30}$$

This can be entered in Equation 7.29, whence

$$\tan^2 \frac{\pi}{1 + (\lambda_1/\lambda_2)} = \frac{y_A y_B - y_C^2}{y_B^2 - (y_A y_C^2 / y_B)} . \tag{7.31}$$

This determines the complete design of the coating. The optical thickness of layer A can be found from Equation 7.30 to be

$$\frac{\lambda_1 \lambda_2}{2(\lambda_1 + \lambda_2)} . \tag{7.32}$$

The only other quantity to be found is the optical thickness of layer C, and we first note that the total optical thickness of the period is $\lambda_0/2$, where λ_0 is the wavelength of the first high-reflectance zone. The optical thicknesses of layers A and B have already been defined as equal, so that the optical thickness of layer C is

$$\frac{\lambda_0}{2} - \frac{2\lambda_1 \lambda_2}{2(\lambda_1 + \lambda_2)} . \tag{7.33}$$

Medium M, introduced as an artificial aid to calculation, disappears and does not figure at all in the results. Any two of the optical admittances y_A, y_B and y_C can be chosen at will. The third one is then found from Equation 7.31.

Thelen gives a large number of examples of multilayers with various zones suppressed. Particularly useful is a multilayer with the second- and third-order zones suppressed. For this,

$$\lambda_1 = \lambda_0/2, \quad \lambda_2 = \lambda_0/3,$$

and all the layers are found to be of equal optical thickness $\lambda_0/10$. Two of the refractive indices of the layers are then chosen, and Equation 7.31 solved for the remaining one. For rapid calculation, Thelen gives a nomogram connecting the three quantities. The transmittance of a multilayer with the second and third orders suppressed is given in Figure 7.18.

Thelen also considered a multilayer in which the second, third, and fourth orders were all suppressed and found the conditions to be as follows:

- Layer thicknesses:

$$A : \lambda_0/12$$

$$B : \lambda_0/12$$

$$C : \lambda_0/6$$

The indices are given by

$$y_B = (y_A y_C)^{1/2}.$$

Figure 7.19 shows the transmittance of a multilayer where the second, third, and fourth orders have been suppressed in this way.

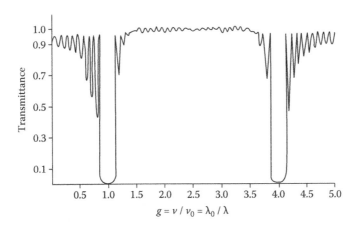

FIGURE 7.18

Calculated transmittance as a function of g of the design: Incident $|(ABCBA)^{10}A|$ substrate with $n_m = 1.50$, $n_0 = 1.00$, $n_A = 1.38$, $n_B = 1.90$ and $n_C = 2.30$. (After A. Thelen, *Journal of the Optical Society of America*, 53, 1266–1270, 1963. With permission of Optical Society of America.)

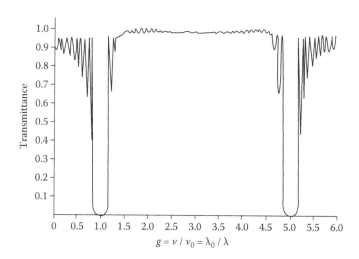

FIGURE 7.19
Calculated transmittance as a function of g of the design:

$$\text{Incident} \mid (\text{AB2CBA})^{10}\,\text{A} \mid \text{substrate}$$

with $n_m = 1.50$, $n_0 = 1.00$, $n_A = 1.38$, $n_B = 1.781$, and $n_C = 2.30$. (After A. Thelen, *Journal of the Optical Society of America*, 53, 1266–1270, 1963. With permission of Optical Society of America.)

A heat-reflecting filter using a combination of stacks, in which the second and third and second, third, and fourth orders have been suppressed, together with the normal quarter-wave stacks, has been designed. The calculated transmittance spectrum is shown in Figure 7.20. Although the production of such coatings is a demanding task, we note that coatings of similar performance are successfully applied to the envelopes of energy-efficient incandescent flood lamps so that emitted heat is redirected to the filament, although the design is not necessarily that shown.

The so-called half-wave hole is a common problem with shortwave pass filters. The visible symptom is a narrow dip in transmittance at around one-half of the fundamental reference wavelength where the layers are quarter waves. The reason for the absence of a second-order peak in structures based on the quarter-wave stack is that the quarter waves simultaneously become half waves and are, together, absentees. Anything that disturbs this relationship causes the missing order to appear as the half-wave hole. We have already commented on that in the discussion on Equation 7.22. We will briefly examine the half-wave hole using a structure of dispersionless materials of indices 1.45 and 2.15, roughly corresponding to silicon dioxide and tantalum pentoxide.

The performance of a classical shortwave pass filter based on a structure $(0.5L\ H\ 0.5L)^{25}$ with a reference wavelength of 1000 nm, and in which the outermost six layers on either side have been refined, is shown in Figure 7.21 along with a modified design to be discussed shortly. When this filter is tilted to an angle of incidence of 40° in the incident medium of air, the result is as shown in Figure 7.22. A severe dip in transmittance has appeared at the left-hand side of the characteristic, roughly corresponding to a wavelength of $\lambda_0/2$. This is the half-wave hole.

We apply a method similar to that of Thelen to the problem of suppressing this hole. In this case, we want to eliminate only one reflecting order, the second. This requires an antireflection coating at every interface between the high- and low-index materials that is effective only at the second-order wavelength. To simplify the manufacturing problems, we shall use the same two materials as are already used in the basic design.

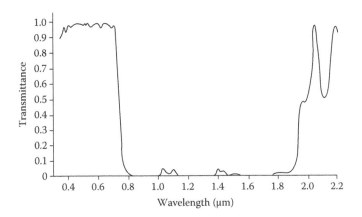

FIGURE 7.20
Calculated transmittance of a triple-stack heat reflector. Design:

$$\text{Incident} \left| \left[1.1 \left(\tfrac{1}{2} AC \tfrac{1}{2} A \right) \right] \left(\tfrac{1}{2} AC \tfrac{1}{2} A \right)^5 \left[1.25 \left(\tfrac{1}{2} AC \tfrac{1}{2} A \right) \right] \right.$$

$$\left[0.57 (ADCDA) \right]^8 \left[0.642 (ADCDA) \right]^8 \tfrac{1}{2} A \left| \text{sub} \right.$$

with $\lambda_0 = 860$ nm, $n_m = 1.50$, $n_M = 1.00$, $n_A = 1.38$, $n_B = 1.781$, $n_C = 2.30$, and $n_D = 1.90$. (After A. Thelen, *Journal of the Optical Society of America*, 53, 1266–1270, 1963. With permission of Optical Society of America.)

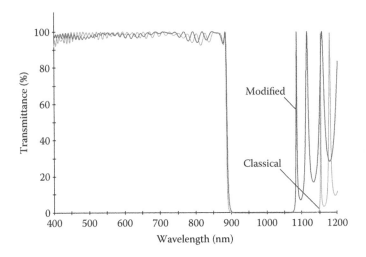

FIGURE 7.21
Two shortwave pass filters at normal incidence. The classical one is based on a 51-layer quarter-wave stack with adjusted outermost layers to reduce ripple. The modified one is the 51-layer stack with inserted antireflection coatings as discussed in the text. The materials used in the designs are dispersionless with indices of refraction 1.45 and 2.15, roughly corresponding to silicon dioxide and tantalum pentoxide.

It is always possible to match two layers with a two-layer coating consisting of the same two materials. The admittance diagram in Figure 7.23 makes that clear. The point on the real axis corresponding to the admittance of the material of an admittance circle is always inside the circle. Thus, each of the two admittance points will be on the periphery of one circle and inside the other so that the two circles must always intersect. Therefore, there will be a continuous path from one admittance to the other and a valid antireflection coating. It can be shown that in this special case,

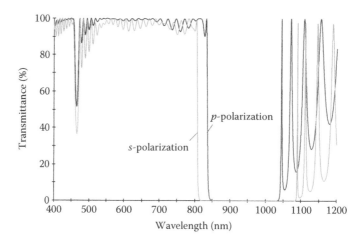

FIGURE 7.22
Classical filter in Figure 7.21 at an angle of incidence of 40° in the air incident medium. The half-wave hole can clearly be seen in the 450–500 nm region.

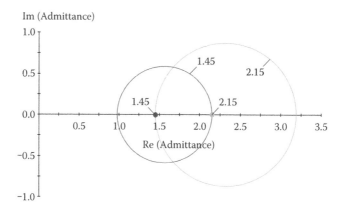

FIGURE 7.23
The geometry of the admittance diagram shows that two admittance circles passing through the admittances of the alternate material will always intersect so that there is a continuous path from one admittance to the other.

the optical thicknesses of the two layers are equal. Clearly, there are two solutions, and we take the thinner.

For these two materials, the optical thicknesses required at the wavelength for the antireflecting action are both 0.08159. At the reference wavelength, the thickness will be just one half of that. The insertion of the antireflection coating implies the removal of an equal optical thickness from each of the layers of the original quarter-wave structure. The starting design is therefore

$$(0.3368L\,0.1632H\,0.1632L\,0.6737H\,0.1632L\,0.1632H\,0.3368L)^{25}$$

Next, the outermost five of the original layers (not the antireflection sections) were refined to give the performance shown as modified in Figure 7.21. The width of the high-reflectance zone has suffered from the changes in the original quarter waves, and so before refining, the edge was adjusted in position to coincide with the edge of the classical design. This implied changing the reference wavelength to 965 nm.

Tilting this filter to 40° in air yields the performance curves shown in Figure 7.24. The half-wave hole has almost disappeared. The residual dip is a consequence of the fact that the antireflection

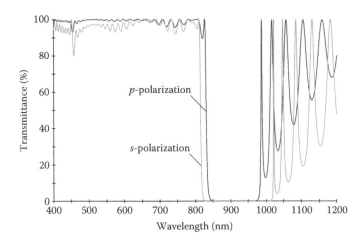

FIGURE 7.24
Performance of the modified design of Figure 7.21 showing the 40° incidence performance. The half-wave hole is much reduced.

coating was designed for normal incidence. It can be further reduced by slightly adjusting the antireflection coating thicknesses, although too much adjustment will begin to cause a half-wave hole to appear at normal incidence.

7.2.9 Reducing the Transmission Zone

The simple quarter-wave multilayer has the even-order high-reflectance bands missing. Sometimes it is useful to have these high-reflectance bands present. The method of the previous section can also be applied to this problem, and the enhancement of the reflectance at the even orders is a relatively simple business. Because it makes the analysis simpler, we assume that the basic period is of the form AB rather than (A/2)B(A/2). Once the basic result is established, it can be easily converted to the form (A/2)B(A/2) if required. The reason that the even-order peaks are suppressed in the ordinary quarter-wave stack is that each of the layers is an integral number of half-waves thick and so are absentee layers. All that is required for a reflectance peak to appear is the destruction of this condition. To achieve this, the thickness of one of the layers must be increased, and the other, decreased, keeping the overall optical thickness constant. The greater the departure from the half-wave condition, the more pronounced the reflectance peak.

Consider the case where reflectance bands are required at λ_0, $\lambda_0/2$, and $\lambda_0/3$, but not necessarily at $\lambda_0/4$. This will be satisfied by making $n_A d_A = n_B d_B/3$ and $n_A d_A = \lambda_0/8$ so that the basic stack becomes either

$$\frac{H}{2} \frac{3L}{2} \frac{H}{2} \frac{3L}{2} \frac{H}{2} \frac{3L}{2} \cdots \frac{3L}{2}$$

or

$$\frac{L}{2} \frac{3H}{2} \frac{L}{2} \frac{3H}{2} \frac{L}{2} \frac{3H}{2} \cdots \frac{3H}{2}.$$

The reflectance peak at $\lambda_0/4$ will be suppressed because the layers at that wavelength have integral half-wave thicknesses.

The method can be used to produce any number of high-reflectance zones. However, it should be noted that the further the thicknesses depart from ideal quarter waves at λ_0, the narrower the first-order reflectance band.

7.2.10 Edge Steepness

We understand very well what is meant by edge steepness, but rather like edge position, quantifying it and comparing measures of it can be problematic. The principal difficulty is its definition that tends to vary from one supplier to another or one application to another. It is a measure of the slope, but the slope itself is not constant. Often the definition will take the form of a difference in the wavelengths corresponding to two different prescribed levels of transmittance normalized by dividing by a wavelength characteristic of the edge position.

In standard longwave and shortwave pass filters, the steepness of edge is, fortunately, not usually a parameter of critical importance. The number of layers necessary to produce the required rejection in the stop band of the filter will generally produce an edge steepness that is quite acceptable. There are, however, some applications where an exceptional degree of edge steepness is required, and then the easiest and most direct way of improving it is to use still more layers in the core of the filter. Increasing the number of layers will tend to cause an apparent increase in the ripple in the passband, because the first minimum in the passband will be brought nearer to the edge and will usually be on a part of the ripple envelope that often increases in width toward the edge. Since the outermost layers, as we have seen, are responsible for the matching that reduces the ripple and as they are most conveniently designed by a refinement process, it is a simple matter to add more layers that can then be further refined.

Filters intended for application in optical telecommunication tend to be subject to very tight specification of exceptionally high performance. Figure 7.25 shows an example of a shortwave pass filter with a steep edge and intended for an application in optical communication. It is

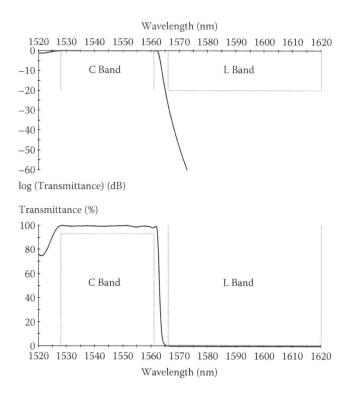

FIGURE 7.25
High-performance shortwave pass filter designed to separate the C and L telecommunication bands. Transmittance must be greater than 93% (−0.3 dB) from 1528 to 1561 nm and less than 1% (−20 dB) from 1566 to 1620 nm. (A. Macleod, *Optical Coatings: Theory, Production and Characterization*, pp. 9–21, Ente per le Nuove Tecnologie, L'Energia e l'Ambiente, Rome, 2001.)

intended to separate the C and L bands of the International Telecommunication Union Grid. The design is

$$\text{Air}|L\,(HL)^4\,[(HL)^{51}H]\,(LH)^5|\text{Glass}$$

where the layers in the square brackets are quarter waves at 1747.5 nm, and those layers outside the square brackets have all been refined to remove passband ripple. There are 122 layers in the design, not unusual in the telecommunication field. The thin-film materials used in the design are Ta_2O_5 ($n = 2.100$) and SiO_2 ($n = 1.442$).

7.3 Notch Filters

A notch filter blocks a band of wavelengths surrounded on both sides by pass regions. It is a combined longwave pass and shortwave pass filter. Other terms sometimes used are *minus filter* and *bandstop filter*.

For a notch filter, we will frequently start with a quarter-wave stack structure. Our first concern is with the width of the notch. The width of the basic high-reflectance zone of our stack is seldom the one that is required by the specification. Fortunately, the required width is usually less than the basic, and there are two principal methods for dealing with this. We know that the width of a quarter-wave stack is a function of the ratios of the indices of the two materials involved. A reduction in that ratio reduces the zone width. This solution depends on the availability of two suitable coating materials, and we shall have more to say about this when we discuss rugate filters later. An alternative and more attractive solution when materials present difficulties is the adjustment of the relative thicknesses of the layers in the basic repeated period. Such adjustment will frequently enhance the reflectance in the second order at $g = 2.0$, but that is worrisome only if the second order intrudes on the zone of high transmittance.

Ripple is our second major concern with notch filters. Figure 7.3 shows that a basic structure will present a rather different equivalent admittance on either side of the high-reflectance zone. This implies that a simple antireflection coating that matches the structure on one side of the zone is unlikely to match it on the other. Matching sequences will therefore be rather more complex than that for a straightforward edge filter.

Figure 7.26 shows an example of the adjustment of relative thickness to give the correct width. The notch is to cover the region from 521 to 579 nm with high transmittance from 400 to 519 nm and from 581 to 800nm. Trial and error yields a formula, at a reference wavelength of 550 nm, of $\text{Air}\,|\,(1.68L\ 0.30H)^{59}\ 1.68L\,|\,\text{Glass}$ with n_H as 2.15, n_L as 1.45, and glass as 1.52, the materials corresponding to Ta_2O_5 and SiO_2 but without dispersion or absorption. Refinement of all layers with targets including the rejection region yielded the performance in Figure 7.26. Outside the target region, the performance is naturally less satisfactory (Figure 7.27), and it is on the shortwave side that the major problems will normally occur.

Similar results will be achieved with starting designs of quarter waves and with complete synthesis.

An interesting technique that has wide implications is known as thickness modulation, already mentioned in Section 6.4 in connection with multiple reflectance peaks. Of course, any design could be described as employing thickness modulation, but the term indicates thicknesses that have a functional relationship, rather than simply varying. The major work on thickness modulation is a book by Perilloux [18], but there are also some recent papers specifically on notch filters [19,20].

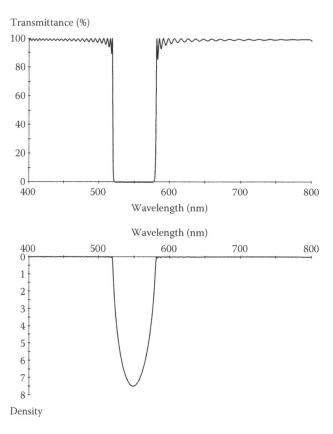

FIGURE 7.26
Design of the notch filter started as Air | $(1.68L\ 0.30H)^{59}\ 1.68L$ | Glass at 550 nm with indices $n_H = 2.15$, $n_L = 1.45$, and $n_{Glass} = 1.52$. It was then refined over the region of 400–800 nm including the rejection region to yield the performance shown.

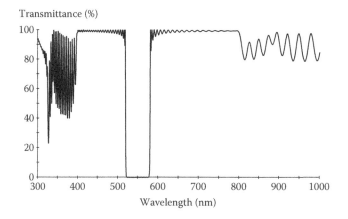

FIGURE 7.27
Notch filter of Figure 7.26 plotted over a wider wavelength region. Outside the target region for refinement, the performance deteriorates.

Thickness modulation is inspired by the index modulation that is at the core of rugate design, and we will deal with rugates later in Section 11.1. But because the classical rugate demands accurately varying refractive index, its attractiveness is reduced by all its accompanying manufacturing and control challenges. In fact, a frequent late stage in rugate design is conversion to discrete layers involving two materials.

The key operation both in rugate design and thickness modulation is the operation known as apodization. Let us illustrate this by an example similar to that in the study by Zhang et al. [20] but based on Figure 7.26. Figure 7.28 shows the basic reflectance peak with no attempt to modify the ripple. The ripple could be said to be at the feet of the peak, and the word *apodization* is derived, through French, from the Greek word *apodos*, meaning "without feet". Apodization involves modulating the structure with a function that gradually weakens the structure toward its extremities. Many different functions can be used: Gaussian, linear, quintic, and sinusoidal are examples. For our illustration, we use a sinusoidal function. The thicknesses of the layers in Figure 7.28 have the same relationship right through the entire coating with the resulting ripple also shown in the figure. We can think of the structure as a repeat of a two-layer period with period thickness of one-half wave at the reference wavelength. Let us retain this relationship at the center of the structure, but as we move toward the peripheries, let us progressively reduce the thickness of high-index material and increase the low index to compensate so that when we eventually arrive at the front and rear surfaces, the high-index material has shrunk to zero. This is the process

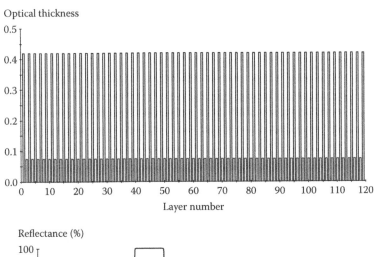

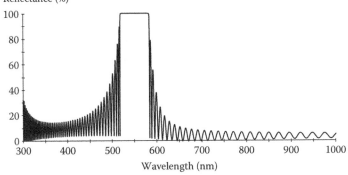

FIGURE 7.28
Basic structure of the notch filter and its reflectance. The thicker films in the upper diagram are the low-index films. The value of A is 0.075.

of apodization. If p is the layer number and q is the total number of layers, then the high-index layer thicknesses will be given by

$$(n_H d_H / \lambda_0)_p = A \sin[\pi(p-1)/(q-1)], \tag{7.34}$$

where A is the optical thickness of the high-index material at the center. The low-index material then has thicknesses given by

$$(n_L d_L / \lambda_0)_p = 0.5 - A \sin[\pi(p-1)/(q-1)]. \tag{7.35}$$

With 119 layers, the structure and performance is as shown in Figure 7.29. A slight improvement in the ripple next to the peak has been achieved by making the first layer a quarter wave instead of a half wave. It then acts as a weak matching layer.

The weakening of the periods results in the reduction of the reflectance and this can best be seen in the density calculation of Figure 7.29. However, the reduction of the ripple is effective over a much wider spectral region if we compare Figures 7.27 and 7.29.

The residual ripple in Figure 7.29 is almost entirely due to a mismatch between the air incident medium and the admittance of the notch structure that is virtually that of the low-index layer. It has not been addressed in this filter but would be a simple matter of adding suitable layers with some refinement.

Multiple notches are more difficult. Rugate structures (Section 11.1) represent a general solution. In this context we look at an elaboration of the single notch technique. Let us examine a requirement for notches at wavelengths of 500 and 750 nm. We use a variant of the technique already described in Section 6.4, "Reflectors with Multiple Peaks," together with the ideas of apodization in the present section.

First of all, we need a structure consisting of a repeated period that presents at both wavelengths a thickness of an integral number of half waves. A little trial and error shows that three half waves at 500 nm gives two half waves at 750 nm. We first look at these periods separately. Let us take 500 nm as our reference wavelength, and then the basic periods are $L\,H\,L\,H\,L\,H$ and $1.5L\,1.5H\,1.5L\,1.5H$ for 500 and 750 nm, respectively. However, rather than quarter waves, we modify the relative thicknesses in the same way as with the single notch so that if we just had the 500 nm peak, the optical thicknesses would be given by

$$(n_L d_L / \lambda_0)_p = 0.5\{1 - A \sin[\pi(p-1)/(q-1)]\} \tag{7.36}$$

and

$$(n_H d_H / \lambda_0)_p = 0.5\{A \sin[\pi(p-1)/(q-1)]\}, \tag{7.37}$$

where p is the layer number, q is the total number of layers, and $0.5A$ is the optical thickness of the high-index material at the center. Keeping 500 nm as λ_0, the relationships for the 750 nm peak on its own would be similar except that the 0.5 in Equations 7.36 and 7.37 would be replaced by 0.75. Now we follow Perilloux [18] and morph the six-layer combination for 500 nm into the four-layer combination for 750 nm. To make it clear, we treat each of the six layers separately.

$$
\begin{aligned}
L_1 &= (0.5 + 0.25k)\{1 - A \sin[\pi(p-1)/(q-1)]\}, \\
H_2 &= (0.5 + 0.25k)\{A \sin[\pi(p-1)/(q-1)]\}, \\
L_3 &= 0.5(1 - k)\{1 - A \sin[\pi(p-1)/(q-1)]\}, \\
H_4 &= 0.5(1 - k)\{A \sin[\pi(p-1)/(q-1)]\}, \\
L_5 &= (0.5 + 0.25k)\{1 - A \sin[\pi(p-1)/(q-1)]\}, \\
H_6 &= (0.5 + 0.25k)\{A \sin[\pi(p-1)/(q-1)]\},
\end{aligned}
\tag{7.38}
$$

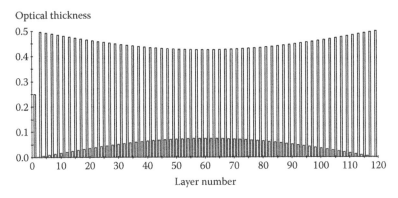

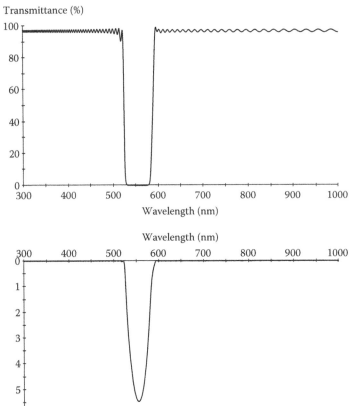

FIGURE 7.29
The process of apodization, using, in this particular case, a sine function, results in a substantial drop in ripple to the extent where virtually no further correction is necessary. The upper diagram shows the variation in thickness of the high- and low index-layers, the high index being the thinner. As before, we use indices $n_H = 2.15$, $n_L = 1.45$, and $n_{Glass} = 1.52$. The substrate is glass and the incident medium is air.

where k, between zero and unity, is the morphing parameter. With 200 layers and A set at 0.2, we find by trial and error a value for k of 0.53 that equalizes the densities of the two peaks. As with the single notch with apodized design, we find that the optimum thickness for layer 1 is 0.31625, exactly half the thickness given by Equation 7.38.

Although the peaks plotted in terms of wavelength in Figure 7.30 appear of quite different width, they are of the same width in terms of frequency as shown in Figure 7.31, where they are plotted in terms of g. There, the adjacent spurious peaks that share the constructive interference at the desired ones can be seen.

The residual ripple is deliberately uncorrected in the two apodized notch designs. It is almost entirely a consequence of the mismatch at the outer surface of the design, where the structure

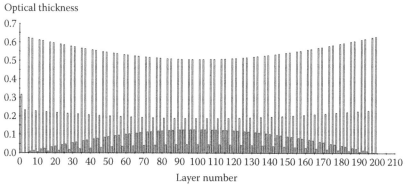

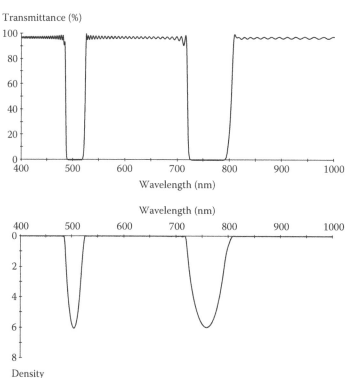

FIGURE 7.30

Apodized double notch filter. The upper figure shows the variation of optical thickness through the filter. The thicknesses increasing toward the peripheries are those of the low-index layers and the decreasing are of those of the high-index layers.

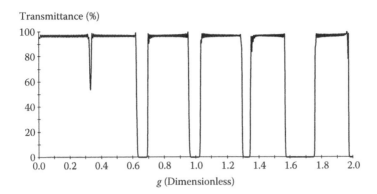

FIGURE 7.31
Apodized double notch filter of Figure 7.30 plotted in terms of g with $\lambda_0 = 500$ nm. The peaks at 1.0 and 0.67 that correspond to those in Figure 7.30 are of the same width.

essentially presents the admittance of the low-index layer while the incident medium is air. This deficiency can be readily corrected by the addition of a simple matching structure, as in other designs of this chapter. Some examples are given by Lyngnes and Kraus [19].

References

1. L. I. Epstein. 1952. The design of optical filters. *Journal of the Optical Society of America* 42:806–810.
2. J. J. Vera. 1964. Some properties of multilayer films with periodic structure. *Optica Acta* 11:315–331.
3. A. Thelen. 1966. Equivalent layers in multilayer filters. *Journal of the Optical Society of America* 56:1533–1538.
4. C. Ufford and P. Baumeister. 1974. Graphical aids in the use of equivalent index in multilayer-filter design. *Journal of the Optical Society of America* 64:329–334.
5. W. T. Welford (writing as W. Weinstein). 1954. Computations in thin film optics. *Vacuum* 4:3–19.
6. R. Jacobsson. 1964. Matching a multilayer stack to a high-refractive-index substrate by means of an inhomogeneous layer. *Journal of the Optical Society of America* 54:422–423.
7. L. Young and E. G. Crystal. 1966. On a dielectric fiber (sic) by Baumeister. *Applied Optics* 5:77–80.
8. J. S. Seeley, H. M. Liddell, and T. C. Chen. 1973. Extraction of Tschebysheff design data for the lowpass dielectric multilayer. *Optica Acta* 20:641–661.
9. J. S. Seeley and S. D. Smith. 1966. High performance blocking filters for the region 1 µ to 20 µ. *Applied Optics* 5:81–85.
10. P. W. Baumeister. 1958. Design of multilayer filters by successive approximations. *Journal of the Optical Society of America* 48:955–958.
11. J. A. Dobrowolski and R. A. Kemp. 1990. Refinement of optical multilayer systems with different optimization procedures. *Applied Optics* 29:2876–2893.
12. J. A. Dobrowolski. 1965. Completely automatic synthesis of optical thin film systems. *Applied Optics* 4:937–946.
13. A. F. Turner and P. W. Baumeister. 1966. Multilayer mirrors with high reflectance over an extended spectral region. *Applied Optics* 5:69–76.
14. L. I. Epstein. 1955. Improvements in heat reflecting filters. *Journal of the Optical Society of America* 45:360–362.
15. A. Thelen. 1963. Multilayer filters with wide transmission bands. *Journal of the Optical Society of America* 53:1266–1270.

16. R. B. Muchmore. 1948. Optimum band width for two layer anti-reflection films. *Journal of the Optical Society of America* 38:20–26.
17. A. Macleod. 2001. Optical coatings for communication technology. In *Optical Coatings: Theory, Production and Characterization*, E. Masetti, D. Ristau, and A. Krasilnikova (eds), pp. 9–21. Rome, Italy: Ente per le Nuove Tecnologie, L'Energia e l' Ambiete (ENEA).
18. B. E. Perilloux. 2002. *Thin Film Design: Modulated Thickness and Other Stopband Design Methods*, vol. TT57. Bellingham, WA: SPIE Optical Engineering Press.
19. O. Lyngnes and J. Kraus. 2014. Design of optical notch filters using apodized thickness modulation. *Applied Optics* 53:A21–A26.
20. J. Zhang, Y. Xie, X. Cheng et al. 2013. Thin-film thickness-modulated designs for optical minus filter. *Applied Optics* 52:5788–5793.

8

Bandpass Filters

A filter possessing a region of transmission bounded on either side by regions of rejection is known as a bandpass filter. For the broadest bandpass filters, the most suitable construction is a combination of longwave pass and shortwave pass filters, already mentioned in Chapter 7. For narrower filters, however, this method is not very successful because of difficulties associated with obtaining both the required precision in positioning and the steepness of edges. Other methods are therefore used, involving a single assembly of thin films to simultaneously produce the pass and rejection bands. The simplest of these is the thin-film Fabry–Perot filter, a development of the interferometer already described in Chapter 6. The spacer layer in the Fabry–Perot etalon acts rather like a resonant cavity and so is usually called a cavity layer. The Fabry–Perot filter then becomes known as a single-cavity filter. The single-cavity filter has a roughly triangular passband shape, and it has been found possible to improve this, by coupling simple single-cavity filters in series, in much the same way as electrical tuned circuits. These coupled arrangements are known as multiple-cavity filters. The terminology has not always been as simple. The cavities are usually a half-wave, or integral multiples of half-waves, thick. Thus, an older term that is still sometimes used is *multiple half-wave filter*. The two-cavity filter was earlier called a double half-wave filter, abbreviated to DHW filter, while the three cavity was called a triple half-wave or THW filter. Another term, less used nowadays, is WADI indicating a wide-band all-dielectric filter.

8.1 Broad Bandpass Filters

Bandpass filters can be very roughly divided into broad bandpass filters and narrow bandpass filters. There is no definite boundary between the two types, and the description of any particular filter usually depends on the application and the filters with which it is being compared. One manufacturer's broad band may be another's narrow. For the purpose of the present work, by broadband filters, we mean filters with bandwidths of perhaps 20% or more, which are made by combining longwave pass and shortwave pass filters. The best arrangement is probably to deposit the two components on opposite sides of a single substrate. To give maximum possible transmission, each edge filter should be designed to match the substrate into the surrounding medium, a procedure already examined in Chapter 7. Such a filter is shown in Figure 8.1.

It is also possible, however, to deposit both components on the same side of the substrate. This was a problem which Epstein [1] examined in his early paper on symmetrical periods. The main difficulty is the combining of the two stacks so that the transmission in the passband is a maximum and so that one stack does not produce transmission peaks in the rejection zone of the other. The transmission in the passband will depend on the matching of the first stack to the substrate, the matching of the second stack to the first, and the matching of the second stack to the surrounding medium. Depending on the equivalent admittances of the various stacks, it may be necessary to insert quarter-wave matching layers or to adopt any of the more involved matching techniques.

In the visible region, with materials such as zinc sulfide and cryolite, or titanium dioxide and silicon dioxide, the combination $[(H/2)L(H/2)]^S$ acts as a good longwave-pass filter with an equivalent admittance at normal incidence and at wavelengths in the pass region not too far

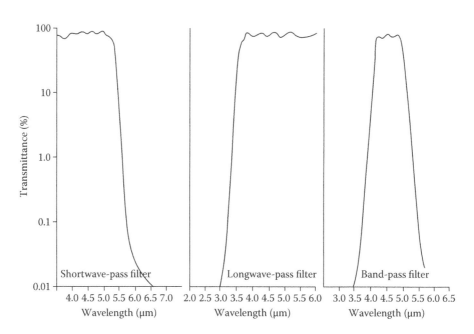

FIGURE 8.1
Construction of a bandpass filter by placing two separate edge filters in series. (Courtesy of Standard Telephones and Cables, Ltd., London.)

removed from the edge of near unity. This can therefore be used next to the air without mismatch. The combination $[(L/2)H(L/2)]^S$ acts as a shortwave-pass filter, with equivalent admittance only a little lower than the first section and can be placed next to it, between it and the substrate, without any matching layers. The mismatch between this second section and the substrate, which in the visible region will be glass of index 1.52, is sufficiently large to require a matching layer. Happily, the $[(H/2)L(H/2)]$ combination with a total phase thickness of 270°, i.e., effectively three quarter waves, has admittance exactly correct for this. The transmittance of the final design is shown in Figure 8.2b with the appropriate admittances of the two sections in Figure 8.2a. Curve A refers to a $[(L/2)H(L/2)]^4$ shortwave-pass section, and B, to a $[(H/2)L(H/2)]^4$ longwave-pass section. The complete design is shown in Table 8.1. The edges of the two sections have been chosen quite arbitrarily and could be moved as required.

To avoid the appearance of transmission peaks in the rejection zones of either component, it is safest to deposit them so that high-reflectance zones do not overlap. The complete rejection band of the shortwave-pass section will always lie over a pass region of the longwave-pass filter, but the higher-order bands should be positioned, if at all possible, clear of the rejection zone of the longwave-pass section. The combination of edge filters of the same type has already been investigated in Chapter 7, and the principles discussed there apply to this present situation. It should also be remembered that although in the normal shortwave-pass filter the second-order reflection peak is missing, a small peak can appear if any thickness errors are present. This can, if superimposed on a rejection zone of the other section, cause the appearance of a transmission peak if the errors are sufficiently pronounced. The expression for maximum transmission is

$$T_{\max} = \frac{T_a T_b}{\left[1 - (R_a R_b)^{1/2}\right]^2},$$

but this holds only if the phase conditions are met.

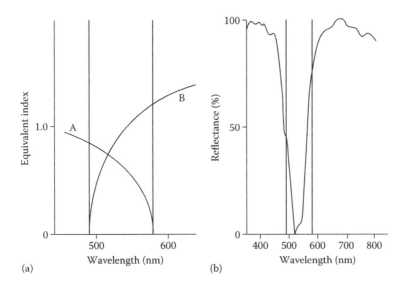

FIGURE 8.2
(a) Equivalent admittances of two stacks made up of symmetrical periods used to form a bandpass filter. A: (0.5LH0.5L);
B: (0.5HL0.5H), where $n_L = 1.38$ and $n_H = 2.30$. (b) Calculated reflectance curve for a bandpass filter. For the complete design
of this filter, made up of two superimposed stacks, one of type A and one of type B, refer to Table 8.1. (After L. I. Epstein,
Journal of the Optical Society of America, 42, 806–810, 1952. With permission of Optical Society of America.)

TABLE 8.1

Epstein Bandpass Design

Layer Index	Phase Thickness of Each Layer Measured at 546 nm (°)	Layer Index	Phase Thickness of Each Layer Measured at 546 nm (°)
1.52	Massive	2.30	55.4
1.38	67.3	1.38	33.9
2.30	134.5	2.30	67.9
1.38	122.7	1.38	67.9
2.30	110.8	2.30	67.9
1.38	110.8	1.38	67.9
2.30	110.8	2.30	67.9
1.38	110.8	1.38	67.9
2.30	110.8	2.30	67.9
1.38	110.8	1.38	67.9
2.30	110.8	2.30	33.9
1.38	110.8	1.00	Massive

Source: L. I. Epstein, *Journal of the Optical Society of America*, 42, 806–810, 1952. With permission of Optical Society of
America.

The Epstein technique is essentially an analytical one and was designed to avoid any involved
and lengthy calculations. We can take advantage of our computer-based design techniques to
create a more advanced design. We combine the idea of thickness modulation (see Sections 6.4 and
7.3 and Perilloux [2]) with computer refinement. To avoid the complication of dispersion, we use
our 2.15 and 1.45 materials, roughly corresponding to Ta_2O_5 and SiO_2 in the design of a bandpass
filter from 500 to 600 nm. The substrate will be glass at 1.52, and the incident medium, air.

We begin with two edge filters of identical design. Each starts with 61 quarter-wave layers, and then the outermost 20 layers on either side are linearly tapered to reduce ripple in much the same way as in the notch filters of the previous chapter. We use a linear taper for simplicity because we are going to refine the thicknesses to complete the design process and the number 20 is not critical. The thicknesses of the first 20 layers are given by

$$\frac{n_L d_L}{\lambda_0} = 0.25\left[1 + \frac{(q-p)}{(q-1)}\right],$$
$$\frac{n_H d_H}{\lambda_0} = 0.25\left[1 - \frac{(q-p)}{(q-1)}\right],$$

(8.1)

where the low-index layer is outermost, q is the number of layers in the taper (20 in this case), and p is the layer number (not greater than q). The final 20 layers are similar but with their order reversed. We use reference wavelengths of 680 and 435 nm and place the two designs in series with the 680 nm version innermost, next to the substrate. Some simple refinements, in which all layers are involved and using targets that cover both the rejection regions and the passband, results in the design and performance in Figure 8.3. Before refinement, the design was reasonably well matched to surrounding media of index 1.52. Changing the incident medium to air upset the matching at the front. The subsequent refinement introduced the ripple in the shortwave and longwave sidebands. The design still shows the basic thickness variation, but some significant changes can be seen in the taper at the left-hand side together with very slight oscillation in the two central quarter-wave sections. The middle layer has also been increased in thickness as part of the ripple correction. Thickness modulation together with refinement is clearly a powerful combination.

8.2 Narrowband Single-Cavity Filters

8.2.1 Metal–Dielectric Single-Cavity Filter

The simplest type of narrowband thin-film filter is based on the Fabry–Perot interferometer discussed in Chapter 6. In its original form, the Fabry–Perot interferometer consists of two identical parallel reflecting surfaces spaced apart at a distance d. In collimated light, the transmission is low for all wavelengths except for a series of very narrow transmission bands spaced at intervals that are constant in terms of wavenumber. This device can be replaced by a complete thin-film assembly consisting of a dielectric layer bounded by two metallic reflecting layers (Figure 8.4). The dielectric layer takes the place of the spacer and is usually known as the cavity layer. Except that the cavity, or spacer, layer now has an index greater than unity, the analysis of the performance of this thin-film filter is exactly the same as for the conventional etalon, but in other respects, there are some significant differences.

While the surfaces of the substrates should have a high degree of polish, their figures need not be worked to the exacting tolerances necessary for etalon plates. Provided the vapor stream in the chamber is uniform, the films will follow the contours of the substrate without exhibiting thickness variations. This implies that it is possible for the thin-film Fabry–Perot filter to be used in a much lower order than the conventional etalon. Indeed, it turns out in practice that lower orders must be used, because the thin-film cavity layers tend to exhibit an increasing roughness with increasing thickness so that their use beyond the fourth or fifth order becomes problematic. The roughness broadens the passband and reduces the peak transmittance, removing any advantage of a very high order. This simple type of filter is known as a metal–dielectric Fabry–Perot, or single cavity, to distinguish it from the all-dielectric equivalent to be described later.

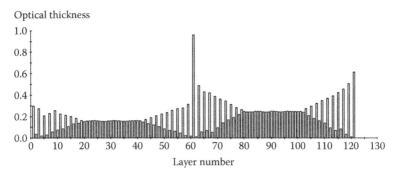

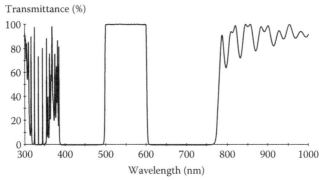

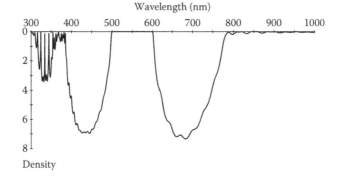

FIGURE 8.3
Thicknesses (odd layers of high index and even of low) of the final design (*top*) and the performance of the filter in transmittance and in density.

It is worthwhile briefly to analyze the performance of the single-cavity filter once again, this time including the effects of phase shift at the reflectors. The starting point for this analysis is Equation 3.20:

$$T_F = \frac{T_a T_b}{\left[1 - (R_a R_b)^{1/2}\right]^2} \frac{1}{1 + F\sin^2\left[\frac{1}{2}(\varphi_a + \varphi_b) - \delta\right]}, \tag{8.2}$$

where

$$F = \frac{4(R_a R_b)^{1/2}}{\left[1 - (R_a R_b)^{1/2}\right]^2}$$

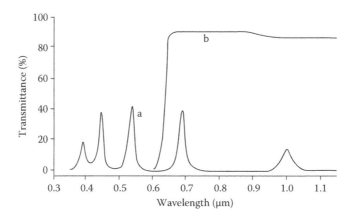

FIGURE 8.4
Characteristics of a metal–dielectric filter for the visible region (curve a). Curve b is the transmittance of an absorption glass filter that can be used for the suppression of the short wavelength sidebands. (Courtesy of Barr and Stroud, Ltd., Suffollk.)

and

$$\delta = \frac{2\pi nd \cos \vartheta}{\lambda}.$$

We have slightly adapted Equation 3.20 by removing the + and − signs on the reflectances. The following analysis is similar to that already performed in Chapter 5, except that here we are including the effects of φ_a and φ_b. The transmittance maxima are given by

$$\frac{2\pi nd \cos \vartheta}{\lambda} - \frac{\varphi_a + \varphi_b}{2} = m\pi, \quad \text{where } m = 0, \pm 1, \pm 2, \pm 3, \ldots. \tag{8.3}$$

The analysis is marginally simpler if we work in terms of wavenumber v instead of wavelength. The positions of the peaks are then given by

$$\frac{1}{\lambda} = v = \frac{m\pi + (\varphi_a + \varphi_b)/2}{2\pi nd \cos \vartheta} = \frac{1}{2nd \cos \vartheta} \left(m + \frac{\varphi_a + \varphi_b}{2\pi} \right). \tag{8.4}$$

Depending on the particular metal, the thickness, the index of the substrate, and the index of the cavity layer, the phase shift on reflection φ will be either in the first or second quadrant. $(\varphi_a + \varphi_b)/(2\pi)$ will therefore be positive between 0 and 1 and, for silver in the middle of the visible region, roughly 0.7. If we take m as the order number, then there will actually be a zeroth order peak. At normal incidence, this peak will be roughly given by $nd/0.35$ so that for this peak to be at λ_0, the cavity layer should have an optical thickness around $0.35\lambda_0$. In the first order, the cavity layer should be around $0.85\lambda_0$ in optical thickness.

The resolving power of the thin-film single-cavity, or Fabry–Perot, filter may be defined in the same way as for the interferometer. As we saw in Chapter 6, a convenient definition is

$$\frac{\text{Peak wavelength}}{\text{Halfwidth of passband}},$$

where the halfwidth is the width of the band measured at half the peak transmittance. Because the halfwidth can be rapidly converted into resolving power, and because it is so easily visualized, it tends to be the parameter of choice to express the performance of all types of narrowband filter, not just the single-cavity filter. Other measures of bandwidth sometimes quoted along with the halfwidth are widths measured at different fractions of peak transmittance, 0.9 or 0.1, for example.

Sometimes, particularly in telecommunication applications, these levels are expressed in decibels below the peak level.

We digress for a moment to explain this different scale. dB is the symbol for decibel, which is a unit of comparison of power levels and, in our case, a comparison of irradiances. The scale is a logarithmic one. The decibel is not included in the SI system, but it is internationally accepted and used. It is based on the bel, a unit that originated at Bell Telephone Laboratories, and is named after Alexander Graham Bell. The bel is defined as the base 10 logarithm of the ratio of the two power levels. A drop in power leads to a negative value. The numbers become a little more convenient when multiplied by 10, and this leads to the decibel scale. The level in decibel of irradiance I_1 compared with irradiance I_0 is, therefore

$$L\,(\mathrm{dB}) = 10 \log_{10}\left(\frac{I_1}{I_0}\right), \tag{8.5}$$

a negative value representing a loss, and a positive value, a gain. The ratio I_1/I_0 can represent a transmittance or reflectance in absolute units. To convert transmittance into decibel, we use

$$T\,(\mathrm{dB}) = 10 \log_{10}[T(\mathrm{abs})]. \tag{8.6}$$

For example, a transmittance of 50% becomes 0.5 in absolute terms, and the value in decibel is 3.01 dB. This is sometimes referred to as an insertion loss of 3.01 dB because it represents a reduction in power. The advantage of the decibel scale is that it is additive. Two components in series, one with an insertion loss of 2 dB and the other of 4 dB would give a total insertion loss of $(2 + 4)$ dB = 6 dB.

It is worthwhile to spend just a little longer on this question of resolving power and resolution. Let the passband be sufficiently narrow, which is the same as F being sufficiently large, so that near a peak, we can replace

$$\frac{\varphi_a + \varphi_b}{2} - \delta \quad \text{by} \quad -m\pi - \Delta\delta$$

and

$$\sin^2\left(\frac{\varphi_a + \varphi_b}{2} - \delta\right) \text{ by } (\Delta\delta)^2,$$

where we are assuming that φ_a and φ_b are constant or vary very much more slowly than δ over the passband.

The half-peak bandwidth, or halfwidth, can be found by noting that at the half-peak transmission points,

$$F\sin^2\left(\frac{\varphi_a + \varphi_b}{2} - \delta\right) = 1.$$

Then, by using the approximation given earlier, this becomes

$$(\Delta\delta_h)^2 = \frac{1}{F},$$

i.e., the halfwidth of the passband,

$$2\Delta\delta_h = 2/F^{1/2}.$$

The finesse is defined as the ratio of the interval between fringes to the fringe halfwidth and is written as $\mathcal{F}$. The change in δ in moving from one fringe to the next is just π, and the finesse is therefore

$$\mathcal{F} = \frac{\pi F^{1/2}}{2}. \tag{8.7}$$

Since $v \propto \delta$, $v_0/\Delta v_h = \delta_0/2\Delta\delta_h$, where v_0 and δ_0 are, respectively, the values of the wavenumber and cavity layer phase thickness associated with the transmission peak, and Δv_h and $2\Delta\delta_h$ are the corresponding values of halfwidth. The ratio of the peak wavenumber to the halfwidth is then given by

$$\frac{v_0}{\Delta v_h} = 7\left(m + \frac{\varphi_a + \varphi_b}{2\pi}\right)$$

for a peak of order m, since

$$\delta_0 = m\pi + \frac{\varphi_a + \varphi_b}{2}.$$

The ratio of peak position to halfwidth expressed in terms of wavenumber is exactly the same in terms of wavelength,

$$\frac{v_0}{\Delta v_h} = \frac{\lambda_0}{\Delta\lambda_h},\tag{8.8}$$

where λ_0 is given by

$$\lambda_0 = \frac{2nd\cos\vartheta}{m + (\varphi_a + \varphi_b)/2\pi},$$

and this was discussed in Chapter 6.

The manufacture of the metal–dielectric filter is straightforward. The main point to watch is that the metallic layers should be evaporated as quickly as possible on to a cold substrate. Good results can be achieved in the visible and near infrared regions with silver and cryolite and in the ultraviolet with aluminum and either magnesium fluoride or cryolite. Wherever possible, the layers should be protected by cementing a cover slip over them as soon as possible after deposition. This also serves to balance the assembly by equalizing the refractive indices of the media outside the metal layers.

Turner [3] quoted some results for metal–dielectric filters constructed for the visible region that may be taken as typical of the performance to be expected. The filters were constructed from silver reflectors and magnesium fluoride cavities. For a first-order cavity, a bandwidth of 13 nm with a peak transmittance of 30% was obtained at a peak wavelength of 531 nm. A similar filter with a second-order cavity gave a bandwidth of 7 nm with peak transmittance of 26% at 535 nm. With metal–dielectric filters, the third order is usually the highest used. Because of scattering in the cavity layer, which becomes increasingly apparent in the fourth and higher orders, any benefit that would otherwise arise from using these orders is largely lost. A typical curve for a metal–dielectric filter for the visible region is shown in Figure 8.4. The particular peak to be used is that at 0.69 µm, which is of the third order. The shortwave sidebands due to the higher-order peaks can be quite easily suppressed by the addition of an absorption glass filter, which can be cemented over the metal–dielectric element to act as a cover glass. Such a filter is also shown in the figure. It is one of a wide range of absorption glasses available for the visible and near infrared and has longwave-pass characteristics. There are, unfortunately, few absorption filters suitable for the suppression of the longwave sidebands. If the detector that is to be used is not sensitive to these longer wavelengths, then no problem exists, and commercial metal–dielectric filters for the visible and near infrared usually possess longwave sidebands beyond the limit of the photocathodes or photographic emulsions or other detectors for this region. If the longwave sideband suppression must be included as part of the filter assembly, then there is an advantage in using metal–dielectric filters in the first order, even though the peak transmittance for a given bandwidth is much lower, since they do not usually possess longwave sidebands. Theoretically, there will always be a peak corresponding to the zero order at very long wavelengths, but this will not usually appear, partly because the substrate will cut off long before the zero order is reached and because the properties of the thin-film materials themselves will radically change. We shall discuss later a special type of metal–dielectric filter, the induced transmission filter, which can be made to have a much higher peak transmittance, although with a rather broader halfwidth, without introducing long-wavelength sidebands, and which is often used as a longwave suppression filter.

Silver does not have an acceptable performance for ultraviolet filters, and aluminum has been found to be the most suitable metal, with magnesium fluoride as the usually preferred dielectric. In the ultraviolet beyond 300 nm, there are few suitable cements (none at all beyond 200 nm), and it is not possible to use cover slips cemented over the layers in the way in which filters for the visible region can be protected. The normal technique, therefore, is to attempt to protect the filter by the addition of an extra dielectric layer between the final metal layer and the atmosphere. These layers are effective in that they slow down the oxidation of the aluminum, which otherwise rapidly takes place and causes a reduction in performance even at quite low pressures. This oxidation has already been referred to in Chapter 5. However, complete stabilization of the filters is very difficult, and slight longwave drifts can occur, as reported by Bates and Bradley [4]. A second function of the final dielectric layer is to act as a reflection-reducing layer at the outermost metal surface and, hence, to increase the transmittance of the filter. This is not a major effect—the problem of improving metal–dielectric filter performance is dealt with later in this chapter—but any technique which helps to improve performance, even marginally, in the ultraviolet, is very welcome. Some performance curves of first-order metal–dielectric single-cavity filters are shown in Figure 8.5.

The formula for transmittance of the single-cavity filter can also be used to determine both the peak transmittance in the presence of absorption in the reflectors and the tolerance that can be allowed in matching the two reflectors. First of all, let the reflectances be equal and let the absorptance be denoted by A, so that

$$R + T + A = 1. \tag{8.9}$$

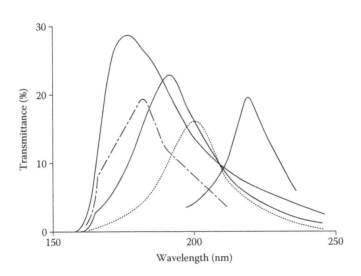

FIGURE 8.5
Experimental transmittance curves of first-order metal–dielectric filters for the far ultraviolet deposited on Spectrosil B substrates. (After B. Bates and D. J. Bradley, *Applied Optics*, 5, 971–975, 1966. With permission of Optical Society of America.)

The peak transmittance will then be given by

$$(T_F)_{peak} = \frac{T^2}{(1 - R)^2}$$

and, using Equation 8.9,

$$(T_F)_{peak} = \frac{1}{(1 + A/T)^2},$$

(8.10)

exactly as for the Fabry–Perot interferometer, which shows that when absorption is present, the value of peak transmission is determined by the ratio A/T.

To estimate the accuracy of matching which is required for the two reflectors, we assume that the absorption is zero. The peak transmission is given by the expression

$$(T_F)_{peak} = \frac{T_a T_b}{\left[1 - (R_a R_b)^{1/2}\right]^2},$$

(8.11)

where the subscripts a and b refer to the two reflectors. Let

$$R_b = R_a - \Delta_a,$$

(8.12)

where Δ_a is the error in matching, so that $T_b = T_a + \Delta_a$. Then we can write

$$(T_F)_{peak} = \frac{T_a(T_a + \Delta_a)}{\left\{1 - [R_a(R_a - \Delta_a)]^{1/2}\right\}^2}$$

$$= \frac{T_a(T_a + \Delta_a)}{\left\{1 - R_a\left[1 - \frac{1}{2}(\Delta_a/R_a) + \ldots\right]\right\}^2}.$$

(8.13)

Now assume that Δ_a is sufficiently small compared with R_a so that we can take only the first two terms of the expansion in Equation 8.13. With some rearrangement, the equation becomes

$$(T_F)_{peak} = \frac{T_a^2}{(1 - R_a)^2} \frac{1 + (\Delta_a/T_a)}{\left[1 + \frac{1}{2}(\Delta_a/T_a)\right]^2}.$$

(8.14)

The first part of the equation is the expression for peak transmission in the absence of any error in the reflectors, while the second part shows how the peak transmission is affected by errors. The second part of the expression is plotted in Figure 8.6, where the abscissa is $T_b/T_a = 1 + \Delta_a/T_a$. Clearly, the single-cavity filter is surprisingly insensitive to errors. Even with reflector transmittance unbalanced by a factor of three, it is still possible to achieve 75% peak transmittance.

8.2.2 All-Dielectric Single-Cavity Filter

In the same way as we found for the conventional Fabry–Perot etalon, if improved performance is to be obtained, then all-dielectric multilayers should replace the metallic reflecting layers.

An all-dielectric filter is shown in diagrammatic form in Figure 8.7. Basically, this is the same as a conventional etalon with dielectric coatings and with a solid thin-film cavity, or spacer, and the observations made for the metal–dielectric filter are also valid. Again, the substrate need not be worked to a high degree of flatness, although the polish must be good, because provided the geometry of the coating machine is adequate, the films will follow any reasonable substrate contours without showing changes in thickness.

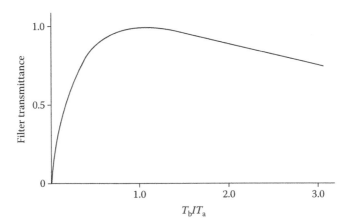

FIGURE 8.6
Theoretical peak transmittance of a single-cavity, or Fabry–Perot, filter with unbalanced reflectors.

FIGURE 8.7
Structure of an all-dielectric single-cavity filter.

The bandwidth of the all-dielectric filter can be calculated as follows. If the reflectance of each of the multilayers is sufficiently high, then

$$F = \frac{4R}{(1-R)^2} \simeq \frac{4}{T^2}$$

and

$$\frac{\lambda_0}{\Delta\lambda_h} = m\mathcal{F} = \frac{m\pi F^{1/2}}{2} \simeq \frac{m\pi}{T}. \tag{8.15}$$

Since the maximum reflectance for a given number of layers will be obtained with a high-index layer outermost, there are really only two cases that need be considered, and these are shown in Figure 8.8. If x is the number of high-index layers in each stack, not counting the cavity layer, then in the case of the high-index cavity, the transmittance of the stack will be given by

$$T = \frac{4n_L^{2x}n_s}{n_H^{2x+1}},$$

where n_s is the substrate index, and in the case of the low-index cavity, by

$$T = \frac{4n_L^{2x-1}n_s}{n_H^{2x}}.$$

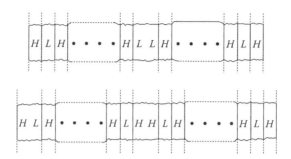

FIGURE 8.8
Structure of the two basic types of all-dielectric single-cavity, or Fabry–Perot, filter.

Substituting these results into the expression for bandwidth, we find, for the high-index cavity,

$$\frac{\Delta\lambda_h}{\lambda_0} = \frac{4n_L^{2x} n_s}{m\pi n_H^{2x+1}},\qquad(8.16)$$

and, for the low-index cavity,

$$\frac{\Delta\lambda_h}{\lambda_0} = \frac{4n_L^{2x-1} n_s}{m\pi n_H^{2x}},\qquad(8.17)$$

where we are adopting the fractional halfwidth $\Delta\lambda_h/\lambda_0$ rather than the resolving power $\lambda_0/\Delta\lambda_h$ as the important parameter. This is customary practice.

In these formulas, we have completely neglected any effect due to the dispersion of phase change on reflection from a multilayer. As we have already noted in Chapter 6, the phase change is not constant. The sense of the variation is such that it increases the rate of variation of $[(\varphi_a + \varphi_b)/2] - \delta$ with wavelength in the formula for transmittance of the single-cavity filter and, hence, reduces the bandwidth and increases the resolving power in Equations 8.16 and 8.17.

Seeley [5] has studied the all-dielectric filter in detail and, by making some approximations in the basic expressions for the filter transmittance, has arrived at formulas for the first-order halfwidths, which, with a little adjustment, become equal to the expressions in Equations 8.16 and 8.17 multiplied by the factor $(n_H - n_L)/n_H$. We can readily extend Seeley's analysis to all-dielectric filters of order m.

We recall that the half-peak points are given by

$$F\sin^2[(2\pi D/\lambda) - \varphi] = 1,\qquad(8.18)$$

where, since the filter is quite symmetrical, we have replaced $(\varphi_a + \varphi_b)/2$ by φ. It is simpler to carry out the analysis in terms of $g = \lambda_0/\lambda = v/v_0$. At the peak of the filter, we have $g = 1.0$. We can assume for small changes Δg in g that

$$2\pi D/\lambda = m\pi(1 + \Delta g)$$

and

$$\varphi = \varphi_0 + \frac{d\varphi}{dg}\Delta g,$$

so that Equation 8.18 becomes

$$F\sin^2\left(m\pi(1 + \Delta g) - \varphi_0 - \frac{d\varphi}{dg}\Delta g\right) = 1.$$

φ_0, as we know, is 0 or π, and so, using the same approximation as before,

$$F\left(m\pi\Delta g - \frac{d\varphi}{dg}\Delta g\right)^2 = 1$$

or

$$\Delta g = F^{-1/2}\left(m\pi - \frac{d\varphi}{dg}\right)^{-1}.$$

The halfwidth is $2\Delta g$ so that

$$2\Delta g = \frac{\Delta v_h}{v_0} = \frac{\Delta\lambda_h}{\lambda_0} = 2F^{-1/2}\left(m\pi - \frac{d\varphi}{dg}\right)^{-1}$$

$$= \frac{2}{m\pi F^{1/2}}\left(1 - \frac{1}{m\pi}\frac{d\varphi}{dg}\right)^{-1}.$$

(8.19)

We now need the quantity $d\varphi/dg$. We use Seeley's technique, but rather than follow him exactly, we choose a slightly more general approach because we shall require the results later. The matrix for a dielectric quarter-wave layer is

$$\begin{bmatrix} \cos\delta & i\sin\delta/\eta \\ i\eta\sin\delta & \cos\delta \end{bmatrix},$$

where, as usual, we are writing η for the optical admittance, which is in free space units. Now, for layers that are almost a quarter wave, we can write

$$\delta = \pi/2 + \varepsilon,$$

where ε is small. Then,

$$\cos\delta \simeq -\varepsilon, \quad \sin\delta \simeq 1,$$

so that the matrix can be written as

$$\begin{bmatrix} -\varepsilon & i/\eta \\ i\eta & -\varepsilon \end{bmatrix}.$$

We limit our analysis to quarter-wave multilayer stacks having high index next to the substrate. There are two cases, even and odd numbers of layers.

8.2.2.1 Case I: Even Number (2x) of Layers

The resultant multilayer matrix is given by

$$\begin{bmatrix} B \\ C \end{bmatrix} = [L][H][L]\ldots[L][H]\begin{bmatrix} 1 \\ \eta_{sub} \end{bmatrix},$$

where η_{sub} is the substrate admittance, and

$$[L] = \begin{bmatrix} -\varepsilon_L & i/\eta_L \\ i\eta_L & -\varepsilon_L \end{bmatrix},$$

$$[L] = \begin{bmatrix} -\varepsilon_H & i/\eta_H \\ i\eta_H & -\varepsilon_H \end{bmatrix}.$$

Then,

$$
\begin{bmatrix} B \\ C \end{bmatrix} = \{[L][H]\}^x \begin{bmatrix} 1 \\ \eta_{\text{sub}} \end{bmatrix}
$$

$$
= \begin{bmatrix} -\left(\dfrac{\eta_H}{\eta_L}\right) & -i\left(\dfrac{\varepsilon_L}{\eta_H} + \dfrac{\varepsilon_H}{\eta_L}\right) \\ -i(\eta_L\varepsilon_H + \eta_H\varepsilon_L) & -\left(\dfrac{\eta_L}{\eta_H}\right) \end{bmatrix}^x \begin{bmatrix} 1 \\ \eta_{\text{sub}} \end{bmatrix}
$$

$$
= \begin{bmatrix} M_{11} & iM_{12} \\ iM_{21} & M_{22} \end{bmatrix} \begin{bmatrix} 1 \\ \eta_{\text{sub}} \end{bmatrix}.
$$

Our problem is to find expressions for M_{11}, M_{12}, M_{21}, and M_{22}. In the evaluation, we neglect all terms of second and higher orders in ε. Terms in ε appearing in M_{11} and M_{22} are of second and higher orders and therefore

$$
M_{11} = (-1)^x \left(\frac{\eta_H}{\eta_L}\right)^x,
$$

$$
M_{22} = (-1)^x \left(\frac{\eta_L}{\eta_H}\right)^x.
$$

M_{12} and M_{21} contain terms of first, third, and higher orders in ε. The first-order terms are

$$
M_{12} = -\left(\frac{\varepsilon_L}{\eta_H} + \frac{\varepsilon_H}{\eta_L}\right)\left(-\frac{\eta_L}{\eta_H}\right)^{x-1} + \left(-\frac{\eta_H}{\eta_L}\right)\left[-\left(\frac{\varepsilon_L}{\eta_H} + \frac{\varepsilon_H}{\eta_L}\right)\right]\left(-\frac{\eta_L}{\eta_H}\right)^{x-2} + \cdots
$$

$$
+ \left(-\frac{\eta_H}{\eta_L}\right)^P\left[-\left(\frac{\varepsilon_L}{\eta_H} + \frac{\varepsilon_H}{\eta_L}\right)\right]\left(-\frac{\eta_L}{\eta_H}\right)^{x-p-1} + \cdots
$$

$$
+ \left(-\frac{\eta_H}{\eta_L}\right)^{x-1}\left[-\left(\frac{\varepsilon_L}{\eta_H} + \frac{\varepsilon_H}{\eta_L}\right)\right]
$$

$$
= (-1)^x\left(\frac{\varepsilon_L}{\eta_H} + \frac{\varepsilon_H}{\eta_L}\right)\left[\left(\frac{\eta_L}{\eta_H}\right)^{x-1} + \left(\frac{\eta_L}{\eta_H}\right)^{x-3} + \cdots + \left(\frac{\eta_H}{\eta_L}\right)^{x-1}\right],
$$

i.e.,

$$
M_{12} = (-1)^x\left(\frac{\varepsilon_L}{\eta_H} + \frac{\varepsilon_H}{\eta_L}\right)\left(\frac{\eta_L}{\eta_H}\right)^{x-1}
$$

$$
\times \left[\left(\frac{\eta_L}{\eta_H}\right)^{2x-2} + \left(\frac{\eta_L}{\eta_H}\right)^{2x-4} + \cdots + \left(\frac{\eta_L}{\eta_H}\right)^2 + 1\right]
$$

$$
= (-1)^x\left(\frac{\varepsilon_L}{\eta_H} + \frac{\varepsilon_H}{\eta_L}\right)\left(\frac{\eta_L}{\eta_H}\right)^{x-1}\left[1 - \left(\frac{\eta_L}{\eta_H}\right)^{2x}\right]\left[1 - \left(\frac{\eta_L}{\eta_H}\right)^2\right]^{-1}
$$

since $(\eta_L/\eta_H) < 1$.

Now, provided x is large enough and (η_L/η_H) is small enough, we can neglect $(\eta_L/\eta_H)^{2x}$ in comparison with 1, and after some adjustment, the expression becomes

$$M_{12} = \frac{(-1)^x \eta_H \eta_L \left(\dfrac{\eta_H}{\eta_L}\right)^x \left(\dfrac{\varepsilon_L}{\eta_H} + \dfrac{\varepsilon_H}{\eta_L}\right)}{(\eta_H^2 - \eta_L^2)}.$$

A similar procedure yields

$$M_{21} = \frac{(-1)^x \eta_H \eta_L \left(\dfrac{\eta_H}{\eta_L}\right)^x (\eta_L \varepsilon_L + \eta_H \varepsilon_H)}{(\eta_H^2 - \eta_L^2)}.$$

8.2.2.2 Case II: Odd Number (2x + 1) of Layers

The resultant matrix is given by

$$\begin{bmatrix} B \\ C \end{bmatrix} = [H][L][H][L] \dots [L][H] \begin{bmatrix} 1 \\ \eta_{sub} \end{bmatrix}$$

$$= [H]\{[L][H]\}^x \begin{bmatrix} 1 \\ \eta_{sub} \end{bmatrix},$$

which we can denote by

$$\begin{bmatrix} N_{11} & iN_{12} \\ iN_{21} & N_{22} \end{bmatrix} \begin{bmatrix} 1 \\ \eta_{sub} \end{bmatrix},$$

and which is simply the previous result multiplied by

$$\begin{bmatrix} -\varepsilon_H & i/\eta_H \\ i\eta_H & -\varepsilon_H \end{bmatrix}.$$

Then,

$$N_{11} = -\varepsilon_H M_{11} - M_{21}/\eta_H = (-1)^{x+1} \left(\frac{\eta_H}{\eta_L}\right)^x \frac{(\varepsilon_L \eta_H \eta_L + \varepsilon_H \eta_H^2)}{(\eta_H^2 - \eta_L^2)},$$

$$N_{12} = -\varepsilon_H M_{12} + M_{22}/\eta_H = (-1)^x \left(\frac{\eta_L}{\eta_H}\right)^x \frac{1}{\eta_H},$$

$$N_{21} = \eta_H M_{11} - \varepsilon_H M_{21} = (-1)^x \left(\frac{\eta_H}{\eta_L}\right)^x \eta_H,$$

$$N_{22} = -\varepsilon_H M_{22} - \eta_H M_{12} = (-1)^{x+1} \left(\frac{\eta_H}{\eta_L}\right)^x \frac{\eta_H^2 \eta_L (\varepsilon_L/\eta_H + \varepsilon_H/\eta_L)}{(\eta_H^2 - \eta_L^2)},$$

where terms in $(\eta_L/\eta_H)^x$ are neglected in comparison with $(\eta_H/\eta_L)^x$.

8.2.2.3 Phase Shift: Case I

We are now able to compute the phase shift on reflection. We take, initially, the index of the incident medium to be η_0. Then,

$$\begin{bmatrix} B \\ C \end{bmatrix} = \begin{bmatrix} M_{11} & iM_{12} \\ iM_{21} & M_{22} \end{bmatrix} \begin{bmatrix} 1 \\ \eta_{sub} \end{bmatrix}$$
$$= \begin{bmatrix} M_{11} + i\eta_{sub}M_{12} \\ \eta_{sub}M_{22} + iM_{21} \end{bmatrix},$$

$$\rho = \frac{\eta_0 B - C}{\eta_0 B + C} = \frac{\eta_0(M_{11} + i\eta_{sub}M_{12}) - \eta_{sub}M_{22} - iM_{21}}{\eta_0(M_{11} + i\eta_{sub}M_{12}) + \eta_{sub}M_{22} + iM_{21}}$$
$$= \frac{(\eta_0 M_{11} - \eta_{sub}M_{22}) + i(\eta_0\eta_{sub}M_{12} - M_{21})}{(\eta_0 M_{11} + \eta_{sub}M_{22}) + i(\eta_0\eta_{sub}M_{12} + M_{21})}, \tag{8.20}$$

$$\tan\varphi = \frac{2\eta_0\eta_{sub}^2 M_{12}M_{22} - 2\eta_0 M_{11}M_{21}}{\eta_0^2 M_{11}^2 - \eta_{sub}^2 M_{22}^2 + \eta_0^2\eta_{sub}^2 M_{12}^2 - M_{21}^2}. \tag{8.21}$$

Inserting the appropriate expressions and once again neglecting terms of second and higher orders in ε and terms in $(\eta_L/\eta_H)^x$, we obtain for φ

$$\tan\varphi = \frac{-2\eta_H\eta_L(\eta_L\varepsilon_H + \eta_H\varepsilon_L)}{\eta_0(\eta_H^2 - \eta_L^2)} \tag{8.22}$$

(for $LH...LHLH \,|\, \eta_m$).

8.2.2.4 Phase Shift: Case II

ρ is given by an expression similar to Equation 7.19, in which M is replaced by N. Then, following the same procedure as for case I, we arrive at

$$\tan\varphi = \frac{-2\eta_0(\eta_L\varepsilon_L + \eta_H\varepsilon_H)}{(\eta_H^2 - \eta_L^2)} \tag{8.23}$$

(for $HLH...LHLH \,|\, \eta_m$).

Equations 8.22 and 8.23 are in a general form that we will need later. For our present purposes, we can introduce some slight simplification.

$$\delta = \frac{2\pi nd}{\lambda} = 2\pi ndv = 2\pi ndv_0(v/v_0) = (\pi/2)g,$$

so that

$$\varepsilon_H = \varepsilon_L = (\pi/2)g - \pi/2 = (\pi/2)(g - 1).$$

Also, when we consider the construction of the Fabry–Perot filters we see that the incident medium in case I will be a high-index cavity layer and a low-index cavity in case II. Thus, for Fabry–Perot filters,

$$\tan\varphi = \frac{-\pi n_L}{(n_H - n_L)}(g - 1)$$

for both cases I and II.

Now, φ is nearly π or 0. Then,

$$\frac{d\varphi}{dg} = \frac{-\pi\eta_L}{(\eta_H - \eta_L)},$$

which is the result obtained by Seeley. This can then be inserted in Equation 8.19 to give

$$\frac{\Delta v_h}{v_0} = \frac{\Delta\lambda_h}{\lambda_0} = \frac{2}{m\pi F^{1/2}}\left(\frac{\eta_H - \eta_L}{\eta_H - \eta_L + \eta_L/m}\right).$$

Then, the expressions for the halfwidth of all-dielectric Fabry–Perot filters of mth order become as follows, which are simply the earlier results multiplied by the factor $(\eta_H - \eta_L)/(\eta_H - \eta_L + \eta_L/m)$:

• High-index cavity:

$$\left(\frac{\Delta\lambda_h}{\lambda}\right)_H = \frac{4\eta_{sub}\eta_L^{2x}}{m\pi\eta_H^{2x+1}}\frac{(\eta_H - \eta_L)}{(\eta_H - \eta_L + \eta_L/m)} \qquad (8.24)$$

• Low-index cavity:

$$\left(\frac{\Delta\lambda_h}{\lambda}\right)_L = \frac{4\eta_{sub}\eta_L^{2x-1}}{m\pi\eta_H^{2x}}\frac{(\eta_H - \eta_L)}{(\eta_H - \eta_L + \eta_L/m)} \qquad (8.25)$$

It should be noted that these results are for first-order reflecting stacks and mth-order cavity. Clearly, the effect of the phase is much greater the closer the two indices are in value and the lower the cavity order m. For the common visible and near infrared materials, zinc sulfide and cryolite, the factor for first-order cavities is equal to 0.43, while for infrared materials such as zinc sulfide and lead telluride, it is greater, around 0.57. Figures 8.9 and 8.10 show the characteristics of typical all-dielectric narrowband single-cavity, or Fabry–Perot, filters.

8.2.2.5 Some Practical Matters

Since the all-dielectric multilayer reflector is effective over a limited range only, transmission sidebands appear on either side of the peak and, in most applications, must be suppressed. In the

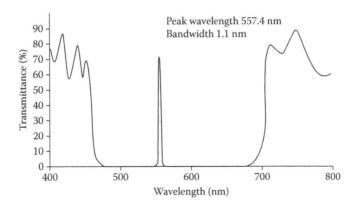

FIGURE 8.9
Measured transmittance of a narrowband all-dielectric filter with unsuppressed sidebands. Zinc sulfide and cryolite were the thin-film materials used. The peak transmittance at around 70% is now known to be a consequence of degradation by adsorption of atmospheric moisture. (Courtesy of Sir Howard Grubb, Parsons & Co. Ltd., Newcastle upon Tyne.)

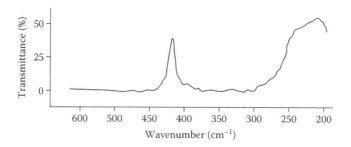

FIGURE 8.10
Measured transmittance of a Fabry–Perot filter for the far infrared. Design: Air | *LHLHHLH* | Ge, with *H* indicating a quarter wave of germanium and *L* of cesium iodide. The rear surface of the substrate is uncoated so that the effective transmittance of the filter is 50%. (Courtesy of Sir Howard Grubb, Parsons & Co. Ltd., Newcastle upon Tyne.)

visible and very near infrared regions, the shortwave sidebands can be very easily removed by adding to the filter a longwave-pass absorption filter, readily available in the form of polished glass disks from a large number of manufacturers. Unfortunately, it is not nearly as easy to obtain shortwave-pass absorption filters, and the rather shallow edges of those that are available considerably tend to reduce the peak transmittance of the filter if the sidebands are effectively suppressed. The best solution to this problem is not to use an absorption type of filter at all, but to employ as a blocking filter a metal–dielectric filter of the type already discussed or, preferably, of the induced transmission multiple-cavity type to be considered shortly. Because metal–dielectric filters used in the first order do not have longwave sidebands, they are very successful in this application. The metal–dielectric blocking filter can, in fact, be deposited over the all-dielectric filter in the same deposition cycle, provided that the layers are controlled using the narrowband filter itself as the test glass—this is known as direct monitoring—but, more frequently, a completely separate metal–dielectric filter is used. The various components that go to make up the final filter are cemented together in one assembly.

In the visible and near infrared regions of the spectrum, materials such as zinc sulfide and cryolite are capable of halfwidths of less than 0.1 nm with useful peak transmittance. Uniformity is, however, a major difficulty for filters of such narrow bandwidths. At 90% of peak points, the single-cavity filter has a width that is one-third of the halfwidth. It is a good guide that the uniformity of the filter should be such that the peak wavelength does not vary by more than one third of the halfwidth over the entire surface of the filter. This means that the effective increase in halfwidth due to the lack of uniformity is kept within some 4.5% of the nominal halfwidth, and the reduction in peak transmittance, to less than 3%. (These figures can be calculated using the expressions derived later for assessing the performance of filters in uncollimated incident light). For filters of less than 0.1 nm halfwidth, this rule implies a variation of not more than 0.03 nm or 0.006% in terms of layer thickness, a very severe requirement even for quite small filters. Halfwidths of 0.3–0.5 nm are less demanding and can be more readily produced provided considerable care is taken. For narrower filters, use is often made of the solid-etalon filters described in Section 8.2.4.

8.2.3 Losses in Single-Cavity Filters

Before we leave the single-cavity filter, we examine the effects of absorption losses in the layers in a manner similar to that already employed in Chapter 6, where we were concerned with quarter-wave stacks. Many researchers have investigated the problem. The following account heavily relies on the work of Hemingway and Lissberger [6], but with slight differences.

We directly apply the method of Chapter 6. There, we recall, we showed that the loss in a weakly absorbing multilayer was given by

$$A = (1 - R) \sum \mathcal{A},$$

where for quarter waves,

$$A = \beta \left(\frac{n}{y_e} + \frac{y_e}{n} \right),$$

$$\beta = \frac{2\pi k d}{\lambda} = \frac{2\pi n d}{\lambda} \frac{k}{n} = \frac{\pi}{2} \frac{k}{n}.$$

y_e is the admittance of the structure on the emergent side of the layer, in free space units; $n - ik$ is the refractive index of the layer; and d is the geometrical thickness. For quarter waves, $nd = \lambda/4$.

The arrangement is shown in Table 8.2, where the admittance y_e is given at each interface and where expressions for either high- or low-index cavities are included. The reflecting stacks are assumed to begin with high-index layers of which there are x per reflector, not counting the cavity layer.

We consider the case of low-index cavities first:

$$\sum \mathcal{A} = \beta_H \left(\frac{n_{sub}}{n_H} + \frac{n_H}{n_{sub}} \right) + \beta_L \left(\frac{n_H^2}{n_L n_{sub}} + \frac{n_L n_{sub}}{n_H^2} \right)$$

$$+ \beta_H \left(\frac{n_L^2 n_{sub}}{n_H^3} + \frac{n_H^3}{n_L^2 n_{sub}} \right) + \beta_L \left(\frac{n_H^4}{n_L^2 n_{sub}} + \frac{n_L^2 n_{sub}}{n_H^4} \right) + \cdots$$

$$+ \beta_L \left(\frac{n_H^{2x-2}}{n_L^{2x-3} n_{sub}} + \frac{n_L^{2x-3} n_{sub}}{n_H^{2x-2}} \right) + \beta_H \left(\frac{n_H^{2x-1}}{n_L^{2x-2} n_{sub}} + \frac{n_L^{2x-2} n_{sub}}{n_H^{2x-1}} \right)$$

$$+ m \left[\beta_L \left(\frac{n_H^{2x}}{n_L^{2x-1} n_{sub}} + \frac{n_L^{2x-1} n_{sub}}{n_H^{2x}} \right) + \beta_H \left(\frac{n_H^{2x}}{n_L^{2x-1} n_{sub}} + \frac{n_L^{2x-1} n_{sub}}{n_H^{2x}} \right) \right]$$

$$+ \beta_H \left(\frac{n_H^{2x-1}}{n_L^{2x-2} n_{sub}} + \frac{n_L^{2x-2} n_{sub}}{n_H^{2x-1}} \right) + \cdots + \beta_H \left(\frac{n_{sub}}{n_H} + \frac{n_H}{n_{sub}} \right),$$

where η_{sub} is the index of the substrate, the final set of terms is a repeat of the first and where the cavity consists of $2m$ quarter waves. By rearranging, we find

$$\sum \mathcal{A} = 2\beta_H \left(\frac{n_{sub}}{n_H} + \frac{n_L^2 n_{sub}}{n_H^3} + \frac{n_L^4 n_{sub}}{n_H^5} + \cdots + \frac{n_L^{2x-2} n_{sub}}{n_H^{2x-1}} \right)$$

$$+ 2\beta_H \left(\frac{n_H}{n_{sub}} + \frac{n_H^3}{n_L^2 n_{sub}} + \frac{n_H^5}{n_L^4 n_{sub}} + \cdots + \frac{n_H^{2x-1}}{n_L^{2x-2} n_{sub}} \right)$$

$$+ 2\beta_L \left(\frac{n_L n_{sub}}{n_H^2} + \frac{n_L^3 n_{sub}}{n_H^4} + \cdots + \frac{n_L^{2x-3} n_{sub}}{n_H^{2x-2}} \right)$$

$$+ 2\beta_L \left(\frac{n_H^2}{n_L n_{sub}} + \frac{n_H^4}{n_L^3 n_{sub}} + \cdots + \frac{n_H^{2x-2}}{n_L^{2x-3} n_{sub}} \right)$$

$$+ 2m\beta_L \left(\frac{n_H^{2x}}{n_L^{2x-1} n_{sub}} + \frac{n_L^{2x-1} n_{sub}}{n_H^{2x}} \right),$$

TABLE 8.2

Admittance y_e at Each Interface

Direction of incidence

$\downarrow$ —————— $n_{sub} = n_0$

n_H —————— n_H^2/n_{sub}

n_L —————— $n_L^2 n_{sub}/n_H^2$

n_H —————— $n_H^4/(n_L^2 n_{sub})$

n_L —————— $n_L^4 n_{sub}/n_H^4$

$\vdots$ —————— $n_H^{2x-}/(n_L^{2x-4} n_{sub})$

n_L —————— $n_L^{2x-2} n_{sub}/n_H^{2x-2}$

n_H —————— $n_H^{2x}/(n_L^{2x-2} n_{sub})$

$\swarrow$ $\qquad\qquad$ $\searrow$

Left branch:

————— $n_H^{2x}/(n_L^{2x-2} n_{sub})$

n_L ————— $n_L^{2x} n_{sub}/n_H^{2x}$

Cavity : $n_L^{2x} n_{sub}/n_H^{2x}$

n_L ————— $n_H^{2x}/(n_L^{2x-2} n_{sub})$

Right branch:

————— $n_H^{2x}/(n_L^{2x-2} n_{sub})$

n_L ————— $n_L^{2x} n_{sub}/n_H^{2x}$

n_H ————— $n_H^{2x+2}/(n_L^{2x} n_{sub})$

Cavity : $n_H^{2x+2}/(n_L^{2x} n_{sub})$

n_L ————— $n_L^{2x} n_{sub}/n_H^{2x}$

n_H ————— $n_H^{2x}/(n_L^{2x-2} n_{sub})$

$\searrow$ $\qquad\qquad$ $\swarrow$

————— $n_H^{2x}/(n_L^{2x-2} n_{sub})$

n_H —————— $n_L^{2x-2} n_{sub}/n_H^{2x-2}$

$\vdots$ —————— $n_H^4/(n_L^2 n_{sub})$

n_H —————— $n_L^2 n_{sub}/n_H^2$

n_L —————— n_H^2/n_{sub}

n_H —————— n_{sub}

where we have combined similar terms due to the two mirrors and where the final term is due to the cavity layer. The first four terms are geometric series, and, therefore, since $(n_L/n_H) < 1$,

$$\sum \mathcal{A} = 2\beta_H \frac{n_{sub}}{n_H} \frac{\left[1 - (n_L/n_H)^{2x}\right]}{\left[1 - (n_L/n_H)^2\right]}$$

$$+ 2\beta_H \frac{n_H^{2x-1}}{n_L^{2x-2} n_{sub}} \frac{\left[1 - (n_L/n_H)^{2x-2}\right]}{\left[1 - (n_L/n_H)^2\right]}$$

$$+ 2\beta_L \frac{n_L n_{sub}}{n_H^2} \frac{\left[1 - (n_L/n_H)^{2x-2}\right]}{\left[1 - (n_L/n_H)^2\right]}$$

$$+ 2\beta_L \frac{n_H^{2x-2}}{n_L^{2x-3} n_{sub}} \frac{\left[1 - (n_L/n_H)^{2x-2}\right]}{\left[1 - (n_L/n_H)^2\right]}$$

$$+ 2m\beta_L \left[\frac{n_H^{2x}}{n_L^{2x-1} n_{sub}} + \frac{n_L^{2x-1} n_{sub}}{n_H^{2x}}\right].$$

(n_L/n_H) will usually be rather less than unity, and x will normally be large, and so we can make the usual approximations and neglect terms such as $(n_L/n_H)^{2x}$ in the numerators and those terms which have (n_m/n_H) as a factor compared with $(n_L/n_m)(n_H/n_L)^{2x-1}$ etc. Then the expression simplifies to

$$\sum \mathcal{A} = 2\beta_H \frac{n_H^{2x-1}}{n_L^{2x-2} n_{sub}} \frac{1}{\left[1 - (n_L/n_H)^2\right]}$$

$$+ 2\beta_L \frac{n_H^{2x-2}}{n_L^{2x-3} n_{sub}} \frac{1}{\left[1 - (n_L/n_H)^2\right]}$$

$$+ 2m\beta_L \frac{n_H^{2x}}{n_L^{2x-1} n_{sub}},$$

but

$$\beta_H = \frac{2\pi n_H d}{\lambda} \frac{k_H}{n_H} = \frac{\pi}{2} \frac{k_H}{n_H},$$

$$\beta_L = \frac{\pi}{2} \frac{k_L}{n_L}.$$

Thus,

$$\sum \mathcal{A} = \frac{\dfrac{\pi k_H n_H^{2x}}{n_{sub} n_L^{2x-2}} + \dfrac{\pi k_L n_H^{2x}}{n_{sub} n_L^{2x-2}}}{(n_H^2 - n_L^2)} + \frac{\pi m k_L n_H^{2x}}{n_L^{2x} n_{sub}}$$

$$= \frac{\pi n_H^{2x}}{n_{sub} n_L^{2x}} \left[\frac{n_L^2 (k_H + k_L)}{(n_H^2 - n_L^2)} + m k_L\right].$$

The absorptance is then given by $A = (1 - R) \sum \mathcal{A}$. If the incident medium has index n_0, then, since the terminating admittance in Table 8.2 is n_{sub},

$$R = \left[\frac{n_0 - n_{sub}}{n_0 + n_{sub}}\right]^2,$$

and, therefore,

$$(1 - R) = \frac{4n_0 n_{sub}}{(n_0 + n_{sub})^2}.$$

The preceding expression for $\sum \mathscr{A}$ should therefore be multiplied by the factor $4n_0 n_{sub}/(n_0 + n_{sub})^2$ to yield the absorptance. However, the filters should be designed so that they are reasonably well matched into the incident medium, and therefore, this factor will be unity, or sufficiently near unity. The absorptance is then given by $\sum \mathscr{A}$. That is,

$$A = \frac{\pi n_H^{2x}}{n_{sub} n_L^{2x}} \left[\frac{n_L^2 (k_H + k_L)}{(n_H^2 - n_L^2)} + m k_L \right] \tag{8.26}$$

for low-index cavities.

For high-index cavities, we work through a similar analysis, and with the same approximations, we arrive at

$$A = \frac{\pi n_H^{2x}}{n_{sub} n_L^{2x}} \left[\frac{n_L^2 k_H + n_H^2 k_L}{(n_H^2 - n_L^2)} + m k_H \right] \tag{8.27}$$

for high-index cavities.

It should be noted that since x is the number of high-index layers, the filter represented by Equation 8.27 will be narrower than that represented by Equation 8.26 for equal x.

A useful set of alternative expressions can be obtained if we substitute Equations 8.24 and 8.25 into Equations 8.26 and 8.27 to give the following:

• High-index cavity:

$$A = 4 \frac{\lambda_0}{\Delta \lambda_h} \frac{\left\{ \left[m + (1 - m)(n_L/n_H)^2 \right] + k_L \right\}}{(n_H + n_L)[m + (1 - m)(n_L/n_H)]} \tag{8.28}$$

• Low-index cavity:

$$A = 4 \frac{\lambda_0}{\Delta \lambda_h} \frac{\left\{ k_L(n_H/n_L) \left[m + (1 - m)(n_L/n_H)^2 \right] + (n_L/n_H)k_H \right\}}{(n_H + n_L)[m + (1 - m)(n_L/n_H)]} \tag{8.29}$$

Figure 8.11 shows the value of A plotted for Fabry–Perot filters with $n_H = 2.35$ and $n_L = 1.35$, typical of zinc sulfide and cryolite. $(\lambda_0/\Delta \lambda_h)$ is taken as 100, and k_H and k_L, as either 0 or 0.0001. The effect of other values of $(\lambda_0/\Delta \lambda_h)$ or k can be estimated by multiplying by an appropriate factor. The approximations are reasonable for $k(\lambda_0/\Delta \lambda_h)$ less than around 0.1.

It is difficult to draw any general conclusions from Figure 8.11 because the results depend on the relative magnitudes of k_H and k_L. However, except in the case of very low k_L, the high-index cavity is to be preferred. There are very good reasons connected with performance when tilted, with energy grasp, and with the manufacture of filters, for choosing high-index cavities rather than low-index cavities.

Small extinction coefficients are difficult to measure, but in the visible and near infrared regions of the spectrum, materials such as zinc sulfide and cryolite, when deposited with great care and attention to detail, clearly have extinction coefficients of significantly less than 10^{-5}, but it is difficult to judge how low because their performance is much more subject to the adsorption of atmospheric moisture, to which we will return in Section 14.6. The coefficients of tantala and silica, preferred nowadays for their great stability, can be less than 10^{-6}. In their regions of transparency, therefore, absorption losses in these high-performance materials do not represent any great problem.

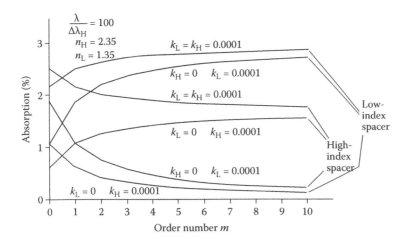

FIGURE 8.11

Value (expressed as a percentage) of the absorptance, as a function of the order number m, of single-cavity filters with $\lambda_0/\Delta\lambda_h$ of 100 and values of extinction coefficients k_H and k_L of 0.0001 or 0. Other values can be accommodated by multiplying by an appropriate factor. n_H is taken as 2.35, and n_L as 1.35. The results are derived from Equations 8.28 and 8.29.

8.2.4 Solid-Etalon Filter

A solid-etalon filter, or, as it is sometimes called, a solid-spacer filter, is a very high-order single-cavity or Fabry–Perot filter in which the cavity consists of an optically worked plate or a cleaved crystal. Thin-film reflectors are deposited on either side of the cavity, or spacer, in the normal way, so that the cavity also acts as the substrate. The problems of uniformity that exist with all-thin-film narrowband filters are avoided, and the thick cavity does not suffer from the increased scattering losses that always seem to accompany the higher-order thin-film cavities. The solid-etalon filter is very much more robust and stable than the conventional air-spaced Fabry–Perot etalon, while the manufacturing difficulties are comparable. The high order implies a small interval between orders, and a conventional thin-film narrowband filter must be used in series with it to eliminate the unwanted orders.

An early account of the use of mica for the construction of filters of this type is that of Dobrowolski [7] who credits Billings with being the first to use mica in this way, achieving halfwidths of 0.3 nm. Dobrowolski obtained rather narrower passbands, and his is the first complete account of the technique. Mica can be readily cleaved to form thin sheets with flat parallel surfaces, but there is a complication due to the natural birefringence of mica, which means that the position of the passband depends on the plane of polarization. This splitting of the passband can be avoided by arranging the thickness of the mica such that it is a half-wave plate, or multiple half-wave, at the required wavelength. If the two refractive indices are n_o and n_e, this implies

$$\frac{2\pi(n_o - n_e)d}{\lambda} = p\pi, \quad p = 0, \pm1, \pm2, \dots .$$

The order of the cavity will then be given by

$$m = \frac{n_o p}{(n_o - n_e)} \text{ or } m = \frac{n_e p}{(n_o - n_e)},$$

depending on the plane of polarization. The difference between these two values is p, but since p is small, the bandwidths will be virtually identical. The separation of orders for large m is approximately given by λ/m. Dobrowolski found that the maximum order separation, corresponding to

$p = 1$, was given by 1.64 nm at 546.1 nm. With such cavities, around 60 μm thick, filters with halfwidths of around 0.1 nm, the narrowest being 0.085 nm, were constructed. Peak transmittance ranged up to 50% for the narrower filters and up to 80% for slightly broader ones with around 0.3 nm halfwidth.

More recent work on solid-etalon filters has concentrated on the use of optically worked materials as cavities. These must be ground and polished so that the faces have the necessary flatness and parallelism. The most complete account so far of the production of such filters is by Austin [8]. Fused silica disks as thin as 50 μm have been produced with the necessary parallelism for halfwidths as narrow as 0.1 nm in the visible region, while thicker disks can give bandwidths as narrow as 0.005 nm. A 50 μm fused silica cavity gives an interval between orders of around 1.4 nm in the visible region, which allows the suppression of unwanted orders to be fairly readily achieved by conventional thin-film narrowband filters.

The process of optical working tends to produce an error in parallelism over the surface of the cavity, which is ultimately independent of its thickness. Let us denote the total range of thickness due to this lack of parallelism and to any deviation from flatness by Δd. This variation in cavity thickness causes the peak wavelength of the filter to vary. We can take an absolute limit for these variations as half the bandwidth of the filter. Then the resultant halfwidth will be increased by just over 10%, and the peak transmittance, reduced by just over 7% (once again using the expressions which we will shortly establish for filter performance in uncollimated light). We can write

$$\Delta\lambda_0/\lambda_0 = \Delta D/D = \Delta d/d \leq 0.5\Delta\lambda_h/\lambda_0,$$

where D is the optical thickness nd of the cavity, $\Delta\lambda_0$ is the error in peak wavelength, and $\Delta\lambda_h$ is the halfwidth. But

$$\text{Resolving power} = \lambda_0/\Delta\lambda_h = m\mathcal{F},$$

and, hence, since

$$D = m\lambda_0/2,$$

$$\mathcal{F} = \frac{0.25\lambda_0}{\Delta D}.$$

Now the attainable ΔD in the visible region is on the order of $\lambda/100$, and this means that the limiting finesse is around 25, independent of the cavity thickness. High resolving power then has to be achieved by the order number m that determines both the cavity thickness $D = m\lambda_0/2$ and the interval between orders λ_0/m. For a halfwidth of 0.01 nm at, say, 500 nm, the resolving power is 50,000. The finesse of 25 implies an order number of 2,000, a cavity optical thickness of 500 μm, and an interval between orders of 0.25 nm. This very restricted range between orders means that it is very difficult to directly carry out sideband blocking by a thin-film filter. Instead, a broader solid-etalon filter can be used with its corresponding greater interval between orders. It, in its turn, can be suppressed by a thin-film filter. For a halfwidth of 0.1 nm, a cavity optical thickness of 50 μm is required, which gives an interval between orders of 2.5 nm.

The temperature coefficient of the peak wavelength change of solid-etalon filters with fused silica cavities is 0.005 nm/°C, and the filters may be finely tuned by altering this temperature.

Candille and Saurel [9] used Mylar foil as the cavity. Their filters were strictly of the multiple-cavity type described later in this chapter. The Mylar acted as a substrate and a high-order cavity. One of the reflectors included a low-order single-cavity filter that served both as blocking filter to eliminate the additional unwanted orders of the Mylar section and as an additional cavity to steepen the sides of the passband. The position of the passband could be altered by varying the tension in the Mylar. The filters were not as narrow as the other solid-etalon filters mentioned, halfwidths of 0.8–1.0 nm being obtained.

Solid-etalon filters have also been constructed for the infrared. Smith and Pidgeon [10] used a polished slab of germanium some 780 µm thick working at around 700 cm^{-1} in the 400th order. Both faces were coated with a quarter wave of zinc sulfide followed by a quarter wave of lead telluride to give a reflectance of 62%, a fringe halfwidth of 0.1 cm^{-1} and an interval between orders of 1.6 cm^{-1}. This particular arrangement was designed so that the lines in the R-branch of the CO_2 spectrum, which are spaced at 1.6 cm^{-1} apart at around 14.5 µm, should be exactly matched by a number of adjacent orders. Order sorting was not, therefore, a problem.

Roche and Title [11] have reported a range of solid-etalon filters for the infrared. These filters are some 13 mm in diameter and have resolving powers in the region of 3×10^4, and the techniques used for their construction are as reported by Austin [8]. For wavelengths equal to or shorter than 3.5 µm, fused silica cavities are quite satisfactory. For longer wavelengths, Yttralox, a combination of yttrium and thorium oxides, was found most satisfactory. With this material, solid-etalon filters were produced, which at 3.334 µm had halfwidths as low as 0.2 nm, and at 4.62 µm, 0.8 nm. At these wavelengths, the attainable finesse was 30–40, and the current limit to the halfwidth that can be achieved is the permissible interval between orders, which determines the arrangement of subsidiary blocking filters.

Floriot et al. [12] made some significant advances in the use of solid-etalon filters for telecommunication applications. These are actually multiple-cavity filters, rather than single-cavity, but it is convenient to refer to the work here. We shall see shortly that the sloping edges of the single-cavity filter can be considerably steepened, so that the response of the filter becomes more rectangular, when multiple cavities are coherently coupled together to present a single, combined interferometric response. The coupling between successive cavity structures is usually accomplished by means of a single quarter-wave layer called the coupling layer. Multiple-cavity filters present challenges in production that are well beyond those of single-cavity filters. The very narrow bandpass filters required for telecommunication applications may require designs of well above 100 discrete layers, sometimes more than 200. If such filters are constructed from solid etalons, then the same reduction in design complexity and greater ease of deposition enjoyed by the single-cavity solid-etalon filters are realized. Of course, there is the difficulty of coupling the cavities together, and this has been accomplished by using air as the coupling material and finely adjusting the thickness using piezoelectric translators. The air-coupling layers can actually have thicknesses considerably in excess of quarter waves and, in the filters described, were arranged at usually slightly more than 20 quarter waves, and to achieve optimum performance the coupling layers, together with the two layers on either side of each coupling layer, were refined in thickness. In practical filters, this refinement process is repeated at the appropriate point in the production of the filters to compensate for any committed errors in deposition. When first produced, the cavities are not necessarily of exactly the required thicknesses. They consist of 10×10 mm silica plates and are measured by examining the interference fringes produced by a tunable laser. They are then adjusted in thickness by depositing a silica film over their surface. The final filter is required to accommodate only the 250 µm beam waist of a Gaussian beam, and so the initial parallelism of 3″ meets the requirements. The problem of the limited free spectral range is partially met by arranging different thicknesses for the cavities so that those peaks close to the fundamental wavelength in the various etalons do not exactly coincide.

A typical example using four cavity structures has the following design, where the symbols indicate quarter-wave thicknesses at the reference wavelength of 1550 nm. C is a quarter wave of the silica cavity material of index 1.44, H is a high-index quarter wave of index 2.09 (Ta_2O_5), L is a low-index quarter wave of index 1.46 (SiO_2), and A represents a quarter wave of air:

$$1.09H(LH)^2 \ 246C \ (HL)^2 \ 1.05H \ 20.66A \ 1.13H(LH)^2 \ 500C \ (HL)^2 \ 1.09H \ 24.79A \ 1.06H \ (LH)^2 \ 412C$$
$$(HL)^2 \ 1.02H \ 31.11A \ 0.73H \ (LH)^2 \ 298C \ (HL)^2 \ 1.27H.$$

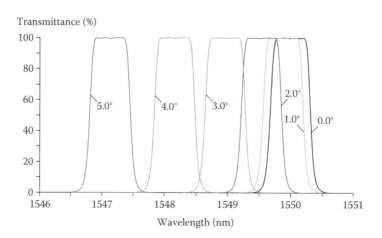

FIGURE 8.12
Calculated performance of the solid-etalon filter, with design as in the text, at normal incidence and at increasing angle of incidence. The ripple performance is exceptional and remains so even at elevated angles of incidence. (After J. Floriot, F. Lemarchand, and M. Lequime, *Applied Optics*, 45, 1349–1355, 2006. With permission of Optical Society of America.)

The incident and emergent media are both air. The calculated performance of this design is shown in Figure 8.12. The ripple performance is exceptionally good. Filters with this level of performance will often exhibit rather poorer performance at oblique incidence where the profile becomes distorted. This solid-etalon filter exhibits virtually no perceptible distortion as evidenced by the figure.

8.2.5 Effect of Varying the Angle of Incidence

As we have seen with other types of thin-film assembly, the performance of an all-dielectric single-cavity filter varies with the angle of incidence, and this effect is particularly important when considering, for instance, the allowable focal ratio of the pencil being passed by the filter or the maximum tilt angle in any application. The variation with the angle of incidence is not altogether a bad thing because the effect can be used to tune filters that are otherwise off the desired wavelength—very important from the manufacturer's point of view because it enables an easing of the otherwise almost impossibly tight production tolerances.

The effect of tilting has been studied by a number of workers, particularly by Dufour and Herpin [13], Lissberger [14], Lissberger and Wilcock [15], and Pidgeon and Smith [16]. For our present purposes, we follow Pidgeon and Smith since their results are in a slightly more suitable form.

8.2.5.1 Simple Tilts in Collimated Light

The phase thickness of a thin film at oblique incidence is $\delta = 2\pi nd \cos \vartheta / \lambda$. This can be interpreted as an apparent optical thickness of $nd \cos \vartheta$ that varies with angle of incidence so that layers seem thinner when tilted. Although the optical admittance changes as the tilt angle varies, in narrowband filters, the predominant effect is the apparent change in thickness that moves the filter passband to shorter wavelengths.

For an ideal single-cavity filter with cavity layer index n^*, where the reflectors have constant phase shift of zero or π regardless of the angle of incidence or wavelength, we can write for the position of peak wavelength in the mth order

$$\frac{2\pi n^* d \cos \vartheta}{\lambda} = \frac{2\pi n^* d}{\lambda_0} \frac{\lambda_0}{\lambda} \cos \vartheta = m\pi,$$

where λ is the new, tilted peak wavelength and λ_0 is the peak at normal incidence. We can replace λ_0/λ by g, which can then be written as $1 + \Delta g$. Then, since $2\pi n \times d/\lambda_0 = m\pi$,

$$1 + \Delta g = \frac{1}{\cos \vartheta},$$

so that

$$\Delta g = \frac{1}{\cos \vartheta} - 1.$$

If the angle of incidence is ϑ_0 in air, then

$$\vartheta = \arcsin(\sin \vartheta_0/n^*),$$

and Δg is given in terms of ϑ_0 and n^*. The effect of tilting then in this ideal filter can be simply estimated from knowledge of the index of the cavity and the angle of incidence. For small angles of incidence, the shift is given by

$$\Delta g = \frac{\Delta v}{v_0} = \frac{\Delta \lambda}{\lambda_0} = \frac{\vartheta_0^2}{2n^{*2}}. \tag{8.30}$$

The index of the cavity n^* determines its sensitivity to tilt; the higher the index, the less the filter is affected.

In the case of a real filter, the reflectors are also affected by the tilting, and so the calculation of the shift in peak wavelength is more involved. It was, however, shown by Pidgeon and Smith [16] that the shift is similar to what would have been obtained from an ideal filter with cavity index n^*, intermediate between the high and low indices of the layers of the filter. n^* is known as the effective index. This concept of the effective index holds good for quite high angles of incidence, up to 20° or 30°, or even higher, depending on the indices of the layers making up the filter.

We can estimate the effective index for the filter by a technique similar to that already used for finding the peak position of metal–dielectrics (Equation 8.4). We retain our assumption of small angle of incidence and small changes in g around the value corresponding to the peak at normal incidence.

The peak position is given, as before, by

$$\sin^2 \left[\frac{2\pi n d \cos \vartheta}{\lambda} - \varphi \right] = 0 \tag{8.31}$$

with, at normal incidence,

$$\sin^2 \left[\frac{2\pi n d}{\lambda_0} - \varphi_0 \right] = 0. \tag{8.32}$$

Now φ_0 is 0 or π, and so Equation 8.32 is satisfied by

$$2\pi n d/\lambda_0 = m\pi, \quad m = 0, 1, 2, \ldots.$$

The analysis is once again easier in terms of g $(= \lambda_0/\lambda = v/v_0)$. Equation 8.31 becomes

$$\sin^2[(2\pi n d/\lambda_0)g \cos \vartheta - \varphi_0 - \Delta\varphi] = 0. \tag{8.33}$$

We write

$$g = 1 + \Delta g \text{ and } \cos \vartheta \simeq 1 - \vartheta^2/2.$$

However, we should work in terms of ϑ_0, the external angle of incidence, which we will assume is referred to free space (if not, then we make the appropriate correction). Then

$$n \sin \vartheta = n_0 \sin \vartheta_0 = \sin \vartheta_0,$$

and, using Equation 8.33,

$$\sin^2 \left[(2\pi n d / \lambda_0) - \varphi_0 + m\pi \Delta g - \left(m\pi \vartheta_0^2 / 2n^2 \right) - \Delta \varphi \right] = 0$$

is the condition for the new peak position. This requires

$$m\pi \Delta g - \left(m\pi \vartheta_0^2 / 2n^2 \right) - \Delta \varphi = 0. \tag{8.34}$$

Now $\Delta \varphi$ is a function of ϑ and Δg, and to evaluate it, we return to Equations 8.21 and 8.22, but, because we are dealing with a perturbation from a reference condition of normal incidence, we replace η by n. The layers in the reflectors are all quarter waves, and so ε is given by

$$\frac{\pi}{2} + \varepsilon = \left(\frac{2\pi n d}{\lambda_0} \right) g \cos \vartheta = \left(\frac{\pi}{2} \right) (1 + \Delta g) \left(1 - \frac{\vartheta^2}{2} \right),$$

but

$$\vartheta = \frac{\vartheta_0}{n},$$

so that

$$\varepsilon = \frac{\pi}{2} \Delta g - \frac{\pi \vartheta_0^2}{4n^2},$$

with n being either n_L or n_H for ε_L or ε_H, respectively.

At this stage, we are forced to consider high-index and low-index cavities separately.

8.2.5.2 Case I: High-Index Cavities

From Equation 8.21, we have, inserting n_H for η_0,

$$\Delta \varphi = -\frac{2n_L^2}{\left(n_H^2 - n_L^2 \right)} \varepsilon_H - \frac{2n_H n_L}{\left(n_H^2 - n_L^2 \right)} \varepsilon_L$$

$$= -\frac{2n_L^2}{\left(n_H^2 - n_L^2 \right)} \left(\frac{\pi}{2} \Delta g - \frac{\pi \vartheta_0^2}{4n_H^2} \right) - \frac{2n_H n_L}{\left(n_H^2 - n_L^2 \right)} \left(\frac{\pi}{2} \Delta g - \frac{\pi \vartheta_0^2}{4n_L^2} \right)$$

$$= -\frac{\pi n_L}{\left(n_H - n_L \right)} \Delta g + \frac{\pi}{2} \frac{\left(n_L^2 + n_H^2 - n_L n_H \right)}{n_H^2 n_L \left(n_H - n_L \right)} \vartheta_0^2,$$

and Equation 8.34 becomes

$$m\pi \Delta g - \frac{m\pi \vartheta_0^2}{2n_H^2} + \frac{\pi n_L}{\left(n_H - n_L \right)} \Delta g - \frac{\pi}{2} \frac{\left(n_L^2 + n_H^2 - n_L n_H \right)}{n_H^2 n_L \left(n_H - n_L \right)} \vartheta_0^2 = 0,$$

giving, after some manipulation and simplification,

$$\Delta g = \frac{1}{n_H^2} \left[\frac{m - 1 - (m - 1)\frac{n_L}{n_H} + \frac{n_H}{n_L}}{m - (m - 1)\frac{n_L}{n_H}} \right] \frac{\vartheta_0^2}{2}.$$

But comparing the expression with Equation 8.30, we find

$$n^{*2} = \frac{n_H^2 \left[m - (m - 1)\frac{n_L}{n_H} \right]}{\left[m - 1 - (m - 1)\frac{n_L}{n_H} + \frac{n_H}{n_L} \right]}$$

or

$$n^* = n_H \left[\frac{m - (m - 1)\frac{n_L}{n_H}}{m - 1 - (m - 1)\frac{n_L}{n_H} + \frac{n_H}{n_L}} \right]^{1/2}. \tag{8.35}$$

For first-order filters,

$$n^* = (n_H n_L)^{1/2}, \tag{8.36}$$

which is the result obtained by Pidgeon and Smith. As $m \to \infty$, then $n^* \to n_H$, as we would expect.

8.2.5.3 Case II: Low-Index Cavities

The analysis is exactly as for case I except that Equation 8.23 is used, and n in Equation 8.34 becomes n_L:

$$n^* = n_L \left[\frac{m - (m - 1)\frac{n_L}{n_H}}{m - m\frac{n_L}{n_H} + \left(\frac{n_L}{n_H}\right)^2} \right]^{1/2}. \tag{8.37}$$

For first-order filters,

$$n^* = \frac{n_L}{\left[1 - \frac{n_L}{n_H} + \left(\frac{n_L}{n_H}\right)^2 \right]^{1/2}}, \tag{8.38}$$

which is, again, the expression given by Pidgeon and Smith, and we note again that as $m \to \infty$, then $n^* \to n_L$.

Typical curves showing how the effective index n^* varies with order number for both low- and high-index cavities are given in Figure 8.13.

Pidgeon and Smith made experimental measurements on narrowband filters for the infrared. The designs in question were

1. $L|Ge|LHLH\ LL\ HLH|$Air,

2. $L|Ge|LHLHL\ HH\ LHLH|$Air,

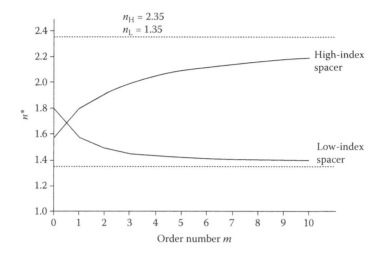

FIGURE 8.13
Effective index n^* plotted against order number m for single-cavity filters constructed of materials such as zinc sulfide, $n = 2.35$, and cryolite, $n = 1.35$. The results were calculated from Equations 8.35 and 8.37. The reason for the curious crossing between the zeroth and first orders is that because of the design, a zero-order low-index cavity becomes a first-order high index and vice versa.

where H represents a quarter-wave thickness of lead telluride and L represents a quarter-wave thickness of zinc sulfide and where the peak wavelength was approximately 15 μm. Calculations of shift were carried out by the approximate method using n^* and by the full matrix method without approximations. The results using n^* matched the accurate calculations up to angles of incidence of 40° to an accuracy representing ±2% change in n^*. The experimental points showed good agreement with the theoretical estimates. Some of the results are shown in Figures 8.14 and 8.15.

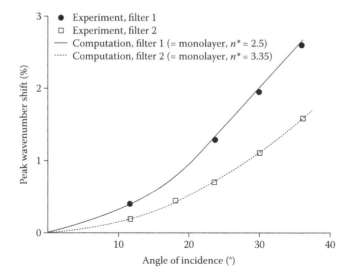

FIGURE 8.14
Shift of peak wavenumber with scanning angle for two single-cavity filters in collimated light. In both cases, the monolayer curves fit the computed curves to ±2% in n. (After C. R. Pidgeon and S. D. Smith, *Journal of the Optical Society of America*, 54, 1459–1466, 1964. With permission of Optical Society of America.)

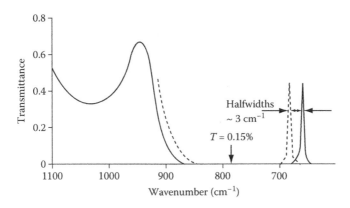

FIGURE 8.15
Measured transmittance of two filters of type 2. Design: Air | HLHL HH LHLHL | Ge substrate | L | Air (H = PbTe; L = ZnS).
(After C. R. Pidgeon and S. D. Smith, *Journal of the Optical Society of America*, 54, 1459–1466, 1964. With permission of Optical Society of America.)

The angle of incidence may be in a medium other than free space, in which case Equation 8.30 becomes

$$\Delta g = \frac{\Delta v}{v_0} = \frac{\Delta \lambda}{\lambda_0} = \frac{n_0{}^2 \vartheta_0^2}{2n^{*2}}, \tag{8.39}$$

where ϑ_0 is measured in radians.

If ϑ_0 is measured in degrees, then

$$\Delta g = \frac{\Delta v}{v_0} = \frac{\Delta \lambda}{\lambda_0} = 1.5 \times 10^{-4} \frac{n_0{}^{20} \vartheta^2}{n^{*2}}. \tag{8.40}$$

8.2.5.4 Effect of an Incident Cone of Light

The analysis can be taken a stage further to arrive at expressions for the degradations of peak transmission and bandwidth, which become apparent when the incident illumination is not perfectly collimated. Essentially, the same results were obtained by Lissberger and Wilcock [15] and by Pidgeon and Smith [16].

It is assumed first of all that in collimated light, the sole effect of tilting a filter is a shift of the characteristic toward shorter wavelengths, or greater wavenumbers, leaving the peak transmittance and bandwidth virtually unchanged. The performance in convergent or divergent light is then given by integrating the transmission curve over a range of angles of incidence. The analysis is simpler in terms of wavenumber or g, rather than wavelength. If v_0 is the wavenumber corresponding to the peak at normal incidence, and v_Θ, to the peak at angle of incidence Θ, then it is plausible that the resultant peak, when all angles of incidence in the cone from 0 to Θ are included, should appear at a wavenumber given by the mean of the extremes mentioned earlier. We shall show shortly that this is indeed the case. The new peak is given by

$$v_{\text{peak}} = v_0 + \frac{1}{2} \Delta v', \tag{8.41}$$

where

$$\Delta v' = v_\Theta - v_0 = v_0 \frac{\Theta^2}{2n^{*2}}.$$

The effective bandwidth of the filter will, of course, appear broader, and since the process is, in effect, a convolution of a function with bandwidth W_0, which is the width of the filter at normal incidence, and another function with bandwidth $\Delta v'$, the change in peak position produced by altering the angle of incidence from 0 to Θ, it seems likely that the resultant bandwidth might be given by the square root of the sum of their squares. This too is indeed the case, as we shall also show.

$$W_\Theta^2 = W_0^2 + (\Delta v')^2.$$

(8.42)

The peak transmission falls and is given by

$$\hat{T}_\Theta = \left(\frac{W_0}{\Delta v'}\right) \arctan\left(\frac{\Delta v'}{W_0}\right).$$

(8.43)

The analysis is as follows.

We consider incident light in the form of a cone with semiangle Θ, that is, a cone of focal ratio $1/(2 \tan \Theta)$. We assume that in collimated light, the effect of tilting the filter is to simply move the characteristic toward shorter wavelengths, leaving the bandwidth and peak transmittance unchanged. We can think of the illumination as in the form of a uniform spherical wavefront with the filter located exactly at the center. Then, the flux at any angle will be proportional to the area subtended on the spherical surface.

For small values of ϑ, the flux incident on the filter is proportional to $\vartheta d\vartheta$. The resultant transmittance of the filter is then given by the total flux transmitted divided by the total flux incident. We can forget about the constants of proportionality because they cancel in this operation.

With that simplification, the total flux incident is proportional to

$$\int_0^\Theta \vartheta d\vartheta = \frac{1}{2}\Theta^2.$$

The total flux transmitted is proportional to

$$\int_0^\Theta \vartheta T(\vartheta) d\vartheta.$$

We can, for small values of ϑ and Δg, set

$$T = \frac{1}{1 + \left\{\dfrac{2}{\Delta g_h}\left(\Delta g - \dfrac{\vartheta_0^2}{2n^{*2}}\right)\right\}^2},$$

where Δg_h is the halfwidth at normal incidence of the filter in units of g. This expression directly follows from the concept of n^*. The transmittance of the filter is then given by

$$\begin{aligned}
T &= \frac{2}{\Theta^2}\int_0^\Theta \frac{\vartheta_0 d\vartheta_0}{1 + \left\{\dfrac{2}{\Delta g_h}\left(\Delta g - \dfrac{\vartheta_0^2}{2n^{*2}}\right)\right\}^2}\\
&= -\frac{2}{\Theta^2}\frac{n^{*2}\Delta g_h}{2}\left[\arctan\left\{\frac{2}{\Delta g_h}\left(\Delta g - \frac{\vartheta_0^2}{2n^{*2}}\right)\right\}\right]_0^\Theta \\
&= \frac{1}{2}\frac{\Delta g_h}{(\Theta^2/2n^{*2})}\left[\arctan\left(2\frac{\Delta g}{\Delta g_h}\right) - \arctan\left\{2\left(\frac{\Delta g}{\Delta g_h} - \frac{\Theta^2}{2n^{*2}}\frac{1}{\Delta g_h}\right)\right\}\right].
\end{aligned}$$

(8.44)

The expression inside the square brackets is the difference of two angles. We can take the tangent of that difference and then the arctangent of the resulting expression to yield an improved form:

$$T = \frac{1}{2} \frac{\Delta g_h}{(\Theta^2/2n^{*2})} \left[\arctan \left(\frac{\left(\frac{2}{\Delta g_h}\right)\left(\frac{\Theta^2}{2n^{*2}}\right)}{1 + \left(\frac{2}{\Delta g_h}\right)^2 \left\{ \Delta g \left(\Delta g - \frac{\Theta^2}{2n^{*2}} \right) \right\}} \right) \right]. \tag{8.45}$$

Differentiating this with respect to Δg, we find zero derivative and, therefore, maximum transmittance at

$$\Delta g = \frac{1}{2} \frac{\Theta^2}{2n^{*2}}.$$

However, $\Theta^2/(2n^{*2})$ is the shift in the position of the peak at angle of incidence Θ. Thus, in a cone of light of semiangle Θ, the peak wavelength of the filter is given by the mean of the value at normal incidence and that at the angle Θ corresponding to Equation 8.41. The value of the peak transmittance is then, from Equation 8.44,

$$T = \frac{1}{2} \frac{\Delta g_h}{(\Theta^2/2n^{*2})} \left[\arctan \left(\frac{(\Theta^2/2n^{*2})}{\Delta g_h} \right) - \arctan \left(-\frac{(\Theta^2/2n^{*2})}{\Delta g_h} \right) \right]$$

$$= \frac{\Delta g_h}{(\Theta^2/2n^{*2})} \arctan \left(\frac{(\Theta^2/2n^{*2})}{\Delta g_h} \right),$$

which corresponds to Equation 8.43.

The half-peak points are given by those values of Δg that yield half the peak transmittance in Equation 8.45. That is,

$$\frac{1}{2} \frac{\Delta g_h}{(\Theta^2/2n^{*2})} \left[\arctan \left(\frac{\left(\frac{2}{\Delta g_h}\right)\left(\frac{\Theta^2}{2n^{*2}}\right)}{1 + \left(\frac{2}{\Delta g_h}\right)^2 \left\{ \Delta g \left(\Delta g - \frac{\Theta^2}{2n^{*2}} \right) \right\}} \right) \right] = \frac{1}{2} \frac{\Delta g_h}{(\Theta^2/2n^{*2})} \arctan \left(\frac{(\Theta^2/2n^{*2})}{\Delta g_h} \right),$$

and this is satisfied by

$$1 + \left(\frac{2}{\Delta g_h}\right)^2 \left\{ \Delta g \left(\Delta g - \frac{\Theta^2}{2n^{*2}} \right) \right\} = 2,$$

i.e.,

$$\Delta g \left(\Delta g - \frac{\Theta^2}{2n^{*2}} \right) - \left(\frac{\Delta g_h}{2}\right)^2 = 0.$$

We are interested in the difference between the roots of the equation, which gives the width of the characteristic

$$(\Delta g_1 - \Delta g_2) = \left[\left(\frac{\Theta^2}{2n^{*2}}\right)^2 + (\Delta g_h)^2 \right]^{1/2},$$

exactly corresponding to Equation 8.42.

Since

$$\arctan x = x - \frac{x^3}{3} + \frac{x^5}{5} - \frac{x^7}{7} + \cdots, \text{ for } |x| \leq 1,$$

for small values of $(\Delta v'/W_0)$, we can write

$$\hat{T}_\Theta = 1 - \frac{1}{3}\left(\frac{\Delta v'}{W_0}\right)^2. \tag{8.46}$$

If *FR* denotes the focal ratio of the incident light, then for values of around 2 to infinity, it is a reasonably good approximation that

$$\Theta = 1/[2FR].$$

Using this, we find another expression for $\Delta v'$ that can be useful:

$$\Delta v' = \frac{v_0}{8n^{*2}(FR)^2}.$$

We can still further extend this analysis to the case of a cone of semiangle Θ incident at an angle other than normal, provided we make some simplifying assumptions. If the angle of incidence of the cone is χ, then the range of angles of incidence will be $\chi \pm \Theta$.

If $\chi < \Theta$, then we can assume that the result is simply that for a normally incident cone of semiangle $\chi + \Theta$.

If $\chi > \Theta$, then we have three frequencies, v_0 corresponding to normal incidence, v_1 corresponding to angle of incidence $\chi - \Theta$, and v_2 corresponding to angle of incidence $\chi + \Theta$. The new filter peak can be assumed to be

$$v_{\text{peak}} = \frac{1}{2}(v_1 + v_2) = \frac{\chi^2 + \Theta^2}{2n^{*2}} v_0 \ (\chi \text{ and } \Theta \text{ in radians}) \tag{8.47}$$

or, alternatively,

$$v_{\text{peak}} = \frac{1.52 \times 10^{-4}(\chi^2 + \Theta^2)}{n^{*2}} v_0 \ (\chi \text{ and } \Theta \text{ in degrees}). \tag{8.48}$$

The halfwidth is

$$\left[W_0^2 + (v_2 - v_1)^2\right]^{1/2}, \tag{8.49}$$

where

$$(v_2 - v_1) = \frac{2\chi\Theta}{n^{*2}} v_0 \ (\chi \text{ and } \Theta \text{ in radians}) \tag{8.50}$$

or

$$(v_2 - v_1) = \frac{6.09 \times 10^{-4}\chi\Theta}{n^{*2}} v_0 \ (\chi \text{ and } \Theta \text{ in degrees}). \tag{8.51}$$

The peak transmittance is

$$\frac{W_0}{(v_2 - v_1)} \arctan\left[\frac{(v_2 - v_1)}{W_0}\right] \simeq 1 - \frac{1}{3}\left[\frac{(v_2 - v_1)}{W_0}\right]^2. \tag{8.52}$$

$(v_2 - v_1)$ is proportional to $\chi\Theta$, and Hernandez [17] found excellent agreement between measurements made on real filters and calculations from these expressions for values of $\chi\Theta$ up to $100°^2$.

We can illustrate the use of these expressions in calculating the performance of a zinc sulfide and cryolite filter for the visible region. We assume that this is a low-index first-order filter with a bandwidth of 1%.

For this filter, we calculate that $n^* = 1.55$. We take 10% reduction in peak transmittance as the limit of what is acceptable. Then, from Equation 8.52,

$$(v_2 - v_1)/W_0 = 0.55,$$

and the increased halfwidth corresponding to this reduction in peak transmittance is

$$(1 + 0.55^2)^{1/2} W_0 = 1.14 W_0$$

or an increase of 14% over the basic width.

At normal incidence, the tolerable cone semiangle is given by

$$1.5 \times 10^{-4} \left(\frac{\Theta^2}{n^{*2}}\right) = \Delta v = 0.55 W_0 = 0.55 \times 0.01 \ (\Theta \text{ in degrees}),$$

i.e.,

$$\Theta = \left[\frac{1.55^2 \times 0.55 \times 0.01}{1.5 \times 10^{-4}}\right]^{1/2} = 9.4°.$$

Such a cone at normal incidence will cause a shift in the position of the peak toward shorter wavelengths or higher frequencies of

$$\frac{1}{2}\frac{\Delta v'}{v_0} = \frac{0.55 \times 0.01}{2} = 0.275\%.$$

Used at oblique incidence in a cone of illumination, we have

$$\frac{6.09 \times 10^{-4}\chi\Theta}{n^{*2}} v_0 = v_2 - v_1 = 0.55 \times 0.01,$$

i.e.,

$$\chi\Theta = \frac{1.55^2 \times 0.55 \times 0.01}{6.09 \times 10^{-4}} = 21.7°^2.$$

This implies that the filter can be used in a cone of semiangle 2° up to an angle of incidence of $21.7°/2° = 10.9°$ or of semiangle 3° up to an angle of incidence of 7° and so on.

One very important result is the shift in peak wavelength in a cone at normal incidence, which indicates that if a filter is to be used at maximum efficiency in such an arrangement, its peak wavelength at normal incidence in collimated light should be slightly longer to compensate for this shift.

8.2.6 Sideband Blocking

There is a disadvantage in the all-dielectric filter: the high-reflectance zone of the reflecting coating is limited in extent, and hence, the rejection zone of the filter is also limited. In the near ultraviolet, visible, and near infrared regions, the transmission sidebands on the shortwave side of the peak can usually be suppressed, or blocked, by an absorption filter with a longwave-pass characteristic in the same way as for metal–dielectric filters. The longwave sidebands are more of a problem. These may be outside the range of the sensitivity of the detector and therefore may not require elimination, but if they are troublesome, then the usual technique for removing them is the addition of a metal–dielectric first-order filter with no longwave sidebands. It is usually very much broader than the narrowband component so that the peak transmittance may be high. The metal–dielectric component is usually added as a separate component, but it can be deposited over the basic Fabry–Perot. Rather than a simple Fabry–Perot filter, a double cavity metal–dielectric is commonly used. Multiple-cavity filters are the next topic of discussion.

8.3 Multiple-Cavity Filters

The transmission curve of the basic all-dielectric single-cavity, or Fabry–Perot, filter is not of ideal shape. It can be shown that one-half of the energy transmitted in any order lies outside the half-width (assuming an even distribution of energy with frequency in the incident beam). A more nearly rectangular curve would be a great improvement. Further, the maximum rejection of the single-cavity filter is completely determined by the halfwidth and the order. The broader filters, therefore, tend to have poor rejection as well as a somewhat unsatisfactory shape.

When tuned electric circuits are coupled, the resultant response curve is rather more rectangular, and the rejection outside the passband, rather greater than a single tuned circuit, and a similar result is found for the single-cavity filter. If two or more of these filters are placed in series, much the same sort of curve is obtained with a much more promising shape. The filters may be either metal–dielectric or all-dielectric, and the basic form is as follows, the outer reflectors being rather weaker than the central one:

|reflector|half-wave cavity|reflector|half-wave cavity|reflector|.

This type of structure has been known by many different names. An old term is double half wave or DHW filter, but the more usual modern terms are either a double-cavity filter or two-cavity filter. Some typical examples of all-dielectric two-cavity filters are shown in Figure 8.16.

Such filters were certainly constructed by A. F. Turner et al.* at Bausch and Lomb in the early 1950s, but the results were published only as quarterly reports in the Fort Belvoir Contract Series over the period of 1950–1968. The earliest filters were of the three-cavity (triple half-wave) type, known at Bausch and Lomb as WADIs (wide-band all-dielectric interference filters) [18]. Two-cavity (double half wave) filters came later but were in routine use at Bausch and Lomb certainly by 1957. They were initially known as TADIs (two-cavity all-dielectric interference filters). The Fort Belvoir Contract Reports make fascinating reading and show just how advanced the work at Bausch and Lomb was at that time. Use was being made of the concept of equivalent admittance for the design of both WADI filters and the edge filters for blocking the sidebands. Multilayer antireflection coatings were also well understood.

8.3.1 Smith's Method

The first complete account in the archival literature of a theory applicable to multiple half-wave filters is due to Smith [19], and it is his method that we follow first.

---

* For example, reports 4, 5, and 6 of Contract DA-44-009-eng-1113 covering the period of January–October 1953. The contract reports were once obtainable from the Engineer Research and Development Laboratories, Fort Belvoir, Virginia, but are now, unfortunately, out of print.

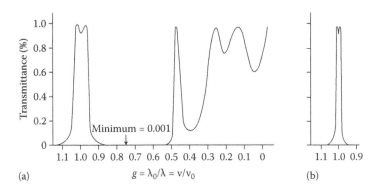

FIGURE 8.16
(a) Computed transmittance of *HLLHLHLLH*. (b) Computed transmittance of *HLHHLHLHHLH*. In both cases $n_H = 4.0$ and $n_L = 1.35$. (After S. D. Smith, *Journal of the Optical Society of America*, 48, 43–50, 1958. With permission of Optical Society of America.)

The reflecting stacks in the classical single-cavity or Fabry–Perot filter have more or less constant reflectance over the passband of the filter. A dispersion of phase change on reflection does, as we have seen, help reduce the bandwidth, but this does so without altering the basic shape of the passband. Smith suggested the idea of using reflectors with much more rapidly varying reflectance to achieve a better shape. The essential expression for the transmittance of the complete filter has already been derived in Section 3.4.3, where we have assumed that $\beta = 0$, that is, no absorption in the cavity layer. From Smith's formula (Equation 3.19),

$$T = \frac{|\tau_a^+|^2 |\tau_b^+|^2}{\left(1 - |\rho_a^-||\rho_b^+|\right)^2} \left[1 + \frac{4|\rho_a^-||\rho_b^+|}{\left(1 - |\rho_a^-||\rho_b^+|\right)^2} \sin^2 \frac{(\varphi_a + \varphi_b - 2\delta)}{2}\right]^{-1}, \tag{8.53}$$

it can be seen that high transmission can be achieved at any wavelength if, and only if, the reflectances on either side of a chosen cavity layer are equal. Of course, there is a phase condition that must be met too, but this can be arranged by choosing the correct cavity thickness to make

$$\left|\frac{\varphi_a + \varphi_b}{2} - \delta\right| = m\pi.$$

In these expressions, the symbols have the same meanings as those given in Figure 3.8.

Smith pointed out the advantage of having reasonably low reflectance in the region around the peak wavelength, which implies that absorption is less effective in limiting the peak transmittance. In the single-cavity filter, low reflectance means wide bandwidth, but Smith limited the bandwidth by arranging for the reflectances to begin to appreciably differ at wavelengths only a little removed from the peak. This is illustrated in Figure 8.17. The figure shows what is probably the simplest type of two-cavity filter with construction *HHLHH*. The *HH* layers are the two cavities, and the *L* layer is a coupling layer. For simplicity, in the following discussion, we shall ignore any substrate. The behavior of the filter is described in terms of the reflectances on either side of one of the two cavities. R_1 is the reflectance of the interface between the high index and the surrounding medium, which we take as air with index unity, and is a constant. R_2 is the reflectance of the assembly on the other side of the cavity and is low at the wavelength at which the cavity is a half wave (i.e., absentee) and rises on either side. The residual reflectance is due to the quarter-wave *L* layer. At wavelengths λ' and λ'', the reflectances R_1 and R_2 are equal, and we would expect to see high transmission if the phase condition is met, which, in fact, it is. The transmittance of the assembly is also shown in the figure, and the shape can be seen to consist of a steep-sided passband with two peaks close together and only a slight dip in transmission between the peaks, much more like the ideal rectangle than the shape of a single-cavity filter.

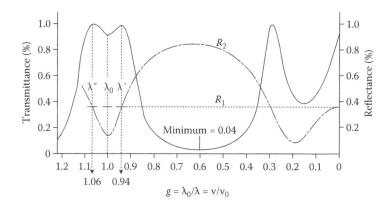

FIGURE 8.17
Computed transmittance of *HHLHH* and explanatory reflectance curves R_1 and R_2 (n_H = 4.0; n_L = 1.35). (After S. D. Smith, *Journal of the Optical Society of America*, 48, 43–50, 1958. With permission of Optical Society of America.)

Smith's formula for the transmittance of a filter can be written as follows:

$$T(\lambda) = T_0(\lambda) \frac{1}{1 + F(\lambda)\sin^2[(\varphi_1 + \varphi_2)/2 - \delta]}, \tag{8.54}$$

where

$$T_0(\lambda) = \frac{(1 - R_1)(1 - R_2)}{\left[1 - (R_1 R_2)^{1/2}\right]^2}, \tag{8.55}$$

$$F(\lambda) = \frac{4(R_1 R_2)^{1/2}}{\left[1 - (R_1 R_2)^{1/2}\right]^2}. \tag{8.56}$$

Both these quantities are now variable since they involve R_2, which varies as in Figure 8.17. The form of the functions is also shown in Figure 8.18. At wavelengths removed from the peak, $T_0(\lambda)$ is low and $F(\lambda)$ is high, the combined effect being to increase the rejection. In the region of the peak,

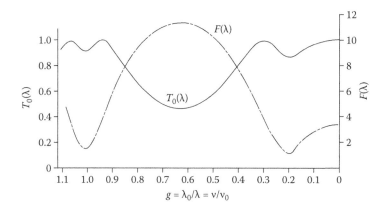

FIGURE 8.18
$T_0(\lambda)$ and $F(\lambda)$ for *HHLHH*. (After S. D. Smith, *Journal of the Optical Society of America*, 48, 43–50, 1958. With permission of Optical Society of America.)

$T_0(\lambda)$ is high, and, just as important, $F(\lambda)$ is low, producing high transmittance that is insensitive to the effects of absorption. As we have shown before, the peak transmittance is dependent on quantity A/T, where A is the absorptance and T the transmittance of the reflecting stacks. Clearly, the greater is T, the higher A can be for the same overall filter transmittance.

The typical double-peaked shape of the two-cavity filter results from the intersection of the R_1 and R_2 curves at two separate points. Two other cases can arise. The curves can intersect at one point only, in which case, the system has a single peak with transmittance that is theoretically unity, provided the phase condition is met, or the curves may never intersect at all, in which case, the system will usually show a single peak of transmittance rather less than unity, the exact magnitude depending on the relative magnitudes of R_1 and R_2 at their closest approach. This third case is to be avoided in design. For the twin-peaked filter, a requirement is that the trough in the center between the two peaks should be shallow, which means that R_1 and R_2 should not be very different at λ_0.

Having examined the simplest type of two-cavity filter, we are in a position to study more complicated ones. What we have to look for is a system of two reflectors, where one of the reflectors remains reasonably constant over the range of interest and where the other should be equal, or nearly equal, to the first over the passband region, but should sharply increase outside the passband. The straightforward single-cavity filter effectively has zero reflectance at the peak wavelength, but the reflectance rapidly rises on either side of the peak. If, then, a simple quarter-wave stack is added to a single-cavity structure, the resultant combination should have the desired property, that is, the reflectance equal or close to that of the simple stack at the center wavelength and sharply increasing on either side. We can therefore use a simple stack as one reflector, with more or less constant reflectance, on one side of the cavity and, on the other side, an exactly similar stack combined with the additional single-cavity filter. This will result in a single-peaked filter if the reflectances will be exactly matched at λ_0. The double-peaked transmission curve will be obtained if the reflectance of the stack plus the single-cavity filter is arranged to be just a little different from the reflectance of the stack by itself. This is the arrangement that is more often used, and it generally occurs naturally because of an extra quarter-wave layer that is inserted in between the stack and the additional single-cavity structure. This avoids the existence of a half-wave layer that would otherwise tend to perturb the form of the reflectance variation remote from the peak. This layer appears as a sort of coupling layer in the filter. Figure 8.19 should make the situation clear.

So far, we have not considered the substrate of the filter, but, of course, it is part of the structure and must be included. The substrate will be on one side of the filter and will alter the reflectance of the system on that side. This change in reflectance can easily be calculated, particularly if the substrate is considered to be on the same side of the cavity, or spacer, as the simple stack. The

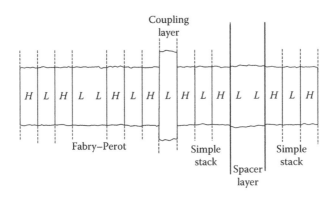

FIGURE 8.19
Construction of a two-cavity filter. The Fabry–Perot or single-cavity filter combined with the adjacent simple stack forms one reflector. The other reflector is the simple stack on the right of the cavity, or spacer, layer.

constant reflectance R_1 of the simple stack will generally be large, and if the substrate index is given by n_m, then the transmittance of the stack on its own $(1 - R_1)$ will become either $(1 - R_1)/n_m$ if the index of the layer next to the substrate is low, or $n_m(1 - R_1)$ if it is high.

This change in reflectance could be considerable, especially if n_m is large, and so the substrate must be taken into account in the design, and it is usually convenient to do this from the beginning of the design process. The substrate can be considered part of the simple stack, and R_1 can be adjusted to include it. Provided the reflectances of the two assemblies on either side of the cavity layer are arranged always to be equal at the appropriate wavelengths, the transmittance of the complete filter will be unity.

As an example, let us consider the case of a filter deposited on a germanium substrate using zinc sulfide for the low-index layers and germanium for the high ones. Let the cavity be of low index and let the reflecting stack on the germanium substrate be represented by $LLHL\,|\,\text{Ge}$, where the LL layer is the cavity. The transmittance of the stack into the cavity layer will be approximately $T_1 = 4n_L^3/n_H^2 n_{Ge}$, which, since the substrate is the same material as the high-index layer, becomes $4n_L^3/n_H^3$. On the other side of the cavity layer, we make a start with the combination $\text{Air}\,|\,HLHLL$, representing the basic reflecting stack, where LL is once again the cavity layer. This has transmission $T_2 = 4n_L^3/n_H^4$, which is $1/n_H$ times T_1. Clearly, this is too unbalanced, and an adjustment to this second stack must be made. If a low-index layer is added next to the air, then the transmittance becomes $T_2 = 4n_L^5/n_H^4$. Since n_L^2 is approximately equal to the index of germanium, the transmittances T_1 and T_2 are now equal and the single-cavity filter can now be added to the second stack to give the desired shape to the reflectance curve. This additional single-cavity filter can take any form, but it is convenient here to use a combination almost exactly the same as the already existing arrangement. The complete design of the filter is then

$$\text{Air}\,|\,HLH\ LL\ HLH\ L\ HLH\ LL\ HL\,|\,\text{Ge},$$

and the performance of the filter is shown in Figure 8.20.

An alternative way of checking whether the filter is going to have high transmittance uses the concept of absentee half-wave layers. The layers in two-cavity filters are usually either of quarter- or half-wave thickness at the center of the passband, as in the filter mentioned earlier, and we can take it as an example to illustrate the method. First, we note that the two cavities are both

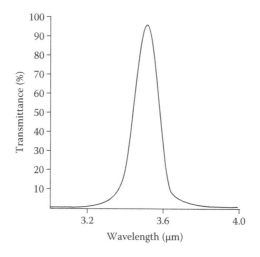

FIGURE 8.20
Computed transmittance of the two-cavity filter. Design: $\text{Air}\,|\,HLH\ LL\ HLH\ L\ HLH\ LL\ HL\,|\,\text{Ge}$. The substrate is germanium ($n = 4.0$); H = germanium ($n = 4.0$), L = zinc sulfide ($n = 2.35$), and the incident medium is air ($n = 1.0$). The reference wavelength is 3.5 μm.

half-wave layers and that they can be eliminated without affecting the transmission. The filter, at the center wavelength, will have the same transmittance as

$$\text{Air}|HLH\ HLH\ L\ HLH\ HL|\text{Ge}.$$

In this, there are two sets of HH layers that can be eliminated in the same way, leaving two sets of LL layers that can be removed in their turn. Almost all the layers in the filter can be eliminated in this way, ultimately leaving

$$\text{Air}|L|\text{Ge}.$$

As we already know, a single quarter wave of zinc sulfide is a good antireflection coating for germanium, and so the transmittance of the filter will be high in the center of the passband.

Knittl [20,21] used an alternative multiple-beam approach to study the design of two-cavity filters. Essentially, he applied a multiple beam summation to the first cavity, the results of which are then used in a multiple beam summation for the second cavity. This yields an expression not unlike Smith's, although slightly more complicated, but with the advantage that it is only the phase that varies across the passband. The magnitude of the reflection and transmission coefficients can be safely assumed constant, and this means that the parameters involving these quantities are also constant. The form of the expression for overall transmittance is then very much easier to manipulate so that the positions and values of maxima and minima in the passband can be readily determined. We shall not deal further with the method here, because it was already well covered by Knittl [21].

Of course, the possible range of designs does not end with the two-cavity, or DHW, filter. Other types of filter exist, involving even more half-wave cavities. An early type of filter, already mentioned, was the WADI devised by Turner, and it consisted of a straightforward single-cavity filter, to either side of which was added a half-wave layer together with several quarter-wave layers. The function of these extra layers was to alter the phase characteristics of the reflectors on either side of the primary cavity layer, so that the passband was broadened and at the same time, the sides became steeper. Similarly, it is possible to repeat the basic single-cavity element used in the two-cavity filter once more to give a three-cavity filter also known as a triple half wave or THW filter, which has a similar bandwidth but steeper sides. WADI and three-cavity, or THW, filters are much the same thing, although the original design philosophy was a little different, and nowadays, the term *three-cavity filter* is more usual. Even more cavity layers may be used yielding multiple-cavity filters. The method we have been using for the analysis of the filters becomes rather cumbersome when many cavities are involved—even the simple method for checking that the transmittance is high in the passband breaks down, for reasons that will be made clear in the next section, where we shall consider a very powerful design method devised by Thelen [22].

8.3.2 Thelen's Method

We have not yet arrived at any ready way of calculating the bandwidth of two- and three-cavity filters. The design method has merely ensured that the transmittance of the filter is high in the passband and that the shape of the transmission curve is steep sided. The bandwidth can be calculated, but to arrive at a prescribed bandwidth in a design involves a degree of trial and error, usually aided by computer. It can indeed be calculated using the formula for transmittance:

$$T = T_0 \frac{1}{1 + F_0 \sin^2 \delta},$$

but this can be very laborious as the phases of the reflectances have to be included in δ. This expression has been very useful in achieving an insight into the basic properties of the multiple-cavity filter, but for systematic design, a method based on the concept of equivalent admittance will be found more useful.

As shown in Chapter 7, any symmetrical assembly of thin films can be replaced by a single layer of equivalent admittance and optical thickness, both of which vary with wavelength, but can be calculated. This concept was used by Thelen [22] in the development of a very powerful systematic design method that predicts all the performance features of the filters including the bandwidth. The basis of the method is the splitting of the multiple-cavity filter into a series of symmetrical periods, the properties of which can be predicted by finding their equivalent admittances. Take, for example, the design we have already examined:

$$\text{Air}|HLHLLHLHLHLHLLHL|\text{Ge.}$$

This can be split up into the following arrangement:

$$\text{Air}|HLHL\ LHLHLHLHL\ LHL|\text{Ge.}$$

The part that determines the properties of the filter is the central section, *LHLHLHLHL*, which is a symmetrical assembly. It can therefore be replaced by a single layer having the usual series of high-reflectance zones, where the admittance is imaginary, and pass zones, where the admittance is real. It turns out that the real region is centered on the wavelength where the layers are quarter waves and with a width that decreases with the number of quarter waves. We are interested in the real region because it represents the passband of the final filter. The symmetrical section must then be matched to the substrate and the surrounding air, and matching layers are added for that purpose on either side. This is the function of the remaining layers of the filter. The condition for perfect matching is easily established because the layers are all of quarter-wave optical thicknesses.

A most useful feature of this design approach is that the central section of the filter can be repeated many times, steepening the edges of the passband and improving the rejection without affecting the bandwidth to any great extent.

In order to make predictions of performance straightforward, Thelen has computed formulas for the bandwidth of the basic sections. We use Thelen's technique here, with some slight modifications, in order to fit in with the pattern of analysis already carried out for the single-cavity filter. In order to include filters of order higher than the first, we write the basic period as

$$H^m LHLHLH...LH^m \text{ or } L^m HLHLHL...HL^m,$$

where there are $2x + 1$ layers, $x + 1$ of the outermost index and x of the other, and m is the order number. We have already mentioned how Seeley [5], in the course of developing expressions for the single-cavity filter, arrived at an approximate formula for the product of the characteristic matrices of quarter-wave layers of alternating high and low indices. Using an approach similar to Seeley's, we can put the characteristic matrix of a quarter-wave layer in the following form:

$$\begin{bmatrix} -\varepsilon & i/n \\ in & -\varepsilon \end{bmatrix}, \tag{8.57}$$

where $\varepsilon = (\pi/2)(g - 1)$, $g = \lambda_0/\lambda$, and we are using n to indicate both refractive index and characteristic admittance. This expression is valid for wavelengths close to that for which the layer is a quarter wave. First, let us consider m to be odd, and write m as $2q + 1$. Then, to the same degree of approximation, the matrix for H^m or L^m is

$$(-1)^q \begin{bmatrix} -m\varepsilon & i/n \\ in & -m\varepsilon \end{bmatrix}.$$

Neglecting terms of second and higher orders in ε, then the product of the $2x - 1$ layers making up the symmetrical period is

$$\begin{bmatrix} M_{11} & iM_{12} \\ iM_{21} & M_{22} \end{bmatrix},$$ (8.58)

where

$$M_{11} = M_{22} = (-1)^{x+2q}(-\varepsilon)\left[m\left(\frac{n_1}{n_2}\right)^x + \left(\frac{n_1}{n_2}\right)^{x-1} + \left(\frac{n_1}{n_2}\right)^{x-2} + \cdots + \left(\frac{n_2}{n_1}\right)^{x-1} + m\left(\frac{n_2}{n_1}\right)^x\right],$$

$$iM_{21} = i(-1)^x/[(n_1/n_2)^x n_1],$$

and

$$iM_{12} = i(-1)^x[(n_2 = n_1)^x/n_1].$$

We immediately see that the ratio of these two quantities is positive and real so that the equivalent admittance is real. We return to that shortly, but there is also the halfwidth to consider. Now it is not easy from these expressions to analytically derive the halfwidth of the final filter. Instead of deriving the halfwidth, therefore, Thelen chose to define the edges of the passband as those wavelengths for which

$$\frac{1}{2}|M_{11} + M_{22}| = 1,$$

or, since $M_{11} = M_{22}$,

$$|M_{11}| = 1.$$

We shall follow Thelen. These points will not be too far removed from the half-peak transmission points, especially if the sides of the passband are steep. Applying this to Equation 8.58, we obtain

$$|M_{11}| = \varepsilon\left[m\left(\frac{n_1}{n_2}\right)^x + \left(\frac{n_1}{n_2}\right)^{x-1} + \left(\frac{n_1}{n_2}\right)^{x-2} + \cdots + \left(\frac{n_2}{n_1}\right)^{x-1} + m\left(\frac{n_2}{n_1}\right)^x\right] = 1.$$ (8.59)

This expression is quite symmetrical in terms of n_1 and n_2. Then if we replace n_1 and n_2 by n_H and n_L, regardless of which is which, we will obtain the same expression:

$$\varepsilon\left[m\left(\frac{n_H}{n_L}\right)^x + \left(\frac{n_H}{n_L}\right)^{x-1} + \left(\frac{n_H}{n_L}\right)^{x-2} + \cdots + \left(\frac{n_L}{n_H}\right)^{x-1} + m\left(\frac{n_L}{n_H}\right)^x\right] = 1,$$

i.e.,

$$\varepsilon\left[(m-1)\left(\frac{n_H}{n_L}\right)^x + (m-1)\left(\frac{n_L}{n_H}\right)^x + \left(\frac{n_H}{n_L}\right)^x\left(\frac{1 - (n_L/n_H)^{x+1}}{1 - (n_L/n_H)}\right)\right] = 1,$$

where we have used the formula for the sum of a geometric series just as in the case of the single-cavity filter. We now neglect terms of power x or higher in (n_L/n_H) to give

$$\varepsilon\left(\frac{n_H}{n_L}\right)^x\left\{(m-1) + \frac{1}{1 - (n_L/n_H)}\right\} = 1,$$

i.e.,

$$\varepsilon = \left(\frac{n_{\mathrm{L}}}{n_{\mathrm{H}}}\right)^{x} \frac{[1 - (n_{\mathrm{L}}/n_{\mathrm{H}})]}{[m - (m-1)(n_{\mathrm{L}}/n_{\mathrm{H}})]}. \tag{8.60}$$

The bandwidth will be given by

$$\left|\frac{\Delta\lambda_{\mathrm{B}}}{\lambda_0}\right| = \left|\frac{\Delta\nu_{\mathrm{B}}}{\nu_0}\right| = 2(g - 1) = \frac{4\varepsilon}{\pi},$$

so that, by slightly manipulating Equation 8.60,

$$\left|\frac{\Delta\lambda_{\mathrm{B}}}{\lambda_0}\right| = \frac{4}{m\pi} \left(\frac{n_{\mathrm{L}}}{n_{\mathrm{H}}}\right)^{x} \frac{(n_{\mathrm{H}} - n_{\mathrm{L}})}{(n_{\mathrm{H}} - n_{\mathrm{L}} + n_{\mathrm{L}}/m)}. \tag{8.61}$$

The equivalent admittance is given by

$$E = \left(\frac{M_{21}}{M_{12}}\right)^{1/2} = \left(\frac{n_1}{n_2}\right)^{x} n_1. \tag{8.62}$$

We recall that m is odd in Equations 8.61 and 8.62, but we shall see in a moment that although Equation 8.61 turns out to be identical for m even, Equation 8.62 must be slightly modified. The case of m even, i.e., $m = 2q$, is arrived at similarly. Here the matrix of H^m or L^m is

$$(-1)^q \begin{bmatrix} 1 & im\varepsilon/n \\ im\varepsilon n & 1 \end{bmatrix},$$

and a similar multiplication, neglecting terms higher than the first in ε, gives

$$\left|\frac{\Delta\lambda_{\mathrm{B}}}{\lambda_0}\right| = \frac{4}{m\pi} \left(\frac{n_{\mathrm{L}}}{n_{\mathrm{H}}}\right)^{x} \frac{(n_{\mathrm{H}} - n_{\mathrm{L}})}{(n_{\mathrm{H}} - n_{\mathrm{L}} + n_{\mathrm{L}}/m)},$$

that is, exactly as Equation 8.61, but

$$E = \left(\frac{M_{21}}{M_{12}}\right)^{1/2} = \left(\frac{n_2}{n_1}\right)^{x-1} n_2$$

for equivalent admittance. This is to be expected since the layers L^m or H^m act as absentees because of the even value of m.

Equation 8.61 should be compared with the single-cavity expressions (Equations 8.24 and 8.25). If we split the multiple-cavity filter into a series of single cavities, then the number of layers in each reflector is half that in the basic symmetrical period. Equations 8.24, 8.25, and 8.61 are therefore consistent.

In order to complete the design, we need to match the basic period to the substrate and the surrounding medium. We first consider the case of first-order filters, and the modifications that have to be made in the case of higher-order filters will become obvious. For a first-order filter then, matching will best be achieved by adding a number of quarter-wave layers to the period. The first layer should have index n_1; the next, n_2; and so on, alternating the indices in the usual manner. The equivalent admittance of the combination of symmetrical period and matching layers will then be

$$\frac{n_1^{2y}}{n_2^{2(y-1)}} \left(\frac{n_2}{n_1}\right)^{x} \frac{1}{n_2} \quad \text{or} \quad \left(\frac{n_2}{n_1}\right)^{2y} n_2 \left(\frac{n_1}{n_2}\right)^{x},$$

where there are y layers of index n_1 and either $(y - 1)$ or y layers of index n_2, respectively. We have also used the fact that the addition of a quarter wave of index n to an assembly of equivalent admittance E alters the admittance presented by the structure to n^2/E, the quarter-wave rule.

This equivalent admittance should be made equal to that of the substrate on the appropriate side and of the surrounding medium on the other. The following discussion should make the method clear.

When we try to apply this formula to the design of multiple half-wave filters, we find to our surprise that quite a number of designs that we have looked at previously, and which seemed satisfactory, do not satisfy the conditions. For example, let us consider the design arrived at in the earlier part of this section:

$$Ge|LH\ LL\ HLH\ L\ HLH\ LL\ HLH|air,$$

where L indicates zinc sulfide of index 2.35, and H, germanium of index 4.0. The central period is $LHLHLHLHL$, which has an equivalent admittance of $n_L{}^5/n_H{}^4$. The LHL combination alters this equivalent admittance to

$$\frac{n_L^4}{n_H^2}\frac{n_H^4}{n_L^5} = \frac{n_H^2}{n_L},$$

which is a gross mismatch to the germanium substrate. The $LHLH$ combination on the other side alters the admittance to

$$\frac{n_H^4}{n_L^4}\frac{n_L^5}{n_H^4} = n_L$$

That, in turn, is not a particularly good match to air.

The explanation of this apparent paradox is that in this particular case, the total filter, taking the phase thickness of the central symmetrical period into account, has unity transmittance because it satisfies Smith's conditions given in the previous section, but over a wide range of wavelengths, pronounced transmission fringes would be seen if the bandwidth of the filter were rather broader than the width of a single fringe. Adding extra periods to the central symmetrical one has the effect of decreasing this fringe width, bringing them closer together. Eventually, given enough symmetrical periods, the width of the fringes becomes less than the filter bandwidth, and they appear as a pronounced ripple superimposed on the passband. This is clearly illustrated in Figure 8.21.

The three-cavity version is still acceptable when an extra L layer is added, but the five-cavity version is quite unusable. The presence or absence of an outermost L layer has no effect on the performance, other than inverting the fringes. The simple method of cancelling out half waves for predicting the passband transmission therefore breaks down, because it ensures only that λ_0 will coincide with a fringe peak.

It is profitable to look at the possible combinations of the two materials that can be made into a filter on germanium and where the center section can be repeated as many times as required. The combinations for up to 11 layers in the center section are given in Table 8.3.

The validity of any of these combinations can be easily tested. Take, for example, the fourth one, with the nine-layer period in the center. Here the equivalent admittance of the symmetrical period is $E = n_H^5/n_L^4$. The $HLHL$ section between the germanium substrate and the center section transforms the admittance into

$$\frac{n_L^4}{n_H^4}\frac{n_H^5}{n_L^4} = n_H,$$

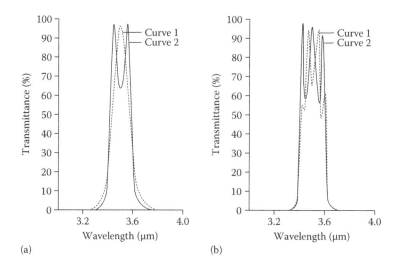

FIGURE 8.21

(a) Curve 1: computed transmittance of the three-cavity filter: Air | *LHLHL* (*LHLHLHLHL*)² *LHL* | Ge. Curve 2: effect of omitting the *L* layer next to the air in the design of curve 1: Air | *HLHL* (*LHLHLHLHL*)² *LHL* | Ge. (b) Computed transmittance of five-cavity filters. Curve 3: Air | *HLHL* (*LHLHLHLHL*)⁴ *LHL* | Ge. Curve 4: same as curve 3 but with an extra *L* layer: Air | *LHLHL* (*LHLHLHLHL*)⁴ *LHL* | Ge. The presence or absence of the *L* layer has little effect on the ripple in the passband. For all curves, *H* = germanium (*n*_H = 4.0) and *L* = zinc sulfide (*n*_L = 2.35).

TABLE 8.3

Suitable Matching Combination for Filters on Germanium

Matching Combination for Air	Symmetrical Period	Matching Combination for Germanium
Air | (already matched)	*LHL*	*L* | Ge
Air | *H*	*HLHLH*	*HL* | Ge
Air | *HL*	*LHLHLHL*	*LHL* | Ge
Air | *HLH*	*HLHLHLHLH*	*HLHL* | Ge
Air | *HLHL*	*LHLHLHLHLHL*	*LHLHL* | Ge

Note: *L*, ZnS; *n*_L = 2.35; *H*, Ge; *n*_H = 4.0.

which is a perfect match for germanium. The matching section at the other end is *HLH*, and this transforms the admittance into

$$\frac{n_H^4}{n_L^2} \frac{n_L^4}{n_H^5} = \frac{n_L^2}{n_H},$$

which, because zinc sulfide is a good antireflection material for germanium, gives a good match for air.

For higher-order filters, the method of designing the matching layers is similar. However, we can choose, if we wish, to add half-wave layers to that part of the matching assembly next to the symmetrical period in order to make the resulting cavity of the same order as the others. For example, the period *HHHLHLHLHLHHH*, based on the fourth example of Table 8.3 with additional outer half waves, can be matched either by Air | *HLH* and *HLHL* | Ge, as shown, or by Air | *HLHHH* and *HHHLHL* | Ge, making all cavities of identical order regardless of the number of periods.

This method then gives the information necessary for the design of multiple half-wave filters. The edge steepness and rejection in the stop bands will determine the number of basic symmetrical periods in any particular case. Usually, because of the approximations in the various formulas, and because the definition used for bandwidth is not necessarily the halfwidth, although not far removed from it, it is advisable to check the design by accurate computation before finalizing it. It may also be advisable to make an estimate of the permissible errors that can be tolerated in the manufacture because it is pointless attempting to achieve a performance beyond the capabilities of the process. The result will just be worse than if a less demanding specification had been attempted. The estimation of manufacturing errors is discussed in Chapter 13. Typical multiple-cavity filters are shown in Figure 8.22.

8.4 Higher Performance in Multiple-Cavity Filters

The curve of Figure 8.22b not only shows the square shape of the passband of a multiple-cavity filter but also illustrates one of the problems inherent in this type of design, the "rabbit's ears," or the rather prominent peaks at either side of the passband. This can become even worse with increasing numbers of periods. Figure 8.23 clearly shows this.

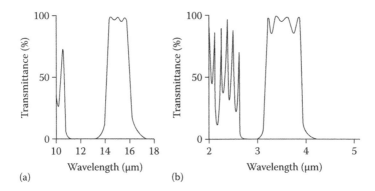

FIGURE 8.22
(a) Transmittance of a multiple half-wave filter. Design: Air | $HHL\ H\ LHHL\ LHHLH$ | Ge with H = PbTe (n = 5.0); L = ZnS (n = 2.35); and λ_0 = 15 μm. (b) Transmittance of a multiple half-wave filter. Design: Air | $HHLH\ LHHLH\ LHHL\ H\ LHHLH$ LHH | silica H = Ge (n = 4.0); L = ZnS (n = 2.35); silica substrate (n = 1.45); and λ_0 = 3.5 μm. (Courtesy of Sir Howard Grubb, Parsons & Co. Ltd., Newcastle upon Tyne.)

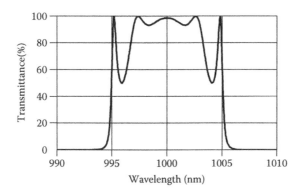

FIGURE 8.23
Multiple-cavity filter with a central core of five symmetrical periods. Design: Glass | $HLHLHLH\ (HLHLHLHLHLHLHLH)^5$ $HLHLHLH$ | Glass, with n_H = 2.35, n_L = 1.35, n_{glass} = 1.52, and λ_0 = 1000 nm. Note the very prominent peaks at the edges of the passband sometimes called rabbit's ears.

The reason for this problem feature is the dispersion of the equivalent admittance of the symmetrical period. In the design approach, this is assumed constant across the passband, but in reality, toward the edges, it considerably varies, tending to either zero or infinity at the passband edges; see Figure 8.24. It is, in fact, exactly the same problem as in edge filters where better ripple suppression near the edge demands a matching system that exhibits similar dispersion. Shifted periods are, however, difficult to arrange in the case of bandpass filters because of the need for ripple suppression at both edges of the passband. However, inspired by the shifted periods technique, we seek a solution, where part of the matching is due to a symmetrical system that has a dispersion of the appropriate form so that its matching remains reasonably good even when the equivalent admittance to be matched to the surrounding media is varying. Any of the symmetrical periods we are dealing with will have an odd number of quarter waves so that the equivalent phase thickness at $g = 1$ will be an odd number of $\pi/2$. This implies that the period could, itself, be used as a simple matching assembly. Since the passband in this type of filter is usually narrow, the matching condition will not vary too much over the passband width except for the dispersion. In order to make use of this possibility, we have to find at least pairs of symmetrical periods that will permit one to be used as a matching assembly for the other. Attempting to find two, or more, periods that have the correct relationship at $g = 1$ for one to match the other to the substrate or incident medium is difficult. If, however, we could find two periods of different width but with the same central admittance, then the broader could make the necessary modification at the edges of the narrower, effectively straightening out the admittance eventually to be matched. This would enable us to complete the design with straightforward matching as in Table 8.3, using a series of quarter-wave layers now perfectly satisfactory over the width of the passband. A solution lies with higher-order periods.

The addition of further half-wave layers to the outside of a symmetrical period does not change its equivalent admittance at the passband center, nor does it change the sense of curvature of the variation of equivalent admittance. Figure 8.25 shows the admittances of $HLHLHLHLHLH$, $HHHLHLHLHLHLHHH$, $HLLLHLHLHLHLHLLLH$, and $HHHLLLLHLHLHLHLHLLLLHHH$. All have the value n_H^6/n_L^5 at $g = 1$, and all exhibit a gradually increasing admittance as the value of g moves away from unity. The wider curves have values of admittance intermediate between the narrower curves and the value that all possess at $g = 1$. All represent an odd number of quarter waves at $g = 1$, but as g varies, the properties of the broader curves remain closer to those of an odd number of quarter waves than do the narrower. These broader curves could therefore be used to match the narrower ones to a notional medium of reasonably constant admittance, in this case, n_H^6/n_L^5. The best one,

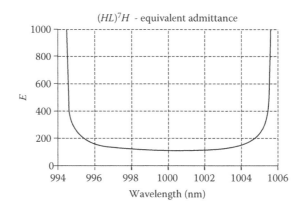

$(HL)^7H$ - equivalent admittance

FIGURE 8.24
Equivalent admittance of $(HL)^7H$ over the potential passband. Note the rapid change near the edges of the passband. This dispersion of equivalent admittance is very difficult to match to an essentially dispersionless medium.

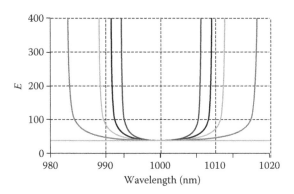

FIGURE 8.25
Equivalent admittances of symmetrical periods from narrower to broader in order *HHHLLLHLHLHLHLLLHHH*, *HHHLHLHLHLHLHHH*, *HLLLHLHLHLHLLLH*, and *HLHLHLHLHLH* with *H* representing characteristic admittance 2.35, and *L*, 1.35. The straight line represents the admittance that all have at $g = 1$.

that is the period closest to the ideal values of the required admittance, can be chosen as the intermediate matching structure. Experience shows that attempts at using more than one of the wider matching systems in series does not often give very good results unless one is very lucky with the matching of the differing dispersion curves. Finally, matching of the now dispersionless notional medium to the incident and emergent media can then be a straightforward matter of a series of quarter waves, as before.

A simple example where we assume equal incident and emergent media uses two of the periods from Figure 8.25, *HLHLHLHLHLH* and *HHHLHLHLHLHLHHH*:

Glass|*HLHLH HLHLHLHLHLH* (*HHHLHLHLHLHLHHH*)^q *HLHLHLHLHLH HLHLH*|Glass.

The characteristic curves of two such filters are shown in Figures 8.26 and 8.27.

We need an expression for the width of such filters. This is principally determined by the highest order periods. If we write the expression for the highest order period as

$$mA \ BABA...B \ mA,$$

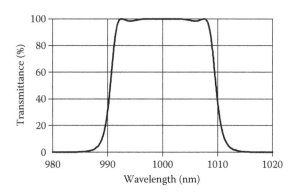

FIGURE 8.26
Multiple-cavity filter similar to that of Figure 8.23 but using periods of increasing order to improve the passband ripple. Design: Glass | *HLHLH HLHLHLHLHLH*(*HHHLHLHLHLHLHHH*)^2 *HLHLHLHLHLH HLHLH* | Glass, with $n_H = 2.35$, $n_L = 1.35$, $n_{glass} = 1.52$, and $\lambda_0 = 1000$ nm. Note the much flatter passband top compared with that in Figure 8.23.

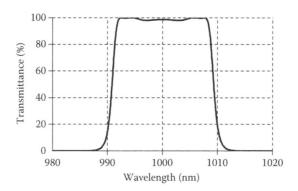

FIGURE 8.27
Multiple-cavity filter similar to that of Figure 8.26 but with three central high-order periods rather than two. Design: Glass |
$HLHLH\ HLHLHLHLHLH\ (HHHLHLHLHLHLHHH)^3\ HLHLHLHLHLH\ HLHLH$ | Glass, with n_H = 2.35, n_L = 1.35, n_{glass} =
1.52, and λ_0 = 1000 nm.

where there are $2x$ +1 layers including the layers mA, then we can show that the bandwidth,
defined in the same way as before, is given by

$$\frac{\Delta\lambda}{\lambda_0} = \frac{4}{m\pi}\left(\frac{n_L}{n_H}\right)^x \frac{(n_H - n_L)}{(n_H - n_L + n_L/m)}.$$ (8.63)

This expression reduces to that already derived if $m = 1$. Using the expression to calculate the
bandwidth of the filters of Figures 8.26 and 8.27, we find 0.018, implying passband edges at 991.1
and 1009.1 nm.

Of course, it will be clear that Figure 8.25 shows a special case, where all the symmetrical periods
have identical equivalent admittance at the reference wavelength. This feature is not absolutely
necessary. It is possible to have matching systems with differing central equivalent admittances as
long as the final combination matches the surrounding media and the dispersion of the admittances gives good ripple performance. Thus, for any given bandwidth, there is usually a very large
number of possible alternative designs. The search for the best performance is often a computer
operation where the definition of best involves a function of merit that includes ripple, bandwidth,
edge steepness, and so on, appropriately weighted.

The design procedures so far have all made use of quarter-wave layers. We shall see later that
this is desirable for reasons connected with their manufacture. Powerful natural deposition error
compensation processes exist in systems of quarter waves but are largely absent for non-quarter-
wave designs.

The designs of Figures 8.26 and 8.27 assume that the incident and emergent media are virtually
identical. This would fit the case where the filter is embedded in a system or, perhaps, a cemented
cover is attached to its outer surface. Because of the symmetry of their structure, the admittance
presented at the reference wavelength to the exterior media is the characteristic admittance of the
coupling layers. In the designs just considered, these are of low index, quite close to the index of
glass. Thus, the match with the surrounding media at the reference wavelength is necessarily good.
There are instances, however, where the incident and emergent media are quite different or when
the index of the coupling layers does not match that of the media quite so well. Recently, driven by
telecommunication applications in the main, there has been a shift to very stable materials such as
silica and tantala and processes such as ion-assisted deposition or ion-beam sputtering that yield
very stable and tough layers. These permit the use of the filter without any protection over the
front surface. This can be a great advantage in well-corrected systems. However, it brings the
complication of the lack of symmetry. Of course, the earlier infrared narrowband filters that we

considered were not symmetrical, but there, the requirements for peak transmittance and lack of ripple were much less demanding.

In such cases it is often sufficient to construct a two-layer antireflection coating of the V-coat type to match the otherwise unmatched medium, normally the air on the side of incidence. This matching coating does not suffer from the same sensitivity to errors of the remainder of the structure, and the normal tolerances for such an antireflection coating apply. An example of such a filter is shown in Figures 8.28 and 8.29. We will return to this filter shortly when we examine tilted performance.

Usually the V-coat approach will be sufficient, but if still greater levels of performance are required then three-, four-, and even five-layer antireflection coatings can be used, between either the filter and incident medium or substrate, or even both. The antireflection coatings are seldom subject to the same tight tolerances as the remainder of the filter.

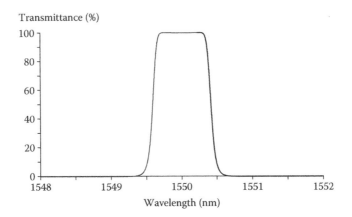

FIGURE 8.28
Performance of a filter of design: Air $|$ [1.2462 L 0.3458H] $(HL)^7$ H $(HL)^{15}$ H $(HHHH$ $(HL)^{15}$ H $HHHH)^2$ $(HL)^{15}$ H $(HL)^7$ H $|$ Glass with $\lambda_0 = 1550$ nm, $n_H = 2.15$ and $n_L = 1.45$ (roughly corresponding to Ta_2O_5 and SiO_2, respectively), $n_{glass} = 1.52$, and $n_{air} = 1.00$. The structure within the square brackets is the antireflection coating of the V-coat type.

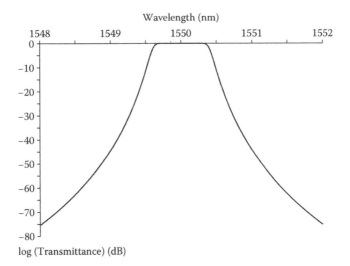

FIGURE 8.29
Performance of the filter of Figure 8.28 plotted on a logarithmic scale. This filter has performance typical of a Dense Wavelength Division Multiplexing beam splitter intended for 200 GHz channel spacing.

As an illustration of the design technique, let us design a filter with a halfwidth approximately 6 nm at a peak wavelength of 1000 nm. We shall use our usual 2.15 and 1.45 film indices with a substrate of 1.52 index and air as the incident medium. We begin by finding the necessary values for x and m. They are listed in Table 8.4, where we recall that there are $2x + 1$ layers in each basic period.

We see that there are two primary possibilities, x of 9 and m of 4 or x of 10 and m of 2. We shall examine both possibilities.

The central period in the case of the first example has 19 layers and is

$$4H \ LHLHLHLH \ L \ HLHLHLHL \ 4H = \ 4H(LH)^9 3H, \tag{8.64}$$

where we have chosen the arrangement using high-index cavity layers to assure a low-index coupling layer. For the intermediate matching, it seems likely that a similar period with $2H$ layers outermost rather than $4H$ would be suitable. Then to complete the basic design, we can use an eight-layer alternate quarter-wave period. We will repeat the basic central period four times as a first attempt. Then finally, we need a matching unit to match the coupling layer to air. This can be a simple V-coat of starting design $1.25L \ 0.4H$ that can be refined to yield an optimum design. Putting all that together, and compacting the period formulas, we have, after our refinement,

$$\text{Air}\Big|1.35L \ 0.3 \ H(HL)^4 2H(LH)^9 \ H\big[4H(LH)^9 3H\big]^4 \ 2H(LH)^9 \ H(LH)^4\Big|1.52. \tag{8.65}$$

This has a total of 127 layers and a physical thickness of 22.12 μm. The performance is illustrated in Figure 8.30.

TABLE 8.4

Bandwidth at 1000 nm as a Function of x and m

	$x = 6$ (nm)	$x = 7$ (nm)	$x = 8$ (nm)	$x = 9$ (nm)	$x = 10$ (nm)
$m = 1$	39.01	26.31	17.74	11.97	8.07
$m = 2$	29.42	19.85	13.38	9.03	6.09
$m = 3$	23.62	15.93	10.75	7.25	4.89
$m = 4$	19.73	13.31	8.98	6.05	4.08
$m = 5$	16.94	11.43	7.71	5.20	3.51
$m = 6$	14.84	10.01	6.75	4.55	3.07

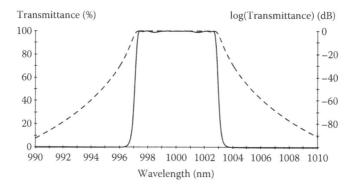

FIGURE 8.30
Performance of the first of the 6 nm filters (Equation 8.65). The full line is the transmittance in percent, and the broken line, in decibels.

The second example has x at 10 and m at 2. The central period therefore has 21 layers with half waves outermost. To yield low-index coupling, these should be of low index, and the structure is

$$2L\,HLHLHLHLHLHLHLHLHLH\,2L = 2L\,(HL)^{10}\,L. \tag{8.66}$$

We peel off the half waves to give the intermediate match of $(HL)^9\,H$, and then the outer matching structure can be $(HL)^4\,H$. Again, putting all this together with our central periods and with a simple V-coat final match gives

$$\text{Air}\big|1.35L\,0.3\,HH(LH)^4\,H(LH)^9\,\big[2L(HL)^{10}L\big]^4\,(HL)^9\,H(HL)^4\,H\big|1.52, \tag{8.67}$$

and the performance is shown in Figure 8.31. There are 137 layers with a total physical thickness of 21.9 μm, slightly less than that of the first filter even though the first has only 127 layers. The first filter is clearly the superior one.

In these two examples, the intermediate matching structure was chosen to be of the same central admittance as that of the basic central period. This is not a necessary constraint, and there are many additional possibilities. The calculations are not difficult, and the design of any possible structure is quite straightforward and especially suitable for computer implementation. It is not surprising that there are various commercial computer tools that will carry out the necessary calculations to produce a range of possible designs that can simply be placed in order of whatever definition of merit is chosen.

It is clear that the design of very high-performance narrowband filters is considerably constrained by the need for structures that can be successfully produced. This has largely meant adherence to quarter waves, and exact multiples of quarter waves, of a limited range of materials. These limitations may be ultimately relaxed as manufacturing techniques advance. Techniques borrowed from the microwave region that ignore such constraints can be harnessed in the optical region to produce outstanding theoretical designs that assume the availability of a continuous range of refractive indices and/or completely unconstrained thicknesses. Baumeister [23–25] has explained the technique.

Baumeister's design technique is based on the parameter standing wave ratio. In an incident medium, the net electric field amplitude varies with position, varying from $\mathcal{E}_1 + \mathcal{E}_2$ to $\mathcal{E}_1 - \mathcal{E}_2$, where $\mathcal{E}_1$ and $\mathcal{E}_2$ are the amplitudes of the two beams. The standing wave ratio is defined as

$$V = \frac{\mathcal{E}_1 + \mathcal{E}_2}{\mathcal{E}_1 - \mathcal{E}_2} = \frac{1 + \sqrt{R}}{1 - \sqrt{R}}, \tag{8.68}$$

where R is the reflectance. An advantage of the standing wave ratio parameter is that it takes a very simple form when the admittance of the system concerned is real. Let the admittance of the

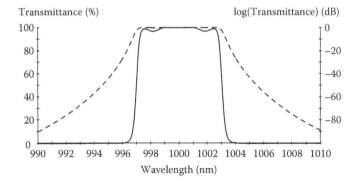

FIGURE 8.31
Performance of the second of the 6 nm filters (Equation 8.67). The full line is the transmittance in percent, and the broken line, in decibels. The ripple performance is not quite as good as that in Figure 8.30.

surface be Y and the characteristic admittance of the incident medium be y_0, both real, then the standing wave ratio is given by

$$V = \frac{y_0}{Y} \text{ or } V = \frac{Y}{y_0}, \text{ whichever is greater than unity.} \tag{8.69}$$

The design of the microwave structure that is the starting design for the operation is based on a series of iris diaphragm reflectors in a waveguide, each of which is adjusted to a given standing wave ratio, assuming that the particular reflector is isolated from the others, and which are spaced apart by what are equivalent to cavity layers. In Baumeister's method, these reflectors are replaced by thin-film structures that are bounded by the materials of the successive cavities. For example, if the indices of two successive cavities are n_L and the cavities are separated by five alternate quarter waves of n_H and n_L, then the standing wave ratio will be n_H^6/n_L^6. To achieve the exact standing wave ratios that are required, the method necessarily makes use of fractional thicknesses or special characteristic admittances. These characteristic admittances can be achieved by introducing symmetrical periods, although these will contain fractional thicknesses. Once the correct standing wave ratios are achieved, the dispersion of the properties of the structures must be taken into account, and a certain amount of manual adjustment, involving such operations as replacing single quarter waves by three quarter waves and altering the order of the cavity layers, is normal.

An example of a two-material 13-cavity design arrived at by this technique is shown in Figure 8.32. This design uses quarter-wave structures for each reflector, but the final layer in each structure is a quarter wave of admittance adjusted to yield the exact required standing wave ratio. The adjusted admittance is then achieved through the use of a three-layer symmetrical period. The final design that uses 346 layers is

Air$|0.241L\,2.499H\,1.241L\,H(L\,H)^3\,2L\,(H\,L)^5\,1.3257H\,0.3329L\,1.3257H\,(L\,H)^5\,2L\,(H\,L)^5\,H$

$1.2077L\,0.5671H\,1.2077L\,H\,(L\,H)^5\,2L\,(H\,L)^6\,1.4654H\,0.0654L\,1.4654H\,(L\,H)^6\,2L\,(H\,L)^6$

$1.4251H\,0.1419L\,1.4251H\,(L\,H)^6\,2L(H\,L)^6\,1.4072H\,0.1759L\,1.4072H\,(L\,H)^6\,2L\,(H\,L)^6$

$1.4001H\,0.1894L\,1.4001H\,(L\,H)^6\,2L\,(H\,L)^6\,1.4001H\,0.1894L\,1.4001H\,(L\,H)^6\,2L\,(H\,L)^6$

$1.4072H\,0.1759L\,1.4072H\,(L\,H)^6\,2L\,(H\,L)^6\,1.4251H\,0.1419L\,1.4251H\,(L\,H)^6\,2L\,(H\,L)^6$

$1.4654H\,0.0654L\,1.4654H\,(L\,H)^6\,2L\,(H\,L)^5\,H\,1.2077L\,0.5671H\,1.2077L\,H\,(L\,H)^5\,2L\,(H\,L)^5$

$1.3257H\,0.3329L\,1.3257H\,(L\,H)^5\,2L\,(H\,L)^4\,H2.628L\,1.818H|$Glass,

where $n_{air} = 1.00$, $n_H = 2.065$, $n_L = 1.47$, and $n_{glass} = 1.50$.

If the designs are constrained to use only quarter waves, or multiple quarter waves, of only two materials, then it is generally no longer possible to match the required standing wave ratios exactly, and so the design operation simply attempts as close a match as possible with the inevitable increase in passband ripple.

8.4.1 Effect of Tilting

A feature of the multiple-cavity designs not so far mentioned is the sensitivity to changes in angle of incidence. Thelen [22] examined this aspect, and for those types which involve symmetrical periods consisting of quarter waves of alternating high and low indices and where the cavities are of the first order, he arrived at exactly the same expressions as those of Pidgeon and Smith for the

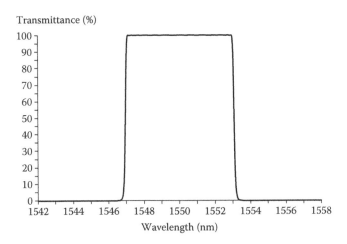

FIGURE 8.32
Performance of the 346-layer filter designed by Baumeister with design as in the text. (After P. Baumeister, *Applied Optics*, 42, 2407–2414, 2003. With permission of Optical Society of America.)

single cavity. As far as angular dependence is concerned, the filter behaves as if it were a single layer with an effective index of

$$n^* = (n_1 n_2)^{1/2},$$

where $n_1 > n_2$, or

$$n^* = \frac{n_1}{\left[1 - (n_1/n_2) + (n_1/n_2)^2\right]^{1/2}},$$

where $n_2 > n_1$.

For higher-order filters, therefore, we should be safe in making use of Equations 8.35 and 8.37.

Figure 8.33 shows the tilted performance of a simple three-cavity narrowband filter of the following design:

$$Air|1.21L\ 0.38H\ (HL)^7 HHHHHH\ (LH)^7\ L\ (HL)^7\ HHHHHH\ (LH)^7\ L(HL)^7\ HHHHHH$$

$$(LH)^7|Glass\ with\ n_{air} = 1.00, n_H = 2.065, n_L = 1.47,\ and\ n_{glass} = 1.50.$$

Unfortunately, there can be complications. The filter of Figure 8.28 uses cavities of different orders to achieve the excellent ripple performance. These cavities move at slightly different rates with changing angle because their effective indices slightly differ. Detuned cavities have a seriously degrading effect on multiple-cavity filters. This can clearly be seen in Figure 8.34. The filter passband shape considerably degrades as angle of incidence increases. As a result, this particular filter is suitable for normal incidence and only small angles on either side of normal.

If we wish to improve tilted performance, we need to bring the effective indices of the cavities closer together. The design of this filter involves high-index cavities, and so the higher-order ones will have higher effective index. Incorporating more low-index material in these cavity structures is the most direct route. Here some trial and error is usually the most straightforward way to proceed. Various tricks are possible, such as changing some single quarter-wave low-index layers to three quarters, but in the case of this particular filter, changing one or more of the high-index half waves in the cavity layers to low-index half waves works reasonably well. The best arrangement appears to be as follows:

$$Air|1.2676L\ 0.3379H\ (HL)^7 H\ (HL)^{15} H\ (HLLH\ (HL)^{15} H\ HLLH)^2\ (HL)^{15} H\ (HL)^7 H|Glass,$$

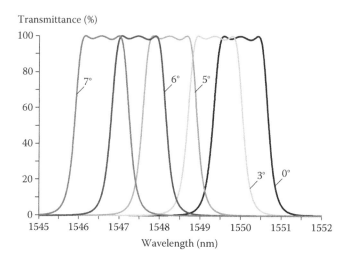

FIGURE 8.33
Performance as a function of the angle of incidence in air of a simple three-cavity narrowband filter where all cavities are exactly the same. There is no perceptible distortion of the characteristic as the angle is increased. By 15° incidence (not shown), polarization splitting is becoming perceptible, but otherwise, the characteristic is still undistorted.

with the values of indices exactly as in Figure 8.28. This gives the tilted performance shown in Figure 8.35.

The design of Figure 8.32 uses cavities that are of the same order and so presents similar tilted performance to that shown in Figure 8.35.

Another advantage to filters that exhibit this insensitivity of shape to tilt angle is their improved inherent insensitivity to a particular type of uniformity error. Deposition thickness monitoring of narrowband filters is usually what is termed *direct*; that is, it takes place on an actual filter that is being produced. This permits the error compensation necessary for successful narrowband filter production. Away from the monitoring area, however, the deposited thicknesses may vary and, in fact, such variation is frequently encouraged in the production of telecom filters. Here, after deposition, a large filter disk is eventually diced into smaller filters, and a spread of peak wavelength yields filters for different channels. It is very difficult to ensure that such varying thicknesses

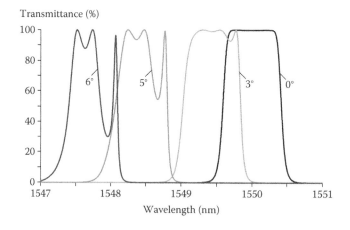

FIGURE 8.34
Performance of the filter of Figure 8.28 at normal incidence (extreme right) and angles of incidence of 3°, 5°, and 6°. Serious distortion is already visible at 3° and increases with angle. Polarization splitting is negligible.

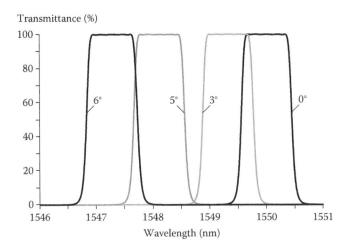

FIGURE 8.35
Tilted performance of the modified design where the cavities now move at similar rates with angle. Note that the effective index of this filter is now very slightly less than that of Figure 8.34. Even at 15° incidence (not shown), the filter characteristic remains undistorted, but there is perceptible polarization splitting.

of high- and low-index films always remain exactly in step. Relative thickness errors have an effect on filter shape similar to that of the changing angle of incidence. The insensitivity of shape to angle of incidence changes therefore implies improved insensitivity also to relative errors in uniformity and increases the yield of usable components. This is illustrated in Figures 8.36 and 8.37, where only the low-index layer thicknesses have been altered.

8.4.2 Losses in Multiple-Cavity Filters

Losses in multiple-cavity filters can be estimated in the same way as for the Fabry–Perot filter. There are so many possible designs that a completely general approach would be very involved. However, we can begin by assuming that the basic symmetrical unit is perfectly matched at either end. The arrangement of admittances through the basic unit will then be as shown in Table 8.5.

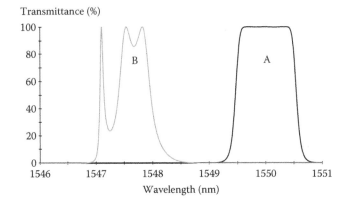

FIGURE 8.36
Filter of Figure 8.28 at normal incidence showing theoretical thicknesses at A and with low-index layer thicknesses reduced by a factor 0.998 at B, the high-index layer thicknesses remaining unchanged. The resulting distortion is clear.

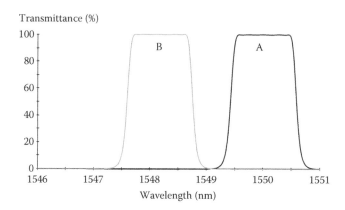

FIGURE 8.37
Filter of Figure 8.35 showing normal incidence performance with theoretical thicknesses at A and with low-index layer thicknesses reduced by a factor 0.998 at B. In both cases, the high-index layers are unchanged. There is virtually no distortion of the passband.

TABLE 8.5

Arrangement of Admittance through Basic Filter Unit

	$n_1{}^x/n_2{}^{x-1}$
n_1	
	$n_2{}^{x-1}/n_1{}^{x-2}$
n_2	
	$n_1{}^{x-2}/n_2{}^{x-3}$
n_1	
	$n_2{}^{x-3}/n_1{}^{x-4}$
$\vdots$	
	$n_1{}^{x-2}/n_2{}^{x-3}$
n_2	
	$n_2{}^{x-1}/n_1{}^{x-2}$
n_1	
	$n_1{}^x/n_2{}^{x-1}$

Then, in the same way as for the single-cavity filter, we can write

$$\sum A = \beta_1\left[\left(\frac{n_1}{n_2}\right)^{x-1} + \left(\frac{n_2}{n_1}\right)^{x-1}\right] + \beta_2\left[\left(\frac{n_1}{n_2}\right)^{x-2} + \left(\frac{n_2}{n_1}\right)^{x-2}\right]$$

$$+\beta_1\left[\left(\frac{n_1}{n_2}\right)^{x-3} + \left(\frac{n_2}{n_1}\right)^{x-3}\right] + \cdots + \beta_2\left[\left(\frac{n_2}{n_1}\right)^{x-2} + \left(\frac{n_1}{n_2}\right)^{x-2}\right]$$

$$\beta_1\left[\left(\frac{n_2}{n_1}\right)^{x-1} + \left(\frac{n_1}{n_2}\right)^{x-1}\right],$$

i.e.,

$$\sum \mathcal{A} = \beta_1 \left\{ \left[\left(\frac{n_1}{n_2}\right)^{x-1} + \left(\frac{n_1}{n_2}\right)^{x-3} + \cdots + \left(\frac{n_2}{n_1}\right)^{x-1} \right] + \left[\left(\frac{n_2}{n_1}\right)^{x-1} + \left(\frac{n_2}{n_1}\right)^{x-3} + \cdots + \left(\frac{n_1}{n_2}\right)^{x-1} \right] \right\}_1$$

$$+ \beta_2 \left\{ \left[\left(\frac{n_2}{n_1}\right)^{x} + \left(\frac{n_2}{n_1}\right)^{x-3} + \cdots + \left(\frac{n_1}{n_2}\right)^{x-1} \right] + \left[\left(\frac{n_1}{n_2}\right)^{x-1} + \left(\frac{n_1}{n_2}\right)^{x-3} + \cdots + \left(\frac{n_2}{n_1}\right)^{x-1} \right] \right\}.$$

We note that the second expression of each pair is the same as the first with inverse order. The layers are quarter waves, and so we can write, as before,

$$\beta_1 = \frac{\pi}{2} \frac{k_1}{n_1} \text{ and } \beta_2 = \frac{\pi}{2} \frac{k_2}{n_2}.$$

Once again, we divide the cases into high- and low-index cavities.

8.4.2.1 Case I: High-Index Cavities

We replace n_1 by n_H, k_1 by k_H, n_2 by n_L, and k_2 by k_L. Then, neglecting, as before, terms in $(n_L/n_H)^x$ compared with unity,

$$\sum \mathcal{A} = \frac{\pi(k_H/n_H)(n_H/n_L)^{x-1}}{1 - (n_L/n_H)^2} + \frac{\pi(k_L/n_L)(n_H/n_L)^{x-2}}{1 - (n_L/n_H)^2}$$

$$= \pi \left(\frac{n_H}{n_L}\right)^{x} \frac{n_L(k_H + k_L)}{(n_H^2 - n_L^2)}.$$

Now Equation 8.61 with $m = 1$ is

$$\left| \frac{\Delta\lambda_B}{\lambda_0} \right| = \frac{4}{\pi} \left(\frac{n_L}{n_H}\right)^{x} \frac{(n_H - n_L)}{n_H},$$

so that

$$\sum \mathcal{A} = 4 \left(\frac{\lambda_0}{\Delta\lambda_B}\right) \frac{n_L(k_H + k_L)}{n_H(n_H + n_L)}.$$

Now, this is the loss of one basic symmetrical unit. If further basic units are added each will have the same loss. In addition, there are the matching stacks at either end of the filter. We will not be far in error if we assume that they add a further loss equal to one of the basic symmetrical units. The total number of units is then equal to the number of cavities. If we denote this by q, then $q = 2$ for a two-cavity filter and so on. We can also assume that $R = 0$ so that the absorption loss becomes

$$A = q\pi \left(\frac{n_H}{n_L}\right)^{x} \frac{n_L(k_H + k_L)}{(n_H^2 - n_L^2)} \tag{8.70}$$

or

$$A = 4q \left(\frac{\lambda_0}{\Delta\lambda_B}\right) \frac{n_L(k_H + k_L)}{n_H(n_H + n_L)}. \tag{8.71}$$

8.4.2.2 Case II: Low-Index Cavities

In the same way, low-index cavities give

$$A = q\pi \left(\frac{n_H}{n_L}\right)^{x} \frac{(n_H^2 k_L + n_L^2 k_H)}{n_H(n_H^2 - n_L^2)} \tag{8.72}$$

or

$$A = 4q \left(\frac{\lambda_0}{\Delta\lambda_B} \right) \left(\frac{n_L}{n_H} \right) \frac{[k_L(n_H/n_L) + k_H(n_L/n_H)]}{(n_H + n_L)}. \tag{8.73}$$

Equations 8.71 and 8.73 are approximately q times the absorption of single-cavity filters with the same halfwidth, a not surprising result.

8.4.3 Further Information

The examples of multiple-cavity filters so far described have been for the visible and infrared, but, of course, they can be designed for any region of the spectrum where suitable thin-film materials exist. Much useful information is given in the book by Cushing [26] and in that by Thelen [27]. An account of filters for the visible and ultraviolet that is still relevant is given by Barr [28]. All-dielectric filters, both of the single-cavity and multiple-cavity types for the near ultraviolet, are described by Neilson and Ring [29], who used combinations of cryolite and lead fluoride and of cryolite and antimony trioxide, the former for the region of 250–320 nm and the latter for 320–400 nm. Apart from the techniques required for the deposition of these materials, the main difference between such filters and those for the visible or infrared is that the values of the high- and low-refractive indices are much closer together, requiring more layers for the same rejection. Neilson and Ring's filters contained basic units of 15–19 layers, in most cases, so that complete two-cavity filters consisted of 31–39 layers. Malherbe [30] has described a lanthanum fluoride and magnesium fluoride filter for 205.5 nm, in which the basic unit had 51 layers (high-index first-order cavity), the full design being $(HL)^{12}H\ H(LH)^{25}H(LH)^{12}$ with a total number of 99 layers, giving a measured bandwidth of 2.5 nm.

8.5 Phase-Dispersion Filter

The phase-dispersion filter predates the successes of the narrowband filters for telecommunication applications and was an attempt to find an approach to the design of narrowband filters, which would avoid some of the manufacturing difficulties inherent in conventional narrowband filters. At the time, the conventional filter was found to be increasingly difficult to manufacture as halfwidths were reduced below around 0.3% of peak wavelength. Attempts to improve the position by using higher-order cavities appeared to be ineffective when the cavity became thicker than perhaps the fourth order because of what was considered to be increased roughness of the cavity. Much more is now known about narrowband filters and the causes of manufacturing difficulties, and those will be dealt with in some detail in a subsequent chapter. A principal problem at the time was moisture-induced drifts, although that was not recognized until somewhat later. Although the phase-dispersion filter was not, as it turned out, the solution to the narrowband filter problem, nevertheless, it does have interesting properties, and the philosophy behind the design is a useful analytical exercise.

As we have seen, the reflecting quarter-wave stack, described in Chapter 6, and a fundamental component of our narrowband filters, shows a significant dispersion of its phase change on reflection that influences the attained halfwidth of the filters. This suggested to Baumeister and Jenkins [31] that it might form the basis for a new type of filter in which the narrow bandwidth would depend almost entirely on this phase dispersion rather than on the high reflectances of the reflecting stacks. They called this type of filter a phase-dispersion filter. It consists quite simply of a single-cavity all-dielectric filter that has, instead of the conventional dielectric quarter-wave stacks on either side of the cavity layer, reflectors consisting of staggered multilayers. Their rapid change in phase causes the bandwidth of the filter and the position of its peak to be much less sensitive to the errors in thickness of the cavity layer than would otherwise be the case.

The results, which they themselves [31] and with Baumeister et al. [32] eventually achieved, were good, although they never quite succeeded in attaining the performance possible in theory. This prompted a study [33] of the influence of errors in any of the layers of a filter on the position of the peak. The idea behind this study was that random errors in both thickness and uniformity in layers other than the cavity might be responsible for the discrepancy between theory and practice. If, in a practical filter, the errors were causing the peak to vary in position over the surface of the filter, then the integrated response would exhibit a rather wider bandwidth and lower transmittance than those of any very small portion of the filter, which might well be attaining the theoretical performance. It seemed possible that there might be a design of filter that could yield the minimum sensitivity to errors and therefore give the minimum possible bandwidth with a given layer roughness.

Giacomo et al.'s findings [33] can be summarized as follows. (The notation in the paper has been slightly altered to agree with that used in this book.) The peak of an all-dielectric multilayer filter is given by

$$\frac{\varphi_a + \varphi_b}{2} - \delta = m\pi, \tag{8.74}$$

where

$$\delta = \frac{2\pi n_c d_c}{\lambda} = 2\pi n_c d_c \nu,$$

d_c being the physical thickness of the cavity layer and the other symbols having their usual meanings.

For a change Δd_i in the ith layer, Δd_j in the jth layer, and Δd_c in the cavity, the corresponding change in the wavenumber of the peak $\Delta\nu$ is given by

$$\sum_i \frac{\partial \varphi_a}{\partial d_i} \Delta d_i + \sum_j \frac{\partial \varphi_b}{\partial d_j} \Delta d_j - 2\frac{\partial \delta}{\partial d_c} \Delta d_c + \left(\frac{\partial \varphi_a}{\partial \nu} + \frac{\partial \varphi_b}{\partial \nu} - 2\frac{\partial \delta}{\partial \nu}\right)\Delta\nu = 0. \tag{8.75}$$

Now,

$$\frac{\partial \delta}{\partial d_c} = 2\pi n_c \nu = \frac{\delta}{d_c} \tag{8.76}$$

and

$$\frac{\partial \delta}{\partial \nu} = 2\pi n_c d_c = \frac{\delta}{\nu}, \tag{8.77}$$

and, since d_i and ν appear in the individual thin-film matrices only in the value of $\delta_i = 2\pi n_i d_i \nu$, then

$$\sum_i \frac{\partial \varphi_a}{\partial d_i} \Delta_0 d_i = \frac{\partial \varphi_a}{\partial \nu} \Delta_0 \nu,$$

and similarly for φ_b, where Δ_0 indicates that the changes in d_i are related by

$$\frac{\Delta_0 d_i}{d_i} = \frac{\Delta_0 \nu}{\nu}.$$

This gives

$$\frac{\partial \varphi_a}{\partial \nu} = \sum_i \left(\frac{\partial \varphi_a}{\partial d_i} \frac{d_i}{\nu}\right), \tag{8.78}$$

which is independent of the particular choice of Δ_0 used to arrive at it. A similar expression holds for φ_b. Using Equations 8.76 through 8.78 in Equation 8.75,

$$\sum_i \left(\frac{\partial \varphi_a}{\partial d_i} \Delta d_i \right) + \sum_j \left(\frac{\partial \varphi_b}{\partial d_j} \Delta d_j \right) - 2\delta \frac{\Delta d_c}{d_c} + \left[\sum_i \left(\frac{\partial \varphi_a}{\partial d_i} d_i \right) + \sum_j \left(\frac{\partial \varphi_b}{\partial d_j} d_j \right) - 2\delta \right] \frac{\Delta v}{v} = 0,$$

i.e.,

$$\frac{\Delta v}{v} = - \frac{\left[\sum_i \left(\frac{\partial \varphi_a}{\partial d_i} d_i \alpha_i \right) + \sum_j \left(\frac{\partial \varphi_b}{\partial d_j} d_j \alpha_j \right) - 2\delta\alpha_c \right]}{\left[\sum_i \left(\frac{\partial \varphi_a}{\partial d_i} d_i \right) + \sum_j \left(\frac{\partial \varphi_b}{\partial d_j} d_j \right) - 2\delta \right]}, \tag{8.79}$$

where

$$\alpha_i = \frac{\Delta d_i}{d_i}, \text{ etc}$$

Now, in a real filter, the fluctuations in thickness, or roughness, will be completely random in character, and in order to deal with the performance of any appreciable area of the filter, we must work in terms of the mean square deviations. Each layer in the assembly can be thought of as being a combination of a large number of thin elementary layers of similar mean thicknesses but which fluctuate in a completely random manner quite independently of each other. The root-mean-square (RMS) variation in the thickness of any layer in the filter can then be considered to be proportional to the square root of its thickness. This can be written as follows:

$$\varepsilon_i = k d_i^{1/2},$$

where k can be assumed to be the same for all layers regardless of thickness. If a_i is the RMS fractional variation of the ith layer, then

$$a_i = \frac{\varepsilon_i}{d_i} = \frac{k}{d_i^{1/2}},$$

where

$$a_i = \overline{\alpha_i^2}.$$

We now define β as being

$$\beta^2 = \overline{\left(\frac{\Delta v}{v} \right)^2}.$$

Then,

$$\beta^2 = \frac{\left[\sum_i \left[\left(\frac{\partial \varphi_a}{\partial d_i} d_i \right)^2 a_i^2 \right] + \sum_j \left[\left(\frac{\partial \varphi_b}{\partial d_j} d_j \right)^2 a_j^2 \right] + 4\delta^2\alpha_c^2 \right]}{\left[\sum_i \left(\frac{\partial \varphi_a}{\partial d_i} d_i \right) + \sum_j \left(\frac{\partial \varphi_b}{\partial d_j} d_j \right) - 2\delta \right]^2},$$

which gives

$$\beta^2 = \frac{k^2 \sum\limits_{k=1}^{q} \left(\frac{1}{d_k} A_k^2 \right)}{\left(\sum\limits_{k=1}^{q} A_k \right)^2}, \tag{8.80}$$

where

$$A_k = \frac{\partial \varphi_a}{\partial d_k} d_k \text{ or } \frac{\partial \varphi_b}{\partial d_k} d_k \text{ or } -2\delta,$$

whichever is appropriate. q is the number of layers in the filter. The expression will be a minimum when

$$\frac{A_k}{d_k} = \frac{A_l}{d_l} = \cdots . \tag{8.81}$$

Then,

$$\beta^2 = k^2/T, \tag{8.82}$$

where T is the total thickness of the filter.

In the general case,

$$\beta \geq k/T^{1/2},$$

and one might hope to attain a limiting resolution of

$$R = T^{1/2}/k. \tag{8.83}$$

The condition written in Equation 8.81 can be developed with the aid of Equation 8.78 into

$$\frac{\partial \varphi_a}{\partial d_k} = \frac{\partial \varphi_b}{\partial d_l} = -4\pi n_c \nu,$$

so that

$$\nu \left(\frac{\partial \varphi_a}{\partial \nu} \right) = \sum_i \left(\frac{\partial \varphi_a}{\partial d_i} d_i \right) = -4\pi n_c d_m \nu,$$

and likewise for reflector b, where d_m = total thickness of the appropriate reflector. This gives

$$\frac{\partial \varphi_a}{\partial \nu} = -4\pi n_c d_m. \tag{8.84}$$

This condition is necessary but not sufficient for the resolution to be a maximum, and it can be used as a preliminary test of the suitability of any particular multilayer reflector that may be employed. The classical quarter-wave stack is very far from satisfying it, but the staggered multilayer is much more promising. In their paper, Giacomo et al. [33] compared a staggered multilayer reflector with a conventional quarter-wave stack. Both reflectors have 15 layers, and the results are quoted for the broadband reflector at 17,000 cm^{-1} and for the conventional reflector at 20,000 cm^{-1}.

Equation 8.84 can be written as

$$\sum_i \left(\frac{\partial \varphi_a}{\partial d_i} d_i \right) = \sum_i \frac{\partial \varphi_a}{\partial \alpha_i} = -4\pi n_c d_m \nu.$$

Now, from Table 8.6,

$$-\sum_i \frac{\partial \varphi_a}{\partial \alpha_i} = 30.662$$

and

$$4\pi n_c d_m \nu = 34.5,$$

so that on the preliminary basis of Equation 8.84, the prospects look extremely good. However, this is not a sufficient condition. We must calculate the actual relationship between β and k and compare it with the theoretical condition given by Equation 8.82. Now

$$A_i = d_i \frac{\partial \varphi}{\partial d_i} = \frac{\partial \varphi}{\partial \alpha_i},$$

which is the last column given for each reflector. This can be used in Equation 8.80 giving for a filter using the broadband reflector

$$\beta = 1.023k,$$

which can be compared with the value obtained in the same way for the conventional quarter-wave stack of Table 8.6:

$$\beta = 1.289k.$$

For a total filter thickness of 2.35 µm, the theoretical minimum value of β is given by Equation 8.82 as

$$\beta = 0.652k$$

(k having units of micrometers raised to 1/2).

TABLE 8.6

Phase Dispersion Filter Reflectors

Layer Number	Broadband Film				Classical Film		
	Thickness d_i (µm)	Index n	$\partial\varphi/\partial d_i$ (µm^{-1})	$\partial\varphi/\partial\alpha_i$	Thickness d_i (µm)	$\partial\varphi/\partial d_i$ (µm^{-1})	$\partial\varphi/\partial\alpha_i$
Substrate		1.52					
1	0.0751	2.30	0.32	0.024	0.0543	0.01	0.001
2	0.1279	1.35	0.60	0.076	0.0926	0.02	0.002
3	0.0751	2.30	1.97	0.148	0.0543	0.05	0.003
4	0.1235	1.35	1.85	0.229	0.0926	0.06	0.005
5	0.0626	2.30	4.75	0.298	0.0543	0.16	0.009
6	0.1299	1.35	4.60	0.597	0.0926	0.16	0.015
7	0.0681	2.30	11.68	0.795	0.0543	0.48	0.026
8	0.0957	1.35	10.63	1.018	0.0926	0.48	0.044
9	0.0566	2.30	30.85	1.746	0.0543	1.39	0.075
10	0.0859	1.35	30.37	2.608	0.0926	1.39	0.128
11	0.0504	2.30	78.33	3.948	0.0543	4.03	0.219
12	0.0805	1.35	62.33	5.019	0.0926	4.03	0.373
13	0.0450	2.30	121.58	5.471	0.0543	11.69	0.635
14	0.0767	1.35	65.41	5.015	0.0926	11.69	1.082
15	0.0450	2.30	81.59	3.672	0.0543	33.92	1.843
Incident medium		1.35					
Σ	1.1978		506.8	30.662	1.0829	69.53	4.460

Source: P. Giacomo, P. W. Baumeister, and F. A. Jenkins, *Proceedings of the Physical Society*, 73, 480–489, 1959, Institute of Physics.

Thus, although the phase-dispersion filter using the reflectors shown in Table 8.6 appears to be promising based on the criterion in Equation 8.84, in the event, its performance is somewhat disappointing. It is, however, certainly better from the point of view of uniform error sensitivity than the straightforward classical filter. So far no design that better meets the condition of Equation 8.81 has been proposed.

Some otherwise unpublished results obtained by Ritchie [34] are shown in Figure 8.38. This filter used zinc sulfide and cryolite as the materials on glass as substrate. Its design is given in Table 8.7. An experimental filter monitored at 1.348 µm gave peaks with the following corresponding bandwidths:

- 1.047 µm; bandwidth, 3.0 nm
- 1.159 µm; bandwidth, 2.5 nm
- 1.282 µm; bandwidth, 4.0 nm

Theoretically, the bandwidths should have been 0.8, 1.7, and 4.6 nm, respectively.

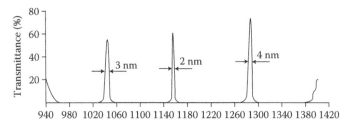

FIGURE 8.38
The measured transmittance of a 35-layer phase-dispersion filter. The design is given in Table 8.7. (After F. S. Ritchie. Unpublished work on Ministry of Technology Contract KX/LSO/C.B.70(a).)

TABLE 8.7

Filter Design by Ritchie

Layer Number	Material	Optical Thickness Units of λ_0
1	ZnS	0.2375
2	Na_3AlF_6	0.2257
3	ZnS	0.2143
4	Na_3AlF_6	0.2036
5	ZnS	0.1934
6	Na_3AlF_6	0.1838
7	ZnS	0.1746
8	Na_3AlF_6	0.1649
9	ZnS	0.1576
10	Na_3AlF_6	0.1498
11	ZnS	0.1423
12	Na_3AlF_6	0.1352
13	ZnS	0.1285
14	Na_3AlF_6	0.1220
15	ZnS	0.1159
16	Na_3AlF_6	0.1101
17	ZnS	0.1046
Cavity	Na_3AlF_6	0.5000

Source: F. S. Ritchie. Unpublished work on Ministry of Technology Contract KX/LSO/C.B.70(a).
Note: These 17 layers are followed by another 17 that are mirror images of the first 17.

8.6 Multiple-Cavity Metal–Dielectric Filters

Metal–dielectric filters are indispensable in suppressing the longwave sidebands of narrowband all-dielectric filters, and as filters in their own right, especially in the extreme shortwave region of the spectrum. Unlike all-dielectric filters, however, they possess the disadvantage of high intrinsic absorption. In single-cavity filters, this means that the passbands must be wide in order to achieve reasonable peak transmission, and the shape is far from ideal. It is possible to combine metal–dielectric elements into multiple-cavity filters that, because of their more rectangular shape, are more satisfactory, but again, losses can be high.

The accurate design procedure for such metal–dielectric filters can be lengthy and tedious, and frequently, they are simply designed by trial and error as they are manufactured. We have already mentioned the metal–dielectric single-cavity filter. These filters may be coupled together by simply depositing them one on top of the other with no coupling layer in between.

We can illustrate this by choosing silver as our metal, which we can give an index of $0.055 - i3.32$ at 550 nm [35]. The thickness of the cavity layer in the single-cavity filter, as we have already noted, should be rather thinner than a half wave at the peak wavelength to allow for the phase changes in reflection at the silver/dielectric interfaces. This phase change varies only slowly with silver thickness when it is thick enough to be useful as a reflector, and we can assume, as a reasonable approximation, that it is equal to the limiting value for infinitely thick material. We can then use Equation 5.6 to calculate the thickness of the cavity. Equation 5.6 calculates for us exactly one half of the filter because it gives the thickness of the dielectric material to yield real admittance with zero phase change at the outer surface of the metal–dielectric combination. Adding a second exactly similar structure with the two dielectric layers facing each other, so that they join to form a single cavity, yields a filter in which the phase condition (Equation 8.3) is satisfied.

Let us choose a cavity of index 1.35, similar to that of cryolite. Then, half the cavity thickness is given by

$$D_c = \frac{1}{4\pi} \arctan\left(\frac{2\beta n_c}{n_c^2 - \alpha^2 - \beta^2}\right), \tag{8.85}$$

where $\alpha - i\beta$ is the index of the metal and n_c is that of the cryolite cavity, and the angle is taken in the first or second quadrant.

With $\alpha - i\beta = 0.055 - i3.32$ and $n_c = 1.35$, we find

$$D_c = 0.18855,$$

so that the cavity thickness should be 0.3771 full waves.

We can choose a metal layer thickness of 35 nm, quite arbitrarily, simply for the sake of illustration. Our single-cavity filter is then as follows:

Glass	Ag	Cryolite	Ag	Glass
	35 nm	0.3771 waves	35 nm	

(the geometrical thickness being quoted for the silver and the optical thickness for the cryolite). And the two-cavity filter is exactly double this structure:

Glass	Ag	Cryolite	Ag	Cryolite	Ag	Glass.
	35 nm	0.3771 waves	70 nm	0.3771 waves	35 nm	

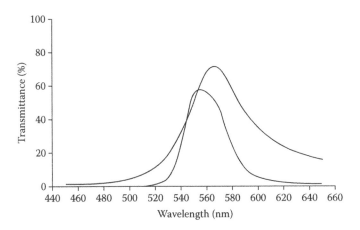

FIGURE 8.39
Transmittance as a function of wavelength of filters of the following designs:

$$\text{Glass}\left|\begin{array}{c|c|c}\text{Ag} & \text{Cryolite} & \text{Ag}\\ 35\,\text{nm} & 0.3771\lambda_0 & 35\,\text{nm}\end{array}\right|\text{Glass}$$

and

$$\text{Glass}\left|\begin{array}{c|c|c|c|c}\text{Ag} & \text{Cryolite} & \text{Ag} & \text{Cryolite} & \text{Ag}\\ 35\,\text{nm} & 0.3771\lambda_0 & 70\,\text{nm} & 0.3771\lambda_0 & 35\,\text{nm}\end{array}\right|\text{Glass.}$$

where $\lambda_0 = 550\,\text{nm}$, $n - ik = 0.055 - 3.32$, $n_{\text{glass}} = 1.52$, and $n_{\text{cryolite}} = 1.35$. Dispersion in the materials has been neglected. The narrower peak corresponds to the two-cavity design.

Transmittance curves of these filters are shown in Figure 8.39. The peaks are slightly displaced from 550 nm because of the approximations inherent in the design procedure.

The single cavity has reasonably good peak transmittance, but its typical triangular shape means that its rejection is quite poor even at wavelengths far from the peak. The two-cavity filter has better shape, but rather poorer peak transmittance. The rejection can be improved by increasing the metal thickness, but at the expense of peak transmittance.

The design approach we have described is quite crude and simply concentrates on ensuring that the peak of the filter is centered near the desired wavelength. Either the peak transmittance and bandwidth are accepted as they are or a new metal thickness is tried. Performance is in no way optimized.

The unsatisfactory nature of this design procedure led Berning and Turner [36] to develop a new technique for the design of metal–dielectric filters, in which the emphasis is on ensuring that maximum transmittance is achieved in the filter passband. For this purpose, they devised the concept of potential transmittance and created a new type of metal–dielectric filter known as the induced transmission filter.

8.6.1 Induced Transmission Filter

Given a certain thickness of metal in a filter, what is the maximum possible peak transmittance? And how can the filter be designed to realize this transmittance? This is the basic problem tackled and solved by Berning and Turner [36]. The development of the technique as given here is based on their approach, but it has been adjusted and adapted to conform more nearly to the general pattern of this book.

The concept of potential transmittance has already been touched on in Section 2.10 and used in the analysis of losses in dielectric multilayers. We recall that the potential transmittance ψ of a layer or assembly of layers is defined as the ratio of the irradiance leaving the rear surface to that actually entering at the front surface, and it represents the transmittance that the layer or assembly of layers would have if the reflectance of the front surface were reduced to zero. Once the parameters of the metal layer are fixed, its potential transmittance is entirely determined by the admittance of the structure at the exit face of the layer. Furthermore, it is possible to determine that particular admittance that gives maximum potential transmittance. To achieve this transmittance once the correct exit admittance is established, it is sufficient to add a coating to the front surface to reduce the reflectance to zero. The maximum potential transmittance is a function of the thickness of the metal layer.

The design procedure is then as follows. The optical constants of the metal layer at the peak wavelength are given. Then the metal layer thickness is chosen and the maximum potential transmittance together with the matching admittance at the exit face of the layer, which is required to produce that level of potential transmittance, is found. Often a minimum acceptable figure for the maximum potential transmittance will exist, and that will put an upper limit on the metal layer thickness. A dielectric assembly to give the correct matching admittance at the rear of the layer, which will usually be directly deposited on the substrate, must then be designed. The filter is then completed by the addition of a dielectric system to match the front surface of the resulting metal–dielectric assembly to the incident medium. Techniques for each of these steps will be developed. The matching admittances for the metal layer rapidly vary with wavelength because of the dispersion of the metal, and so the dielectric stacks that have difficulty in matching this dispersion are efficient in matching over a limited region only, outside which their performance falls off rapidly. It is this rapid fall in performance that defines the limits of the passband of the filter.

Before we can proceed further, we require some analytical expressions for the potential transmittance and for the matching admittance. This leads to some lengthy and involved analysis that is not difficult but dismayingly tedious.

8.6.1.1 Potential Transmittance

We limit the analysis to an assembly in which there is only one absorbing layer, the metal. The potential transmittance is then related to the matrix for the assembly, as shown in Chapter 2. For the film in question, we have

$$\begin{bmatrix} B_i \\ C_i \end{bmatrix} = [M] \begin{bmatrix} 1 \\ Y_e \end{bmatrix},$$

where $[M]$ is the characteristic matrix of the metal layer and Y_e is the admittance of the terminating structure. The potential transmittance ψ is then given by

$$\psi = \frac{\mathrm{Re}(Y_e)}{\mathrm{Re}(B_i C_i^*)}. \tag{8.86}$$

Let the exit admittance be given by $X + iZ$. Then

$$\begin{bmatrix} B_i \\ C_i \end{bmatrix} = \begin{bmatrix} \cos\delta & (i\sin\delta)/y \\ iy\sin\delta & \cos\delta \end{bmatrix} \begin{bmatrix} 1 \\ X + iZ \end{bmatrix},$$

where

$$\delta = 2\pi(n - ik)d/\lambda = (2\pi nd)/\lambda - i(2\pi kd)/\lambda$$

$$= \alpha - i\beta,$$

$$\alpha = (2\pi nd)/\lambda,$$

$$\beta = (2\pi kd)/\lambda.$$

If free space units are used, then

$$y = n - ik.$$

Now,

$$(B_i C_i^*) = [\cos \delta + i(\sin \delta / y)(X + iZ)][iy \sin \delta + \cos \delta (X + iZ)]^*$$

$$= [\cos \delta + i(\sin \delta / y)(X + iZ)][-iy^* \sin \delta^* + \cos \delta^* (X - iZ)]$$

$$= -iy^* \cos \delta \sin \delta^* + \frac{\sin \delta \sin \delta^* y^{*2}(X + iZ)}{yy^*}$$

$$+ \cos \delta \cos \delta^* (X - iZ) + \frac{i \sin \delta \cos \delta^* y^* (X + iZ)(X - iZ)}{yy^*}.$$

We require the real part of this, and we take each term in turn:

$$-iy^* \cos \delta \sin \delta^* = -i(n + ik)(\cos \alpha \cosh \beta + i \sin \alpha \sinh \beta)$$

$$\times (\sin \alpha \cosh \beta + i \cos \alpha \sinh \beta),$$

and the real part of this, after a little manipulation, is

$$\text{Re}(-iy^* \cos \delta \sin \delta^*) = n \sinh \beta \cosh \beta + k \cos \alpha \sin \alpha.$$

Similarly,

$$\text{Re}\left(\frac{\sin \delta \sin \delta^* y^{*2}(X + iZ)}{yy^*}\right) = \frac{X(n^2 - k^2) - 2nkZ}{(n^2 + k^2)}$$

$$\times (\sin^2\alpha\cosh^2\beta + \cos^2\alpha\sinh^2\beta),$$

$$\text{Re}[\cos \delta \cos \delta^* (X - iZ)] = X(\cos^2\alpha\cosh^2\beta + \sin^2\alpha\sinh^2\beta),$$

$$\text{Re}\left(\frac{i \sin \delta \cos \delta^* y^*(X + iZ)(X - iZ)}{yy^*}\right) = \frac{X^2 + Z^2}{(n^2 + k^2)}(n \sinh \beta \cosh \beta - k \sin \alpha \cos \alpha).$$

The potential transmittance ψ is then given by the ratio of the net output and the net input, that is, $X/\text{Re}(BC^*)$. A slightly more friendly expression is the inverse of this:

$$\frac{1}{\psi} = \left[\frac{(n^2 - k^2 - 2nkZ/X)}{(n^2 + k^2)}(\sin^2\alpha\cosh^2\beta + \cos^2\alpha\sinh^2\beta)\right.$$

$$+ (\cos^2\alpha\cosh^2\beta + \sin^2\alpha\sinh^2\beta) \tag{8.87}$$

$$+ (1/X)(n \sinh \beta \cosh \beta + k \cos \alpha \sin \alpha)$$

$$\left. + \frac{X^2 + Z^2}{X(n^2 + k^2)}(n \sinh \beta \cosh \beta - k \cos \alpha \sin \alpha)\right].$$

8.6.1.2 Optimum Exit Admittance

Next, we find the optimum values of X and Z. From Equation 8.87,

$$\frac{1}{\psi} = \left[\frac{p(n^2 - k^2 - 2nkZ/X)}{(n^2 + k^2)} + q + \frac{r}{X} + \frac{s(X^2 + Z^2)}{X(n^2 + k^2)}\right], \tag{8.88}$$

where p, q, r, and s are shorthand for the corresponding expressions in Equation 8.87. ψ is always positive and is a well-behaved function, so for an extremum in ψ, we have an extremum in $1/1\psi$ of the opposite sense. We can safely write that the optimum exit admittance is given by

$$\frac{\partial}{\partial X}\left(\frac{1}{\psi}\right) = 0 \text{ and } \frac{\partial}{\partial Z}\left(\frac{1}{\psi}\right) = 0,$$

i.e.,

$$\frac{p \cdot 2nkZ}{X^2(n^2+k^2)} - \frac{r}{X^2} + \frac{s}{(n^2+k^2)} - \frac{sZ^2}{X^2(n^2+k^2)} = 0 \tag{8.89}$$

and

$$\frac{p(-2nk)}{X(n^2+k^2)} + \frac{2sZ}{X(n^2+k^2)} = 0. \tag{8.90}$$

From Equation 8.90,

$$Z = \frac{nkp}{s},$$

and, substituting in Equation 8.89,

$$X^2 = \frac{r(n^2+k^2)}{s} - \frac{n^2k^2p^2}{s^2}.$$

Then, inserting the appropriate expressions for p, r, and s, from Equations 8.87 and 8.88,

$$X = \left[\frac{(n^2+k^2)(n\sinh\beta\cosh\beta + k\sin\alpha\cos\alpha)}{(n\sinh\beta\cosh\beta - k\sin\alpha\cos\alpha)}\right.$$
$$\left. - \frac{n^2k^2\left(\sin^2\alpha\cosh^2\beta + \cos^2\alpha\sinh^2\beta\right)^2}{(n\sinh\beta\cosh\beta - k\sin\alpha\cos\alpha)^2}\right]^{\frac{1}{2}}, \tag{8.91}$$

$$Z = \frac{nk\left(\sin^2\alpha\cosh^2\beta + \cos^2\alpha\sinh^2\beta\right)}{(n\sinh\beta\cosh\beta - k\sin\alpha\cos\alpha)}. \tag{8.92}$$

We note that for β, large $X \rightarrow n$ and $Z \rightarrow k$, that is,

$$Y_e \rightarrow (n + ik) = (n - ik)^*.$$

8.6.1.3 Maximum Potential Transmittance

The maximum potential transmittance can then be found by substituting the values of X and Z, calculated by Equations 8.91 and 8.92, into Equation 8.87. All these calculations are best performed by computer or calculator, and so there is little advantage in developing a separate analytical solution for maximum potential transmittance.

8.6.1.4 Matching Stack

We have to devise an assembly of dielectric layers that, when deposited on the substrate, will have an equivalent admittance of

$$Y = X + iZ.$$

This is diagrammatically illustrated in Figure 8.40, where a substrate of admittance $(n_s - ik_s)$ has an assembly of dielectric layers, terminating such that the final equivalent admittance is $(X + iZ)$. Now, the dielectric layer circles are executed in a clockwise direction always. If we therefore reflect the diagram in the x axis and then reverse the direction of the arrows, we get exactly the same set of circles—that is, the layer thicknesses are exactly the same—but the order is reversed (it was ABC and is now CBA), and they match a starting admittance of $X - iZ$, i.e., the complex conjugate of $(X + iZ)$, into a terminal admittance of $(n_s + ik_s)$, i.e., the complex conjugate of the substrate index. In our filters, the substrate will have real admittance, i.e., $k_s = 0$, and it is a more straightforward problem to match $(X - iZ)$ into n_s than n_s into $(X + iZ)$.

There is an infinite number of possible solutions, but the simplest involves adding a phase-matching dielectric layer to change the admittance $(X - iZ)$ into a real value and then to add a series of quarter waves to match the resultant real admittance into the substrate. We will illustrate the technique shortly with several examples. At the moment, we recall that the necessary analysis was carried out in Section 5.1.2. There we showed that a film of optical thickness D given by

$$D = \frac{1}{4\pi} \arctan\left[\frac{2Zn_f}{(n_f^2 - X^2 - Z^2)}\right],$$ (8.93)

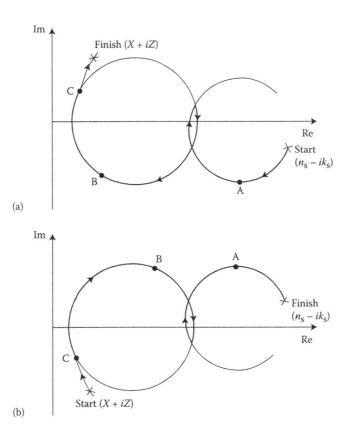

FIGURE 8.40

(a) Sketch of the admittance diagram of an arbitrary dielectric assembly of layers matching a starting admittance of $(n_s - ik_s)$ to the final admittance of $(X + iZ)$. (b) The curves of (a) reflected in the real axis and with the directions of the arrows reversed. This is now a multilayer identical to (a) but in the opposite order and connecting an admittance of $(X - iZ)$ (i.e., $(X + iZ)^*$) to one of $(n_s + ik_s)$ (i.e., $(n_s - ik_s)^*$).

(where the tangent is taken in the first or second quadrant) will convert an admittance $(X - iZ)$ into a real admittance of value

$$\mu = \frac{2Xn_f^2}{\left(X^2 + Z^2 + n_f^2\right) + \left[\left(X^2 + Z^2 + n_f^2\right)^2 - 4X^2n_f^2\right]^{1/2}}. \tag{8.94}$$

n_f, the characteristic admittance of the film, can be high or low, but μ will always be lower than the substrate admittance (except in very unlikely cases) because it is the first intersection of the locus of n_f with the real axis, which is given by Equations 8.93 and 8.94. Since the admittance of the substrate will always be greater than unity, the quarter-wave stack to match μ to n_s should start with a quarter wave of low index. Alternate high- and low-index layers follow, the precise number being found by trial and error.

In order to complete the design, we need to know the equivalent admittance at the front surface of the metal layer, and then we construct a matching stack to match it to the incident medium.

8.6.1.5 Front Surface Equivalent Admittance

If the admittance of the structure at the exit surface of the metal layer is the optimum value $(X + iZ)$ given by Equations 8.91 and 8.92, then it can be shown that the equivalent admittance, which is presented by the front surface of the metal layer, is simply the complex conjugate $(X - iZ)$. The analytical proof of this requires a great deal of patience, although it is not particularly difficult. Instead, let us use a logical justification.

Consider a filter consisting of a single metal layer matched on either side to the surrounding media by dielectric stacks. Let the transmittance of the assembly be equal to the maximum potential transmittance and let the admittance of the structure at the rear of the metal layer be the optimum admittance $(X + iZ)$. Let the equivalent admittance at the front surface be $(\xi + i\eta)$, and let this be perfectly matched to the incident medium. Now we know that the transmittance is the same regardless of the direction of incidence. Let us turn the filter around, therefore, so that the transmitted light proceeds in the opposite direction. The transmittance of the assembly must be the maximum potential transmittance once again. The admittance of the structure at what was earlier the input, but is now the new exit face of the metal layer, must therefore be $(X + iZ)$. But since the layers are dielectric and the medium is of real admittance, this must also be the complex conjugate of $(\xi + i\eta)$, that is, $(\xi - i\eta)$. $(\xi + i\eta)$ must therefore be $(X - iZ)$, which is what we set out to prove.

The procedure for matching the front surface to the incident medium is therefore exactly the same as that for the rear surface, and indeed, if the incident medium is identical to the rear exit medium, as in a cemented filter assembly, then the front dielectric section can be an exact repetition of the rear.

8.6.2 Examples of Filter Designs

We can now attempt some filter designs. We choose the same material, silver, as we did for the metal–dielectric filters earlier. Once again, arbitrarily, we select a thickness of 70 nm. We retain the wavelength as 550 nm, at which the optical constants of silver can be taken as $0.055 - i3.32$ [35].

The filter is to use dielectric materials of indices 1.35 and 2.35, corresponding to cryolite and zinc sulfide, respectively. The substrate is glass, $n = 1.52$, and the filter will be protected by a cemented cover slip so that we can also use $n = 1.52$ for the incident medium:

$$\alpha = 2\pi nd/\lambda = 0.04398,$$

$$\beta = 2\pi kd/\lambda = 2.6549.$$

And from Equations 8.91 and 8.92, we find the optical admittance

$$X + iZ = 0.4572 + i(3.4693).$$

Substituting this in Equation 8.87 gives

$$\psi = 80.50\%.$$

We can choose to have either a high- or a low-index phase-matching layer. Let us choose first a low index, and from Equation 8.93, we obtain an optical thickness for the 1.35 index layer of 0.19174 full waves. Equation 8.94 yields a value of 0.05934 for μ that must be matched to the substrate index of 1.52. We start with a low-index quarter wave and simply work through the sequence of possible admittances:

$$\frac{n_L^2}{\mu}, \ \frac{n_H^2\mu}{n_L^2}, \ \frac{n_L^4}{n_H^2\mu}, \ \frac{n_H^4\mu}{n_L^4}, \text{ etc.,}$$

until we find one sufficiently close to 1.52. The best arrangement in this case involves three layers of each type.

$$\frac{n_H^6\mu}{n_L^6} = 1.6511,$$

equivalent to a loss of 0.2% at the interface with the substrate.

The structure so far is then

$$|\text{Ag}|L'' \, LHLHLH|\text{Glass},\qquad\qquad (8.95)$$

with $L'' = 0.19174$ full waves. This can be combined with the following L layer into a single layer $L' = 0.25 + 0.19174 = 0.44174$ full waves, i.e.,

$$|\text{Ag}|L' \, HLHLH|\text{Glass}.$$

Since the medium is identical to the substrate, then the matching assembly at the front will be exactly the same as that at the rear, so that the complete design is

$$\text{Glass}|HLHLH \, L' \, \text{Ag} \, L' \, HLHLH|\text{Glass},$$

with the following:

- Ag: 70 nm (physical thickness)
- L': 0.44174 full waves (optical thickness)
- H, L: 0.25 full waves
- λ_0: 550 nm

The performance of this design is shown in Figure 8.41. The dispersion of the silver has not been taken into account to give a clearer idea of the intrinsic characteristics. The peak is indeed centered at 550 nm with transmittance virtually that predicted.

A high-index matching layer can be handled in exactly the same way. For an index of 2.35, Equation 8.93 yields an optical thickness of 0.1561 and Equation 8.94 gives a value of 0.1426 for μ. Again, the matching quarter-wave stack should start with a low-index layer. There are two possible arrangements, each with H' representing 0.1561 full waves:

$$\text{Ag}|H' \, LHLH|\text{Glass},$$

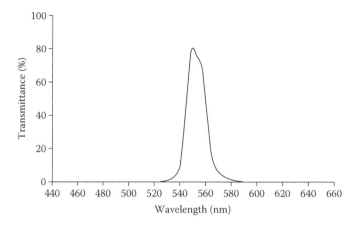

FIGURE 8.41
Calculated performance of the design: Glass | $HLHLH\ L'$ Ag $L'\ HLHLH$ | Glass, where $n_{Glass} = 1.52$, Ag = 70 nm (physical thickness) of index $0.055 - i3.32$, $H = 0.25\lambda_0$ (optical thickness) of index 2.35, $L = 0.25\lambda_0$ (optical thickness) of index 1.35, $L' = 0.4417\lambda_0$ (optical thickness) of index 1.35, and $\lambda_0 = 550$ nm. Dispersion has been neglected.

with $n_H^4 \mu / n_L^4 = 1.310$, i.e., a loss of 0.6% at the glass interface, or

$$Ag|H'\ LHLHL|Glass,$$

with $n_L^6 / n_{LH}^4 \mu = 1.392$, representing a loss of 0.2% at the glass interface.

We choose the second alternative, and the full design can then be written as

$$Glass|LHLHL\ H'\ Ag\ H'\ LHLHL|Glass,$$

with the following:

- Ag: 70 nm (physical thickness)
- H': 0.1561 full waves (optical thickness)
- H, L: 0.25 full waves

The performance of this design is shown in Figure 8.42, where, again, the dispersion of silver has not been taken into account. Peak transmission is virtually as predicted.

When, however, we plot, over an extended wavelength region, the performance of any of these designs, including the metal–dielectric filters from the early part of this chapter, we find that the performance at longer wavelengths appears disappointing. One example, the low-index matched induced transmission filter, is shown in Figure 8.43. In the case of the filters from the early part of this chapter, the rise is smoother, but is of a similar order of magnitude. The reason for the rise is, in fact, our assumption of zero dispersion. This means that β is reduced as λ increases. α is always quite small, and the performance of the metal layers is principally determined by β. Silver, however, over the visible and near infrared, shows an increase in k, which roughly corresponds to the increase in λ, so that k/λ is roughly constant (to within around ±20%) over the region of 400 nm to well beyond 2.0 μm. In fact, this behavior is common to most high-performance metals. The dispersion completely alters the picture and is the reason why the first-order metal–dielectric filters do not show longwave sidebands.

Taking dispersion into account, the performance of the induced transmission filter considerably improves and is shown in Figure 8.44. Close examination of the rejection, however (Figure 8.45), shows that it is not particularly high, being between 0.01% and 0.1% transmittance over most of the range with an increase to 0.15% in the vicinity of 850 nm. This level of rejection can be

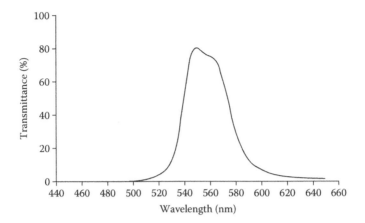

FIGURE 8.42
Calculated performance neglecting dispersion of the design: Glass | *LHLHL H'* Ag *H' LHLHL* | Glass, where n_{Glass} = 1.52, Ag = 70 nm (physical thickness) of index 0.055 − i3.32, H = 0.25λ_0 (optical thickness) of index 2.35, L = 0.25λ_0 (optical thickness) of index 1.35, H' = 0.1561λ_0 (optical thickness) of index 2.35, and λ_0 = 550 nm.

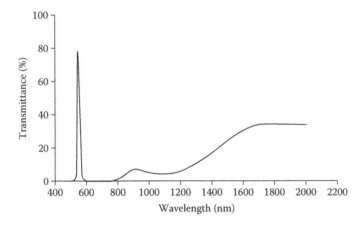

FIGURE 8.43
Design of Figure 8.41 computed over a wider spectral region neglecting dispersion.

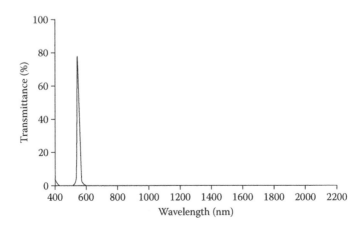

FIGURE 8.44
Design of Figure 8.41 computed this time including dispersion. The rise in transmittance at longer wavelengths has vanished, but there is now obvious transmittance at 400 nm.

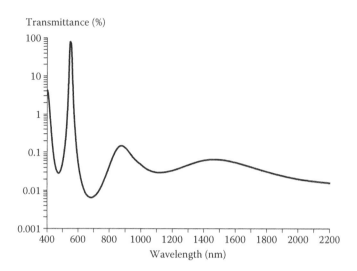

FIGURE 8.45
Filter of Figure 8.41 with performance displayed on a logarithmic scale.

acceptable in some applications, and the induced transmission filter represents a very useful, inexpensive general-purpose filter. The dispersion that improves the performance on the longwave side of the peak degrades it on the shortwave side, and to complete the filter, it is normal to add a longwave-pass absorption glass filter cemented to the induced transmission component.

To improve the rejection of the basic filter, it is necessary to add further metal layers. The simplest arrangement is to have these extra metal layers of exactly the same thickness as the first. The potential transmittance of the complete filter will then be the product of the potential transmittances of the individual layers. The terminal admittances for all the metal layers can be arranged to be optimum, quite simply, giving optimum performance for the filter. All that is required is a dielectric layer in between the metal layers which is twice the thickness given by Equation 8.93 for the first matching layer. We can see why this is so, by imagining a matching stack on the substrate overcoated with the first metal layer. Since its terminal admittance will be optimum, the input admittance will be the complex conjugate, as we have already discussed. The addition of the thickness given by Equation 8.93 renders the admittance real; that is, the admittance locus has reached the real axis. The addition of a further identical thickness must give an equivalent input admittance that is the complex conjugate of the metal input admittance and, hence, is equal to the optimum admittance. This can be repeated as often as desired.

Returning to our example, a two-metal layer induced transmission filter will have peak transmittance, if perfectly matched, of $\psi^2 = (0.80501)^2$, that is, 64.8%; a three-metal layer filter should have $\psi^3 = (0.80501)^3$, that is, 52.17% and so on.

The designs, based on the low-index matching layer version, are then, from Equation 8.95,

$$\text{Glass}|HLHLHL\,L''\,Ag\,L''\,L''\,Ag\,L''\,LHLHLH|\text{Glass}$$

$$= \text{Glass}|HLHLH\,L'\,Ag\,L'''\,Ag\,L'\,HLHLH|\text{Glass},$$

where

$$L' = 0.25 + 0.19174 = 0.44174 \text{ full waves,}$$

$$L'' = 0.19174 \text{ full waves,}$$

$$L''' = 2 \times 0.19174 = 0.38348 \text{ full waves,}$$

$$Ag = 70 \text{ nm,}$$

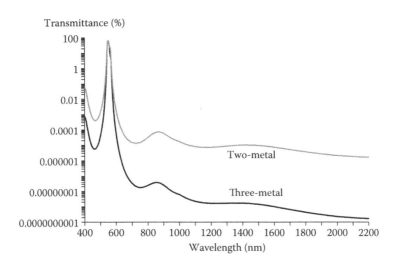

FIGURE 8.46
Performance on a logarithmic scale of the filters of the two-metal design (Equation 8.96) and the three-metal design (Equation 8.97). The rejection is several orders of magnitude better than that of the single-metal design of Figure 8.45.

and

$$\text{Glass}|HLHLH\,L'\,Ag\,L'''\,Ag\,L'''\,Ag\,L'\,HLHLH|\text{Glass}.$$

The performance of these two filters on a logarithmic scale is plotted in Figure 8.46, and a considerable improvement is evident.

Unfortunately, these designs, although they do have the peak transmittance predicted, possess a poor passband shape, in that, it has a hump or shoulder on the longwave side. The hump can be eliminated by adding an extra half-wave layer to those layers with label L''', i.e.,

$$\text{Glass}|HLHLH\,L'\,Ag\,L^{IV}\,Ag\,L'\,HLHLH|\text{Glass}$$

and

$$\text{Glass}|HLHLH\,L'\,Ag\,L^{IV}\,Ag\,L^{IV}\,Ag\,L'\,HLHLH|\text{Glass}, \tag{8.99}$$

where

$$L^{IV} = 0.5 + 0.38348 = 0.88348 \text{ full waves.}$$

Figure 8.47 shows the form of designs in Equations 8.96 and 8.97, and the hump can clearly be seen together with the improved shape of designs in Equations 8.98 and 8.99.

Dispersion was not included in the computation of Figure 8.47. The rejection over an extended region, including the effects of dispersion, is plotted in Figure 8.48. Unfortunately, the modified designs in Equations 8.98 and 8.99 act as metal–dielectric–metal (M-D-M is a frequently used shorthand notation for such a filter) and metal–dielectric–metal–dielectric–metal (M-D-M-D-M) filters at approximately 1100 nm, which gives a very narrow leak, rising to around 0.15% in the former and 0.05% in the latter. Elsewhere, the rejection is excellent, of the order of 0.0001% at 900 nm and 0.000015% at 1.5 μm for the former and 0.0000001% at 900 nm and 3×10^{-9}% at 1.5 μm for the latter.

If the leak is unimportant, then the filters can be used as they are, with the addition of a longwave-pass filter of the absorption type as before. For the suppression of all-dielectric filter sidebands, it is better to use the filters of the types indicated in Equations 8.96 and 8.97 since the

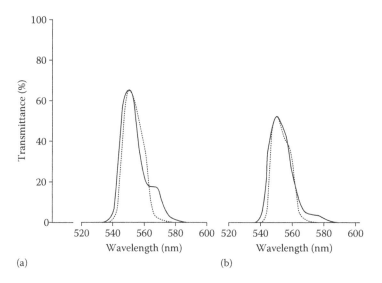

FIGURE 8.47
Performance, neglecting dispersion, of (a) two-metal-layer designs and (b) three-metal-layer designs of induced transmission filter. The full curves denote Equations 8.96 and 8.97, and there is a spurious shoulder on the longwave side of the peak in each case. This can be eliminated by the addition of half-wave decoupling layers as the dashed lines show. They are derived from Equations 7.89 and 7.90, respectively.

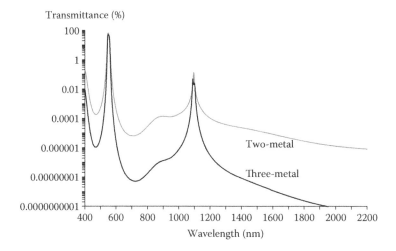

FIGURE 8.48
The extended performance of the filters of the design shown in Equations 8.98 and 8.99. The spikes at 1050 nm occur because the thick dielectric structures between the metal layers act as rather poor cavities.

shape of the sides of the passband is relatively unimportant. The rejection of these filters is excellent, and, of course, the leak is missing.

If shape as well as rejection is important, then it is possible with a slight alteration of the design to achieve a better shape at the expense of a slight loss in transmittance. We retain the 70 nm thickness of silver at a reference wavelength of 550 nm as in Figure 8.47a but use high-index phase matching layers to which we can conveniently assign the index of 2.35 that we have been using. We note that this is not critical. The three-layer core of the filter to give maximum potential transmittance of 80.5% is shown in Figure 8.49. The silver layer swings round clockwise cutting the real axis at point A that is almost at its center. We observe that a high-index quarter-wave layer

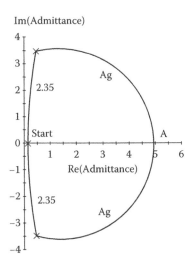

FIGURE 8.49
Admittance diagram of the central three-layer core of the induced transmission filter. The 70 nm thick silver layer cuts the real axis at A, very close to the center of the layer.

starting at point A, while not closing the gap completely between A and 1.52, the admittance of the surrounding media, would nevertheless considerably reduce it. Thus, we can imagine a structure that follows exactly the admittance locus but starts at A and passes around the locus twice to also end at A. Since the locus is exactly that to achieve the maximum potential transmittance of the 70 nm thickness of silver, and since this new structure has twice the total thickness of silver exactly coinciding with the optimum locus, the potential transmittance will be the square of the single 70 nm thickness value—that is, 64.8%—even though there are now three silver layers, two of thickness 35 nm and one of 70 nm. But the locus is not so far connected to the surrounding media. We therefore add a quarter wave of high0index (2.35) material to both the front and the back of the structure. The structure is now of the form D-M-D-M-D-M-D. If this were an exact match, then the design would be complete, and the peak transmittance, 64.8%, but we know that the high-index layers have indices a little too low for a perfect match. To complete the design, therefore, we gently refine the dielectric layers but retain the thicknesses of the metal layers. Because of the symmetry, it is useful to retain the equality of the two outer layers and of the two inner dielectric layers.

The final design is

$$1.52|0.991H|\text{Ag 35 nm}|11.231H|\text{Ag 70 nm}|11.231H|\text{Ag 35 nm}|0.991H|1.52, \qquad (8.100)$$

and the peak transmittance is 63.9%.

The filter performance is shown in Figures 8.50 and 8.51. It is rather broader than the two-metal design of Figure 8.48, but the rejection now shows no surprises. Of course, this design does not yield the maximum possible transmittance for the two outer silver layers. They are not correctly positioned in the admittance diagram for that. However, they do almost equal the maximum transmittance possible if the two outer silver layers were combined into one.

The bandwidth of the filters is not an easy quantity to analytically predict, and the most straightforward approach is to simply compute the filter profile.

Berning and Turner [36] suggested a figure of merit indicating the potential usefulness of a metal in induced transmission filters as the ratio k/n. The higher this ratio, the better should be the performance of the completed filter. We saw in Chapter 3 that another useful indicator of potential loss in a material is the product nk. The smaller this quantity, the lower the level of loss.

Induced transmission filters for the visible region having only one single metal layer are relatively straightforward to manufacture. The thickness of the metal layer is sensitive to the actual

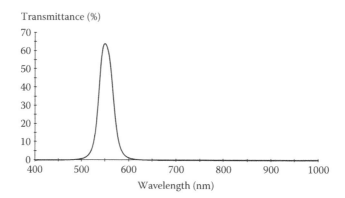

FIGURE 8.50
Calculated performance of the metal dielectric filter.

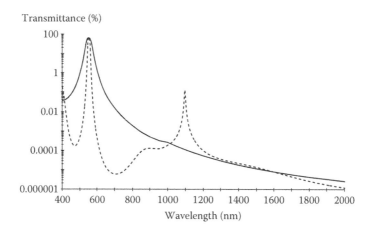

FIGURE 8.51
Comparison of the filter of Figure 8.50 (*full line*) with that of Figure 8.48 with two metal layers (*broken line*).

optical constants both of the metal and of the dielectrics. If necessary, it can be adjusted by a little trial and error. If the metal layer is less than optimum in thickness, the effect will be a broadening of the passband and a rise in peak transmittance at the expense of an increase in background transmittance remote from the peak. A splitting of the passband will also eventually become noticeable with the appearance of two separate peaks if the thickness is further reduced. If, on the other hand, the silver layer is made too thick, the effect will be a narrowing of the peak with a reduction of peak transmittance. The best results are usually obtained with a compromise thickness where the peak is still single in shape but where any further reduction in silver thickness would cause the splitting to appear. A good approximation in practice, which can be used as a first attempt at a filter, is to deposit the first dielectric stack and to measure the transmittance. The silver layer can then be deposited using a fresh monitor glass so that the optical density is twice that of the dielectric stack. The second cavity and stack can then be added on yet another fresh monitor. A measurement of the transmittance of the complete filter will quickly indicate which way the thickness of the silver layer should be altered in order to optimize the design. Usually, one or two tests are sufficient to establish the best parameters. If, after this optimizing, the background rejection remote from the peak is found to be unsatisfactory, then not enough silver is being used. As the thickness was chosen to be optimum for the two dielectric sections, a pair of quarter-wave

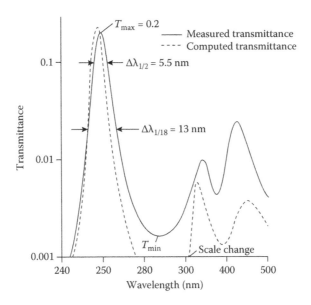

FIGURE 8.52
Computed and measured transmittance (scale of 0–1.0) of an induced transmission filter for the ultraviolet. Design: Air |
HLHLHLH 1.76*L* Al 1.76*L HLHLHLH* | Quartz, where *H* = PbF$_2$ (n_H = 2.0) and *L* = Na$_3$AlF$_6$ (n_L = 1.36). The physical
thickness of the aluminum layer is 40 nm and λ_0 = 253.6 nm. (After P. W. Baumeister, V. R. Costich, and S. C. Pieper, *Applied
Optics* 4, 911–913, 1965. With permission of Optical Society of America.)

layers should be added to each in the design, and the trial-and-error optimization, repeated. This
will also narrow the bandwidth, but this is usually preferable to high background transmission.

In the ultraviolet, the available metals do not have as high a performance as, for instance, silver
in the visible, and it is very important, therefore, to ensure that the design of a filter is optimized as
far as possible; otherwise, a very inferior performance will result. An important paper in this field
is that by Baumeister et al. [37]. Aluminum is the metal commonly used for this region and
measured, and computed results obtained by these workers for filters with aluminum layers are
shown in Figure 8.52. The performance achieved is satisfactory, and the agreement between
practical and theoretical curves is good.

Induced transmission filters have been the subject of considerable study by many workers.
Metal–dielectric multilayers were reviewed by MacDonald [38]. A useful account of induced
transmission filters was given by Lissberger [39]. Multiple-cavity induced transmission filters
were described by Maier [40]. An alternative design technique for metal–dielectric filters involving
symmetrical periods was published by Macleod [41]. Symmetrical periods for metal–dielectric
filter design were also used by McKenney [42] and Landau and Lissberger [43].

8.7 Measured Filter Performance

Not a great deal has been published on the measured performance of actual filters, and the main
source of information for a prospective user is always the literature issued by manufacturers. The
performance of current production filters tends to improve all the time, so that inevitably, such
information does not remain up to date for long. Two papers [44,45] quote the results of a number
of tests on commercial filters, and although they were written some time ago, they will still be
found useful sources of information.

Blifford examined the performance of the products of four different manufacturers, covering the region of 300–1000 nm. The variation of peak wavelength with the angle of incidence was found to be similar to the relationship already established in Equation 8.30. Unfortunately, information on the design and materials is lacking, so that the expression for the effective index cannot be checked. The sensitivities to tilt varied from $P = 0.22$ to $P = 0.51$, where P corresponds to the quantity $1/n^{*2}$ in Equations 8.30, 8.39, and 8.40. Blifford suggests that an average value of 0.35 for P, that is, an effective index of 1.7, would probably be a useful value to assume in any case where no other data were available. Changes in peak transmittance with angle of incidence were found, but were not consistent from one filter to another and should apparently always be measured for each individual filter. Possibly, the variations in the effect are due to the absorption filters that are used for sideband suppression and which, because they do not show any shift in edge wavelength with angle of incidence, may cut into the passband of the interference section at large angles of incidence. In most cases examined, the change in peak transmittance was less than 10% for angles of 5–10°.

The variation in peak transmittance over the surface of the filter was also measured in a few cases. For a typical filter with a peak wavelength of 500 nm and a bandwidth not explicitly mentioned, but probably 2.1 nm (from information given elsewhere in the paper), the extremes of peak transmission were 54% and 60%. This is, in fact, one aspect of a variation of peak wavelength, bandwidth, and peak transmittance that frequently occurs, although the magnitude can range from very small to very large. The cause is principally the adsorption of water vapor from the atmosphere before a cover slip can be cemented over the layers, and it is dealt with in greater detail in Chapter 14. Infrared filters appear to suffer less from this defect than visible and near infrared filters. We note that the more modern and stable tantala/silica filters deposited by one of the energetic processes mostly eliminate this adsorption problem.

Another parameter measured by Blifford was the variation of peak wavelength with temperature. The variation of the temperature from −60°C to +60°C resulted in changes of peak wavelength from +0.01 to +0.03 nm/°C. The relationship was found to be linear over the whole of this temperature range with little, if any, change in the passband shape and peak transmittance. In most cases, the temperature coefficients of bandwidth and peak transmittance were found to be less than $0.01°C^{-1}$. Filters for the visible region were also the subject of a detailed study by Pelletier et al. [46]. The shift with temperature for any filter is a function of the coefficients of optical thickness change with temperature, depending on the design of the filter and especially on the material used for the cavities. Measurements made on different filter designs yielded the following coefficients of optical thickness for the individual layer materials:

- Zinc sulfide: $(4.8 \pm 1.0) \times 10^{-5}°C^{-1}$
- Cryolite: $(3.1 \pm 0.7) \times 10^{-5}°C^{-1}$

Hysteresis is frequently found with temperature cycling of narrowband filters over an extended temperature range. The hysteresis is particularly pronounced when the filters are uncemented and when they are heated toward 100°C. It is usually confined to the first temperature cycle. It normally takes the form of a shift of peak wavelength toward shorter wavelengths and is caused by the desorption of water, discussed again in Chapter 14.

Our ideas of the effects of temperature on optical coatings were completely revolutionized by an important paper by Takahashi [47]. Films that have been deposited by the energetic processes (Chapter 13) usually exhibit lower temperature coefficients than those that have been thermally evaporated, even when the effects of moisture desorption and adsorption are discounted. This is at first sight a quite surprising result. But the explanation appears to lie in the microstructure. The lateral thermal expansion of the loosely packed columns in the thermally evaporated films appears to enhance the drifts due to temperature changes. In the energetically deposited films, the material is virtually bulk-like, in that there are no voids in between any residual columns, and so the material exhibits bulk-like properties. In the tantala/silica combination, it is also amorphous. The change in characteristics with a change in temperature now corresponds to what would be expected from bulk materials. Indeed, Takahashi [47] showed that for multiple-cavity narrowband filters, and surely for other filter types, once the design and materials are chosen, the expansion coefficient of the substrate dominates the behavior and can even change the sense of the induced spectral shift. The strain induced in the coating by the differential lateral expansion and contraction of substrate and coating is translated by Poisson's ratio into a swelling or contraction normal to the film surfaces. The indices also change through strain birefringence. If, as usual, the expansion coefficients of film materials are higher than those of the substrate, the birefringence together with the film swelling increases the optical thickness and, with increasing temperature, moves the characteristic to a longer wavelength. Increasing the expansion coefficient of the substrate reduces the longwave shift. As a result of this modeling and improved understanding, temperature coefficients of peak wavelength shift at 1550 nm of 3 pm/°C (pm is picometer, i.e., 0.001 nm, so that this figure represents 0.0002%/°C) are routinely achieved in energetically deposited tantala/silica filters for communication purposes, and shifts significantly lower than 1 pm/°C, even virtually 0, are possible. We shall return to this model in more detail in Section 16.6.

An effect of a different kind, although related, is the subject of a contribution by Title [48] and Title et al. [49]. A permanent shift of a filter characteristic toward shorter wavelengths amounting to a few tenths of nanometers accompanied by a distortion of passband shape was produced by a high level of illumination. The filters were for the Hα wavelength (656.3 nm), and the changes were interpreted as due to a shift in the properties of the zinc sulfide material, the fundamental nature of the shift being unknown. Zinc sulfide can be transformed into zinc oxide by the action of ultraviolet light, especially in the presence of moisture, and the shifts that were observed could have possibly been caused by such a mechanism.

The possibility of variations in filter properties both over the surface of the filter and as a function of time, temperature, and illumination level should clearly be borne in mind in the designing of apparatus incorporating filters.

A useful survey that compares the performance achievable from different types of narrowband filters was the subject of a report by Baumeister [50].

Baker and Yen [44] studied the variations in the properties of infrared filters with angle of incidence and temperature and reported both theoretical and experimental results. The accurate calculation of the effects of changes in the angle of incidence yielded a variation of peak wavelength of the expected form, but no significant variation of bandwidth for angles of incidence of up to 50°. They also calculated that the peak transmittance and the shape of the passband should remain unchanged for angles up to 45°. For angles above 50°, both the shape and the peak transmittance gradually deteriorated. The calculations were confirmed by measurements on real filters. The effects of varying temperatures were also investigated both theoretically and practically. As in the case of the shorter wavelength filters examined by Blifford, they measured a shift toward longer wavelengths with increasing temperature. For temperatures down to liquid helium, the filters show little loss of peak transmittance or variation of characteristic passband shape. However, serious losses in transmittance occurred above 50°C. Although not mentioned in the paper, this is probably due to the use of germanium, either as substrate or as one of the layer materials. Germanium always exhibits a marked fall in transmittance at elevated temperatures above 50°C. Baker and Yen make the point that filters designed to be least sensitive to variations in the angle of incidence are usually most sensitive to temperature and vice versa. Again, this is likely to be due to the use of germanium as a high-index cavity. The temperature coefficients of peak wavelength that they quote vary from +0.0035 to +0.0125%/°C. Unfortunately, neither the materials used in the filters nor the designs are quoted in the paper, but it is likely that the figures will apply to many interference filters for the infrared.

Similar measurements of the temperature shift of infrared filters were made at Grubb Parsons. The materials used were zinc sulfide and lead telluride, and the filters that had first-order high-index cavities gave temperature coefficients of peak wavelength of −0.0135%/°C. These filters were long-wavelength narrowband filters for use in a satellite-borne selective chopper radiometer for atmospheric temperature sounding. A negative temperature coefficient is usual with filters having lead telluride as one of the layer materials. This negative coefficient in lead telluride is especially useful as it tends to compensate for the positive coefficient in zinc sulfide, and Evans et al. [51] succeeded in designing and constructing filters using lead telluride, which have essentially zero temperature coefficient. A more recent and very detailed study of the temperature sensitivity of infrared filters containing lead telluride is that by Stolberg-Rohr and Hawkins [52]. This paper contains much useful information.

References

1. L. I. Epstein. 1952. The design of optical filters. *Journal of the Optical Society of America* 42:806–810.
2. B. E. Perilloux. 2002. *Thin Film Design: Modulated Thickness and Other Stopband Design Methods*, vol. TT57. Bellingham, WA: SPIE Optical Engineering Press.
3. A. F. Turner. 1950. Some current developments in multilayer optical films. *Journal de Physique et le Radium* 11:443–460.
4. B. Bates and D. J. Bradley. 1966. Interference filters for the far ultraviolet (1700A to 2400A). *Applied Optics* 5:971–975.
5. J. S. Seeley. 1964. Resolving power of multilayer filters. *Journal of the Optical Society of America* 54:342–346.
6. D. J. Hemingway and P. H. Lissberger. 1973. Properties of weakly absorbing multilayer systems in terms of the concept of potential transmittance. *Optica Acta* 20:85–96.
7. J. A. Dobrowolski. 1959. Mica interference filters with transmission bands of very narrow half-widths. *Journal of the Optical Society of America* 49:794–806.
8. R. R. Austin. 1972. The use of solid etalon devices as narrow band interference filters. *Optical Engineering* 11:65–69.
9. M. Candille and J. M. Saurel. 1974. Réalisation de filtres "double onde" a bandes passantes très étroites sur supports en matière plastique (mylar). *Optica Acta* 21:947–962.
10. S. D. Smith and C. R. Pidgeon. 1963. Application of multiple beam interferometric methods to the study of CO_2 emission at 15µm. *Mémoires de la Société Royale de Science de Liège 5ième Serie* 9:336–349.
11. A. E. Roche and A. M. Title. 1974. Tilt tunable ultra narrow-band filters for high resolution photometry. *Applied Optics* 14:765–770.
12. J. Floriot, F. Lemarchand, and M. Lequime. 2006. Solid-spaced filters: An alternative for narrow-bandpass applications. *Applied Optics* 45:1349–1355.
13. C. Dufour and A. Herpin. 1954. Applications des methodes matricielles au calcul d'ensembles complexes de couches minces alternees. *Optica Acta* 1:1–8.
14. P. H. Lissberger. 1959. Properties of all-dielectric filters: I. A new method of calculation. *Journal of the Optical Society of America* 49:121–125.
15. P. H. Lissberger and W. L. Wilcock. 1959. Properties of all-dielectric interference filters: II. Filters in parallel beams of light incident obliquely and in convergent beams. *Journal of the Optical Society of America* 49:126–130.
16. C. R. Pidgeon and S. D. Smith. 1964. Resolving power of multilayer filters in nonparallel light. *Journal of the Optical Society of America* 54:1459–1466.
17. G. Hernandez. 1974. Analytical description of a Fabry Perot spectrometer. 3: Off-axis behaviour and interference filters. *Applied Optics* 13:2654–2661.
18. A. F. Turner. 1952. Wide passband multilayer filters. *Journal of the Optical Society of America* 42:878(a).
19. S. D. Smith. 1958. Design of multilayer filters by considering two effective interfaces. *Journal of the Optical Society of America* 48:43–50.
20. Z. Knittl. 1965. Dielektrische Interferenzfilter mit rechteckigem Maximum. In *Proceedings of the Colloquium on Thin Films*, pp. 153–161, Budapest.

21. Z. Knittl. 1976. *Optics of Thin Films.* London, Sydney, New York and Toronto: John Wiley & Sons; Prague: SNTL.

22. A. Thelen. 1966. Equivalent layers in multilayer filters. *Journal of the Optical Society of America* 56:1533–1538.

23. P. Baumeister. 2001. Design of a coarse WDM bandpass filter using the Thelen bandpass design method. *Optics Express* 9:652–657.

24. P. Baumeister. 2003. Application of microwave technology to design an optical multilayer bandpass filter. *Applied Optics* 42:2407–2414.

25. P. W. Baumeister. 1982. Use of microwave prototype filters to design multilayer dielectric bandpass filters. *Applied Optics* 21:2965–2967.

26. D. H. Cushing. 2011. *Enhanced Optical Filter Design.* Bellingham, WA: SPIE Press.

27. A. Thelen. 1988. *Design of Optical Interference Coatings.* First ed. New York: McGraw-Hill.

28. E. E. Barr. 1974. Visible and ultraviolet bandpass filters. *Proceedings of SPIE* 50:87–118.

29. R. G. T. Neilson and J. Ring. 1967. Interference filters for the near ultra-violet. *Journal de Physique* 28: C2,270–C2,275.

30. A. Malherbe. 1974. Interference filters for the far ultraviolet. *Applied Optics* 13:1275–1276.

31. P. W. Baumeister and F. A. Jenkins. 1957. Dispersion of the phase change for dielectric multilayers: Application to the interference filter. *Journal of the Optical Society of America* 47:57–61.

32. P. W. Baumeister, F. A. Jenkins, and M. A. Jeppesen. 1959. Characteristics of the phase-dispersion interference filter. *Journal of the Optical Society of America* 49:1188–1190.

33. P. Giacomo, P. W. Baumeister, and F. A. Jenkins. 1959. On the limiting bandwidth of interference filters. *Proceedings of the Physical Society* 73:480–489.

34. F. S. Ritchie. Unpublished work on Ministry of Technology Contract KX/LSO/C.B.70(a).

35. G. Hass and L. Hadley. 1972. Optical constants of metals. In *American Institute of Physics Handbook*, D. E. Gray (ed.), pp. 6.124–6.156. New York: McGraw-Hill.

36. P. H. Berning and A. F. Turner. 1957. Induced transmission in absorbing films applied to band pass filter design. *Journal of the Optical Society of America* 47:230–239.

37. P. W. Baumeister, V. R. Costich, and S. C. Pieper. 1965. Bandpass filters for the ultraviolet. *Applied Optics* 4:911–913.

38. J. MacDonald. 1971. *Metal-Dielectric Multilayers.* London: Adam Hilger.

39. P. H. Lissberger. 1981. Coatings with induced transmission. *Applied Optics* 20:95–104.

40. R. L. Maier. 1967. 2M interference filters for the ultraviolet. *Thin Solid Films* 1:31–37.

41. H. A. Macleod. 1978. A new approach to the design of metal-dielectric thin-film optical coatings. *Optica Acta* 25:93–106.

42. D. B. McKenney. 1969. Ultraviolet interference filters with metal-dielectric stacks. PhD Thesis. Tucson, AZ: University of Arizona.

43. B. V. Landau and P. H. Lissberger. 1972. Theory of induced-transmission filters in terms of the concept of equivalent layers. *Journal of the Optical Society of America* 62:1258–1264.

44. M. L. Baker and V. L. Yen. 1967. Effects of the variation of angle of incidence and temperature on infrared filter characteristics. *Applied Optics* 6:1343–1351.

45. I. H. Blifford. 1966. Factors affecting the performance of commercial interference filters. *Applied Optics* 5:105–111.

46. E. Pelletier, P. Roche, and L. Bertrand. 1974. On the limiting bandwidth of interference filters: influence of temperature during production. *Optica Acta* 21:927–946.

47. H. Takashashi. 1995. Temperature stability of thin-film narrow-bandpass filters produced by ion-assisted deposition. *Applied Optics* 34:667–675.

48. A. M. Title. 1974. Drift in interference filters: 2. Radiation effects. *Applied Optics* 13:2680–2684.

49. A. M. Title, T. P. Pope, and J. P. Andelin. 1974. Drift in interference filters. *Part 1. Applied Optics* 13:2675–2679.

50. P. W. Baumeister. 1973 March. Thin films and interferometry. *Applied Optics* 12:1993–1994.

51. C. S. Evans, R. Hunneman, J. S. Seeley et al. 1976. Filters for ν2 band of CO_2: Monitoring and control of layer deposition. *Applied Optics* 15:2736–2745.

52. T. Stolberg-Rohr and G. J. Hawkins. 2015. Spectral design of temperature-invariant narrow bandpass filters for the mid-infrared. *Optics Express* 23:580–596.

9

Tilted Dielectric Coatings

9.1 Introduction

We have already seen in Chapter 2 that the characteristics of coatings change when they are tilted with respct to the incident illumination, and the degree of change depends on the angle of incidence. We have also studied the shifts that are induced in narrowband filters. Narrowband filters are a simple case because the tilt angle is usually small, and we can assume that the major effect is in the phase thickness of the layers, which is equally affected for each plane of polarization. For larger tilts, however, the admittances are also affected, and then the performance for each plane of polarization differs. Some important applications take advantage of the difference in the performance between one plane and the other, which can be controlled to some extent, making the construction of phase retarders and polarizers possible. On the other hand, the differences in performance can create problems, and although it is impossible to completely cancel the effects, there are ways of modifying it so that a more acceptable performance may be achieved. With purely dielectric layers, tilting is certainly a complication, but when metal layers are added, the behavior is still more complicated and can even be rather strange. In this chapter, therefore, we concentrate on the dielectric case. Metals will be included in the following chapter.

We will begin with the effect on phase thickness and then consider the addition of tilting effects to the admittance diagram, which allows us to qualitatively explain the behavior of many different types of tilted coatings and which involves a slight modification to the traditional form of the tilted admittances.

Polarization splitting, a marked difference between p- and s-polarized performance parameters, is a common consequence of tilting. The splitting can be encouraged, to yield polarizing components, or discouraged, to suppress, at least partially, polarization sensitivity. Some of the material in this chapter has already been mentioned and discussed in earlier chapters, but there are advantages in attempting a consistent and connected account.

9.2 Tilted Thicknesses

The phase thickness δ is one of the two parameters that characterize the thin film, the other being the (tilted) admittance. At normal incidence, and for a dielectric film, the phase thickness is given by

$$\delta = \frac{2\pi n d}{\lambda}, \tag{9.1}$$

where the symbols have their usual meaning. At oblique incidence where the propagation angle within the film is ϑ, the phase thickness becomes

$$\delta = \frac{2\pi n d \cos \vartheta}{\lambda}, \tag{9.2}$$

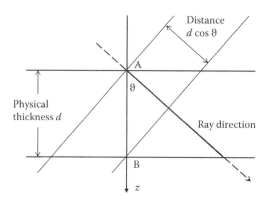

FIGURE 9.1
The physical thickness of the film is d. A and B are the reference points between which the phase change will be calculated. The two planes of constant phase that cut through A and B are shown, and the corresponding distance along the direction of propagation of the ray is $d \cos \vartheta$. The related change in phase is therefore $2\pi n d \cos \vartheta / \lambda$.

and it is clear that the effect of the tilting has been to reduce the value of δ. This can be a somewhat surprising result at first sight because the slant depth of the film is clearly greater, but we are dealing with the change in phase of the light on its passage through the film. In Chapter 2, we saw that the important change was along the z-axis rather than along the direction of the wave. This is such an important result that perhaps some further explanation is in order.

When we carry out an interference calculation, we are adding the electric and magnetic fields of the various waves involved in the interference. These must be added at exactly the same point; otherwise, the sum is invalid. Since the point where we accomplish this sum is on the interface on the side of incidence and since the plane wave solution is independent of the values of x and y, we can most conveniently choose the point where the z-axis crosses the interface as the calculation point. We can call this point A. The light undergoes a double traversal of the film, and to correctly calculate the change in phase, we must also choose a reference point on the rear surface of the film that we can call B. The phase change is the that due to the traversal AB together with that due to the return traversal BA. We can save ourselves a deal of trouble if we make the choice of B such that the traversals are symmetrical. This results in the arrangement shown in Figure 9.1. The distance measured along the direction of propagation between the planes of constant phase that intersect reference points A and B is $d \cos \vartheta$, clearly less than the physical thickness of the film. This must be multiplied by $2\pi n / \lambda$ to yield the phase thickness δ.

We recall that the phase thickness of an isotropic film is insensitive to polarization.

9.3 Modified Admittances

The form of the admittances of a film illuminated at oblique incidence are given in Chapter 2 and have already been used in considering the performance of some coatings including narrowband filters. For dielectric materials, they are

$$\eta_s = y \cos \vartheta, \tag{9.3}$$

$$\eta_p = \frac{y^2}{\eta_s} = \frac{y}{\cos \vartheta}, \tag{9.4}$$

where

$$n_0 \sin \vartheta_0 = n_1 \sin \vartheta_1 = n_2 \sin \vartheta_2 = \ldots, \tag{9.5}$$

and n, y, and ϑ are the values appropriate to the particular material, the subscript 0 indicating, as usual, the incident medium.

The calculation of multilayer properties at angles of incidence other than normal involves the use of the preceding expressions instead of those for normal incidence. It should be emphasized that the appropriate tilted values are to be adopted for the incident medium and substrate as well as for the films. This does not present any problem for our computer calculations, which can accommodate this without difficulty, but our sense of reflectance, and our appreciation of the consequent properties of our coating, is also dependent on the admittance of the incident medium and that this changes every time we alter the angle of incidence or change to the alternate polarization complicates our understanding. In the admittance diagram, our isoreflectance and isophase contours depend on the admittance of the incident medium and on the polarization. We therefore need new sets of values for s-polarization and quite different sets for p-polarization each time the angle of incidence is changed. This, too, hinders our understanding. Fortunately, we can considerably ease the situation, although complexity remains and we cannot completely cure it.

It has been shown by Thelen [1] that the properties of a multilayer are unaffected if all the admittances are multiplied or divided by a constant factor, and indeed, it is usual to divide the admittances by Y, the admittance of free space, so that the normal incidence admittance is numerically equal to the refractive index. We now propose an additional correction to the admittances, the division of s-polarized admittances, and the multiplication of p-polarized admittances, by $\cos \vartheta_0$. This has the effect of preserving, for both s- and p-polarizations, the admittance of the incident medium at its normal incidence value, regardless of the angle of incidence, and means that our appreciation of reflectance and the isoreflectance and isophase contours of the admittance diagram retain their normal incidence values whatever the angle of incidence or plane of polarization. We can simply call these admittances the modified admittances, and the expressions for them become for dielectric materials

$$\eta_s = \frac{y \cos \vartheta}{\cos \vartheta_0} \tag{9.6}$$

and

$$\eta_p = \frac{y \cos \vartheta_0}{\cos \vartheta}. \tag{9.7}$$

The values of reflectance, transmittance, absorptance, and phase changes on either transmission or reflection are completely unchanged by the adoption of these values for the admittances. Since the expressions involve $\cos \vartheta_0$ and $\cos \vartheta$, which are connected by the admittance of the incident medium, then the dependence of the modified admittances on the index of the incident medium will be somewhat different from the unmodified, traditional ones. Nevertheless, we shall see that this does carry some advantages. Note, however, that we do not make any change to the phase thickness.

In dielectric materials, provided that $n_0 \sin \vartheta_0$ is less than n, the film index, the two values for the modified admittances are real and positive. If, however, n_0 is greater than n, then there is a real value of ϑ_0 at which $n_0 \sin \vartheta_0$ is equal to n. This angle is known as the critical angle (see Chapter 2), and for angles of incidence greater than this value, we need a different approach to the admittances.

Let us consider air of index unity as the incident medium, at least initially. We recall that all transparent thin-film materials have refractive index greater than unity. In Figure 9.2, the modified admittance is shown for a number of thin-film materials as a function of angle of incidence. The p-admittances of all materials cross line $n = 1$ at the value known as the Brewster angle for which

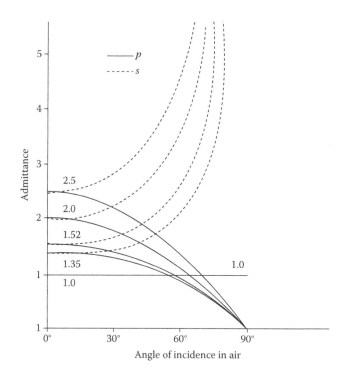

FIGURE 9.2

Modified p- and s-admittances (i.e., including the extra factor of $\cos\vartheta_0$) of materials of indices 1.0, 1.35, 1.52, 2.0, and 2.5 for an incident medium of index 1.0.

the single-surface p-reflectance is zero. The s-admittances all increase away from line $n = 1$, so that the single-surface s-reflectance simply increases with the angle of incidence. There can be, of course, Brewster angles for some pairs of the lower index materials at still higher angles of incidence, difficult to see in the diagram. Since all these materials are dielectric with indices higher than that of the incident medium, their modified optical thickness is real, and although a correction has to be made for the effect of angle of incidence, quarter- and half-wave layers can be produced at nonnormal incidence just as readily as at normal. It cannot be too greatly emphasized that although the optical thickness changes with angle of incidence, it does not vary with the plane of polarization.

It is possible to make several deductions directly from Figure 9.2. The first is that for any given pair of indices, the ratio of the s-admittances increases with angle of incidence, while that for p-admittances reduces. Since the width of the high-reflectance zone of a quarter-wave stack decreases with decreasing ratio of these admittances, the width will be less for p-polarized light than that for s-polarized. As we shall shortly see, this effect is used in a useful type of polarizer. The splitting of the admittance of dielectric layers also means that there is a relative phase shift between p- and s-polarized lights reflected from a high-reflectance coating when the layers depart from quarter-waves. This effect can be used in the design of phase retarders, and we will include a brief account of them. The diagram also helps us consider the implications of antireflection coatings for high angles of incidence. A frequent requirement is an antireflection coating for a crown glass of index around 1.52. For a perfect single-layer coating, we should have a quarter wave of material of optical admittance equal to the square root of the product of the admittances of the glass and the incident medium. At normal incidence in air, there is, of course, no sufficiently robust material with index as low as 1.23. For greater angles of incidence, the s-polarized reflectance increases still further from its normal incidence value, and the admittance required for a perfect single-layer antireflection coating remains outside the range of practical materials, corresponding to still lower indices of refraction. The p-polarized behavior is, however, completely different, and in the range from approximately 50° to 70°, the admittance required for the anti-reflection coating is within the range of what is possible. No coating is required, of course, at the Brewster angle. For angles greater than the Brewster angle, the index required is *greater* than that for the glass. Note, however, that the Brewster condition implies two solutions to the antireflection coating for *p*-polarization. The other solution usually corresponds to a very low unobtainable index. Antireflection coatings for high angles of incidence will also be discussed shortly.

The behavior of dielectric materials when the incident medium is of a higher index (one that is within the range of available thin-film materials) is somewhat more complicated. Figure 9.3 illustrates the variation of the modified admittances when the incident medium is a glass of index 1.52. There is the familiar splitting of the *s*- and *p*-polarized admittances, increasing, as before, with angle of incidence. For indices lower than that of the glass, it is possible to reach the critical angle, and at that point, the admittances reach either zero or infinity and disappear from the diagram. We will return later to the behavior beyond the critical angle. A further very important feature is that while for indices higher than that of the incident medium, the *p*-polarized modified admittance falls with angle of incidence; for indices lower than the incident medium, the *p*-polarized modified admittance rises. All cut the incident medium admittance at the Brewster angle, but now a new phenomenon is apparent. The *p*-admittance curves for materials of index lower than that of the incident medium intersect the curves corresponding to higher indices. They exhibit a Brewster angle. An immediate deduction is that a quarter-wave stack, composed of such pairs of materials, will simply behave, at the angle of incidence corresponding to the point of intersection, as a thick slab of material. Provided the admittances of substrate, thin films and incident medium are not too greatly different, the *p*-reflectance will be low. The ratio of the *s*-admittances is large, because their splitting increases with angle of incidence, and so the corresponding *s*-reflectance is high, and the

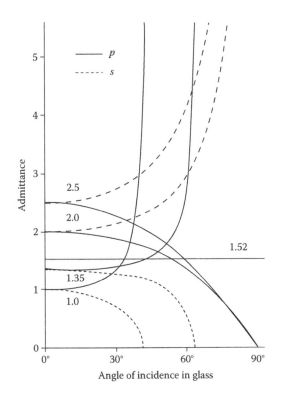

FIGURE 9.3

Modified *p*- and *s*-admittances (i.e., including the extra factor of $\cos \vartheta_0$) of materials of indices 1.0, 1.35, 1.52, 2.0, and 2.5 for an incident medium of index 1.52.

width of the high-reflectance zone, large. This is the basic principle of the MacNeille [2] polarizing beam splitter that we analyze in a later section. The range of useful angles of incidence will partly depend on the rate at which the curves of p-polarized admittance diverge on either side of the intersection, and this can be estimated from the diagram.

Apart from the polarization-splitting of the admittance, the behavior of dielectric layers at angles of incidence less than critical is reasonably straightforward and does not involve difficulties of a more severe order than what exist at normal incidence. The behavior becomes rather more complicated, and interesting, with angles beyond critical, and then, when metal films are introduced, however, the behavior becomes still stranger especially when metal and dielectric films are combined. We are leaving metal films for the next chapter.

We already know that the admittance locus of a dielectric layer at normal incidence is a circle centered on the real axis. Tilted dielectric layers at angles of incidence less than critical still have circular loci that can be calculated from the tilted admittances in exactly the same way. Provided the modified admittances are used in constructing the loci, then the isoreflectance and isophase circles on the admittance diagram will remain exactly the same as at normal incidence for both p- and s-polarizations.

9.4 Polarizers and Analyzers

Polarizers and analyzers are essentially the same components used in different roles. A polarizer converts an unpolarized ray into, usually, a linearly polarized one, while an analyzer is a polarizer that is used to identify the orientation of an already linearly polarized ray. We concentrate on linear polarizers and analyzers, but we note that there are devices, usually involving a combination of components, which can manipulate polarization in more complicated ways. One example is the conversion of unpolarized light into a circularly polarized ray of a given handedness.

The extinction ratio, that is, the ratio of the irradiance of the undesired mode of polarization to that of the desired (although extinction ratio is sometimes defined as the reciprocal), is a guide to the performance of a polarizer, but the extinction ratio takes no account of the level of throughput of the device, and so we must supplement it by some measure of efficiency, usually the transmittance, or the reflectance if appropriate. Another older, and rather less useful, performance parameter is polarizing efficiency defined as $(I_{desired} - I_{unwanted})/(I_{desired} + I_{unwanted})$ and usually expressed as a percentage.

High-performance polarizers depend on birefringent crystals for their operation. In a common arrangement, the crystal presents to the incident light a refractive index that depends on the orientation of the polarization. Unpolarized light incident on the crystal penetrates it to reach a tilted surface. The surface tilt is arranged so that for the greater of the two refractive indices, the surface is slightly beyond critical angle, and for the lower index, slightly below critical. Then that portion of the incident light with appropriate orientation of polarization is totally internally reflected, while the other portion is largely transmitted. The transmitted light is then returned to its original direction by a second similar prism separated from the first by a suitable gap, often simply of air. In a different arrangement, the two principal planes of polarization emerge in slightly different directions, and a second prism assures that only one mode finally emerges. The unwanted light is usually absorbed at a blackened surface. For crystal polarizers, an extinction ratio of 10^{-6} is readily attainable.

Crystal polarizers are limited in size by the availability of suitable crystals. Thin-film polarizers are limited in size by the dimensions of the coating machines. The thin-film polarizer also presents advantages in cost. But there are limitations in performance associated with the thin-film polarizers, limitations that will be examined in greater detail in a later chapter (Section 16.2.4), and so the thin-film polarizer is not, in general, a replacement for the crystal polarizer. Except in a few cases, it should be considered as a different component with different applications.

9.4.1 Plate Polarizer

The width of the high-reflectance zone of a quarter-wave stack is a function of the ratio of the admittances of the two materials involved. This ratio varies with the angle of incidence and is different for s- and p-polarizations. We recall that

$$\eta_s = n\cos\vartheta \quad \text{while} \quad \eta_p = n/\cos\vartheta$$

so that

$$\eta_{Hs}/\eta_{Ls} = \cos\vartheta_H/\cos\vartheta_L$$

and

$$\eta_{Hp}/\eta_{Lp} = \cos\vartheta_L/\cos\vartheta_H,$$

whence

$$\frac{(\eta_H/\eta_L)_s}{(\eta_H/\eta_L)_p} = \frac{\cos^2\vartheta_H}{\cos^2\vartheta_L}. \tag{9.8}$$

The factor $\cos^2\vartheta H/\cos^2\vartheta L$ is always less than unity so that the width of the high-reflectance zone for p-polarized light is always less than that for s-polarized light. Within the region outside the p-polarized but inside the s-polarized high-reflectance zone, the transmittance is low for s-polarized light but high for p-polarized so that the component acts as a polarizer. The region is quite narrow, so that such a polarizer will not operate over a wide wavelength range, but for single wavelengths, such as a laser line, it can be very effective. To complete the design of the component, it is necessary to reduce the ripple in transmission for p-polarized light, and this can be performed using any of the techniques of Chapter 7, and it is the performance right at the edge of the pass region that is important. Computer refinement of the few outermost layers will usually work well. It is normal to use the component as a longwave-pass filter because this involves thinner layers and less material than would a shortwave-pass filter. The rear surface of the component requires an antireflection coating for p-polarized light. We can omit this altogether if the component is used at the Brewster angle. The design of such a polarizer, using shifted periods as matching structures, is described by Songer [3], who gives the design shown in Figure 9.4. Plate polarizers are used in preference to the prism, or MacNeille, type when high powers are concerned. Plate polarizers arranged at the Brewster angle for the rear surface are commonly used to clean up the polarization of high-power lasers.

Virtually any coating possessing a sharp edge between transmission and reflection can potentially be used as a polarizer. It has been suggested that narrowband filters have advantages over simple quarter-wave stacks as the basis of plate polarizer coatings, because the monitoring of the component during deposition is a more straightforward procedure [4]. With such polarizers, high power can be more problematic because of the greater field magnification inherent in the cavity structures.

The plate polarizer reflects what it does not transmit. Thus, it can be considered to be also a polarizing beam splitter. The p-performance in terms of extinction ratio is largely a matter of the total number of layers, but there is always a difficulty in achieving precisely 100% transmittance for p-polarization. Ripple reduction will usually leave a residual reflectance of at least a fraction of a percentage. Thus, although an extinction ratio of 10^{-6} or better is readily achievable for p-polarization, it is difficult to reach an extinction ratio of even 10^{-3} for s-polarization.

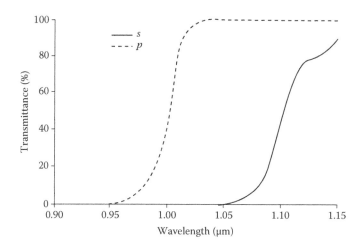

FIGURE 9.4
Characteristic curve of a plate polarizer for 1.06 μm. Design: Air | $(0.5H'\,L'\,0.5H')^3\,(0.5H''\,L''\,0.5H'')^8\,(0.5H'\,L'\,0.5H')^3$ | Glass, where $H' = 1.010H$, $L' = 1.146L$, $H'' = 1.076H$, and $L'' = 1.220L$ and with $n_H = 2.25$, $n_L = 1.45$, $\lambda_0 = 0.9$ μm, and $\vartheta_0 = 56.5°$. The solid line indicates s-polarization, and the dashed line, p-polarization. (After L. Songer, *Optical Spectra*, 12, 45–50, 1978.)

9.4.2 Cube Polarizers

An advantage of the polarizer immersed in a prism is that the effective angle of incidence can be very high—much higher than if the incident medium were air. This enhances the polarization splitting and gives broader regions of a high degree of polarization than could be the case with air as the incident medium. Thus, there is an advantage in using an immersed design, provided the incident power is not too high.

Netterfield [5] considered the advantages of a cube polarizer when only a very limited range of wavelengths is required. A symmetrical dielectric structure where odd and even layers have identical thickness, although not necessarily the same for odd as for even, will, when incident and emergent media are of equal index, present at least one wavelength where the transmittance is 100%. Tilting this structure separates the p and s-performances so that their maxima do not coincide. If the structure is a quarter-wave stack, or similar, then the high-transmission wavelength will be at the edge of the high-reflectance zone where the broader s-characteristic implies that s-transmittance will be low when the p-transmittance is maximum. Such a polarizer needs no matching, and the high angle of incidence assures low s-transmittance with a modest number of layers. The prism structure assures large polarization splitting, helping the performance, and the remaining problem is any mismatch between the prism and the cement holding the assembly together, which can cause the appearance of unwanted fringes. Thus, the most important feature of Netterfield's approach is to make sure that the glass is chosen to match as accurately as possible the index of the cement. Then any suitable high and low-index pairs of layers can be chosen. The remainder of the design is a simple computer exercise.

9.4.3 Brewster Angle Polarizing Beam Splitter

For polarizers with much wider useful regions, we look to the Brewster phenomenon. This type of beam splitter was first constructed by Mary Banning [6] at the request of S. M. MacNeille, the inventor of the device [2], which is frequently known as a MacNeille polarizer.

The principle of the device is that it is always possible to find an angle of incidence so that the Brewster condition for an interface between two materials of differing refractive index is satisfied. When this is so, the reflectance for the p-plane of polarization vanishes. The s-polarized light is partially reflected and transmitted. To increase the s-reflectance, retaining the p-transmittance at or

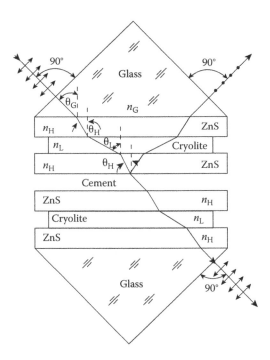

FIGURE 9.5
Schematic diagram of a polarizing beam splitter. (After M. Banning, *Journal of the Optical Society of America*, 37, 792–797, 1947. With permission of Optical Society of America.)

very near unity, the two materials may then be made into a multilayer reflecting stack. In the simplest case, the layer thickness should be quarter-wave optical thicknesses at the appropriate angle of incidence. When the Brewster angle for normal thin-film materials is calculated, it is found to be greater than 90° when referred to air as the incident medium. This presents a problem that is solved by building the multilayer filter into a glass prism so that the light can be incident on the multilayer at an angle greater than is possible with air as incident medium. Figure 9.5 shows the arrangement of the first Brewster angle polarizers ever made. Banning's original coating consisted of three layers, probably because of practical difficulties at that time. Two coatings were therefore prepared, one on the hypotenuse of each prism making up the cube, as shown in Figure 9.5. The modern approach takes advantage of the improvements in deposition techniques and places all the layers on one prism hypotenuse that is then cemented to the second, uncoated, prism. Any number of layers can be used in the coating. We suggest a quarter-wave stack structure for simplicity, but it will be obvious that the Brewster condition exists whatever the thickness of the individual layers, and a tapered broad-band stack will exhibit the same suppression of the *p*-reflectance as does the quarter-wave structure. The Brewster angle polarizer is certainly not limited to the quarter-wave stack design. For example, broader performance can be achieved with tapered stacks.

Let the incident medium be of index n_0. Then the angle of incidence in the incident medium in order to achieve the Brewster angle in the two materials n_1 and n_2 will be given by

$$\sin^2\vartheta_0 = \frac{n_1^2}{n_0^2}\sin^2\vartheta_1$$

$$= \frac{n_1^2}{n_0^2}\frac{\tan^2\vartheta_1}{1+\tan^2\vartheta_1} \tag{9.9}$$

$$= \frac{n_1^2 n_2^2}{n_0^2\left(n_1^2 + n_2^2\right)},$$

so that

$$n_0 = \frac{n_1 n_2}{\sin \vartheta_0 \left(n_1^2 + n_2^2\right)^{1/2}}.$$

(9.10)

Typical values of n_1 and n_2 require values of n_0 rather greater than unity for real values of ϑ_0. This implies that a polarizer based on this principle must be immersed in a medium of elevated index. In the most usual arrangement, the device is made up of two isosceles right-angle prisms, at least one having a coated hypotenuse, cemented into a cube. Note that it is impossible that the prism material should exhibit an admittance at the design angle equal to that of the two film materials. This would imply that a Brewster condition could apply to three different materials simultaneously. Thus, there will always be a mismatch, however slight, between the film structure and the prisms leading to a slight residual ripple. The ripple is normally very small but, for ultimate performance, can be reduced by a suitably designed simple antireflecting coating. The antireflection coating can use one of the materials already in the polarizing structure, but since both have exactly the same tilted admittance, the matching also requires a third material that is different from either of the two existing ones.

Banning [6] adopted zinc sulfide ($n = 2.30$) and cryolite ($n = 1.25$) as the materials, the cryolite being evaporated at rather high pressure to attain a very low index. To obtain an angle of incidence of 45° and, hence, a cube prism, she used glass of index 1.55. Banning correctly pointed out that the dispersion of the film materials would imply a variation in the value of n_0 from Equation 9.10. In order to preserve the angle of incidence and, hence, the optimum polarization, the glass of the prism should have the correct dispersion.

The *Abbe number* (or sometimes the V-number) is defined as

$$V = \frac{n_D - 1}{n_F - n_C},$$

(9.11)

where the subscripts D, F, and C indicate that the indices are measured at the corresponding Fraunhofer lines at 589.2, 486.1, and 656.3 nm, respectively. Note that the Abbe number is greater the smaller the degree of dispersion. A high Abbe number implies low dispersion. In the visible region, low-index films have usually quite small dispersion that can be neglected. The high-index films show much greater dispersion. If we assume, therefore, that only one film material has perceptible dispersion and, further, that the denominator of Equation 9.11 can be replaced by a differential, then we can treat n_0 and n_1 in Equation 9.10 as the variables and differentiate it to give

$$\Delta n_0 = \frac{\Delta n_1}{\sin \vartheta_0} \left[\frac{n_2}{\left(n_1^2 + n_2^2\right)^{1/2}} - \frac{n_1^2 n_2}{\left(n_1^2 + n_2^2\right)^{3/2}} \right].$$

(9.12)

We can then, using Equation 9.9, replace $\sin \vartheta_0$ to yield

$$\frac{n_0 - 1}{\Delta n_0} = V_0 = V_1 \frac{(n_0 - 1)}{(n_1 - 1)} \frac{\left(n_1^2 + n_2^2\right) n_1}{n_0 n_2^2},$$

(9.13)

where V_0 and V_1 can actually be derived from Equation 9.11 so that they are the correct Abbe numbers of the glass and high-index material n_1, respectively.

We will use the expressions to design a cube polarizing beam splitter of high performance. We will assume tantalum oxide and silicon oxide materials and a design wavelength of 510 nm. The assumed optical constants are shown in Figure 9.6.

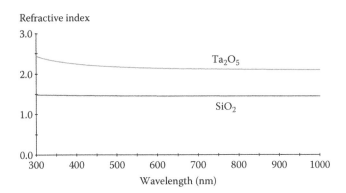

FIGURE 9.6
Refractive indices of the thin-film materials used in the design of the Brewster polarizer.

Using Equation 9.10 and values of 2.168 and 1.462 for the indices at 510 nm for Ta_2O_5 and SiO_2, we find 1.714 as the value of the ideal index of the glass. The dispersion of the SiO_2 is negligible, and the Abbe number we estimate for the Ta_2O_5 is 21.6. Equation 9.13 then suggests an Abbe number for the glass of 53.4. N-LAK 8 from the Schott catalog has index of 1.720 and an Abbe number of 53.8, both of which are closer to the requirements than we can usually expect.

The remainder of the design is straightforward. We create a quarter-wave stack with, in this case, the high-index layers outermost because they appear to give a slightly better match to the surrounding glass with its slightly higher than required index. Best performance is achieved with thicknesses that are tuned to the appropriate angle of incidence. There is no point in using more layers than necessary because the aperture of the polarizer fixes its limiting properties. We shall discuss this in greater detail in Chapter 16 (Section 16.2.4) where some performance limits are discussed. The design is then

$$\text{N-LAK8}|(1.208H\ 1.802L)^{10}\ 1.208H|\text{N-LAK8}, \tag{9.14}$$

where we assume that any cement layer has an index identical to that of the glass. Computer calculations show that at 510 nm, the extinction ratio is an impressive 2.3×10^{-7}, and the degree of polarization in transmission, 99.999954%. We have to correctly use at least five significant figures after the decimal point to state the degree of polarization. This is why it is not as useful as an extinction ratio.

We can also analytically derive the degree of polarization at the center wavelength.

$$R = \left[\frac{\eta_G - (\eta_H^2/\eta_G)(\eta_H/\eta_L)^{q-1}}{\eta_G + (\eta_H^2/\eta_G)(\eta_H/\eta_L)^{q-1}}\right]^2, \tag{9.15}$$

where q is the number of layers, and we are assuming q to be odd.

	For s-waves:	For p-waves:
	$\eta_G = n_G \cos\vartheta_G$	$\eta_G = n_G/\cos\vartheta_G$
	$\eta_H = n_H \cos\vartheta_H$	$\eta_H = n_H/\cos\vartheta_H$
	$\eta_L = n_L \cos\vartheta_L$	$\eta_L = n_L/\cos\vartheta_L$

Now, for p-waves, by the condition we have imposed, $\eta_H = \eta_L$ and

$$
\begin{aligned}
R_p &= \left[\frac{\eta_G - (\eta_H^2/\eta_G)}{\eta_G + (\eta_H^2/\eta_G)} \right]^2 \\
&= \left[\left(\frac{n_G^2 \cos^2 \vartheta_H}{n_H^2 \cos^2 \vartheta_G} - 1 \right) \bigg/ \left(\frac{n_G^2 \cos^2 \vartheta_H}{n_H^2 \cos^2 \vartheta_G} + 1 \right) \right]^2 .
\end{aligned}
\tag{9.16}
$$

Similarly,

$$
R_s = \left[\frac{n_G^2 \cos^2 \vartheta_G - n_H^2 \cos^2 \vartheta_H \left(\dfrac{n_H \cos \vartheta_H}{n_L \cos \vartheta_L} \right)^{q-1}}{n_G^2 \cos^2 \vartheta_G + n_H^2 \cos^2 \vartheta_H \left(\dfrac{n_H \cos \vartheta_H}{n_L \cos \vartheta_L} \right)^{q-1}} \right]^2 .
\tag{9.17}
$$

Now,

$$
\frac{n_H \cos \vartheta_L}{n_L \cos \vartheta_H} = 1,
$$

so that

$$
\frac{n_H \cos \vartheta_H}{n_L \cos \vartheta_L} = \frac{n_H^2}{n_L^2}
$$

and

$$
R_s = \left[\frac{n_G^2 \cos^2 \vartheta_G - n_H^2 \cos^2 \vartheta_H \left(\dfrac{n_H}{n_L} \right)^{2(q-1)}}{n_G^2 \cos^2 \vartheta_G + n_H^2 \cos^2 \vartheta_H \left(\dfrac{n_H}{n_L} \right)^{2(q-1)}} \right]^2 .
\tag{9.18}
$$

The degree of polarization in transmission is given by

$$
P_T = \frac{T_p - T_s}{T_p + T_s} = \frac{1 - R_p - 1 + R_s}{1 - R_p + 1 - R_s} = \frac{R_s - R_p}{2 - R_p - R_s},
\tag{9.19}
$$

and in reflection by

$$
P_R = \frac{R_s - R_p}{R_s + R_p}.
\tag{9.20}
$$

It can be seen that in general, for a small number of layers, the polarization in reflection is better than the polarization in transmission, but for a large number of layers, it is inferior to that in transmission.

The very great advantage that this type of polarizing beam splitter has over the other polarizers, such as the pile of plates, is its wide spectral range coupled with a large physical aperture. Unfortunately, it does suffer from a limited angular field, particularly at the center of its range, partly because the Brewster condition is met exactly only at the design angle. As the angle of incidence moves away from this condition, a residual reflectance peak for p-polarization gradually appears in the center of the range. This hardly affects the p-polarized extinction ratio, but the throughput is reduced. As an example, we can consider our Figure 9.7 polarizer. The performance at angles of incidence off normal is shown in Figure 9.8. Skew rays present a further difficulty. Polarization performance is measured with reference to the s- and p-directions associated with the principal plane of incidence containing the axial ray. A skew ray possesses a plane of incidence that is rotated

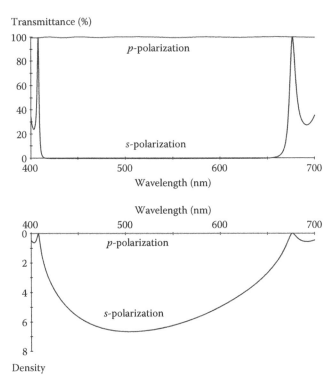

FIGURE 9.7
Transmittance and density in transmission of the Brewster angle polarizer.

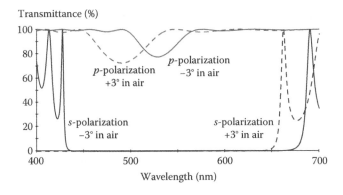

FIGURE 9.8
Performance of the polarizer of Figure 9.7 at incident angles on the prism entrance surface of ±3° in air. The positive angle increases the angle of incidence on the hypotenuse. The negative angle decreases it.

with respect to the principal plane. Thus, the s- and p-planes for skew rays are not quite those of the axial ray, and although the s-polarized transmittance can be very low, there can be a component of the p-polarized light, parallel to the axial s-direction and representing leakage that can be significant. In fact, given the apex angle of a cone of incident illumination, there is a limit to the degree of polarization that cannot be exceeded even by adding further layers to the design. We return to this question later in Chapter 16. The width of the region of high-polarization performance can, however, be extended by adding further reflecting structures or tapering the design in the manner of the extended zone reflectors.

The leakage can be reduced by moving to higher angles of incidence. Li and Dobrowolski [7] describe such a polarizer where one of the materials is operating beyond the critical angle. This also considerably improves the possible p-polarized performance that can be obtained. The disadvantage is that more glass is required than for smaller incidence angles.

A detailed study of the polarizing prism was carried out by Clapham [8].

Provided it can be accepted that the s-performance cannot be quite the equal of the p-performance, the thin-film polarizer can also serve as a polarizing beam splitter.

9.5 Nonpolarizing Coatings

The design of coatings that avoid polarization problems is a much more difficult task than that of polarizer design, and there is no completely effective method. The changes in the phase thickness of the layers and in the optical admittances are fundamental and cannot be avoided. The best we can hope to do, therefore, is to arrange a sequence of layers to give as far as possible similar performance for p- and for s-polarizations. Clearly, the wider the range of either the angle of incidence or the wavelength, the more difficult the task. The techniques currently available operate only over very restricted ranges of wavelength and angle of incidence (effectively over a very narrow range of angles). There is a small body of published work, but the principal analytical techniques that we shall use here heavily rely on techniques devised by Thelen [9,10].

9.5.1 Edge Filters at Intermediate Angle of Incidence

This section is entirely based on an important paper by Thelen [10]. However, the expressions found in the original paper have been altered in order to make the notation consistent with the remainder of this book. Care should be taken, therefore, in reading the original paper. In particular, the x found in the original is defined here in a slightly different way.

At angles of incidence not so severe that the p-reflectance suffers, the principal effect of operating edge filters at oblique incidence is the splitting between the two planes of polarization. This limits the edge steepness that can be achieved for unpolarized light. Edge filters with quite limited pass regions can be constructed from bandpass filters, but because bandpass filters also suffer from polarization splitting, the bandwidth for s-polarized light shrinking, and that for p-polarized light expanding, they still present the same problem. However, there is a technique that can displace the passbands of a bandpass filter to make one pair of edges coincide. This results in an edge filter of rather limited extent, which, for a given angle of incidence, has no polarization splitting. The position of the peak of a bandpass filter can be considered to be a function of both the cavity thickness and the phase shift of the reflecting stacks on either side. At oblique incidence, the relative phase shift between s- and p-polarized lights reflected by the stacks can be adjusted by adding or removing material. This alters the relative positions of the peaks of the passbands for the two planes of polarization, and if the adjustment is correctly made, it can make a pair of edges coincide. This, of course, is for one angle of incidence only. As the angle of incidence moves away from the design value, the splitting will gradually reappear.

Rather than apply this technique exactly as we have just described it, we instead adapt the techniques for the design of multiple-cavity filters based on symmetrical periods. Let us take a typical multiple-cavity filter design:

$$\text{Incident medium|matching (symmetrical stack)}^q\text{matching|substrate.}$$

The symmetrical stack that forms the basis of this filter can be represented as a single matrix with the same form as that of a single film, as we have already seen in Chapter 8. The limits of the passband are given by those wavelengths for which the diagonal terms of the matrix are unity and the off-diagonal terms are zero. That is, if the matrix is given by

$$
\begin{bmatrix} N_{11} & iN_{12} \\ iN_{21} & N_{22} \end{bmatrix},
$$

then the edges of the passband are given by

$$
N_{11} = N_{22} = \pm 1.
$$

The design procedure simply ensures that this condition is satisfied for the appropriate angle of incidence.

We can consider the symmetrical period as a quarter-wave stack of $2x + 1$ layers with two identical additional layers, one on either side:

$$
fBABAB \ldots AfB,
$$

where A and B indicate quarter-wave layers and f is a correction factor to be applied to the quarter-wave thickness to yield the thicknesses of the detuned outer layers. We can write the overall matrix as $fB\,M\,fB$, where $M = ABAB...A$, giving the product

$$
\begin{bmatrix} \cos\alpha & i\sin\alpha/\eta_B \\ i\eta_B\sin\alpha & \cos\alpha \end{bmatrix} \begin{bmatrix} M_{11} & iM_{12} \\ iM_{21} & M_{11} \end{bmatrix} \begin{bmatrix} \cos\alpha & i\sin\alpha/\eta_B \\ i\eta_B\sin\alpha & \cos\alpha \end{bmatrix}.
$$

Then, N_{11} is given by

$$
N_{11} = N_{22} = M_{11}\cos 2\alpha - 0.5(M_{12}\eta_B + M_{21}/\eta_B)\sin 2\alpha = \pm 1 \tag{9.21}
$$

for the edge of the zone for each plane of polarization. This must be simultaneously satisfied for both planes of polarization for the edges of the passbands to coincide. In fact, symmetrical periods made up of thicknesses other than quarter waves can also be used, when some trial and error will be required to satisfy Equation 9.21. A computer can be of considerable help. For quarter-wave stacks, we seek assistance in the expressions derived in Chapter 8 for a narrowband filter design. We use the analysis starting with Equation 8.58, with $m = 1$ and $q = 0$, giving

$$
M_{11} = (M_{22} =)(-1)^x(-\varepsilon)[(\eta_A/\eta_B)^x + \cdots + (\eta_B/\eta_A)^x],
$$

$$
iM_{21} = i(-1)^x/[(\eta_A/\eta_B)^x\eta_A], \tag{9.22}
$$

$$
iM_{12} = i(-1)^x[(\eta_B/\eta_A)^x/\eta_A],
$$

ε indicating, as before, a small departure from a quarter wave, that is, $\varepsilon = (\pi/2)(g-1)$. Note that $2x + 1$ is now the number of layers in the inner stack. The total number of layers, including the detuned ones, is $2x + 3$. Now, using exactly the same procedure as in Chapter 8, we can write expressions for the coefficients in Equation 9.21 as

$$
M_{11} = \frac{(-1)^x(-\varepsilon)(\eta_H/\eta_L)^x}{(1 - \eta_L/\eta_H)}
$$

$$
= (-1)^x(-\varepsilon)P
$$

and

$$
0.5(M_{12}\eta_B + M_{21}/\eta_B) = 0.5(-1)^x[(\eta_B/\eta_A)^{x+1} + (\eta_A/\eta_B)^{x+1}]
$$

$$
= (-1)^x Q,
$$

where

$$P = \frac{(\eta_H/\eta_L)^x}{(1 - \eta_L/\eta_H)} \quad \text{and} \quad Q = 0.5(\eta_H/\eta_L)^{x+1}.$$

Then the two equations become

$$\pm 1 = \varepsilon P_p \cos 2\alpha + Q_p \sin 2\alpha,$$
$$\pm 1 = \varepsilon P_s \cos 2\alpha + Q_s \sin 2\alpha,$$

$$(9.23)$$

and give for α and ε

$$\sin 2\alpha = \pm \frac{P_s - P_p}{\left(P_s Q_p - P_p Q_s\right)},$$

$$(9.24)$$

$$\varepsilon = \frac{\pm 1 - Q_p \sin 2\alpha}{P_p \cos 2\alpha}.$$

$$(9.25)$$

Now,

$$\varepsilon = (\pi/2)(1 - g), \quad \text{where} \quad g = \lambda_0/\lambda,$$

$$\alpha = (\pi/2)(\lambda_R/\lambda) = (\pi/2)(\lambda_R/\lambda_0)g = (\pi/2)fg,$$

so that

$$f = \alpha/(\pi g/2) = \alpha/[(\pi/2) - \varepsilon].$$

$$(9.26)$$

Two values for f will be obtained. Usually, the larger corresponds to a shortwave-pass filter, and the smaller, to a longwave-pass filter.

There are some important points about the particular values of α and ε, which are best discussed within the framework of a numerical example. Let us attempt the design of a longwave-pass filter at 45° in air having a symmetrical period of

$$fL\,HLHLHLH\,fL,$$

where H represents an index of 2.35 and L of 1.35. The inner stack has seven layers, which corresponds to $2x + 1$, so that x in this example is 3. We will use the modified admittances that for this combination are as follows (the subscripts S and A referring to the substrate [$n = 1.52$] and to air [$n = 1.00$], respectively):

$$\eta_{Hs} = 3.1694, \quad \eta_{Ls} = 1.6264,$$
$$\eta_{Ss} = 1.9028, \quad \eta_{As} = 1.0000,$$
$$\eta_{Hp} = 1.7425, \quad \eta_{Lp} = 1.1206,$$
$$\eta_{Sp} = 1.2142, \quad \eta_{Ap} = 1.0000.$$

Then,

$$P_s = 15.201, \quad P_p = 10.535,$$
$$Q_s = 7.211, \quad Q_p = 2.923,$$

giving $\sin \alpha = \pm 0.1480$.

Now, the outer tuning layers in their unperturbed state will be quarter waves, and so the two solutions we look for will be near $2\alpha = \pi$, that is, in the second and third quadrants. We continue to keep the results in the correct order and find

$$2\alpha = \pi \pm 0.1485 = 3.2901 \text{ or } 2.9931.$$

Then, in both cases, $\cos 2\alpha = 0.9890$, and so

$$\varepsilon = \pm(1 + 2.923 \times 0.148)/(-10.535 \times 0.9890) = \pm(-0.1375),$$

whence

$$f = (3.2901/2)/[(\pi/2) - 0.1375] = 1.148,$$

with

$$g = 1 - (2 \times 0.1375)/\pi = 0.9125$$

and

$$f = (2.9931/2)/[(\pi/2) + 0.1375] = 0.876,$$

with

$$g = 1 + (2 \times 0.1375)/\pi = 1.088.$$

We take the second of these, which will correspond to a longwave-pass filter. We now need to consider the matching requirements. Since we are attempting to obtain coincident edges for both planes of polarization in an edge filter of limited passband extent, we will interest ourselves in having good performance right at the edge of the passband with little regard for performance further away. We use the method based on symmetrical periods. The basic period is

$$0.876L \, HLHLHLH \, 0.876L,$$

with H and L quarter-waves of indices 2.35 and 1.35, respectively, and tuned for 45°. The calculation of the equivalent admittances for the symmetrical period gives the values for s- and p-polarizations shown in Table 9.1. (Again, they are modified admittances.) We will arrange matching at $g = 1.08$. Adding an $HLHL$ combination to the period with the L layer next to it yields admittances of 0.9625 for p-polarization and 1.416 for s-polarization. The media we have to match have modified admittances of 1.0 for air and 1.214 for glass for p-polarization and 1.0 and 1.903 for

TABLE 9.1

Equivalent Admittances and Phase Thicknesses of the Symmetrical Period ($0.876L \, HLHLHLH \, 0.876L$)

	s-Polarization		p-Polarization	
g	E (Modified)	γ/π	E (Modified)	γ/π
1.04	Imaginary values		0.1946	4.4372
1.05	0.0949	4.2955	0.2018	4.4372
1.06	0.1190	4.4454	0.1993	4.5884
1.07	0.1202	4.5786	0.1861	4.6652
1.08	0.0982	4.7211	0.1588	4.7486
1.09	Imaginary values		0.1049	4.8530
1.10	Imaginary values		Imaginary values	

Note: L and H indicate quarter waves at 45° angle of incidence of indices 1.35 and 2.35, respectively.

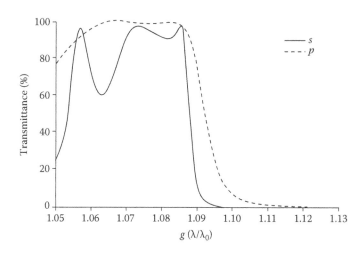

FIGURE 9.9
Calculated performance of a polarization-free edge filter designed for use at 45° in air using the method of Thelen [10]. The multilayer structure is given in the text. The solid curve indicates s-polarization, and the dashed curve, p-polarization.

s-polarization. As an initial attempt, therefore, this matching is probably adequate. Since the matching is to be at $g = 1.08$, the thicknesses of the four layers in the matching assemblies must be corrected by the factor 1.0/1.08. To complete the design, we need to make sure that all layers are tuned for 45°, which means multiplying their effective thicknesses for 45° by the factor $1/\cos \vartheta$. The final design with all thicknesses quoted as their normal incidence values is then

$$\text{Air} | (0.971\,H\,1.087\,L)^2 (1.028\,L(1.049\,H\,1.174\,L)^3\,1.049\,H\,1.028\,L)^q\,(1.087\,L\,0.971\,H)^2 | \text{Glass} .$$

The performance with $q = 4$ is shown in Figure 9.9. Since the p-admittances are less effective than the s-admittances in achieving high reflectance, the steepness of the edge for s-polarization is somewhat greater, and so the two edges coincide at their upper ends. The adjustment of the factor f can move this point of coincidence up and down the edges. Thelen gives many examples of designs including some that are based on symmetrical periods containing thicknesses other than quarter waves.

9.5.2 Reflecting Coatings at Very High Angles of Incidence

Reflecting coatings at very high angles of incidence suffer catastrophic reductions in reflectance for p-polarization. This is especially true for coatings that are embedded in glass, such as cube beam splitters, and we have already seen how they are well suited to making good polarizers. The admittances for p-polarized light are not favorable for high reflectance, and so to increase the p-reflectance, we must use a large number of layers—many more than is usual at normal incidence. The s-reflectance must also be considerably reduced at the same time; otherwise, it will vastly exceed what is possible for p-polarization. The technique we use here is based on yet another method originated by Thelen [9]. A number of authors have studied the problem. For a detailed account of the use of symmetrical periods in the design of reflecting coatings for oblique incidence, the paper by Knittl and Houserkova [11] should be consulted.

We consider a quarter-wave stack. The surface admittance of such a stack on a substrate is given at normal incidence by

$$Y = \frac{y_1^2 y_3^2 y_5^2 \cdots y_m}{y_2^2 y_4^2 y_6^2 \cdots} , \tag{9.27}$$

with y_m in the numerator, as shown, if the number of layers is even, or in the denominator, if odd. The reflectance is

$$R = \left[\frac{y_0 - Y}{y_0 + Y}\right]^2$$

in the normal way. Now, if the stack of quarter waves is considered to be tilted, with the thicknesses tuned to the particular angle of incidence, the expression for reflectance will be similar except that the appropriate tilted admittances must be used. Here we will use the modified admittances so that y_0 will remain the same. Then, Y becomes

$$Y = \frac{\eta_1^2 \eta_3^2 \eta_5^2 \cdots \eta_m}{\eta_2^2 \eta_4^2 \eta_6^2 \cdots}, \tag{9.28}$$

and in order for the reflectances for p- and s-polarizations to be equal, the modified admittances Y for p- and s-polarizations must be equal or, alternatively,

$$\frac{y_0 - Y_p}{y_0 + Y_p} = \frac{Y_s - y_0}{Y_s + y_0}.$$

However, this second condition reduces to $Y_s Y_p = y_0^2$, and since $\eta_p \eta_s = y^2$, then this is equivalent to a set of quarter waves that would yield zero reflectance if quarter waves at normal incidence, not an encouraging result if we are looking for high reflectance.

If we write Δ_1 for (η_{1p}/η_{1s}), and similarly for the other layers, then the first condition is

$$\frac{\Delta_1^2 \Delta_3^2 \Delta_5^2 \cdots \Delta_m}{\Delta_2^2 \Delta_4^2 \Delta_6^2} = 1. \tag{9.29}$$

(Note that Thelen's paper does not use modified admittances and so includes the incident medium in the formula.) The procedure then is to attempt to find a combination of materials such that condition in Equation 9.29 is satisfied, and the value of admittance is such that the required reflectance is achieved. This is a matter of trial and error.

An example may help make the method clear. Table 9.2 gives some figures for modified admittances in glass ($n = 1.52$) and at an angle of incidence of 45°. In this particular case, the substrate and incident medium are of the same material. There is a number of possible arrangements, but the most straightforward is to find three materials H, L and Q, Q, being of intermediate index, such that

$$\Delta_H \Delta_L = \Delta_Q^2. \tag{9.30}$$

Then the multilayer structure can be ...HQLQHQLQHQLQ... so that the form of admittance is

$$Y = \frac{\eta_H^2 \eta_L^2 \eta_H^2 \cdots}{\eta_Q^2 \eta_Q^2 \eta_Q^2 \cdots}. \tag{9.31}$$

The number of layers can then be chosen so that the required reflectance is achieved. The substrate does not appear in Equation 9.31 because, as already mentioned, it is of the same material as the incident medium, and therefore, $\Delta_m = 1$. When the substrate is of a different material, there may be a slight residual mismatch, but practical difficulties will usually make the achievement of an exact match difficult. Theoretically, it is always possible to remove the residual mismatch by adding an extra section that matches the substrate to the incident medium.

From Table 9.2, we see that a set of layers giving an approximate match at 45° has indices of 1.35, 2.25, and 1.57. For this combination,

$$\frac{\Delta_H \Delta_L}{\Delta_Q^2} = \frac{1.3656 \times 0.6478}{0.941^2} = 0.999.$$

TABLE 9.2

Values Using Thelen's Method

n_f	$1/\cos\vartheta$	η_p	η_s	$\Delta\ (=\eta_p/\eta_s)$
1.35	1.6526	1.5776	1.1553	1.3656
1.38	1.5943	1.5558	1.2241	1.2710
1.45	1.4898	1.5275	1.3765	1.1097
1.52	1.4142	1.5200	1.5200	1.0000
1.57	1.3719	1.5230	1.6185	0.9410
1.65	1.3180	1.5377	1.7705	0.8685
1.70	1.2907	1.5515	1.8627	0.8330
1.75	1.2672	1.5680	1.9531	0.8028
1.80	1.2466	1.5867	2.0419	0.7771
1.85	1.2286	1.6072	2.1295	0.7548
1.90	1.2127	1.6292	2.2158	0.7353
1.95	1.1985	1.6525	2.3010	0.7182
2.00	1.1858	1.6770	2.3853	0.7030
2.05	1.1744	1.7023	2.4687	0.6895
2.10	1.1640	1.7285	2.5514	0.6775
2.15	1.1546	1.7554	2.6334	0.6666
2.20	1.1461	1.7829	2.7147	0.6568
2.25	1.1383	1.8110	2.7955	0.6478
2.30	1.1311	1.8396	2.8757	0.6397
2.35	1.1245	1.8686	2.9554	0.6323
2.40	1.1184	1.8980	3.0347	0.6254

Note: Modified admittances: incident medium index = 1.52; angle of incidence = 45°.

The *p*-admittance increase due to one four-layer period of that type is

$$\frac{\eta_{Hp}^2\eta_{Lp}^2}{\eta_{Qp}^4} = \frac{1.811^2 \times 1.578^2}{1.523^2} = 1.518.$$

Eight periods give a value of 28.2, that is, a reflectance of 87% for 32 layers. The particular arrangement of *H*, *L*, and *Q* layers is flexible as long as *H* or *L* are odd and *Q* is even. The performance of a coating to this design is shown in Figure 9.10. The basic period is four-quarter-waves thick. High-reflectance zones exist wherever the basic period is an integral number of half-waves thick. Since in this case we have four quarter waves, we expect extra high-reflectance zones at *g* = 0.5 and *g* = 1.5. The peak at *g* = 0.5 (i.e., λ = 2 × 510 = 1020 nm) is visible at the long wavelength end of the diagram.

The examination of the modified admittances for the materials shows how the coating does yield the desired performance. Each second pair of layers tends to reduce the *s*-reflectance of the preceding pair but slightly to increase the *p*-reflectance. To achieve high reflectance, large numbers of layers are needed. Angular sensitivity is quite high, and there is little that can be done to improve it.

9.5.3 Edge Filters at Very High Angles of Incidence

It is possible to adapt the treatment of the previous section to design edge filters for use at high angles of incidence. We cannot expect to be able to do this with a modest number of layers. Let us illustrate the method by using the example we have just calculated. Figure 9.10 shows the performance. We wish to use this component as a longwave-pass filter and, hence, to eliminate the

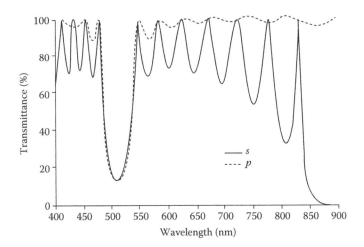

FIGURE 9.10
Calculated performance of a polarization-free reflector at an angle of incidence of 45o in glass. The coating was designed using the method of Thelen [9]. Design: Glass | $(1.38H\ 1.372Q\ 1.653L\ 1.372Q)^8$ | Glass, with $n_H = 2.25$, $n_Q = 1.57$, $n_L = 1.35$, $n_{Glass} = 1.52$, and $\lambda_0 = 510$ nm. The solid line indicates s-polarization, and the dashed line, p-polarization.

ripple on the longwave side of the peak. However, the rejection is not good enough. We need to use more layers. An arrangement with 20 repeats gives lower than 0.1% over most of the rejection region, but then there is the ripple. We therefore choose 24 repeats and use the outermost eight layers on either side as the basis for our matching. Computer refinement is the simplest way to achieve this. For this number of layers, we need a somewhat more accurate estimate of the layer thicknesses. The final design, before refinement, is then

$$\text{Glass}|(1.13827H\,1.37187Q\,1.65262L\,1.37187Q)^{24}|\text{Glass}.$$

The performance after refinement is shown in Figure 9.11.

Shortwave-pass filters or filters with different materials can be designed in the same way. Such designs are fairly sensitive to materials and to the angle of incidence.

The width of the rejection region is very limited and less than the width of the split region in a conventional two-material edge filter. Provided the requirements are not too severe, adding a

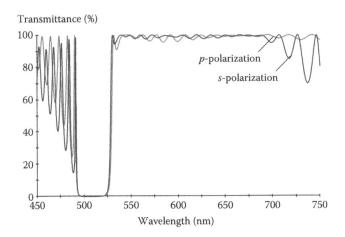

FIGURE 9.11
Calculated performance of a polarization-free edge filter at an angle of incidence of 45°o in glass. Design: Glass | $(1.13827H\ 1.37187Q\ 1.65262L\ 1.37187Q)^{24}$ | Glass, with $n_H = 2.35$, $n_Q = 1.57$, $n_L = 1.35$, $n_{Glass} = 1.52$, and $\lambda_0 = 510$ nm.

glass absorption filter is probably the most practical way of dealing with the problem. Otherwise, a second interference component at a slightly shorter wavelength will be additionally required.

9.6 Antireflection Coatings

Antireflection coatings at high angles of incidence are rather more difficult than the design of coatings for normal incidence. Some simplification occurs when only one plane of polarization has to be considered. Then it is a case of converting the admittances into tilted or modified optical admittances at the appropriate angle of incidence and then using these values to design coatings in much the same way as for normal incidence. The major complication is that the range of admittances available is different from the range at normal incidence and, especially in the case of s-polarization, less favorable. We will briefly consider the problem of antireflection coatings for one polarization first and then treat both polarizations where there is the additional problem of a completely different set of admittances for each polarization.

In order to simplify the discussion of design, we will assume an angle of incidence of 60° in air with a substrate of index 1.5 and possible film indices of 1.3, 1.4, 1.5, . . ., 2.5. Real designs will be based on available indices, will therefore be more constrained, and may require more layers. The modified admittances with values of $\Delta(= \eta_p/\eta_s)$ are given in Table 9.3.

9.6.1 p-Polarization Only

At 60°, the modified p-admittance of the substrate is only 0.9186, giving a single-surface reflectance for p-polarized light of less than 0.2%, acceptable for most purposes. The angle of incidence of 60° is only just greater than the Brewster angle. If still lower reflectance is required, then a single quarter-wave of admittance given by $(0.9186 \times 1.0000)^{1/2}$, that is, 0.9584, is required. This corresponds from Table 9.3 to an index of just over 1.6, that is, *greater* than the index of the substrate. As the angle of incidence increases, still further from 60°, the required index will become still greater. Eventually, at

TABLE 9.3

Values of Δ at 60° for a Range of Indices

n_f	$1/\cos \vartheta$	η_p	η_s	$\Delta (= \eta_p/\eta_s)$
1.00	2.0000	1.0000	1.0000	1.0000
1.30	1.3409	0.8716	1.9391	0.4495
1.40	1.2727	0.8909	2.2000	0.4050
1.50	1.2247	0.9186	2.4495	0.3750
1.60	1.1893	0.9514	2.6907	0.3536
1.70	1.1621	0.9878	2.9258	0.3376
1.80	1.1407	1.0266	3.1560	0.3253
1.90	1.1235	1.0673	3.3823	0.3156
2.00	1.1094	1.1094	3.6056	0.3077
2.10	1.0977	1.1526	3.8262	0.3012
2.20	1.0878	1.1966	4.0448	0.2958
2.30	1.0794	1.2414	4.2615	0.2913
2.40	1.0722	1.2867	4.4766	0.2874
2.50	1.0660	1.3325	4.6904	0.2841

Note: Modified admittances: incident medium index = 1.00; angle of incidence = 60°.

very high angles of incidence indeed, the required single-layer index will be greater than the highest index available, and at that stage, designs based on combinations such as Air | HL | Glass will be required with quarter-wave thicknesses at the appropriate angle of incidence. Such coatings operate over a very small range of angles of incidence only and are very difficult to produce because of the required accuracy. If at all possible, it is better to avoid such designs altogether by redesigning the optical system.

9.6.2 *s*-Polarization Only

The modified *s*-admittance for the substrate is 2.449, and the required single-layer admittance for perfect antireflection is $(2.4495 \times 1.0000)^{1/2}$ or 1.5650, well below the available range. The problem is akin to that at normal incidence where we do not have materials of sufficiently low index. Here the solution is similar. We begin by raising the admittance of the substrate to an acceptable level by adding a quarter-wave of higher admittance. In this case, a layer of index 1.9 or admittance of 3.3823 is convenient and gives a resultant admittance of $3.3823^2/2.449$ or 4.6713 that requires a quarter wave of admittance $(4.6713 \times 1.0000)^{1/2}$ or 2.1613 to complete the design. This corresponds most nearly to an index of 1.4 and admittance of 2.2000, and the residual reflectance with such a combination is 0.03%, a considerable improvement over the 17.7% reflectance of the uncoated substrate. We cannot expect that such a coating will have a broad characteristic, and Figure 9.12 confirms it. A small improvement can be made by adding a high-admittance half-wave layer between the two quarter waves or a low-admittance half wave next to the substrate. The latter is also shown in the figure. In terms of normal incidence thicknesses, the two designs are

$$\text{Air} | 1.273L\ 1.123H | \text{Glass}$$

and

$$\text{Air} | 1.273L\ 1.123H\ 2.682A | \text{Glass},$$

where L, H, and A indicate quarter waves at normal incidence of films of index 1.4, 1.9, and 1.3, respectively. The *p*-reflectance of these designs is very high, and they are definitely suitable for *s*-polarization only.

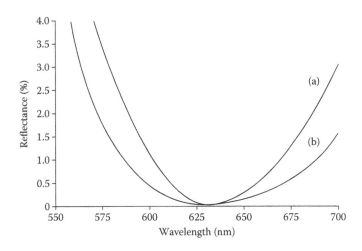

FIGURE 9.12
Antireflection coatings for *s*-polarized light at an angle of incidence of 60° in air. (a) Air | 1.273L 1.123H | Glass and (b) Air | 1.273L 1.123H 2.682A | Glass, with $n_L = 1.4$, $n_H = 1.9$, $n_A = 1.3$, $n_{Glass} = 1.5$, and $\lambda_0 = 632.8$ nm.

Again, it is better wherever possible to avoid the necessity for such antireflection coatings by rearranging the optical design of the instrument so that s-polarized light is reflected and p-polarized light is transmitted.

9.6.3 s- and p-Polarizations Together

The task of assuring low reflectance for both s- and p-polarized lights with discrete layers is very difficult and should only be attempted as a last and usually very expensive resort. It is possible to arrive at designs that are effective over a narrow wavelength region, and one such technique is included here. Again, we use the range of indices given in Table 9.3 and design a coating to give low s- and p-reflectances on a substrate of index 1.5 in air.

We use quarter-wave layer thicknesses only and a design technique similar to the procedure we have already used for high-reflectance coatings but with an additional condition that the admittance of both substrate and coating for both p- and s-polarizations should be unity to match the incident medium. This implies

$$\frac{\Delta_1^2 \Delta_3^2 \Delta_5^2 \cdots \Delta_m}{\Delta_2^2 \Delta_4^2 \Delta_6^2} = 1 \qquad (9.32)$$

and

$$Y = \frac{\eta_{1s}^2 \eta_{3s}^2 \eta_{5s}^2 \cdots \eta_{m,s}}{\eta_{2s}^2 \eta_{4s}^2 \eta_{6s}^2 \cdots} = 1. \qquad (9.33)$$

Equation 9.32 ensures that the p-reflectance equals the s-reflectance, and Equation 9.33 ensures that the s-reflectance and, therefore, the p-reflectance, is zero. From Table 9.3, the starting values are $\Delta_m = 0.3750$ and $\eta_m = 2.4495$. Trial and error shows that with the addition of one single quarter-wave layer, the best result corresponds to an index of 1.3, for which $\Delta_1^2/\Delta_m = 0.4495^2/0.3750 = 0.5387$ and $\eta_{1s}^2/\eta_m = 1.9391^2/2.4495 = 1.5350$. Other combinations give values that are farther from unity in each case. When adopting a quarter wave of index 1.3 as the first layer of the coating, we need a further combination of layers that will provide a correction factor of 1.3624 in Δ and of 0.8071 in η_s. An additional single layer will not do, but two-layer combinations of a high-index layer followed by a low-index layer can be found that will correct Δ but that are inadequate in terms of η_s. The two-layer combination that comes nearest to satisfying the requirements is a layer of index 1.8 followed by one of index 1.3, making the design so far:

$$\text{Air}|n = 1.3|n = 1.8|n = 1.3|\text{Glass.}$$

This has an overall Δ of $(0.4495^2 \times 0.4495^2)/(0.3253^2 \times 0.375) = 1.0288$ and a η_s of $(1.9391^2 \times 0.9391^2)/(3.1560^2 \times 2.4495) = 0.5795$. However, the combination of index 2.5 followed by 1.4 gives approximately the same correction for Δ but a different correction for η_s. This gives the opportunity of using both combinations in a four-layer arrangement to adjust the value of η_s without altering Δ. The correction factor for Δ is given by $(0.4495^2 \times 0.2841^2)/(0.4050^2 \times 0.3253^2) = 0.9396$ and for η_s by $(1.9391^2 \times 4.6904^2)/(2.2000^2 \times 3.1560^2) = 1.7159$. This then yields an overall value for Δ of $0.9396 \times 1.0288 = 0.9667$ and for η_s of $1.7159 \times 0.5795 = 0.9944$. The seven layers can be put in various orders without altering the reflectance at the reference wavelength. All that is required is that the 1.3 and 2.5 indices should be odd, and the 1.4 and 1.8 indices, even. Here we put them in descending value of index from the substrate so that the final design is

$$\text{Air}|1.3409L\,1.2727A\,1.3409L\,1.1407B\,1.3409L\,1.1407B\,1.066H|\text{Glass,}$$

with $n_L = 1.30$, $n_A = 1.40$, $n_B = 1.80$, and $n_H = 2.50$.

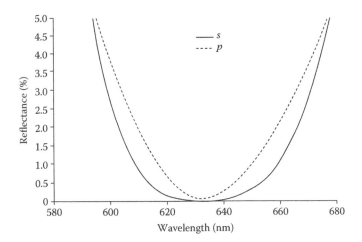

FIGURE 9.13
Calculated performance of an antireflection coating for glass to have low reflectance for both *p*- and *s*-polarizations at an angle of incidence of 60° in air. The solid line indicates *s*-polarization, and the dashed line, *p*-polarization. λ_0 = 632.8 nm, and the design is given in the text.

The calculated performance of this coating for a reference wavelength of 632.8 nm is shown in Figure 9.13. As we might have suspected, the width of the zone of low reflectance is narrow. An alternative design arrived at in the same way but for a substrate of index 1.52 and a range of film indices from 1.35 to 2.40 uses 10 layers:

Air|1.3036L 1.1748A 1.3036L 1.1748A 1.3036L 1.1407B 1.0722H 1.1235C 1.0722H 1.1235C|Glass,

with n_L = 1.35, n_A = 1.65, n_B = 1.80, n_C = 1.90, n_H = 2.40, n_{Glass} = 1.52, and n_{air} = 1.00. The performance is similar to that of Figure 9.13.

The most successful designs for equal *p* and *s*-performances in the last two sections have illustrated the usefulness of employing more than two materials. It is fairly easy to see why. The phase thickness at oblique incidence is exactly the same for both polarizations, and with only two materials, there is no flexibility in the properties of the interfaces. An interface of material *A* to material *B* has virtually the same properties as material *B* to material *A*, except for a 180° difference in the phase shifts. Altering the polarization-sensitive properties independently is therefore virtually impossible. A slight increase in flexibility can be obtained by the use of three materials, with three different interface properties. There is a price to pay in that the properties of coatings at oblique incidence exhibit considerable sensitivity to the precise values of refractive index.

Refinement and synthesis are valid techniques for design at oblique incidence, but they cannot find a performance that does not exist in parameter space. The designs they produce using discrete layers therefore tend to be complex and disappointing in their performance level. Figure 9.14 shows the reflectance of a glass surface in air over 55° incidence to 65°. An antireflection coating for *p*-polarization is hardly necessary over the angular range, but if the surface is to be antireflected for both polarizations, then the coating must reduce the *s*-reflectance while avoiding any too great increase in the *p*-reflectance. Again, three materials are helpful (here we use indices of 1.45, 1.65, and 2.40), but the synthesized design is complex, consisting of 42 layers with no discernible pattern, and the reflectance is rising toward 1.5% at 65°. Figures 9.14 through 9.17 show details of the layer sequence and the performance.

The ideal antireflection coating is, of course, the inhomogeneous layer, already discussed in Chapter 4. The properties of the coating are not very sensitive to the profile of index as long as it is smooth and gradual. It acts as an efficient antireflection coating as long as its total phase thickness is greater than one-half wavelength. Thus, when tilted, although the profile slightly changes, as long as the angle of incidence is below that at which the phase thickness would become less than

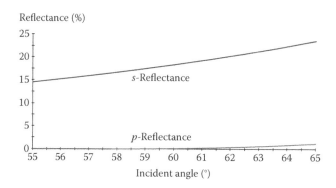

FIGURE 9.14
Reflectance as a function of angle of incidence of a glass surface of index 1.52 in air. The Brewster angle is at 56.7°.

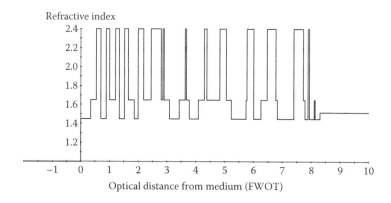

FIGURE 9.15
Index distribution of the 42-layer synthesized antireflection coating for the Figure 9.14 glass intended for the region of 530–570 nm and for angles of incidence of 55–65° in air. The horizontal scale is in terms of full wave optical thickness measured from the front surface of the coating. The substrate is on the right.

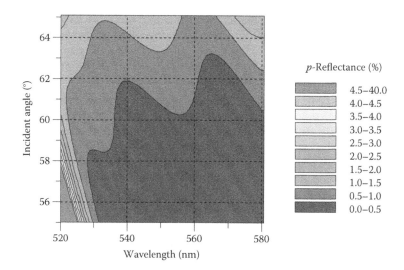

FIGURE 9.16
Reflectance for *p*-polarization of the 42-layer antireflection coating on glass in an incident medium of air. Over the required range of 530–570 nm and angles of incidence from 55° to 65°, the reflectance is less than 1.5%. The index of glass is 1.52 and those of the materials are 1.45, 1.65, and 2.40.

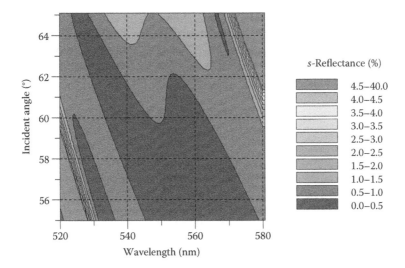

FIGURE 9.17
Reflectance for *s*-polarization of the 42-layer antireflection coating on glass in an incident medium of air. Over the required range of 530–570 nm and angles of incidence from 55° to 65°, the reflectance is less than 1.5%. The index of glass is 1.52 and those of the materials are 1.45, 1.65, and 2.40.

one-half wavelength, the antireflecting properties will be maintained. Unfortunately, the difficulty of finding sufficiently low refractive indices for the outermost parts of the inhomogeneous layer still exists at oblique incidence. For this reason, a structural gradation where the structural units are roughly conical or pyramidal in shape is preferred. Such a coating, already described in Chapter 4, is inevitably weaker than a solid film, but there are many applications, the inner surfaces of compound lenses, for example, where sufficient protection from environmental disturbance exists. The main design consideration is that the coating should still be greater than a half wave in optical thickness at the greatest angle of incidence involved.

9.7 Retarders

In many applications, the polarization state of a ray is of great importance. Optical surfaces, with or without coatings, modify these properties. Knowledge of the nature of the modification is necessary for any investigation of the initial polarization. Also, the measurement of the modification is an important surface and coating characterization tool. Deliberate manipulation of the polarization is possible with specially designed coatings. Before we can examine and quantify these properties, we need to define some terms and conventions.

9.7.1 Ellipsometric Parameters and Relative Retardation

We assume a completely polarized beam defined by two orthogonal components of electric field that exhibit complete coherence. Then, if z is the direction of propagation, the ellipsometric parameters ψ (psi) and Δ (delta) are defined as

$$\tan \psi = \left| \frac{\mathcal{E}_x}{\mathcal{E}_y} \right| \quad \text{and} \quad \Delta = \varphi_x - \varphi_y. \tag{9.34}$$

Δ is also known as the relative retardation or retardance. Note that ψ and Δ depend on the choice of the reference axes. Note also that this Δ should not be confused with that in Equation 9.29 and the following ones. When reflection at a surface is concerned, it is convenient to make the reference axes coincide with the p- and s-directions. Then, we can introduce the idea of ψ and Δ that are properties of the surface, or of the coating. We can define them as

$$\tan \psi = \left| \frac{\rho_p}{\rho_s} \right| \quad \text{or} \quad \left| \frac{\tau_p}{\tau_s} \right| \quad \text{and} \quad \Delta = \varphi_p - \varphi_s. \tag{9.35}$$

However, there is a mild problem. The convention needs a consistent handedness of the reference axes together with the propagation direction, that is, p- and s- directions and the direction of propagation, *in that order*. The thin-film convention, detailed in Chapter 2, is right-handed in incidence and transmission but *left-handed* in reflection. To keep the complications to a minimum, the ellipsometric convention then flips the p-direction in the reflected ray. This creates problems at normal incidence where there is no plane of incidence and where there can be no orientational dependence in the behavior of linearly polarized light. The thin-film community, therefore, retains the Chapter 2 convention for positive directions and simply amends the definition of Δ to be

$$\Delta = \varphi_p - \varphi_s \quad \text{in transmission,}$$
$$\Delta = \varphi_p - \varphi_s \pm \pi \quad \text{in reflection.} \tag{9.36}$$

This also helps remove a complication at normal incidence. Circularly polarized light that is reflected from the surface of a coating at normal incidence remains circularly polarized, but it changes its handedness. This implies a relative retardance, or Δ, of π, or 180°, which is fortunately in accordance with Equation 9.36. The value of ψ is therefore 45°, and Δ, 180°, or π. Any pair of orthogonal axes making up a right-handed set with the direction of propagation can therefore be used without requiring any change in the values of ψ and Δ.

A reflection at normal incidence is sometimes referred to as a half-wave plate or a half-wave retarder. This is an unsatisfactory way of referring to it because it does not behave as a normal half-wave retarder. We cannot use it to rotate a plane of polarization, for example. We shall deal with retarders in more detail shortly.

9.7.2 Series of Coated Surfaces

A knowledge of the absolute phase of a light ray demands an exact knowledge of the path length traversed by the ray. In an optical system, this is not usually known or even constant to the necessary precision. For the understanding of the effects on the polarization of the light, the absolute phase is not required but only the difference in phase between the reference components. Thus, unless there is some optical activity that would introduce a differential phase shift between the components, we normally ignore the phase change that results from the passage between components and simply include the changes that take place at the surfaces. Since it is the difference in phase that is important, we can simply use Δ. Should we need separate phases for each polarization, we can adopt the net Δ for the final p-polarized component and zero for s. A system where the surface normals are coplanar is relatively simple in its calculation, but a more complicated system where the normals may have any orientation requires a conversion in the reference directions at each interface. There is no ambiguity, but there is considerable complication in the calculations.

At oblique incidence, ψ and Δ are initially defined with respect to the plane of incidence. Provided the plane of incidence for any subsequent reflection or transmission does not change, that is, the normals are coplanar, then combining the effect of multiple surfaces is straightforward. Since the p-direction and s-direction are completely independent, then, ignoring the phase shift between elements, we can write

$$\rho_p = \rho_{p1} \cdot \rho_{p2} \cdot \rho_{p3} \cdot \cdots,$$
$$\rho_s = \rho_{s1} \cdot \rho_{s2} \cdot \rho_{s3} \cdot \cdots, \tag{9.37}$$

where each ρ may be replaced by a τ for a transmission and where we must use the ellipsometric convention for ρ_p. Then,

$$\tan \psi = \tan \psi_1 \cdot \tan \psi_2 \cdot \tan \psi_2 \cdot \cdots,$$
$$\Delta = \Delta_1 + \Delta_2 + \Delta_3 + \cdots. \tag{9.38}$$

The combination becomes much more complicated if the planes of incidence are not coincident. Then it is not particularly helpful to think in terms of ψ and Δ. Since the plane of incidence always contains the ray, we will be dealing with a rotation of each fresh plane of incidence about the ray direction. This can be handled as an application of the Jones matrices [12–14] for rotation. First, we need to choose reference directions for the input electric field of the wave, and these can conveniently be the p and s-directions of the beam as it emerges from the previous element. Then, let ϑ be the angle of rotation of the new set of p and s-directions around the direction of propagation. The calculation can be written as

$$\begin{bmatrix} \mathcal{E}_{p'} \\ \mathcal{E}_{s'} \end{bmatrix} = \begin{bmatrix} \rho_{p'} & 0 \\ 0 & \rho_{s'} \end{bmatrix} \begin{bmatrix} \cos \vartheta & \sin \vartheta \\ -\sin \vartheta & \cos \vartheta \end{bmatrix} \begin{bmatrix} \mathcal{E}_p \\ \mathcal{E}_s \end{bmatrix}. \tag{9.39}$$

The components $\mathcal{E}_{p'}$ and $\mathcal{E}_{s'}$ are referred to the p and s-directions of the final emergent beam. Once again, the ellipsometric convention is used for $\rho_{p'}$.

The Jones matrix approach involves keeping track of the reference directions and their rotations. This can be quite difficult and include awkward subsidiary calculations, especially when transmission rather than reflection is involved. An alternative and somewhat more attractive approach involves setting up a set of three-dimensional reference axes and referring everything to them. Vector analysis is the most straightforward way of handling the calculations. Once the calculation framework is set up, it is then necessary only to specify the directions of the various surface normals, the materials, and coatings and whether or not reflection or transmission is involved.

The most important prerequisite for such calculations is a clear head.

9.7.3 Retarders

Phase retarders introduce a relative phase shift between two orthogonal planes of polarization and are characterized by their relative retardance Δ. Most commonly, these are plates of birefringent materials with different indices of refraction for each polarization. If these indices are given by n_a and n_b, then, ignoring any effect of the surfaces of the plate, the relative phase shift that is introduced is $2\pi(n_a - n_b)d/\lambda$, where d is the physical thickness of the plate. Usually, the direction in the plate that corresponds to the electric field direction for the lower refractive index is called the fast axis. This helps to determine the sense of the relative phase shift, but the terminology is not well organized. The birefringent retarder is a very straightforward component that is easy to use and possesses the great advantage of preserving the direction of the light beam. There are limitations, however. Retarders of large size may be impossible, or prohibitively expensive. The surfaces have the usual reflection loss and may need antireflection coatings to minimize them. Then there is the dependence on $1/\lambda$. Achromatic retarders, therefore, are more complicated, generally using two or more different materials. Thin-film retarders can avoid some of these problems.

Isotropic materials show no polarization-sensitive effects at normal incidence. Once they are tilted, there is a difference in the properties for s- and p-polarized lights, and this is the basis for their application in retarders.

The phase thickness of a layer is the same for both s- and p-polarizations. For a device to operate in transmission, the reflectance should be low, and the light then tends to be transmitted with almost identical phase shift for both polarizations. Only in narrowband filters where the light is stored in cavities do we find any very significant differences between the polarizations, and such components are rather more useful as polarizers than as retarders. Thus, useful thin-film retarders are almost invariably reflecting devices. Since the polarization state of the light can be perturbed by a simple difference in reflectance even when there is no difference in the relative phase, these devices should have p- and s-reflectances that are as closely equal as possible. This normally implies making the reflectances as high as possible.

Before we consider the design of thin-film retarders, we need to consider our conventions. We are exclusively dealing with tilted reflectances and with changes in the polarization state of the light. It is therefore convenient to use the ellipsometric parameters ψ and Δ of the coating or device (Equations 9.35 and 9.36) to describe the performance. Since we will attempt to have no influence on polarization from reflectance differences, ψ will normally be 45°.

We should briefly consider the implications of a value of ψ that differs from 45°. We imagine linearly polarized light incident with polarization direction at 45° to the plane of incidence. This gives equal amplitude of both s- and p-polarizations. Then on suffering a reflection characterized by ψ, the new plane of polarization will be rotated to angle ψ with respect to the direction of s-polarization. In fact, errors in ψ always result in a rotation of the plane of polarization of linearly polarized light. The largest effect is produced when the incident light has polarization direction at 45° to the plane of incidence. The ellipsometric convention measures angles with respect to the s-direction (this is in accordance with the convention for ψ), and so the rotation in that case will be $\varepsilon = (45° - \psi)$. Circularly polarized light will become slightly elliptical, as if the original linearly polarized light used to generate it were rotated through ε away from the ideal 45° with the axis of a quarter-wave plate. We will use ψ as a characteristic of our thin-film retarders.

9.7.4 Simple Retarders

The simplest form of this type of retarder is not strictly a thin-film system at all. Below the critical angle, the ψ and Δ associated with reflection at a simple dielectric surface are of little interest. ψ varies from 45° to 0 and back to 45°, while Δ starts at 180° and flips to 0 as the angle of incidence ϑ passes through the Brewster angle and rises either to grazing incidence or the critical angle, whichever comes first. Beyond the critical angle, which requires an incident medium of greater refractive index than the emergent, the situation changes. Then ψ remains fixed at 45° and Δ rises from 0 to a maximum in the first quadrant and then falls back to 0. The behavior is illustrated in Figure 9.18, where the reflection is at an inner surface of glass. There are two angles of incidence at which the retardance is exactly 45°. With two such reflections, a total Δ of 90°, that is, a quarter-wave retardance, can be obtained. A device that employs this effect is known as a Fresnel rhomb (Figure 9.19).

Although we delay our discussion of metals and other absorbing materials until the following chapter, it is convenient to include the mention of the use of a high-performance metal layer as a very simple retarder. We shall see that as the angle of incidence changes, the admittance (modified) of a high-performance metal layer is given to a good degree of accuracy by $\eta_s = (n - ik)/\cos \vartheta_0$ and $\eta_p = (n - ik)\cos \vartheta_0$. As the angle of incidence increases, therefore, the s- and p-admittances move along a straight line from $(n - ik)$ at normal incidence toward the origin for p-polarization and away from the origin for s-polarization. The phase shifts at normal incidence are equal and then move gradually apart, the p-phase shift reducing to 0 at grazing incidence and the s-phase shift increasing to 180° at grazing incidence. The retardance Δ is then 180° at normal incidence and

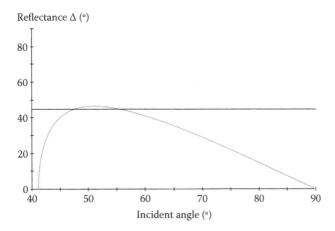

FIGURE 9.18
Variation of Δ with angle of incidence (°) for total internal reflection at a glass surface (index of 1.52). Δ is 45° for angles of 47.6° and 55.4°.

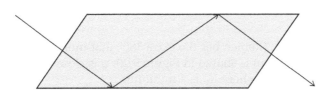

FIGURE 9.19
Sketch of a Fresnel rhomb. The double reflection inside the device gives a total retardance Δ of 90°.

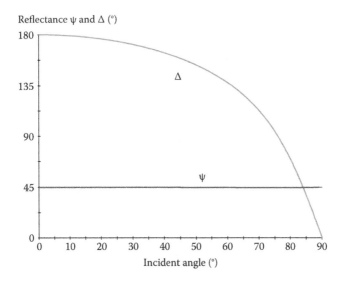

FIGURE 9.20
Calculated retardance Δ and ψ for an opaque silver layer at 600 nm, where the optical constants are (0.06 − i3.75).

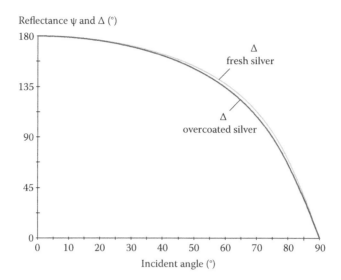

FIGURE 9.21
Effect of a thin dielectric layer over the silver of Figure 9.20. A 3 nm thick layer of index 1.70 slightly reduces the retardation. The original silver upper curve is compared with an overcoat of 3 nm of material of index 1.70. The two curves are almost identical on this scale, but there is a reduction of around 4° in the center.

falls to 0 at grazing incidence. (Remember the extra 180° that must be included in Δ.) The calculated behavior of silver at 600 nm is shown in Figure 9.20. ψ is close to 45° falling to around 44.5° near the pseudo-Brewster angle where R_p is a minimum.

The effect of a thin dielectric overcoat, like a layer of tarnish, for example, is to slightly reduce the retardation. Figure 9.21 shows the effect of 3 nm of a material of index 1.70. The p-polarization locus falls faster than the s-polarization, and so the p-phase shift moves more rapidly into the first quadrant. This implies a reduction in the value of Δ, and at its maximum, there is a reduction of around 4°.

Of course, the retardance varies with wavelength as well as angle of incidence, but this simple retarder can be quite useful, especially soon after deposition in a reasonably benign laboratory atmosphere.

The retardance of a simple surface beyond critical can be altered by a thin-film overcoat. For example, the Fresnel rhomb is almost achromatic in performance, but the dispersion of the glass causes the retardance to gradually increase with decrease in wavelength. A further disadvantage of the Fresnel rhomb is its sensitivity to angle of incidence changes. The performance of the Fresnel rhomb can be considerably improved in both these directions by the addition of a thin-film coating to both surfaces of the rhomb. King [15] has manufactured Fresnel rhombs exhibiting a phase retardance varying by less than 0.4° over the wavelength range of 330–600 nm. These were made from hard crown glass with one surface coated with magnesium fluoride that is 20 nm thick. Then, Lostis [16] gave an early example of the considerable modification possible in the retardance of a prism hypotenuse by the addition of a single thin film.

We begin with 45° as the angle of incidence, glass ($n = 1.52$) as the prism material, and air as the outermost medium, beyond the hypotenuse. This incidence is beyond the critical angle. To achieve a qualitative understanding, we start by imagining that a dielectric film is added, at oblique incidence, to the hypotenuse of the prism where the light is totally reflected. Provided that there are no losses, the light will still be totally reflected, but the value of Δ will have changed. We can understand the change by considering a simple hypothetical case where the base of the prism has $\varphi_s = \varphi_p = 0$. This means that the admittance is at the origin. We imagine that the film is of sufficiently high index and that its p- and s-admittances are real, but, of course, of different values. If the

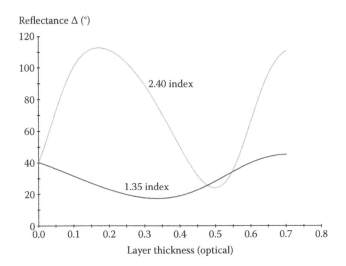

FIGURE 9.22
Retardance, or Δ, produced by total internal reflection at 45° in a prism of index 1.52 when a layer of index 2.40 or 1.35 is added to the outer surface of the hypotenuse. Air is the outer medium. A thickness of around 0.08 full waves (measured at normal incidence) of index 2.40 gives a retardance of 90° in one reflection.

index is high, compared with glass, then the s-admittance will be high, and the p-admittance low (note that these are the *modified* admittances). A low-index film will have the p-admittance higher than the s-admittance. Figure 9.3 should make this clear. Eighth-wave thicknesses arrive at η_p and η_s on the imaginary axis while the boundary between the fourth and third quadrants is the point y_0 in between these two values. The addition of an eighth wave then takes the loci for p- and s-polarizations to points that correspond to different phase shifts on reflection, and for a high-index film, the p-admittance will be lagging behind the s-admittance. Subsequently, as the thickness becomes a quarter wave, the s- and p-admittances are in step and the retardance is again 180°. For the second quarter wave, the situation is reversed and the retardance rises above 180°. For a low-index film, the p- and s-admittances are reversed, and so the variation of retardance will tend to be opposite to that of the high-index film, although the low-index film will exhibit tighter cycles because of the larger $\cos\vartheta$ term. The extent of the polarization splitting in Figure 9.3 is also an indication of the maximum value of Δ attainable.

Figure 9.22 shows the accurately calculated variation of retardance of a high-index film ($n = 2.40$) and low-index film ($n = 1.35$). A thin (around 0.08 full waves for 2.40) layer of high admittance can clearly transform the retardance so that it is 90° in just one reflection at 45° incidence. This is illustrated in Figure 9.23, where although there is a variation with wavelength, it is much less than that of a simple crystal plate.

9.7.5 Multilayer Retarders at One Wavelength

There has been a number of applications, where reflecting coatings have been required, which introduced specified phase retardances between s- and p-polarizations. In particular, certain types of high-power laser resonators have required coatings that introduce a 90° phase shift between s- and p-polarizations at an angle of incidence of 45°. Coatings designed and manufactured for this purpose have generally been designed for wavelengths in the infrared and have taken the form of silver films with a multilayer dielectric overcoat.

The first published designs were due to Southwell [17,18], who used a computer synthesis technique. Then, Apfel [19,20] devised an analytical approach that we follow here. The principle of the operation of the coatings is that an added dielectric layer will not affect the reflectance of a

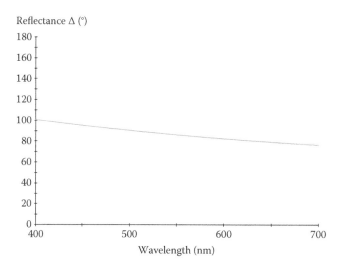

FIGURE 9.23
Variation with wavelength of the retardance at 45° incidence of the design: Glass ($n = 1.52$) | $0.062\lambda_0$ $n = 2.40$ | air with $\lambda_0 = 510$ nm.

system that already has a reflectance of unity. It will simply alter the phase change on reflection. When the component is used at oblique incidence, the alteration in phase will be different for each plane of polarization. By adding layers in the correct sequence, eventually any desired phase difference between p- and s-polarizations for a single specified angle of incidence and wavelength can be achieved. In practice, a silver layer is used as the basic reflecting coating, and although this has reflectance slightly less than unity, in the infrared, it is high enough for it to be possible to neglect any error that might otherwise be introduced. It is, of course, not necessary to use a metal layer as starting reflector. A dielectric stack would be equally effective but would simply have more layers.

The basis of Apfel's method is a plot of phase retardance, denoted by Apfel as D, against the average phase shift A as a function of thickness of added layer of a given index. Apfel defines D as the difference between p- and s-phase shifts. To be consistent, we will rather use the normal sign convention for Δ so that we replace Apfel's D by Δ and adjust his treatment accordingly.

We will use the modified admittances so that $\eta_0 = y_0$. The starting point of the treatment is a reflector with a reflectance of unity, that is, a surface with imaginary admittance. Let this imaginary admittance be $i\beta$ for p-polarization and $-i\beta$ for s-polarization. Then for this surface,

$$|\rho_p|e^{i\varphi_p} = e^{i\varphi_p} = \frac{(\eta_0 - i\beta)}{(\eta_0 + i\beta)} = \frac{(y_0 - i\beta)}{(y_0 + i\beta)},$$
(9.40)

i.e.,

$$\tan\left(\frac{\varphi_p}{2}\right) = \frac{-\beta}{y_0}$$
(9.41)

and

$$\tan\left(\frac{\varphi_s}{2}\right) = \frac{\beta}{y_0}.$$
(9.42)

These two values are numerically identical but inverted in sign. They can be represented by $\pm\xi$ when

$$\beta = -y_0 \tan\left(\frac{\xi}{2}\right).$$
(9.43)

From the form of the admittance diagram, we can see that for β positive, φ_p should be in the third or fourth quadrant, and, hence $\tan(\varphi_p/2)$ is negative, confirming the sign in Equation 9.43. Now let us add a film of admittance η_1 and phase thickness δ_1 to the substrate:

$$\begin{bmatrix} B \\ C \end{bmatrix} = \begin{bmatrix} \cos\delta_1 & i(\sin\delta_1)/\eta_1 \\ i\eta_1\sin\delta_1 & \cos\delta_1 \end{bmatrix} \begin{bmatrix} 1 \\ i\beta \end{bmatrix}$$

$$= \begin{bmatrix} \cos\delta_1 - (\beta/\eta_1)\sin\delta_1 \\ i(\eta_1\sin\delta_1 + \beta\cos\delta_1) \end{bmatrix}. \tag{9.44}$$

The phase shift φ_p is now given, from Equation 9.44, as

$$\tan\left(\varphi_p/2\right) = \frac{-(\eta_1\sin\delta_1 + \beta\cos\delta_1)}{y_0[\cos\delta_1 - (\beta/\eta_1)\sin\delta_1]}. \tag{9.45}$$

While that for φ_s becomes

$$\tan(\varphi_s/2) = \frac{-(\eta_1\sin\delta_1 - \beta\cos\delta_1)}{y_0[\cos\delta_1 + (\beta/\eta_1)\sin\delta_1]}, \tag{9.46}$$

with appropriate adjustments to Equations 9.45 and 9.46 should $\beta \rightarrow \infty$.

To draw a Δ–A curve, we choose a starting point given by $\Delta = 2\xi \pm 180°$ and $A = 0$ and plot the difference in phase against the average phase all calculated from Equations 9.45 and 9.46. Different values of ξ yield a family of curves. There is, however, a slight problem connected with principal range. If this is limited to $360°$, then there are two values of Δ for every possible value of A. Apfel's plots show only one of the solutions. It is better to permit Δ to range from $-360°$ to $+360°$ so as to separate the two solutions. Then, a certain amount of manipulation of the results into the correct range is sometimes necessary for the results to cleanly plot. This family of curves can have a scale of thickness marked along them, in the manner of Figure 9.24. Thickness increases with movement toward the left. Note that as curves disappear off the left-hand side of the diagram, they reappear at the right-hand side and the same with the top and the bottom of the diagram.

The curves now make it possible to determine the phase retardation produced by any combination of thicknesses of layers of the given dielectric materials that are added to any substrate of unity reflectance. Since the curves for different materials do not coincide, it is possible to reach any point of the diagram simply by moving from one set of curves to the other in succession.

The method is illustrated in Figure 9.25, where we design a phase retarder to have a retardance of $180°$. The start point is a glass/air interface beyond critical. The layers follow the appropriate curves and, as can be seen, terminate at the extrema. This assures the most efficient design. The final retardance in this case is $172°$, and the design is optimum for four layers. To do better will require yet another layer. The performance as a function of wavelength is shown in Figure 9.26.

Computer refinement and synthesis are powerful techniques that work very well with retarders. A useful aspect of this graphical technique is that it does very clearly indicate the least number of layers that will be required and their approximate thicknesses. Once this information is available, the automatic design processes can then be used to great effect.

9.7.6 Multilayer Retarders for a Range of Wavelengths

There is no body of published knowledge on analytical design techniques for thin-film retarders over a wide range of wavelengths. Fortunately, as mentioned earlier, the existence of powerful computer programs for refinement and synthesis with the capability of targeting retardance makes the design of such systems relatively straightforward. As emphasized in the previous section, the use of reflection rather than transmission is strongly advised, and if possible, operating

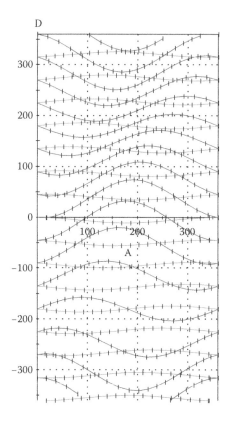

FIGURE 9.24
Δ–A plot for films of index 1.45 (*flatter curves*) and 2.15 (*steeper curves*) over the hypotenuse of a glass prism of index 1.52, internally illuminated at angle of incidence 45°. The tick marks are spaced at intervals of one tenth of a quarter wave at the appropriate angle of propagation.

on a surface with incidence beyond critical, with its natural total reflection, considerably reduces the difficulties in, and the complexity, of the design.

A typical design for a 90° retarder is given in Table 9.4 with performance plotted in Figure 9.27. For simplicity, no dispersion in the materials was assumed. The complexity of the design depends very much on the tolerances required. Over the range of 400–700 nm, the variation in Δ is largely within ±0.5°, slightly rising at the very ends of the range. To reduce this variation, it is sufficient to apply more layers.

The advantage of the use of prisms beyond the critical angle is that whatever thicknesses are used for the dielectric layers, provided they have very low losses, then the reflectance will always be total. In addition, at the very high angles of propagation in the film materials, the large splitting between the *p*- and *s*-admittances also greatly helps in simplifying the design. At angles that are less than critical, these advantages are lacking. Otherwise, the design operation is similar. We need high reflectance, and we can achieve that by a dielectric or metal–dielectric assembly. This must have sufficient layers so that the additional retardance-trimming layers do not cause too great a fall in reflectance.

Since it usually makes little difference whether the relative retardance is positive or negative, the best choice is to have the targets represent as small as possible a *reduction* from the starting values. This makes the use of the natural tendency of the dielectric overcoating layers.

It cannot be expected that such retarders could have performance equivalent to the totally internally reflecting retarders. The performance is generally inferior and, unless starting with a high-performance metal, the number of layers large.

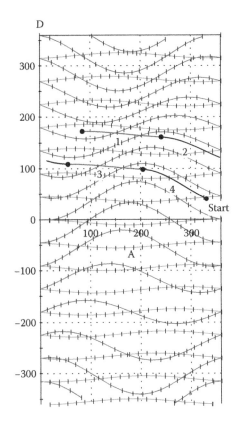

FIGURE 9.25
Design of a retarder to give 180° retardance at one wavelength and consisting of four layers with final design of 1.52 |
1.2533L 0.8950H 1.1935L 0.6902H | Air, with $\lambda_0 = 600$ nm, $n_H = 2.15$, and $n_L = 1.45$. The angle of incidence in the incident
medium of index 1.52 is 45°. The final retardance is 172°. To do better will require another layer.

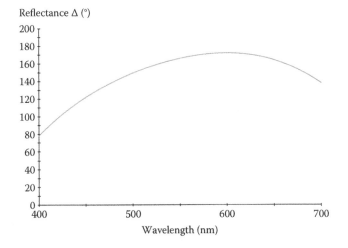

FIGURE 9.26
Retardance as a function of wavelength, at 45° in glass, of the design of Figure 9.25. At 600 nm, the design wavelength,
it is 172°.

TABLE 9.4

90° Retarded Design

Layer	Material	Refractive Index	Thickness (Optical)
Medium	1.52	1.5200	
1	2.40	2.4000	0.06390
2	1.38	1.3800	0.13084
3	2.40	2.4000	0.08567
Substrate	Air	1.0000	

Note: Reference wavelength 510 nm; angle of incidence: 45°.

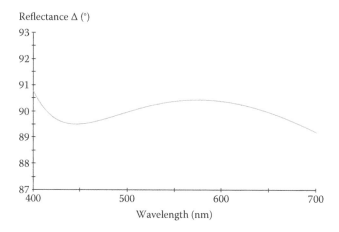

FIGURE 9.27
Retardance Δ of a three-layer system on the base of a prism of index 1.52 with internal incident angle 45°. The design of the coating is given in Table 9.4.

An example of the design of a 90° retarder that operates at 45° in air and uses a completely dielectric approach is given in Table 9.5 with performance in terms of both retardance and ψ in Figures 9.28 and 9.29. The basic reflector is a 21-layer quarter-wave stack at a reference wavelength of 600 nm and tuned to 45°. It is constructed from material with indices of 1.45 and 2.40. The target specification of 90° is defined over 550–650 nm. The eventual variation in retardance is clearly rather larger than in the internally reflected cases, and the range of wavelengths over which the performance is achieved, rather less. The overcoat that assures the retardance has 12 layers. The value of ψ for this coating is shown in Figure 9.29, and any errors from this source are clearly well below those from the achieved values of Δ.

9.8 Optical Tunnel Filters

At an earlier stage in the development of narrowband filters, a main barrier to their construction was the fabrication of reflecting stacks of sufficiently low loss, and it appeared that the phenomenon of frustrated total internal reflection might offer some hope as a possible solution. This phenomenon has been known for some time. If light is incident on a boundary beyond the critical angle, it will normally be completely reflected. However, the incident light does in fact penetrate a

TABLE 9.5

90° Retarded for Air Incident Medium

Layer	Refractive Index	Thickness (Optical)
Medium	1.0000	
1	1.4500	0.185990
2	2.4000	0.360087
3	1.4500	0.408571
4	2.4000	0.364333
5	1.4500	0.374153
6	2.4000	0.344197
7	1.4500	0.380938
8	2.4000	0.329630
9	1.4500	0.344621
10	2.4000	0.291818
11	1.4500	0.368586
12	2.4000	0.079470
13	2.4000	0.261612
14	1.4500	0.286358
15	2.4000	0.261612
Layers 16 to 31 are repeats of 14 and 15		
32	1.4500	0.286358
33	2.4000	0.261612
Substrate	1.5200	

Note: Reference wavelength: 600 nm; angle of incidence in air: 45°.

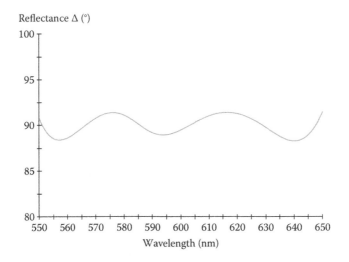

FIGURE 9.28

Retardance Δ at 45° in air from a dielectric reflector of materials 2.40 and 1.45 arranged as a basic reflector for 45° with outer correcting layers. See Table 9.5 for the design.

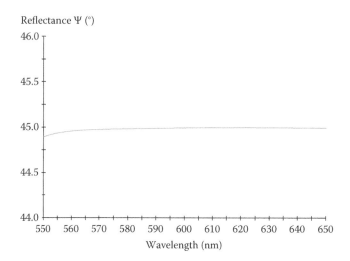

FIGURE 9.29
Value of ψ for the retarder of Figure 9.28 and Table 9.5.

short distance into the second medium, where it exponentially decays. Provided the second medium is somewhat thicker than a wavelength or so, the decay will be more or less complete, and the reflectance, unity. If, on the other hand, the second medium is made extremely thin, then the decay may not be complete when the wave meets the boundary with the third medium. Then, if in the third medium the angle of propagation is no longer greater than critical, a proportion of the incident light will appear as a progressive wave and the reflectance at the first boundary will be something short of total. This, as Baumeister [21] pointed out, is very similar to the behavior of fundamental particles in tunneling through a potential barrier, and he has used the term *optical tunneling* to describe the phenomenon. The most important feature of the effect, as far as the thin-film filter is concerned, is that the frustrated total reflection can be adjusted to any desired value, simply by varying the thickness of the frustrating layer between the first and third media.

The method of constructing a filter using this effect is very similar to the polarizing beam splitter (Section 9.4.2). The hypotenuse of a prism is first coated with a frustrating layer of lower index so that the light will be incident at an angle greater than critical. This is a function of the prism angle, refractive index, and the refractive index of the frustrating layer. Next follows the cavity layer that must necessarily be of higher index, so that a real angle of propagation will exist. This, in turn, is followed by yet another frustrating layer. The whole is then cemented into a prism block by adding a second prism. The angle at which light is incident on the diagonal face must be greater than the angle ψ given by

$$\sin \psi = n_F / n_G,$$

where n_F is the index of the frustrating layer and n_G is the index of the glass of the prism. For $n_F = 1.35$ and $n_G = 1.52$, we find $\psi = 63°$, a quite appreciable angle. Usually, glass of rather higher index, nearer 1.7, is used to reduce the angle as far as possible.

Although at first sight the optical tunnel or frustrated total reflectance (FTR) filter appears most attractive and simple, there are some considerable disadvantages. First, there is an enormous shift in peak wavelength between the two planes of polarization. Typical figures quoted are on the order of 100 nm in the visible region, the peak corresponding to *p*-polarization being at a shorter wavelength. This large polarization splitting is due to the large angle of incidence at which the device must be used. Although the phase thickness of the cavity does not vary with polarization, the phase changes at the interface do vary. Another effect of this large angle is that the angle sensitivity of the filter is extremely large. Shifts of 5 nm/° have been calculated [21].

Added to these disadvantages is the fact that the attempts made to produce FTR filters have been very disappointing in their results, the performance appearing to fall far short of what was theoretically expected. It seemed that the difficulties inherent in the construction of the FTR filter were at least as great as those involved in the conventional Fabry–Perot filter. Because of this, interest in the FTR filter has been mainly theoretical, and the filter does not appear to have ever been in commercial production.

The theory is given in great detail by Baumeister [21], who also surveyed the quite extensive literature at the time. Not only did he cover the FTR filter, but he also pointed out that as far as the theory is concerned, the frustrating layer, or, as he renamed it, the tunnel layer, behaves exactly as a loss-free metal layer. This implies that all sorts of filters including ones similar to induced transmission filters are possible using tunnel layers, although strictly, the induced transmission filter is one that maximizes transmittance when there is loss in the metal layer. Designs for a number of these are included in the paper. One conclusion that Baumeister reaches is that there appears to be no practical application for the tunnel layer filter of the induced transmission and FTR single-cavity types. However, he does mention the possibility of a longwave-pass filter constructed from an assembly of many tunnel layers separated by dielectric layers with the advantage of a limitless rejection zone on the shortwave side of the edge. Even with this type of filter, there are some disadvantages that could be serious. The characteristics of the filter near the edge suffer from strong polarization splitting. This could be overcome by adding a conventional edge filter to the assembly at the front face of the prism. However, the second disadvantage is rather more serious: the appearance of passbands in the stop region when the filter is tilted in the direction so as to make the angle of incidence more nearly normal. Curves given by Baumeister show a small transmission spike appearing even with a tilt of only 1° internal, or 2.7° external, with respect to the design value.

Despite the pessimistic conclusion in the Baumeister paper, the concept was revitalized by Li and Dobrowolski [7] in the construction of highly efficient thin-film polarizers. The difficulties of very low admittance contrast between high- and low-index materials for p-polarization at angles around 45° in glass disappear at high angles of incidence when one of the materials is operating beyond critical as a tunnel layer. The design concepts are rather more complex; they do not involve the Brewster condition, but computers are not at all inhibited by such complexity, and so automatic refinement and synthesis are particularly useful methods.

References

1. A. Thelen. 1966. Equivalent layers in multilayer filters. *Journal of the Optical Society of America* 56:1533–1538.
2. S. M. MacNeille. 1946. *Beam splitter*. US Patent, 2,403,731.
3. L. Songer. 1978. The design and fabrication of a thin film polarizer. *Optical Spectra* 12(10):45–50.
4. D. Blanc, P. H. Lissberger, and A. Roy. 1979. The design, preparation and optical measurement of thin film polarizers. *Thin Solid Films* 57:191–198.
5. R. P. Netterfield. 1977. Practical thin-film polarizing beam splitters. *Optica Acta* 24:69–79.
6. M. Banning. 1947. Practical methods of making and using multilayer filters. *Journal of the Optical Society of America* 37:792–797.
7. L. Li and J. A. Dobrowolski. 2000. High-performance thin-film polarizing beam splitter operating at angles greater than the critical angle. *Applied Optics* 39:2754–2771.
8. P. B. Clapham. 1969. *The Preparation of Thin Film Polarizers*. OP. Met. 7. Teddington, UK: National Physical Laboratory.
9. A. Thelen. 1976. Nonpolarizing interference films inside a glass cube. *Applied Optics* 15:2983–2985.
10. A. Thelen. 1981. Nonpolarizing edge filters. *Journal of the Optical Society of America* 71:309–314.
11. Z. Knittl and H. Houserkova. 1982. Equivalent layers in oblique incidence: The problem of unsplit admittances and depolarization of partial reflectors. *Applied Optics* 11:2055–2068.

12. R. C. Jones. 1941. A new calculus for the treatment of optical systems: I. Description and discussion of the calculus. *Journal of the Optical Society of America* 31:488–493.

13. J. Henry Hurwitz and R. C. Jones. 1941. A new calculus for the treatment of optical systems: II. Proof of three general equivalence theorems. *Journal of the Optical Society of America* 31:493–499.

14. R. C. Jones. 1941. A new calculus for the treatment of optical systems: III. The Sohnke theory of optical activity. *Journal of the Optical Society of America* 31:500–503.

15. R. J. King. 1966. Quarter wave retardation systems based on the Fresnel rhomb. *Journal of Scientific Instruments* 43:617–622.

16. M. P. Lostis. 1957. Etude et realisation d'une lame demi-onde en utilisant les proprietes des couches minces. *Journal de Physique et le Radium* 18:51S–52S.

17. W. H. Southwell. 1979. Multilayer coatings producing 90° phase change. *Applied Optics* 18:1875.

18. W. H. Southwell. 1980. Multilayer coating design achieving a broadband 90° phase shift. *Applied Optics* 19:2688–2692.

19. J. H. Apfel. 1981. Graphical method to design multilayer phase retarders. *Applied Optics* 20:1024–1029.

20. J. H. Apfel. 1984. Graphical method to design internal reflection phase retarders. *Applied Optics* 23:1178–1183.

21. P. W. Baumeister. 1967. Optical tunneling and its application to optical filters. *Applied Optics* 6:897–905.

10

More on Tilted Coatings

10.1 Introduction

In Chapter 9, we limited ourselves largely to dielectric layers. Now, in this chapter, we extend our investigation to include absorbing layers and especially metals, and we pay particular attention to the critical angle and beyond. The complication is that instead of a real n, the refractive index, we now have $(n - ik)$, the complex form.

We have already very briefly associated metals with the admittance diagram in Chapter 3. We will now examine metallic admittance loci in more detail. Also, we have not faced the problems of Snell's Law with complex arguments, necessary when we are dealing with tilted absorbing layers. There are some, at first sight, strange effects that occur with dielectric-coated metallic reflectors. Under certain conditions and at reasonably high angles of incidence, sharp absorption bands can exist for one plane of polarization. This can create difficulties with dielectric-overcoated reflectors such as protected silver. Metal layers beyond the critical angle are still more interesting. They can exhibit resonance-like features that are useful in many ways, particularly in sensing. We begin by returning to the tilted admittances, but this time including absorption.

10.2 Modified Admittances and the Tilted Admittance Diagram

We recall the form of the admittances and the phase thickness of a dielectric material at oblique incidence:

$$\delta = \frac{2\pi nd \cos \vartheta}{\lambda},$$ (10.1)

$$\eta_s = y \cos \vartheta,$$ (10.2)

$$\eta_p = \frac{y^2}{\eta_s} = \frac{y}{\cos \vartheta},$$ (10.3)

where

$$n_0 \sin \vartheta_0 = n_1 \sin \vartheta_1 = n_2 \sin \vartheta_2 = \dots,$$ (10.4)

and n, y, d, and ϑ are the values appropriate to the particular material, the subscript 0 indicating, as usual, the incident medium.

We recall from Chapter 2 that the problem presented when the values of n and y are permitted to be complex is solved by adopting a slightly different form for Equations 10.1 through 10.3. The incident medium retains its real characteristics so that it is free from absorption, but the index of the absorbing material is complex, and so the relevant ϑ, the propagation angle, must be permitted to become complex to support the complex $\sin\vartheta$. A complex angle is difficult to visualize. However, we can avoid any interpretation of the complex angle by noting that it is actually $\cos\vartheta$ that we need, and $\cos^2\vartheta$ is just $1 - \sin^2\vartheta$. Then the arrangement when absorbing materials are concerned becomes

$$\delta = \frac{2\pi d\left(n^2 - k^2 - n_0^2\sin^2\vartheta_0 - 2ink\right)^{1/2}}{\lambda}, \tag{10.5}$$

where physical considerations tell us that the square root must be taken in the fourth quadrant. Also,

$$\eta_s = \left(n^2 - k^2 - n_0^2\sin^2\vartheta_0 - 2ink\right)^{1/2}, \tag{10.6}$$

so that

$$\eta_p = \frac{(n - ik)^2}{\eta_s}, \tag{10.7}$$

where we are working, as usual, in free space units.

The two sets of relationships are completely compatible. Equations 10.1 through 10.3 are better stated as Equations 10.5 through 10.7 when $\cos\vartheta$ is permitted to become complex.

In Chapter 9, we saw the advantages for the admittance diagram of introducing the modified admittances, normalized so that the incident medium, necessarily dielectric, always exhibited its normal incidence value. By applying the same technique to our absorbing materials, we obtain the following for their modified admittances:

$$\eta_s = \frac{\left(n^2 - k^2 - n_0^2\sin^2\vartheta_0 - 2ink\right)^{1/2}}{\cos\vartheta_0} \tag{10.8}$$

and

$$\eta_p = \frac{(n - ik)^2}{\eta_s}, \tag{10.9}$$

the fourth quadrant solution still being the correct one for the square root.

We recall that provided the modified admittances are used in constructing the loci, then the isoreflectance and isophase circles on the admittance diagram will remain exactly the same as at normal incidence for both p- and s-polarizations.

The admittance locus of a metal layer is a little more complicated than that of a dielectric. For a lossless metal in which the refractive index, and hence the optical admittance, is purely imaginary and given by $-ik$, the loci are a set of circles with centers on the real axis and passing through points ik and $-ik$ on the imaginary axis. Figure 10.1 shows the typical form. The circles, like the dielectric ones, are traced out clockwise starting on ik and ending on $-ik$. Real metallic layers do depart from this ideal model, but if the metal is of high performance, i.e., if the ratio k/n is high, then the loci are similar to the perfect case. It is as if the diagram were slightly rotated about the origin so that the points where all circles intersect are $(-n, k)$ and $(n, -k)$, respectively, although the circles can never reach point $(-n, k)$ since admittance loci are constrained to the first and second quadrants of the admittance diagram. Figure 10.2 shows a set of optical admittance loci calculated for silver, $n - ik = 0.075 - i3.41$ (value at 546 nm from Berning and Turner [1]) demonstrating this typical behavior, although the departure from ideal can be seen as very small. The direction of the loci is perhaps now better described as terminating on $(n, -k)$, because although most are still

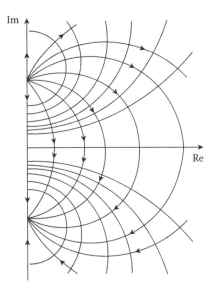

FIGURE 10.1
Admittance loci for an ideal metal with admittance $-ik$. The loci begin at point ik and terminate on at $-ik$. Equithickness contours are also shown at no fixed intervals. Similar loci are obtained for s-polarized FTR layers. For p-polarized FTR layers, the shape of the loci is similar, but they are traced in the opposite direction.

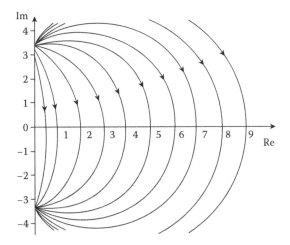

FIGURE 10.2
Admittance loci for silver at normal incidence in the visible region. The value assumed for the optical constants at 546 nm is, from Berning and Turner [1], $0.075 - i3.41$.

described in a clockwise direction, it is possible for some very truncated loci now to come from the left of the end point, and some of these are strictly counterclockwise in direction. As we depart more and more from perfection with the k/n ratio falling, the circles gradually transform into spirals. We will omit metals not of high optical quality with their spiral loci terminating at $(n, -k)$ from the discussion in this chapter. Now we need to consider what happens to the loci at oblique incidence.

The phase thickness at normal incidence is

$$\delta = \frac{2\pi(n - ik)d}{\lambda},$$
(10.10)

dominated by the imaginary part. At oblique incidence, it becomes

$$\delta = \frac{2\pi\left(n^2 - k^2 - n_0^2\sin^2\vartheta_0 - 2ink\right)^{1/2}d}{\lambda},$$
(10.11)

still in the fourth quadrant. Since $n_0 \sin \vartheta_0$ is normally small compared with k, it has little effect on the phase thickness. It slightly reduces the real part and increases the imaginary part, but the effect is small, and the behavior is essentially similar to that at normal incidence. At an angle of incidence of 80° in air, for example, the phase thickness of silver changes from $2\pi(0.075 - i3.41)d/\lambda$ to $2\pi(0.00721 - i3.549)d/\lambda$. The change in the modified admittance is therefore mainly due to the $\cos \vartheta_0$ term. The ratio of real to imaginary parts remains virtually the same, and the p-admittance simply moves toward the origin (both real and imaginary parts reduced), and the s-admittance, away from the origin. Thus, the principal effect for high-performance metal layers with tilt is an expansion of the circular loci for s-polarization and a contraction for p-polarization. The basic form remains the same. We shall see shortly that our explanation must be modified when we move beyond the critical angle.

10.3 Application of the Admittance Diagram

The shift in the modified optical admittance does make it easier to see that the phase shift on reflection from a massive metal will vary with angle of incidence. For silver, or most other high-performance metals, at normal incidence, the phase shift will be in the second quadrant. As the angle of incidence increases, the movement of the p-polarized admittance toward the origin implies that the p-polarized phase shift moves toward the first quadrant. We know that isoreflectance circles interest the real axis in two points with product y_0^2. We also know, therefore, that the isoreflectance circle in Figure 10.3, ACB, which is tangent to the metal admittance line at point C, must have $OA \cdot OB = y_0^2$. Geometry tells us that OC^2 is equal to $OA \cdot OB$ and must be given by y_0^2. The circle centered on the origin and passing through the point y_0 on the real axis divides the lower quadrant of the admittance diagram into reflectance phase shifts in the first and second quadrants, as shown in the figure and must pass through C. Point C therefore marks a minimum p-reflectance and a phase shift on reflection of 90°. Because the behavior somewhat resembles the Brewster phenomenon with dielectric surfaces, the angle is known as the pseudo-Brewster angle. At the pseudo-Brewster angle, the phase shift for s-reflectance is not quite 180°, and so the angle of incidence has to slightly increase for the difference in phase shift between the two polarizations to become 90°. This slightly greater angle is called the principal angle. We can see that a reasonable approximation for the pseudo-Brewster angle is given by $\cos \vartheta_0 = y_0/(n^2 + k^2)^{1/2}$. For silver with optical constants $(0.125 - i3.339)$ in air [2], this expression gives about 72.6°. Accurate calculation makes it nearer 70°. Figure 10.4 shows the calculated reflectance and retardance Δ of an unprotected silver surface at 550 nm as a function of the angle of incidence in air. Note that the principal angle is just slightly greater than the pseudo-Brewster angle, as predicted by the admittance locus.

Now we examine what happens when a metal layer is overcoated with a dielectric layer. The arrangement is schematically sketched in Figure 10.5. Provided the admittance η_f of the dielectric layer is less than $(\eta_g\eta_g{}^*)^{1/2}$, where η_g is the admittance of the metal layer, the admittance locus will loop outside the line joining the origin to the starting point, as in the diagram. For dielectric layers having admittance greater than that of the incident medium, the reflectance falls with increasing

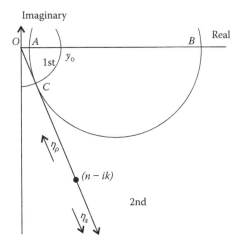

FIGURE 10.3
Variation of the modified admittance of a typical high-performance metal as the angle of incidence is increased. The diagram exaggerates the distance of point $(n - ik)$ from the imaginary axis to make the relationships clearer. At normal incidence, both s- and p-polarizations have admittance $(n - ik)$. As ϑ_0 increases, $\cos\vartheta_0$ gradually drops from unity at normal incidence to zero at grazing incidence. The modified η_p therefore drops toward the origin while η_s moves away. Geometry tells us that a minimum p-reflectance occurs at point C, where the phase shift on reflection crosses the boundary between second and first quadrants.

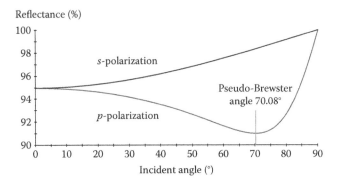

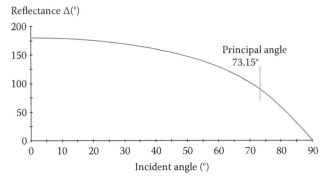

FIGURE 10.4
Reflectance and retardance of a silver surface at 550 nm. Optical constants from Palik [2] are $0.125 - i3.339$.

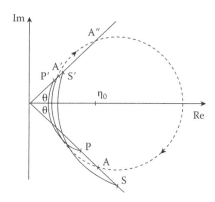

FIGURE 10.5
Schematic diagram of a dielectric overcoat on a metal surface. At normal incidence, the metal admittance is at point A. A' represents a quarter-wave thickness of material, while A" represents the point at which the reflectance returns to the starting value. The lowest reflectance is given by the intersection with the real axis between points A and A'. When tilted, the p-locus is given by PP', and the s-locus, by SS'.

thickness while the locus is in the fourth quadrant of the admittance diagram. At the intersection with the real axis, the reflectance is a minimum. It then begins to rise, but at quarter-wave point A' given by η_f^2/η_g, it is still below the reflectance of the bare metal. Only at point A" does the reflectance return to its initial level. The drop in reflectance for silver is usually slight, but for aluminum, it can be catastrophic. Silver is therefore usually overcoated with a quarter-wave protective layer, but aluminum, especially if luminous reflectance is what is required, with a half-wave one, as explained in Chapter 5. The higher the index of the overcoat, the lower the reflectance at the minimum, and so low-index protecting layers such as silica are much to be preferred.

As the metal–dielectric combination is tilted, the modified p-admittance of the metal slides toward the origin, with a drop in reflectance, while the s-admittance moves away from the origin with a rise in reflectance. The dielectric layer shows a drop in admittance for p-polarized light and an increase for s-polarized. For dielectric coatings that are near a quarter wave, these changes tend to compensate, and, indeed, in silver slightly overcompensate, for the changes in reflectance of the bare metal. The p-reflectance of the overcoated metal tends to be slightly higher than the s-reflectance.

Eventually, for very high angles of incidence, the p-polarized admittance of the dielectric layer falls below the admittance of the incident medium, and now the fourth quadrant portion of the locus represents increasing reflectance. This means that the dielectric overcoating, when thin, instead of reducing the reflectance of the metal, actually slightly enhances it. Thus, although it depends on the final thickness of the dielectric layer, the p-reflectance will tend to be high. For s-polarized light, the admittance of the dielectric layer tends to infinity as the angle of incidence tends to 90°. The locus of the dielectric overcoat therefore tends more and more toward a vertical line. As the admittance of the metal moves away from the origin, its projection in the real axis moves further to the right, eventually crossing the incident medium admittance and continuing toward infinity. There must therefore be an angle of incidence, which is very high, where the locus of the dielectric overcoat will intersect the real axis at the admittance of the incident medium. If the thickness is chosen so that the locus terminates at this point, then the reflectance of the metal–dielectric combination will be zero. This will occur for one particular value of angle of incidence and for a precise value of the dielectric layer thickness, and the dip in reflectance will show a rapid variation with angle of incidence. Such behavior, for s-polarized light, of a metal overcoated with a thin dielectric layer was predicted by Nevière and Vincent [3] from a quite different analysis based on a Brewster absorption phenomenon in a lossy waveguide used just under its cutoff thickness. Since the modified admittance for s-polarized light increases with angle of incidence only in the case where its refractive index is greater than that of the incident medium, this is a necessary condition for the observation of the effect. It turns out that there is a similar resonance for *p*-polarization. The increased flexibility given by two dielectric layers deposited on a metal has been used to advantage in the design of reflection polarizers [4]. For a more detailed analysis of a dielectric overcoated metal including the effects due to *p*-polarization, see Macleod [5].

A different but related phenomenon, already briefly alluded to in Chapter 5, was discovered by Pellicori [6] and theoretically investigated by Cox et al. [7,8] in connection with an infrared mirror of protected aluminum with a protective overcoat of silicon dioxide. We can understand the behavior with the help of an admittance diagram. The silicon dioxide is heavily absorbing in the region beyond 8 µm. At a wavelength of just over 8 µm, *n* and *k* have values of around 0.4 and 0.3, respectively. At normal incidence, the admittance loci of the silicon dioxide are spirals that end on the characteristic admittance of the silicon dioxide and are described in a clockwise manner, in much the same way as the silver loci already discussed. At nonnormal incidence, the *s*-polarized admittance and the phase factor for the layer remain in the fourth quadrant, and so the behavior of the silicon oxide is similar to that at normal incidence. The *p*-polarized admittance, however, moves toward the first quadrant and enters it at an angle of incidence of around 40°. The behavior of such a material, where the phase thickness is in the fourth quadrant but the optical admittance is in the first, is different from normal materials in that the spirals are now traced out counterclockwise, rather than clockwise. The admittance of aluminum at 8.1 µm is around $18.35 - i55.75$, and for *p*-polarized light at an angle of incidence of 60°, the modified admittance becomes $9.176 - i27.87$. The dielectric locus sweeps down toward the real axis, as in Figure 10.6 and, in a thickness of 150 nm, terminates near point $(1, 0)$, so that the reflectance is near zero. Calculated performance at long wavelengths of a protected aluminum reflector is shown in Figure 10.7.

This behavior is quite unlike the normal behavior to be expected with lossless dielectric overcoats that have refractive index greater than that of the incident medium. However, we shall see that it does have a certain similarity with one of the techniques for generating surface electromagnetic waves, which we shall be dealing with shortly, where the coupling medium is a dielectric

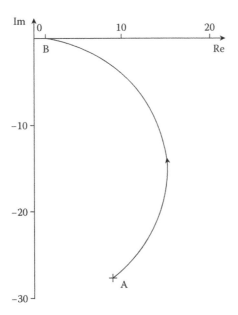

FIGURE 10.6
p-Polarized admittance locus for 150 nm thickness of SiO_2 $(0.39 - i0.29)$, on aluminum, $(18.35 - i55.75)$, at an angle of incidence of 60°. A is the point corresponding to the modified admittance of aluminum, and the counterclockwise curvature of the spiral locus carries it into the region of low reflectance.

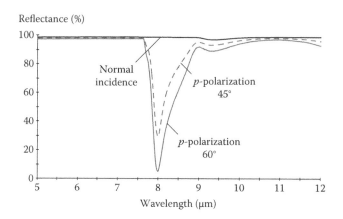

FIGURE 10.7
Calculated reflectance of an aluminum reflector protected by 150 nm thickness of SiO_2 at normal and oblique incidences. Reflectance for s-polarization is not shown but is virtually indistinguishable from the normal incidence curve. (Material data taken from E. D. Palik, ed. 1985. *Handbook of Optical Constants of Solids I*. Orlando, FL: Academic Press.)

layer of index lower than that of the incident medium, and where the angle of incidence is beyond the critical angle.

We now turn back to dielectric materials and investigate what happens when angles of incidence exceed the critical angle. The critical angle and angles beyond it are discussed in Chapter 2, but Equations 10.8, 10.9, and 10.11 are also the relevant equations when we have $k = 0$ and $n_0 \sin \vartheta_0 > n$. The phase thickness at normal incidence $2\pi nd/\lambda$ becomes, from Equation 10.11,

$$\delta = \frac{2\pi\left(n^2 - n_0^2\sin^2\vartheta_0\right)^{1/2}d}{\lambda},$$

i.e.,

$$\delta = -i\frac{2\pi\left(n_0^2\sin^2\vartheta_0 - n^2\right)^{1/2}d}{\lambda} \qquad (10.12)$$

at oblique incidence, where, again, the fourth rather than second quadrant solution is correct. The modified admittances are then

$$\eta_s = -i\frac{\left(n_0^2\sin^2\vartheta_0 - n^2\right)^{1/2}}{\cos\vartheta_0},$$

$$\eta_p = \frac{n^2}{\eta_s}. \qquad (10.13)$$

Since, beyond critical, η_s is negative imaginary, η_p must be positive imaginary. The behavior of the modified admittance is diagrammatically shown in Figure 10.8. For a thin film of material used beyond the critical angle, then the s-polarized behavior is indistinguishable from that of an ideal metal. We have a set of circles centered on the real axis, described clockwise and ending on point η_s on the negative imaginary axis. For p-polarized light, the behavior is, in one important respect, different. Here, the combination of negative imaginary phase thickness and positive imaginary admittance inverts the way in which the circles are described, so that although they are still centered on the origin, they are counterclockwise and terminate at η_p on the positive imaginary axis. This behavior plays a significant part in what follows. We assume a beam of light incident on the hypotenuse of a prism beyond the critical angle. Simply for plotting some of the following figures, we assume a value for the index of the incident medium of 1.52.

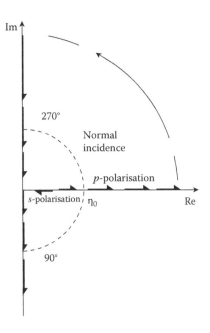

FIGURE 10.8
Variation of the s-polarized and p-polarized modified admittances of free space with respect to an incident medium of higher index. η_0 is the incident admittance. The s-admittance falls along the real axis until zero at the critical angle, and then it turns along the negative direction of the imaginary axis tending to negative imaginary infinity as the angle of incidence tends to 90°. The p-admittance rises along the real axis, passing the point η_0 at the Brewster angle, becoming infinite at the critical angle, switching over to positive imaginary infinity and then sliding down the imaginary axis tending to zero as the angle of incidence tends to 90°.

For an uncoated hypotenuse, the second medium is air of refractive index unity. The modified admittance for p-polarized light is positive imaginary and, as ϑ_0 increases, falls down the imaginary axis toward the origin. The reflectance is unity, and Figure 10.8 shows that the phase shift varies from 180° through the third and fourth quadrants toward 360° or 0°. The s-polarized reflectance is likewise unity, but the admittance is negative imaginary and falls from zero to infinity along the imaginary axis so that the s-polarized phase shift increases with ϑ_0 from zero, through the first and second quadrants toward 180°. Since the incident medium has an admittance of 1.52, the circle separating the first and second quadrants and the third and fourth quadrants, which has its center the origin, has a radius of 1.52.

Now let a thin film be added to the hypotenuse. Since we are treating our glass prism as the incident medium, we should treat the surrounding air as the substrate. Thus, the starting admittance for the film is on the imaginary axis. Provided the thin film has no losses, then the admittance of the film–substrate combination must remain on the imaginary axis. If the film admittance is imaginary, the combination admittance will simply move toward the film admittance. If, however, the film admittance is real, the admittance of the combination will move along the imaginary axis in a positive direction, returning to the starting point every half wave at the given angle of incidence. The lower the modified admittance, the slower the locus moves near the origin and the faster at points far removed from the origin. The variation of phase change between the fourth quadrant and the start of the first quadrant is therefore slower, while that between the third and second quadrants is faster than for a higher admittance. Thus, there is a wide range of possibilities for varying the relative phase shifts for p- and s-polarizations by choosing an overcoat of higher or lower index and varying the thickness or even by adding additional layers [9–11].

Given that the starting point is on the axis, then the only way in which the admittance can be made to leave it is by an absorbing layer. We turn to the set of metal loci (Figure 10.2), and we can see that for a range of values of starting admittance on the imaginary axis, the metal loci loop around, away from the axis, to cut the real axis. Although Figure 10.2 shows the behavior of metal layers for an incident medium of unity at normal incidence, the tilted behavior for an incident admittance of 1.52 is quite similar. Figure 10.9 shows the illuminating arrangement and the *p*-polarized loci. For a very narrow range of starting values, the metal locus cuts the real axis near the incident admittance, and if the metal thickness is such that the locus terminates there, then the reflectance of the combination will be low. For one particular angle of incidence and metal thickness, the reflectance will be zero. It should not be too much of a surprise to find that the condition is very sensitive to the angle of incidence. Since the admittance of the metal varies much more slowly than that of the air substrate, the zero reflectance condition will no longer hold, even for quite small tilts. One can think of it in terms of a rather long lever with a fulcrum and a forced movement very close to it so that the end of the lever considerably magnifies the movement. This very narrow drop in reflectance to a very low value, which has all the hallmarks of a sharp resonance, can be interpreted as the generation of a surface plasma wave, or plasmon, on the metal film. This coupling arrangement, due to Kretschmann and Raether [12], cannot operate for *s*-polarized light without modification. The admittance of the substrate for *s*-polarization is now on the negative part of the imaginary axis, and therefore, any metal that is deposited will simply move the admittance of the combination toward the admittance of the bulk metal.

Our explanation of the phenomenon concentrates on the electromagnetic behavior. An alternative theoretical route treats the free electrons in the metal, which with the positively charged fixed atoms, form a kind of neutral plasma. The electrons are found to collectively oscillate in the manner of a plasma wave, and their collective oscillations can be treated as a quasiparticle subject to quantum mechanical laws and known as a Plasmon. There are bulk plasmons, but those that concern us are a surface phenomenon where the quasiparticle is known as a surface plasmon. The surface plasmon can couple to an electromagnetic wave, the combination also acting as a quasiparticle known as a polariton. To distinguish this particular class of polariton from other similar combinations, it is usually known as a surface plasmon polariton. Our optical interest is in the electromagnetic character of this phenomenon, and although the correct term for the total entity is

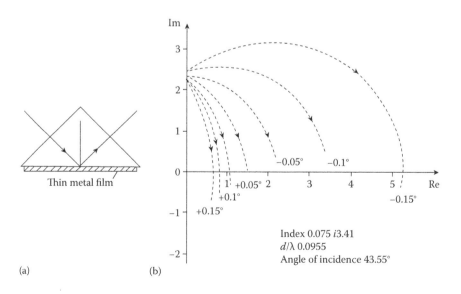

(a)   (b)

FIGURE 10.9

(a) Coupling to a surface plasmon. (After E. Kretschmann and H. Raether, *Zeitschrift für Naturforschung*, 23A, 2135–2136, 1968.) (b) *p*-Polarized admittance locus corresponding to the arrangement in (a). The solid curve corresponds to the optimum angle of incidence and thickness of metal (silver) film. The dashed curves correspond to changes in the angle of incidence as marked on each curve.

surface plasmon polariton, it is normally shortened to surface plasmon. We continue with the electromagnetic aspects.

An alternative coupling arrangement, devised by Otto [13] predating the Kretchmann arrangement and essentially launching the optical studies, involves the excitation of surface plasmons through an evanescent wave in an FTR layer. We recall that the admittance locus for *p*-polarization of a layer used beyond the critical angle is a circle described in a counterclockwise direction. This means that such a layer can be used to couple into a massive metal. Here the metal acts as the substrate, with a starting admittance in the fourth quadrant of the admittance diagram. For *p*-polarized light, the dielectric FTR layer has a circular locus that cuts the real axis. Clearly, then for the correct angle of incidence and dielectric layer thickness, the reflectance can be made zero. Surface plasmons and their applications were extensively reviewed by Raether [14]. Abelès [15] included an account of the optical features of such effects in his review of the optical properties of very thin films.

Now let us return to the first case of coupling and let us examine what happens when a thin layer is deposited over the metal next to the surrounding air. The starting admittance is, as before, on the imaginary axis, but now the dielectric layer modifies that position, so that the starting point for the metal locus is changed. Because the metal loci at the imaginary axis are closely clustered together, almost intersecting, a small change in the starting point produces an enormous change in the locus and, hence, in the point at which it cuts the real axis, leading to a substantial change in reflectance (Figure 10.10). This very large change that a thin external dielectric film makes to the internal reflectance of the metal film has been used in the study of contaminant films adsorbed on metal surfaces. Film thicknesses of a few angstroms have been detected in this way. Provided that the film is very thin, then an additional tilt of the system will be sufficient to pull the intersection of the metal locus with the real axis back to the incident admittance, and so the effect can be interpreted as a shift in the resonance rather than a damping.

This result helps us devise a method for exciting a similar resonance with *s*-polarized light. The essential problem is the starting point on the negative imaginary axis, which means that the subsequent metal locus remains within the fourth quadrant, never crossing the real axis to make it possible to have zero reflectance. The addition of a dielectric layer between the metal surface and the surrounding air can move the starting point for the metal on to the positive part of the imaginary axis so that the coated metal locus can cut the real axis for *s*-polarized light in just the same way as the uncoated metal in *p*-polarized light. Moreover, for both *p*- and *s*-polarized lights,

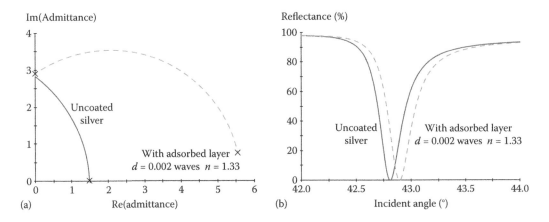

FIGURE 10.10
(a) Effect of a thin adsorbed layer on the surface plasmon resonance in silver at 632.8 nm. The solid line is the optimum while the dashed line is the change in the metal locus due to the adsorbed layer. (b) Calculated reflectance as a function of angle of incidence with and without the adsorbed layer.

the low reflectance will be repeated for each additional half-wave dielectric layer that is added. This behavior was used by Greenland and Billington [16] for the monitoring of optical layers intended as cavity layers for metal–dielectric interference filters. The operation of the cavities for inducing absorption devised by Harrick and Turner [17], although designed on the basis of a different approach, can also be explained this way.

Under certain conditions, it is also possible to arrange the dielectric overcoat such that an s-polarized resonance and a p-polarized resonance should be sufficiently close together in angle to make it possible to measure both together [18,19]. The electric field is still high at the outermost surface, and so the sensitivity to a thin added layer of material is undiminished. The added s-polarization resonance yields additional information on the structure of the detected material. Figure 10.11 shows typical resonances. The outer medium is water, which is common in biochemical investigations. The rugged nature of the outer SiO_2 layer implies that the detector can be cleaned and reused, an additional advantage over bare metal.

The long-range surface plasmon is an interesting and potentially useful phenomenon [20,21]. Once again, the light must be p-polarized for the effect to be possible. A thinner metal layer than in the classical case is employed, and it is surrounded by dielectric media that are beyond the critical angle so that they support evanescent waves. A resonance that is exceedingly narrow is induced in the reflectance of the assembly (Figure 10.12). This implies lower losses, and hence greater range, than the classical surface plasmon. We use, once again, an infinite plane wave to excite the resonance. We can understand it by using an admittance diagram.

Figure 10.13 shows the admittance diagram at resonance. The massive material SiO_2 behind the silver is effectively the substrate for the system. Since it is beyond critical, its p-admittance is positive imaginary, and so the starting point for the locus is on the positive limb of the imaginary axis. The silver locus loops around in a huge circle and ends such that the starting point for the subsequent SiO_2 locus is below the real axis and close to the imaginary axis. The final (front) layer is also SiO_2 and beyond critical. Its p-admittance is therefore also positive imaginary. This implies a counterclockwise direction for its admittance locus that loops around to terminate at the admittance of the incident medium, in this case, 1.8. The enormous length of the admittance locus implies a rapid change in the conditions as the angle of incidence changes, and hence, a very narrow dip in the p-reflectance of the system and, hence, a lower loss for the plasmon.

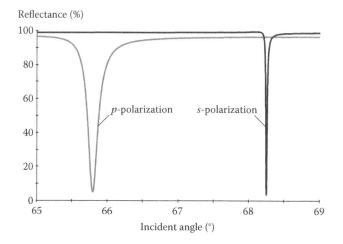

FIGURE 10.11

Resonances at 632.8 nm for both p- and s-polarizations achieved by overcoating the silver layer with a thick low-index layer (SiO_2). The outer medium is water rather than air, and the thickness of silver is 51.1 nm, and of SiO_2, 502.3 nm. The incident medium is borosilicate glass. Note that both resonances are considerably narrower than that in Figure 10.10.

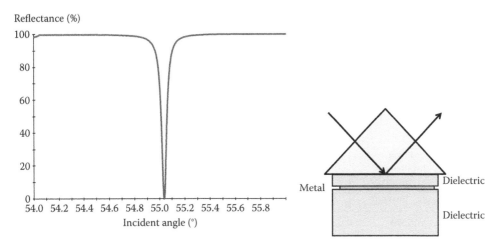

FIGURE 10.12
Long-range surface plasmon resonance. As before, the resonance exists for p-polarized light. It is shown here for an incident medium of index 1.8 with a thin silver film, which is 20.64 nm thick, surrounded by SiO_2 that is 1037 nm thick in front and massive behind. The wavelength is 632.8 nm.

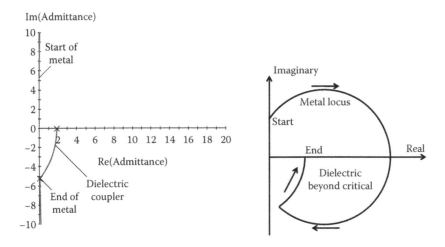

FIGURE 10.13
p-Polarized admittance locus associated with the long-range surface plasmon. The angle of incidence in the 1.8 index incident material is 55.035°. The massive material SiO_2 behind the silver is beyond critical, and therefore, the starting point for the locus is on the positive limb of the imaginary axis. The silver locus loops around and ends close to the imaginary axis and below the real axis so that the subsequent SiO_2 locus that is beyond critical loops around to terminate at the admittance of the incident medium, in this case, 1.8. Note that the modified admittances are used. Note further that the loci of dielectric materials beyond critical for p-polarized light are circles described *counterclockwise*. An actual locus is on the left while that on the right is meant for explanation.

The total electric field amplitude is shown in Figure 10.14. The field in the metal layer is extremely small explaining the low loss. The field is largely, but not completely, perpendicular to the metal surfaces.

An unexpected transparency of metal layers surrounded by dielectric layers beyond the critical angle (Figure 10.15) was described by Dragila et al. [22], who explained it by coupled plasma waves on either side of the metal layer. This is virtually the same as the long-range plasmon except that now we are dealing with enhanced transmission rather than enhanced plasmon range. The problem [23,24] can be readily treated as an induced transmission filter, and the transmittance,

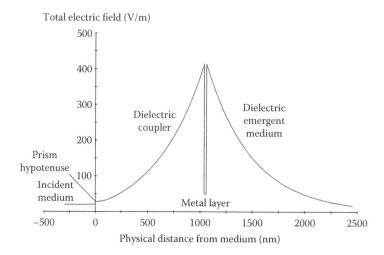

FIGURE 10.14

Total electric field amplitude through the system at resonance. Note that the electric field in the metal, that is, silver, layer is exceedingly low, and so the loss is smaller than that for the conventional one-sided surface plasmon. The field is plotted in volts per meter, assuming 1 W/m² input power density.

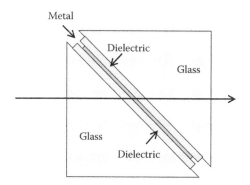

FIGURE 10.15

Aarrangement of the metal layer (silver) surrounded by two dielectric layers operating beyond the critical angle described by Dragila et al. [22].

maximized, all in a simple three-layer, dielectric–metal–dielectric design. Maximum potential transmittance from any given absorbing layer is achieved when the admittance at the rear surface is an optimum value that solely depends on the geometric thickness (that is, d/λ) of the layer and its optical constants. The theory is dealt with in Chapter 8. This optimum admittance is in the first quadrant and can be calculated from the analytical expressions already given. Here we have identical incident and exit media. The p-polarized locus for the low-index material next to the substrate sweeps around in a counterclockwise manner in the first quadrant from the admittance of the exit medium toward the imaginary axis, where it will eventually intersect it in the point that corresponds to the modified tilted admittance. Over the range of angles beyond critical, the p-admittance varies from infinity to zero along the positive limb of the imaginary axis. The metal optimum admittance varies more slowly, and it is straightforward to identify the appropriate angle for which the optimum admittance for a given metal thickness lies on the dielectric locus. This fixes the dielectric layer thickness, and the remainder of the design is then the addition of a layer in front of the metal exactly the same as that behind.

Figure 10.16 shows the admittance diagrams for a filter of this type, and Figure 10.17, the performance. The incident and exit media are high-index glass ($n = 1.8$); the wavelength is taken as 632.8 nm; and the thickness of the metal layer silver, as 80 nm with optical constants (0.0666 – i4.0452). The low-index material is silica ($n = 1.45$), and both first and third layers have a physical thickness of 314 nm. Once the wavelength and thickness of the silver, and the index of the matching dielectric layers, are chosen, the thickness of the dielectric layers and angle of incidence are then fixed.

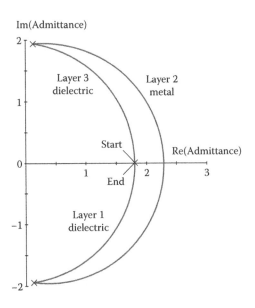

FIGURE 10.16
Admittance diagrams for the induced transmission filter described in the text. The counterclockwise p-loci of the dielectric layers are in opposition to the clockwise rotation of the silver loci. Design: 1.8 | 1.45, 344 nm, (0.0666 – i4.0452), 80 nm, 1.45, 344 m | 1.8, $\lambda = 632.8$ nm, $\vartheta_0 = 59.44°$.

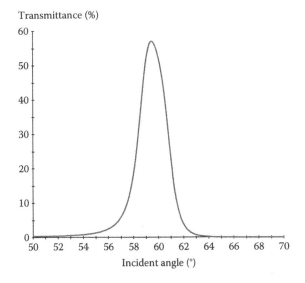

FIGURE 10.17
p-Polarized performance of the filter of Figure 10.16 as a function of angle of incidence at $\lambda = 632.8$ nm. The peak transmittance is the maximum possible from a silver layer of this thickness at this angle of incidence.

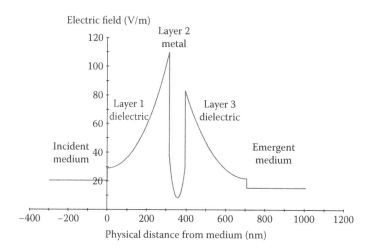

FIGURE 10.18
Distribution of total electric field amplitude through the filter of Figures 10.16 and 10.17. Note the similarity with the long-range plasmon distribution in Figure 10.14.

Note the similarity between the loci of the metal layer followed by the dielectric layer in Figure 10.16, that is, layers 2 and 3, and the corresponding layers in Figure 10.13. In fact, the mechanism that assures the high transparency of the metal layer in this filter is virtually the same as the one that ensures the low loss in the long-range surface plasmon as the electric field in Figure 10.18 suggests. An almost identical mechanism is responsible for the curious induced transparency of metal films pierced with regular arrays of small holes [25,26]. Bonod et al. [27] explain the phenomenon as a scattering of the light by the array of holes into a surface plasmon on the side of incidence. This induces a similar plasmon on the other side of the metal that is then scattered by the holes back into the direction of incidence. Indeed a diffraction grating on either side is calculated to perform in the same way. The electric field in the metal is reduced and, hence, the losses. The great advantage of this arrangement is that it can operate at normal incidence. The maximum possible transmittance is that predicted by the theory of induced transmission in Chapter 8.

10.4 Alternative Approach

We have seen how we can use our normal thin-film calculation techniques to explain the resonances that we associate with the concept of a surface plasmon, or more correctly, a surface plasmon polariton. What exactly is a surface plasmon, and where did the idea originate? The optical resonance is completely explicable within our classical electromagnetic theory of optical effects in thin films, and indeed, it was observed by various workers, largely in passing, before plasmons were recognized. Plasmon, however, suggests something to do with a plasma, and indeed, the modern thread of the surface plasmon story started with the first paper of a series by David Bohm and David Pines [28] on oscillations of free electrons in a metal. The ultimate objective was to explain the losses suffered by a fast charged particle impacting the metal. The mobile electrons together with their supporting matrix of fixed positively charged particles can be thought of as a kind of plasma, and this plasma responds to the impinging particle in collective oscillations that dissipate energy. The motion is quantized and, by suitable transformation,s can be represented as a quasiparticle that eventually came to be known as a plasmon. Then a little later, it was shown [29] that there are two kinds of collective motion, one involving electrons throughout the bulk of the metal, not the one that interests us, and the other involving the surface electrons and

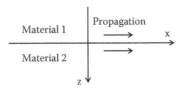

FIGURE 10.19
Surface and reference axes.

becoming known as a surface plasmon. The surface plasmon can be coupled with an electro-magnetic wave pinned to the metal surface and traveling along it, together with the surface plasmon. The combination is correctly known as a surface plasmon polariton, but in thin-film optics, the term is usually shortened to *surface plasmon*. The electromagnetic disturbance must be *p*-polarized, and it exponentially decays away from the surface both into the dielectric and into the metal. We shall use our normal theory to analyze it. The treatment that follows is similar to and relies on that in Macleod [30].

We start with a flat featureless plane separating two materials with labels 1 and 2 (Figure 10.19). A full general analysis is extremely tedious, and so we will use some special knowledge in arranging the problem. We assume propagation in the positive *x*-direction, and we know that the fields will exponentially decay away from the surface into the two materials. We know that any component of electric field in the *y*-direction is ineffective, and so we will confine the electric field to the *x–z* plane. We therefore have *x*- and *z*-components of electric field, and *y*-components of magnetic field only, that is, amplitudes $\mathcal{E}_x$, $\mathcal{E}_z$, and $\mathcal{H}_y$. In what follows, we will use the normal thin-film sign conventions (our conventions) rather than the solid state ones and that implies that some of the relationships will have signs differing from those usual in solid state textbooks.

We can represent the wave components in the two materials as plane waves having the following form:

$$
\begin{aligned}
E &= \mathcal{E}\exp[i(\omega t - \kappa_x x - \kappa_z z)], \\
H &= \mathcal{H}\exp[i(\omega t - \kappa_x x - \kappa_z z)],
\end{aligned}
\tag{10.14}
$$

where we will add subscripts 1 and 2 to distinguish between the parts in each material and where any relative phases are included in the complex amplitudes. There are certain boundary conditions that we can apply. The boundary is defined as $z = 0$. Frequency is invariant, and therefore, the phase factors in Equation 10.14 are identically equal at the boundary provided

$$
\kappa_x = \kappa_{x1} = \kappa_{x2}.
\tag{10.15}
$$

This directly leads to Snell's law. Additionally, there are the continuity of the tangential components of electric and magnetic fields and the normal components of the electrical displacement. Using the components that we have identified, these conditions become

$$
\begin{aligned}
\mathcal{H}_y &= \mathcal{H}_{y1} = \mathcal{H}_{y2}, \\
\mathcal{E}_x &= \mathcal{E}_{x1} = \mathcal{E}_{x2}, \\
\varepsilon_1 \mathcal{E}_{z1} &= \varepsilon_2 \mathcal{E}_{z2},
\end{aligned}
\tag{10.16}
$$

where ε is the appropriate permittivity. Then, from Maxwell's equations, we have two vector relationships:

$$
\begin{aligned}
\nabla \times \mathbf{H} &= \frac{\partial}{\partial t}(\varepsilon \mathbf{E}), \\
\nabla \times \mathbf{E} &= -\frac{\partial}{\partial t}(\mu_0 \mathbf{H}),
\end{aligned}
\tag{10.17}
$$

where we have used the condition that at optical frequencies, $\mu = \mu_0$. With some manipulation, this yields

$$\kappa_z \mathcal{H}_y = \omega \varepsilon \mathcal{E}_x,$$

$$\kappa_x \mathcal{H}_y = -\omega \varepsilon \mathcal{E}_z, \tag{10.18}$$

$$\kappa_x \mathcal{E}_z - \kappa_z \mathcal{E}_x = -\omega \mu_0 \mathcal{H}_y$$

for both materials 1 and 2. We have nine variables: $\mathcal{E}_{x1}$, $\mathcal{E}_{z1}$, $\mathcal{H}_{y1}$, $\mathcal{E}_{x2}$, $\mathcal{E}_{z2}$, $\mathcal{H}_{y2}$, κ_x, κ_{z1}, and κ_{z2}. Equation 10.18 has six relationships (three for each of the materials), and Equation 10.16 has three, and so it is possible to solve for all nine variables. In fact, what we particularly want is to eliminate the amplitudes and find values for κ_x, κ_{z1}, and κ_{z2}.

We recall that $\varepsilon = \varepsilon_0 \varepsilon_r$, where ε_r is the relative permittivity and that the velocity of light in free space is $c = 1/\sqrt{(\mu_0 \varepsilon_0)}$. With that, we eliminate the various amplitudes to give the three relationships:

$$\frac{\kappa_{z1}}{\varepsilon_{r1}} = \frac{\kappa_{z2}}{\varepsilon_{r2}},$$

$$\kappa_{z1}^2 + \kappa_x^2 = \frac{\omega^2 \varepsilon_{r1}}{c2}, \tag{10.19}$$

$$\kappa_{z2}^2 + \kappa_x^2 = \frac{\omega^2 \varepsilon_{r2}}{c^2}.$$

These can be reduced to

$$\kappa_{z1} = \frac{\omega \varepsilon_{r1}}{c(\varepsilon_{r1} + \varepsilon_{r2})^{1/2}},$$

$$\kappa_{z2} = \frac{\omega \varepsilon_{r2}}{c(\varepsilon_{r1} + \varepsilon_{r2})^{1/2}}, \tag{10.20}$$

$$\kappa_x = \frac{\omega}{c} \left\{ \frac{\varepsilon_{r1} \varepsilon_{r2}}{\varepsilon_{r1} + \varepsilon_{r2}} \right\}^{1/2}.$$

For the wave to be locked to the surface, it should exponentially decay in each direction away from the surface. This implies that both κ_{z1} and κ_{z2} should have large imaginary parts, κ_{z2} being primarily negative imaginary, and κ_{z1}, positive imaginary. Then κ_x should have a positive real part and as small as possible imaginary part so that we have propagation along the x-axis with reasonable range. All this implies that ε_{r1} and ε_{r2} should be largely real with opposite signs, and the magnitude of the negative one should be greater than that of the positive.

We recall that

$$\varepsilon_r = (n - ik)^2. \tag{10.21}$$

Let us assume that ε_{r2} is largely negative real. This implies that k_2 should be large, and n_2, small. Then, for ε_{r1} to be positive real but smaller in magnitude than ε_{r2}, k_1 should be vanishingly small and n_1 be less than k_2. In other words, material 2 should be a metal of reasonably high optical performance and material 1 should be a dielectric preferably of low index. Then, from Equations 10.20 and 10.21,

$$\kappa_{z1} = \frac{\omega}{c} \cdot \frac{n_1^2}{\left\{ n_1^2 + (n_2 - ik_2)^2 \right\}^{1/2}} \approx \frac{\omega}{c} \cdot \frac{n_1^2}{\left\{ n_1^2 - k_2^2 \right\}^{1/2}},$$

$$\tag{10.22}$$

$$\kappa_{z2} = \frac{\omega}{c} \cdot \frac{(n_2 - ik_2)^2}{\left\{ n_1^2 + (n_2 - ik_2)^2 \right\}^{1/2}} \approx \frac{\omega}{c} \cdot \frac{-k_2^2}{\left\{ n_1^2 - k_2^2 \right\}^{1/2}},$$

where we assume k_1 as zero and we take the negative imaginary root in the denominator.

$$\kappa_x = \frac{\omega}{c} \cdot \left[\frac{n_1^2 (n_2 - ik_2)^2}{n_1^2 + (n_2 - ik_2)^2} \right]^{1/2}$$

$$= \frac{\omega}{c} \cdot \left[\frac{n_1^2 (n_2^2 - k_2^2) - i2n_1^2 n_2 k_2}{n_1^2 + n_2^2 - k_2^2 - i2n_2 k_2} \right]^{1/2} \qquad (10.23)$$

$$= \frac{\omega}{c} \cdot \left[\frac{n_1^2 (n_2^2 - k_2^2)}{n_1^2 + n_2^2 - k_2^2} \right]^{1/2} \left[\frac{1 - i2n_1^2 n_2 k_2 / \{ n_1^2 (n_2^2 - k_2^2) \}}{1 - i2n_2 k_2 / \{ n_1^2 + n_2^2 - k_2^2 \}} \right]^{1/2}.$$

Now we apply the binomial theorem to the second factor in Equation10.23, taking the first two terms only, to give

$$\kappa_x = \frac{\omega}{c} \cdot \left[\frac{n_1^2 (n_2^2 - k_2^2)}{n_1^2 + n_2^2 - k_2^2} \right]^{1/2} \left[1 - in_1^2 n_2 k_2 / \{ (n_2^2 - k_2^2)(n_1^2 + n_2^2 - k_2^2) \} \right]. \qquad (10.24)$$

The factor in front will be positive since k_2 is much larger than the other parameters. The second term in the right-hand square bracket is negative for the same reason and, therefore, indicates attenuation. The absorption coefficient will be twice the imaginary part of κ_x since the irradiance is proportional to the square of the field amplitude. Thus,

$$\alpha_x = \frac{\omega}{c} \cdot \left[\frac{n_1^2 (n_2^2 - k_2^2)}{n_1^2 + n_2^2 - k_2^2} \right]^{1/2} \cdot \left[\frac{2n_1^2 n_2 k_2}{(n_2^2 - k_2^2)(n_1^2 + n_2^2 - k_2^2)} \right]. \qquad (10.25)$$

The propagation length of the plasmon is the distance traveled where the power drops to $1/e$ times its original value. This is $1/\alpha_x$.

Because there are always two possible square roots with opposite sign, another solution of Equation 10.24 is possible, but it simply represents a surface plasmon propagating in the opposite direction.

A perfect metal where the electrons are completely free, that is, there is no damping whatsoever, and where the electrons, since they are unbound, have no resonant frequency, has a relative permittivity that is completely real. At high frequencies, it is unity; in other words, the frequency is too high for the electrons to respond at all. Then as the frequency reduces, the relative permittivity drops toward zero and passes through it at what is known as the plasma frequency. Below that frequency, the relative permittivity is negative. Above the plasma frequency, therefore, the metal acts like a dielectric with index increasing with frequency toward unity while below the plasma frequency, it acts like a perfect metal with zero index and with extinction coefficient increasing with wavelength. Real high-performance metals are more complicated, but they do have somewhat similar behavior. Although n and k are not zero, k dominates below the plasma frequency, rising with wavelength, but n is not zero. In the vicinity of the plasma frequency, the k and n curves cross, so that the real part of the permittivity (Equation 10.21) passes through zero. All this is a bulk rather than a surface effect. Now let us return to our surface wave. The denominator in both Equations 10.24 and 10.25 contains the factor $(n_1^2 + n_2^2 - k_2^2)$. As the frequency increases toward the crossing point, that is, the wavelength is reduced, the extinction coefficient of the metal k_2 falls and the refractive index n_2 rises, but n_1 remains positive, and so the factor becomes zero at a frequency a little below that for the crossing. This frequency is usually known as the surface plasmon frequency. The real part of κ_x and α_x should both become infinite according to Equations 10.24 and 10.25, and clearly, the surface plasmon frequency reduces with increasing n_1. Equations 10.24 and 10.25 are, however, approximations and when the more accurate expressions (Equation 10.20) are used, the values of real κ_x and of α_x do not actually rise to infinity, but rather to a maximum that is at a still slightly lower frequency for the real part of κ_x but very slightly higher for α_x. In the next section, we shall use expressions Equations 10.20 and 10.21 for the calculations.

10.5 Calculated Values for Silver

We take silver and free space (or air) as an example. We use the optical constants of silver shown in Figure 10.20. The long-wave performance is quite similar to what is expected from a perfect metal. The short-wave performance is a little more complicated, but we particularly look at the region from 300 to 1000 nm.

Figure 10.21 shows the calculated value of the factor $(n_1^2 + n_2^2 - k_2^2)$, assuming the metal, with optical constants of Figure 10.20, are in contact with a vacuum with n_1 of 1.00. The zero crossing occurs at 336.9 nm shown as the limiting wavelength in Figure 10.22.

The relationship between the angular frequency ω and the magnitude of the wave vector κ, (strictly, the real part of κ) often called the wavenumber, is known as the dispersion relationship. ω is traditionally plotted against κ, but such plots are unusual in the optical coating field, where the free space wavelength is more frequently the independent variable. Figure 10.22 has therefore been plotted in a different way, showing the actual wavelength against the free space wavelength. The actual wavelength is given by

$$\lambda_{actual} = \frac{2\pi}{\kappa}. \tag{10.26}$$

The plasmon wavelength is always less than the free space wavelength, but the difference reduces toward the infrared. The absorption coefficient is shown in Figure 10.23. There is a rapid increase in absorptance toward shorter wavelengths. The propagation length of the surface plasmon is the distance over which the power drops by $1/e$. This is the inverse of the absorption coefficient.

The dispersion relationship in Figure 10.22 shows that the surface plasmon wavelength is less than that of an electromagnetic wave propagating in the adjacent dielectric medium, and whatever the direction of the light with respect to the surface, the interval between the intersections of the planes of constant phase with the surface cannot be less than the wavelength. Thus, a free electromagnetic wave cannot couple to a surface plasmon at a smooth, featureless surface. As we have already seen in this chapter, coupling cannot only be deliberately arranged through an evanescent

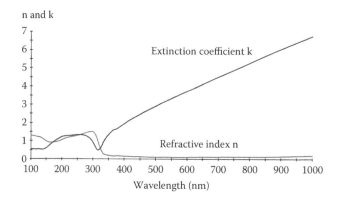

FIGURE 10.20
Optical constants of silver. (From E. D. Palik, ed., *Handbook of Optical Constants of Solids I*, Academic Press, Orlando, FL, 1985.)

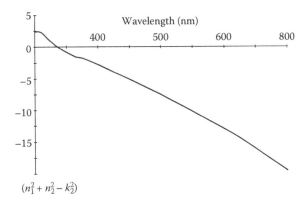

FIGURE 10.21
Factor $(n_1^2 + n_2^2 - k_2^2)$ calculated from the values in Figure 10.20 together with n_1 as 1.00. The zero crossing is at 336.9 nm.

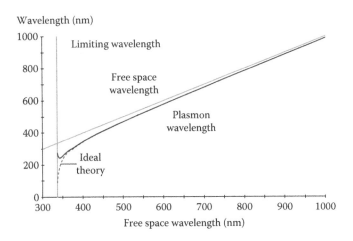

FIGURE 10.22
Dispersion relation of the surface plasmon on a silver surface in a vacuum. Here the relationship has been plotted in a way that is more meaningful in the optical coating field. The dielectric over the silver in this case is vacuum, and in that medium, the light has the free space wavelength, identical to the horizontal scale. This is everywhere greater than the plasmon wavelength.

wave or an almost evanescent wave in the metal, but it can also take place at a surface that departs from perfect smoothness. A rough surface behaves as a kind of perturbed diffraction grating that presents a disturbance that can be represented as a range of surface frequencies that scatter the light in different directions as with a grating but less regularly. With such a scattering range, it is easy to see that there may well be a component scattered along the surface with the correct frequency that excites the appropriate plasmon. The process can proceed in either direction. An incident electromagnetic wave can couple into the plasmon in the way described, but equally, a plasmon can scatter some of its energy into an emitted electromagnetic wave.

We can examine the surface profile for evidence of periodicities by constructing an autocorrelation function that shows the relative strength of the periodic component as a function of the period. This will usually show a gradual drop in strength with increasing period length, indicating that the scattering will increase in efficiency with reducing wavelength and increasing frequency. We can expect, therefore, a gradual increase in the coupling into a surface plasmon with reducing wavelength and increasing frequency, until it approaches the surface plasmon frequency beyond which the coupling will vanish.

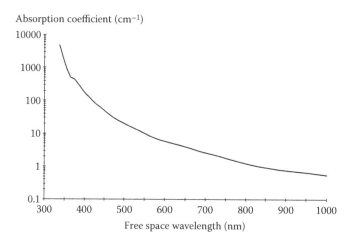

FIGURE 10.23
Absorption coefficient of the surface plasmon. It becomes very large as it approaches the surface plasmon wavelength.

This effect can be observed in metallic reflectors especially of silver. At the short wavelength end of their range of high reflectance, there can often be a dip that cannot be explained in terms of the variation of their optical constants. Further, it was found by Stanford and Bennett [31] that a protecting layer enhanced this dip in reflectance that moved toward longer wavelengths and increased in depth as the refractive index of the overcoat increased. This is nowadays a well-known effect that is due either to a roughness of the metal surface or a scattering within the protecting layer, or a combination of both. Exact quantitative calculation is difficult, but we can readily understand the trends by returning to our theory.

The surface plasmon wavelength that marks the upper limit of the effect (that is, strictly, the surface plasmon polariton effect) is approximately the point at which the factor $(n_1^2 + n_2^2 - k_2^2)$ becomes zero. $n_2^2 - k_2^2$ becomes increasingly negative with increasing wavelength, and thus, the cutoff wavelength increases with increasing n_1. Further, as demonstrated in Figure 10.24, the absorption coefficient rises with increasing dielectric index. Although energy can scatter back out of the surface plasmon, the enormously increased absorption coefficient at the shortwave end of the scale is responsible for significant losses. For high performance from a protected or enhanced metal reflector near its shortwave limit, very smooth surfaces and low scattering from the dielectric materials are required.

There are other interesting beneficial effects. In some recent devices, advantage has been taken of deliberate scattering into and out of surface plasmons. We shall mention just two of these. An array of small holes in a thin metal layer acts as a two-dimensional grating that scatters incident light of the correct wavelength into a surface plasmon on the surface of the metal. This couples into a similar plasmon on the other surface, such that the decaying electric fields of each interfere to reduce the internal field in the metal and, hence, the total losses. The plasmon on the other side is scattered by the array of holes back into a progressive electromagnetic wave. The performance is similar to that of an induced transmission filter but without the dielectric multilayers on either side of the metal. The effect was discovered by Ebbesen et al. [26] and was explained in detail by Bonod et al. [27]. Then, a group [32] led by Fu created a device based on a corrugated metal layer that again, by similar coupling, results in a typical induced transmission filter performance, again without the need for any additional matching multilayers. An advantage of this latter device is that it can be represented by an equivalent circuit that permits it to be theoretically treated in a similar way to a thin film.

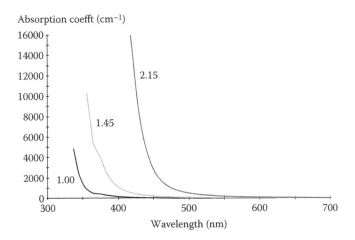

FIGURE 10.24
Variation of the plasmon absorption coefficient with wavelength, and with different dielectric indices, showing the increase with index of the dielectric medium. The increase is very large at the shortwave end of the scale and particularly so with the highest dielectric index. The curves terminate at the appropriate plasmon cutoff wavelength.

References

1. P. H. Berning and A. F. Turner. 1957. Induced transmission in absorbing films applied to band pass filter design. *Journal of the Optical Society of America* 47:230–239.
2. E. D. Palik, ed. 1985. *Handbook of Optical Constants of Solids I.* Orlando, FL: Academic Press.
3. M. Nevière and P. Vincent. 1980. Brewster phenomena in a lossy waveguide used just under the cut-off thickness. *Journal d'Optique* 11:153–159.
4. M. Ruiz-Urbieta, E. M. Sparrow, and P. D. Parikh. 1975. Two-film reflection polarizers: Theory and application. *Applied Optics* 14:486–492.
5. A. Macleod. 2010. Oblique incidence and dielectric-coated metals. *Society of Vacuum Coaters Bulletin* (Spring):24–29.
6. S. F. Pellicori. 1974. *Private communication* (Santa Barbara Research Center, Goleta, CA.)
7. J. T. Cox, G. Hass, and W. R. Hunter. 1975. Infrared reflectance of silicon oxide and magnesium fluoride protected aluminum mirrors at various angles of incidence from 8 μm to 12 μm. *Applied Optics* 14:1247–1250.
8. J. T. Cox and G. Hass. 1978. Protected Al mirrors with high reflectance in the 8-12-μm region from normal to high angles of incidence. *Applied Optics* 17:2125–2126.
9. P. B. Clapham, M. J. Downs, and R. J. King. 1969. Some applications of thin films to polarization devices. *Applied Optics* 8:1965–1974.
10. M. P. Lostis. 1957. Etude et realisation d'une lame demi-onde en utilisant les proprietes des couches minces. *Journal de Physique et le Radium* 18:51S–52S.
11. K. Rabinovitch and G. Toker. 1994. Polarization effects in optical thin films. *Proceedings of SPIE* 2253:89–102.
12. E. Kretschmann and H. Raether. 1968. Radiative decay of non-radiative surface plasmons excited by light. *Zeitschrift für Naturforschung* 23A:2135–2136.
13. A. Otto. 1968. Excitation of non-radiative surface plasma waves in silver by the method of frustrated total reflection. *Zeitschrift für Physik* 216:398–410.
14. H. Raether. 1977. Surface plasma oscillations and their applications. In *Physics of Thin Films*, G. Hass and M. H. Francombe (eds), pp. 145–261. New York: Academic Press.
15. F. Abelès. 1976. Optical properties of very thin films. *Thin Solid Films* 34:291–302.

16. K. M. Greenland and C. Billington. 1950. The construction of interference filters for the transmission of specified wavelengths. *Journal de Physique et le Radium* 11:418–421.
17. J. Harrick and A. F. Turner. 1970. A thin film optical cavity to induce absorption of thermal emission. *Applied Optics* 9:2111–2114.
18. Z. Salamon, H. A. Macleod, and G. Tollin. 1997. Coupled plasmon-waveguide resonators: A new spectroscopic tool for probing proteolipid film structure and properties. *Biophysical Journal* 73:2791–2797.
19. Z. Salamon, G. Tollin, H. A. Macleod et al. 1998. Thin layer surface resonators: A new spectroscopic tool for probing dielectric film structure and properties. In *41st Annual Technical Conference Proceedings*, pp. 238–242, Society of Vacuum Coaters, Boston, MA.
20. A. E. Craig, G. A. Olson, and D. Sarid. 1983. Experimental observation of the long-range surface-plasmon polariton. *Optics Letters* 8:380–382.
21. J. C. Quail, J. G. Rako, and H. J. Simon. 1983. Long-range surface-plasmon modes in silver and aluminum films. *Optics Letters* 8:377–379.
22. R. Dragila, B. Lutherda, and S. Vukovic. 1985. High transparency of classically opaque metallic films. *Physical Review Letters* 55:1117–1120.
23. H. A. Macleod. 1992. Unconventional coatings. In *Tutorials in Optics*, D. T. Moore (ed.), pp. 121–135. Washington, DC: Optical Society of America.
24. A. Macleod and C. Clark. 2001. Evanescent waves and some of their applications. In *44th Annual Technical Conference Proceedings*, pp. 75–80, Society of Vacuum Coaters, Philadelphia, PA.
25. T. W. Ebbesen, H. F. Ghaemi, T. Thio et al. 1999. *Sub-wavelength aperture arrays with enhanced light transmission.* US Patent, 5,973,316.
26. T. W. Ebbesen, H. J. Lezec, H. F. Ghaemi et al. 1998. Extraordinary optical transmission through subwavelength hole arrays. *Nature* 391:667–669.
27. N. Bonod, S. Enoch, L. Li et al. 2003. Resonant optical transmission through thin metallic films with and without holes. *Optics Express* 11:482–490.
28. D. Bohm and D. Pines. 1951. A collective description of electron interactions: I. Magnetic interactions. *Physical Review* 82:625–634.
29. R. H. Ritchie. 1957. Plasma losses by fast electrons in thin films. *Physical Review* 106:874–881.
30. A. Macleod. 2011. Surface plasmons: Fundamentals Part 1. *Society of Vacuum Coaters Bulletin* (Spring): 24–29.
31. J. L. Stanford and H. E. Bennett. 1969. Enhancement of surface plasma resonance absorption in mirrors by overcoating with dielectrics. *Applied Optics* 8:2556–2557.
32. L. Fu, H. Schweizer, T. Weiss et al. 2009. Optical properties of metallic meanders. *Journal of the Optical Society of America B* 26:B111–B119.

11

Other Topics: From Rugate Filters to Photonic Crystals

The field of optical coatings is exceedingly broad. The earlier chapters contain descriptions and explanations of thin film structures that could be readily classified. Here, in this chapter, we look at some further coatings that do not quite fit into the classifications so far. That should not be taken as any indication of relative importance.

11.1 Rugate Filters

The term *rugate* is derived from biology, where the meaning is essentially "corrugated." It was adopted to describe a structure exhibiting a regular cyclic variation of refractive index, resembling a sine or cosine wave. Such structures have the property of reflecting a narrow spectral region and transmitting all others. They exhibit properties similar to a quarter-wave stack but without the higher-order reflection bands. Thus, they are notch filters and particularly useful in removing bright spectral lines from weaker continua. Many of their applications involve laser sources, and they are especially relevant in the field of laser protection. This strict original definition of rugate has been somewhat relaxed, so that the term is also sometimes used for any layer system in which there is a deliberate attempt to induce an inhomogeneity whether or not it is of a cyclic kind.

We begin with the strict interpretation of rugate. It can easily be shown that all the beams, emerging from the front surface of a multilayer constructed from a series of quarter-wave layers of alternate high and low indices, are exactly in phase, and we know that this leads to high reflectance but limited in width in terms of wavelength or frequency because the constructive interference condition applies only close to the wavelength for which the layers are exact quarter waves. Outside the zone of high reflectance, it is the transmittance that is high. The quarter-wave stack, therefore, acts as a notch filter, and we have already considered it in that role. The lower the ratio of the high- to low-refractive index at the interfaces, the lower the amplitude reflection coefficients, and the greater the number of beams required to achieve a given reflectance. The rate at which the interference condition decays with change in wavelength determines the width of the high reflectance zone. Smaller index contrast implies more beams, faster decay of the constructive interference, and, hence, narrower reflectance zones. A narrow zone of high reflectance in turn implies a large number of layers of low-index contrast. All this is considered in greater detail in Chapters 6 and 7.

A limitation of systems made up of discrete dielectric layers is that a change in wavelength does not change the amplitude of the beams, except for the usually slight changes due to dispersion. The same beams with the same amplitudes exist over a wide spectral region. It is impossible to distinguish between the interference effect between two beams with phase difference φ and the same two beams with phase difference $\varphi \pm 2m\pi$, where m is an integer. Thus, in the case of the quarter-wave stack, the interference condition that exists at wavelength λ_0 also exists at wavelengths $\lambda_0/3$, $\lambda_0/5$, $\lambda_0/7$, and so on, leading to the higher-order reflectance zones that can sometimes limit its usefulness as a notch filter. A typical characteristic curve plotted in terms of g, that is, λ_0/λ, is shown in Figure 11.1.

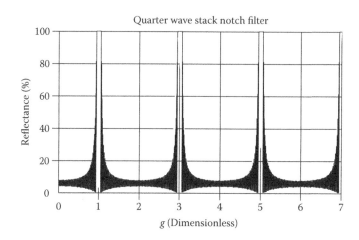

FIGURE 11.1
Typical characteristic of a quarter-wave stack used as a notch filter showing the higher orders at g of 3, 5, and 7. The fringes in the pass regions are so tightly packed, they cannot be distinguished.

The higher orders may not present any problem in many applications, and for these, the discrete layer design will be quite satisfactory. For those others where the peaks are a problem, we do need to suppress them. Some discrete layer approaches to suppress just one or two peaks are already discussed in Chapter 7. They have their origins in the interference between beams reflected at all the interfaces. In other words, their origin is distributed throughout the multilayer. We therefore need a distributed solution. We need to retain the beams at the fundamental peak at $g = 1.0$, but we must remove them at all other integral values of g. An antireflection coating that does not affect the performance at $g = 1$, but that operates at values of g greater than unity, is required for each interface. An inhomogeneous layer is such an antireflection coating.

We shall return shortly to the derivation of performance of such systems. For the moment, let us accept the two possible profiles for inhomogeneous layers shown in Figure 11.2. If we assume that the layers have an optical thickness of one quarter wavelength at $g = 1.0$, then the performances in terms of reflectance against g are those shown in Figure 11.3.

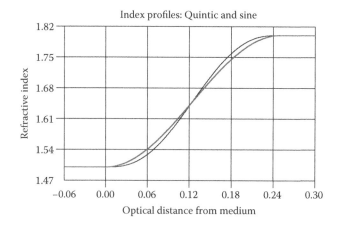

FIGURE 11.2
Inhomogeneous layer profiles rising from 1.50 to 1.80. The layers are one quarter of a wavelength in optical thickness, and the profile of refractive index follows a sine law (shallower curve: thick line) or a fifth-order polynomial (steeper curve: thin line) with zero first and second derivatives with respect to thickness at the end points.

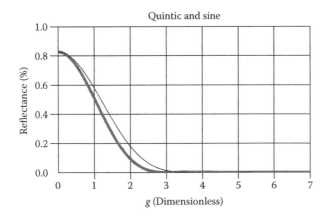

FIGURE 11.3
Reflectance against g for the inhomogeneous layers shown in Figure 11.2. The sine law variation is steeper than the fifth-order polynomial so the curve of reflectance (left-hand curve: thick line) drops faster, but the fifth-order polynomial (right-hand curve: thin line) gives lower reflectance at greater values of g.

This antireflection coating must now be inserted at each interface in the discrete layer coating. Figure 11.4 shows the resulting profile. The coating now has a sinusoidal variation of index throughout and is known as a rugate structure because of the smooth cyclic variation.

The new variation of reflectance is shown in Figure 11.5. Note the small residual peak at $g = 2.0$. This is due to the failure of the sinusoidal variation of refractive index to act completely like the absentee half-wave layers of the discrete design. The slight residual reflectance change accumulates in a coating with a large number of layers and gives the slight perturbation from the regular fringe pattern that appears elsewhere. Southwell [1] pointed out that an inhomogeneous layer based on an exponential sine does act as an absentee layer at even values of g even though its profile is almost indistinguishable from that of a sine function.

The inhomogeneous antireflection coating is a very robust one from the point of view of errors. There is an insensitivity to the actual profile of index. As long as the thickness at a given wavelength is greater than roughly a half wave, then the reflectance at that wavelength should be

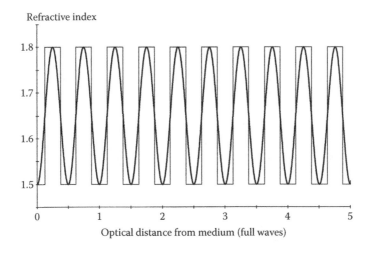

FIGURE 11.4
Result of replacing each discrete interface (square plot: thin line) by one graded to have a sine profile (rounded plot: thick line). This gives the rugate structure.

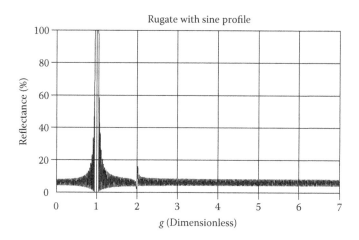

FIGURE 11.5
Reflectance curve of the rugate filter. The variation of index is shown in Figure 11.4 except that the filter actually calculated had the equivalent of 64 discrete layers.

very low. Thus, even quite large errors in the profile of a rugate filter are not normally serious unless they are systematic and lead to a change in the pitch of the cycle. Such errors tend to broaden the fundamental peak. Quite severe errors are required before the higher-order peaks begin to return. This has useful implications for the manufacturing of such filters.

Yet a further advantage of the rugate structure is that it is very easy to design rugates for multiple wavelengths. It is simply necessary to add the sinusoidal variations around the mean index. An example of this will be given shortly in Section 11.1.2.

11.1.1 Apodization

We have already mentioned apodization as an effective way of reducing ripple in Chapter 7 when discussing thickness modulation. It is just as powerful in the context of index modulation. Apodization involves a gradual reduction in the index swing from the center to the outer edges of the structure so that the modulation varies from full at the center to zero at the outer edges. The process is not very sensitive to the form of the variation as long as it is smooth. Sinusoidal variation is suitably effective. The rugate then presents its mean index to the surrounding media. If these media have the identical index, then the ripple largely disappears. If the index of an adjacent medium differs from the mean, then the addition of a suitable broadband antireflecting structure will be necessary.

A simple example of an apodized single-peak rugate is shown in Figures 11.6 and 11.7. The apodizing function is a simple half cycle of a sine wave. The matching structures have profiles corresponding to half cycles of a sine wave, but from −90° to +90°, designed to be of half-wave thickness at the longest wavelength of the performance curve. The index variation is achieved by varying the packing density with the thin film material having an index of 2.00 and voids of 1.45. Multiline rugates can likewise be apodized by suitable apodizing functions.

11.1.2 Discrete Layer Replacements

The control of the deposition of rugate filters is a rather more involved task than for a simple discrete layer quarter-wave stack. In discrete layer deposition, it is optical thickness that has always been the object of the closed loop control system. Refractive index has been considered to be characteristic of the particular material being deposited, and so the control of that aspect of the layers has been open loop. The deposition methods have concentrated on the control of source

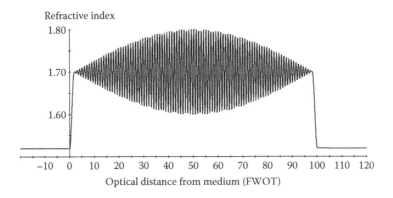

FIGURE 11.6
Design of a single-peak rugate consisting of three regions, an inhomogeneous match to the incident medium of 1.52 index; the apodized rugate; and, finally, an inhomogeneous match to the substrate, also of 1.52 index. The horizontal scale is in terms of full waves at 1000 nm. The total optical thickness is 100 μm.

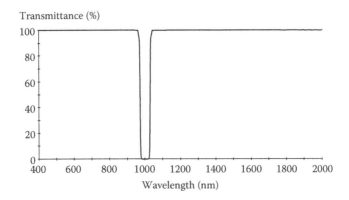

FIGURE 11.7
Performance of the single-peak rugate with design as in Figure 11.6.

temperature, rate of deposition, and so on. The rugate filter represents a greater challenge because there is no natural material that yields the desired profile of refractive index. It must be engineered. Compositional changes are necessary, and in the true rugate filter, these changes should be smooth. This tends to imply some form of active index control.

The absence of the need for direct index control, however, makes discrete layers very attractive. Although they are not strictly true rugates, nevertheless, it is possible to create discrete layer structures that have, up to a point, similar properties. To replace a rugate structure by a discrete layer structure, we can imagine slicing a rugate period into a large number of thin layers of equal optical thickness. Each thin slice has an inhomogeneous index profile, but we can convert it into a homogeneous index that simply has the central value. This gives a staircase profile of index. In fact, and we return to this point later in this section, the calculation of the properties of rugate filters with arbitrary profile is normally carried out in this way with the thicknesses chosen to be so thin that further subdivision makes no changes to the results. Here we use rather thicker slices.

Figure 11.8 shows the profile of a rugate filter that has been converted in this way. The steps are arranged so that in each rugate cycle, there are 10 of them. This means that at the reflectance peak where the rugate cycle is one half-wave thick, the individual discrete layers are just one twentieth of a wave thick. As long as the individual layers are thin compared with a quarter wave, then the discrete version works well. However, as the wavelength reduces, the phase thickness of the individual layers increases and eventually becomes much thicker compared with a wavelength.

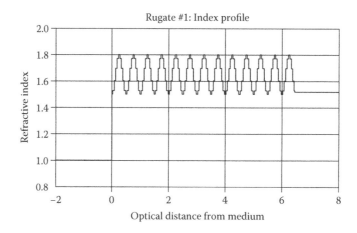

FIGURE 11.8
Profile of a rugate filter with a cycle consisting of 10 discrete layers rather than a continuously varying profile.

However, the behavior of the system does not just simply deteriorate but is quite regular and understandable. At a value of g of zero, the layers are effectively of zero phase thickness, and so the reflectance of the system is that of the uncoated substrate. At $g = 1.0$, the rugate cycle is now a half wave and the reflectance is high. As g increases, the cycle, at first, retains its antireflecting properties and the higher-order peaks are suppressed. Now let us jump to the case where g is large enough for the layers to be of half-wave thickness. Here we have absentee layers, and the reflectance is that of the uncoated substrate. At this value of g, we still have exactly the same beams taking part in the interference as at all other values of g. The phase shifts between them, however, are exactly the same as at $g = 0$, except that in every case, there is an additional path difference of a wavelength, that is, $360°$, which is indistinguishable from zero. Furthermore, as we now reduce g from this value, we find the same interference pattern as a function of the reduction in g that we find as a function of the increase in g from zero in the normal way. Thus, if we have 10 equal steps or discrete layers making up the rugate cycle with a fundamental peak at $g = 1$, then there will be a similar peak at $g = 9$. A cycle made up of 4 layers will have a further peak at $g = 3$ and so on. Figure 11.9 illustrates this for the rugate of Figure 11.8. Figure 11.10 shows similar performance for a rugate with a four-layer

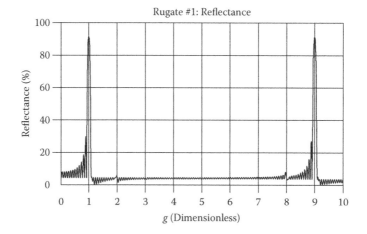

FIGURE 11.9
Performance of the rugate of Figure 11.8 as a function of g showing the harmonic peak at $g = 9.0$. Note the subtle differences in the low reflectance performance from $g = 0$ to $g = 2$ and from $g = 8$ to $g = 10$. This is due [1] to the use in Figure 11.8 of a half-cycle that is the mirror image of the alternate half cycle only if the outer layers are half the thickness of the others.

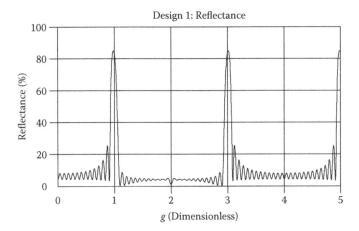

FIGURE 11.10
Performance of a rugate similar to that of Figure 11.8 except that the cycle is made up of four discrete layers of equal thickness. The harmonic peak appears now at $g = 3.0$.

cycle. In this case, the harmonics begin at $g = 3$, and so the sole peak that is eliminated is at $g = 2$. This may not appear to be any different from a two-layer cycle, but in fact, the extra layers help suppress the half-wave-hole peak that appears at $g = 2$ when the coating based on the two-layer cycle is tilted.

Southwell [1] indicated that the slight lack of symmetry in the result in Figure 11.9 is a consequence of the use of a set of sublayers of identical thickness such that there are no two adjacent sublayers with the same index. This effectively makes the rugate period symmetrical only if the two outermost layers are considered to be half the thickness of the others. A rearrangement where the outermost sublayers have the same index and the full sublayer thickness, implying a merging of the innermost layer pair and the ending layer of each cycle with the starting of the next, gives a perfectly symmetrical performance.

An alternative technique for the replacement of the continuous variation with a series of discrete layers uses two materials with fixed indices of refraction. One of the indices must be equal to or less than the lowest in the rugate structure, and the other, equal to or greater than the highest. The method uses the properties of the characteristic matrices of the films. There are two variants. The first uses the result that the matrix of any symmetrical arrangement of layers, absorbing and inhomogeneous layers included, can be replaced by the characteristic matrix of a single equivalent homogeneous layer [2,3]. This equivalence is dealt with more fully in Chapter 4 and is a purely analytical relationship and certainly not physical, but it is valid wherever the properties involve only the characteristic matrices. This relationship can be reversed, so that the homogeneous film matrix can be replaced by the matrix of a symmetrical combination of layers. Since the eventual result involves identical matrices, properties such as reflectance and transmittance at one particular angle of incidence and wavelength are unchanged when the equivalent sequences are interchanged. One of the most useful aspects of this relationship is the replacement of a layer of intermediate index by a symmetrical combination of layers of given high and low indices. At one angle of incidence and one wavelength, this equivalence completely holds for any property that can be calculated using the characteristic matrices. For the equivalence to be retained exactly with changes in wavelength demands a particular dispersion of the indices of the replacement layers. This implies that when real layers are involved with their natural dispersion, the equivalence becomes gradually poorer as the wavelength changes, especially as the wavelength decreases. The equivalence does strictly not extend to changes in the angle of incidence, although the deterioration is not usually very rapid. The second variant uses an approximate method based on pairs of layers. When both members of a layer pair are thin compared with a wavelength, then the characteristic matrix of the combination of the two layers is equivalent to that of a single layer of intermediate index [4]. Again, this relationship is not valid for changes in angle of incidence, and it becomes poorer as the wavelength decreases. Both variants can take the staircase approximation to the rugate cycle and convert it into an equivalent series of alternate high- and low-index layers of differing thicknesses. We illustrate the method by using the second variant, the two-layer approximation. The product of two thin-film matrices can be considerably simplified if the layers are thin compared with the wavelength. Following Southwell [4], let a layer be dielectric and of physical thickness d, refractive index n, and admittance $y = n$ free space units, then

$$
\begin{bmatrix} \cos \delta & \dfrac{i \sin \delta}{y} \\ iy \sin \delta & \cos \delta \end{bmatrix} \rightarrow \begin{bmatrix} 1 & i\dfrac{2\pi}{\lambda}d \\ i\dfrac{2\pi}{\lambda}n^2 d & 1 \end{bmatrix}.
\tag{11.1}
$$

The product of two thin film matrices becomes

$$
\begin{bmatrix} \cos \delta_1 & \dfrac{i \sin \delta_1}{y_1} \\ iy_1 \sin \delta_1 & \cos \delta_1 \end{bmatrix} \begin{bmatrix} \cos \delta_2 & \dfrac{i \sin \delta_2}{y_2} \\ iy_2 \sin \delta_2 & \cos \delta_2 \end{bmatrix} \rightarrow \begin{bmatrix} 1 & i\dfrac{2\pi}{\lambda}d_1 \\ i\dfrac{2\pi}{\lambda}n_1^2 d_1 & 1 \end{bmatrix} \begin{bmatrix} 1 & i\dfrac{2\pi}{\lambda}d_2 \\ i\dfrac{2\pi}{\lambda}n_2^2 d_2 & 1 \end{bmatrix}
$$

$$
= \begin{bmatrix} 1 & i\dfrac{2\pi}{\lambda}(d_1 + d_2) \\ i\dfrac{2\pi}{\lambda}(n_1^2 d_1 + n_2^2 d_2) & 1 \end{bmatrix},
\tag{11.2}
$$

where we are neglecting terms in $d_1 d_2$. Then, by comparing Equation 11.2 with Equation 11.1, we find

$$
d = d_1 + d_2,
$$

$$
n^2 = \frac{n_1^2 d_1 + n_2^2 d_2}{d_1 + d_2}.
$$

Now, knowing n_1 and n_2 and specifying n and d, the equivalent parameters of the two-layer combination, we have for d_1 and d_2

$$
d_1 = \left(\frac{n^2 - n_2^2}{n_1^2 - n_2^2} \right) d,
\tag{11.3}
$$

$$
d_2 = d - d_1.
$$

Note that the total physical thickness remains constant so that the total physical thickness of the rugate to be replaced is preserved. The order of the layers is unimportant too. Interchanging the two layers makes no difference to the equivalent parameters of the combination. The important consideration is that n must lie between n_1 and n_2; otherwise, negative thicknesses would result.

Figure 11.11 shows a single cycle that has been replaced in this way. (There is an extra layer at the end that is strictly the first layer of a following cycle.) The performance of a rugate filter based on 14 of these cycles in series is shown in Figure 11.12.

The important point about these calculations is that a discrete layer approximation to a rugate filter can give performance that is nevertheless acceptable. The range of transparency of the materials is rarely greater than the clear ranges shown in Figures 11.9 and 11.11. Many of the techniques for coating production lend themselves much better to the construction of discrete layer systems than to the creation of smoothly varying index profiles.

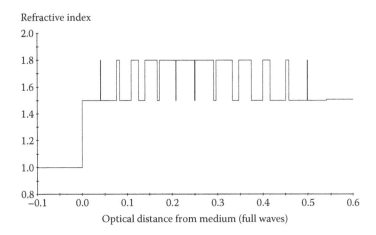

FIGURE 11.11
Twenty-two-layer representation of a single half-wave rugate cycle. The layers are either of high (1.8) index or low (1.5), and their thicknesses are varied so that the overall effect is similar to the smooth variation of the classical rugate.

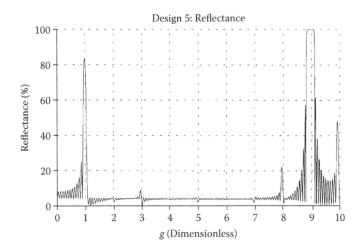

FIGURE 11.12
Performance of the rugate of Figure 11.11. The performance has characteristics similar to those of the stepped version from which it was derived.

As an example, we show the index variation of an apodized three-line rugate in Figure 11.13 and the performance in Figure 11.14. The design is a simple combination of three suitable sine-wave variations apodized by modulating by a half cycle of a sine function. The material used in the design was of index 2.1, and its packing density was varied to give the index variation shown with inhomogeneous matching at either side to match to the surrounding media of index 1.52. Note the total thickness of 80 μm.

The design was then replaced with a series of thin discrete layers of alternating indices of 2.1 and 1.45 to give the design of Figure 11.15 and the performance in Figure 11.16. There are 2000 layers in the replacement structure. Obviously, the construction of such rugates is not a simple task. Nevertheless, rugates consisting of still greater number of layers, some over 4000, have been successfully constructed [5].

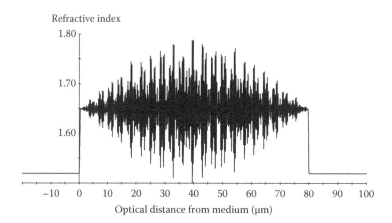

FIGURE 11.13
Index variation in an apodized three-line rugate.

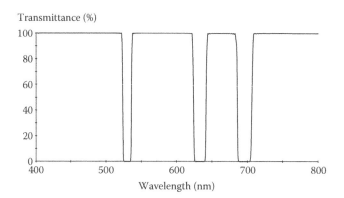

FIGURE 11.14
Calculated performance of the design of Figure 11.13. The minima represent a density of greater than 5.0.

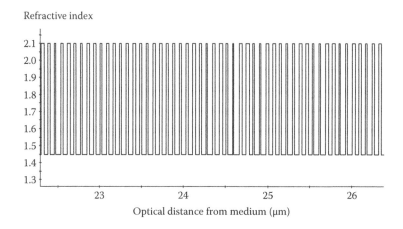

FIGURE 11.15
Section of the 2000-layer replacement for the rugate consisting of a series of thin layers of indices 2.1 and 1.45.

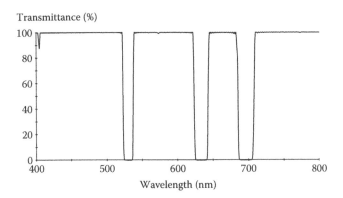

FIGURE 11.16
Performance of the two-material discrete film replacement for the rugate shown in Figure 11.15. The performance is virtually identical with that of Figure 11.12 except for a very small artifact at close to 400 nm.

11.1.3 Fourier Technique

We now turn to the broader meaning of rugate as presenting any continuous variation of index and consider the theoretical problems in more detail. Figure 11.17 shows a representation of an inhomogeneous layer that is linking two media. The optical admittance y is plotted against the optical thickness z. The accurate calculation of such layers involves slicing them into sufficiently thin homogeneous sublayers and then using the normal calculation techniques. The slices should be rather thinner than a quarter wave at the shortest wavelength in the calculation. It is rarely necessary to use less than one tenth of a wave. To test the adequacy of the approximation, the layers can be made still thinner, and the calculation, repeated. A completely unchanged performance is an indication that the approximation is satisfactory. For the design of such structures, it is usual to employ an approximate technique based on what is essentially an application of the vector method.

This technique was pioneered by Dobrowolski and Lowe [6], who credited several earlier workers with the origins of the technique. If the performance is to be calculated at the plane denoted by $z = 0$, then the vector derived from the step at the plane z will be given by

$$\rho \exp(-i2\delta) = \frac{\Delta y}{2y} \exp(-i2\kappa z),$$

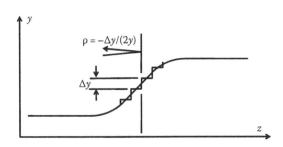

FIGURE 11.17
To derive an expression for the performance of a dielectric inhomogeneous layer, we first divide the layer into a series of separate steps. These steps are chosen close enough so that closer spacing still yields an unchanged result. Each step has an amplitude reflection coefficient of $-\Delta y/(2y)$.

where κ, the wavenumber, is given by $2\pi/\lambda$, λ being the free space wavelength. If we represent twice the optical thickness z by x, then we can write the sum of all the various vectors as

$$\sum \frac{\Delta y}{2y} \exp(-i\kappa x). \tag{11.4}$$

In the simple vector method, this sum is simply equal to the amplitude reflection coefficient. However, when many such vectors are involved with a quite thick inhomogeneous structure, a correction may be made that represents a better approximation. The conversion of the sum of Equation 11.4 to an integral then yields

$$\int_{-\infty}^{\infty} \frac{dy}{dx} \cdot \frac{1}{2y} \cdot \exp(-i\kappa x) \cdot dx = Q(\kappa) \cdot \exp[i\varphi(\kappa)], \tag{11.5}$$

connecting a function of performance with a function of the distribution of characteristic admittance through a Fourier integral expression. This may be inverted so that the distribution of y may be calculated from the distribution of performance. Q is a function of performance $\kappa = 2\pi/\lambda$, and x is twice the optical path. $\varphi(\kappa)$ is a phase factor that must be an odd function to ensure that $y(\kappa)$ (that is also numerically $n(\kappa)$) is real. Although multiple beam effects are neglected, a judicious choice of Q can reduce the errors that arise from this approximation. Note that the Equation 11.5 is frequently written with a positive argument for the exponential. This is simply a consequence of the particular sign convention that is used.

Functions that have been proposed and used for Q include the following (the first represents the simple amplitude reflection coefficient):

$$Q = \sqrt{R},$$

$$Q = \sqrt{\frac{R}{T}},$$

$$Q = \sqrt{\frac{1}{2}\left(\frac{1}{T} - T\right)}, \tag{11.6}$$

$$Q = \sqrt{\frac{1}{\sqrt{T}} - \sqrt{T}}.$$

For a more complete and detailed treatment, see Dobrowolski and Lowe [6] and Bovard [7].

The great advantage of this approach is the analytical connection in either direction of a function of design with a function of performance. If we know the performance, we can find a design and vice versa. Disadvantages are that the technique is approximate and that considerable skill and experience are required in the choice of the appropriate Q function and phase factor φ. Although the resulting design is a continuously varying admittance profile, it can be converted into a discrete layer design, the thicknesses being chosen thin enough not to affect performance at the shortest wavelength of interest.

11.2 Ultrafast Coatings

Traditionally, coating designers have been able to rely on the steady-state nature of the effects they seek to produce. There are now laser systems, known as ultrafast, capable of generating pulses of light that are short enough for transient response to become significant. A normal high reflector consisting of a quarter-wave stack might be some 25 quarter waves in thickness. At a wavelength of 1 μm, this implies a trip length for light travelling from the front to the rear of the coating and back again of 12.5 μm or a trip time of around 42 fs (1 fs is 1/1000 ps). Pulses that are around 50 fs in length are now common, and there are still shorter pulses, some 5 fs or so in length. It is clear that the transient response of coatings must now be considered important in such applications, but the effects, in fact, can be significant even with pulses of some two or three orders of magnitude longer. The idea that coating properties should have an influence on short pulses and that they might be engineered to have prescribed effects is not new. It is, however, only recently that the field has expanded and the technology advanced to the stage where the application is becoming of major importance. The filters used in digital telecommunication tend to be very complicated, and especially in narrowband filters used for channel separation, they exhibit similar effects to the coatings used for ultrafast pulses. Thus, although the thrust of this section is ultrafast effects, the theory and behavior also apply in telecommunication applications.

A short pulse can be thought of as an envelope over a carrier. The carrier contains the phase information associated with the pulse, and it travels at what is known as the *phase velocity*. The energy is obviously associated with the envelope that travels at what is known as the *group velocity*. In the presence of dispersion, the group velocity and the phase velocity are different, normal dispersion making the phase velocity greater. Thus, the carrier appears to run through the pulse envelope. The form of a short pulse with Gaussian envelope is shown in Figure 11.18.

The pulse may also be visualized in a different way, as a collection of monochromatic component waves with a continuous distribution of frequencies over a given band. The coherent combination of these monochromatic waves yields the envelope and carrier of the alternative model. Both models are entirely equivalent, and if we wish, we can pass from one to the other by way of a Fourier transform.

Pulse envelopes frequently have a Gaussian shape [8,9]. For simplicity, we can look at the temporal variation at the origin of our coordinates, $z = 0$ and then, if the peak of the pulse corresponds to $t = 0$,

$$F(t) = Ae^{-\frac{t^2}{2\mu^2}}, \tag{11.7}$$

where μ has the dimension of time. The Fourier transform gives the frequency distribution, and it is also a Gaussian function,

$$G(\omega) = Be^{-\frac{\mu^2(\omega-\omega_0)^2}{2}}. \tag{11.8}$$

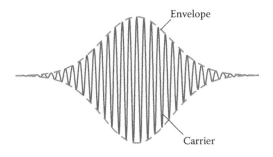

Envelope

Carrier

FIGURE 11.18
Short Gaussian-shaped pulse consisting of an envelope over a carrier of constant frequency. The carrier phase may move faster than the pulse when it will appear to run through the envelope as it travels.

If the time between the half-maximum points is τ and the width of the pulse (angular) frequency distribution at half-maximum is $\Delta\omega$, then

$$\tau \cdot \Delta\omega = 4 \cdot \log_e 2.$$

Note that both these quantities are functions of μ. For example,

$$\tau = 2\sqrt{\log_e 2} \cdot \mu.$$

The center of the pulse is the point where all the component waves are exactly of identical phase. If all the component waves travel at the speed of light in vacuo, then the phase coincidence will also travel at that speed and the center of the pulse will move with it. Similarly, if all waves slow down equally, then the pulse will slow down to the same extent, but will otherwise be unchanged.

The relative phase of the carrier within the pulse is set by the value of phase where all the component waves coincide. If the phase of the waves is zero, then the carrier will have a peak exactly at the peak of the pulse. We can find the position of the pulse peak at any time by a simple procedure.

The pulse can be considered to be made up of monochromatic component waves. As these propagate, the phase relationships between them will change, but if the pulse shape is unaltered as it propagates, then at any particular time, there must be a distance along the path where the phase is identical for all the component waves, and this must correspond to the pulse center. We use the normal thin film convention of $(\omega \tau - \kappa z)$ in the phase factor, where $\kappa = 2\pi n/\lambda$ and with λ the free-space wavelength. We write the component wave phase at distance z and time t as $\varphi = \varphi_0 + \Delta\varphi$. Then for coincidence of all component phases, $\Delta\varphi$ must be zero.

This condition is

$$(\omega_0 + \Delta\omega)t - (\kappa_0 + \Delta\kappa)z = \varphi_0 + \Delta\varphi,$$

$$\omega_0 t - \kappa_0 z = \varphi_0,$$

$$\Delta\varphi = 0 = (\Delta\omega)t - (\Delta\kappa)z, \tag{11.9}$$

$$z = \frac{\Delta\omega}{\Delta\kappa}t = v_g t.$$

The quantity $\Delta\omega/\Delta\kappa$ is known as the group velocity v_g, and clearly, it must remain constant if position z is to be the same for all the component waves, and the shape of the pulse, unchanged.

An alternative visualization involves a simple diagram (Figure 11.19). We plot the z-direction horizontally, and ω, vertically. We sketch the bundle of component waves making up the pulse as

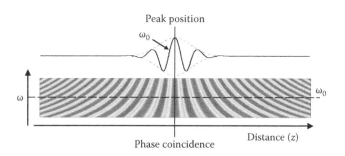

FIGURE 11.19
Sketch showing the component waves of the pulse as a continuous distribution of horizontal lines along the direction of propagation and with their relative phases marked as contours across them. The pulse peak coincides with the position where the phase of all the components is exactly equal.

a continuous set of lines through the appropriate values of ω and parallel to the z-axis. We mark contours of constant φ on the lines. Provided there is one contour that runs normally across the lines, so that there is an exact phase coincidence, then the pulse peak will be positioned there and the pulse shape will be unchanged.

In a nondispersive medium, the phase at the peak will be zero because all the component waves will be travelling at identical velocity even though it may be less than the velocity in free space. In a dispersive medium, the component waves travel at different velocities according to the particular value of refractive index. Provided the variation in velocities still permits a phase coincidence somewhere, then the pulse will appear there and will be unchanged in shape, although the phase of the carrier wave will be altered. It is clear from Equation 11.9 that the critical condition is for the group velocity to remain constant across the frequency spectrum of the pulse.

In a dispersive medium, the refractive index changes with frequency. We can calculate the group velocity in terms of this change. We write n as a function of ω as $n(\omega)$:

$$\kappa = \frac{2\pi n(\omega)}{\lambda} = \frac{\omega n(\omega)}{c},$$

$$\frac{d\kappa}{d\omega} = \frac{n(\omega)}{c} + \frac{\omega}{c} \cdot \frac{dn(\omega)}{d\omega},$$

$$v_g = \frac{c}{n(\omega) + \omega \cdot \dfrac{dn(\omega)}{d\omega}}. \tag{11.10}$$

In a medium with normal dispersion, this is not constant.

There is thus no guarantee that the group velocity should be constant with changing frequency. If the second derivative of κ with respect to ω is nonzero, then there can be no phase coincidence, and the pulse will be perturbed. Again, we can consider the operation in two different equivalent ways. If we limit ourselves to the second derivative, then we can write the expression for the phase of an arbitrary component wave as

$$(\omega_0 + \Delta\omega) \cdot t - \left(\kappa_0 + \Delta\omega \cdot \left.\frac{d\kappa}{d\omega}\right|_0 + \frac{1}{2}(\Delta\omega)^2 \cdot \left.\frac{d^2\kappa}{d\omega^2}\right|_0 \right) \cdot z = \varphi + \Delta\varphi, \tag{11.11}$$

and we can immediately identify a problem. The third term in the coefficient of z is even in $\Delta\omega$, and so it cannot be compensated for by the other terms. We must therefore split the frequency distribution of the pulse into two parts, one with positive $\Delta\omega$ and the other with negative $\Delta\omega$, and look at each separately. In each case, we ensure that the value of $\Delta\varphi$ is zero. This gives two equations instead of the usual one. We keep the value of z the same in each and introduce a different time t representing the interval in time between the pulse centers that correspond to each part of the split distribution. If the spectral width of the split distribution were halved, then each component pulse would have twice the basic pulse width. As a crude correction for this effect, therefore, we treat the $\Delta\omega$ in the following expressions as the width of the frequency distribution of the basic initial pulse:

$$\Delta\omega \cdot t_1 - \Delta\omega \cdot \left.\frac{d\kappa}{d\omega}\right|_0 \cdot z - \frac{1}{2}(\Delta\omega)^2 \cdot \left.\frac{d^2\kappa}{d\omega^2}\right|_0 \cdot z = 0,$$

$$-\Delta\omega \cdot t_2 - \Delta\omega \cdot \left.\frac{d\kappa}{d\omega}\right|_0 \cdot z - \frac{1}{2}(\Delta\omega)^2 \cdot \left.\frac{d^2\kappa}{d\omega^2}\right|_0 \cdot z = 0.$$

Then, since $\dfrac{d}{d\omega}\left(\dfrac{d\kappa}{d\omega}\right) = -\dfrac{1}{v_g^2} \cdot \dfrac{d}{d\omega}(v_g),$

$$\Delta t = (t_1 - t_2) = -\Delta\omega \cdot \frac{d^2\kappa}{d\omega^2} \cdot z = \Delta\omega \cdot \frac{dv_g}{d\omega} \cdot \frac{1}{v_g^2} \cdot z, \tag{11.12}$$

and the result (Equation 11.12) actually corresponds to that of a much more strict derivation using Gaussian pulses. For very short pulses that are considerably broadened, Equation 11.12 gives the width of the broadened pulse. For longer pulses or smaller broadening, it is a little more complicated.

Alternatively, we can use the diagram to see the way in which the phase coincidences are affected by the variation of group velocity. Figure 11.20 shows the modified arrangement of the various component waves and their contours of equal phase. The phase broadening itself causes a widening of the pulses corresponding to each band of frequencies, and so there is a still greater broadening as the pulse propagates. The carrier in each of the pulses differs in frequency. The change in the carrier frequency through the width of the resultant pulse in any real case is gradual rather than abrupt and is known by the term *chirp*. The pulse is broadened and chirped.

The effect, because it is due to a change in the group velocity across the frequency range of the pulse, is usually known as group velocity dispersion (often abbreviated to GVD). Similar effects occur in waveguides and optical fibers. GVD is measured in units of (time)^2/(unit length) and is given by

$$\mathrm{GVD} = \frac{d^2\kappa}{d\omega^2}\bigg|_0 . \tag{11.13}$$

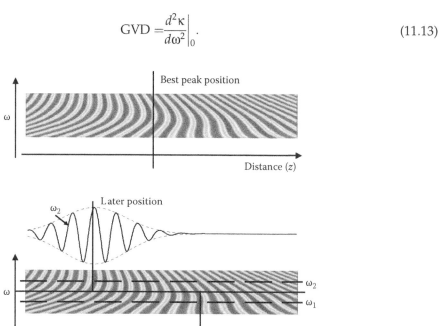

FIGURE 11.20

Pulse frequency distribution in the upper figure is notionally split into two parts in the lower. Each of these parts can be looked on as representing a component pulse with its own center position. Since the group velocity is different for the two component pulses, they separate such that one lags behind the other and the combined pulse is broadened. The carrier of the combined pulse shifts from, in this figure, a lower value at the front to a greater value at the rear. This shift in carrier frequency through the pulse is known as a *chirp*. The pulse is said to be broadened and chirped.

If the original pulse is of Gaussian shape as in Equation 11.7, then if we write

$$\tau_g^2 = \left.\frac{d^2\kappa}{d\omega^2}\right|_0 \cdot z,$$

it can be shown [8] that the new pulse width is given by

$$\tau_{new} = \tau \left[1 + \frac{\tau_g^4}{\mu^4}\right]^{\frac{1}{2}}. \tag{11.14}$$

All these effects are linear, and so they can be undone by a similar but opposite effect. Further, the order in which the effects occur is unimportant. A dispersive broadening may be canceled by an opposite dispersion.

A pulse, consisting of an envelope over a carrier, may be subjected to a modification, by passing through a crystal modulator, for example, in which the phase of the carrier is gradually varied throughout the length of the pulse. If this variation is a linear function of time, then the effect is just as though the frequency of the carrier had been changed. There is little other effect. However, if the phase is changed as a quadratic function of time, then it is as though the frequency of the carrier were gradually shifted throughout the length of the pulse [9]:

$$\cos(\omega t + at^2) = \cos[(\omega + at)t]. \tag{11.15}$$

The pulse has frequency $(\omega + at)$ and, in other words, is chirped. This chirped pulse appears indistinguishable from a short pulse that has been dispersion broadened, except that the apparent dispersion can be opposite in sign to normal dispersion. The pulse can then be subjected to the action of a dispersive medium where there is significant GVD. Provided this dispersion is of the correct magnitude and sense, then it will undo the artificially induced effect in the pulse, leaving it considerably narrowed. Various components have been used for this purpose but the flexibility of optical coatings makes them particularly attractive in this application [10–13].

Optical coatings affect both the amplitude and the phase of incident light. They can therefore, in principle, make the kinds of adjustments to incident light that we have been considering. They have an advantage over dispersive systems, in that the correction is made immediately. We must first consider the nature of the effect that thin-film coatings have on the pulse.

Amplitude reduction over part of the range of frequencies leads to pulse broadening because the narrower the frequency spectrum, the broader the pulse. We therefore limit ourselves to the consideration of those systems that have flat performance in terms of either transmittance or reflectance and that make adjustments to the phase.

The sign convention is important. We use the normal thin-film convention.

The coordinate system has its origin at the surface where the reflection is said to be taking place and the phase shift is measured at that surface. The electric field retains its incident positive direction. An incident wave, say, $\mathcal{E}\cos(\omega t - \kappa z + \varphi_{inc})$, suffers a phase change φ_{ref} at surface $z = 0$. The electric field at that surface for the reflected beam therefore becomes $\mathcal{E}\cos(\omega t + \kappa z + \varphi_{inc} + \varphi_{ref})$, where the positive sign before κz indicates propagation in the opposite direction to that of incidence. The returned beam is now propagating along the negative direction of the z-axis. We can avoid the sign change in z if we introduce the idea of the total path traveled by the wave that we denote by x, which always increases as the wave propagates and is along the positive direction of the z-axis before reflection and along the negative direction after reflection. (Note the temptation when using the alternative phase factor convention of $(\kappa z - \omega t)$ to reverse the direction of the wave by incorrectly writing $(\kappa z + \omega t)$, reversing the direction of time rather than, correctly, $(-\kappa z - \omega t)$, reversing the propagation direction.)

The expression for the wave now becomes

$$\mathcal{E}\cos(\omega t - \kappa x + \varphi_{inc} + \varphi_{ref}), \tag{11.16}$$

where x is always positive for increasing propagation length.

Now let us examine the effects of the various phase angles on the pulse and its components. We take Equation 11.10, and we rewrite the left-hand side to include a change of phase on reflection. Then,

$$
\omega_0 t - \kappa x + \Delta\omega \cdot t - \Delta\omega \frac{d\kappa}{d\omega}\bigg|_{\omega_0} x - \frac{1}{2}(\Delta\omega)^2 \frac{d^2\kappa}{d\omega^2}\bigg|_{\omega_0} x + \varphi_0 + \Delta\omega \frac{d\varphi}{d\omega}\bigg|_{\omega_0} + \frac{1}{2}(\Delta\omega)^2 \frac{d^2\varphi}{d\omega^2}\bigg|_{\omega_0}
$$

$$
= (\omega_0 t - \kappa L) + \Delta\omega\left\{ t - \left(\frac{d\kappa}{d\omega}\bigg|_{\omega_0} x - \frac{d\varphi}{d\omega}\bigg|_{\omega_0} \right) \right\} - \frac{1}{2}(\Delta\omega)^2 \left(\frac{d^2\kappa}{d\omega^2}\bigg|_{\omega_0} x - \frac{d^2\varphi}{d\omega^2}\bigg|_{\omega_0} \right).
$$

(11.17)

The quantity $-(d\varphi/d\omega)$ has units of time, and we can identify it as equivalent in its effect to the group delay (GD) due to dispersion, and it is therefore known as the group delay, sometimes abbreviated to GD. The next term, $-(d^2\varphi/d\omega^2)$, has an effect equivalent to the GVD. Since the negative first derivative is known as GD, this second derivative is known as group delay dispersion, abbreviated to GDD, and has units of (time)^2. Although we have said little about it here, the third derivative is sometimes called the third-order dispersion, with units of (time)^3, and abbreviated to TOD. TOD is usually small, but if it is significant, it can adversely affect the shape of the pulse. The GDD is particularly important because it can be adjusted in sign and can therefore be used to offset the effects of GVD and to operate on chirped pulses.

For most simple reflectors, φ increases with wavelength. This is the case with the classical quarter-wave stacks. φ slowly increases with λ, the rate of change being a minimum at the central wavelength, and the greater the index contrast in the layers, the slower the change. An outer low-index layer actually still further reduces the rate of change. The calculated GDD for a quarter-wave stack is shown in Figure 11.21. The outermost layer in this case is of high-refractive index. Although a low-index layer outermost leads to a slight gain, it gives an antinode of electric field at the outer surface and may therefore be undesirable. It is obvious that the calculated GDD for quarter-wave stacks will normally be very small, and so it is a particularly safe type of reflector to use with short pulses.

Transparent optical materials with normal dispersion show a refractive index n that reduces as the wavelength increases. We can write for κ

$$
\kappa = \frac{2\pi n}{\lambda} = \frac{n\omega}{c},
$$

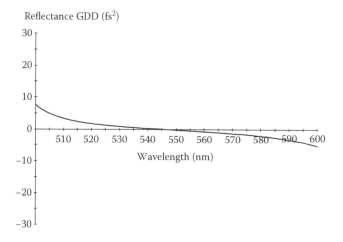

FIGURE 11.21
Calculated GDD for a 23-layer classical quarter-wave stack of titanium dioxide and silicon dioxide, the titanium dioxide outermost. Reference wavelength is 550 nm. The effect is clearly quite small, and this is normal for quarter-wave stacks in general. Variation in the total number of layers makes little difference.

c being, as usual, the velocity of light in vacuo. The variation of n in the region of normal dispersion can usually be represented as a Cauchy expression:

$$n = A + \frac{B}{\lambda^2} + \frac{C}{\lambda^4},$$

so that since $\lambda = 2\pi c/\omega$,

$$\kappa = \frac{n\omega}{c} = \frac{A\omega}{c} + \frac{B\omega^3}{4\pi^2 c^3} + \frac{C\omega^5}{16\pi^4 c^5},$$

and the GVD is given by

$$\frac{d^2\kappa}{d\omega^2} = \frac{6B\omega}{4\pi^2 c^3} + \frac{20C\omega^3}{16\pi^4 c^5}$$
$$= \frac{1}{2\pi c^2}\left(\frac{6B}{\lambda} + \frac{20C}{\lambda^3}\right). \tag{11.18}$$

Typical values for B might be on the order of 10^4 nm^{-2}, and for C, 10^9 nm^{-4}. c is approximately 300 nm/fs. Thus, Equation 11.18 becomes on the order of 10^{-4} fs^2/nm or 100 fs^2/mm.

Figure 11.22 shows the GVD calculated from the manufacturer's data for SK7 glass [14]. Clearly, for very short pulses of a few femtosecond length, the propagation of even a very few millimeters in a dispersive medium can degrade the pulse.

Rather longer pulses with length measured in picoseconds are used in telecommunications. Here, propagation in media such as optical fibers can be over lengths measured in kilometers. Now the effect of GVD and TOD can be sufficiently great to affect even these much longer pulses.

The net GDD is given by

$$\left(\frac{d^2\kappa}{d\omega^2}\bigg|_0 L - \frac{d^2\varphi}{d\omega^2}\bigg|_0\right). \tag{11.19}$$

Straightforward quarter-wave stacks show small GDD implying that although useful in reflecting short pulses without distortion, they are not useful as they stand for the correction of the effects of GVD. The principle of coatings for the correction of the dispersive effects is that light may penetrate into them to a rapidly varying extent and therefore show rapid phase dispersion, which, in turn, is translated into the high GDD that is required. Broadband reflectors with extended zones exhibit this effect and, incidentally, may have a considerable broadening effect when used as

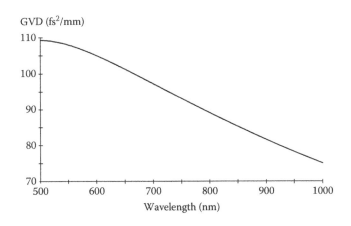

FIGURE 11.22
GVD in femtoseconds squared per milimeter for SK7 glass [14] calculated from the manufacturer's data.

TABLE 11.1

Design of Chirped Reflector

Layer	Material	Optical Thickness	Layer	Material	Optical Thickness
Medium	Air	Massive			
1	TiO_2	0.048	13	TiO_2	0.282
2	SiO_2	0.239	14	SiO_2	0.285
3	TiO_2	0.336	15	TiO_2	0.275
4	SiO_2	0.208	16	SiO_2	0.291
5	TiO_2	0.231	17	TiO_2	0.306
6	SiO_2	0.197	18	SiO_2	0.324
7	TiO_2	0.225	19	TiO_2	0.362
8	SiO_2	0.292	20	SiO_2	0.320
9	TiO_2	0.292	21	TiO_2	0.355
10	SiO_2	0.287	22	SiO_2	0.323
11	TiO_2	0.279	23	TiO_2	0.273
12	SiO_2	0.288	Substrate	Glass	Massive

Source: Thin Film Center Inc., Tucson, AZ.
Note: $\lambda_0 = 700$ nm.

simple reflectors. They are, however, useful for operating on chirped pulses [12,13], and because they often have a structure that exhibits a gradual tapering of layer thickness through the structure, they are often known as chirped mirrors. Table 11.1 and Figures 11.23 and 11.24 show the details of the design and calculated performance of a simple example of such a coating with a GDD of -30 fs^2 over the region of 750–900 nm. This is an example of a design arrived at purely by synthesis with no starting information other than the materials silica and titania that were to be used. Szipöcs et al. [12] and Szipöcs and Köházi-Kis [15] give more detailed accounts of a more systematic approach to the design of such chirped mirrors.

If the depth of penetration D is a smooth function of wavelength, then a constant, negative GDD $-|K|$ demands an optical depth of penetration D given by

$$D = \frac{nd}{\lambda_0} = \frac{\pi c^2 |K|}{2\lambda_0} \cdot \left[\frac{1}{\lambda_a} - \frac{1}{\lambda_b} \right], \tag{11.20}$$

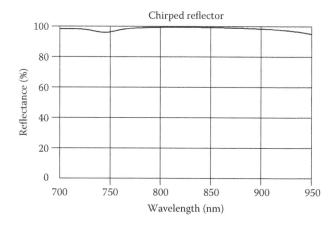

FIGURE 11.23
Calculated reflectance of the coating of Table 11.1.

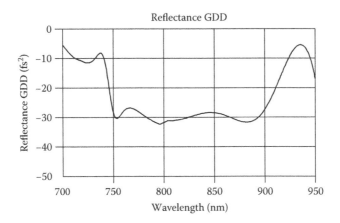

FIGURE 11.24
Calculated GDD of the coating of Table 11.1

where the range of wavelengths is λ_a to λ_b. This can give an idea of the thickness of coating required, although it can be made a little thinner by the use of a cavity structure.

Figures 11.25 and 11.26 show the construction and performance of a more complicated dispersive coating for a GDD of $-500\ \mathrm{fs}^2$. It consists of a reflector over which a variable-depth section has been placed to give the necessary phase characteristic.

The chirped mirror is less useful for compensating for the GVD of a longer pulse passing through a considerable thickness of optical material. Here the requirement is a much narrower spectral range but a much higher GDD. Some way of increasing the magnitude of the negative values of GDD of an optical coating is required. The addition of a weak cavity to the front of the quarter-wave stack has been shown to be one fairly successful way of achieving this result, provided the wavelength region is limited; that is, the pulse is reasonably long. The presence of the cavity can be thought of as causing a variable GD that has a maximum at resonance and falls away on either side. Such an arrangement is usually known as a Gires–Tournois interferometer after the originators [16,17]. The weak cavity does not reduce the reflectance too much, but the result is a very rapid change of phase on reflection that leads to the desired effect.

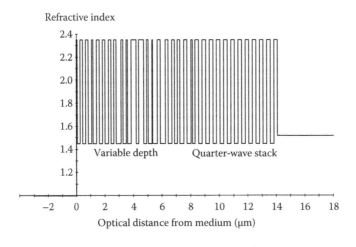

FIGURE 11.25
Design of a dispersive coating for a GDD of $-500\ \mathrm{fs}^2$. It consists of a quarter-wave stack reflector with a variable-depth section in front.

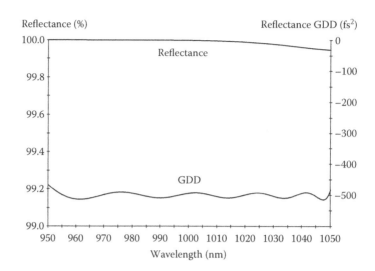

FIGURE 11.26
Performance of the dispersive coating of Figure 11.25.

Figure 11.27 shows the dispersive parameters of a simple Gires–Tournois interferometer of design

$$\text{Air}|(HL)^6HH(LH)^6L(HL)^9H|\text{Glass,} \qquad (11.21)$$

using tantalum pentoxide and silicon dioxide as materials. This type of interferometer is particularly useful for compensating for TOD in optical fibers [18]. Its characteristic with the large central lobe and much less pronounced secondary lobes makes it particularly suitable for mounting in a cascade of reflecting components, each with slightly shifted resonances so as to broaden the resultant characteristic.

In some versions of the interferometer, the final quarter-wave stack is largely replaced by a silver film. This saves many layers. The same considerations that apply to enhanced metal reflectors to avoid reducing rather than increasing the reflectance apply in this case.

The Gires–Tournois interferometer leads us to a reasonably straightforward design procedure. We start with a reflector that can conveniently be a quarter-wave stack. Then, at the front of the reflector, we build a variable depth reflector that produces the necessary phase variation without destroying the reflectance. This was the technique used for dispersive coating in Figures 11.25 and 11.26. This design began with the creation of the quarter-wave stack reflector, to which was added by synthesis a reflector of varying penetration depth. All this assured high reflectance over the 950–1050 nm range with the achievement of a GDD of some −500 fs². The thickness variation through the design suggests that the process has actually introduced some cavities that introduce an enhanced phase shift compared with a simple taper. The reflectance is perhaps higher than is strictly necessary depending on the application. If so, then a few layers can be taken from the quarter-wave stack section.

A major problem with coatings of this type is their acute error sensitivity. The deposited version of the coating of Figures 11.25 and 11.26 will show inevitably larger oscillations in GDD. Fortunately, the performance is additive, and the normal practice is to have several dispersive components in series so that the ripple in one can be canceled out by that in another. Pervak et al. [19] demonstrated a clever way of using mirrors produced in the same deposition run at different angles of incidence so as to cancel the ripple by shifting the peaks of one mirror into the troughs of another. The GDD can also be considerably increased by the use of components in series. Pervak

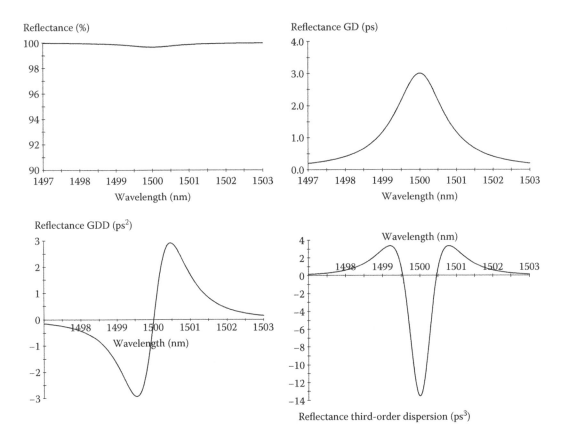

FIGURE 11.27
Various dispersive parameters of the Gires–Tournois interferometer of design given in Equation 11.21. These are several orders of magnitude greater than those of the chirped coating but over much narrower spectral regions and are appropriate for telecom pulses rather than ultrafast.

et al. [20] described a system for pulse amplification, where the pulse is first stretched to reduce peak power in the amplifier and then shortened by 52 successive reflections at dispersive mirrors. This is an area that is very fast moving.

11.3 Glare Suppression Filters and Coatings

Glare is a term that is extensive in its coverage. What we mean by the term in this context is specular reflection of illumination from a bright source that enters the eyes and masks a usually weaker, desired visual image. Much of the earlier work reported here was connected with cathode ray tube displays, and although these displays are fast disappearing, we include the coatings because they are interesting applications, and we can still learn from them. Furthermore, even the latest displays can suffer from glare problems.

Polarizing sunglasses represent an early example of glare reduction. Sunlight reflected by water or silica sand is a common source of glare. When the sun is at an angle that makes the glare a problem, the reflection is usually in the vertical plane and at or near the Brewster angle so that the reflected light is principally s-polarized. A person who is upright will receive this glare light as primarily linearly horizontally polarized, and it can therefore be virtually eliminated by a suitably oriented polarizer.

This solution depends on reflection in the vicinity of the Brewster angle and is not available for the now common glare caused by unsuitable lighting where visual display units are concerned. In this case, the signal light from within the device is masked by specularly reflected ambient light from its surface. The orientation of the plane of incidence can enormously vary, and the glare can be reflected at angles that are near normal. A solution that has been much used in electronic instruments consists of a circular polarizer inserted before the display. Specular reflection at near normal incidence reverses the handedness of the circularly polarized glare light that has already passed through the polarizer on its inward journey and makes it impossible for it to pass through it again on the outward journey. This works well when the specular reflectance of the outer surface of the polarizer is appreciably less than that of the underlying display. In other cases, the reflectance must be reduced by the application of an antireflection coating. Since the circular polarizer protects against glare from its own rear surface, the antireflection coating is required over the front surface only.

Later, it was found that a quite satisfactory reduction in glare could be achieved by replacing the circular polarizer by a simple neutral density filter such as a sheet of absorbing glass or plastic. Specular reflectance from the filter is eliminated by antireflection coatings at the front and back (Figure 11.28). Glare light then passes through the filter twice while signal light passes through only once. This nominally reduces the glare to signal ratio by a factor equal to the transmittance of the filter. However, the brightness of the display can be raised to compensate, and so a typical glare reduction is equal to the square of the transmittance. A transmittance of 50% then reduces the glare by a factor of 4, a quite acceptable figure.

The glare reduction filter of this type is a separate component that is fitted, at a late stage, as an accessory to the display unit. The absorbing glare reduction component can be included in the antireflection coating itself. The simplest way of achieving this is to replace the normal completely transparent high-index materials by high-index absorbing materials. The most common arrangement takes the four-layer high-efficiency antireflection coating and replaces the usual zirconia or titania with indium tin oxide (ITO) (Figure 11.29). A good antireflection coating that is completely transparent reduces the glare by 50%. Normally, it is arranged to have a certain amount of absorption that acts to reduce the glare still further. Figure 11.30 shows a calculated characteristic that uses ITO data from Gibbons et al. [21]. The overall transmittance of the coating is around 90%, and so the glare is further reduced by a factor of 0.8.

To enhance the absorption still further and increase the glare reduction, materials that are still more absorbing may be used. Transition metal nitrides, such as titanium nitride, are one possibility [22]. Wolfe [23] used layers of silver and nickel to increase the absorption and at the same time assure electrical conductance so that the coating could also serve as a radio-frequency screen. Silver was incorporated in the form of a subsystem consisting of around 8 nm of silver surrounded by 1.2–2.0 nm of $NiCrN_x$ that was, in turn, surrounded by some 20–30 nm of SiN_x or $SiZrN_x$. An outer layer of SiO_2 then completed the coating. Alternatively, a layer of nickel, perhaps 6–9 nm thick surrounded by protecting layers of SiN_x to protect it from oxidation, was found to be satisfactory. Coatings involving these materials could be made to have transmittances in the range of 30–80%.

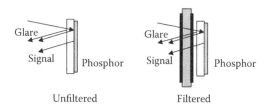

FIGURE 11.28
Principle of an external antiglare filter. Glare light passes through the filter twice while signal light passes through only once. Although the illustration imagines a cathode ray tube type of display, the problem and solution exist with other current types also.

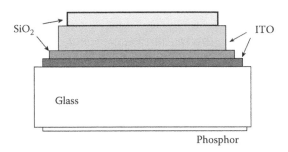

FIGURE 11.29
Glare reduction filter applied to the face of a display. The high-index material is made both absorbing and conducting.

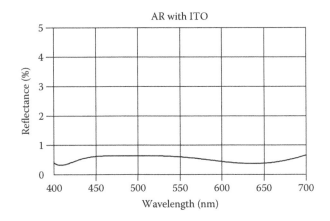

FIGURE 11.30
Response of a four-layer antireflection coating using silica and ITO. The ITO constants are taken from Gibbons et al. [21].

An ingenious family of two-layer coatings for glare reduction has been proposed. Early development was carried out by a group at the Asahi Glass Company Ltd. in Japan [24], who termed the coating *ARAS*. A further description is given by Ishikawa and Lippey [25]. Absorbing two-layer coatings are also discussed in detail by Zheng et al. [26]. At the shortest wavelength, the coating can be considered to essentially consist of a typical V-coat with a thin high-index layer next to the substrate and a rather thicker low-index layer outermost. For simplicity, the substrate in this description is transparent, but this is not a necessary condition. Now let the wavelength move to a longer value. The physical thicknesses of the layers will remain constant, but in the absence of dispersion, both optical thicknesses will become smaller fractions of the wavelength, and so the admittance loci will shrink. Now imagine that as the wavelength changes, the reflectance of the coating remains at zero. The outermost low-index layer can be considered to be a normal dielectric material, like silica, and so it will exhibit negligible dispersion. The end point is firmly fixed at unity on the real axis, the admittance of the incident medium, and so since the locus is shorter, the starting point moves around the existing circle. Similarly, if the index of the high-index inner layer remains constant and the starting point is firmly fixed at the admittance of the substrate on the real axis, the end point will move around the high admittance circle and a gap will open up in the locus so that it is no longer valid. Now let the optical constants of the inner layer, the high-index layer, be completely adjustable. By adjusting both the index of refraction and the extinction coefficient, the end point of the locus can be swept over a quite large area of the admittance diagram. The gap in the admittance locus can be closed so that it becomes valid, and the reflectance remains at zero. By arranging for appropriate smooth variations in both n and k, the reflectance can be retained at zero over the entire visible region.

TABLE 11.2

Optical Constants of Tungsten Doped Titanium Nitride

Wavelength	Refractive Index	Extinction Coefficient
405.00	2.5	0.7
510.00	1.8	1.3
632.80	1.2	1.7

Source: H. Ishikawa and B. Lippey. 1996. Two layer broad band antireflection coating. In Tenth International Conference on Vacuum Web Coating, pp. 221–233. Bakish Materials Corporation, Englewood, NJ.

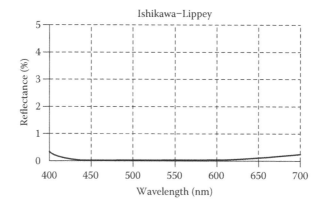

FIGURE 11.31
Performance calculated for design Air | SiO$_2$: 80 nm | TiN$_x$W$_y$: 10 nm | Glass. (Calculation parameters from H. Ishikawa and B. Lippey. 1996. Two layer broad band antireflection coating. In Tenth International Conference on Vacuum Web Coating, pp. 221–233. Bakish Materials Corporation, Englewood, NJ.)

The properties of tungsten-doped titanium nitride are very close to ideal. Measured values extracted from the study by Ishikawa and Lippey (estimating from their graph) are given in Table 11.2. The thicknesses of the tungsten-doped titanium nitride film and the silica film were 10 and 80 nm, respectively.

We use a cubic spline interpolation to smooth the constants given in the table and then calculate the performance assuming a normal dispersive index for the glass substrate to give Figure 11.31. This is impressive.

The calculated transmittance of the coating is shown in Figure 11.32. It is surprisingly neutral and will contribute to a satisfactory reduction in glare. Although no figures are given, the coating also reduces emissions from the display unit.

The admittance diagram in Figure 11.33 clearly shows the way in which the dispersion of the optical constants of the absorbing layer holds the termination of the locus in the vicinity of the incident medium admittance and keeps the reflectance low.

11.4 Some Coatings Involving Metal Layers

11.4.1 Electrode Films for Schottky Barrier Photodiodes

A simple diode photodetector consists of a metal layer deposited over a semiconductor forming a Schottky barrier. High quantum efficiency can be achieved. The incident light passes through the

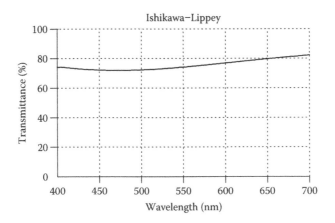

FIGURE 11.32
Calculated transmittance of the coating of Figure 11.31.

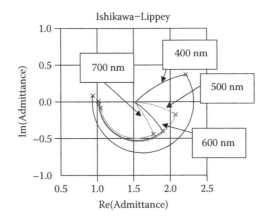

FIGURE 11.33
Admittance locus of the antireflection coating of Figure 11.31 showing how the dispersion of the optical constants of the layer next to the substrate compensates for the shortening of the locus as the wavelength increases.

metal layer into the depletion layer of the diode where it creates electron–hole pairs. The metal contact layer must transmit the incident light, and since it has intrinsically high reflectance, it must be coated to reduce its reflection loss. We give here a very simple approach to the design of a combination of electrode and antireflection coating. A number of workers [27–29] have made contributions in this area, with probably the most complete account of an analytical approach being that of Schneider [29].

The substrate for the thin films is the semiconducting part of the diode, and it is fixed in its optical admittance. The metal layer goes directly over the semiconductor (in some arrangements, there is a very thin insulating layer that has negligible optical interference effect), and so the potential transmittance is fixed entirely by the thickness of the metal. All that can be done to maximize the actual transmittance is to simply reduce the reflectance to zero.

We take as an example a gold electrode layer deposited on silicon. We assume a wavelength of 700 nm and optical constants of $0.131 - i3.842$ for gold and $3.92 - i0.05$ for silicon [30]. The optical constants of silver and copper are quite similar to those of gold at this wavelength, and the results apply almost equally well to these two alternative metals. The admittance locus of a single gold film on silicon is shown in Figure 11.34. An antireflection coating must bridge the gap between the appropriate point on the metal locus to point $(1,0)$, corresponding to the admittance of air. We can

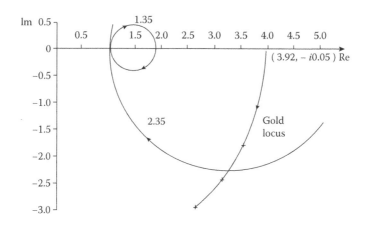

FIGURE 11.34
Admittance diagram showing some of the factors in the antireflection of a metal electrode layer in a photodiode. The optical constants of gold are assumed to be $(0.131 - i3.842)$ at a wavelength λ_0 of 700 nm. The gold is deposited on silicon with optical constants $(3.92 - i0.05)$. The crosses on the gold locus mark thickness increments of $0.005\lambda_0$, i.e., 3.5 nm. Also shown are loci corresponding to dielectric layers of indices 1.35 and 2.35 that terminate at point 1.00.

assume that the maximum and minimum values of the dielectric layer admittance available for antireflection coating are 2.35 and 1.35, respectively. Using these values, we can add to the admittance diagram two circles that pass through point (1,0) and correspond to the admittance loci of dielectric materials of characteristic admittances 2.35 and 1.35, respectively. These loci define the limits of a region in the complex plane. Provided a metal locus ends within this region, then it will be possible to find a dielectric overcoat of admittance between 1.35 and 2.35 that, when the thickness is correctly chosen, will reduce the reflectance to zero. It is clear from the diagram that the thicker the metal film, the higher must be the admittance of the antireflection coating. Once the metal locus extends beyond this region, a single dielectric layer can no longer be used and a multilayer coating (or a single absorbing layer, although it would reduce transmittance and so would not be very useful in this particular application) becomes necessary. We have already considered multilayer coatings in the section on induced transmission filters. Here we limit ourselves to a single layer and take the highest available index of 2.35.

The remaining task in the design is then to find the thicknesses of metal and dielectric corresponding to the trajectories between the substrate and the point of intersection and between the point of intersection and point (1,0) in Figure 11.34. The points marked along the metal locus correspond to intervals of $0.005\lambda_0$ in geometrical thickness, that is, to the thickness intervals of 3.5 nm. Visual estimation suggests a value of $0.013\lambda_0$ for the thickness to the point of intersection. A more accurate calculation gives $0.0133\lambda_0$, that is, a thickness of 9.3 nm. The dielectric layer has an optical thickness of somewhere between an eighth and a quarter wave, and accurate calculation yields $0.186\lambda_0$.

The calculated performance of this coating is shown in Figure 11.35. Of course, the thickness of the metal film is rather small, and it is unlikely that the values of optical constants measured on thicker films would apply without correction, but the form of the curve and the basic principles of the coating are as discussed here.

11.4.2 Spectrally Selective Coatings for Photothermal Solar Energy Conversion

Coatings for application in the field of solar energy represent a complete subject in their own right. They were discussed in detail by Hahn and Seraphin [31]. Here we simply consider a limited range of coatings based on antireflection coatings over metal layers that have much in common with the electrode film of the previous section.

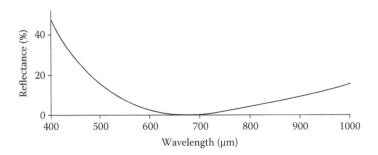

FIGURE 11.35
Calculated transmittance, including dispersion, of the gold electrode film and antireflection coating designed in the text.

Solar absorbers that operate at elevated temperatures can lose heat by radiation unless steps are taken to reduce their emittance in the infrared. Yet to efficiently operate, they must have high solar absorptance in the visible and near infrared. Optimum results are obtained from an absorbing coating that exhibits a sharp transition from absorbing to reflecting at a wavelength in the near infrared that varies with the operating temperature of the absorber. One way of constructing such a coating is to start with a thick metal film or a metal substrate and apply an antireflection coating that is efficient over the visible but which becomes ineffective in the infrared, so that at longer wavelengths, the reflectance is high and the thermal emittance, as a result, is low. Fortunately, we are simply interested in a reduction of reflectance. Transmittance is unimportant. The energy that is not absorbed in the coating is absorbed in the substrate. Thus, the antireflection coating can include absorbing layers.

A useful approach to the design is the use of a semiconducting layer over a metal. The semiconductor becomes transparent in the infrared beyond the intrinsic edge, and so in that region, the reflectance of the underlying metal predominates. In the visible and near infrared, the absorption in the semiconductor is sufficient to suppress the metallic reflectance, and to complete the design, it is sufficient to add an antireflection coating to reduce the reflectance of the front face of the semiconductor. Since the metal is to dominate the infrared performance, either the semiconductor layer must be relatively thin in the infrared or the metal must have sufficiently high k/n to be only slightly affected by the high index of the semiconductor in its transparent region. From the point of view of optical constants, silver is therefore the most favorable metal, but it suffers from a lack of stability at elevated temperatures that cause it to agglomerate so that its optical constants are shifted and its reflectance is reduced. Hahn and Seraphin (see Hahn and Seraphin [31] for a readily available summary and more detailed references) have developed coatings in which the silver is stabilized by layers of chromium oxide (Cr_2O_3), which act as diffusion inhibitors. The silicon films are produced by chemical vapor deposition, in which the silicon–hydrogen bonds in silane gas flowing over the substrate are broken by elevated substrate temperature and, as a result, silicon deposits. Adding oxygen or nitrogen to the gas stream gives an antireflection coating of silicon oxide or nitride that can be graded in composition by the continuous variation of gas–stream composition. Such coatings can withstand temperatures in excess of 600°C without degradation.

The design of such coatings is straightforward. First of all, the thickness of silicon must be such that the visible absorption is sufficiently high to mask the underlying silver but not so thick that interference effects reduce reflectance and increase emittance in the infrared. In the visible region, the light that enters the silicon layer and is reflected from the silver at the rear surface should be sufficiently attenuated that only a very small proportion ever reemerges. We can assume that the attenuation of this light depends on the law of the form $\exp(-4\pi kd/\lambda)$, and for the entire round trip from the front surface to the rear of film and back again to the front surface, we should have a value roughly in the range of $0.01 – 0.05$. Let us choose a design wavelength of 500 nm in the first instance at which silicon in thin-film form has optical constants of $4.3 – i0.74$ [30]. Then for exp

$(-4\pi kd/\lambda)$ to be 0.05, the value of d must be 160 nm. Since this is for the entire round trip, the film thickness should be half this value or 80 nm. An antireflection coating must then be added to reduce the visible reflectance of the front surface of the silicon layer. Since we have reduced the interference effects to a low level, the front surface will be similar to bulk silicon with optical constants characteristic of the material. Seraphin and his colleagues used a graded-index film of silicon nitride and silicon dioxide, but for simplicity, we assume here a homogeneous film of roughly 2.0 admittance and a quarter-wave thick at 500 nm. We can take zirconium dioxide with its characteristic admittance of 2.07 as an example. The performance of the complete coating is shown in Figure 11.36. The extra dip at 600 nm is a result of the thickness of the silicon. The silicon admittance locus spirals around, converging on the optical constants. At 600 nm, the spiral is somewhat shorter, but the end point is passing through a region where the zirconium oxide layer can act as a reasonably efficient antireflection coating once again, and so the dip appears. The silver begins to assert itself at around 700 nm in this design. We can shift the reflectance through to a longer wavelength, say, 750 nm, by carrying out a completely similar procedure, but this time using $4.17 - i0.37$ for the optical constants. Now a double-pass reduction of 0.05 leads to a round-trip thickness of 480 nm, representing a film thickness of 240 nm. The performance is also shown in Figure 11.36. In both traces, the optical constants of silicon and silver were derived from Hass and Hadley [30].

An alternative arrangement makes use of metal layers as part of an antireflection coating for silver. The great problem in designing an antireflection coating for a high-efficiency metal using entirely dielectric layers is that the admittance where the locus of the first dielectric layer, that is, the layer next to the metal, first cuts the real axis is far from point $(1,0)$, where we want to terminate the coating, and with each pair of subsequent quarter-waves, we can modify that admittance by only $(n_H/n_L)^2$. Many quarter waves are needed, as we have seen with the induced transmission filters. A metal layer, on the other hand, follows a different trajectory from a dielectric layer, cutting across dielectric loci, and can be used to bridge the gap between the large radius circle of the dielectric next to the metal and a dielectric locus that terminates at $(1,0)$.

The metal locus itself can be arranged to pass through $(1,0)$, but the extra dielectric layer is capable of giving a slightly broader characteristic and some protection to the metal layer. Silver could be used as the matching metal, but its high k/n ratio leads to rather narrow spikelike characteristics even with the terminating dielectric layer, and a metal with rather greater losses is better. We use chromium here as an illustration with aluminum oxide as dielectric. These materials

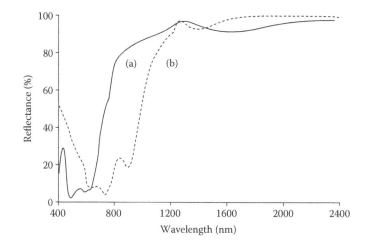

FIGURE 11.36

Calculated performance including dispersion of solar absorber coatings consisting of antireflected silicon over silver. Designs:Further details are given in the text.

have figured in published coatings (see Hahn and Seraphin [31] for further details). We choose a wavelength of 500 nm for the design, and the optical constants we assume for our materials are silver: $0.05 - i2.87$; aluminum oxide: 1.67; and chromium; $2.86 - i4.11$. Again the optical constants of the metals were obtained from Hass and Hadley [30] with interpolation if necessary. An admittance diagram of a coating of design

$$\text{Air} \left| \begin{array}{c} \text{Al}_2\text{O}_3 \\ 0.1841\lambda_0 \end{array} \right| \begin{array}{c} \text{Cr} \\ 7.5\,\text{nm} \end{array} \left| \begin{array}{c} \text{Al}_2\text{O}_3 \\ 0.184\lambda_0 \end{array} \right| \text{Ag}(\lambda_0 = 500\,\text{nm})$$

is shown in Figure 11.37. The chromium locus bridges the gap between the two dielectric layers. Because of its rather lower k/n ratio than silver, its trajectory is flatter and the entire characteristic is less sensitive to wavelength changes. The arrangement helps keep the final end point of the coating in the vicinity of (1,0) as the loci increase or decrease in length with changing wavelength or g.

No attempt was made to refine this design, although clearly, because of the wide range of possible thickness combinations that would lead to zero reflectance at the design wavelength, there must be a scope for performance improvement by refinement. The characteristic of the coating is shown in Figure 11.38. The reflectance minimum can be shifted to longer wavelengths by repeating the design process with appropriate values of the optical constants. This gives the desired zero but then at shorter wavelengths, where the dielectric loci are departing further and further from ideal and the chromium layer is unable to bridge the gap between them, a peak of high reflectance is obtained. At still shorter wavelengths, there is a second-order minimum where the dielectric layers make a complete revolution and are once again in the vicinity of the correct position. For the ideal values we have used in these calculations, the central peak of high reflectance is very high indeed. Practical coatings also show this double minimum (see Hahn and Seraphin [31]), but the central maximum is very much less prominent, the most likely explanation being that the layers in practice have much greater losses than we have assumed. In particular, the

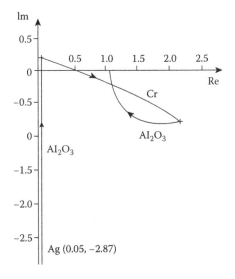

FIGURE 11.37
Admittance diagram at λ_0 of an absorber coating of design

$$\text{Air} \left| \begin{array}{c} \text{Al}_2\text{O}_3 \\ 0.1841\lambda_0 \end{array} \right| \begin{array}{c} \text{Cr} \\ 7.5\,\text{nm} \end{array} \left| \begin{array}{c} \text{Al}_2\text{O}_3 \\ 0.184\lambda_0 \end{array} \right| \text{Ag}(\lambda_0 = 500\,\text{nm}).$$

See the text for an explanation.

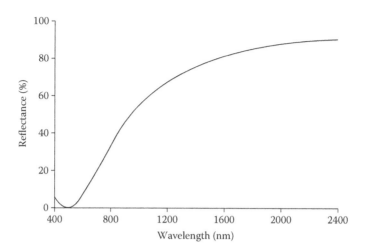

FIGURE 11.38
Calculated performance, including dispersion, of the absorber coating of Figure 11.37.

thin chromium layers are unlikely to have ideal optical constants. High losses would make the loci spiral in toward the center of the diagram and reduce the wavelength sensitivity.

The major problems associated with such coatings are not their design but the necessary high-temperature stability. Spectrally selective solar absorbers are only economically viable when they are used to produce high temperatures, and indeed, it is only at high temperatures that they offer an advantage over the more conventional spectrally flat black absorbing surfaces that can be produced very much more cheaply. They are used under vacuum to eliminate gas conduction heat losses, and so the major degradation mechanism is diffusion within the coatings. Silver is particularly prone to agglomerate at high temperatures, and much development effort has resulted in the incorporation of thin diffusion barriers such as chromium oxide that inhibit diffusion and agglomeration of the components without affecting the optical properties. The achievements in terms of lifetime at high temperatures are impressive. Further details are found in Hahn and Seraphin [31].

Some particularly interesting nine-layer selective solar absorber coatings consisting of three or four different materials, some metallic, are given by Kennedy [32].

11.4.3 Heat-Reflecting Metal–Dielectric Coatings

There are several applications where a cheap and simple heat-reflecting filter would be valuable. For example, a normal, spectrally flat solar absorber can be combined with such a filter so that the combination acts as a spectrally selective absorber. It is possible to construct a very simple bandpass filter that has the desired characteristics from a single metal layer surrounded by two dielectric matching layers [33–36]. The filter is similar in some respects to the induced transmission filter, although the maximum potential transmittance that is theoretically possible cannot usually be achieved. One design technique uses the admittance diagram, and we can illustrate it with an example in which we consider a glass substrate and an incident medium of air or vacuum. Silver, with optical constants of $0.06 - i3.75$ at 600 nm, can serve as metal, and we assume a dielectric layer material of index 2.35. Zinc sulfide, which has such an index, has been used in this application, but the most durable and stable coatings are ones incorporating a refractory oxide. Figure 11.39 shows an admittance diagram in which one dielectric locus begins at the substrate and a second terminates at (1,0), corresponding to the incident medium. If the complete coating is to have zero

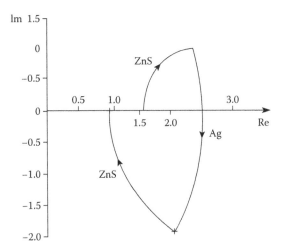

FIGURE 11.39
Admittance diagram of a metal–dielectric heat-reflecting filter. The diagram shows the locus at a wavelength of 600 nm of a ZnS | Ag | ZnS combination deposited on glass.

reflectance, then the remaining layers must bridge the gap between these two loci. Once again, it is easy to see that a metal layer can do this and that the particular optical constants of the metal are unimportant. They will simply somewhat alter the points of intersection with the two loci. The loci shown approximately correspond to the thickest silver film that will still give zero reflectance. Increasing the silver thickness without sacrificing the zero reflectance requires that the indices of the two dielectric layers be increased. A small increase in thickness of metal without a gross alteration in the design could be achieved by the insertion of a low-index quarter-wave layer next to the substrate to move the starting point of the next high-index dielectric layer, the upper one in the admittance diagram, further along the real axis toward the origin. The new locus would be outside the existing one demanding a thicker metal matching layer. In the absence of such a low-index layer, the final three-layer design is

$$
\begin{array}{|c|c|c|c|c|}
\text{Air} & \begin{array}{c}\text{ZnS}\\ 2.35\\ 0.146\lambda_0\end{array} & \begin{array}{c}\text{Ag}\\ 0.06-i3.75\\ 0.25\,\text{nm}\end{array} & \begin{array}{c}\text{ZnS}\\ 2.35\\ 0.141\lambda_0\end{array} & \begin{array}{c}\text{Glass}\\ \\ 1.52\end{array}
\end{array}\quad \lambda_0 = 600\,\text{nm}
$$

with performance shown in Figure 11.40. The steep fall toward the infrared is partly due to the drop in efficiency of the matching, but an inspection of the admittance diagram quickly reveals that the reduction in length of each locus accompanying an increase in wavelength should not by itself change the reflectance grossly. Metals, however, show large dispersion, the nature of which is a k that increases in step with λ into the infrared. This keeps the value of $(2\pi k d/\lambda)$ high, and that, together with the increasing k, lengthens the locus through the infrared, inducing a considerable increase in reflectance. The coating could be based on virtually any metal with high infrared reflectance and high-index dielectric material. Gold and bismuth oxide have been successfully used [36].

Silver tends to oxidize if the coating is exposed to the air, and it is normal to surround it with thin diffusion barrier layers. These can be nickel, chromium, titanium, and so on, and it is likely that these protecting metals oxidize in situ so that they expand, and the consequent high packing density makes them an effective barrier. These thin oxide layers have very little optical effect.

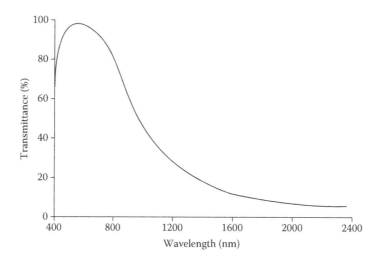

FIGURE 11.40
Transmittance, calculated with dispersion included, of the heat-reflecting coating of Figure 11.39. Details of the design are given in the text.

11.5 Gain in Optical Coatings

Little has been written on the subject of gain in optical coatings. This may be because the topic has not so far been of any great practical importance, but there are some signs in recent research papers that this situation may change. There are some interesting, even fascinating, effects that are connected with gain, and so we include some notes on it here. More details can be found in Macleod [37–39] and Macleod and Clark [40]. Gain calculations are handled through the extinction coefficient in the same way as absorption, except that the coefficient has an opposite sign. We shall write the complex index with gain as $(n + ik)$ in this section, where a positive value will be substituted for k, although we shall usually refer to the presence of gain as k being negative. We hope that this will make it a little clearer than writing $(n - ik)$ with a negative value substituted for k.

A particularly useful way of visualizing the difference between absorption and gain is to plot an admittance locus. A locus with a value of k representing a material with loss is a gradually collapsing spiral that eventually, if enough material is present, reaches point $(n - ik)$ (Figure 11.41). The spiral for a layer with negative k, or gain, opens out. If the material is thick enough, the spiral eventually reaches the imaginary axis and will actually pass through it to the other side. At that point, it reverses direction and spirals inward toward point $-(n + ik)$ (Figure 11.42). The value of k used in Figure 11.42 is intentionally very large to clearly show the spiral. In a real case, the value would be much lower, and the spiral, much slower.

In any real case, gain is a consequence of a population inversion maintained by some process of pumping, and there is a depletion rate that is a function of irradiance. The gain will eventually stabilize at a level where the depletion is just canceled by the pumping. Thus, in any real case, k will be constant only for very low power levels. We will not include such effects in this short discussion. We are effectively assuming that power levels are very low. The effect of the gain is to increase the transmittance and reflectance that now, because of the gain, can exceed unity. Figure 11.43 shows the behavior of both transmittance and reflectance in a system consisting of 4000 nm thick film with optical constants $(1.5 + i0.1)$ with incident medium of air and emergent of BK 7 glass. The incident medium of air implies a much higher reflection loss than the emergent medium of BK 7 glass. This makes the transmitted output rather larger than the reflected output. Both are

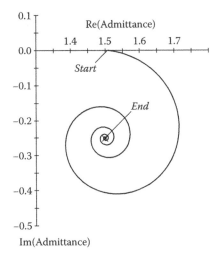

FIGURE 11.41
The locus of a thick absorbing layer deposited over glass. The end point corresponds to $(n - ik)$ that, in this figure, has a value of $(1.5 - i2.5)$.

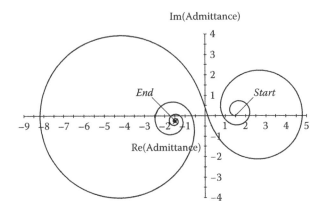

FIGURE 11.42
Locus of a material with negative k, or gain. The spiral reverses direction as it enters the left-hand side of the admittance diagram and gradually collapses toward point $-(n + ik)$. The reflectance in the left-hand side of the diagram is everywhere greater than 100% and is infinity at the point $-y_0$.

considerably greater than 100%. Again, for simplicity and to clearly show the effects, we are assuming a very large gain.

An adjustment of both gain and wavelength will yield a reflectance of infinity, that is, will result in oscillation of the system. The power output will determine the ultimate level of extinction coefficient. If the reflectance is high enough, that is, the termination point of the locus is close to $-y_0$, then spontaneous emission in the film will be enough to start the oscillation. Note that the normal thin-film calculations, those that we are using, do always assume an input from the incident medium. As the reflectance rises, this input becomes smaller and smaller with respect to the output, but the calculations themselves do not take spontaneous emission into account.

If the reflectance at the output side (0.11%; the emergent medium is BK 7) is removed by a perfect antireflection coating, then the reflectance becomes that of the front surface, and the device

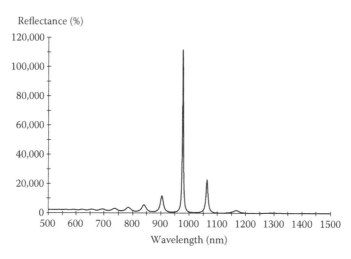

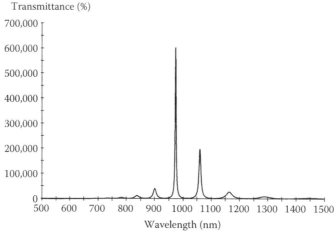

FIGURE 11.43
Reflectance (*upper*) and transmittance (*lower*) of a 4000 nm thick film with optical constants $(1.5 + i0.1)$ with incident medium of air and emergent of BK 7 glass. The unbalanced residual reflectances are responsible for the lack of symmetry in the results.

acts as an ideal amplifier (Figure 11.44). Again, the gain is set enormously high so that the effect can be clearly seen. In this case, however, any small irregularity such as a small inhomogeneity, residual reflectance, or spontaneous emission could affect the stability of the device as evidenced by a sudden expansion of the admittance locus. These need to be taken into account in any real amplifier so that such instability is prevented. An important requirement is that the output reflectance should be sufficiently low.

Added dielectric layers in the left-hand side of the complex plane behave almost in the normal way. The loci of dielectrics are still circular and centered on the real axis, and the quarter-wave rule holds. The difference is that the loci are described counterclockwise instead of clockwise. If the starting point for a dielectric layer is on the left of the imaginary axis, then the entire locus, and that of any following dielectrics, will remain on the left-hand side.

The matching of the rear of the gain layer to the eventual emergent medium presents some interesting aspects. In the case of the rear surface of an absorbing substrate, maximum throughput

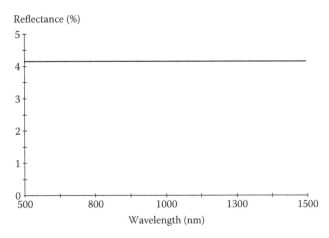

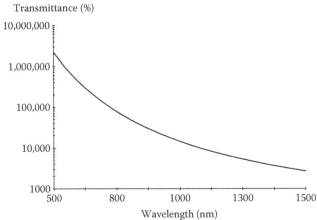

FIGURE 11.44
Reflectance and transmittance of the system of Figure 11.43 with the output side perfectly matched to the substrate.

is the criterion for the optimum matching system. Then, a useful definition of transmittance [41] is suggested in Chapter 2, Section 2.3.4:

$$T = \frac{4y_0 y_0^* \, \mathrm{Re}(Y)}{\mathrm{Re}(y_0)[(y_0 + Y)(y_0 + Y)^*]},$$ (11.22)

where y_0 is the absorbing incident medium admittance and Y is the admittance of the rear surface. This transmittance is maximized when Y is the complex conjugate of the incident admittance y_0. Provided the matching system is without loss, it can readily be shown that if the direction of the light is reversed, then this coating will be an exact antireflection coating for the absorbing material. The same definition of transmittance could be used for the gain material. Maximum transmittance would be obtained with an exit admittance, the complex conjugate of the characteristic admittance. In an amplification application, however, this is not a useful definition.

In an amplifier, it is most important to avoid any standing wave in the gain medium or, alternatively that there should be no counterpropagating wave. The existence of such a condition is indicated by an opening admittance spiral. To prevent the spiral from opening, the exit admittance for the gain medium must be equal to the characteristic admittance of the gain medium, not the complex conjugate. If this condition is satisfied, then the locus will be a point, and the reflectance of the front surface, simply that of the bulk material. An antireflection coating at the front surface will follow the normal rules. Note that the matching applies to the given value of gain only. A gain change will disturb the matching, and the spiral will begin to open.

11.5.1 Oblique Incidence including Beyond Critical

Modified, tilted admittances were introduced in Chapter 9, where the expressions for absorbing materials were given in Equations 9.10 and 9.11. These are repeated in the following equations:

$$\eta_s = \frac{\sqrt{n^2 - k^2 - n_0^2 \sin^2 \vartheta_0 - 2ink}}{\cos \vartheta_0} \quad \text{(fourth quadrant)}, \tag{11.23}$$

$$\eta_p = \frac{(n - ik)^2}{\eta_s}. \tag{11.24}$$

We recall that the modified admittances are normalized so that the incident medium retains its normal incidence value making the admittance diagram easier to interpret. These expressions are, of course, also valid for a completely dielectric layer, that is, where k is zero when they have a particularly simple form. As long as n is greater than $n_0 \sin \vartheta_0$, then η_s is positive real, and so must be η_p. If n is greater than n_0, then η_s rises toward infinity as ϑ_0 tends to 90°, and consequently, η_p falls toward zero. Matters are a little more complicated when n is less than n_0. Then as the angle of incidence increases, η_s falls toward zero, while η_p rises toward infinity. These values are reached at the critical angle when $n_0 \sin \vartheta_0$ becomes equal to n. Beyond the critical angle, η_s is negative imaginary and slides down the negative limb of the imaginary axis toward infinity as ϑ_0 tends to 90°. At the critical angle, η_p switches to imaginary infinity and moves down the imaginary axis toward zero as ϑ_0 increases. While the admittances are on the real axis, the phase shift on reflection is either zero or 180°. Beyond the critical angle, when the admittances are on the imaginary axis, the reflectance becomes 100%, usually referred to as total internal reflection, and the phase shift for s-polarization moves through the first quadrant into the second, while for p-polarization, through the third quadrant into the fourth. All this is indicated in Figure 10.8.

An absorbing material changes the results only slightly. The value of η_s moves off the real or imaginary axis into the fourth quadrant of the complex plane, and η_p, into the first quadrant (Figure 11.45). Total internal reflection is replaced by high reflectance, and the phase shift on reflection is slightly modified (Figure 11.46). There is a slight rounding of the corners that occur with the lossless material (thin lines), but otherwise, there is virtually no change. Strictly, there is now no critical angle, but we can neglect small values of k in calculating an angle that we can designate as critical simply as reference. This is roughly the angle on the loci of the tilted admittances in Figure 11.45, nearest the origin for s-polarization and furthest from the origin for p-polarization.

There is, however, a drastic change when gain is involved. Now k in Equation 11.23 is negative and the quantity under the square root sign in the first or second quadrant. The two possible roots must therefore be in the first or third quadrant. Physical conditions, discussed shortly, require that the chosen solution must be the first quadrant one. Figure 11.47 shows the variation of tilted admittance as a function of angle of incidence when the index of the incident medium is greater than that of the emergent medium. Total internal reflection does not quite occur, similar to the case of slightly absorbing emergent media. Comparison with Figure 11.45, however, shows that the admittances have jumped into different quadrants. Figure 11.47 is virtually an inverted Figure 11.45.

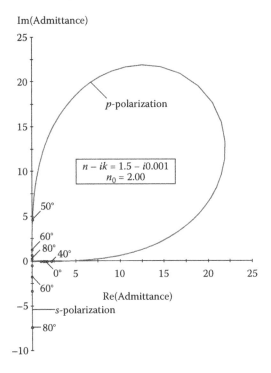

FIGURE 11.45
When the material is slightly absorbing, the tilted admittances move off the axes into the first quadrant for *p*-polarization and fourth quadrant for *s*-polarization.

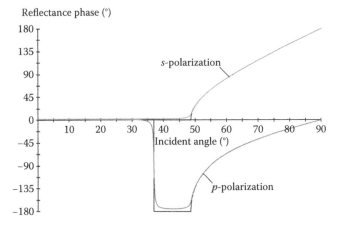

FIGURE 11.46
Form of the phase shift on reflection for a lossless dielectric and for a slightly absorbing material. The thin lines with the square corners indicate the lossless results.

The phase behavior shows a large jump that accompanies even the slightest gain. Below the critical angle, there is virtually no difference between gain, loss, and lossless. The behavior abruptly changes as soon as the critical angle is exceeded (Figure 11.48). This curious effect was remarked on by Siegman [42], but it never seems to have been observed, and for reasons that will shortly be explained, it probably will never be.

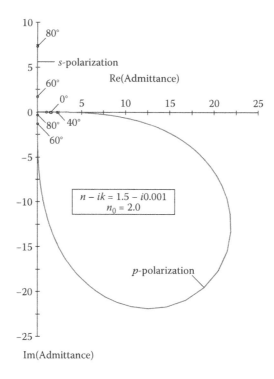

FIGURE 11.47
When the material has slight gain, the tilted admittances move off the real and imaginary axes, but unlike the case of slight absorption, they jump into opposite quadrants. The s-admittance is in the first quadrant, and the p-admittance, in the fourth.

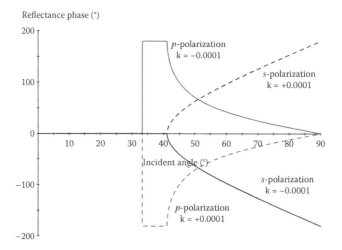

FIGURE 11.48
The phase shift on reflection accompanying a change from absorption to gain shows the flip in phase that occurs beyond the critical angle. The incident medium is glass, and the index of the emergent medium is unity with an extinction coefficient that can be either +0.0001 (absorption) or −0.0001 (gain).

To understand what is happening, we need to think of the complete expression for the wave, which, in a gain medium, is given by

$$E = \mathcal{E} \exp\left[i\left\{\omega t - \frac{2\pi}{\lambda}\left(n_0 \sin \vartheta_0 x + [\xi + i\zeta]z\right)\right\}\right]$$

$$= \mathcal{E} \exp\left(\frac{2\pi\zeta z}{\lambda}\right) \exp\left[i\left\{\omega t - \frac{2\pi}{\lambda}\left(n_0 \sin \vartheta_0 x + \xi z\right)\right\}\right].$$

(11.25)

Here ξ and ζ are both positive. The direction of propagation of phase of the wave is given by the coefficients $n_0 \sin \vartheta_0$ and ξ, while the increase in amplitude of the wave is governed by ζ and is purely along the z-direction. Such a wave where planes of constant amplitude and constant phase are no longer parallel is known as inhomogeneous. ξ is small beyond critical, and the wave skims close to the interface at an angle of $\arctan[\xi/(n_0 \sin \vartheta_0)]$. The angle is smaller with smaller gain. Since it is an infinite plane wave, it can travel enormous distances in the presence of small gain. The propagation angle effect compensates the gain effect, and so we have an increasing exponential into the emergent medium. For zero gain and no absorption, the wave becomes parallel to the interface, but for an absorbing emergent medium, it again is inhomogeneous and inclined to the surface, but this time with a decreasing exponential into the emergent medium. The phase jump and the switch in admittance quadrant can thus be readily understood.

11.5.2 Evanescent Gain

There has been some controversy over what is sometimes known as evanescent gain. The primary application is in amplifying fibers where the cladding possesses gain. The reflection at the core-cladding boundary amplifies the pulses that are directed along the core, so that experimentally, there is certainly gain, but there has been discussion about the mechanism and whether or not there could be gain below the critical angle [42–47]. There is no consensus. Some arguments deny gain anywhere, but there is a recognizable bias toward the existence of gain beyond critical but not below critical. From the thin-film point of view, this is similar to reflection beyond critical at the hypotenuse of a prism where gain is introduced into the outside medium.

It is easy to see that the result at normal incidence is

$$R = \left[\frac{(n_0 - n)^2 + k^2}{(n_0 + n)^2 + k^2}\right],$$

(11.26)

where the sign of k has no influence. At oblique incidence, the reflectance is also readily calculated by our normal expressions and yields the usual

$$R = \left[\frac{\eta_0 - \eta}{\eta_0 + \eta}\right]\left[\frac{\eta_0 - \eta}{\eta_0 + \eta}\right]^*,$$

(11.27)

where, again, it is clear that the sign of k is without influence so that whether or not the k represents absorption or gain, the result is exactly the same. This applies both before and beyond critical and gives the result shown in Figure 11.49, suggesting the complete absence of gain in reflection at any incidence. This, however, neglects the question of stability.

The admittance locus of a gain medium is shown in Figure 11.42 and is in the form of an opening spiral at the right-hand side of the complex plane becoming converging on entering the left-hand side of the plane to eventually converge on point $-(n + ik)$. Point $(n + ik)$ on the right-hand side is, therefore, an unstable point. Any slight perturbation, even if hardly detectable, will cause the spiral to open, and if enough material remains between the instability and the surface, the spiral will terminate on the stable point $-(n + ik)$. This, of course, is at normal incidence, but the behavior

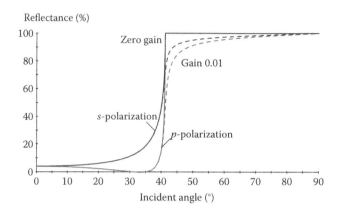

FIGURE 11.49
Internal reflectance of a boundary between glass of index 1.52 and air and a gain medium with optical constants of 1.00 + i0.01. The gain curve is indistinguishable from the result with optical constants 1.00 − i0.01, that is exactly the same amount of absorption.

will be quite similar at oblique incidence with the appropriate tilted admittances. Thus, there are two limiting solutions, an unstable one that is our normal reflectance without gain, as in Equations 11.26 and 11.27, and a stable one that is the inverse of the unstable solution and, therefore, shows gain everywhere, although much smaller and reducing with increasing angle beyond critical. The various theoretical studies terminated their gain media in various ways and, as a result, represent a mixture of stable and unstable results. The bias toward no gain below critical but gain above is almost certainly related to the rate of opening and closing of the spiral loci, very slow below critical but exceedingly fast above [37,39]. Thus, although the ultimate stable effect beyond critical is much smaller in magnitude (Figure 11.50), the total thickness of gain material required beyond a perturbation to reach stability is orders of magnitude less than that required below critical [37].

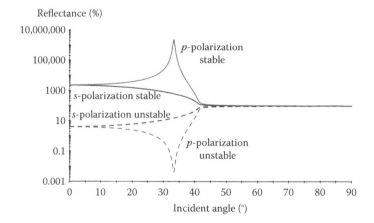

FIGURE 11.50
Two solutions for reflectance including both s- and p-polarizations. The p-polarization peak corresponds to the Brewster angle, although because of the finite k-value, it does not yield exactly zero reflectance. The results below 100% are unstable in the sense that any slight perturbation in the semiinfinite gain medium will cause the admittance spiral to begin to expand. If both the perturbation and the film thickness to the front surface are both great enough, then the stable limit will be reached.

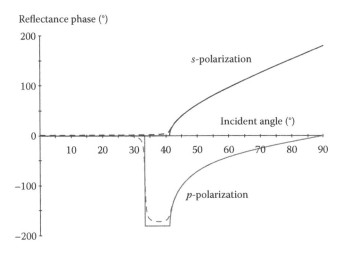

FIGURE 11.51
The phase shift on internal reflection when the admittance locus terminates at $-(n + ik)$ (*broken curves*) is very similar to the phase shift for a semiinfinite material with optical constants $(n - ik)$ (*full curves*). Beyond the critical angle, the curves with and without gain virtually coincide, and there is no phase flip. Incident medium index is 1.52, and the gain medium has constants $(1.00 + i0.01)$.

The phase change on reflection associated with the stable solution is shown in Figure 11.51. There is now no phase flip beyond critical. The flip is associated with the unstable solution and, as a result, will likely never be observed.

11.6 Perfect Absorbers

There is considerable current interest in enhanced absorption in thin layers of absorbing materials [48–54]. Descriptive terms such as perfect absorption, coherent perfect absorption, and critical coupling, among others, are used. Many of the papers are in other than optics journals, and the treatment is sometimes remote from normal thin-film theory. In this section, we attempt an explanation of the design of such components within the framework of the approaches to thin-film optical theory used in this book. In what follows, to keep the discussion as straightforward as possible, we consider normal incidence only, but extension to oblique incidence will be reasonably clear. We follow the treatment in Macleod [39,55,56].

Perfect absorption lacks a rigorous definition in the literature. The expression is sometimes used to describe a component that neither reflects nor transmits. Since many of the substrates are opaque, all that is required for such absorption is a perfect antireflection coating. Here we adopt a more rigorous definition of 100% absorptance within one layer only of one input ray from one side of the coating.

Then there is the coherent perfect absorber that can receive input from both sides. In one configuration, the input is only from the nominal incident medium, and the ray is largely reflected. In a second configuration, rays are incident from both sides, and each is totally absorbed. This therefore acts as a kind of switch exhibiting a reflectance that can be turned off. The component completely operates by absorption. No nonlinear material is involved.

We start with the perfect single-sided absorber. We begin with potential absorptance $\mathcal{A}$ of a layer, already shown to be related to the potential transmittance ψ by

$$\mathcal{A} = 1 - \psi = 1 - \frac{T}{(1-R)} = 1 - \frac{\text{Re}(Y_e)}{\text{Re}(BC^*)}, \quad (11.28)$$

where B and C have their usual meanings and Y_e is the exit admittance at the rear. We note that Re(BC^*) is twice the net irradiance entering the front surface. The absorptance A is

$$A = (1-R)\mathcal{A} = (1-R)\left[1 - \frac{\text{Re}(Y_e)}{\text{Re}(BC^*)}\right]. \quad (11.29)$$

Since we must have some input, Re(BC^*) must be finite, and so for 100% absorptance, both R and Re(Y_e) must be zero, implying

$$\frac{C}{B} = y_0, \quad Y_e = \text{imaginary}. \quad (11.30)$$

The conditions are interdependent and require a precise value of Y_e.

We will take iron at a wavelength of 1000 nm and normal incidence as absorbing layer and, for the demonstration, will assume optical constants of $(3.208 - i4.185)$. The admittance at the rear surface of the iron film Y_e must be imaginary so that the admittance locus of the iron film must start at the imaginary axis, but also to assure zero reflectance, the termination of the locus must be exactly at the admittance of the incident medium, implying that both the starting point for the locus and the iron thickness must be chosen exactly to satisfy the termination condition. To determine the necessary parameter values, we plot a reversed locus. This is an admittance locus that moves in reverse from the desired end point to find the desired start point. Note that it is not the normal locus of a reversed design but uses, instead, negative values of thickness. The necessary locus is shown in Figure 11.52, and the absorptance with the exactly optimum conditions, in Figure 11.53.

Of course, we have to achieve the imaginary exit admittance. We need to start with a structure presenting 100% reflectance with a phase adjusting layer between it and the metal and, at normal incidence, we can approach this but never exactly achieve it. We note in passing that it does become possible beyond the critical angle. The nominal 100% reflectance can involve dielectric or a mixture of dielectric and metal layers. Once the approach to 100% is close enough, then the surface admittance can be adjusted by adding the dielectric phase adjusting layer of correct thickness so as to move the end point of the locus up the imaginary axis, or as close to it as we can get. A quite useful performance can be achieved with a silver reflector and an overcoating dielectric phase

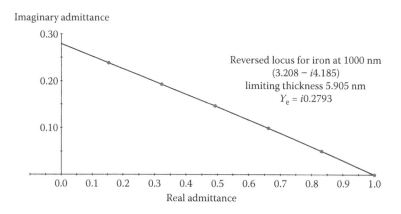

FIGURE 11.52

Reversed locus for iron at 1000 nm. This marks the locus of the exit admittances for a layer of the limiting thickness (5.905 nm) to terminate at 1.00. The intercept with the imaginary axis is at 0.2793. The dots on the locus are 1nm apart.

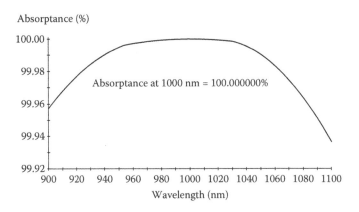

FIGURE 11.53
Absorptance of the iron layer given the exact optimum conditions from Figure 11.52.

adjusting layer, preferably of low index to be as near the imaginary axis as possible. Since the end point is not quite on the imaginary axis, the thickness of the iron film should be very slightly reduced accordingly. Gentle refinement can quickly find the optimum.

The term *coherent perfect absorber* has been used to describe a device where reflectance at a front surface can be switched to zero by the application of light to the rear surface of exactly the same frequency and a given amplitude and relative phase [49,57]. The technique that has been used for their design is time reversed lasing.

The technique makes use of the reversible nature of light. Linear systems have the property of time-reversal symmetry. If we reverse time, then the system will simply run backward. We can think of an absorbing medium supporting a positive-going wave in our usual convention, that is, with the phase term written as $(\omega t - \kappa z)$ as

$$\mathcal{E}\exp\left[i\left(\omega t - \frac{2\pi(n - ik)z}{\lambda}\right)\right] \rightarrow \mathcal{E}\exp\left[i\left(-\omega t - \frac{2\pi(n - ik)z}{\lambda}\right)\right], \tag{11.31}$$

where on the right-hand side, we have reversed the sign of time. We can rewrite this wave as

$$\mathcal{E}\exp\left[i\left(-\omega t - \frac{2\pi(n - ik)z}{\lambda}\right)\right] = \mathcal{E}\exp\left[i\left(\frac{2\pi(n - ik)(-z)}{\lambda} - \omega t\right)\right], \tag{11.32}$$

and we see that this time reversal can be interpreted as a progressive wave expressed in the alternative convention of $(\kappa z - \omega t)$ but that is moving in the opposite direction, that is, the negative z-direction. However, the convention for the complex index in this alternative convention is that an absorbing film should be represented as $(n + ik)$ rather than $(n - ik)$, and this new wave expression is clearly representing gain rather than absorption. Had the original wave been traveling through a gain medium, then the flip in time would have changed the gain to absorption. The time-reversal symmetry is then completely assured if we replace gain with equivalent absorption and vice versa. In time reversed lasing, we start with a gain medium outputting a coherent beam from each side with no externally incident beam involved. If we now apply the time reversal technique, we have incident beams that are simply the coherent output beams reversed in direction with no reflected beams and a gain medium that is now switched to absorption. This arrangement is sometimes termed *critical coupling* or *coherent perfect absorption*, and the objective is new optical switching devices perhaps based on semiconductors. The absorbing layer is normally outermost with no overcoat, and the linearity condition is satisfied by considering the system at the lasing threshold where the gain is unmodified. However, time reversal is not necessary for the understanding and optimization of the absorption, and we will look at it as a straightforward optical thin-film problem.

The normal expression for the performance calculation of a structure of thin films is

$$
\begin{bmatrix} B \\ C \end{bmatrix} = \left\{ \prod_{j=1}^{q} \begin{bmatrix} \cos \delta_j & \dfrac{i \sin \delta_j}{y_j} \\ iy_j \sin \delta_j & \cos \delta_j \end{bmatrix} \right\} \begin{bmatrix} 1 \\ y_m \end{bmatrix},
\tag{11.33}
$$

where y_m is the surface admittance of the emergent surface and C/B of the front surface. The elements of the final column matrix indicate an electric field amplitude of unity that emerges into the emergent medium with no light propagating in the opposite direction. For light to emerge from the substrate into the coating, we reverse the handedness of the fields so that y_m becomes $-y_m$. In no way should this be considered as a negative index. It is simply a reversal of the normal direction of the light. Then, to complete the design, C/B should be equal to y_0, the admittance of the incident medium. B is then the complex electric field amplitude, including phase, of the input ray that corresponds to unity amplitude and zero phase entering from the rear. An admittance locus is the most straightforward way of visualizing and arranging the correct structure.

We continue with iron at 1000 nm, and as usual, we start at the final interface. Since this is of negative admittance, it must be located on the left-hand side of the imaginary axis. To cross from the left-hand to the right-hand side, where the termination point should be y_0, an absorbing layer is necessary. Dielectric layers cannot cross the imaginary axis.

It is very unlikely that the metal locus on its own can start at $-y_m$ and terminate exactly at y_0, and so some matching structure has to be devised. This can be either behind or in front of the absorbing layer, but in this case, we will place it behind, next to the substrate. The usual technique of starting from both ends to find a point of intersection permitting a continuous locus can be used. A reversed locus of the metal from y_0 must then intersect a forward dielectric locus starting at $-y_m$. In this case, we are fortunate in that the reversed metal locus does just intersect a dielectric locus of 2.40 index, roughly corresponding to TiO$_2$. We choose the thicker of two close solutions to yield slightly higher reflectance in the unswitched case. The final switched locus with complete absorption of both beams is shown in Figure 11.54. The ratio of switched to switching power is only 0.374, but the unswitched reflectance is 52.17% (Figure 11.55). Although, in the other direction, the ratio is much more favorable, the unswitched reflectance (not shown) would be just only over 11%. We emphasize that this is simply an illustration of the method, and no attempt has been made to

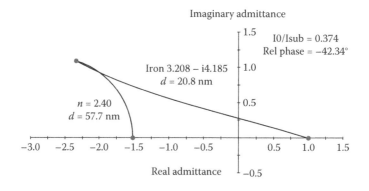

FIGURE 11.54
Final switched locus at 1000 nm with input from both sides and zero reflectance on either side. The optical constants and thicknesses are indicated against the appropriate loci. In this configuration, the switching power is roughly three times the switched power.

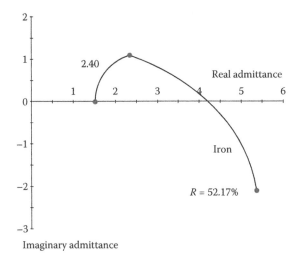

FIGURE 11.55
The unswitched admittance locus showing a reflectance of 52.17% at 1000 nm.

optimize its performance beyond assuring complete absorption of both switched and switching beams. It is, in fact, an interesting application of the mixed Poynting vector (Section 2.13).

11.7 Photonic Crystals

Recently, we have started to see the term *photonic crystal* appearing in the optical thin-film literature. The term was first coined by Yablonovitch, and more information is found in Yablonovitch [58,59]. This section attempts to explain, from the point of view of optical coating practitioners, the concept of photonic crystal and its application to optical coatings.

11.7.1 What Is a Photonic Crystal?

The band structure associated with electrons in crystals is a consequence of their wave nature. The electron wavelength is of the same order as the lattice spacing, and the consequence is an interference phenomenon that is related to the electron wavelengths and, hence, their energies. The principal effect is a forbidden energy zone, called the bandgap, between the energy possessed by bound valence electrons and free electrons, and this gap has enormous influence on the optical properties of the crystal. For example, an interaction between an optical photon and a bound electron is not possible if the electron would acquire energy that would place it in the forbidden zone. In such a case, the photon would simply be transmitted by the crystal. If the photon were to have higher energy, sufficient to shift the electron energy into the allowed band beyond the bandgap, the conduction band, then it would be absorbed. This explains the high infrared transmittance and high visible absorptance of the semiconductors. The light, however, has a very large wavelength compared with the lattice spacing, and so the lattice has no direct effect. The interaction with the light is indirect. The lattice influences the electrons, and the electrons then affect the photons. All of this involves three dimensions, and it is normal to assume that the crystal medium is infinite in extent.

For the lattice to directly affect the photons, its spacing should be comparable with the wavelength of light. Thus, the scale has to be several orders of magnitude greater than that of a crystal. We are led to the idea of a regular three-dimensional assembly of blocks of one refractive index in a matrix of a different one with spacing chosen to induce interference effects in light propagating through it. As we would expect, such an assembly exhibits a pattern of regions where propagation is forbidden and regions where it is permitted. By analogy, with normal crystals and their electronic properties, this much larger structure is called a *photonic crystal*, and the forbidden propagation regions, as *photonic bandgaps* (PBGs).

There are two major problems associated with three-dimensional photonic crystals. The first is the calculation of the properties, and the second, fabrication. Both problems can be eased by reducing the number of dimensions.

11.7.2 Two-Dimensional Photonic Crystals

Two-dimensional photonic crystals can be constructed by the pulling process used for the construction of optical fibers. Here the matrix is the normal dielectric glass material used for the fiber while the scattering features are cylindrical holes running parallel to the axis. This can be effectively converted into an optical fiber light guide by removing the central hole so that the light propagates along its path and is contained by the reflection from the surrounding photonic crystal structure. Since the process effectively removes an "atom" from the lattice, this absence of the hole is called a *defect* even though it is an intentional feature of the structure. Defect is used frequently in this sense in photonic crystal terminology.

11.7.3 One-Dimensional Photonic Crystals

A reduction in the number of dimensions to one represents a considerable decrease in complexity and allows rapid development of photonic crystal theory. A one-dimensional photonic crystal is a multilayer optical coating under a different name. Unfortunately, this appears to have been not completely understood in the early photonic crystal studies. Also the language and the approach adopted owe much to solid-state theory. The problems and their proposed solutions are generally described in this different language. Adding confusion is that many of the quite fundamental results in traditional thin-film theory were developed some time ago. The flood of recent literature is such that older papers are less likely to be recognized. As a result, earlier one-dimensional photonic crystal publications would sometimes repeat results already well known and understood in the field of optical coatings. More recently, the close links between optical coatings and photonic crystals have become recognized. In fact, Yablonovitch, the originator, has recommended reserving the term *photonic crystal* for the two-and three-dimensional structures, which are truly different from optical coatings [59]. Unfortunately, the term is still being used for one-dimensional structures, and an accompanying side effect is that some developments that are of more importance to the thin-film community are nevertheless presented as developments in photonic crystals. This section is an attempt to bridge the gap.

A term that is often used is *photonic bandgap*, with the acronym PBG. In one dimension, a PBG is simply a high reflectance zone. We know that any repeated dielectric structure will exhibit a performance in terms of wavelength that will consist of well-defined reflecting regions, where the reflectance simply rises steadily as a function of the number of repeats of the structure and regions where the transmittance is high but where there is ripple oscillating between two well-defined envelopes. As the number of repeats increases, the density of the ripple oscillations increases, but the ripple continues to follow exactly the same envelopes. The high-reflectance regions are the PBGs, and the coating is known as a PBG structure.

One of the early assumptions about the one-dimensional photonic crystal was that it would be completely impossible for it to support a forbidden gap over a range of angles from 0 to 90° or, in optical coating terms, that the reflectance could remain high from normal incidence to an incidence of 90°. No good proof of this assumption appears to exist, and so it is difficult to be certain of its origin. Photonic crystal theory, by analogy with solid-state crystal theory, tends to treat infinite assemblies. It is natural for a photonic crystal theorist, therefore, to be thinking in terms of propagation within the crystal, whereas the thin-film theorist is well aware of restrictions on the number of layers and of the presence of an incident and emergent medium. If propagation is visualized as within the crystal, then it is clear that propagation could laterally be along the layers and not across them, and the resistance to propagation would disappear. Also, it was recognized that there would be a particular propagation direction that would correspond to the Brewster condition between the two materials of the crystal, and this would mean a collapse of the band-gap, or reflecting zone, for p-polarized light. The fact that it is possible, when the incident medium is air, and there is sufficient contrast between the high index and low index of the materials to construct a reflector that, for a limited range of wavelengths, can exhibit high reflectance for both p- and s-polarizations over all angles of incidence from 0 to 90° came, therefore, as a surprise to the photonic crystal community but not to the thin-film community. The effect was called either the *omnidirectional reflector* [60] or, sometimes, the *perfect mirror*.

A quarter-wave stack is the basic dielectric reflector. As the multilayer is tilted to greater angles of incidence, the characteristic moves to a shorter wavelength. But some wavelengths remain within the shifted high-reflectance zone. An omnidirectional mirror is one where there is at least one wavelength that remains within the high reflectance zone right up to grazing incidence. To construct such a reflector, we know that we should do two things. We should make the high-reflectance zone as wide as possible and move as little as possible with changing incidence. To make the zone wide, we choose two materials with very different refractive indices, and to minimize the high-reflectance zone movement, we give one of the materials a maximum refractive index. One other thing that we must do is to use sufficient layers to realize the high reflectance over the whole range.

The limits of the high-reflectance zone of a repeated two-layer structure is given by the expression

$$\cos \delta_A \cos \delta_B - \frac{1}{2} \left(\frac{y_B}{y_A} + \frac{y_A}{y_B} \right) \sin \delta_A \sin \delta_B = \pm 1, \tag{11.34}$$

where the quantities have their usual meaning. At normal incidence and for a quarter-wave stack, where $2\delta_A$ and δ_B are equal, the expression is easily inverted to give the high-reflectance zone edges and width. At oblique incidence it is more complicated and not analytically friendly. But it is not difficult to arrange a numerical method on a computer, and once we have the calculation programmed, we can readily and almost painlessly calculate any particular case. Here we have chosen to use materials of indices 1.45 and 2.5 that are typical of the visible and near infrared regions.

The high-reflectance zone edges could be plotted against ϑ_0, the angle of incidence in the incident medium. But a more convenient quantity is $n_0 \sin \vartheta_0$ because this is the basis for Snell's Law and makes the diagram much more general. Also, we can make $n_0 \sin \vartheta_0$ negative for p-polarization [60], which has no effect on the results but makes the plots easier to read. Figure 11.56 shows the edges of the high-reflectance zone for the first three orders. At normal incidence, the orders at $g = 2.0$ and $g = 4$ are missing because all the layers are absentees. However, as the layers are detuned by tilting, a reflectance zone appears. This is the well-known *half-wave hole* phenomenon that is a major problem in shortwave pass filters. The Brewster angle between the film materials is over to the left where the p-reflectance zone boundaries cross.

The range of $n_0 \sin \vartheta_0$ for an incident medium of air runs from zero to unity, and so, on the figure, if we include both p- and s-polarizations, it runs from −1.0 to + 1.0. It is fairly easy to see that in the first order, the reflectance will stay within the high-reflectance zone for values of g from 1.1012 to

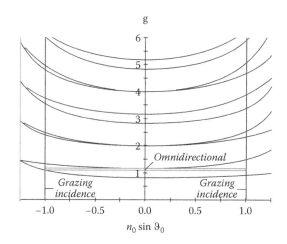

FIGURE 11.56
Limits of the high-reflectance regions plotted in terms of g against $n_0 \sin \vartheta_0$. The right-hand half of the diagram shows results for s-polarization, and the left-hand, for p-polarization. The inner vertical lines mark grazing incidence in air. The materials have indices of 1.45 and 2.5. The outer line on the extreme left marks the Brewster angle between the layer materials. The omnidirectional region is marked.

1.1713. This is a wavelength range of 854–908 nm for a reference wavelength of 1000 nm. Figure 11.57 shows the performance of a 31-layer structure, $\text{Air} \,|\, (0.5H \, L \, 0.5H)^{15} \,|\, \text{Glass}$, with a reference wavelength of 1000 nm, calculated for normal incidence (black) and for p-polarization at 85° incidence (gray). The reflectance clearly remains high over the 854–908 nm region. The curves for s-polarization (not shown) are wider. Between 85° and grazing incidence 90°, the reflectance simply rises everywhere as is normal at grazing incidence.

In the study of electrons in solids, much importance is attached to diagrams of energy against momentum. If the x–z plane is the plane of incidence with z normal and x parallel to the surface, then the complete phase factor of a wave propagating obliquely in the plane of incidence is

$$\exp[i\{\omega t - (\kappa_x x + \kappa_z z)\}], \tag{11.35}$$

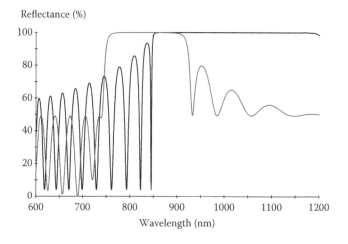

FIGURE 11.57
Reflectance of a 31-layer structure, $\text{Air} \,|\, 0.5H \, L \, 0.5H \,|\, \text{Glass}$ with indices H: 2.5, L: 1.45 and glass: 1.52. Black curve: normal incidence; gray: p-polarization at 85°. The reflectance clearly remains high over the region of 854–908 nm.

where ω, the angular frequency, is analogous to energy, and κ, the wavenumber, to momentum. In thin-film optics, it is the z-component that interests us and we ignore the x-component. Here, however, we choose the x-component. κ_x can be written as

$$\kappa_x = \frac{2\pi n_0 \sin \vartheta_0}{\lambda} = \frac{2\pi}{\lambda_0} \cdot g \cdot n_0 \sin \vartheta_0, \tag{11.36}$$

and, ω as

$$\omega = \frac{2\pi c}{\lambda} = \frac{2\pi c}{\lambda_0} \cdot g, \tag{11.37}$$

where c is the free space velocity of light. We can normalize the expressions by removing the constant factors $2\pi/\lambda_0$ and $2\pi c/\lambda_0$, and this gives us a plot of g against $g \cdot n_0 \sin \vartheta_0$, that is the same as the curves of Figure 11.56, but with the horizontal scale multiplied by g. Vertical lines of constant $n_0 \sin \vartheta_0$ now become lines emanating from the origin with slope $1/n_0 \sin \vartheta_0$, sometimes called *light lines*. If n_0 varies with frequency, that is, if the incident medium is dispersive, then the lines will become curves. Figure 11.58 shows the new style. Light lines at grazing incidence for air incident medium are shown as two tilted straight lines passing through the origin. A similar plot is given by Fink [60]. Here we have used parameters more familiar to the thin-film community.

The PBG structure can have what is termed a *defect*. The defect consists of extra or missing material. The most common form of defect is a missing quarter-wave layer in a quarter-wave stack. The result is a half-wave layer surrounded by reflectors, in other words, a single-cavity coating. If the structure around the defect is symmetrical, then the performance of the device exhibits a narrow transmission band, a single-cavity filter. If the structure is not symmetrical, then the result will depend on the degree of asymmetry. In the case where the asymmetry is pronounced, we have a Gire–Tournois interferometer [16], that is, a device exhibiting reflectance with a resonant GD. Structures with multiple defects become multiple-cavity filters, or multiple-peak filters, depending on the details of the structure.

Certain other terms associated with electron behavior in crystals are sometimes used to describe optical behavior of thin-film structures. *Bloch waves* is a term used in the band structure of solids. In optics, its meaning is associated with the equivalent properties of the multilayer. In particular, the dispersion relation

$$\cos(K\Lambda) = \cos\left(\frac{\omega n_1 d_1}{c}\right) \cos\left(\frac{\omega n_2 d_2}{c}\right) - \frac{1}{2}\left(\frac{n_2}{n_1} + \frac{n_1}{n_2}\right) \sin\left(\frac{\omega n_1 d_1}{c}\right) \sin\left(\frac{\omega n_2 d_2}{c}\right), \tag{11.38}$$

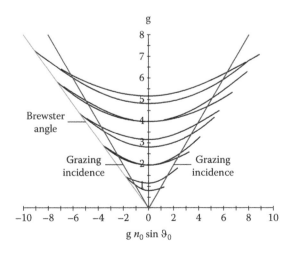

FIGURE 11.58
Band limits plotted in the style of energy against momentum.

where $\Lambda = d_1 + d_2$, can be identified as the expression for the equivalent phase thickness. K is known as the *Bloch wavenumber*. The forbidden bands or the bandgaps are those regions where K becomes imaginary.

Quantum tunneling is another term associated with PBG structures. Quantum tunneling is essentially another term for the effects predicted by induced transmission theory.

Although the one-dimensional photonic crystal is strictly a multilayer optical coating, nevertheless, the expression of its properties from a different point of view can help bring a greater depth of understanding to the field.

References

1. W. H. Southwell. 1998. Rugate filter structures. Private communication (Rockwell Science Center, Thousand Oaks, CA).
2. L. I. Epstein. 1952. The design of optical filters. *Journal of the Optical Society of America* 42:806–810.
3. L. I. Epstein. 1955. Improvements in heat reflecting filters. *Journal of the Optical Society of America* 45:360–362.
4. W. H. Southwell. 1985. Coating design using very thin high- and low-index layers. *Applied Optics* 24:457–460.
5. K. D. Hendrix, C. A. Hulse, G. J. Ockenfuss et al. 2008. Demonstration of narrowband notch and multi-notch filters. *Proceedings of SPIE* 7067:706702–1 to 706702–14.
6. J. A. Dobrowolski and D. Lowe. 1978. Optical thin film synthesis program based on the use of Fourier transforms. *Applied Optics* 17:3039–3050.
7. B. G. Bovard. 1993. Rugate filter theory: An overview. *Applied Optics* 32:5427–5442.
8. B. E. A. Saleh and M. C. Teich. 1991. *Fundamentals of Photonics.* First ed. New York: John Wiley & Sons.
9. A. Yariv and P. Yeh. 1984. *Optical Waves in Crystals.* First ed. New York: John Wiley & Sons.
10. K. Ferencz and R. Szipöcs. 1993. Recent developments of laser optical coatings in Hungary. *Optical Engineering* 32:2525–2538.
11. A. Stingl, C. Spielmann, F. Krausz et al. 1994. Generation of 11-fs pulses from a Ti:sapphire laser without the use of prisms. *Optics Letters* 19:204–206.
12. R. Szipöcs, K. Ferencz, C. Spielmann et al. 1994. Chirped multilayer coatings for broadband dispersion control in femtosecond lasers. *Optics Letters* 19:201–203.
13. R. Szipöcs and F. Krausz. 1998. Dispersive dielectric mirror. US Patent, 5,734,503.
14. Schott. 1992. *Schott Optical Glass.* Duryea, PA: Schott Glass Technologies.
15. R. Szipöcs and A. Köházi-Kis. 1997. Theory and design of chirped dielectric laser mirrors. *Applied Physics B* 65:115–135.
16. F. Gires and P. Tournois. 1964. Interféromètre utilisable pour la compression d'impulsions lumineuses modulées en fréquence. *Comptes Rendus de l'Academie de Science* 258:6112–6115.
17. J. Kuhl and J. Heppner. 1986. Compression of femtosecond optical pulses with dielectric multilayer interferometers. *IEEE Transactions on Quantum Electronics* QE-22:182–185.
18. M. Jablonski, Y. Takushima, and K. Kikuchi. 2001. The realization of all-pass filters for third-order dispersion compensation in ultrafast optical fiber transmission systems. *Journal of Lightwave Technology* 19:1194–1205.
19. V. Pervak, I. Ahmad, M. K. Trubetskov et al. 2009. Double-angle multilayer mirrors with smooth dispersion characteristics. *Optics Express* 17:7943–7951.
20. V. Pervak, I. Ahmad, S. A. Trushin et al. 2009. Chirped-pulse amplification of laser pulses with dispersive mirrors. *Optics Express* 17:19204–19212.
21. K. P. Gibbons, C. K. Carniglia, R. E. Laird et al. 1997. ITO coatings for display applications. In *40th Annual Technical Conference*, pp. 216–220. Society of Vacuum Coaters, New Orleans, LA.
22. E. J. Bjornard. 1992. Electrically-conductive, light attenuating antireflection coating. US Patent, 5,091,244.

23. J. Wolfe. 1995. Anti-static, anti-reflection coatings using various metal layers. In *38th Annual Technical Conference*, pp. 272–275. Society of Vacuum Coaters, Chicago.

24. T. Oyama and Y. Katayama. 1997. Light absorptive antireflector. US Patent, 5,691,044.

25. H. Ishikawa and B. Lippey. 1996. Two layer broad band antireflection coating. In *Tenth International Conference on Vacuum Web Coating*, pp. 221–233. Bakish Materials Corporation, Englewood, NJ.

26. Y. Zheng, K. Kikuchi, M. Yamasaki et al. 1997. Two-layer wideband antireflection coating with an absorbing layer. *Applied Optics* 36:6335–6339.

27. H. J. Hovel. 1976. Transparency of thin metal films on semiconductor substrates. *Journal of Applied Physics* 47:4968–4970.

28. Y. C. M. Yeh, F. P. Ernest, and R. J. Stirn. 1976. Practical antireflection coating for metal-semiconductor solar cells. *Journal of Applied Physics* 47:4107–4112.

29. M. V. Schneider. 1966. Schottky barrier photodiodes with antireflection coating. *Bell System Technical Journal* 45:1611–1638.

30. G. Hass and L. Hadley. 1972. Optical constants of metals. In *American Institute of Physics Handbook*, D. E. Gray (ed.), pp. 6.124–6.156. New York: McGraw-Hill.

31. R. E. Hahn and B. O. Seraphin. 1978. Spectrally selective surfaces for photothermal solar energy conversion. *Physics of Thin Films* 10:1–69.

32. C. E. Kennedy. 2014. High temperature solar selective coatings. US Patent, 8,893,711 B2.

33. J. C. C. Fan and F. J. Bachner. 1976. Transparent heat mirrors for solar energy applications. *Applied Optics* 15:1012–1017.

34. J. C. C. Fan, F. J. Bachner, G. H. Foley et al. 1974. Transparent heat-mirror films of $TiO_2/Ag/TiO_2$ for solar energy collection and radiation insulation. *Applied Physics Letters* 25:693–695.

35. B. Bhargava, R. Bhattacharya, and V. V. Shah. 1977. A broad band (visible) heat reflecting mirror. *Thin Solid Films* 40:L9–L11.

36. L. Holland and G. Siddall. 1958. Heat-reflecting windows using gold and bismuth oxide films. *British Journal of Applied Physics* 9:359–361.

37. A. Macleod. 2011. Gain in optical coatings: Part 1. *Society of Vacuum Coaters Bulletin* (Fall):22–27.

38. A. Macleod. 2012. Gain in optical coatings: Part 2. *Society of Vacuum Coaters Bulletin* (Spring):20–25.

39. A. Macleod. 2015. Some aspects of absorption and gain. *Proceedings of SPIE* 9627:96270T-1–96270T-8.

40. A. Macleod and C. Clark. 2012. Gain in optical coatings. *Proceedings of SPIE* 8486:848612-1–848612-10.

41. H. A. Macleod. 1995. Antireflection coatings on absorbing substrates. In *38th Annual Technical Conference*, pp. 172–175. Society of Vacuum Coaters, Chicago.

42. A. Siegman. 2010. Fresnel reflection, Lenserf reflection, and evanescent gain. *Optics and Photonics News* 21 (1):38–45. Optical Society of America, Washington, DC.

43. P. R. Callary and C. K. Carniglia. 1976. Internal reflection from an amplifying layer. *Journal of the Optical Society of America* 66:775–779.

44. G. N. Romanov and S. S. Shakhidzhanov. 1972. Amplification of electromagnetic field in total internal reflection from a region of inverted population. *Pis'ma v Zhurnal Eksperimental' noi i Teoreticheskoi Fiziki (Journal of Experimental and Theoretical Physics Letters)* 16:209–211.

45. K. J. Willis, J. B. Schneider, and S. C. Hagness. 2008. Amplified total internal reflection: Theory, analysis, and demonstration of existence via FDTD. *Optics Express* 16:1903–1914.

46. F. V. Ignatovich and V. K. Ignatovich. 2010. On Fresnel reflection and evanescent gain. *Optics and Photonics News* 21 (5):6–7. Optical Society of America, Washington, DC.

47. F. V. Ignatovich and V. K. Ignatovich. 2011. *On Fresnel Reflection and Evanescent Gain*. Available from http://textbookphysics.org/TIR%20OPN/TIR%20100212.pdf.

48. J. R. Tischler, M. S. Bradley, and V. Bulovi. 2006. Critically coupled resonators in vertical geometry using a planar mirror and a 5 nm thick absorbing film. *Optics Letters* 31:2045–2047.

49. Y. D. Chong, L. Ge, H. Cao et al. 2010. Coherent perfect absorbers: Time-reversed lasers. *Physical Review Letters* 105:053901.

50. M. A. Kats, D. Sharma, J. Lin et al. 2012. Ultra-thin perfect absorber employing a tunable phase change material. *Applied Physics Letters* 101:221101–1 to 221101-5.

51. M. A. Kats, R. Blanchard, P. Genevet et al. 2013. Nanometre optical coatings based on strong interference effects in highly absorbing media. *Nature Materials* 12:20–24.

52. M. A. Badsha, Y. C. Jun, and C. K. Hwangbo. 2014. Admittance matching analysis of perfect absorption in unpatterned thin films. *Optics Communications* 332:206–213.

53. T. Y. Kim, M. A. Badsha, J. Yoon et al. 2016. General strategy for broadband coherent perfect absorption and multi-wavelength all-optical switching based on epsilon-near-zero multilayer films. *Scientific Reports* 6:22941.

54. J. Yoon, M. Zhou, M. A. Badsha et al. 2015. Broadband epsilon-near-zero perfect absorption in the near-infrared. *Scientific Reports* 5:12788.

55. A. Macleod. 2013. Optical absorption. Part 1. *Society of Vacuum Coaters Bulletin* 13 (Summer):24–30.

56. A. Macleod. 2013. Optical absorption. Part 2. *Society of Vacuum Coaters Bulletin* 13 (Fall):28–33.

57. W. Wan, Y. Chong, L. Ge et al. 2011. Time-reversed lasing and interferometric control of absorption. *Science* 331:889–892.

58. E. Yablonovitch. 1993. Photonic band-gap structures. *Journal of the Optical Society of America B* 10:283–297.

59. E. Yablonovitch. 2007. Photonic crystals: What's in a name? *Optics and Photonics News* 19 (March):12–13. Optical Society of America, Washington, DC.

60. Y. Fink, J. N. Winn, S. Fan et al. 1998. A dielectric omnidirectional reflector. *Science* 282:1679–1682.

12

Color in Optical Coatings

12.1 Introduction

Anyone who works with optical coatings knows that they can present exceedingly attractive colors. These colors originate in interference effects that enhance reflectance or transmittance in certain parts of the visible spectrum and inhibit it in others. Although colors occur with both transmitted and reflected light, it has long been observed that the most vivid effects are usually to be found in reflection. In the same way that coatings can be designed to have desired spectral properties, they can also be designed to present desired colors. This is a little more complicated than the usual design processes, because of the subjective nature of color itself. Color is strictly a human response to a luminous stimulus. The response varies with the individual observer. Since the coating is not luminous in itself, a source of light is also required before the colors can be observed. We avoid the obvious difficulties inherent in the variations between individual observers and light sources by using theoretical standards that represent more or less well the average properties and permit us to remove the subjective nature of the problem. The design goals can then be presented in unambiguous terms, but there remain some complications. In order to observe the color, there must be an acceptable level of reflected or transmitted light. This, in turn, has a major influence on what can be achieved. A frequent further requirement is that the coatings should be sufficiently simple for large-scale production at reasonable cost. This chapter is a quite abbreviated account of certain aspects of color that are particularly relevant to optical coatings.

12.2 Color Definition

Color is a subjective, human response to the spectral quality of light [1]. The human eye has two different types of receptors, known as cones and rods. Cones predominate in the center of the retina while rods are more plentiful toward the periphery. The cones look very much like the rods but are very slightly conical, and it is they that are sensitive to color. The rods respond to weaker levels of illumination and do not provide any color information. In the normal eye, there are three different types of cones distinguished by their spectral responses. We subjectively describe the sensations they produce as being red, or green, or blue. There is overlap of the different responses, especially in the case of green and red. It appears that around 95% of the world's population can closely agree on the quality of the response to any particular color, and it is this agreement that allows us to devise objective definitions of color.

We are unable to directly measure the color response, and so, as with other subjective phenomena, we must content ourselves with comparative measurements involving certain agreed standards. These standards may take, and have taken, different forms, such as sets of colored objects such as tiles, but the now generally accepted system is based on the use of three standard sources of illumination, one emitting light to produce a red sensation, one green, and one blue.

Since the eye operates by converting the light into a combination of red, green, and blue responses, we can represent any color as a combination of red, green, and blue stimuli derived from standard sources, or primary stimuli as they are more properly called, in a process known as color matching. The measure of the color, then, is the three amounts of light that must be delivered from the primary stimuli and combined, in order to produce exactly the same response as the color to be measured. Fortunately, the human color response is such that we can treat the three amounts of standard light as three independent color coordinates. Since there is a high degree of agreement on the results of such color matching experiments, we can convert the results of large numbers of subjective tests into a representative objective measure.

The concept of color purity is one that we can readily understand, and we can accept that the purest color response would be derived from a single spectral element or line. We are quite used to the representation of illumination in terms of its spectrum. If we know the color coordinates of every spectral line, that is, the amounts of the standard light in each, then, because the process is linear, we can simply add the appropriate amounts to derive the coordinates of the particular color concerned. This is the basis of the objective measure of color. The sets of coordinates for each spectral line are known as the color matching functions.

Although there are different systems of color representation and specification in existence in different countries and professions, the most significant, and now almost universal, is the system defined by the Commission Internationale de l'Eclairage (CIE). This organization was established in 1913 and is the body internationally agreed as setting standards for colorimetry and photometry. In 1931, the CIE defined a Standard Observer consisting of three color matching functions based on three fixed primary color stimuli, the color matching functions being normalized sets of values of relative amounts of the primary stimuli required to reproduce the stimuli of all the lines of the visible spectrum, the lines having equal radiant power.

As might be expected, it turns out that it is impossible to match all the lines of the spectrum with mixtures of any three practical primary stimuli. Negative amounts are necessary. Negative amounts can be accommodated in actual experiments by adding the appropriate amount of the primary stimulus to the color to be matched, but this complicates the color specification to an undesirable extent. The solution adopted by the CIE is exceptionally clever. Since the color matching functions depend on the nature of the primary stimuli, these were defined so as to require only positive values in color matching. As a result, although the stimuli roughly correspond to red, green, and blue, they have a color purity well beyond that of any single spectral line. They are completely unattainable practically, but it is not necessary that they be constructed. Their purpose is to simply define the color matching functions. So that we can refer to the primary stimuli, they have been given quite simple names, X, Y, and Z, corresponding to red, green, and blue, respectively, but avoiding any residue of a subjective description. The corresponding color matching functions are named $\bar{x}(\lambda)$, $\bar{y}(\lambda)$, and $\bar{z}(\lambda)$, respectively. However, the CIE went still further. The flexibility in the definition of the primary stimuli permitted the arrangement that the areas of the color matching functions, plotted against a linear scale of wavelength, should be identical. Still further, the $\bar{y}(\lambda)$ color matching function was made identical to the standard photopic response curve of the human eye. This last is of great importance to thin-film practitioners. The color matching functions are illustrated in Figure 12.1. We emphasize that these are color matching functions, and although obviously ultimately related to the spectra sensitivity of the cone receptors, they do not represent their spectral responses.

The cone receptors in the human eye are responsible for the response to normal levels of illumination and, of course, for the color response. The photopic response curve of the human eye is the relative response of the cones to monochromatic light of identical power but varying wavelength. It is a measure of the spectral variation of the apparent luminosity, or brightness, of the light. The correspondence of the $\bar{y}(\lambda)$ color matching function and the standard photopic response function means that we can readily derive luminous properties from our color calculations.

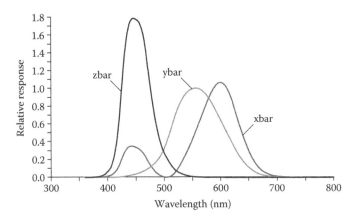

FIGURE 12.1
Color matching functions (1931) defined by the CIE. The ybar curve corresponds to the standard photopic response curve of the human eye.

Optical coatings are not self-luminous. A source of illumination is necessary for us to be able to perceive their color. Obviously, the spectral variation of the illumination will affect that of the light reflected or transmitted by the coating, that is, the color of the coating. The specification of the source of illumination, then, is a necessary component of the definition of coating color. Standard definitions of sources of illumination are therefore included in the CIE system. The CIE definitions give special meanings to the terms *sources* and *illuminants*. A source is a practical device that can be constructed and used in an actual measurement of color. An illuminant is a theoretical distribution of relative output to be used purely in calculation. Some sources correspond sufficiently well to illuminants to be defined as equivalent, but there are illuminants that have no exact corresponding source. In this chapter, we will be concerned more with illuminants than sources.

There are many different defined illuminants in the CIE system, but of particular interest in optical coatings are Standard Illuminants A, E, and D_{65}. Illuminant A represents black body radiation at a temperature of 2856 K, and the corresponding source is a gas-filled coiled tungsten filament lamp, operating at a correlated color temperature of 2856 K. *Correlated color temperature* implies a close relationship between the source color and that of a black body at that temperature. E is the equal energy illuminant emitting equal power per wavelength unit over the visible spectrum. This illuminant can be considered as that assumed in all normal calculations of spectral coating characteristics. The corresponding source does not exist in practice. D_{65} is a representation of daylight (but not direct sunlight) with a correlated color temperature of 6504 K. Unfortunately, there is no artificial source that exactly matches this, and so the CIE has developed a technique for quantifying the usefulness of artificial sources in representing D_{65} and other artificial daylight sources. The spectral distributions of illuminants A, E, and D_{65} are illustrated in Figure 12.2. The spectral distributions are normalized so that the relative output is 100 at 560 nm. D_{65} is, in fact, one of a series of standard daylight illuminants with designations D_{xx}, where $xx00$ K is the nominal correlated color temperature.

All color calculations start with three basic color coordinates, known as the tristimulus values, denoted by X, Y, and Z. Let the spectral output of the source of illumination be $S(\lambda)$ and the response of the coating be $R(\lambda)$. Then,

$$X = 100 \frac{\int_\lambda S(\lambda)R(\lambda)\bar{x}(\lambda)d\lambda}{\int_\lambda S(\lambda)\bar{y}(\lambda)d\lambda}, \tag{12.1}$$

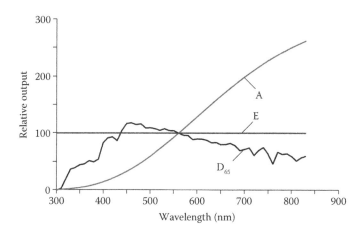

FIGURE 12.2
Relative outputs of illuminants A, E, and D$_{65}$.

$$Y = 100 \frac{\displaystyle\int_\lambda S(\lambda)R(\lambda)\bar{y}(\lambda)d\lambda}{\displaystyle\int_\lambda S(\lambda)\bar{y}(\lambda)d\lambda},$$
(12.2)

$$Z = 100 \frac{\displaystyle\int_\lambda S(\lambda)R(\lambda)\bar{z}(\lambda)d\lambda}{\displaystyle\int_\lambda S(\lambda)\bar{y}(\lambda)d\lambda},$$
(12.3)

where the factor 100 is omitted should $R(\lambda)$ be given in percent rather than absolute terms.

Because of the way in which $\bar{y}$ (λ) is defined, Y is also the luminous reflectance or luminous transmittance, in percent, depending on the nature of the response R. It is therefore also known as the *luminance factor*.

X, Y, and Z are components of a vector in three-dimensional space making it difficult to visualize the color. Two dimensions are easier. We can retain the relationship between the tristimulus values but reduce the number of necessary dimensions by normalizing the components so that their sum is always unity. This leads to what are known as the *chromaticity coordinates*, x and y, defined as

$$x = \frac{X}{X + Y + Z},$$
(12.4)

$$y = \frac{Y}{X + Y + Z}.$$
(12.5)

The chromaticity coordinates define points in a rectangular plot known as a chromaticity diagram.

We can take as an example a quite simple decorative coating consisting of a titanium oxide layer some 300 nm thick deposited over an opaque titanium metal foil or layer. The variation of reflectance as a function of wavelength is shown in Figure 12.3. We can deduce from the form of the curve that the color of the coating will be green.

The results of calculating the chromaticity coordinates of the coating, using the three illuminants from Figure 12.2, are shown in Figure 12.4. We would describe a color as white if it were to introduce no further spectral variation in the color of the reflected light so that the reflected light has exactly the color of the source. In fact, the human eye adapts to the color of the source of illumination. A white paper looks white, even when the general illumination has a blue or yellow cast to it. The point corresponding to the particular illuminant in the chromaticity diagram is therefore referred to as the white point, and the appropriate white points are plotted along with the coating in Figure 12.4. Note that because of the way in which the parameters are defined, the white point corresponding to illuminant E has coordinates (1/3, 1/3). A spectrum line stimulates the purest possible color response, and the locus of the chromaticity coordinates of the lines of the spectrum is shown on the diagram with the appropriate wavelengths marked off on it. Purple is not a spectral color. The line joining the two ends of the spectrum line is the locus of the purest purple colors, known as the purple line.

It is difficult to make any deductions about the color of the coating from the chromaticity coordinates on their own until we see how they compare with the coordinates of the illuminants, or white points. Then we can see that the chromaticity coordinates represent, in each case, a move toward the green part of the spectrum locus.

We can quantify this procedure by drawing a straight line from the white point, through the color coordinates of the sample, to meet the spectrum locus as in Figure 12.5. The wavelength that corresponds to the point of intersection is then known as the dominant wavelength. The color associated with this dominant wavelength is then of the same quality as that of the sample. In fact, we can think of the sample color as being a mixture of the spectrum line color and white. The purity of the resulting color is less, the greater the proportion of white light that must be added. This then leads to a definition of color purity. There are two definitions of purity in common use. Colorimetric purity is an expression of purity based on the idea just expressed, the fraction of the total luminance of the color that is due to the monochromatic stimulus. Excitation purity is a simpler definition that is the fraction of the total distance from the white point to the spectrum line that corresponds to the color stimulus in question. Excitation purity roughly corresponds to what is often termed the *saturation of the color*.

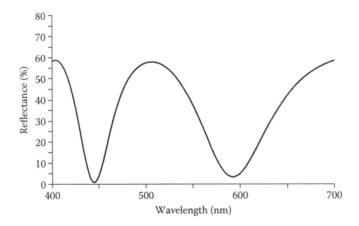

FIGURE 12.3
Spectral variation of reflectance of a titanium oxide layer (300 nm thick) over titanium metal.

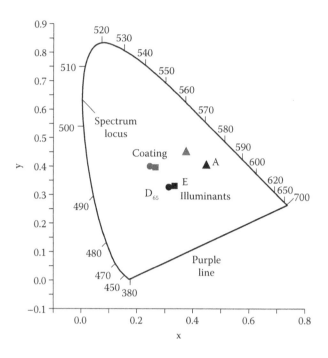

FIGURE 12.4
Chromaticity coordinates of the coating of Figure 12.3 calculated with illuminants D_{65}, E, and A. The locus of the lines of the spectrum is shown and marks the purest colors. The scale around it is wavelength in nanometers. Purple is not a spectral color, and the purest purples lie on the line labeled purple. The color white corresponds to the particular illuminant, and so the corresponding point is also known as the white point.

Both expressions of purity can be reduced to measurements taken from the chromaticity diagram. If we indicate the chromaticity of the white point by (x_w, y_w), of the spectral line by (x_λ, y_λ), and of the point in question by (x, y), then the two measures of purity are given by

$$p_e = \frac{x - x_w}{x_\lambda - x_w} = \frac{y - y_w}{y_\lambda - y_w},$$
$$p_c = \frac{y_\lambda}{y} \cdot \frac{x - x_w}{x_\lambda - x_w} = \frac{y_\lambda}{y} \cdot \frac{y - y_w}{y_\lambda - y_w},$$

(12.6)

where the particular formula to be used is the one that gives least rounding error.

If the coating chromaticity is beneath the white point, it will be of a purple hue, and the line from the white point through the color stimulus will intersect the purple line rather than the spectrum locus. In that case, there is no dominant wavelength. The line, produced backward, will however intersect the spectrum locus, and that point of intersection is known as the complementary wavelength. All chromaticities will have one or other of dominant wavelength or complementary wavelength. Some blue and some red chromaticities will have both.

The expressions for purity in Equation 12.6 need to be extended to accommodate color stimuli without a dominant wavelength. In the case of excitation purity, the coordinates used for (x_λ, y_λ) are those of the point of intersection with the purple line. In the case of colorimetric purity, (x_λ, y_λ) are replaced by the coordinates of the complementary wavelength.

Purity and dominant or complementary wavelength are rarely used as a specification for color, but rather as an aid to the appreciation of the color.

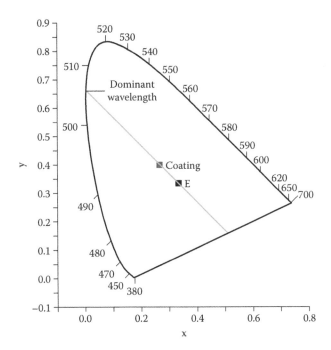

FIGURE 12.5
Dominant wavelength is found by drawing a line from the white point, through the color stimulus in question to meet the spectrum locus. The wavelength corresponding to the point of intersection is the dominant wavelength.

12.3 1964 Supplementary Colorimetric Observer

Human color vision is a complicated process, in which the brain and eye are both involved and the perception of color is found to depend not only just on the spectral content of the stimulus but also on the size of the retinal area that is stimulated or, more usefully for interpretation, on the angular subtense of the area. The experiments that were carried out to establish the parameters of the 1931 Standard Observer used fields that subtended 2°. Experiments demonstrated the validity of the functions for fields subtending from 1° to 4° but larger fields exhibit an increasing difference. For this reason, in 1964, the CIE defined a set of color matching functions for fields subtending 10° at the eye. The associated color matching functions are usually written as $\bar{x}_{10}(\lambda)$, $\bar{y}_{10}(\lambda)$, and $\bar{z}_{10}(\lambda)$. This supplementary observer was arranged so that the equal energy stimulus still has the value 1/3 for its chromaticity coordinates. Figure 12.6 compares the two sets of color matching functions.

12.4 Metamerism

It is clear from the nature of the definition of the tristimulus values that it is possible for color stimuli with quite different radiant power distributions to have identical tristimulus values. Such stimuli are said to be metamers, the adjective being *metameric* and the concept as *metamerism*.

In the case of optical coatings (and indeed of any nonself-luminous stimulus), the spectral distribution of the source of illumination must be included. Since this will vary from one type of source to another, a pair of optical coatings that are metamers for one type of illumination will not

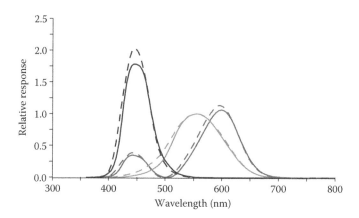

FIGURE 12.6
Supplementary color matching functions (1964) (*broken lines*) compared with the 1931 functions (*solid lines*).

be so for another. Only if the spectral response of two coatings is identical will they exhibit the same color response under all possible qualities of illumination. Such color matches are said to be nonmetameric.

Metamerism is important in the design of coatings that are required to have a particular color response. The illuminant under which the response is to be achieved must be specified. In the case of pairs of coatings that are to be always matched in their color qualities, then their spectral responses must be equal.

12.5 Other Color Spaces

The X, Y, and Z tristimulus values unambiguously define a color, and the derived chromaticity coordinates allow us to plot color information in two dimensions. How closely do we have to match a particular set of tristimulus values so that the perceived colors should be the same or within some acceptable error? Here the chromaticity diagram is of less use. It makes the appreciation of color straightforward, but the distances between points are not simply related to the perceived color differences. This leads to the idea of a uniform color space, where the linear distance between two points is directly proportional to the perceived color difference. There is still progress to be made in this area, but two color spaces that go a long way toward this concept of a uniform color space are recommended by the CIE. These were both defined in 1976. The first is the CIE ($L^*u^*v^*$) space, and the second, the CIE ($L^*a^*b^*$) space, frequently referred to as CIELUV and CIELAB, respectively. Both are derived from the tristimulus values, but the expressions are quite complicated. In both, the coordinates are plotted with rectangular axes in three dimensions, and the perceived color difference is proportional to the linear distance between points.

The quantities L^*, u^*, and v^* are defined by the following expressions:

$$L^* = 116\left(\tfrac{Y}{Y_w}\right)^{\frac{1}{3}} - 16 \quad \text{for } \tfrac{Y}{Y_w} > 0.008856,$$
$$L^* = 903.3\tfrac{Y}{Y_w} \qquad \text{for } \tfrac{Y}{Y_w} \leq 0.008856,$$

(12.7)

$$u^* = 13L^*(u' - u'_w),$$
$$v^* = 13L^*(v' - v'_w),$$

(12.8)

where

$$u' = \frac{4X}{X+15Y+3Z} \quad \text{and} \quad u'_w = \frac{4X_w}{X_w+15Y_w+3Z_w},$$
$$v' = \frac{9Y}{X+15Y+3Z} \quad \text{and} \quad v'_w = \frac{9Y_w}{X_w+15Y_w+3Z_w},$$

(12.9)

and the subscript w indicates the attributes of the white point.

The quantity L^*, in the $(L^*a^*b^*)$ space, is defined exactly as in the $(L^*u^*v^*)$ space. Then a^* and b^* are given by

$$a^* = 500\left[f\left(\frac{X}{X_w}\right) - f\left(\frac{Y}{Y_w}\right)\right],$$
$$b^* = 200\left[f\left(\frac{Y}{Y_w}\right) - f\left(\frac{Z}{Z_w}\right)\right],$$

(12.10)

where

$$f(\varphi) = \varphi^{\frac{1}{3}} \qquad \text{if } \varphi > 0.008856,$$
$$f(\varphi) = 7.787\varphi + \frac{16}{116} \quad \text{if } \varphi \le 0.008856,$$

(12.11)

φ indicating X/X_w, Y/Y_w, etc.

The advantage of these spaces is the closer agreement between perceived color difference and linear distance, but because the coordinates u^* and v^* and a^* and b^* involve the value of L^*, a plot of u^* against v^* or of a^* against b^* does not yield unique chromaticity coordinates. This makes it difficult to appreciate the color, and so these color spaces do not diminish the importance of the chromaticity diagram.

These spaces are very frequently used for the specification of a required color.

12.6 Hue and Chroma

The terms *hue* and *chroma* are used in descriptions of color. Hue is essentially what we mean when we commonly say color. Red, green, yellow, blue, and purple are all hues. Chroma is much the same as purity. In other words, chroma indicates how closely the color response matches that of a monochromatic stimulus. The most common use of the terms is in the Munsell system of color, which uses a color space with essentially cylindrical coordinates. There are three attributes of any color, chroma, hue, and value that we can associate with the cylindrical coordinates ρ, ϑ, and z. We can roughly visualize the Munsell system as a vertical cylinder. The value of the color, what we refer to as lightness, is the vertical position on the cylinder axis. The chroma is the distance toward the periphery of the cylinder, and the hue is the intercept on the circumference of the cylinder that is marked off as a ring of the colors red, yellow, green, blue, and purple. In the actual Munsell system, the periphery of the figure somewhat departs from a perfect cylinder, but otherwise, the coordinates are as described.

The CIE recognizes the usefulness of such coordinates in that they are closely related to color perception, and so the 1976 $(L^*u^*v^*)$ space and $(L^*a^*b^*)$ space include the possibility of calculating what are called correlates of lightness, chroma and hue. L^* is the correlate of lightness.

The correlates of chroma are

$$C_{uv}^* = \left[(u^*)^2 + (v^*)^2\right]^{\frac{1}{2}},$$
$$C_{ab}^* = \left[(a^*)^2 + (b^*)^2\right]^{\frac{1}{2}}.$$

(12.12)

The hue correlate is defined as an angle in degrees by

$$h_{uv} = \arctan\left(\tfrac{v^*}{u^*}\right),$$
$$h_{ab} = \arctan\left(\tfrac{b^*}{a^*}\right),$$

(12.13)

where the signs of u^* and v^* or a^* and b^* are kept separate so that the quadrant of the angle can be unambiguously assigned. Note that these are not in any way a conversion to the Munsell system. They are simply cylindrical coordinates in the appropriate CIE space. Because they are so closely related to color perception, they are of great usefulness in specifying tolerances for color in coating design.

12.7 Brightness and Optimal Stimuli

Decorative coatings modify the spectral quality of illumination to produce a particular color sensation. Although it is not a universal rule, most decorative coatings operate in reflection and for the sake of simplicity, we shall assume reflection in what follows, although the results will be equally applicable to transmitting coatings.

Color purity is a straightforward concept that we can readily appreciate, but there is another important attribute, usually referred to as brightness. Brightness is easily understood, but perhaps slightly more difficult to define. It is an attribute that describes the tendency of a stimulus to appear more or less intense than another. A brighter stimulus emits more light. It appears that our perception of brightness follows a 1/3 power law, hence the 1/3 power in the $L^*a^*b^*$ and $L^*u^*v^*$ color space definitions, but for the purposes of this section, we shall concentrate on the luminance factor, that is, the Y tristimulus value. With any given illuminant, the larger Y, the larger the brightness.

Let us imagine that we have a decorative coating that is completely white in appearance and has a luminance factor of 100. Now let us gradually modify the spectral profile of the coating so that the reflected light is gradually reduced in spectral bandwidth. As the bandwidth is reduced, the reflected color gains in purity until it reaches a single spectral element. This gain in purity must be accompanied by a reduction in the amount of light that is reflected, that is, the brightness of the coating. Thus, there is a relationship between color and brightness or, better, luminance factor. Indeed, we can use the luminance factor to compare the quality of two coatings with similar chromaticity coordinates. This directly leads us to the idea of an optimal color stimulus that exhibits the greatest possible luminance factor for that set of chromaticity coordinates.

As might be guessed, optimal color stimuli have a perfectly rectangular spectral response. The response is uniformly 100% between two edge wavelengths outside of which the response is uniformly zero. Alternatively, it can be zero within a band and uniformly 100% outside. It is perhaps instinctively easier to sense the validity of this than to prove it. Given the form of this optimal stimulus, it becomes possible to calculate the maximum luminance factor for any given chromaticity and to plot the locus in the chromaticity diagram of the optimal stimuli for a given value of luminance factor, which forms a closed curve in the chromaticity diagram. This locus will depend, of course, on the particular illuminant, but such loci have been calculated and are

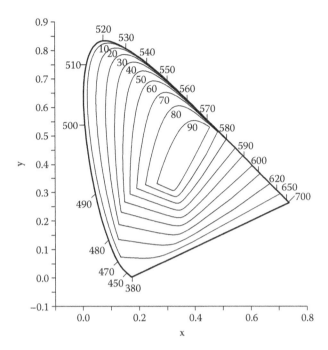

FIGURE 12.7
Chromaticity loci of optimal object–color stimuli for D_{65} illuminant and values of Y from 10 to 90.

tabulated. See, for example, Wyszecki and Styles [1]. Figure 12.7 shows loci calculated for the D_{65} illuminant.

Now that we unambiguously know the maximum possible luminance factor for any given chromaticity, we can use it in an objective assessment of the quality of any given decorative coating. Probably the most useful approach is to plot, in the chromaticity diagram, the chromaticity concerned, together with the locus of a constant maximum luminance factor corresponding to the luminance factor of the coating.

Let us use the diagram to assess the quality of a simple coating. Figure 12.8 shows the performance of a simple 9-layer optical coating with design as listed in the figure caption. Its chromaticity coordinates for the D_{65} illuminant and 1931 Standard Observer are plotted in Figure 12.9. The luminance factor Y is calculated as 54.85, and the locus of the optimal stimuli corresponding to this luminance factor is plotted in the diagram. The coating is clearly not quite optimum, principally because the reflectance is only 90% rather than 100%, but the gap between current performance and optimum is small.

Knowledge of the optimum luminance factor helps us avoid unrealistic targets for color performance in the design of optical coatings.

12.8 Colored Fringes

Colored fringes of constant thickness can often be seen in simple film systems such as oil on water or protective lacquer over metal and the like. Since they are localized in the film, they are very easy to see. They are sometimes known as Newton's rings because the first very detailed study of such

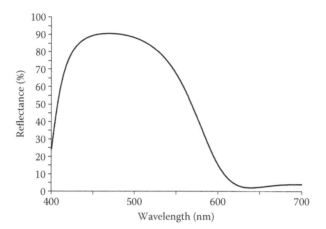

FIGURE 12.8
Reflectance over the visible region of a simple coating consisting of the following design: Air | $(0.5H\ L\ 0.5H)^4$ | Glass, with $\lambda_0 = 500$ nm. H represents a quarter wave of Ta_2O_5 and L of SiO_2.

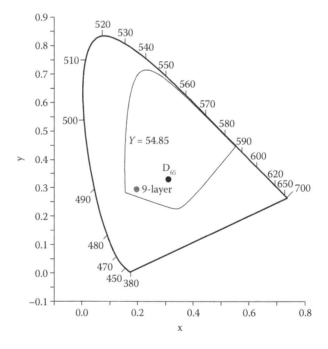

FIGURE 12.9
The dot labeled "9-layer" represents the calculated chromaticity of the coating of Figure 12.8 with illuminant D_{65}. The dot to the right is the white point. The luminance factor of the coating is calculated to be 54.85, and the locus of the optimal stimuli for that value is also plotted in the diagram. The coating is only a little short of the possible optimum.

fringes was carried out by Isaac Newton and described in his *Opticks* [2]. The appearance of such fringes depends very much on both viewing conditions and the source of illumination.

The idea of coherence was introduced in Chapter 2 along with coherence length, a simple quantitative parameter that can be used to predict the appearance or nonappearance of interference effects. Coherence and coherence length should be thought of as properties of the entire system. Coherence length is the path difference where interference just disappears. Of course, one could argue about the exact point where interference effects disappear, but a very useful and much used expression for coherence length is $\lambda^2/\Delta\lambda$, where $\Delta\lambda$ is the limiting bandwidth of the system and can be a property of the light source, the receiver, or the intervening components. Fringes in a film tend to disappear when a double traversal of the film becomes greater than the applicable coherence length.

When we use broadband sources of illumination, such as daylight, the spectral bandwidth of the source is very large, and it is the spectral response of the cones in the retina that possesses the limiting bandwidth. We can take the bandwidth of a cone as roughly 100 nm that would, at, say, 550 nm, yield a coherence length of $550^2/100$ nm, that is, just over 3 μm. This implies that roughly slightly greater than 1.5 μm would be the limiting film optical thickness. However, there are other sources of illumination. Fluorescent lights are mainly based on a mercury discharge, and despite the broadening of the lines in the discharge and the effect of the phosphor that coats the inside of the discharge vessel, there are still strong lines in the output spectrum. We can illustrate this with the CIE F10 illuminant that represents a daylight fluorescent source (Figure 12.10). A significant component of the fluorescent light is in the form of narrow spikes that are some 10 nm wide. This gives a coherence length at 550 nm of around 30 μm and, since this is very much greater than that of the human eye, will be the dominant determinant of the coherence. Thus, we can expect, in such fluorescent light, to perceive fringes in films up to around 15 μm in optical thickness.

Let us use a very simple example as an illustration. We assume oil of refractive index 1.5 over water that has a refractive index of approximately 1.33. We will assume that the physical thickness of the oil varies from 0 to 5 μm, and using the 1931 Standard Observer, we will calculate the appearance of the fringes in the oil in both D_{65} and F10 illuminants. The appearance of the fringes will be affected to some extent by the reproduction process, but the comparison will be valid. This is shown in Figure 12.11. The oil with its refractive index of 1.5 shows fringes vanishing at a physical thickness of 1 μm, that is, an optical thickness of 1.5 μm, just as predicted, while in F10 illumination, the fringes are still clearly visible at a physical thickness of 5 μm, that is, an optical thickness of 7.5 μm.

This effect is often seen in decorative metal objects that are protected by transparent lacquer, such as door handles and faucets.

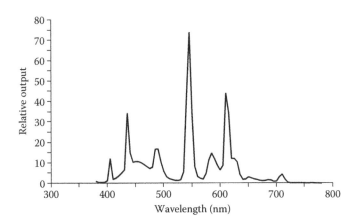

FIGURE 12.10
Relative output of the CIE F10 illuminant representing a daylight fluorescent source. Apart from the pronounced spikes in the output, the blue content is slightly less than that in the D_{65} illuminant.

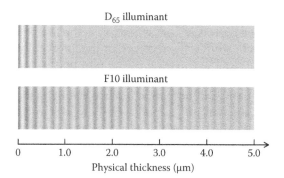

FIGURE 12.11
Appearance of fringes in an oil film ($n = 1.5$) over water ($n \approx 1.33$).

References

1. G. Wyszecki and W. S. Stiles. 1982. *Color Science.* 2nd ed. New York: John Wiley & Sons.
2. Isaac Newton. 1704. *Opticks: Or a Treatise of the Reflections, Refractions, Inflections and Colours of Light.* London: The Royal Society.

13

Production Methods

In this chapter, we shall briefly deal with the fundamental processes and the machines that are used for optical thin-film deposition. We shall also discuss the control of layer thickness and include an examination of some tolerancing methods.

13.1 Deposition of Thin Films

There is a considerable number of processes that can be and are used for the deposition of optical coatings. The commonest take place under vacuum and can be classified as physical vapor deposition (sometimes abbreviated to PVD). In these processes, the thin film directly condenses in the solid phase from the vapor. The word *physical*, as distinct from *chemical*, does not imply the complete absence of chemical parameters in the formation of the film. Chemical reactions are in fact involved, but the term *chemical vapor deposition* (sometimes abbreviated to CVD) is reserved for a family of techniques where the growing film is formed by a chemical reaction between precursors so that the growing film substantially differs in composition and properties from the starting materials.

The physical vapor deposition processes can be classified in various ways, but the most useful classifications for our purposes are based on the methods used for producing the vapor and on the energy that is involved in the deposition and growth of the films. Vacuum, or thermal, evaporation has been the principal physical vapor deposition process for years, and because of its simplicity, its flexibility, and its relatively low cost, and because of the enormous number of existing deposition systems, it is likely to continue so for some considerable time. It is, however, clear that it possesses major shortcomings, especially in respect of the microstructure of the films, and particularly for high-performance specialized coatings, alternative processes, such as sputtering, are being adopted. In thermal evaporation, the material to be deposited, the evaporant, is simply heated to a temperature at which it vaporizes. The vapor then condenses as a solid film on the substrates, which are maintained at temperatures below the freezing point of the evaporant, although they are usually heated above room temperature. Molecules travel virtually in straight lines between source and substrate, and the laws governing the thickness of deposit are similar to the laws that govern illumination. In sputtering, the vapor is produced by bombarding a target with energetic particles, mostly ions, so that the atoms and molecules of the target are ejected from it. Such vapor particles have much more energy than the products of thermal evaporation, and this energy has considerable influence on the condensation and film-growth processes. In particular, the films are usually much more compact and solid. In other variants of physical vapor deposition, the condensation of thermally evaporated material is supplied with additional energy by direct bombardment by energetic particles. Such processes, together with sputtering, are collectively known as the energetic processes.

Although physical vapor deposition is the predominant class of deposition processes in optical coatings, the application of chemical vapor deposition is gradually increasing. The chemical reactions between the starting materials, the precursors, to form the material of the coating may be triggered in various ways, but the most common is probably by means of an electrically induced plasma in the active vapor. Such processes are collectively known as plasma enhanced. Chemical vapor deposition is complementary to, rather than a direct competitor of, physical vapor deposition. Physical vapor deposition is an exceedingly flexible process in terms of materials, substrate shape, and coating type. Chemical vapor deposition is somewhat less so because the reactor where the reaction and deposition take place usually has to be designed, or at least modified, to fit the particular product. It has been used with great success in the production of large numbers of similar components. It is also capable of the deposition of films that present a challenge for physical vapor deposition. The boundary between the two classes of process is rather blurred.

In Chapter 1, we saw how the subject could be said to begin with Fraunhofer's preparing of thin films by the chemical etching of glass and by deposition from solution. These and similar methods have been used to some extent in optical thin-film work. Other techniques that, at different stages in the development of the subject, have been, and are still sometimes, employed include anodic oxidation of aluminum to form a protective coating and the spraying of material onto a surface either in solution or in the form of a substance that can be chemically converted into the desired material later. Even the substance itself is sometimes sprayed on, possibly after vaporization in a hot flame. The polymerization of monomers deposited on surfaces by condensation or from solution is also used occasionally. The extrusion of self-supporting thin-film multilayers is yet another technique.

It is impossible to cover everything, or even anything, to the depth it deserves. There is a number of books that specifically deal with processes. Useful works include those by Vossen and Kern [1, 2] and Glocker and Shah [3]. We shall primarily deal with physical vapor deposition and especially with thermal evaporation since that is still the staple process.

13.1.1 A Word about Pressure Units

The standard SI unit of pressure is the pascal. In much of the earlier optical coating literature, the unit of pressure used was the torr, defined as $1/760$ atmosphere and intended to be equivalent to a millimeter of mercury. Since an atmosphere is 101,325 Pa, 1 Torr equals $101,325/760$ Pa, that is, 133.3224 Pa. Another unit, allowed although not recommended by the SI system, is the bar, defined as 100 kPa, so that the torr is 1.333224 mbar, while the millibar is 100 Pa. The closeness of the torr and millibar has led to a slight preference in the optical coating field for the millibar in any conversion from the old unit. In what follows, we will normally refer to the pressure in pascal but include the value in millibar.

13.1.2 Thermal Evaporation

In thermal evaporation, the vapor is simply produced by heating the material, known as the evaporant. Because of the reduced pressure in the chamber, the vapor is given off in an even stream, the molecules appearing to travel in straight lines so that any variation in the thickness of the film that is formed is smooth and principally depends on the position and orientation of the substrate with respect to the vapor source. The properties of the film are broadly similar to those of the bulk material, although, as we shall see, there are important differences in the detailed microstructure. Precautions that have to be taken to ensure good film quality include scrupulous cleanliness of the substrate surface, near normal incidence of the vapor stream and, sometimes, heating the substrate to temperatures of 200–300°C (or even higher, depending on the material) before commencing deposition. The evaporation is carried out in a sealed chamber that is evacuated to a pressure usually of the order of 10^{-3} Pa (10^{-5} mbar). The materials to be deposited are melted within the chamber, using one of a number of possible techniques that will be described. The complete machine consists of the chamber together with the necessary pumps, pressure gauges, power supplies for supplying the energy necessary to melt the evaporant, monitoring equipment for the measurement of the thin-film thickness during the process, substrate holding jigs, substrate heaters, and the controls. Modern thin-film coating machines are shown in Figures 13.1 and 13.2. We shall see in the next section that a powerful technique for improving coating quality in thermal evaporation is what is known as ion-assisted deposition. This consists of thermal evaporation accompanied by bombardment of the growing film by energetic ions to compact it by the impulsive effective of transferred momentum. The machines of Figures 13.1 and 13.2 incorporate suitable sources for the generation of these energetic ions.

In order to evaporate the material, it must be contained in some kind of crucible, and it must be heated until molten with a sufficiently high vapor pressure, unless it sublimes. There is a number

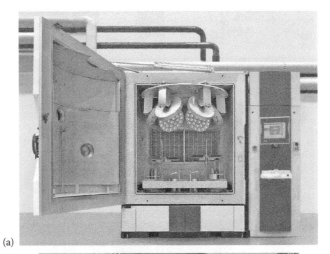

(a)

(b)

FIGURE 13.1
Thin-film coating machines. These are known as box coaters because the chamber is fabricated in the form of a box with a front door, rather than as a bell jar on a baseplate. They are normally designed to be mounted on the wall of a clean room so that loading and unloading of substrates can take place inside the clean room, while servicing of the equipment is accomplished in the gray area behind the clean room wall. (a) The SyrusPro 1510 machine manufactured by Leybold Optics showing a planetary substrate carrier. The front surface heaters, usual with planetary systems, can be seen at the foot of the port to the pumping system at the rear. (Courtesy of Leybold Optics GmbH, Alzenau, Germany.) (b) The OTFC 1800 machine manufactured by Optorun. (Courtesy of Optorun Co. Ltd., Kawagoe, Japan.)

(a)

(b)

FIGURE 13.2
CES Series continuous vacuum thin film coater. The operation is completely automatic. A continuous supply of jigs carrying preloaded substrates are heated under vacuum before passing into the coating chamber. Once coated, they pass back out of the system, and fresh jigs take their place. (a) The coating chamber. Some of the transport and heating chambers can be seen at the top of the photograph. (b) The interior of the coating chamber showing two electron beam evaporation sources with automatic feed mechanisms for tablets on the right and granules on the left. (Courtesy of Shincron Co. Ltd., Yokohama, Japan.)

of ways of achieving this. The simplest method is to make use of a crucible of refractory metal that also acts as a heater when an electric current is passed through it. The crucibles are usually elongated in shape with flat contact areas at either ends so that they look like, and are commonly referred to as, boats. Electrodes within the machine, which are insulated from the structure, act as both terminals and supports. The resistance of the boats is low, and high currents, several hundred amps at low voltages, are required to heat them. Considerable power is used that heats everything including the electrodes, and, especially to protect their sealing rings, the electrodes are normally water-cooled. Figure 13.3 shows a baseplate of an older bell jar type of machine complete with a set of electrodes, and Figure 13.4 shows a molybdenum boat, mounted between electrodes in a similar machine, being charged with material. Tantalum, molybdenum, and tungsten are all suitable for the manufacture of boats, tantalum and molybdenum being easily bent and formed, tungsten much less so. Rather less often, in special applications, platinum may be used. A wide range of materials can be evaporated from tantalum, and of these materials, it is probably that most

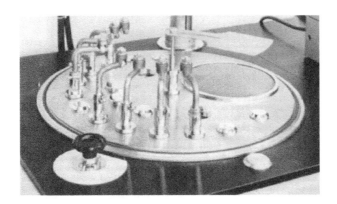

FIGURE 13.3
Baseplate of a thin-film coating machine showing the electrodes and the shutter used for terminating the layers. (Courtesy of Balzers AG, Balzers, Liechtenstein.)

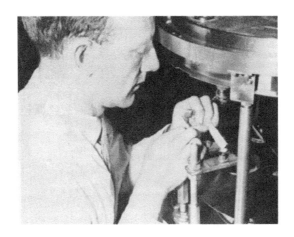

FIGURE 13.4
Molybdenum boat, mounted between electrodes in an Edwards E19E machine, being charged with material (Courtesy of Edwards High Vacuum, Crawley, UK.)

frequently used. However, some materials react with it (ceric oxide for example) or with molybdenum and require the less reactive but rather more difficult tungsten.

Considerable skill is required in the manufacture of tungsten boats, and they will usually be obtained in final form from specialized manufacturers. If it is necessary, perhaps in an emergency, to construct such a boat from tungsten strip, it should be heated to red heat before bending; otherwise, it will crack. Only the simplest of shapes can be attempted. Certain evaporants react even with tungsten. In some cases, a protective liner of alumina can be added, or an alumina crucible surrounded by a tungsten heater can even be used. In other cases, such as aluminum, the reaction is not very fast, and a tungsten wire helix is a satisfactory source. The aluminum, which wets the tungsten, forms droplets along the helix that has its axis horizontal. The area of tungsten in contact with the aluminum for a given evaporation rate is somewhat less, and the thickness of the wire, somewhat greater, than that for a boat, so that the tungsten is dissolved away more slowly and a greater proportion can be removed before failure. Even though the tungsten is dissolved in the aluminum, it appears not to contaminate the deposited film. Different types of boat are shown in Figure 13.5.

Materials such as zinc sulfide or silicon monoxide, which sublime at not too high a temperature, can be heated in a crucible of alumina, or even fused silica, by radiation from above. A tungsten

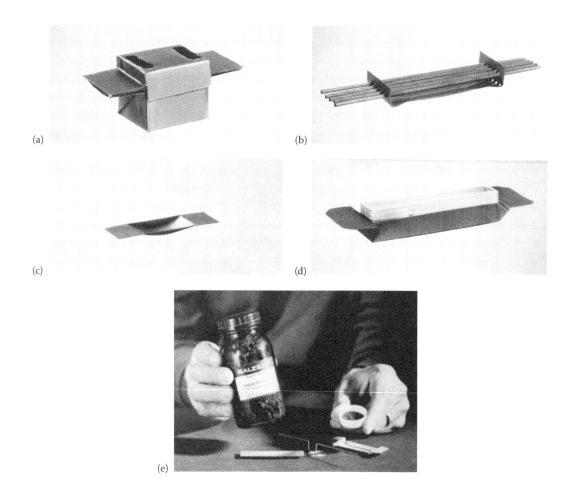

FIGURE 13.5
Various evaporation sources. (a) Tantalum box source (660 A, 1695 W for 1600°C). (b) Tungsten source for large quantities of metals such as aluminum, silver, and gold (475 A, 1400 W for 1800°C). (c) Tungsten boat (325 A, 565 W for 1800°C). (d) Aluminum oxide crucible with molybdenum heater. (e) Aluminum oxide crucible with tungsten filament. Two tungsten boats can also be seen. (Courtesy of Balzers AG, Balzers, Liechtenstein.)

spiral just above the surface of the material can produce enough heat to vaporize it. This means that the hottest part of the material is the evaporating surface, and so the material is much less prone to spitting. One example of such a source is shown in Figure 13.5—the crucible is being held in the hand and the spiral is on the table. A development of this type of source is the *howitzer* source that is shown in Figure 13.6, which is particularly useful for zinc sulfide in the infrared as the capacity can be very great [4].

Germanium is an example of a material that reacts even with alumina. The reaction is not particularly fast, but the germanium films become contaminated and show higher longwave infrared absorption than is usual. Graphite has been found to be a useful boat material in this case. Supplied in rod form for use as furnace heating elements, it can be easily machined into almost any desired shape. Copper, graphite, or one of the refractory metals should be used to make the contacts to the graphite boats. At the high temperatures involved, steel and graphite interact so that the former tends to melt and pit badly and is therefore quite unsuitable.

A form of heating which avoids many of the difficulties associated with directly and indirectly heated boats is electron beam heating, and this is now the preferred technique for most materials, especially the refractory oxides. In this method, the evaporant is contained in a suitable crucible, or hearth, of electrically conducting material and is bombarded with a beam of electrons to heat and

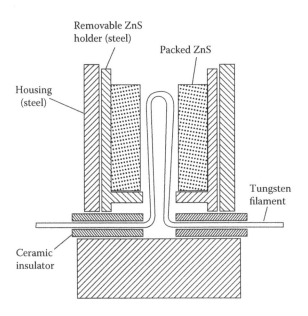

FIGURE 13.6

Howitzer—a source for evaporating large quantities of ZnS at high deposition rates. The removable ZnS holder shown as steel can also be made of fused silica or alumina, and the hairpin filament can be replaced by a tungsten helix. (After J. T. Cox and G. Hass, *Journal of the Optical Society of America*, 48, 677–680, 1958. With permission of Optical Society of America.)

vaporize it. The portion of the evaporant that is heated is in the center of the exposed surface, and there is a reasonably long thermal conduction path through the material to the hearth that can therefore be held at a rather lower temperature than the melting temperature of the evaporant, without prohibitive heat loss. This means that the reaction between the evaporant and the hearth can be inhibited, and the hearth is normally water-cooled to maintain its low temperature. Sometimes liners of various materials, often tungsten or tantalum, are used between charge and crucible. These limit the thermal conductivity so that the charge can be melted with lower beam power, but care must be taken to avoid contamination. Copper, because of its high thermal conductivity, is the preferred hearth material. The electrons are emitted by a hot filament, normally tungsten, and are attracted to the evaporant by a potential usually between 6 and 10 kV. Various types of electrodes and forms of focusing have been used at different times, but the arrangement that has now been almost universally adopted is what is known as the bent-beam type of gun. The hearth is at ground potential and the filament is negative with respect to it. The filament and electrodes are placed under the hearth, well out of reach of the emitted evaporant. There is usually a plate at filament potential situated close to the filament with a beam-defining slit through which the electrons pass, and this is closely followed by the anode at the same potential as the hearth and incorporating a slightly larger slit so that the beam passes through it. The beam is bent around through rather greater than a semicircle by a magnetic field and focused on the material in the hearth. This avoids the problems of early electron beam systems that had filaments in the line of sight of the hearth and, hence, considerably shortened life due to reactions with the evaporant. Supplementary magnetic fields derived from coils allow the position of the spot to be varied so that the mean can be placed in the center of the hearth and a raster can be described that increases the area of heated material. This reduces the temperature necessary to maintain the same rate of deposition, improves the efficiency of use of the material in the crucible, and makes the electron beam source more stable. Typical electron beam sources of this type are shown in Figure 13.7.

The electron beam source is particularly useful for materials that react with boats or require very high evaporation temperatures, or both. Even in quite small sources, beam currents of perhaps 1 A

FIGURE 13.7
Electron beam sources of the bent-beam type. The water-cooled crucibles in these examples have several pockets that can be rotated into position at the focus of the electron beam that issues from the slot below the opening in the top of the gun. The sides of the gun are the pole pieces of the focusing and deflecting magnet. Fine-tuning of the field is accomplished by the adjustable tabs on the upper parts of the sources. (Courtesy of Telemark, Battle Ground, Washington, DC.)

at voltages of around 10 kV can be achieved and refractory oxides such as aluminum oxide, zirconium oxide and hafnium oxide, and reactive semiconductors such as germanium and silicon, can be readily evaporated. Furthermore, materials that can be evaporated quite satisfactorily by a directly heated boat can be evaporated still more easily by electron beam, and so the tendency is to use electron beam sources, once they are installed, for virtually all materials. To improve their flexibility, they can be constructed with multiple pockets in the hearth, as in Figure 13.7, so that the same source can handle several different materials in a single coating cycle. Of course, the capacity of each individual pocket in a multiple-pocket version is usually rather less than that of the single-pocket version of the same source. In addition, it is not currently possible to maintain the alternative crucibles at near evaporation temperatures implying a delay between layers as the source is brought up to temperature. For large-scale production, therefore, or for coatings for the infrared, it is normal to use two or more single-pocket sources.

In the operation of an electron beam source, not all electrons necessarily disappear into the evaporant material in the crucible. Reflected electrons and secondary electrons can exist and may sometimes influence the growing film. Sometimes a grounded plate is fixed behind the source to attempt to trap such electrons. Bangert and Pfefferkorn [5] suggested that in the case particularly of zinc sulfide, the effect of the electrons is actually beneficial. Not much is known about this aspect of the use of electron beam sources.

The temperature of the substrate also plays a part in determining the properties of the condensed films. Higher substrate temperatures lead to denser and more stable films (see Chapter 14). For refractory oxides, temperatures are frequently around 300°C or slightly higher. Usually, it is the consistency of temperature from one coating run to the next that is of greater importance than the absolute level, although Ritchie [6], working in the far infrared beyond 12 μm, found substrate temperature to be of critical importance and devised ways of controlling it to within 2°C of the experimentally determined optimum. Substrates are often of low thermal conductivity and are mounted on rotating jigs to ensure uniformity of film thickness so that the measurement of the absolute temperature of the substrates is difficult. The heating is usually by means of radiant elements placed a short distance behind the substrates or by tungsten halogen lamps placed so that they illuminate the front surfaces of the substrates, the latter method gaining in popularity. For planetary arrangements of substrate carriers (see Figure 13.1), it is difficult to arrange rear-surface heating, and so front surface heaters are normal. Measurement is most often carried out by placing a thermocouple just in front of the substrate carrier. This will not accurately measure substrate temperature but will give an indication of the constancy of process conditions; frequently, this is the most important characteristic. An improvement can be obtained by embedding the thermocouple in a block of material of the same type as the substrates. Thermocouples have been placed on the rotating jig, and the signal led out through silver slip rings, but even in this case, the temperature of the front surface of the substrates is still not necessarily known to any high degree of accuracy, especially if they are of material of low thermal conductivity such as glass or silica. Results that are rather more accurate are achievable with substrates of germanium or silicon, frequently used in the infrared. A more consistent technique that is becoming more common is the use of an infrared remote sensing thermometer that detects infrared radiation from the hot substrates. Usually mounted outside the chamber, this views the substrates through an infrared-transmitting window. The absolute calibration of the device depends on the emittance of the substrate. This varies less for substrates such as glass and dielectric coatings for the visible region than for infrared components. Again, consistency from one run to the next is of prime importance. One tends to find more accurate and complex arrangements in the research laboratory, where very small numbers of samples are the norm and large batches of coatings are rare.

Usually, metals should be deposited at low substrate temperatures to avoid scatter—particularly important in metal–dielectric filters and in ultraviolet-reflecting coatings, although there is an exception to this rule of thumb in the cases of rhodium and platinum, both of which give substantially better results when deposited hot [7,8]. There are difficulties in refrigerating substrates, and substrate temperatures below ambient encourage thicker adsorbed gas layers that inhibit the condensation of the films and cause contamination. Thus, it is not normal to operate with substrate temperatures below ambient, at which adequate results are obtained. The softer dielectric materials, such as zinc sulfide and cryolite, can also be deposited at room temperature (except, as we shall see, if zinc sulfide is to be used in the infrared). The harder dielectric materials, however, usually require elevated substrate temperatures, often 200–300°C. These materials include ceric oxide, magnesium fluoride, and titanium dioxide. Some of the semiconductors for the infrared must be similarly treated. Frequently, optimum mechanical properties demand deposition at a temperature that is different from that for optimum optical properties, and a compromise that depends on the particular application is necessary. Some further details are given when individual materials are discussed.

Figure 13.2 is significant for a number of reasons. The exposure of the deposition chamber to the atmosphere every time fresh substrates for coating are to be introduced enormously disturbs the coating environment. The walls of the chamber adsorb considerable quantities of atmospheric gases, particularly water vapor, and the subsequent outgassing during evacuation is a major contributor to the time to reach final coating conditions, and even then, there are difficulties in determining whether or not the conditions are finally stable. The machine in Figure 13.2 uses what is known as a load-lock system, the term *load lock likely* being derived from a lock on a river or canal. In such a system, the chamber is maintained under deposition conditions. Substrates are introduced into a separate load chamber, isolated from the deposition chamber by a gate valve. The gate valve is opened only when the load chamber has reached the correct vacuum conditions when the substrates are transferred to the deposition chamber. The machine of Figure 13.2 takes the concept still further in that the substrates are thermally soaked until the precise deposition conditions are reached, and only then are they transferred. Immediately after coating, they are transferred out. The deposition chamber is then used only for deposition. In the normal way, the opening of the deposition chamber to atmosphere affords the opportunity of recharging sources, and so ways of doing that remotely under vacuum also had to be devised.

One of the biggest problems in optical coating manufacture is the cleaning of the deposition chamber. Particularly in thermal evaporation, the material in vapor form goes everywhere there is a line of sight from the source and coats everything. There can also be some deposition of material in the shadows of the chamber fittings even, sometimes, on the rear surfaces of the substrates. It is normal, therefore, to fit shields in the chamber to minimize deposition on any parts that would be difficult to clean later, such as the actual walls of the chamber. These shields can be removed from time to time and cleaned outside the deposition room. To still further ease the problems of cleaning, it is common practice to add disposable sheets of aluminum foil to screen even the shields. The cleaning operation can be mechanical; bead blasting is common or chemical. It is good practice to bake the shields after cleaning. This not only removes any traces of trapped fluid but also tends to dislodge any loosely bound particles that remain and could cause problems later in the deposition process. The cleaning operation is not only a technical problem. The dust and fragments of coating material that can be produced in cleaning can be hazardous, and it is important that operators be protected. Largely because of the health issues, there are also all kinds of legal requirements that vary from one region to another.

13.1.3 Energetic Processes

The energetic processes, as the name suggests, are ones that involve energies rather greater than thermal. Thin films deposited by thermal evaporation have a pronounced columnar structure that is a major cause of coating instability and drift. This is discussed later in Chapter 14. The idea behind the energetic processes is to disrupt the columnar structure with its accompanying voids by supplying extra energy, and this does work well. Some of the energetic processes are old ones that have always involved extra energy and are now recognized as having certain advantages because of it. Although we describe the processes as energetic, it has been shown that momentum transfer is the important feature.

Sputtering is an old process that actually predates thermal evaporation, although its application was largely later. Momentum transfer from incident energetic ions is used to eject atoms and molecules from a target into the vapor phase. The kinetic energy and momentum of the ejected particles are high, and so the growing film is subjected to a much greater impulse each time a fresh particle arrives, which disrupts the void and columnar structure. In the conventional form of sputtering, the target is metallic so that it conducts, and the bombarding ions are derived from a direct current (DC) discharge near the target. This discharge may be confined by crossed electric and magnetic fields when it is known as magnetron sputtering, and this is a frequent way of applying the process in optical coating. DC planar magnetron targets are most common; Figure 13.8 shows a schematic form of such a target. The great advantage of magnetron sputtering is the much longer path length of the electrons so that the discharge can be maintained at a considerably lower pressure (0.3 Pa or 0.3×10^{-2} mbar, for example) than is required compared with conventional sputtering in the absence of the magnetic field. A further advantage is the lower supply voltage of some hundreds of volts rather than kilovolts.

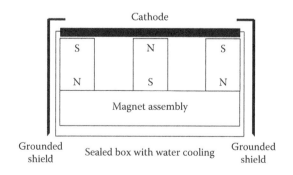

FIGURE 13.8
Schematic representation of a planar magnetron source. The target or cathode is connected to the negative supply. The structure of the coating machine including the grounded shield is the positive side of the supply. Electrons leaving the cathode surface move outward but are turned into a cycloidal path by the field of the magnets. The polarity of the magnets is unimportant as long as they are arranged with the outer poles opposite to the inner as shown.

There are, however, some disadvantages. The arrangement of magnets concentrates the discharge in the region between the pole pieces, and the erosion of the target is greatest there, while other areas of the target show negligible erosion. With long rectangular targets, the appearance of the eroded region is not unlike the shape of a racetrack, a term often used to describe it. Target utilization is therefore not good, and so used targets are usually recovered rather than scrapped. Since the targets in DC magnetron sputtering are metallic, a process of reactive sputtering must be used to produce oxides or nitrides, and the sputtering gas is therefore usually a mixture of a noble gas such as argon and oxygen or nitrogen. This reactive gas also reacts with the target to produce a skin of oxide or nitride and the skin tends to build up in the less eroded regions. These patches of insulating film have a high capacitance (capacitance is proportional to the inverse of the thickness), and as they are bombarded, they gradually charge, and because their capacitance is high, they store a considerable charge. They are not particularly reliable as capacitors and tend to break down in a sudden and violent discharge that is essentially an arc. This arcing usually produces molten droplets of material that are often embedded in the film. In the worst case, the discharge can actually seriously damage the target. The insulating skin also modifies the electrical properties of the sputtering system, inducing a hysteresis that complicates the control. Such effects are particularly severe with silicon targets, and silicon oxide is the sole low-index material really suitable for sputtering. The problem is often called target poisoning.

There are several current solutions to the target-poisoning problem. The target surface may be moved with respect to the magnets so that the region of high erosion moves over the surface and scrubs off the insulating material. In the usual embodiment, the target is made in cylindrical form and rotated about a longitudinal axis around the magnets and inside the grounded shield. Then there are special power supplies that regularly reverse the polarity of the supply so as to discharge the capacitors. Because the capacitance is high, the time constant is quite long, and the discharging can be performed at a relatively low frequency. A more recent solution, and perhaps the most popular in precision optics, involves twin magnetron targets connected to opposite poles of a midfrequency power supply (Figure 13.9). The targets are now alternately the anode and cathode of the system. This discharges the effective capacitors before they can cause damage, and because of the long time constant, the frequency can be as low as 10 kHz, although the more usual frequency is around 40 kHz. This is sometimes termed *midfrequency sputtering*, or twin, or dual magnetron sputtering. This arrangement of alternating anode and cathode solves another problem. In normal single-target sputtering, the chamber structure is usually the anode of the supply, although sometimes there may be an additional rod that functions as an anode. The insulating film also gradually covers this structure, making the anode less and less effective with

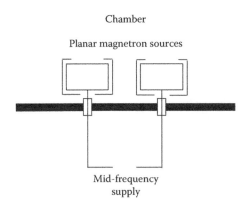

FIGURE 13.9
Dual magnetron arrangement in which two magnetron targets are connected to a midfrequency power supply so that each is alternately anode and cathode. The arrangement avoids the charging problems of reactive DC sputtering without the complications of RF sputtering.

serious implications especially for the control of the process. The effect is known as the disappearing anode. Since the midfrequency arrangement provides the anode that is constantly scrubbed with every alternate half cycle, the technique solves the disappearing anode problem along with the poisoning. Usually, the double magnetrons are planar, but the process has also been used with rotating magnetrons.

An alternative process places the magnetron source inside a shroud where it can be operated in argon and thus avoid poisoning. The material escapes through a large aperture above the source in the center of the shroud. Outside the shroud in the main chamber, the material coats the substrates, but the growing film is also bombarded with a beam of oxygen or nitrogen ions in the manner of ion-assisted deposition, described shortly. Enormous quantities of gas enter the deposition chamber, and to remove the gas very fast, high capacity pumps are used. The films that grow are amorphous of very high packing density. This process is known as microplasma, and most of what is publicly known about the process comes from an issued patent [9]. An advantage of the process appears to be that the geometry of the coating chamber can be similar to that for thermal evaporation. Presumably, the increased positional stability of the magnetron sources is a further advantage over a similar evaporation geometry.

A rather different approach [10,11] uses intermittent deposition on a cylindrical work holder but completely extends the plasma around the cylinder so that the efficiency of conversion of the metal to the oxide or nitride is exceedingly high This is achieved by arranging a number, usually four, of magnetron sources around the periphery of the cylinder, only the one with the appropriate material operating at any particular time. The magnets within the targets are arranged so that a target with a north magnetic pole outermost is always next to one with a south pole outermost. Thus, as we move around the cylinder, we have north–south–north for the first magnetron, then south–north–south for the next, and so on. This forms a kind of magnetic bottle, and the plasma spreads out all around the work holder. Very high-quality coatings are reported. The process is termed *closed-field magnetron sputtering*. A machine is shown in Figure 13.10.

There are other solutions. The oxidation or nitriding may take place remote from the deposition. This requires that only a small amount of material be deposited then treated, then more deposited and then treated, and so on. The process is implemented by placing the substrates on a cylindrical drum that is then rapidly and continuously rotated past a linear magnetron-sputtering source, often midfrequency, then past an ion source and round to the magnetron targets again. The idea originated in the community that coats tools [12], and thermal evaporation was the method

FIGURE 13.10
CFM450 closed-field magnetron coating system. (Courtesy of Applied Multilayers, Coalville, UK.)

uppermost in the minds of the authors for the deposition of the material. Nevertheless, the idea of intermittent deposition can readily be applied to other types of deposition and particularly to the sputtering of optical coatings. An early version of such a process is known as metamode, short for metal mode and is the subject of an issued patent [13].

A fairly recent innovation in optical coatings is the load-lock system for the introduction of substrates into the coating chamber, already mentioned in connection with the machine of Figure 13.2. Since sputtering sources do not need to be renewed after every deposition cycle, they especially lend themselves to load-lock design. Figure 13.11 shows a small machine that uses intermittent deposition by sputtering, but in this case, the deposition chamber is isolated from the outside world by a load-lock system. This particular machine was originally designed for small batch ophthalmic applications, but it has been demonstrated to be also very successful in other types of coating [14].

A machine that goes still further in that the loading system is not only through a load lock but is also robotically handled is shown in Figure 13.12. Here, the input and output sides of the machine is designed to be situated in a clean room while the chamber is in a gray area where planned maintenance can be performed without exposing any of the other parts of the machine. The deposition is intermittently interspersed with bombardment by oxygen ions from an radio-frequency (RF) plasma source. The work holder is a rapidly rotating (250 revolutions per minute) flat circular plate rather than a cylinder.

The insulating skin that forms over the metallic cathode has a relatively high capacitance, and this permits the midfrequency sputtering approach. The surface of a purely dielectric target has very low capacitance because the thickness of dielectric material is now very large. The small

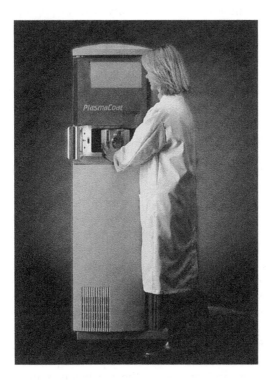

FIGURE 13.11
Plasmacoat is a small machine, not only originally intended principally for the coating of spectacle lenses but also suitable for small batches of other types of coating. The process is one of reactive sputtering, and the operation is entirely automatic. The coating chamber is permanently under vacuum. For loading, the substrate carrier drops down into the loading chamber leaving the coating chamber sealed off. The carrier can then be loaded through the access door. Once substrates are loaded, the access door closes and the substrate carrier moves upward back into the deposition chamber. (Courtesy of Applied Multilayers, Coalville, UK.)

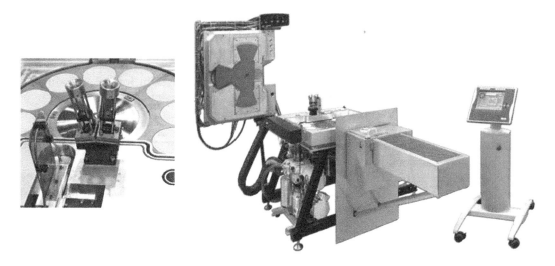

FIGURE 13.12
Helios magnetron sputtering system of Leybold Optics incorporates the intermittent deposition technique in a chamber that is never opened to atmosphere except for servicing. The substrates are fed into the coating chamber robotically through a load-lock system. The loading side is situated in a clean room, and the coating chamber, in a gray room. The substrate carrier is shown on the left with an optional optical monitoring system. (Courtesy of Leybold Optics GmbH, Alzenau, Germany.)

capacitance has a very short time constant, and any discharging operation must take place at elevated frequency, that is, at RF. RF sputtering does avoid the problems of an insulating target. Although it has been much used in other areas of thin-film deposition, it has not been as popular in optical coatings, possibly partly because of additional requirements of screening and matching and possibly because although it is an excellent process capable of very high-quality films, it seems not to present a sufficiently compelling advantage over other deposition techniques. Nevertheless, in applications where speed is less important than quality, it has been found to be remarkably reliable and stable, to the extent where even quite complex coatings can be controlled entirely by power, gas pressure, and time with no ongoing layer thickness measurement whatsoever [15]. Even here, however, we find similar stability with the lower-frequency forms of sputtering [16].

High-power impulse magnetron sputtering, with a somewhat variable acronym but usually HIPIMS [17], is a quite recent innovation. As the name suggests, the process is pulsed at very high power so that the sputtered material is essentially a highly ionized metal. The resulting dense and smooth films exhibit excellent adhesion. Heating of the targets and of the substrates is kept within bounds by the very short, a few hundred microseconds, pulse duration and the relatively low pulse frequency of normally less than 1 kHz. The process has not so far made any significant impact on optical coatings, but there are encouraging results particularly with titanium dioxide that, sputtered from a ceramic $TiO_{1.8}$ target, gave an index of 2.48 at 550 nm [18].

A completely different approach to sputtering involves the use of a broad-beam ion source that forms a kind of separate enclosed vessel, to generate the ions that are then extracted and directed toward the target (Figures 13.13 and 13.14). This is known as ion-beam sputtering [19]. It is capable of a very high degree of film purity, and the lowest published losses in multilayer reflecting coatings (1 ppm or less) have been achieved with this process [20,21]. This is the process used for mirrors in gravitational wave receivers. Since the ion beam is usually neutralized by adding electrons, charging problems with insulating targets can be avoided, and the process is as useful for insulating materials as for conductors, although metal targets that produce fewer scattering particles are preferred for highest quality. Ion-beam sputtering is slow compared with most other processes, and it is not yet able to cope with deposition over very large areas. It has, however, made big inroads into the area of precision optics. The reason for this is partly the evident success

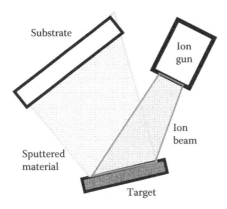

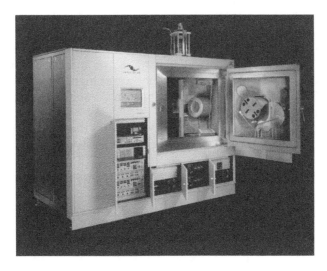

FIGURE 13.14
Spector ion-beam sputtering system for the production of high-quality optical coatings especially narrowband filters for dense wavelength division multiplexing. (Courtesy of Ion Tech, Inc., Fort Collins, CO.)

of the process and partly the resolution of a protracted patent dispute involving ion-beam sputtering, but there was also an element of chance. During the 1990s, it had become a significant technique for the production of dense wavelength division multiplexing filters, and when that market contracted, other uses had to be found for the equipment that now happily demonstrated considerable success in the construction of other types of high-quality coatings. Sometimes a second ion beam is directed at the growing film for stress control when the process is known as dual ion-beam sputtering, although only one beam is actually used for the sputtering.

Not all materials are suitable for sputtering. In particular, fluorides present considerable difficulties because of preferential sputtering of fluorine atoms. The film is then fluorine deficient and optically absorbing. The fluorine vacancies can be filled with oxygen—there is usually plenty of oxygen around—that removes the absorption, at least at longer wavelengths, but the film becomes an oxyfluoride with altered (usually raised) index of refraction and frequently poorer environmental resistance. Various techniques have been proposed for the successful sputtering of fluorides, mostly involving some method of replacing the lost fluorine, but that they have not so far been generally adopted suggests that difficulties may still remain. The most promising report is a recent one by Ode [22], in which several important fluorides for the 193 nm wavelength were successfully deposited by reactive single ion-beam sputtering with xenon and using nitrogen trifluoride (NF_3) as the reactive gas. No heating was employed, and so the deposition temperatures were not higher than 40°C and usually below 30°C. Nitrogen trifluoride is rather less of a hazard than fluorine and is compatible with stainless steel and even with some plastics. It is already used in a number of applications in the semiconductor industry, and so a body of knowledge and experience in its use does exist. Successful materials so far reported are GaF_3, AlF_3, and LaF_3, all with very low extinction coefficients at 193 nm, the extinction coefficient of AlF_3 being as low as 0.0000083.

In reactive low-voltage ion plating [23,24], a high-current beam of low voltage electrons is directed into the region above the hearth in an electron beam source (Figure 13.15). This results in a very high degree of ionization of evaporant material, usually a metal or suboxide, ensuring that the melt should be conducting. Reactive gases, oxygen or nitrogen, fed separately into the chamber, are also highly ionized. There is a complete circuit from electron source to electron beam source to low-voltage high-current power supply and back to electron beam source, and it is completely isolated from the remainder of the structure. The substrate carrier is also electrically isolated. There are many electrons, and they are very mobile, and so the isolated substrates acquire a charge that is negative with respect to the electron beam source. This attracts the positive ions from the source so that they arrive at the film surface with additional momentum that is transferred to the film and compact it. Films are tough, hard, dense, and usually amorphous. Because of the very efficient reaction with the additional gas, they are of high optical quality. Recent versions of the process include the possibility of a deliberate bias on the substrate carrier rather than simply leaving it to the accumulation of electrons.

Ion-assisted deposition (Figure 13.16) is an energetic process that has the great advantage that it is easy to implement in traditional thermal equipment. It consists of thermal evaporation to which

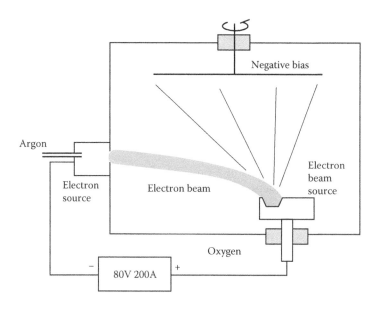

Low-voltage ion plating

FIGURE 13.15

Low-voltage ion plating process. The negative bias on the electrically isolated substrates is acquired from the free electrons in the chamber. (See, for example, Pulker and Guenther [24].) (After H. K. Pulker and K. H. Guenther, *Thin Films for Optical Systems*, F. R. Flory (ed.), pp. 91–115, Marcel Dekker, New York, 1995.)

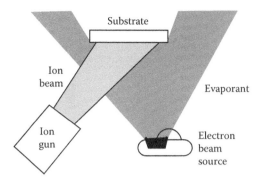

FIGURE 13.16
Addition of ion bombardment of the growing film transforms conventional thermal evaporation into ion-assisted deposition.

has been added bombardment of the growing film with a neutral beam of energetic ions (that is, a beam of positive ions injected with sufficient electrons). All that is required to put it into operation in a conventional machine is therefore the addition of an ion source. The most common types of ion sources for this purpose are broad-beam, often with extraction grids. Much of the earlier published research and reported successes employed the Kaufman, or gridded type of, ion gun. In that, the source of electrons is a hot filament, and the extraction system consists of two closely aligned grids, the inner floating and acquiring the potential of the discharge so that it confines it within the gun, and the second applying a field to draw the positive ions out of the discharge chamber through the apertures in the inner grid. The beam of ions is neutralized outside the discharge chamber by adding electrons, often from a hot filament, immersed in the beam to avoid space charge limitation or, more commonly today, from a separate hollow cathode electron emitter. The Kaufman source has an advantage in that the ion energy is well controlled, an advantage in research, but the grids are fragile and easily misaligned or damaged, and so some effort has been put into the development of sources that do not require extraction grids, and they are being used in increasing numbers in production. For further information, see Bovard [25] and Fulton [26].

A variant of ion-assisted deposition termed *effective physical vapor deposition* has been described and works especially well with sensitive plastic substrates [27]. The principle of this method is to separate the deposition by thermal evaporation from the bombardment simply by a mild rearrangement of the geometry of the machine. The result is a tenacious, tough coating of very high quality especially in terms of residual scattering. Both evaporation and bombardment are continuous from the point of view of the sources but pulsed as far as the substrates are concerned, which rotate on a normal calotte through the deposition and bombardment areas in turn. High rates of deposition without the danger of thermal distortion can readily be achieved.

The ionized plasma-assisted deposition process includes features of both ion-assisted deposition and low-voltage ion plating. It makes use of what is known as an advanced plasma source (Figure 13.17) [28–30]. The source, which is insulated from the chamber and floats in potential, is of simple construction. A central indirectly heated cathode is made of lanthanum hexaboride. This lies along the axis of a vertical cylinder that is the anode. A noble gas, usually argon, is introduced into the source. The cylinder contains a solenoid that produces an axial magnetic field. The crossed electric and magnetic fields make the electrons move in cycloids with the usual increase in path length and degree of ionization, so that an intense plasma is produced in the source. The fields do not axially confine the plasma, and so it escapes from the source into the chamber. There, the electrons, which are very mobile, preferentially escape to the chamber structure, leaving the plasma positively charged without the need for isolated substrate holders. The deposition sources are thermal, usually electron beam, and they emit evaporant into the plasma where it gains energy and is

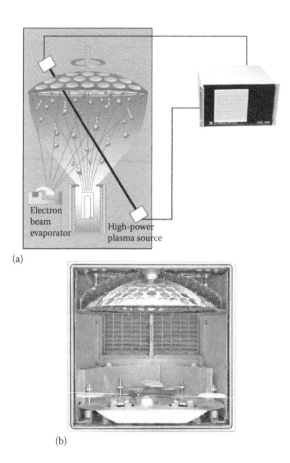

(a)

(b)

FIGURE 13.17

Advanced plasma source. (a) Diagram of the advanced plasma source and the arrangement of the machine for plasma ion-assisted deposition. The sketch shows also the application of a system of optical monitoring (OMS 5000) that takes its signals directly from one of the batch substrates. The optical properties of the substrate are sampled every time it passes through the beam. (b) Photograph of the interior of a SyrusPro 1110 system showing two boat sources toward the front and two electron beam sources toward the rear together with a central advanced plasma source. The single-rotation calotte permits rear surface heating, and the shroud that can be seen above the rear half of the calotte carries the heater elements on its lower-surface electron beam sources and just slightly to the right of the center the cylindrical advanced plasma source. (Courtesy of Leybold Optics GmbH, Alzenau, Germany.)

partially ionized. The evaporant then condenses on the growing film with additional energy, as in ion plating, and is simultaneously bombarded by ions from the plasma as in ion-assisted deposition. For reactive processes, the reacting gas is not fed into the source, but into the plasma as it leaves the source. The process has been very successful in the production of narrowband filters for dense wavelength division multiplexing and, since the contraction of the telecom market, in precision optics in general. More recent plasma sources are powered at RF [31] with impressive results particularly in terms of long-term stability.

Another interesting process that has some potential for optics is cathodic arc evaporation [32]. In this process, a high-current and low-voltage arc is struck between an electrode and a conducting cathode. Since the arc has very small cross-sectional area, the point of impingement on the target can reach temperatures as high as 15,000°C so that material leaves the cathode highly ionized and with very high energy. Promising results for TiO_2 such as an index of 2.735 at 633 nm have been demonstrated [32]. The principal problem is the ejection of molten droplets of material that must not be permitted to reach the growing film. For optical purposes, therefore, the evaporated material can be passed through a filter. An effective one relies on the charged nature of the evaporant and magnetically alters its direction to separate it from the unwanted particles [33]. Ion bombardment can be added to the process when it has been termed *ion-assisted filtered cathodic arc evaporation* [33].

A recent development of some importance is known as radical assisted sputtering or RAS [34]. Here the configuration is also of a vertical drum rotating past dual magnetron sources this time, but the interaction is arranged to be with atomic oxygen or, in some cases, nitrogen. There is essentially no bombardment. The reactivity of the atomic oxygen is so high that bombardment to assure implantation is not necessary and, in fact, is intentionally suppressed. The films are consequently of very high quality. Although this is not strictly an energetic process, nevertheless, the films have consistent high packing density and all the attributes of an energetic process. The machines are configured with load locks so that the chamber is exposed to the atmosphere only for maintenance. Again, the stability of the process has been demonstrated in the monitoring of film thickness purely by timing. This is a particularly fast process, and so batch costs are reported as attractively low, and the machine is capable of large batches. A machine is shown in Figure 13.18.

The improvements brought by the energetic processes are achieved at comparatively low substrate temperatures, which helps with the difficult coating of plastic substrates. It has been theoretically demonstrated by advanced computer modeling [35,36] that the major effects are due to the additional momentum of the molecules, either supplied by collisions with the incoming energetic ions or derived from the additional kinetic energy of the evaporant. Experimental evidence exists [37,38] that shows correlation of the effects with momentum rather than energy of the bombarding ions. Major benefits of these processes are the increased packing density of the films, making them more bulk-like and, hence, increasing their ruggedness; the improved adhesion resulting from a mixing of materials at the interfaces between layers; and a reduction of the sometimes quite high tensile stress in the layers, although the stress can become sufficiently compressive that thin substrates require stress compensation to avoid distortion. The increase in packing density also reduces the moisture sensitivity and can actually eliminate it altogether [39]. The increased packing density also improves the stability of the films in other ways. Magnesium

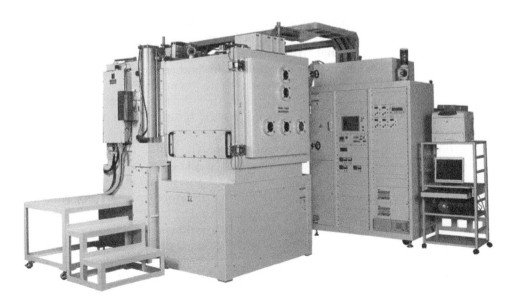

FIGURE 13.18
Machine for implementing the RAS process. The front door opens into the load chamber. The actuator for the gate valve into the deposition chamber is visible on the left-hand wall and, behind it, a cover giving access to the rear of one of the magnetron sources. (Courtesy of Shincron Company Ltd., Yokohama, Japan.)

fluoride films can be only very lightly bombarded without decomposition, but even then, they resist high-temperature oxidation better, for example [40]. The hardness and corrosion resistance of metal films, especially with dielectric overcoats [41], is improved by ion-assisted deposition, but the optical properties tend to be slightly adversely affected, possibly by the implantation of a small fraction of the bombarding ions [42]. The increased reactivity of the bombarding ions permits the deposition of compounds, such as nitrides [43], that are difficult or impossible by normal vacuum evaporation.

13.1.4 Other Processes

Physical vapor deposition processes are those most often used for the production of optical coatings. However, in the electronic device field, chemical vapor deposition is a principal method for thin-film deposition, and there is increasing interest in it for optical purposes, usually with regard to very special requirements.

Chemical vapor deposition differs from physical vapor deposition in that the film material is produced by a reaction among components of the vapor that surrounds the substrates. The reaction may be induced by the temperature of the substrates themselves, when the process is the classical thermal chemical vapor deposition or, and this is more usual in the optical field, it may be a plasma-induced process.

Usually the components, the reactants or precursors, will be introduced into a carrier gas that is permitted to flow through the system. This ensures a constant supply of the reactants to the growing interface and allows sufficient dilution so that the reaction is not so fast as to overwhelm the film growth. In this classical form of chemical vapor deposition, great problems are created by reactions that are too efficient. A reaction that rapidly proceeds tends to produce a film that is poorly packed and poorly adherent. The term *snow* is often used to describe it. The reactions must therefore be quite weak, and this means that impurities that have strong reactions can play havoc with the process and severely limit the possible range of processes.

Because of all the difficulties, the classical thermal chemical vapor deposition process is not often used for optical coatings. Instead, pulsed processes have been largely adopted. Material added to a thin film is assimilated provided it is not immobilized by material deposited over it before it has had time to relax into favorable positions. The problem is not really the strength of the reaction, but rather the large amount of material that arrives in a given time. Earlier material is buried under the later material and cannot relax to a state of equilibrium, and snow is the result. If an efficient reaction can be made to deliver material at a correct rate, then the film will be dense. It is the overall rate of deposition that determines the microstructure. Pulsing the reaction gives the control of rate that is required. The pulsing can most conveniently be achieved when a plasma-assisted process is involved [44].

A related process that is sometimes called plasma polymerization, and sometimes plasma-enhanced (or induced) chemical vapor deposition or PECVD [45–47], is used to deposit dense organic layers with stable optical properties over curved and irregular surfaces with good uniformity. Plasma polymerization is quite unlike normal polymerization where monomers are linked into chains of repeat units. The plasma is characterized by energetic electrons that break the reactants into active fragments, and these fragments link with each other to form the deposited film. Some of this combination may take place in the gaseous phase forming clusters that may deposit on the growing film or may be broken into fragments again by the plasma. Strong binding occurs so that the deposited film is tough and hard and dense. It is not strictly polymeric and contains free radicals that may combine with any oxygen that is also present. The mechanical properties can range from plastic to elastic and glass-like. Because the films are insulating, in fact, they are used as capacitor dielectrics in some applications, RF discharges are usual for this process. The speed of deposition can be very high, up to 1 μm/min, although rates of one-tenth to one-hundredth of this are more common.

The process has been used for some time in the semiconductor industry to deposit silicon dioxide. The normal precursor is tetraethoxysilane (TEOS) together with oxygen, but the substrate temperature is usually quite high, at least 250°C, much higher than can be possible for plastic substrates. When the temperature is reduced to permit coating of plastic substrates, the film composition becomes much more complicated. Apart from the silicon oxide content, they include, for example, silanol that results from reactions involving residual water vapor. There are, in fact, many silicone compounds that can be and have been used as precursors in the PECVD of such silica-rich films. The feature that they tend to share is a backbone of alternate silicon and oxygen atoms. Apart from TEOS already mentioned, other suitable compounds include hexamethyldisiloxane, tetramethoxysilane, methyltrimethoxysilane, and trimethylmethoxysilane. As might be expected, they are toxic, although their toxicity varies. The makeup of the precursors determines the character of the film to a large extent. With organic silicone compounds or silanes present in the gas along with oxygen, the coatings are particularly tough and resistant to abrasion and form the basis for a number of different hard coats. The name *hard coat* is normally given to an initial layer over a plastic substrate that acts as a transition between the organic plastic and an overlying essentially inorganic optical coating. Fluorine compounds give films that have very low friction and are hydrophobic and are frequently used as the outermost antismudge layer in an antireflection coating. The precise details of the precursors are difficult to obtain. They are considered part of the expertise of the process.

A technique that has some features in common with chemical vapor deposition is atomic layer deposition or ALD. In atomic layer deposition, the film molecules are assembled from their components on the actual surface to be coated. The substrates are placed in a reactor, and a precursor in the gaseous phase is entered. This precursor is chosen so that one of the components is chemisorbed on the surface of the substrate. Ideally, this process saturates when complete cover of the surface by a monolayer is achieved, although, in practice, there may not be a complete monolayer, or there may be some multilayer coverage. At saturation, the reactor is purged. A second cycle then involves a precursor for the second component of the molecules. This second component is now chemisorbed over the first, and the same process of saturation ideally ensures the completion of a single monolayer of the compound over the surface, although in practice, there may be some small departure from perfection. The reactor is purged again and the first component is reintroduced to produce the first part of a second molecular layer, and the process repeats itself many times to build up the required thickness of film. As might be expected, the process is slower than either conventional chemical vapor deposition or the physical vapor deposition processes we have considered. The various steps of the process must be activated, and that can be accomplished in a number of ways including simply heating the substrates, or supplying the energy in the form of ultraviolet radiation, or through a plasma that will be excited at RF to avoid charging effects over the dielectric materials involved. The basic principle is that the surface controls the growth, not the reaction. Thus, despite the low rate of deposition, the process has the great intrinsic advantage of depositing uniformly thick films over even quite complex substrates together with films of very high quality with easily controllable thicknesses because of the stability of the incremental thickness in each cycle.

Titanium dioxide and aluminum oxide have been used very successfully in optical coatings. For aluminum oxide, the precursors can be trimethylaluminum (TMA or $Al(CH_3)_3$) and water, alternately pulsed into the reactor. For titanium dioxide, tetrakis(ethylmethylamido)titanium (TEMAT or $Ti[N(CH_3)C_2H_5]_4$) as one of the precursors and water as another are possible.

The process has proven itself in many different applications outside optics, especially in the semiconductor industry and has succeeded in the manufacture of telecom-quality narrowband filters. The process is beginning to penetrate the optical thin-film industry, but the preferred lower refractive index material for optical purposes is still aluminum oxide. Although there are processes for silicon dioxide, they have not been as easy to implement nor as popular. Supplementing the aluminum oxide with silicon dioxide would make the acceptance of the process in optical coatings in general much more likely.

There are signs that this situation is changing. A very encouraging report [48] describes the deposition of an eight-layer broadband antireflection coating using HfO_2 and SiO_2 as the materials. The HfO_2 high-index material was derived from tetrakis[dimethylamido]hafnium and the SiO_2 from tris[dimethylamino]silane (3DMAS) and, in both cases, plasma activated oxygen. The study also included the investigation of single SiO_2 films with several different precursors, all with success, but with the best results from the point of view of film solidity being achieved with a commercial precursor AP-LTO®330, a product of Air Products and Chemicals, Inc., of Allentown, Pennsylvania. This was used with O_3 as the oxygen source and substrate temperature of 200°C.

The features of uniformity over complex substrates, high-quality essentially pinhole-free films, and accurate control over film thickness make the process very attractive. Figure 13.19 shows a typical machine.

There are many other techniques for the deposition of optical coatings. Probably the most important of these is the sol–gel process. The name *sol–gel* refers to those processes that involve a solution that undergoes a transition of the sol–gel type; that is, a solution is transformed into a gel. The common form of the sol–gel process starts with a metal alkoxide. This organometallic compound is hydrolyzed when it is mixed with water in an appropriate mutual solvent. The solution is usually made slightly acidic to control the rates of reaction and to help the formation of a polymeric material with linear molecules. The result is a gradual transition to an oxide polymer with liquid-filled pores. This gel can be deposited over the surface of an optical component by dipping. The coating is then heat treated to remove the liquid in the pores and to densify it;

FIGURE 13.19
TFS 500 system for atomic layer deposition capable of optical coatings. (Courtesy of Beneq Oy, Vantaa, Finland, http://www.beneq.com.)

the higher the temperature to which it is raised, the denser is the film. By treating the gel film at temperatures as high as 1000°C, complete densification is achieved. Lower temperatures give partial densification, but by 600°C, the film is already largely impermeable. Typical materials are TEOS (tetraethylorthosilicate, $Si(OC_2H_5)_4$) for eventual films consisting of silica and titanium tetraethoxide ($Ti(OC_2H_5)_4$) for films of titanium oxide. These materials are dissolved in ethanol and then hydrolyzed by adding a little distilled water. In the case of the titanium compound, the rate of hydrolyzation is much faster. Thus, nitric acid is added to control the transformation, and so the solution is made rather weaker.

There are quite considerable difficulties in producing multilayer coatings by the sol–gel process, and so the principal applications have been in high-durability antireflection coatings of a few layers. The process has not competed with vacuum deposition in the production of more complex multilayers.

Interest in the sol–gel process enormously increased when it was discovered that sol–gel-deposited antireflection coatings had an exceptionally high laser damage threshold [49]. The technique is therefore much used in producing antireflection coatings for components in the very large lasers for fusion experiments. These coatings are unbaked and quite porous; otherwise, the refractive index would not be suitable for single-layer antireflection coatings for low-index materials. In uncontrolled environments, such porous coatings take up moisture and other contaminants, and their index tends to vary over a period, and their performance, fall. Regular coatings must be baked at high temperature. However, the environment of the large lasers is tightly controlled, and the fragility of the unbaked coatings can be viewed as an advantage if they have to be removed to permit recoating of the component.

13.2 Baking

A final stage of the manufacturing process for optical coatings that is seldom discussed is that of baking. This is probably the one aspect of coating production that might still be referred to as an art rather than a science. Baking normally consists of heating the coated component in air at temperatures of usually between 100°C and 300°C for a period of perhaps several hours.

A common reaction in most coating departments to a batch of coatings that exhibit less than acceptable properties is to bake the coatings in air for a time to simply see if their properties improve. They frequently do. There is no doubt that such treatment can improve the properties of the coatings in several respects.

Coated substrates that are to be used as laser mirrors cemented to laser tubes are almost invariably baked before mounting because it is believed that this increases their stability. Such treatment certainly does reduce the drift that may occur at the early stages of laser operation, but the reason for this is obscure. It might be at least partially related to the desorption of moisture.

Frequently, baking causes the absorption in the layers to fall. This may be simply a case of improved oxidation. We know that baking of titanium suboxides in air improves their transmittance and reduces their absorptance [50]. High-quality films are frequently amorphous, and prolonged baking may induce an undesired slow amorphous-to-crystalline transition in such films. This process may compete with the oxidation process so that an optimum period of baking may result. This may be one reason why details of baking are frequently considered proprietary.

Much more work is required on the whole matter of baking and consequent filter stability before all becomes completely clear, but the oven is already an indispensable apparatus in virtually all coating shops.

We return to the matter of baking or annealing in more detail in Chapter 14.

13.3 A Word about Materials

Probably the most difficult aspect of thin-film coating and filter production is that of materials. These are not always satisfactory, and there are still problems associated with their stability, and we discuss aspects of materials in Chapter 14. Once the materials have been chosen, and their properties are known, the thin-film designer, using the methods discussed earlier, can usually produce a design to meet a given specification. Given suitable materials, an acceptable design, and a useful process, there are still further difficulties to be overcome in the construction of a practical filter. One other important topic is substrate preparation, and that is considered in Section 13.5. The two most important remaining factors are, first, controlling the uniformity of layer thickness over the area of the substrate and, second, controlling the overall thickness of each layer. The lack of uniformity causes a shift of characteristic wavelength over the surface of the filter, without necessarily affecting the performance in other ways, while thickness errors usually cause a reduction in performance. The magnitude of the errors that can be tolerated will vary from one design to another, and the estimation of this is dealt with briefly. Much of the remainder of this chapter is concerned with the general problem of minimizing these two sources of error.

13.4 Uniformity

In the evaporation process, it is usual to maintain the pressure within the chamber sufficiently low to ensure that the molecules in the stream of evaporant will travel in straight lines until they collide with a surface. In order to calculate the thickness distribution in a machine, the assumption is usually made that every molecule of evaporant sticks where it lands. This assumption is not always strictly correct, but it does allow uniformity calculations that are sufficiently accurate for most purposes. The distribution of thickness is then calculated in exactly the same way as the intensity of illumination in an optical calculation. All that is required to enable the thickness to be estimated is the knowledge of the distribution of evaporant from the source.

Holland and Steckelmacher [51] published an early and detailed account of techniques for the prediction of layer thickness and uniformity and established the theory that is essentially still used in uniformity predictions. Their expressions were later extended by Behrndt [52]. Holland and Steckelmacher divided sources into two broad types: those that have even distribution in all directions and can be likened to a point source and those that have a distribution similar to that from a flat surface, the intensity falling off as the cosine of the angle between the direction concerned and the normal to the surface. The expressions for the distribution of material emitted from the two types of source are as follows:

- For the point source:

$$dM = [m/(4\pi)]d\omega$$

- For the directed surface source:

$$dM = [m/\pi](\cos\varphi)d\omega,$$

where m is the total mass of material emitted from the source in all directions and dM is the amount passing through solid angle $d\omega$ (at angle φ to the normal to the surface in the case of the second type of source).

If the material is being deposited on the surface element dS of the substrate that has its normal at angle ϑ to the direction of the source from the element, then the amount that will condense on the surface will be given by the following:

- For the point source:

$$dM = \left(\frac{m}{4\pi}\right)\left(\frac{\cos\vartheta}{r^2}\right)dS$$

- For the directed surface source:

$$dM = \left(\frac{m}{\pi}\right)\left(\frac{\cos\varphi\cos\vartheta}{r^2}\right)dS$$

In order to estimate the thickness t of the deposit, we need to know the density of the film. If this is denoted by μ, then the thickness will be as follows:

- For the point source:

$$t = \left(\frac{m}{4\pi\mu}\right)\left(\frac{\cos\vartheta}{r^2}\right)$$

- For the directed surface source:

$$t = \left(\frac{m}{\pi\mu}\right)\left(\frac{\cos\varphi\cos\vartheta}{r^2}\right)$$

These are the basic equations used by Holland and Steckelmacher for estimating the thickness in uniformity calculations.

13.4.1 Flat Plate

The simplest case is that of a flat plate held directly above and parallel to the source. Here the angle φ is equal to the angle ϑ, and the thickness is as follows:

- For the point source:

$$t = \left(\frac{m}{4\pi\mu}\right)\left(\frac{\cos\vartheta}{r^2}\right) = \frac{mh}{4\pi\mu(h^2+\rho^2)^{3/2}}$$

- For the directed surface source:

$$t = \left(\frac{m}{\pi\mu}\right)\left(\frac{\cos^2\vartheta}{r^2}\right) = \frac{mh^2}{\pi\mu(h^2+\rho^2)^2}$$

with notation as in Figure 13.20. These expressions simplify, for the point source, to

$$\frac{t}{t_0} = \frac{1}{\left[1+(\rho/h)^2\right]^{3/2}}$$

and, for the directed surface source, to

$$\frac{t}{t_0} = \frac{1}{\left[1+(\rho/h)^2\right]^2}.$$

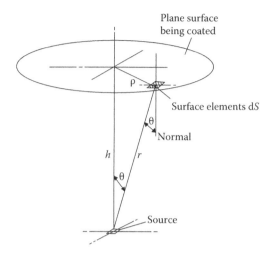

FIGURE 13.20
Diagram showing the geometry of the evaporation from a central source onto a parallel plane surface.

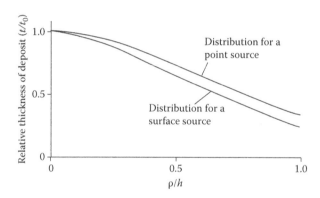

FIGURE 13.21
Film thickness distribution on a stationary substrate from a central source.

They are plotted in Figure 13.21. t_0 is the thickness immediately above the source, where $\rho = 0$. In neither case is the uniformity at all good. Clearly, the geometry is not suitable for any very accurate work unless the substrate is extremely small and in the center of the machine.

13.4.2 Spherical Surface

A slightly better arrangement that can sometimes be used is a spherical geometry where the substrates lie on the surface of a sphere. A point source will give uniform thickness of deposit on the inside surface of a sphere when the source is situated at the center. It can be shown that the directed surface source will similarly give uniform distribution when it is made part of the surface itself. In fact, it was the evenness of the coating within a sphere that led Knudsen [53] to first propose the cosine law for thin-film deposition. The method is often used in machines for simple antireflecting of components such as lenses where the uniformity need not be better than, say, 10% of the layer thickness at the center of the component. However, for precise work, this uniformity is still not adequate.

A higher degree of uniformity involves rotation of the substrate carrier, which we shall now consider.

13.4.3 Rotating Substrates

The situation in Figure 13.22 is as if, in Figure 13.20, the surface for coating were rotated about a normal now displaced at distance R from the source. As the surface rotates, the thickness deposited at any point will be equal to the average of the thickness that would be deposited on a stationary substrate around a ring centered on the axis of rotation, provided that the number of revolutions during the deposition is always sufficiently great to make the amount deposited in an incomplete revolution a sufficiently small proportion of the total thickness. By choosing the correct distance between source and axis of rotation, the uniformity can be made vastly superior to that for stationary substrates.

We shall first consider the directed surface source. Figure 13.22 shows the situation. The calculation is similar to that for the flat plate with a central source. Here we stop the plate and calculate the mean thickness around the circle containing the point in question and centered on the axis of rotation. The radius of the circle is ρ, and if we define any point P on the circle by the angle ψ, then the thickness at the point is given by

$$t = \frac{m}{\pi\mu} \cdot \frac{h^2}{(h^2 + \rho^2 + R^2 - 2\rho R \cos \psi)^2},$$

where r, the distance from the source to the point, is given by

$$r = (h^2 + \rho^2 + R^2 - 2\rho R \cos \psi)^{1/2}.$$

Then, taking the mean of the thickness around the circle, we have for the thickness of the deposit in the rotating case

$$t = \frac{m}{\pi\mu} \cdot \frac{1}{2\pi} \int_0^{2\pi} \frac{h^2 d\psi}{(h^2 + \rho^2 + R^2 - 2\rho R \cos \psi)^2}.$$

Now the integral $\int_0^{2\pi} d\psi/(1 - a \cos \psi)^2$ can be evaluated by contour integration to give

$$\int_0^{2\pi} \frac{d\psi}{(1 - a \cos \psi)^2} = \frac{2\pi}{(1 - a^2)^{3/2}},$$

so that the expression for thickness becomes

$$t = \frac{m}{\pi\mu} \cdot \frac{h^2}{(h^2 + \rho^2 + R^2)^2} \cdot \frac{1}{\left[1 - \{2\rho R/(h^2 + \rho^2 + R^2)\}^2\right]^{3/2}},$$

$$\frac{t}{t_0} = \frac{(1 + R^2/h^2)^2 (1 + \rho^2/h^2 + R^2/h^2)}{[1 + \rho^2/h^2 + R^2/h^2 - 2(\rho/h)(R/h)]^{3/2}[1 + \rho^2/h^2 + R^2/h^2 + 2(\rho/h)(R/h)]^{3/2}},$$

where t/t_0 is, as before, the ratio of the thickness at the radius in question to that at the center of the substrate holder.

Figure 13.23 shows this function plotted for several different dimensions typical of medium-sized coating machines. The distribution is vastly superior to that when the substrates are stationary. For one particular combination of dimensions, that corresponding to around $R = 7$, the distribution is reasonably even over the central part (radius: 4.0) of the machine. This is similar to

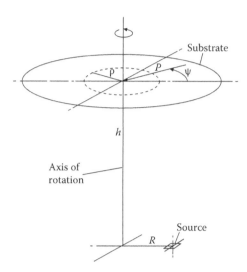

FIGURE 13.22
Diagram showing the geometry of the evaporation from a stationary offset source onto a rotating substrate.

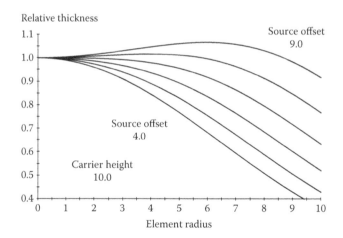

FIGURE 13.23
Theoretical film thickness distribution on substrates rotated about the center of the machine for values of source offset from 4.0 to 9.0 at intervals of 1.0. The height of the flat substrate holder above the source is 10.0. The sources are assumed to be small surfaces parallel to the substrates and with a cosine distribution.

the arrangement used in the production of narrowband filters where the uniformity must necessarily be very good. If the uniformity is not quite so important, where rather broader filters or perhaps antireflection coatings are concerned, then the sources can be moved outward, allowing a larger area to be coated at the expense of a slight reduction in uniformity.

A similar expression is found for a point source, but this time involving elliptic integrals. The thickness at the point P, assuming that the substrate does not rotate, is given by

$$t = \frac{m}{4\pi\mu} \cdot \frac{h^2}{\left(h^2 + \rho^2 + R^2 - 2\rho R \cos\psi\right)^{3/2}},$$

and in the presence of rotation, the thickness at any point around the ring of radius ρ will be the mean of the expression, i.e.,

$$t = \frac{m}{4\pi\mu} \cdot \frac{1}{2\pi} \cdot \int_0^{2\pi} \frac{h d\psi}{(h^2 + \rho^2 + R^2 - 2\rho R \cos\psi)^{3/2}}$$

$$= \frac{m}{4\pi^2\mu} \cdot \int_0^{\pi} \frac{h d\psi}{(h^2 + \rho^2 + R^2 - 2\rho R \cos\psi)^{3/2}} \cdot$$

Now let $(\pi - \psi)/2 = \gamma$, then $d\psi = -2d\gamma$, and the expression for thickness becomes

$$t = \frac{m}{4\pi^2\mu} \cdot \int_{\pi/2}^0 \frac{-h d\gamma}{\left[h^2 + (R + \rho)^2 - 4\rho R \sin^2\gamma\right]^{3/2}},$$

which can be written as

$$t = \frac{m}{4\pi^2\mu} \cdot \frac{h}{\left[h^2 + (R + \rho)^2\right]^{3/2}} \cdot \int_{\pi/2}^0 \frac{d\gamma}{\left\{1 - \left[4\rho R / \{h^2 + (R + \rho)^2\}\right]\sin^2\gamma\right\}^{3/2}} \cdot$$

Now the integral in this expression is a standard form:

$$\frac{1}{(1 - k^2)} E(k, \alpha) = \int_0^{\alpha} \frac{d\gamma}{\left(1 - k^2\sin^2\gamma\right)^{3/2}},$$

where $E(k, \alpha)$ is an elliptic integral of the second kind and is a tabulated function [54]. The expression for thickness then becomes

$$t = \frac{hm}{4\pi^2\mu} \cdot \frac{E(k, \pi/2)}{\left[h^2 + (R + \rho)^2\right]^{1/2}\left[h^2 + (R - \rho)^2\right]},$$

where

$$k = 4\rho R / \left[h^2 + (R + \rho)^2\right].$$

Curves of this expression are given by Holland and Steckelmacher [51], and the shape is actually very similar to that for the directed surface source.

Almost all the sources used in the production of thin-film filters, especially the boat type, give distributions similar to the directed surface source. Holland and Steckelmacher also describe some experiments that they carried out to determine this point. Keay and Lissberger [55] have studied the distribution from a howitzer source loaded with zinc sulfide, and it appears that this is somewhere in between the point source and the directed surface source, probably due to scattering in the evaporant stream immediately above the heater where the pressure is high. The cloud of vapor that forms seems to act to some extent as a secondary point source. This behavior of the howitzer probably depends to a considerable extent on the material being evaporated. Graper [56] studied the distribution of evaporant from an electron beam source and has found that this is somewhat more directional than the directed surface source. Its distribution can be described by a $\cos^x \vartheta$ law, where x is somewhere between 1 and 3 and depends on the power input and on the amount of material in the hearth. Using zinc sulfide and cryolite, Richmond [57] found that the distribution from an electron beam source was best represented by a law of the form $\cos^{2.1} \vartheta$.

Normally, in calculating the distribution to be expected from a particular geometry, we assume that we are using directed surface sources, and then, when setting up a machine for the first time,

FIGURE 13.24
Photograph showing the interior of a machine with a domed calotte. (Courtesy of Balzers AG, Balzers, Liechtenstein.)

the sources are placed at the theoretically best positions. The first few runs soon show whether any further adjustments are necessary, and if they are, they are usually very slight and can be made by trial and error. Once the best positions are found, it is important to ensure that the sources are always accurately set to reproduce them. Care should be taken to make sure that the angular alignment is correct. A source at the correct geometrical position but tilted away from the correct direction will give uniformity errors just as much as if it were laterally displaced.

Where uniformity must be good over as large an area as possible containing many rather smaller substrates, it is possible to use a combination of a spherical surface and rotating plate. A domed work holder, or calotte, is rotated about its center with the sources offset beneath it so that they are approximately on the surface of the sphere, with slight adjustments made during setting up. This gives very good results over a much larger area than would be possible with the simple rotating flat plate. Figure 13.24 shows the interior of a machine that uses this arrangement.

When still improved uniformity is required, it is possible to achieve it by what is known as a planetary geometry. In this arrangement, the substrates are held in a number of small carriers that rotate not only about the center of the machine, but also about their own individual centers at much greater speed, so that they execute many revolutions for each single revolution of the carrier as a whole. This carries a stage further the averaging process that occurs with the simple rotating carrier. A planetary system can be seen in the open machine in Figure 13.1a.

13.4.4 Use of Masks

It is possible to make corrections to distribution by careful use of masks. In their simplest form, they are stationary and are placed just in front of the substrates that rotate on a single carrier about a single axis. The masks are cut so that they modify the radial distribution of thickness. Theoretical calculations give dimensions for masks of approximately the correct shape, which can then be trimmed according to experimental results to arrive at the final form. For a number of reasons, it is normal to leave the central monitor glass uncorrected. It is difficult to correct the central part of the chamber where the mask width tends to zero, and, in any case, the monitor is usually stationary. Furthermore, in some monitoring arrangements, there is an advantage in having more material on the monitor than on the batch.

An additional degree of freedom was introduced by Ramsay et al. [58] in the form of a rotating mask. For a large flat substrate approaching the dimensions of the machine, there is little other than simple rotation that can be done, in terms of the carrier, to improve uniformity. Planetary arrangements require much more room. Stationary masks are of some help, but they are somewhat sensitive to the characteristic of the sources and are not therefore sufficiently stable for a very high degree of uniformity. A much more stable arrangement, which has been shown capable of uniformities of the order of 0.1% over areas of around 200 mm diameter, involves rotating the mask about a vertical axis at a rotational speed considerably in excess of that of the substrate carrier. This effectively corrects the angular distribution of the source that can then be positioned at the center of the machine. The mask rotation axis is usually placed very near the source and positioned so that the line drawn from the source through the mask center intersects the perimeter of the substrate carrier. In practice, the axis of rotation and the rotating shutter are close to the source position, and slight adjustment of the axis can be made for trimming purposes. It has been found to be an exceptionally stable arrangement.

13.5 Substrate Preparation

Before a substrate can be coated, it must be cleaned. The forces that hold films together and to the substrate are all short-range interatomic and intermolecular forces. These forces are extremely powerful, but their short range means that we can think of each atomic layer as being bound to the neighboring layers only and being little affected by material further removed from it. Thus, the adhesion of a thin film to the substrate critically depends on conditions at the substrate surface. Even a monomolecular layer of a contaminant on the surface can change the force of adhesion by orders of magnitude. The condensation of evaporant, too, is just as sensitive to surface conditions that can completely alter the characteristics of the subsequent layers. Substrate cleaning so that the condensing material attaches itself to the substrate and not an intervening layer of contaminant is therefore of paramount importance.

The typical symptoms of an inadequately cleaned substrate are a mottled, oily appearance of the coating, usually coupled with poor adhesion and optical performance. This can also be caused by such defects in the machine as backstreaming of oil from the pumps. When these symptoms appear, it is usually advisable to extend any subsequent improvements in cleaning techniques to the machine as well.

A good account of various cleaning methods is given by Holland [59]. A more recent account is that of Mattox [60]. The best cleaning process will depend very much on the nature of the contamination that must be removed, and although it may seem self-evident, in all cleaning operations, it is essential to avoid contaminating the surface rather than cleaning it. For laboratory work, when the substrates are reasonably clean to start with (microscope slide glass is usually in this condition), then for most purposes, it will be found sufficient to thoroughly wash the substrates in detergent and warm water (not household detergent that sometimes has additives that cause smears to appear on the finished films), to thoroughly rinse them in running warm water (in areas where tap water is fairly pure, hot tap water will often be found adequate), and then to thoroughly and immediately dry them with a clean towel or soft paper tissue or, better still, to blow them dry with a jet of clean dry nitrogen. The substrates should never be allowed to dry themselves or stains will certainly occur, which are usually impossible to remove. Substrates should be handled as little as possible after cleaning and, since they never remain clean for long, immediately placed in the coating machine, and the coating operation, started. Wax or grease will probably require treatment with an alcohol such as isopropyl, perhaps rubbing the surface with a clean fresh cotton swab soaked in the alcohol and then flooding the surface with the liquid. Care must be taken to ensure that the alcohol is clean. A bottle of alcohol available to all in a laboratory seldom remains clean for long, and a better arrangement is to keep it under lock and key and to allow the alcohol into the laboratory in wash bottles that emit the alcohol when squeezed.

This basic cleaning procedure can be modified and supplemented in various ways, especially if large numbers of substrates are to be automatically handled. Ultrasonic scrubbing in detergent solution or in alcohol is a very useful technique, although prolonged ultrasonic exposure is to be avoided since it can eventually cause surface damage. It is important that the substrates be kept wet right through the cleaning procedure until they are dried as the final stage. Vapor cleaning is frequently used for final drying. The substrates are exposed to the vapor of alcohol or other degreasing agents, so that initially, it condenses and runs off, taking any residual contamination or the remains of the agent from the previous cleaning stage with it. The substrates gradually reach the temperature of the vapor. Then no further condensation takes place, and the substrates can be withdrawn perfectly dry. Since the agent is condensing from the vapor phase, it is in an extremely pure form. An alternative end to the cleaning process is a rinse in deionized water followed by drying in a blast of dry, filtered nitrogen.

It is very difficult to see marks on the surface of the substrate with the naked eye. Dust can be picked up by oblique illumination, but wax and grease cannot. An old test for assessing the quality of a cleaning process is to breathe on one of the substrates so that moisture condenses on it in a thin layer. This tends to magnify the effects of any residue. The moisture acts in almost exactly the same way as a condensing film since the condensation pattern depends on the surface conditions. A surface examined in this way is said to exhibit a good or bad *breath figure*. A contaminated surface gives a smeared pattern, while a clean surface is completely even. Since even this step can introduce slight residual contamination, it is better used only on a sample as an indication of the condition of the batch.

Once the substrates are in the chamber, and they should always be loaded as soon as possible after cleaning, they can be given a final clean by a glow discharge. The equipment for this, which consists of a high-voltage supply, preferably DC, together with the necessary lead-in electrodes, is fitted as standard in most machines. At a suitable pressure, which will vary with the particular geometry of the electrodes but which will usually be around 6 Pa (0.06 mbar), a glow discharge is struck, and provided the geometry is correct, the surface of the substrates is bombarded with positive ions. This effectively removes any light residual contamination, although gross contamination will persist. It is not certain whether the cleaning action actually arises from a form of sputtering or whether the glow discharge is merely a convenient way of raising the temperature of the surfaces so that contaminants are baked off. Generally, the glow discharge is limited in duration to 5 or perhaps 10 minutes. It has been suggested that although glow discharge cleaning does remove grease, it does encourage dust particles; for coatings where minimum dust is required, such as high-performance laser mirrors, glow discharge cleaning is frequently omitted. Lee [61] found that the omission of glow discharge cleaning led to a very great increase in the incidence of moisture penetration patches in his films and consequently to a fall in the performance of his filters.

The evaporation of the first layer should begin as soon as possible after the glow discharge has stopped. Cox and Hass [4] used a discharge current of 80 mA and a voltage of 5000 V for 5 minutes to clean substrates before coating them with zinc sulfide and found that the time between finishing the discharge and starting the evaporation should be not greater than 3 minutes. If the time was allowed to exceed 5 minutes, then the quality of the films, especially their adhesion, deteriorated.

If, as sometimes happens, a filter is left for a period, say, overnight, in an uncompleted state, it will often be found advisable to carry out a short period of glow discharge cleaning before starting to evaporate the remaining layers.

In the energetic processes such as ion-assisted deposition, the substrate can be readily bombarded by the ion beam before coating. This is a very effective cleaning method. Care should be taken, however, not to prolong this bombardment. The substrate surface is usually sputtered along with the contaminants. Since this vigorous cleaning process can be carried out at deposition temperature and pressure, the deposition can immediately start when the surface is considered clean, even without any pause in the bombardment.

13.6 Thickness Monitoring and Control

Given suitable materials, clean substrates, and a machine with substrate holder geometry to give the required distribution accuracy, the main remaining problem is that of controlling the deposition of the layers so that they have the characteristics required by the coating or filter design. Of course, many properties are required, but refractive index and optical thickness are the most important. There is no satisfactory way, at present, of measuring the refractive index of that portion of a film actually being deposited. Such measurements can be made later, but for closed loop control, dynamic measurements are required. Normal practice is therefore simply to control, as far as possible, those deposition parameters that would affect refractive index so that the index produced for any given material, or mixture of materials, is consistent. This procedure, while it usually gives satisfactory results, is not ideal and is used simply because, at the present time, there is no better way. Fortunately, the energetic processes, especially sputtering, do exhibit very good stability in terms of refractive index and other properties.

Film thickness can more readily be measured and, therefore, controlled. The simplest systems display a signal to a machine operator who is responsible for interpreting it and assessing the correct instant to terminate deposition. At the other end of the scale, there are completely automatic systems in which operator judgment plays no part and in which even operator intervention is rarely required. The term *monitoring* strictly means keeping track of a parameter, but in the thin-film community, it is understood as including both measurement and control.

There are many ways in which the thickness can be measured. All that is necessary is to find a parameter that varies in a suitable fashion with thickness and to devise a way of monitoring this parameter during deposition. Thus, parameters such as mass, electrical resistance, optical density, reflectance, and transmittance have all been used. Of all the methods, those most frequently employed involve optical measurements of reflectance or transmittance, or the measurement of total deposited mass by the quartz crystal microbalance or, in some very stable processes, time of deposition.

The question of the best method for the monitoring of thin films is, of course, inseparable from that of how accurately the layers must be controlled. This second question is a surprisingly difficult one to answer. Indeed, it is impossible to separate the two questions: the tolerances that can be allowed and the method used for monitoring are closely related, and one cannot be considered in depth independently of the other. For convenience, however, we will consider some of the more common arrangements for monitoring, including only the most rudimentary ideas of accuracy, and then, at a later stage, consider the question of tolerances along with some of the more advanced ideas of monitoring and its various classifications.

13.6.1 Optical Monitoring Techniques

Optical monitoring systems consist of a light source of some description illuminating a test substrate that may or may not be one of the filters in the batch and a detector analyzing the reflected or transmitted light. From the results of that analysis, the deposition of the layer is stopped as far as possible at the correct point. Usually, so that the layer may be stopped as sharply as possible, the machine is fitted with a shutter that can be inserted in front of the operating source. This is a much more satisfactory method than merely turning off the supply to the sources that always take a finite time to stop emitting. Such a shutter can be seen in Figure 13.3.

Almost all the early researchers in the field used the eye as the detector, and the thicknesses of the films were determined by assessing their color appearance in white light. In many cases, they were concerned with simple single-layer coatings such as single-layer antireflection coatings, which are not at all susceptible to errors. When the layer is of the correct thickness for visible light, the color reflected from the surface in white light has a magenta tint, owing to the reduction of the reflectance in the green. The visual method is quite adequate for this purpose and is still being widely used. A very clear account of the method is given by Mary Banning [62], who compiled Table 13.1.

In the production of other types of filters where the errors of the visual method would be too large, other methods must be used. The simplest appears to be the use of a receiver in the measurement of the variation of transmittance or reflectance at a single wavelength, often known as single-point monitoring. It seems that many researchers adopted such a method virtually simultaneously. An early paper by Polster [63] describes a photoelectric method that is basically the same as that used most often today. Holland [59] credits Dufour with an early (1948) version of such a system including a chip changer. In 1950, Billings [64] mentions a multilayer technique that

TABLE 13.1

Banning's Colors for Thickness Monitoring

Color Change for		Optical Thickness for Green Light
ZnS	Na_3AlF_6	
Bluish white	Yellow	
↓	↓	
White	Magenta	$\lambda/4$
↓	↓	
Yellow	Blue	
↓	↓	
Magenta	White	$\lambda/2$
↓	↓	
Blue	Yellow	
↓	↓	
Greenish white	Magenta	$3\lambda/4$
↓	↓	
Yellow	Blue	
↓	↓	
Magenta	Greenish white	λ
↓	↓	
Blue	Yellow	
↓	↓	
Green	Magenta	$5\lambda/4$

Source: M. Banning, *Journal of the Optical Society of America*, 37, 792–797, 1947. With permission of Optical Society of America.

had already been operated for some years by Fogelsanger at Evaporated Metal Films. We saw in Chapter 2 that if the film is without absorption, then its reflectance and transmittance measured at any one wavelength will vary with thickness in a cyclic manner, similar to a sine wave, although, for the higher indices, the waves will be more flattened at their tops. The turning values correspond to those wavelengths for which the optical thickness of the film is an integral number of quarter wavelengths, the reflectance being equal to that of the substrate when the number is even and a maximum amount removed from the reflectance of the substrate when the number is odd. Figure 13.25 illustrates the behavior of films of different values of refractive index. This affords the means for measurement. If the detector in the system is made highly selective, for example, by putting a narrow filter in front of it, then the measured reflectance or transmittance will vary in this cyclic way, and the film may be monitored to an integral number of quarter waves by counting the number of turning points passed through in the course of the deposition. A typical arrangement to perform this operation is shown in Figure 13.26. The filter may be an interference filter or, more flexible, an adjustable prism or grating monochromator.

Consider the deposition of a high-reflectance multilayer stack where all the layers are quarter waves. Let the monitoring wavelength be the wavelength for which all the layers are one quarter-wavelength thick. The reflectance of the test piece will vary as shown in Figure 13.27 [65]. The example shown is typical of a reflecting stack for the visible region. The reflectance can be seen to increase during the deposition of the first layer, of high index, to a maximum where the deposition is terminated. During the second layer, the reflectance falls to a minimum where the second layer is terminated. The third layer increases the reflectance once again, and the fourth layer reduces it. This behavior is superimposed on a trend toward a reflectance of unity so that the variable part of the signal becomes a gradually smaller part of the total. This puts a limit on the number of layers that can be monitored in reflectance in this way to around four, when a fresh monitoring substrate should be inserted. In transmission monitoring, this effect does not exist, and although the signal level falls, the variable part of the signal remains a sufficiently large part of the whole. The only problem is the overall trend of the signal toward zero, so that eventually, it will become too small in comparison with the noise in the system. With reasonable optics and a photomultiplier detector, the number of layers that may be dealt with in this way is around 21. After this stage, the noise often becomes too great.

Frequently, automatic methods of detection of the layer end point are used. Automatic methods, however, are not universally employed, and operator control of the machine is still an important

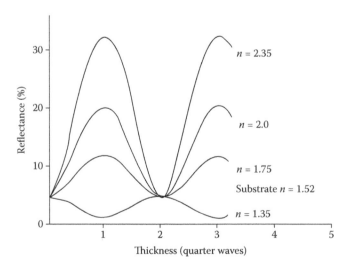

FIGURE 13.25
Curves showing the variation with thickness of the reflectance of several films with different refractive indices.

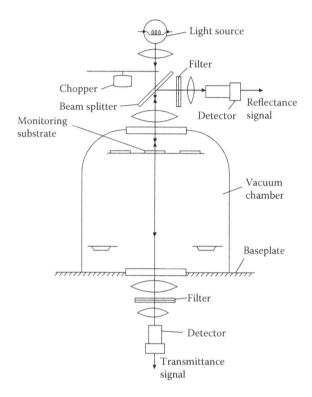

FIGURE 13.26
Possible arrangement of a monitoring system for reflectance and transmittance measurements.

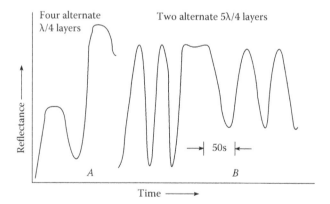

FIGURE 13.27
Record taken from a pen recorder of the reflectance of a monitor glass during film deposition. (After D. L. Perry, *Applied Optics*, 4, 987–991, 1965. With permission of Optical Society of America.)

technique. For the greatest accuracy in control by operator, the output of the detector should be presented on a display unit or perhaps a chart recorder, making it easier to determine the turning values. With such an arrangement, a trained operator can readily terminate the layers to an accuracy on the monitoring substrate of rather better than 5%, depending on the index of the film. With care and attention, 2% is a good estimate. Of course, as we shall see, this does not necessarily mean that the actual thickness of the filters in the batch will be as accurate. Other sources of error operate to introduce differences between the monitor and the batch.

To improve the signal-to-noise ratio it is usual to chop the light before it enters the machine, partly because the evaporation process produces a great deal of light during the heating of the boats, but mainly because, at the signal levels encountered, the electronic noise without some filtering would be impossibly great. The chopper is best placed immediately after the source of light but before the machine, and the filter, after the machine. This arrangement reduces the stray light to a greater extent than would placing either the filter before the machine or the chopper after it. It is, of course, always advisable to limit as far as possible the total light incident on the detector, partly because unchopped radiation can push the detector into a nonlinear region and partly because it can damage the device especially if it is a photomultiplier. If a filter rather than a monochromator is used, then great care should be taken to ensure that the sidebands are particularly well suppressed. Photomultipliers and other detectors have characteristics that can considerably vary with wavelength, and if the monitoring wavelength lies in a rather insensitive region compared with the peak sensitivity, then small leaks in the more sensitive region, which might not be very noticeable in the characteristic curve of the filter, can cause considerable difficulties from stray light, even giving spurious signals of similar or greater magnitude than the true signal. Prism or grating monochromators are often safer for this work, besides being considerably more flexible.

The technique in which the layer termination is at an extremum of the signal is sometimes called turning-value monitoring or turning-point monitoring. We can investigate the errors likely to arise in this type of monitoring as follows. Suppose that in the monitoring of a single quarter-wave layer there is an error γ in the value of reflectance at the termination point.

$$\gamma = \Delta R/R.$$

This will give rise to a corresponding error φ in the phase thickness of the layer δ, where

$$\delta = (\pi/2) - \varphi.$$

The result is symmetrical around the turning value, and so we can assume an undershoot rather than overshoot to give a marginally simpler analysis. Because of the nature of the characteristic reflectance curve of the single layer, the error in phase thickness will be rather greater in proportion than the original error in reflectance. The surface admittance of the layer will be given by the characteristic matrix:

$$\begin{bmatrix} \cos\delta & (i\sin\delta)/y \\ iy\sin\delta & \cos\delta \end{bmatrix} \begin{bmatrix} 1 \\ y_{sub} \end{bmatrix},$$

where

$$\cos\delta = \sin\varphi \quad \text{and} \quad \sin\delta = \cos\varphi.$$

This gives

$$Y = \frac{\sin\varphi + i(y_{sub}\cos\varphi)/y}{y_{sub}\sin\varphi + iy\cos\varphi},$$

where the symbols have their usual meaning. Introducing approximations for $\sin\varphi$ and $\cos\varphi$ up to and including powers of the second order, we have

$$Y = \frac{\varphi + i(y_{sub}/y)(1-\varphi^2/2)}{y_{sub}\varphi + iy(1 - \varphi^2/2)},$$

and the reflectance of the monitor in vacuo will be given by

$$R = \left| \frac{(y_{sub} - 1)\varphi + i(y - y_{sub}/y)(1 - \varphi^2/2)}{(y_{sub} + 1)\varphi + i(y + y_{sub}/y)(1 - \varphi^2/2)} \right|^2,$$

which simplifies to

$$R = \frac{(y - y_{sub}/y)^2}{(y + y_{sub}/y)^2} \left\{ 1 + \frac{4y_{sub}\left(1 - y^2 + y_{sub}^2 - y_{sub}^2/y^2\right)}{\left(y^2 - y_{sub}^2/y^2\right)^2} \varphi^2 \right\}. \tag{13.1}$$

The values of γ and φ are related as follows:

$$\gamma = \frac{4y_{sub}\left(1 - y^2 + y_{sub}^2 - y_{sub}^2/y^2\right)}{\left(y^2 - y_{sub}^2/y^2\right)^2} \varphi^2 = \sigma\varphi^2, \tag{13.2}$$

since the first factor in Equation 13.1 is just the reflectance when γ and φ are both zero.

Now, in most cases, it will be very difficult to determine the reflectance at the turning value to better than 1% of the true value. In many cases, especially where there is noise, it will not be possible even to do as well as this. However, assuming this value for γ, the expression for the error in the layer thickness becomes

$$\pm 0.01 = \sigma\varphi^2,$$

where the sign $\pm$ is taken to agree with $\sigma\varphi^2$ and depends on whether or not the turning value is a maximum or a minimum. If the error is expressed in terms of a quarter-wave thickness, equivalent to $\pi/2$ radians, the expression becomes

$$\text{Error} = \frac{\varphi}{\pi/2} = \frac{0.1}{(\pi/2)\sigma^{1/2}}. \tag{13.3}$$

A typical case is the monitoring of a quarter wave of zinc sulfide on a glass substrate where $y = 2.35$ and $y_m = 1.52$. Substituting these values in Equation 13.2 and using it in Equation 13.3, the fractional error in the quarter wave becomes 0.08. This is a colossal error compared with the original error in reflectance and illustrates the basic lack of accuracy inherent in this method.

In the infrared, it is often possible to use wavelengths for monitoring that are shorter than the wavelengths of the desired filter peaks by a factor of perhaps 2 or even 4. This improves the basic accuracy by the same factor. For layers similar to that considered earlier, the errors would then be 0.04 or 0.02. These errors are on the limit of permissible errors, and it is clear that this simple system of monitoring is not really adequate for any but the simplest of designs.

What makes the method particularly difficult to apply is that it is only the portion of the signal before the turning point that is available to the operator, who therefore has to anticipate the turning value, and the fact that trained machine operators can achieve the theoretical figures for accuracy says much for their skill.

An alternative method, inherently more accurate, involves the termination of the layer at a point remote from a turning value where the signal changes much more rapidly. This consists of the prediction of the reflectance of the monitoring substrate when the layer is of the correct thickness and then the termination of the deposition at that point. One disadvantage is that the reflectance of the monitor, or its transmittance, is not an easy quantity to measure absolutely, because of calibration drifts during the process, due partly to such causes as the gradual coating of the machine windows—almost impossible to avoid. Another is that whereas with turning value monitoring, it is often possible to use just one single monitor, on which all the layers can be deposited, so that it becomes an exact replica of the other filters in the batch; in this alternative method, the prediction of the reflectances used as termination values is very difficult if only one monitor is used, because small errors in early layers affect the shape of the curve for later layers. However, we shall see later in this section that this single-point method is still used effectively—although with some innovations.

Some of these difficulties may be avoided by using a separate monitor for each and every layer. To avoid the errors due to any shift in calibration that may occur in changing from one monitor to the next or in the coating of the machine windows, it is wise if it is at all possible to choose the parameters of the system so that the layer is thicker than a quarter wave at the monitoring wavelength. This ensures that the termination point of the layer is beyond at least the first turning value, which can therefore be used as a calibration check. It will also be found necessary to set up the reflectance scale for each fresh monitoring substrate, and the initial uncoated reflectance that will be accurately known can be used for this. Because a large number of monitor glasses is required, special monitor changers have been designed and are commercially available. These will accommodate stacks of 40 or more glasses. The low-index material may have rather poor contrast on the monitor substrates, and a frequent variant of this method is the deposition of two layers, high index followed by low index, on each monitor substrate. The word *chip* has become the almost universal term for a monitor glass and chip changer for the monitor glass changer.

The principal objection that most researchers almost instinctively feel toward this system is that the monitor is no longer an exact replica of the batch of filters. This is to some extent a valid objection. The layer being deposited on an otherwise uncoated substrate is condensing on top of what may be quite a different structure from the partially finished filters of the batch. Behrndt and Doughty [66] noticed a definite measurable difference between layers deposited on top of an already existing structure and those deposited on fresh substrates. They compared the deposition of zinc sulfide shown by a crystal monitor (this special type of monitor will be discussed shortly), which already had a number of layers on it, with the layer going down on a fresh glass substrate, and they found that the layer began to immediately grow on the crystal when the source was uncovered, but the optical monitor took some time to register any deposition. The difference could amount to several tens of nanometers before the rates became equal. This, they decided, was due to the finite time for nuclei to form on the fresh glass surface and the rather small probability of sticking of the zinc sulfide until the nuclei were well and truly formed. Once the film started to grow, all the molecules reaching the surface would stick. On the crystal where a film already existed, not necessarily of zinc sulfide, nucleation sites were already there and the film started to grow immediately. The sticking coefficient of a material on a fresh monitor surface falls with rising vapor pressure, and zinc sulfide has a particularly large vapor pressure. Similar trouble was not experienced with thorium fluoride that has a much lower vapor pressure. Behrndt and Doughty found that the problem could be solved by providing nucleation sites on the clean monitor slides by precoating them with thorium fluoride that has a refractive index very close to that of glass. Some 20 nm or so of thorium fluoride was found to be sufficient and did not affect the monitoring of zinc sulfide deposited on top. (Since thorium fluoride is radioactive and somewhat out of favor, a different low-index fluoride would be advisable.) This effect becomes greater, the greater the surface temperature of the monitor. By changing the type of evaporation source to an electron beam unit, which produced less radiant heat for the same evaporation rate, it was found possible to operate at monitor temperatures low enough to cause the effect to disappear.

The same authors also remarked on an effect well known in thin-film optics. Thick substrates tend to have layers condensing on them that are thicker than those on thin substrates in the same or similar positions in the machine. In the case cited by the authors, the thin substrates were around 0.040 in. thick, while the thick ones were around half an inch thick. The difference in coating thickness was sufficient to shift the reflectance turning values by some 40–50 nm at 632.8 nm. This was qualitatively shown to be due to the difference in temperature between the two substrates. The thicker substrates took longer to heat up than the thin ones. The heating in this particular case was almost entirely due to radiation from the sources, and again when electron beam sources were introduced, the effect was considerably reduced.

The accuracy of the monitoring process can be greatly improved if a system devised by Giacomo and Jacquinot [67], and usually known as the *maximètre*, is employed. This involves the measurement of the derivative of the reflectance versus wavelength curve of the monitor. At points where the reflectance is a turning value, the derivative of the reflectance with respect to wavelength is zero and is rapidly changing from a positive to a negative value in the case of a maximum and vice versa in the case of a minimum. The original apparatus consisted of a monochromator with a small vibrating mirror before the slits on the exit side so that a small spectral interval was sinusoidally scanned. The output signal from the detector consisted of a steady DC component, representing the mean reflectance, or transmittance, over the interval, a component of the same frequency as the scanning mirror representing the first derivative of the reflectance against wavelength; a component of twice the scanning frequency, representing the second derivative of the reflectance; and so on. A slight complication is the variation in the sensitivity of the system with wavelength that appears as a change in the reflectance signal and, hence, the derivative, unless it is compensated for. In their arrangement, Giacomo and Jacquinot produced an intermediate image of the spectrum within the monochromator, and a razor blade positioned along it made a linear correction to the intensity over a sufficiently wide region and was found to be accurate enough. A more usual technique today would be to make a correction electronically. The accuracy claimed for this system is a few tenths of a nanometer, typically 0.2–0.3 nm, and this is certainly achieved. A problem is that the layers may be insufficiently stable themselves to retain optical thicknesses to this accuracy, especially when exposed to the atmosphere.

A method, similar in some respects, but with some definite advantages in interpretation, was devised by Ring [68] and Lissberger and Ring [69]. It consists of measuring the reflectance or transmittance at two wavelengths and finding the difference. In the original system, a monochromator was used, containing a chopping system that switched the output of the monochromator from one wavelength to another and back again. The alternating current signal from the detector was a measure of the difference. Since the two wavelengths could be placed virtually anywhere within the region of sensitivity of the detector, the method had greater flexibility than the Giacomo and Jacquinot system. Greatest contrast in the two reflectance signals as a layer was being deposited could be obtained by placing the two wavelengths at the points of greatest opposite slope in the characteristic of the thin-film structure at the appropriate stage. When the signals at the two wavelengths were equal, the output of the system passed through a null and, if displayed on a chart recorder, made detection of the terminal point of a particular layer, usually indicated by the null, particularly easy to detect.

More recently, the ideas inherent in these systems have been extended to broad spectral regions. Although the principles of these more modern methods are not new, it is the advances in detectors and in electronics and data analysis that have made them practical. Many of the systems have been developed in industry and have not been frequently published. In the cases of those that have been written up, detailed descriptions of the precise way in which they are used have often been lacking. Usually the technique involves a comparison between the spectral characteristic that is actually obtained at any instant and that required at the instance of termination of the particular layer. In the earlier systems, this was visually carried out by displaying both curves on a monitor. This works well when there is a close match between predicted and measured performance, but frequently, errors in earlier layers, and changes in the characteristics of layers from what is expected, cause the actual curves to differ to a greater or lesser extent from the predictions. In these circumstances, there can be great difficulty in visually assessing the correct moment to terminate a layer. The most recent systems are therefore usually linked to a computer that calculates a figure of merit that can either be displayed to a machine operator or, better still, used in the completely automatic termination of layers.

Details of scanning monochromator systems have been published by a number of authors. An early description of such a system is that of Hiraga et al. [70], where the scanning was carried out by a rotating helical slit assembly. Borgogno et al. [71] and Flory et al. [72] in Marseille developed two such systems. The first uses a stepping motor to rotate a grating and scan the system over a wide wavelength region; the second uses a holographic grating with a flat spectrum plane in which is situated a silicon photodiode array detector that can be electronically scanned. Sullivan et al. [73–75] had great success in implementing a completely automatic system of monitoring including error compensation. A good description is also given by Starke et al. [76]. A study of broad-band optical monitoring from the deep ultraviolet to the near infrared is that of Ristau et al. [77].

From time to time, ellipsometric monitoring has been examined. This has the advantage that an assessment of how stable is the refractive index of the deposited layer can readily be made. The principal disadvantage is the increased difficulty of implementing such a monitoring system that requires a high angle of incidence on the measuring chip. Ellipsometric monitoring was used to advantage by Dligatch and Netterfield [78].

It should not be thought that such advances in monitoring have superseded the direct single-point system where all the layer are monitored on one single substrate that forms one of the batch under construction. Figure 13.17a shows a sketch of a SyrusPro machine (Leybold Optics GmbH—now Bühler Leybold Optics) where all the layers are monitored using one single wavelength, which is changed from layer to layer, with termination points that are, in general, on the side of the oscillatory monitoring curve, rather than at the turning values. The single, chosen, monitoring substrate is in the midst of the batch and passes through the monitoring beam once per substrate carrier rotation. This sampling of the monitoring signal is, however, sufficiently accurate to allow a high degree of control. There is a large advantage in that the chosen substrate is much more representative of the coating batch than a centrally placed separate monitoring substrate [79]. Especially for such systems, the accompanying computation can contain an element of prediction so that the instant of termination can be estimated, avoiding the errors that would occur if termination could take place only at passage of the monitor through the measuring beam [76]. It is also thus possible to compensate for the finite time it can take to terminate a layer deposition.

13.6.2 Quartz Crystal Monitor

The normal modes of mechanical vibration of a quartz crystal have very high Q and can be transformed into electric signals by the piezoelectric properties of the quartz and vice versa. The crystal therefore acts as a very efficient tuned circuit that can be coupled into an electrical oscillator by adding appropriate electrodes. Any disturbance of its mechanical properties will cause a change in its resonant frequency. Such a disturbance might be an alteration of the temperature of the crystal or its mass. The principle of monitoring by the quartz crystal microbalance (as it is called) is to expose the crystal to the evaporant stream and to measure the change in frequency as the film deposits on its face and changes the total mass. In some arrangements, the resonant frequency of the crystal is compared with that of a standard outside the machine, and the difference in frequency is measured; in others, the number of vibrations in a given time interval is digitally measured. Usually, the frequency shift will be internally converted into a measure of film thickness using film constants fed in by the operator. Since the signal from the quartz crystal monitor constantly changes in the same direction, it can be used more easily in automatic systems than optical signals.

The mechanical vibrational modes of a slice of quartz crystal are very complicated. It has been found possible to limit the possible modes and the coupling between them by cutting the slice with respect to the axes of the crystal in a particular way, by proportioning the dimensions of the slice correctly and by supporting the crystal in its holder in the correct way. Quartz crystal vibrational modes also vary with temperature, some having positive temperature coefficient and some negative, and it has been found possible to cut the slice in such a way that modes which have opposite temperature dependence are intentionally coupled so that the combined effect is a resonant frequency independent of temperature over a limited temperature range. The usual cut of crystal used in thin-film monitors is the AT cut. This is cut from a slice oriented so that it contains the x-axis of the crystal and is at an angle of $35°15'$ to the z-axis. The mode of vibration is a high-frequency shear mode (Figure 13.28), and the temperature coefficient is small over the range of $-40°C$ to $+90°C$, of the order of $\pm10^{-6}°C^{-1}$ or slightly greater. The coefficient changes sign several times throughout the range so that the total fractional change in frequency over the complete range is only around 5×10^{-5}. Usually, the frequency chosen is around 5 MHz or sometimes 6 MHz, although the range could be anything from 0.5 to 50 or 100 MHz. Generally, the temperature of the crystal must be limited to below 120°C (otherwise, the temperature coefficient becomes excessively large), so it may not always be possible to keep it at the same temperature as the other substrates in the machine and this may be of importance in some applications.

As the thickness of the evaporant builds up, the frequency of the crystal falls and the reduction in frequency is proportional both to the square of the resonant frequency and to the mass of the film deposited. Accuracy depends on a wide range of factors, many related more to the stability of the installation than to the crystal itself, but a good starting rule of thumb for a typical arrangement is that the measurement of mass thickness can be readily carried out to an accuracy of perhaps 2%, adequate for most optical filters. Unfortunately, the sensitivity of the crystal decreases with increasing build up of mass, and the total amount of material that can be deposited before the crystal must be cleaned is limited. With existing crystals, this makes them less useful for multilayer work, especially in the infrared, where a single crystal would have difficulty in accommodating a complete filter. One way around this problem is to place a screen over the crystal that cuts down the material reaching it to a fraction of that reaching the substrates in the batch. This, of course, reduces the accuracy of the system. A more satisfactory solution is to use a multiple crystal head that can automatically change crystals when one is exhausted.

Because the crystal measures mass and not optical thickness, it must be calibrated separately for each material used. The calibration depends on the acoustical properties not only of the crystal but also of the deposited material. Any mismatch in the acoustic impedances for the particular shear mode at a boundary acts in much the same way as a mismatch in optical admittance and perturbs the natural frequency of the combined crystal and film. This must be taken into account in the calculations, and most commercial crystal monitors include an input for the acoustic impedance values. These values are not always well known for thin-film materials, which creates one problem. There are further problems in crystals that are controlling the deposition of a multilayer coating. The discontinuity in acoustic impedance now occurs at every interface and complicates the calculations to a completely impracticable extent. Thus, the control of the deposition of a multilayer coating on one single crystal carries an additional inaccuracy because of the difficulty of allowing for these discontinuities. This has led to the practice of a separate crystal for each material, since that involves only one discontinuity and so only one calibration for each crystal.

Because of these difficulties, some recent crystal monitors employ multiple modes of vibration. The redundancy in the mass of added material, which is necessarily identical for each mode,

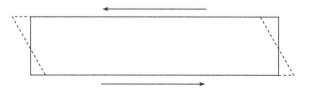

FIGURE 13.28
Quartz crystal operating in shear.

permits an estimate of the acoustical impedance mismatch and, therefore, avoids the need for a user-supplied value. This technique can even be extended to the corrections necessary in the deposition of a multilayer on one crystal [80].

There can be considerable advantages in the use of quartz crystal monitors. Since the output moves in a constant direction and does not reverse, it is more readily accommodated by automatic control systems. Further, the crystal does not need optical windows with their attendant difficulties of maintenance and screening from the evaporant. Alignment is simpler than that for optical monitors, although the requirements for dimensional stability are just as severe. In recent years, there have been developments in the use of multiple-crystal sensors distributed around the chamber able to sense changes in the plume of material from the sources and make appropriate corrections to the monitoring calculations. The automatic corrections for acoustic impedance remove the associated uncertainty. With such improvements, the results that can be achieved by pure-crystal monitoring are excellent.

In the case of narrowband filters, the optical monitoring is successful because of a built-in error compensation process. This makes it difficult for the crystal monitor to achieve the same yield. For processes where error compensation is necessary to achieve the optical performance, optical monitoring is preferred. In those cases, the crystal monitoring is usually still employed, but for source and rate control sensing rather than for primary monitoring.

A useful set of instructions and tips on the quartz crystal monitor will be found in a paper by Riegert [81], which deals much more fully with the topics mentioned earlier and, despite its age, is still valid. Manufacturers' manuals also include good information.

13.6.3 Monitoring by Deposition Time

The stability of the sputtering deposition process renders very consistent the thickness of material added in each increment of the incremental processes described in Chapter 11. This makes it possible to control layer thicknesses by time or, alternatively, by the number of rotations of the substrate drum.

Spencer [14] describes successful experiments in the time control of thickness in a small ophthalmic coater with sputtering sources of ZrO_2 and SiO_2 in the construction of optical coatings including a single-cavity narrowband filter. Pervak et al. [16] with pure time control successfully produced narrowband notch filters using the Helios machine of Figure 13.12. Gibson et al. [11] reported run-to-run reproducibility of $\pm0.3\%$ in the characteristics of edge filters of TiO_2 and SiO_2 for the visible region produced by the closed-field magnetron sputtering system of Figure 13.10. Figure 13.29 illustrates the stability of time monitoring in the production using the RAS process

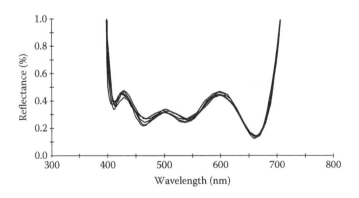

FIGURE 13.29
Performance of a 10-layer antireflection coating monitored by time. There are results from five successive production runs superimposed. The materials are SiO_2 and Si_3N_4, and deposition was by the RAS process (see Chapter 11). (Courtesy of Shincron Company Ltd., Yokohama, Japan.)

(Figure 13.18) of successive batches of antireflection coatings with 10-layer design, using SiO_2 and Si_3N_4.

A variant of time control has long been used in in-line sputtering systems where substrates move through the system at a fixed rate. Short-term stability of the process has to be good, but there can be slow drifts over long periods. Often the thickness of the deposited layers will be measured and the results will be used in the gradual adjustment of sputtering power so that the long-term stability of the process is assured.

13.7 Tolerances

The question of how accurately we must control the thickness of layers in the deposition of a given multilayer is surprisingly difficult to answer and has attracted a great deal of attention over the years. Nowadays, we immediately think of the computer when we wish to carry out numerical studies, but this is a relatively recent innovation. The earlier studies lacked this luxury and so were greatly influenced by the need to limit the volume of calculation. Nevertheless, the results were, and still are, of value.

One of the earliest approaches to the assessment of errors permissible in multilayers was devised by Heavens [82], who used an approximate method based on the alternative matrix formulation in Section 3.4.2. His method, mainly useful when calculations must be performed, consisted of a technique for fairly simply recalculating the performance of a multilayer with a small error in thickness in one of the layers. He showed that the final reflectance of a quarter-wave stack is scarcely affected by a 5% error in any one of the layers.

Lissberger [83] and Lissberger and Wilcock [84] developed a method for calculating the performance of a multilayer involving the reflectances at the interfaces. In multilayers made up of quarter waves, the expressions took on a fairly simple form that permitted the effects of small errors, in any or all of the layers, on the phase change caused in the light reflected by the multilayer to be estimated. Lissberger's results, applied to the all-dielectric single-cavity (Fabry–Perot) filter, show that the most critical layer is the cavity layer. The layers on either side of the cavity layer are the next most sensitive, and the remainder of the layers, progressively less sensitive the further they are from the cavity.

We have already mentioned in Chapter 8 the paper by Giacomo et al. [85], where they examined the effects on the performance of narrowband filters of local variations in thickness, or roughness, of the films. This involved the study of the influence of thickness variations in any layer on the peak frequency of the complete filter. The treatment was similar in some respects to that of Lissberger. For the conventional single-cavity filter, layers at the center had the greatest effect. If all layers were assumed equally rough, the design least affected by roughness would have all the layers of equal sensitivity, and attempts were made to find such a design. A phase-dispersion filter gave rather better results than the simple conventional single cavity, but still fell short of ideal.

Baumeister [86] introduced the concept of sensitivity of filter performance to changes in the thickness of any particular layer. The method involved the plotting of sensitivity curves over the whole range of useful performance of a filter, curves which indicated the magnitude of performance changes due to errors in any one layer. His conclusions concerning a quarter-wave stack were that the central layer is the most sensitive, and the outermost layers, least sensitive. An interesting feature of these sensitivity curves for the quarter-wave stack is that the sensitivity is greatest nearest the edge wavelength. This is confirmed in practice with edge filters, where errors usually produce more pronounced dips near the edge of the transmission zone than appear in the theoretical design.

Smiley and Stuart [87] adopted a different approach using an analog computer. There were some difficulties involved in devising an analogue computer, but once constructed, it possessed the advantage at the time that any of the parameters of the thin-film assembly could be easily varied. A particular filter, which they examined, was

$$Air|4H\,L\,4H|Air,$$

with $n_H = 5.00$ and $n_L = 1.54$. This is a multiple-cavity filter of simple design. Errors in one of the $4H$ layers and in the L layer were investigated separately. They found that errors greater than 1% in one $4H$ layer had a serious effect; errors of 5%, for example, caused a drop in peak transmittance to 70%, and errors of 10%, a drop to 50%, together with considerable degradation in the shape of the passband. Errors of up to 10% in the L layer had virtually no effect on either the shape of the passband or the peak transmittance. This is absolutely in line with what we would nowadays expect from a multiple-cavity filter.

An investigation was performed by Heather Liddell as part of a study reported by Smith and Seeley [88] into some effects of errors in the monitoring of infrared single-cavity filters of designs:

$$Air|HLHL\,HH\,LHLHL|Substrate$$

and

$$Air|HL\,HH\,LHL|Substrate.$$

A computer program to calculate the reflectance of a multilayer at any stage during deposition was used. Monitoring was assumed to be at or near a frequency of four times the peak frequency (i.e., a quarter of the desired peak wavelength) of the completed filter. It was shown that if all layers were monitored on one single substrate, then, provided the form of the reflectance curve during deposition was predicted, and it was possible to terminate layers at reflectances other than turning values, there could be an advantage in choosing a monitoring frequency slightly removed from four times peak frequency. If no corrections were made for previous errors, then a distinct tendency for errors to accumulate in even-order monitoring (that is, monitoring frequency an even integer times peak frequency) was noted.

The major problem in tolerancing is that real errors cannot be treated as small, that is to say that first-order approximations are unrealistic. The error in one layer interacts with the errors in other layers, and it is not realistic to treat them as though their effects can be calculated in isolation and then linearly combined.

In recent years, the most satisfactory approach for dealing with the effects of errors and the magnitude of permissible tolerances has been found to be the use of Monte Carlo techniques. In this method, the performance of the filter is calculated, first with no errors and then a number of times with errors introduced in all the layers. In the original form of the technique, introduced by Ritchie [6], the errors are thickness errors and completely random and uncorrelated. They belong to the same infinite population, taken as normal with prescribed mean and standard deviation. The performance curves of the filter without errors and of the various runs with errors are calculated. Although statistical analyses of the results can be made, it is almost always sufficient to simply plot the various performance curves together, when visual assessment of the effects of errors of the appropriate magnitude can be made. The method really provides a set of traces that reproduce, as far as possible, what would actually be achieved in a succession of real production batches suffering from errors of the chosen magnitude. The characteristics of the infinite normal population can be varied, and the procedure, repeated. It is sufficient to calculate some 8 or perhaps 10 curves for a set of error parameters. The level of error at which a satisfactory process yield would be achieved can then be readily determined. In the earliest version of the technique, the various errors were manually drawn from random number tables and converted into members of a normal population using a table of area under the error curve. (The procedure is described in textbooks of statistics—see Yule and Kendall [89], for example.) Later versions of the technique simply generate the random errors by computer. Although the errors are usually drawn from a normal population, the type of population has little effect on the order of the results. Normal distributions are convenient to program, and since there is no strong reason for not using them and because errors made up of a number of uncorrelated effects are well represented by normal distributions, most error analyses do make use of them.

The level of permissible errors depends to some extent on the index contrast in the filter. Figure 13.30 shows some examples of plots where the errors are simple independent thickness errors of zero mean. From these and similar results, we find that the thickness errors that can be tolerated in

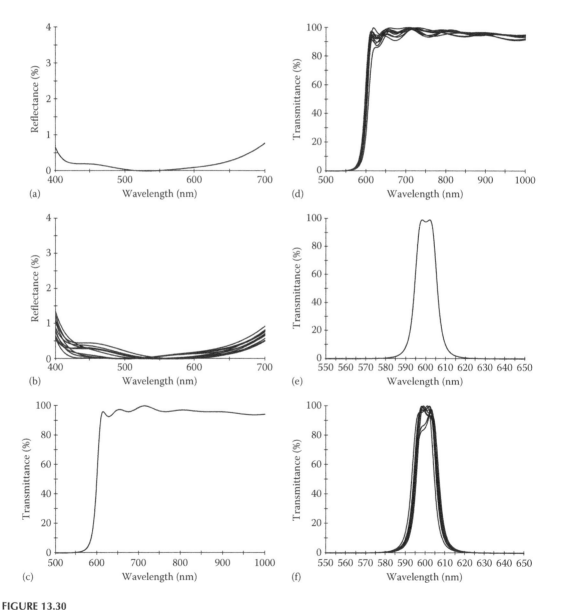

FIGURE 13.30
Effect of random in layer thickness on the performance of thin-film filters. (a) Four-layer antireflection coating with no errors, (b) 4-layer antireflection coating with thickness errors of 2% standard deviation. (c) Longwave pass filter with no errors. (d) Longwave pass filter with thickness errors of 2% standard deviation. (e) Two-cavity filter with no errors. (f) Two-cavity filter with thickness errors of 0.5% standard deviation.

simple edge filters and antireflection coatings are normally around 2% standard deviation. This correlates quite well with the accuracy usually achievable by normal optical or quartz crystal monitoring. Narrowband filters require rather better accuracy when random errors in thickness are involved. The two-cavity filter of Figure 13.30 is showing unacceptable passband distortion with random thickness errors as small as 0.5% standard deviation. This filter has a roughly 2% halfwidth. For narrower filters or filters with greater number of cavities, the tolerances must be still tighter. In a single-cavity filter, the main effect of random errors is a peak wavelength shift, the shape of the passband being scarcely affected even by errors as large as 10%. The standard deviation of the scatter in peak wavelength is slightly less than the standard deviation of the layer thickness errors so that some averaging process is operating, although the orders of magnitude are the same.

A system of monitoring in which the thickness errors in different layers are uncorrelated requires that each layer should be controlled independently of the others. In this type of monitoring, therefore, we cannot expect high precision in the centering of narrowband single-cavity filters and we foresee great difficulties in being able to produce narrowband multiple-cavity filters at all.

This monitoring arrangement is what we have called indirect. Systems where each layer is controlled on a separate monitoring chip are of this type. There are difficulties with the monitoring of low-index layers on a fresh glass substrate because of the small changes in transmittance or reflectance, and so the monitoring chips are usually changed after a low-index layer and before a high index, two or four layers per chip being normal. Sometimes these layers will be monitored to turning values. More frequently, what is sometimes called level monitoring will be used. Here the layer reflectance or transmittance signal is terminated at a point removed from the turning value where the signal is still changing, leading to an inherently greater accuracy. This approach involves what is really an absolute measurement of reflectance or transmittance, and so the termination point is frequently chosen to be after a turning value rather than before, so that the extremum can be used as a calibration. This usually implies a shorter wavelength for monitoring or the introduction of a geometrical difference between batch and monitor, placing the monitor nearer the source or placing masks in front of the batch.

Narrowband filters are not normally monitored in this way. Instead, all the layers are monitored on the same substrate, usually the actual filter being produced, a system known as direct monitoring. At the peak wavelength of the filter, the layers should all be quarter waves or half waves, and so we can expect a signal that reaches an extremum at each termination point. The accuracy cannot therefore be particularly high for any individual layer, and at first sight, it would appear that the achievable accuracy should be far short of what must be required. Since each layer is being deposited over all previous layers on the monitor substrate, then there is an interaction between the errors in any layer and those in the previous layers, and not included in the tolerancing calculation described earlier. We really require a technique that models the actual process as far as possible, and this is a quite straightforward computing operation. Each layer is simply considered to be deposited on a surface of optical admittance corresponding to that of the multilayer that precedes it, rather than on a completely fresh substrate. The results of such a simulation are shown in Figure 13.31, taken from Macleod [90], which demonstrates the powerful error compensation mechanism that has been found to exist. The compensation has also been independently and simultaneously confirmed by Bousquet et al. [91]. Its nature is perhaps best explained by the use of an admittance diagram.

Figure 13.32 shows such a diagram drawn for two quarter waves. Since both the isoreflectance contours (see Section 3.2) and the individual layer loci are circles centered on the real axis, the turning values must always occur at the intersections of the loci with the real axis, regardless of what has been deposited earlier. At the termination point of each layer, there is the possibility of restoring the phase to zero or to π. As far as any individual layer is concerned, it is principally the overshoot or undershoot of the previous layer that affects it. If the previous layer is too thick, the current one will tend to be thinner to compensate for this, and vice versa. Of course, it is impossible

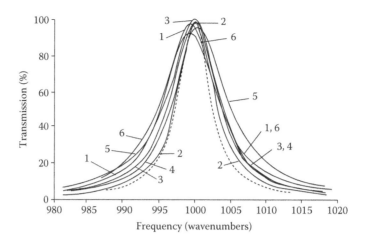

FIGURE 13.31
Effect of 1% standard deviation reflectance error on the performance of the Fabry–Perot filter: Air | *HLHL HH LHLH* | Ge. The substrate is germanium ($n = 4.0$), L represents a quarter wave of ZnS ($n = 2.3$), and H, a quarter wave of PbTe ($n = 5.4$). The monitoring is of the first order. The dashed curve is the performance with no errors. (After H. A. Macleod, *Optica Acta*, 19, 1–28, 1972. With permission of Taylor & Francis Ltd.)

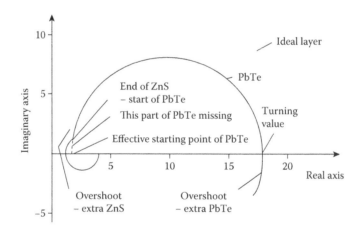

FIGURE 13.32
Admittance locus of the first two layers of the filter in Figure 13.31 when there is an overshoot in the first layer of around one-eighth wave optical thickness. (After H. A. Macleod, *Optica Acta*, 19, 1–28, 1972. With permission of Taylor & Francis Ltd.)

to completely cancel all effects of an error in a layer. The process is actually transforming the thickness errors into errors in reflectance at each stage since the loci will be slightly displaced from their theoretical position. This is not a serious error. As can be guessed from the shape of the diagram, the reflectance error is a second-order effect. Since the phase is self-corrected each time a layer is deposited, the peak wavelength of the filter will remain at the desired value, that of the monitoring wavelength. The remaining error, the residual one in reflectance, is then translated into changes in peak transmittance and halfwidth. Since the reflectance change is always a reduction, the bandwidth of an actual filter is invariably wider than theoretical. The peak transmittance falls to the extent that the reflectances on either side of the spacer layer are unbalanced. This is usually a quite small effect and the reduction in peak transmittance is generally much less important that the increase in bandwidth.

In this monitoring arrangement, thickness errors in any individual layer are a combination of a compensation of the error in the previous layer together with the error committed in the layer itself. The magnitude of the thickness errors can be quite misleading in interpreting whether or not the filter can be successfully made. In Figure 13.31, for example, thickness errors of the order of 50% occur in some layers, and yet the filter characteristics are all useful ones.

The important characteristic is actually the error in reflectance or transmittance in determining the turning values, and it is possible to develop theoretical expressions relating the reflectance or transmittance errors to the reduction in performance of the final filter [90]. This analysis includes an assessment of the sensitivity of each layer to errors that indicate those layers where the greatest care in monitoring should be exercised. These can be different from the thickness sensitivity that Lissberger [83] and Lissberger and Wilcock [84] already mentioned. With high-index cavity layers, greatest sensitivity is found in the low-index layers following the cavity, while with low-index cavities, the cavity itself has the highest sensitivity. A feature of this analysis is that it demonstrates that for any particular error magnitude, there is a point where improved halfwidth does not result from an increase in the number of layers because the effect of errors is increasing more rapidly than the theoretical decrease in bandwidth. Then it is necessary to move to second- and higher-order cavity layers if decreased bandwidth is to result. This corresponds to what is found in practice. The error analysis also demonstrates that from the point of view of monitoring, high-index cavities are to be preferred over low-index cavities. We have already seen in Chapter 8 that high-index cavities give decreased angular sensitivity and greater tuning range. However, for filters in the visible region where the absorption loss in the high-index layers is greater than that in the low-index layers, low-index cavity layers are more common and the same is true for telecom filters in the near infrared.

Formulas that permit the calculation of the errors in reflectance, in halfwidth and in peak transmittance as a function of the magnitude of the random errors in determining the turning values, exist [90], but for most purposes, a computer simulation will suffice. It should be noted that the compensation is most effective only for the first order. Second-order monitoring, that is, monitoring at the wavelength for which the layers are all half waves, is not effective in preserving the peak wavelength. We can understand this because the admittance diagram is quite different, and so the compensation is of a different nature. Likewise, third-order monitoring is not as effective as first-order monitoring, and although the scatter in peak wavelength is less than that obtained with second-order monitoring, it is, nevertheless, quite large.

Multiple-cavity filters are similar in behavior, but there are some complications. The coupling layers in between the various Fabry–Perot sections of the filter turn out to be particularly sensitive to errors in a rather peculiar way. Preliminary examination of the admittance diagram for the various layers of a multiple-cavity filter and even the standard error analysis do not immediately reveal any marked difference in terms of error sensitivity between these layers and those of single-cavity filters. Closer investigation shows that there is always one transition from one layer to the next occurring at or near to the central coupling layer where a thickness error is compensated for by an error of the same rather than the opposite sense [92]. The condition is sketched in Figure 13.33. An increase in thickness in the first layer results in an increase in thickness of the subsequent layer and vice versa. This condition must occur once between each pair of cavities. The net result is an increase or decrease in the relative spacing of the cavities causing the appearance of a multiple-peaked characteristic curve. The peaks become more pronounced, the greater the relative error in spacing. One of the peaks always corresponds to the normal control wavelength and is close to the theoretical trans-mittance. The other peaks (one for a two-cavity, two for a three-cavity, and so on) can appear on either side of the main peak depending on the nature of the particular errors. This false compensation can be destroyed if the second of the two layers concerned can be controlled independently of the others, either on a separate monitor plate or by a quartz crystal monitor, or even by simple timing. It is essential that it should also be deposited on the regular monitor as well, so that the compensation of the full filter should not be destroyed [92].

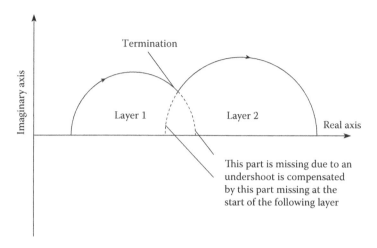

FIGURE 13.33
Error compensation when the admittance circles are on the same side of the real axis. (After H. A. Macleod and D. Richmond, *Optica Acta*, 21, 429–443, 1974. With permission of Taylor & Francis Ltd.)

Another problem surfaced in the production of narrowband filters for applications in telecommunications. Figure 13.34 shows the specification for a dense wavelength division multiplexing filter together with the effect on a suitable design of 10 sets of independent random thickness errors drawn from an infinite normal population with standard deviation of 0.003%. This perturbs the performance of the filter to the limit of what is acceptable by the specification. We have already seen in this chapter that turning value monitoring carries with it the automatic compensation of thickness errors, but in the case especially of these high-performance filters, random thickness errors also have an implication. The uniformity of the filter will be assured by rotating the substrate about its axis above an offset source. These multiplexing and demultiplexing filters are normally quite small, often 1.4 mm^2 in size, and so normal practice is to make a rather larger filter on a disk and to dice it into smaller units after coating. Those parts of the disk that are displaced from the center of rotation will suffer a variable deposition rate as the disk rotates, but each complete rotation will have an incremental thickness added that is equivalent to that at the center or deliberately adjusted so that the filters cut from the disk will span a number of communication channels. However, it is impossible to assure that each layer will correspond to an

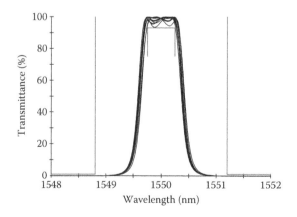

FIGURE 13.34
Effect of random errors in layer thickness of standard deviation 0.003%. The perturbed performance is just within the specification indicated by the straight lines.

exact whole number of rotations. There will therefore be a random error in termination of each layer corresponding to the final fractional turn that will be larger, the further the element is from the center of rotation. This random error must not exceed the already established 0.003%. We can assume for the sake of argument that the largest radius is such that the typical error from this effect is 25% of the thickness that would be deposited in a full turn. Then the total number of turns necessary for this error to not exceed our 0.003% figure is 0.25/0.00003, that is, 8333 complete rotations. Assuming the deposition of a single quarter wave takes some 5 minutes, then the required rotational speed of the disk is 1700 revolutions per minute. Such rotational rates, and even higher, are quite typical of machines for the production of telecommunication-quality filters.

Pelletier et al. [93] theoretically studied the behavior of the maximètre types of monitoring systems in the production of narrowband filters. They conclude that, as we would expect, the accuracy of the system in the production of single layers is very much better than a single-wavelength system. In the monitoring of narrowband filters all on one substrate, there is a compensation process operating like the turning value method, but it is more complex in operation. For very small errors in most layers, the system works adequately, but for large errors in most layers or small errors in certain critical layers, the errors accumulate in such a way as to cause a drastic broadening of the bandwidth of a single-cavity filter or complete collapse of a multiple-cavity filter. Pelletier has introduced two concepts to describe this behavior. Accuracy represents the error that will be committed in any particular layer without reference to the multilayer system as a whole. Stability represents the way in which the errors accumulate as the multilayer deposition proceeds. The accuracy of the maximètre is excellent and greater than in the turning value method, but the stability in the control of narrowband filters is very poor, and it can easily become completely unstable. Subsidiary measurements are therefore required to ensure stability if advantage is to be taken of the very great accuracy that is possible. Narrowband filters and their monitoring systems have been surveyed by Macleod [94].

The concepts of accuracy and stability and the discovery that the one does not ensure the other imply that different measurements may be necessary to ensure that both are simultaneously assured. This leads to the idea of broadband monitoring in which simultaneous measurements are made at a large number of wavelengths over a wide spectral region and a merit function representing the difference between actual and desired signals is computed. The merit function can then be used as a monitoring signal, and layer deposition, terminated when the merit function reaches a minimum. Although perfect deposition should ensure a minimum of zero in the figure of merit, inevitable errors in layer index and homogeneity will perturb the result. The accuracy and stability of such a broadband system in the monitoring of certain components such as beam splitters has been investigated by computer simulation [95], and evidence, found for useful error compensation. Apart from very qualitative justification as discussed earlier, no theory for such compensation yet exists, and it may operate only in quite specific cases.

Extensions of broadband monitoring to a system that would reoptimize, on the basis of errors measured in earlier layers, those layers of a design yet to be deposited are possible, but there are also considerable dangers. The most important aspect of such processes is that the errors should be correctly characterized. They may be errors in thickness, optical constants, or both, and if incorrectly characterized, the results can be rather worse than in the absence of compensation [74,75]. Such techniques should not be thought of as removing the need for stable reproducible materials.

As computing power has increased, so has the ability to model the production process. Such modeling, almost invariably of the Monte Carlo type, allows the study of errors and tolerances in an almost completely realistic way. Some results are described by Macleod [96] and Clark and Macleod [97]. The technique was termed *computational manufacturing* by Tikhonravov and Trubetskov [98] and Tikhonravov et al. [99,100].

We can illustrate the method with a simple example of a longwave pass filter for the visible and near infrared. The filter consists of a total of 31 layers of silica and tantala, arranged as a core of quarter waves bounded on each side by four layers that have been refined to act as ripple-reducing matching systems. The theoretical performance is shown in Figure 13.35.

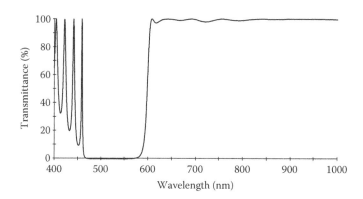

FIGURE 13.35
Theoretical performance of the longwave pass filter used in the simulation exercise.

For this exercise, we will limit our model to include signal noise only. We assume transmittance monitoring, and Figure 13.36 shows the signal for the first two layers of the simulation illustrating the level of noise that we will assume as 0.4% standard deviation in terms of transmittance.

Our first task is to decide on the interpretation of the monitoring signal. Because of the difficulty of maintaining accurate calibration of reflectance or transmittance, common practice in optical monitoring is to arrange that the termination point should, wherever possible, follow a signal extremum. The termination level is then specified as a prescribed signal increase or reduction following the extremum. Thus, the monitoring system must be capable of recognizing an extremum and then detecting the necessary overshoot. We shall use in our simulation a straightforward technique that is mirrored in many commercial systems. The noisy signal will present many extrema, and we will select the correct one by introducing the equivalent of mechanical backlash. The system will accept an extremum once the signal has reversed in sense by a prescribed amount that must be greater than the total noise perturbations of the signal. The result will be a delay in recognition that will increase with the level of noise. Once the extremum has been registered in this way, the prescribed overshoot will be applied to the accepted extremum to derive the termination level. The layer will then be terminated as soon as this level is reached or, immediately, if it is actually less than the backlash. Figure 13.37 illustrates the method. Noise implies that the detection of an extremum will be late, but termination at a prescribed level will tend to be early.

Next we need to decide on a monitoring procedure. We will, as is common practice, use multiple chips, and initially, we choose four layers per chip. Ideally, the first layer on a chip should be of high index to maximize the signals, and happily, in this case, the final layer in the design, that is, the first layer to be deposited, is of high index. The reference wavelength for the design is 523 nm, but for better signals, we move the monitoring wavelength to 450 nm, choose a spectral bandwidth

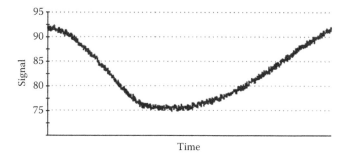

FIGURE 13.36
Simulated monitoring signal for the first two layers of the longwave pass filter shown as if on a strip chart. (Courtesy of Thin Film Center Inc., Tucson, AZ.)

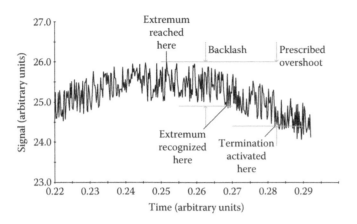

FIGURE 13.37
Noise on the signal implies an enormous numbers of local extrema. The occurrence of the correct extremum will be recognized when the signal has reversed a prescribed amount that must be greater than the total excursion due to the noise. Once the extremum is recognized, the prescribed overshoot will be applied to arrive at the correct termination level for the signal.

of 10 nm, and increase the tooling factor for both materials in the machine to 1.1, that is, 10% more material must be deposited on the chip for the correct thickness on the part. This should ensure as far as possible monitoring signals that exhibit extrema before termination. Beyond 450 nm, we are climbing the dispersion curve for tantala, and we are running out of energy without a special source of illumination.

For such exercises, 10 separate runs of the Monte Carlo model are usually sufficient. The results of the first 10 runs of this particular model are shown in Figure 13.38. The pass region is poor, and the yield quite, unacceptable. The advantage of simulation is that the designs that were achieved are all available for inspection. We can quite quickly determine that it is the first chip with the first four layers to be deposited that is the problem. The second pair of layers is suffering from errors accumulated from the first pair. Replacing the first chip by two chips, each with only two layers, results in a successful monitoring arrangement as shown in Figure 13.39. It is difficult to detect it in Figure 13.39, but the inherent shortening of the layers due to noise in level monitoring has slightly moved the edge to the short wave. This can readily be corrected in practice by increasing the monitoring wavelength.

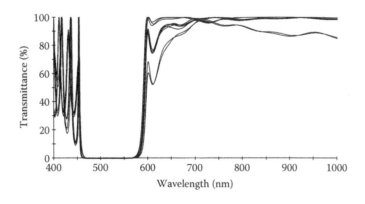

FIGURE 13.38
First attempt at simulating the construction of the filter results in a poor performance in the pass region. (Courtesy of Thin Film Center Inc., Tucson, AZ.)

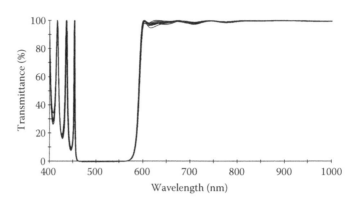

FIGURE 13.39
Changing the first four-layer chip into two two-layer chips solves the problem. The 10 runs now show satisfactory performance. (Courtesy of Thin Film Center Inc., Tucson, AZ.)

This is, admittedly, a quite simple example. Effects such as varying tooling factors, changing temperature, noise varying with time, and so on can all be readily simulated. This makes it easy for the design process to include an element of simulation.

Quartz crystal monitoring, in which the mass rather than optical thickness is measured, seems unlikely to possess powerful compensation. Yet the simulation of a simple broadband system for antireflection coatings comparing optical monitoring with quartz crystal gave results which indicate that the quartz crystal is in no way inferior [96]. The relative merits of quartz crystal and optical monitoring form a subject of almost constant debate, and published results for quartz crystal are impressive [101,102]. It is clear that narrowband filters, if they are to be controlled in peak wavelength, do require direct optical monitoring, but quartz crystal monitoring is suitable for most other filter types. The general opinion, based to some extent on instinct, is that quartz crystal monitoring is most suitable for the production of successive batches of identical components. For single runs of varying coating types, optical monitoring normally appears to be preferred. Optical monitoring is also preferred in applications such as filters for the far infrared, where very large thicknesses of materials are deposited in each coating run.

13.8 Performance Envelopes

We mentioned earlier the problem with sensitivity analysis where the errors that can be accommodated with first-order theory are normally too small to be typical of errors committed in practice. Performance envelopes are not so limited. An interesting and useful property of any dielectric layer in a thin-film structure is that whatever the thickness of the layer, the transmittance and reflectance of the entire coating remain within clearly defined upper and lower bounds [103, 104] known as the performance envelopes. The performance envelopes are valid even in the presence of absorption in the other layers. Since the envelopes in transmission are simpler analytically than those in reflection, we concentrate on transmission. A useful addition to performance envelopes is the round-trip phase change in the chosen layer that allows us to predict not only where the actual transmittance characteristic will touch the upper and/or lower envelopes, but how these points of tangency will move with varying thickness of the layer. Thus, the envelopes go well beyond simple sensitivity curves.

The transmittance of the system defined in Figure 13.40 is given by the usual multiple-beam summation:

$$T = \frac{T_a T_b}{\left[1 - (R_a R_b)^{1/2}\right]^2} \cdot \frac{1}{\left[1 + \frac{4(R_a R_b)^{1/2}}{\left[1 - (R_a R_b)^{1/2}\right]^2} \sin^2\left(\frac{\varphi_a + \varphi_b}{2} - \delta\right)\right]}. \tag{13.4}$$

The $\sin^2$ term oscillates between zero and unity, each of these values being associated with an extremum, the maximum corresponding to zero and the minimum to unity, so that the two envelopes are

$$T = \frac{T_a T_b}{\left[1 \mp (R_a R_b)^{1/2}\right]^2}, \tag{13.5}$$

with the minus corresponding to the maximum, and the plus, to the minimum envelopes. The round-trip phase change is given by

$$\gamma = \varphi_a + \varphi_b - 2\delta, \tag{13.6}$$

and a value of zero, or integral multiple of 360°, yields the maximum, and of ±180°, or odd multiple thereof, the minimum.

Figures 13.41 and 13.42 show a quite simple case of a niobium pentoxide film on a glass substrate. The envelopes are shown in Figure 13.41 and then, with the fringes superimposed, in Figure 13.42. An increase in film thickness will increase the magnitude of the negative round-trip phase change, that is, will move it downward. That will move the zero intersections to the right so that the fringe peaks will occur at longer wavelengths, and vice versa for reduced thickness. The greater the slope of the phase change curve, the smaller the relative movement, that is, the less is the susceptibility to errors.

A more complicated example is afforded by a multiple-cavity filter (Figure 13.43). The matching is a simple single layer accounting for the slight dip in the center of the passband. The envelopes associated with the first cavity layer are shown in Figure 13.44. The slope of the round-trip phase change is quite low, as evidenced by the truncated phase scale, implying a large movement of the tangent point with the upper envelope with even a small change in thickness. The lower envelope

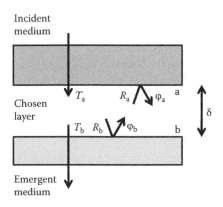

FIGURE 13.40
Two structures, a and b, surround the chosen dielectric layer. T and R indicate transmittance and reflectance, respectively; φ, the phase shift on reflection; and δ, the phase thickness.

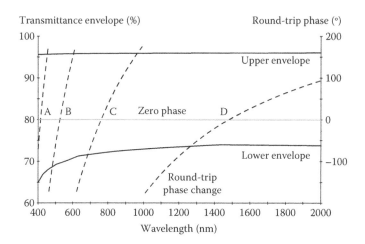

FIGURE 13.41
Envelopes and round-trip phase change for a single layer of Nb_2O_5 three quarters of a wave in optical thickness at 1000 nm on glass. A to D mark the zero crossings where the transmittance will touch the upper envelope.

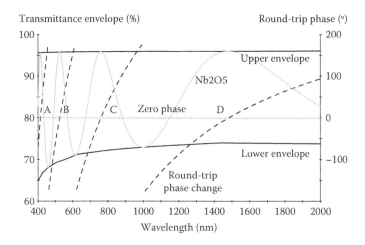

FIGURE 13.42
Response of the Nb_2O_5 film has been added to the plot of the envelopes from Figure 13.41, and the extrema are as predicted.

is virtually on top of the horizontal scale so that except for the tangent point, the filter characteristic will rapidly drop.

The situation with the coupling layers is still more interesting. Figure 13.45 shows that the lower envelope reaches up almost to touch the upper envelope in the center of the characteristic. This is typical of coupling layer envelopes. The slope of the phase curve is quite high so the error sensitivity is reasonably low, but with largish errors, the characteristic of the filter drapes itself over the lower envelope to give the quite characteristic shape shown in Figure 13.46. When such a shape is exhibited by a multiple-cavity filter, it is a sign that the coupling layer or layers are suspect.

For further examples of the use of envelopes in the study of error sensitivity in coatings see Macleod [104].

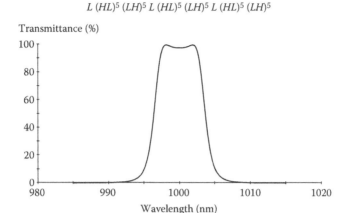

$L\ (HL)^5\ (LH)^5\ L\ (HL)^5\ (LH)^5\ L\ (HL)^5\ (LH)^5$

FIGURE 13.43
Transmittance of the multiple-cavity filter created to demonstrate performance envelopes.

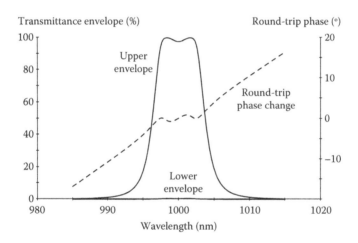

FIGURE 13.44
Envelopes and round-trip phase change for the first cavity layer, layer 11, in the multiple-cavity filter. Note the short scale for the phase implying a very low slope and hence high error sensitivity. The lower envelope is hardly visible at the foot of the plot.

13.9 Reverse Engineering

Reverse engineering has a broad connotation. The reverse engineering that is the subject of this abbreviated section is the attempt to understand what went wrong with the construction of a desired coating. It is a problem with multiple possible answers, and so the more information that can be consulted, the greater the possibility of finding the one correct answer. Keeping good records can be tedious and time consuming, but in times of trouble, they are one of the best possible resources. Also the person immediately in charge of the machine will frequently have a good idea of the source of the problem.

Errors in manufacture can be divided into systematic errors and random errors, and a first step in reverse engineering is to determine which classification applies. Random errors cannot be cured by systematic changes in deposition parameters, which will simply shift the distribution.

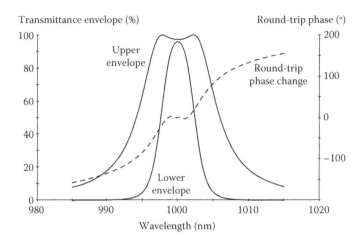

FIGURE 13.45
Performance envelopes and round-trip phase change for the first of the coupling layers, layer 21. Coupling layers always show the curious lower envelope reaching up almost to the upper envelope. Whatever the error in the layer thickness, and it is not sensitive, the characteristic drapes itself over the lower envelope.

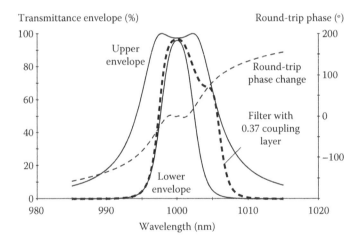

FIGURE 13.46
Multiple-cavity filter with an error in thickness in the coupling layer 21, the thickness of which is now 0.37 rather than 0.25, showing the characteristic shape associated with such an error.

Tighter control is necessary, and it is difficult to determine such a classification from just one result. A frequent random problem is a process that has been successfully running but is now not. Cleaning the machine, using new source materials, and similar attempts at a cure are ineffective. The coatings exhibit greater instability than normal, or poorer hardness or environmental resistance, or similar disappointing performance. In your author's experience, such problems are very often the result of some small water leak. Machines generally have extensive water cooling, and a pinhole leak can have disastrous consequences. Part of the trouble is that such leaks tend to seal themselves under vacuum with a small plug of ice so that pumping performance that is usually checked at the start of a production run appears satisfactory. When sources are activated, however, and the machine heats up, the ice can melt and water can be injected into the process. Such leaks are difficult to find, but obvious places should just have seals replaced. The rotating hearth of an electron beam source will usually have a sealing ring under it to block any outlet of cooling water, for example.

A frequent operation in reverse engineering is the determination of the actual thicknesses of the films in a coating. We know what they should be, but the performance suggests the presence of errors. The process is akin to a normal design operation with a starting design equal to that which was attempted and the required performance that which was measured but with the difference that there is only one correct answer. There are some other differences. Our information on the performance of the coating is a measurement. Absolute accuracy is beyond us. Good spectrometric measurements are difficult to achieve beyond an accuracy of 0.1% absolute, particularly in reflection, and experience shows that any spectrometer available to all comers in a laboratory will rarely achieve such accuracy. The first requirement is therefore to have measurements of the highest possible precision. Once we have precise measurements, then we can go ahead with the reverse engineering process. Constraints are important, and experience is especially helpful. Also a range of results at different angles of incidence as well as wavelength can be particularly useful. Goldstein [105] suggested a useful preliminary theoretical approach that he terms a *Gedankenspektrum method*. The starting design in question is perturbed by the application of random errors that are nevertheless known. The reverse engineering process now attempts to find the correct solution that can be identified. This can be repeated many times, and the range of measurements, extended until the theoretical results are satisfactory. Then the actual reverse engineering can proceed.

An interesting study by Gao et al. [106] describes the reverse engineering of a number of different coatings ranging from single layers to a 22-layer structure where many different scans at different angles of incidence helped guide the procedure to the likely answer. The authors estimate accuracies of 1 nm in thickness and 10^{-3} in layer index achieved for each layer in a 22-layer coating.

References

1. J. L. Vossen and W. Kern. 1978. *Thin Film Processes*. New York: Academic Press.
2. J. L. Vossen and W. Kern. 1991. *Thin Film Processes II*. San Diego, CA: Academic Press.
3. D. A. Glocker and S. I. Shah. 1995. *Handbook of Thin Film Process Technology*. Bristol, UK: Institute of Physics.
4. J. T. Cox and G. Hass. 1958. Antireflection coatings for germanium and silicon in the infrared. *Journal of the Optical Society of America* 48:677–680.
5. H. Bangert and H. Pfefferkorn. 1980. Condensation and stability of ZnS thin films on glass substrates. *Applied Optics* 19:3878–3879.
6. F. S. Ritchie. 1970. *Multilayer filters for the infrared region 10–100 microns*. PhD Thesis. Reading, UK: University of Reading.
7. J. K. Coulter, G. Hass, and J. B. Ramsay. 1973. Optical constants and reflectance and transmittance of evaporated rhodium films in the visible. *Journal of the Optical Society of America* 63:1149–1153.
8. G. Hass and E. Ritter. 1967. Optical film materials and their applications. *Journal of Vacuum Science and Technology* 4:71–79.
9. M. A. Scobey. 1996. *Low pressure reactive magnetron sputtering apparatus and method*. US Patent, 5,525,199.
10. D. R. Gibson, I. Brinkley, E. M. Waddell et al. 2008. Closed field magnetron sputtering: New generation sputtering process for optical coatings. *Proceedings of SPIE* 7101:710108.
11. D. R. Gibson, I. T. Brinkley, E. M. Waddell et al. 2008. Closed field magnetron sputter deposition of carbides and nitrides for optical applications. In *51st Annual Technical Conference Proceedings*, pp. 487–491, Society of Vacuum Coaters, Chicago.
12. S. Schiller, U. Heisig, and K. Goedicke. 1975. Alternating ion plating—A method of high-rate ion vapor deposition. *Journal of Vacuum Science and Technology* 12:858–864.

13. M. A. Scobey, R. I. Seddon, J. W. Seeser et al. 1989. Magnetron sputtering apparatus and process. US Patent, 4,851,095.
14. A. G. Spencer. 1997. Precision optical coatings from express ophthalmic coaters. In *40th Annual Technical Conference Proceedings*, pp. 259–265, Society of Vacuum Coaters, New Orleans, LA.
15. F. Placido. 1998. Radio frequency sputtering of optical coatings including rugate filters. Private communication (University of the West of Scotland, Paisley).
16. V. Pervak, A. V. Tikhonravov, M. K. Trubetskov et al. 2007. Band filters: Two-material technology versus rugate. *Applied Optics* 46:1190–1193.
17. P. J. Kelly and J. W. Bradley. 2009. Pulsed magnetron sputtering—Process overview and applications. *Journal of Optoelectronics and Advanced Materials* 11:1101–1107.
18. K. Sarakinos, J. Alami, and M. Wuttig. 2007. Process characteristics and film properties upon growth of TiO_x films by high power pulsed magnetron sputtering. *Journal of Physics D* 40:2108–2114.
19. D. T. Wei, H. R. Kaufman, and C.-C. Lee. 1995. Ion beam sputtering. In *Thin Films for Optical Systems*, F. R. Flory (ed.), pp. 133–201. New York: Marcel Dekker.
20. R. Lalezari, G. Rempe, R. J. Thompson et al. 1992. Measurement of ultralow losses in dielectric mirrors. In *Topical Meeting on Optical Interference Coatings*, pp. 331–333, Optical Society of America, Tucson, AZ.
21. J. M. Mackowski, L. Pinard, L. Dognin et al. 1998. Different approaches to improve the wavefront of low-loss mirrors used in the VIRGO gravitational wave antenna. In *Optical Interference Coatings*, pp. 18–20, Optical Society of America, Washington, DC.
22. A. Ode. 2014. Ion beam sputtering of fluoride thin films for 193 nm applications. *Applied Optics* 53:A330–A333.
23. H. K. Pulker, M. Bühler, and R. Hora. 1986. Optical films deposited by a reactive ion plating process. *Proceedings of SPIE* 678:110–114.
24. H. K. Pulker and K. H. Guenther. 1995. Reactive physical vapor deposition processes. In *Thin Films for Optical Systems*, F. R. Flory (ed.), pp. 91–115. New York: Marcel Dekker.
25. B. G. Bovard. 1995. Ion-assisted deposition. In *Thin Films for Optical Systems*, F. R. Flory (ed.), pp. 117–132. New York: Marcel Dekker.
26. M. L. Fulton. 1994. Applications of ion-assisted-deposition using a gridless end-Hall ion source for volume manufacturing of thin-film optical filters. *Proceedings of SPIE* 2253:374–393.
27. S. Samori, T. Shimizu, T. Watanabe et al. 2013. High-rate AR coating for plastic substrate by EPD. In *Optical Interference Coatings*, Optical Society of America, Whistler, BC.
28. K. Matl, W. Klug, and A. Zöller. 1991. Ion-assisted deposition with a new plasma source. *Materials Science and Engineering* A140:523–527.
29. S. Pongratz and A. Zöller. 1992. Plasma ion-assisted evaporative deposition of surface layers. *Annual Review of Materials Science* 22:279–295.
30. A. Zöller, R. Götzelmann, K. Matl et al. 1996. Temperature-stable bandpass filters deposited with plasma ion-assisted deposition. *Applied Optics* 35:5609–5612.
31. H. Hagedorn, M. Klosch, H. Reus et al. 2008. Plasma ion-assisted deposition with radio frequency powered plasma sources. *Proceedings of SPIE* 7101:710109–1 to 710109–6.
32. P. J. Martin, R. P. Netterfield, T. J. Kinder et al. 1991. Deposition of TiN, TiC, and TiO_2 films by filtered arc evaporation. *Surface and Coatings Technology* 49:239–243.
33. M. L. Fulton. 1999. New Ion-assisted Filtered Cathodic Arc Deposition (IFCAD) technology for producing advanced thin-films on temperature-sensitive substrates. *Proceedings of SPIE* 3789:29–37.
34. S. Matsumoto, K. Kikuchi, M. Yamasaki et al. 2001. Method of forming a thin film of a composite metal compound and the apparatus for carrying out the method. US Patent, 6,207,536 B1.
35. K.-H. Müller. 1986. Monte Carlo calculation for structural modifications in ion-assisted thin film deposition due to thermal spikes. *Journal of Vacuum Science and Technology A* 4:184–188.
36. K.-H. Müller. 1988. Models for microstructure evolution during optical thin film growth. *Proceedings of SPIE* 821:36–44.
37. J. D. Targove, L. J. Lingg, and H. A. Macleod. 1988. Verification of momentum transfer as the dominant densifying mechanism in ion-assisted deposition. In *Optical Interference Coatings*, pp. 268–271, Optical Society of America, Tucson, AZ.
38. J. D. Targove and H. A. Macleod. 1988. Verification of momentum transfer as the dominant densifying mechanism in ion-assisted deposition. *Applied Optics* 27:3779–3781.

39. P. J. Martin, H. A. Macleod, R. P. Netterfield et al. 1983. Ion-beam-assisted deposition of thin films. *Applied Optics* 22:178–184.
40. M. J. Messerly. 1987. Ion-beam analysis of optical coatings. PhD Dissertation. Tucson, AZ: University of Arizona.
41. W. G. Sainty, R. P. Netterfield, and P. J. Martin. 1984. Protective dielectric coatings produced by ion-assisted deposition. *Applied Optics* 23:1116–1119.
42. C. K. Hwangbo, L. J. Lingg, J. P. Lehan et al. 1989. Ion-assisted deposition of thermally evaporated Ag and Al films. *Applied Optics* 28:2769–2778.
43. C. K. Hwangbo, L. J. Lingg, J. P. Lehan et al. 1989. Reactive ion-assisted deposition of aluminum oxynitride thin films. *Applied Optics* 28:2779–2784.
44. J. Segner. 1995. Plasma impulse chemical vapor deposition. In *Thin Films for Optical Systems*, F. R. Flory (ed.), pp. 203–229. New York: Marcel Dekker.
45. R. Hora and C. Wohlrab. 1993. Plasma polymerization: A new technology for functional coatings on plastics. In *36th Annual Technical Conference*, pp. 51–55, Society of Vacuum Coaters, Albuquerque, NM.
46. W. Möhl, U. Lange, and V. Pacquet. 1994. Optical coatings on plastic lenses by PICVD-technique. *Proceedings of SPIE* 2253:486–491.
47. C. Wohlrab and M. Hofer. 1995. Plasma polymerization of optical coatings on organic substrates: Equipment and processes. In *38th Annual Technical Conference*, pp. 222–230, Society of Vacuum Coaters, Albuquerque, NM.
48. K. Pfeiffer, S. Shestaeva, A. Bingel et al. 2016. Comparative study of ALD SiO_2 thin films for optical applications. *Optical Materials Express* 6:660–670.
49. I. M. Thomas. 1993. Sol-gel coatings for high power laser optics: Past present and future. *Proceedings of SPIE* 2114:232–243.
50. G. Hass. 1952. Preparation, properties and optical applications of thin films of titanium dioxide. *Vacuum* 2:331–345.
51. L. Holland and W. Steckelmacher. 1952. The distribution of thin films condensed on surfaces by the vacuum evaporation method. *Vacuum* 2:346–364.
52. K. H. Behrndt. 1963. Thickness uniformity on rotating substrates. In *Transactions of the 10th AVS National Vacuum Symposium*, pp. 379–384. London: Macmillan.
53. M. Knudsen. 1915. Das Cosinusgesetz in der kinetischen Gastheorie. *Annalen der Physik, 4th Series* 48:1113–1121.
54. E. Jancke and F. Emde. 1952. *Tables of Higher Functions*. Fifth ed. Leipzig: Teubner.
55. D. Keay and P. H. Lissberger. 1967. Application of the concept of effective refractive index to the measurement of thickness distributions of dielectric films. *Applied Optics* 6:727–730.
56. E. B. Graper. 1973. Distribution and apparent source geometry of electron-beam heated evaporation sources. *Journal of Vacuum Science and Technology* 10:100–103.
57. D. Richmond. 1976. Thin film narrow band optical filters. PhD Thesis. Newcastle upon Tyne: Northumbria University.
58. J. V. Ramsay, R. P. Netterfield, and E. G. V. Mugridge. 1974. Large-area uniform evaporated thin films. *Vacuum* 24:337–340.
59. L. Holland. 1956. *Vacuum Deposition of Thin Films*. London: Chapman & Hall.
60. D. M. Mattox. 1978. Surface cleaning in thin film technology. *Thin Solid Films* 53:81–96.
61. C. C. Lee. 1983. Moisture adsorption and optical instability in thin film coatings. PhD Dissertation. Tucson, AZ: University of Arizona.
62. M. Banning. 1947. Practical methods of making and using multilayer filters. *Journal of the Optical Society of America* 37:792–797.
63. H. D. Polster. 1952. A symmetrical all-dielectric interference filter. *Journal of the Optical Society of America* 42:21–25.
64. B. H. Billings. 1950. A birefringent frustrated total reflection filter. *Journal of the Optical Society of America* 40:471–476.
65. D. L. Perry. 1965. Low loss multilayer dielectric mirrors. *Applied Optics* 4:987–991.
66. K. H. Behrndt and D. W. Doughty. 1966. Fabrication of multilayer dielectric films. *Journal of Vacuum Science and Technology* 3:264–272.

67. P. Giacomo and P. Jacquinot. 1952. Localisation précise d'un maximum ou d'un minimum de transmission en fonction de la longeur d'onde: Application à la préparation des couches minces. *Journal de Physique et le Radium* 13:59A–64A.
68. J. Ring. 1957. PhD Thesis. Manchester: University of Manchester.
69. P. H. Lissberger and J. Ring. 1955. Improved methods for producing interference filters. *Optica Acta* 2: 42–46.
70. R. Hiraga, N. Sugawara, S. Ogura et al. 1974. Measurement of spectral characteristics of optical thin film by rapid scanning spectrophotometer. *Japanese Journal of Applied Physics* 13 (Supplement 2-1): 689–692.
71. J. P. Borgogno, P. Bousquet, F. Flory et al. 1981. Inhomogeneity in films: Limitation of the accuracy of optical monitoring of thin films. *Applied Optics* 20:90–94.
72. F. Flory, B. Schmitt, E. Pelletier et al. 1983. Interpretation of wide band scans of growing optical thin films in terms of layer microstructure. *Proceedings of SPIE* 401:109–116.
73. B. T. Sullivan and J. A. Dobrowolski. 1992. Optical multilayer coatings produced with automatic deposition error compensation. In *Optical Interference Coatings*, pp. 278–279, Optical Society of America, Tucson, AZ.
74. B. T. Sullivan and J. A. Dobrowolski. 1992. Deposition error compensation for optical multilayer coatings: 1. Theoretical description. *Applied Optics* 31:3821–3835.
75. B. T. Sullivan and J. A. Dobrowolski. 1993. Deposition error compensation for optical multilayer coatings: II. Experimental results—Sputtering system. *Applied Optics* 32:2351–2360.
76. K. Starke, T. Grosz, M. Lappschies et al. 2000. Rapid prototyping of optical thin film filters. *Proceedings of SPIE* 4094:83–92.
77. D. Ristau, H. Ehlers, T. Gross et al. 2006. Optical broadband monitoring of conventional and ion processes. *Applied Optics* 45:1495–1501.
78. S. Dligatch and R. P. Netterfield. 2007. In-situ monitoring and deposition control of a broadband multilayer dichroic filter. In *Optical Interference Coatings*, pp. 1–3, WC4. Optical Society of America, Washington, DC.
79. A. Zoeller, M. Boos, R. Goetzelmann et al. 2005. Substantial progress in optical monitoring by intermittent measurement technique. *Proceedings of SPIE* 5963:59630D 1–9.
80. A. Wajid. 1992. Measuring and controlling deposition on a piezoelectric monitor crystal. US Patent, 5,112,642.
81. R. P. Riegert. 1968. Optimum usage of quartz crystal monitor based devices. In *IVth International Vacuum Congress*, pp. 527–530, Institute of Physics and the Physical Society, Manchester.
82. O. S. Heavens. 1954. All-dielectric high-reflecting layers. *Journal of the Optical Society of America* 44:371–373.
83. P. H. Lissberger. 1959. Properties of all-dielectric filters: I. A new method of calculation. *Journal of the Optical Society of America* 49:121–125.
84. P. H. Lissberger and W. L. Wilcock. 1959. Properties of all-dielectric interference filters: II. Filters in parallel beams of light incident obliquely and in convergent beams. *Journal of the Optical Society of America* 49:126–130.
85. P. Giacomo, P. W. Baumeister, and F. A. Jenkins. 1959. On the limiting bandwidth of interference filters. *Proceedings of the Physical Society* 73:480–489.
86. P. W. Baumeister. 1962. Methods of altering the characteristics of a multilayer stack. *Journal of the Optical Society of America* 52:1149–1152.
87. V. N. Smiley and F. E. Stuart. 1963. Analysis of infrared interference filters by means of an analog computer. *Journal of the Optical Society of America* 53:1078–1083.
88. S. D. Smith and J. S. Seeley. 1968. *Multilayer Filters for the Region 0.8 to 100 Microns*. Bedford, MA: Air Force Cambridge Research Laboratories.
89. G. U. Yule and M. G. Kendall. 1958. *An Introduction to the Theory of Statistics*. Fourteenth ed. London: Charles Griffin & Co.
90. H. A. Macleod. 1972. Turning value monitoring of narrow-band all-dielectric thin-film optical filters. *Optica Acta* 19:1–28.
91. P. Bousquet, A. Fornier, R. Kowalczyk et al. 1972. Optical filters: Monitoring process allowing the autocorrection of thickness errors. *Thin Solid Films* 13:285–290.
92. H. A. Macleod and D. Richmond. 1974. The effect of errors in the optical monitoring of narrow-band all-dielectric thin film optical filters. *Optica Acta* 21:429–443.

93. E. Pelletier, R. Kowalczyk, and A. Fornier. 1973. Influence du procédé de contrôle sur les tolérances de réalisation des filtres interférentiels à bande étroite. *Optica Acta* 20:509–526.
94. H. A. Macleod. 1976. Thin film narrow band optical filters. *Thin Solid Films* 34:335–342.
95. B. Vidal, A. Fornier, and E. Pelletier. 1978. Optical monitoring of nonquarterwave multilayer optical filters. *Applied Optics* 17:1038–1047.
96. H. A. Macleod. 1981. Monitoring of optical coatings. *Applied Optics* 20:82–89.
97. C. Clark and H. A. Macleod. 1997. Errors and tolerances in optical coatings. In *40th Annual Technical Conference Proceedings*, pp. 274–279, Society of Vacuum Coaters: New Orleans, LA.
98. A. V. Tikhonravov and M. K. Trubetskov. 2005. Computational manufacturing as a bridge between design and production. *Applied Optics* 44:6877–6884.
99. A. V. Tikhonravov, M. K. Trubetskov, and T. V. Amotchkina. 2006. Investigation of the effect of accumulation of thickness errors in optical coating production by broadband optical monitoring. *Applied Optics* 45:7026–7034.
100. A. V. Tikhonravov, M. K. Trubetskov, and T. V. Amotchkina. 2006. Statistical approach to choosing a strategy of monochromatic monitoring of optical coating production. *Applied Optics* 45:7863–7870.
101. H. K. Pulker. 1978. Coating production: New ideas at a time of demand. *Optical Spectra* 12:43–46.
102. C. J. vd. Laan and H. J. Frankena. 1977. Monitoring of optical thin films using a quartz crystal monitor. *Vacuum* 27:391–397.
103. A. Macleod and C. Clark. 2000. Envelopes in optical coating design. In *43rd Annual Technical Conference Proceedings*, pp. 197–202, Society of Vacuum Coaters, Denver, CO.
104. A. Macleod. 2010. Performance envelopes. *Society of Vacuum Coaters Bulletin* (Fall):22–26.
105. F. T. Goldstein. 2008. "Gedankenspektrum" methods in optical coatings. *Proceedings of SPIE* 7101: 710105–1 to 710104–6.
106. L. Gao, F. Lemarchand, and M. Lequime. 2012. Exploitation of multiple incidences spectrometric measurements for thin film reverse engineering. *Optics Express* 20:15734–15751.

14

Material Properties

Much of this chapter is concerned with the properties of materials, ways of measuring them, and some examples of the results of the measurements of the important parameters. Probably the most important properties from the thin-film point of view are given in the following list, although the order is not that of relative importance, which will vary from one application to another:

1. Optical properties such as refractive index and region of transparency
2. Method that must be used for the production of the material in thin-film form
3. Mechanical properties of thin films, such as hardness or resistance to abrasion, and the magnitude of any built-in stresses
4. Chemical properties such as solubility and resistance to attack by the atmosphere and compatibility with other materials
5. Toxicity
6. Price and availability
7. Other properties that may be important in particular applications, for example, electrical conductivity or dielectric constant

Item 7 is not one on which we comment further here. On the question of price and availability (item 6), also little can be said. The situation is changing all the time. Note, however, that price is of secondary importance to suitability. The cost of a failed batch of coatings is very great compared with the price of the source materials. Many companies are able to offer a wide range of materials completely ready for thin-film production, together with all the necessary information on the techniques that should be used.

14.1 Properties of Common Materials

So far, little has been said about the actual properties of the more useful materials employed in thin-film work. The list that follows is far from exhaustive, but gives the more important properties of some commonly used materials. Thin-film properties are very dependent on process conditions, and so the same material may exhibit a range of properties even varying from one machine to another similar one. What is particularly important is that, whatever the achieved properties, they should be sufficiently stable from one production run to another. Random fluctuations can be solved only by tighter control. Published figures therefore tend to be a guide to, rather than a precise indication of, film properties.

The material probably used more than any other in thin-film work is magnesium fluoride. This has an index of approximately 1.38 or 1.39 in the visible region (see Figure 14.1) and is extensively used in the antireflection of lenses. In the simplest case, this is a single layer. Early workers used fluorite (CaF_2), but this was found to be rather soft and vulnerable and was subsequently replaced by magnesium fluoride. Magnesium fluoride can be evaporated from a tantalum or molybdenum boat, and the best results are obtained when the substrate is hot at a temperature of

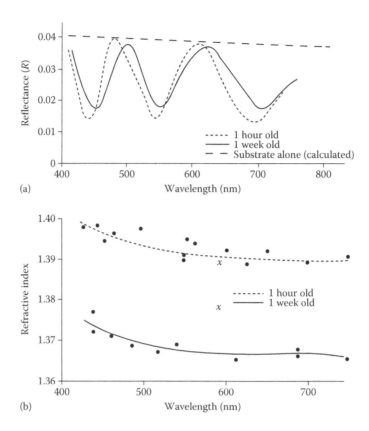

FIGURE 14.1
Refractive index of magnesium fluoride films. (a) The reflectance of a single film. (b) The reflectance result transforms into refractive index. The curves are formed by the results from many films. x denotes bulk indices of the crystalline solid. (After J. F. Hall, Jr. and W. F. C. Ferguson, *Journal of the Optical Society of America*, 45, 74–75, 1955. With permission of Optical Society of America.)

some 200–300°C. When magnesium fluoride is evaporated, trouble can sometimes be experienced through spitting and flying out of material from the boat. This is thought to be caused by thin coatings of magnesium oxide round the grains of magnesium fluoride in the evaporant. Magnesium oxide has a rather higher melting point than magnesium fluoride, and the grains tend to explode once they have reached a certain temperature. It is important, therefore, to use a reasonably pure grade of material, preferably one specifically intended for thin-film deposition and, especially, to protect it from atmospheric moisture.

Magnesium fluoride suffers, as do many of the fluorides, from rather high tensile stress. A thin film of infinite area, even if highly stressed, does not exert any shear stress across the interface with the substrate or neighboring material. However, in a film of limited area, the shear stress rises to a maximum at the edge, where delamination often begins. In single films of magnesium fluoride, the shear stress is not usually dangerously high, but in multilayers containing many magnesium fluoride layers, such as high reflectors, the total strain energy and consequent shear loading can become high enough for spontaneous destruction of the coating to occur. Thus, magnesium fluoride is not recommended for use in structures containing many layers.

Probably the easiest materials of all to handle are zinc sulfide and cryolite. They have a good refractive index contrast in the visible, the index of zinc sulfide being around 2.35 and that of cryolite being around 1.35. Both materials sublime rather than melt and can be deposited from a tantalum or molybdenum boat or else from a howitzer (shown in Chapter 13, Figure 13.6). Although these materials are not particularly robust, they are so easy to handle that they are very much used, especially in the construction of multilayer filters for the visible and near infrared that can subsequently be protected by a cemented cover slip. The substrates need not be heated for the deposition of the materials when intended for the visible region. Zinc sulfide is also a particularly useful material in the infrared out to about 25 μm. In the infrared, however, the substrates must be heated for best performance. The conditions are given by Cox and Hass [1], who state the best conditions to be on substrates that have been heated to around 150°C and cleaned with an effective glow discharge just prior to the evaporation and certainly not more than 5 minutes beforehand. Films produced under these conditions will withstand several hours of boiling in 5% salt water, exposure to humid atmospheres, and cleaning with detergent and cotton wool.

A trick, which has sometimes been used with zinc sulfide to improve its durability, is bombardment of the growing film with electrons. This can be achieved by positioning a negatively biased hot filament, somewhere near the substrate carrier, in such a way that the filament is shielded from the arriving evaporant, but is in the line of sight of the substrates. This process is still not entirely understood, but it has been suggested [2] that an important factor is the modification of the crystal structure of the zinc sulfide layers by electron bombardment. Resistively heated boats produce a mixture of the cubic zinc blende and the hexagonal wurtzite structure, while electron beam sources purely produce the zinc blende modification. The hexagonal form is a high-temperature modification, which, it is suspected, will tend to transform into the lower-temperature cubic modification, particularly when water vapor is present, a transformation accompanied by a weakening of adhesion and even delamination. Deliberate electron bombardment of growing zinc sulfide films from boat sources results in films with entirely cubic structure and with the improved stability expected from that structure.

For more durable films in the visible region, use can be made of a range of refractory oxide layers. More of these are available for the role of high-index layer than that for low-index index.

Cerium dioxide is a high-index material that is not now as commonly used as it once was. It can be evaporated from a tungsten boat (it strongly reacts with molybdenum, producing dense white powdery coatings that completely cover the inside of the system). The procedure to be followed is given by Hass et al. [3]. Unless the material is one of the types especially prepared for vacuum evaporation, it should first be fired in air at a temperature of around 700–800°C. If this procedure is not followed, the films will have a lower refractive index. Even with these precautions, cerium dioxide is an awkward material to handle. It tends to form inhomogeneous layers, and the index varies throughout the evaporation cycle as the material in the tungsten boat is used up. It is therefore difficult to achieve a very high performance from cerium dioxide layers, in terms of maximum transmission from a filter or from an antireflection coating, and its chief use tended to be in the production of high-reflectance coatings, for high-power lasers, for example, where high reflectance coupled with low loss was the primary requirement, and transmission in the pass region, not as important.

Titanium dioxide is nowadays preferred over cerium oxide and is probably one of the most common high-index materials for the visible and near infrared. It has the advantage of the highest index of any of the transparent high-index materials. It is extremely robust but has a rather high melting point of 1925°C, which makes it very difficult to directly evaporate from a boat source. Tungsten boats are most useful. One of the most successful early methods [4] was the initial evaporation of pure titanium metal, which is then subsequently oxidized in air by heating it to temperatures of 400–500°C. To obtain the highest possible index, it is important to evaporate the titanium metal as quickly as possible at as low a pressure as possible so that little oxygen is dissolved in the film. On oxidation in air, indices of around 2.65 can be attained. If the deposit is partially oxidized beforehand, the index is usually rather lower, on the order of 2.25. Other early methods involved the reaction between atmospheric moisture and titanium tetrachloride. Titanium dioxide forms on a hot surface introduced into the vapor of hot titanium tetrachloride in the presence of atmospheric moisture. Best results on glass are obtained when the temperature of the glass is maintained at around 200°C.

Both of these methods are useful for single layers but are almost impossibly complicated where multilayers are required. More modern alternative methods involve what is known as reactive deposition using either evaporation from electron beam sources or sputtering.

Reactive evaporation was developed as a useful process in the early 1950s, Auwärter [5] and Vogt [6] in Europe and Brinsmaid et al. [7] in the United States being major contributors. The problem with the direct evaporation of titanium dioxide is that the very high temperatures that are required cause the titanium dioxide to be reduced, so that absorption appears in the film. It was found that the reduced titanium oxide can be reoxidized to titanium dioxide during the deposition by ensuring that there is sufficient oxygen present in the atmosphere within the chamber. It appears that the oxidation actually takes place on the surface of the substrate rather than in the vapor stream, and the pressure of the residual atmosphere of oxygen must be arranged to be high enough for the necessary number of oxygen molecules to collide with the substrate surface. If the pressure is too high, then the film becomes porous and soft. There is therefore a range of pressures over which the process works best, usually 5×10^{-3} to 3×10^{-2} Pa (5×10^{-5} to 3×10^{-4} mbar). However, it is not possible to give hard and fast figures because they vary from machine to machine and depend on the particular evaporation conditions such as substrate temperature and speed of evaporation. The conditions must therefore be established by trial and error in each process. A suboxide is normally used as starting material. There are two reasons for this. The suboxide usually melts at a lower temperature than the dioxide or the metal and so is useful when a tungsten boat must be used. However, the reduction of the oxide in melting and vaporizing has been mentioned. This causes the composition of the vapor to vary unless the evaporation is what is known as congruent, that is, the composition of the vapor is the same as the composition of the material in the source. Experimental evidence shows that reasonably congruent evaporation is obtained when the composition is near either Ti_2O_3 or Ti_3O_5 [8]. It is usual to use a starting material that has one or other of these compositions. The evaporation should proceed slowly enough to ensure that complete oxidization takes place. This means that several minutes should be allowed for a thickness corresponding to a quarter wave in the visible region. Provided the rate of evaporation is kept substantially constant, then the refractive index of the film can be as high as 2.45 in the visible region. The titanium dioxide remains transparent throughout the visible, the absorption in the ultraviolet becoming intense at around 350 nm.

Titanium oxide is also used with success in sputtering processes. Sputtering is the process of bombardment of the material to be deposited with high-energy positive ions so that molecules are ejected and deposited on the substrate. Reactive sputtering is the same process except that the gas in the chamber is one that can and does react with the material as it is sputtered. Usually this gas is oxygen, and in this case, it reacts with the titanium to produce titanium dioxide without requiring any subsequent oxidation. The problems of poisoning of the sputtering cathodes and the various solutions have already been mentioned in connection with reactive sputtering. The rotating cylindrical magnetron and the midfrequency double magnetron are two current solutions.

The most complete account of the properties of titanium dioxide, and the way in which they depend on deposition conditions, is that of Pulker et al. [9]. The behavior is exceedingly complicated, and the results depend on starting material, oxygen pressure, rate of deposition, and substrate temperature. The evaporation of Ti_3O_5 as the starting material gave more consistent results than were obtained with other possible starting materials. With other forms of titanium oxide, the composition varied as the material was depleted, tending in each case towards Ti_3O_5.

Apfel [10] pointed out the conflict between high optical properties and durability. Optical absorption falls as the substrate temperature is reduced and the residual gas pressure is raised. At the same time, the durability of the layers is adversely affected, and a compromise, which depends on the actual application, is usually necessary. Substrate temperatures between 200°C and 300°C are usually satisfactory, with gas pressures around 1.3×10^{-2} Pa (1.3×10^{-4} mbar or 10^{-4} Torr in the paper).

The low-index material that is normally used in conjunction with titanium dioxide is silicon dioxide (silica). Indeed, there is virtually no other choice among the oxides. The usual current method for the evaporation of silicon dioxide uses an electron beam source. Chunks of silica or machined plates are used as source material, and a slight background pressure of oxygen may sometimes be used. The silicon oxide forms amorphous layers that are dense and resistant. As with most materials, a high substrate temperature during deposition is an advantage.

The high melting temperature of silica makes it difficult to evaporate it directly from heated boats. However, it is possible to use a reactive method [5,7] that avoids this problem. Silicon monoxide is a convenient starting material, which, in its own right, is a useful material for the infrared. The silicon monoxide can be readily evaporated from a tantalum boat or, as the material sublimes rather than melts, a howitzer source (Figure 13.6) or a baffled box source (Figure 13.5a). Provided that there is sufficient oxygen present, the silicon monoxide will oxidize to a form mostly Si_2O_3 that has a refractive index of 1.52–1.55 and exhibits excellent transmission from just on the longwave side of 300 nm out to 8 µm [11].

An interesting effect involving the ultraviolet irradiation of films of Si_2O_3 has been reported [12]. With ultraviolet power density corresponding to a 435 W quartz envelope Hanovia lamp at a distance of 20 cm, the refractive index of the film, after 5 five hours of exposure, drops to 1.48 (at 540 nm). This change in refractive index appears to be due to an alteration in the structure of the film, rather than in the composition, that remains Si_2O_3. At the same time as the reduction in refractive index, an improvement in the ultraviolet transmission is observed, the films becoming transparent to beyond 200 nm. Longer exposure to ultraviolet, around 150 hours, does eventually alter the composition of the films to SiO_2. These changes appear to be permanent. Si_2O_3 is a particularly useful material for protecting aluminum mirrors, and this method of improvement by ultraviolet irradiation opens the way to greatly improved mirrors for the quartz ultraviolet. The effect was studied in some detail by Mickelsen [13], who proposed an explanation involving electron traps.

Heitmann [14] made considerable improvements to the reactive process by ionizing the oxygen in a small discharge tube through which the gas is admitted to the coating chamber. The degree of ionization is not high, but the reactivity of the oxygen is enormously improved, and the titanium oxide and silicon oxide films produced in this way have appreciably less absorption than those deposited by the conventional reactive process. The silicon oxide films show infrared absorption bands characteristic of the SiO form rather than the more usual Si_2O_3. The technique was further improved by Ebert [15] and his colleagues, who developed a more efficient hollow cathode ion source and extended the method to materials such as beryllium oxide, with useful transmittance in the ultraviolet.

Other materials found useful in thin films are the oxides and fluorides of a number of the lanthanides or rare earths. Ceric oxide [3], although possibly strictly not a rare earth, has already been mentioned. Cerium fluoride forms very stable films of index 1.63 at 550 nm when evaporated from a tungsten boat.

Similarly, the oxides of lanthanum, praseodymium and yttrium, and their fluorides, form excellent layers when evaporated from tungsten boats. A good account of their properties is given by Hass et al. [16]. The properties of the rare earth oxides have been shown [17] to have improved transparency, especially in the ultraviolet, when electron beam evaporation is used.

A detailed study of the fluorides of the lanthanides and their usefulness in the extreme ultraviolet (in fact, there is little else that can be used in that region) was performed by Lingg [18] and Lingg et al. [19].

Then there is a number of other hard oxide materials that were extremely difficult to evaporate until the advent of the high-power electron beam gun and so were used only relatively infrequently, if at all. Zirconium dioxide [17,20] is a very tough, hard material which has good transparency from around 350 nm to some 10 µm. It tends to give inhomogeneous layers, the degree of inhomogeneity principally depending on the substrate temperature. Hafnium oxide [17,21] has good transparency to around 235 nm and an index around 2.0 at 300 nm, so that it is a good high-index material for that region. It is a preferred high-index material for high power laser mirrors. Both yttrium and hafnium oxide have been found to be good protecting layers for aluminum in the 8–12 μm region [22,23], which avoid the drop in reflectance at high angles of incidence associated with SiO_2 and with Al_2O_3.

Possibly due to their increased use in multiplexing and demultiplexing filters during the telecommunications surge of the late 1990s and early 2000s, tantalum pentoxide and niobium pentoxide have become exceedingly popular. Both materials are suitable for deposition by a wide range of processes, from thermal evaporation through ion-assisted deposition, to sputtering, and exhibit very low losses in their regions of transparency [24–27]. Under the bombardment inherent in the energetic processes, both materials become amorphous, and so films can exhibit exceptionally low scattering losses. Tantalum is used in so many applications that its supply can be a little problematic, and thus, tantalum pentoxide can be rather expensive. Niobium pentoxide has very similar properties, and although its shortwave properties are a little inferior to those of tantalum pentoxide, its use is rapidly increasing. Their stable and predictable behavior has meant that these two materials have largely replaced titanium dioxide as the principal high-index material for precision coatings in the visible and near infrared. They behave very well with silicon dioxide as the accompanying low-index material.

In the infrared, many more possibilities are available. Semiconductors all exhibit a sudden transition from opacity to transparency at a certain wavelength known as the intrinsic edge. This wavelength corresponds to the energy gap between the filled valence band of electrons and the empty conduction band. At wavelengths shorter than this gap, photons are absorbed in the material because they are able to transfer their energy to the electrons in the filled valence band by lifting them into the empty conduction band. At wavelengths longer than this value, the photon energy is not sufficient, and apart from a little free carrier absorption, there is no mechanism for absorbing the energy, and the material appears transparent until the lattice vibration bands at rather long wavelengths are encountered. For the more common semiconductors, silicon and germanium, the intrinsic edge wavelengths are 1.1 μm and 1.65 μm, respectively. Thus, both of these materials are potentially useful in the infrared. A great advantage that they possess is their high refractive index, 3.5 for silicon and 4.0 for germanium.

Silicon, however, is not at all easy to evaporate because it strongly reacts with any crucible material, and almost the only way of dealing with it in thermal evaporation is to use an electron gun with a water-cooled crucible so that the cold silicon in contact with the crucible walls acts as its own container. The high thermal conductivity of silicon makes it necessary to use high power. Sputtering is a viable process, and in fact, most large-area silicon dioxide coatings are produced by the reactive sputtering of silicon from magnetron targets. The poisoning problem in reactive sputtering and its solutions have already been mentioned. Germanium, on the other hand, is a most useful material and straightforward techniques have been devised to handle it. Tungsten boats can be used provided that the total thickness of material to be deposited is not too great, say, 2 or 3 μm, because germanium does react with tungsten. Molybdenum boats have been used with greater success [21]. A quite satisfactory method is to use a crucible made from graphite and heated directly or indirectly when the germanium films obtained are extremely pure and free from absorption. Again, the method of choice nowadays is the electron beam source when the hearth material can be graphite or water-cooled copper.

There are other semiconductors of use as follows. Tellurium [28,29] has an index of 5.1 at 5 μm and good transmission from 3.5 μm to at least 12 μm and can be easily evaporated from a tantalum boat. Lead telluride [30–38] has an even higher index of around 5.5 with good transmission from 3.4 μm out to beyond 20 μm. A tantalum boat is the most suitable source. Care must be taken to not overheat the material; the temperature should be just enough to cause the evaporation to proceed, otherwise some alteration in the composition of the film will take place, causing an increase in free-carrier absorption and consequent falloff in longwave transparency. The substrates should be heated, best results being obtained with temperatures around 250°C, but as this will be too great for the low-index film, which is usually zinc sulfide, a compromise temperature that is rather lower, usually around 150°C, is often used for both materials. One difficulty with lead telluride is the ease with which it can be upset by impurities that cause free-carrier absorption. It is extremely important to use pure grades of material, and this applies to the accompanying zinc sulfide as well as the lead telluride, especially if the material is to be used at the longwave end of its transparent region. Lead telluride also appears to be incompatible with a number of other materials, particularly some of the halides, presumably because the material diffuses into the lead telluride generating free carriers. An annealing process which can in certain circumstances improve the transmission of otherwise absorbing films of lead telluride in the region beyond 12 μm is described by Evans and Seeley [33].

Lead telluride can in some circumstances behave in a curious way immediately after deposition [30,31]. The optical thickness of the material is observed to grow during a period of around 15 minutes while the layer is still under vacuum. Typical gains in optical thickness of a half-wave layer are on the order of 0.007 full waves, although in any particular case, it considerably varies and can often be zero. The reasons for this behavior are not clear, but the layers do not exhibit any further instability, once they have ceased growing. It is simply a matter of allowing for this behavior in the monitoring process.

A wide range of low-index materials is used in the infrared. Zinc sulfide [1,39] in comparison with the high-index semiconductors has a relatively low index. If an electron beam source is not available, then zinc sulfide should be deposited from a tantalum boat or, better still, a howitzer, on freshly cleaned substrates by a glow discharge and held at temperatures of around 150°C, if the maximum durability is to be obtained. Zinc sulfide films so treated will withstand boiling for several hours in 5% salt solution, cleaning with cotton wool, and exposure to moist air, without damage [1]. Silicon monoxide is another possibility [1,40]. It can also be deposited from a tantalum boat or a howitzer. The deposition rate should be fast, and the pressure, low, on the order of 1.3×10^{-3} Pa (1.3×10^{-5} mbar) or less if possible. The refractive index is around 1.85 at 1 μm and falls to 1.6 at 7 μm. A strong absorption band prevents the use of the material beyond 8 μm. Thorium fluoride, unfortunately radioactive, has been much used in the past, although it is less in favor nowadays because of its radioactivity. It is still used in high-power coatings for the CO_2 laser because a completely suitable replacement in that application has not yet been found. Then there are many other materials, such as fluorides of lead, lanthanum, barium, cerium, and oxides such as titanium, yttrium, hafnium, and cerium.

The nitrides of silicon and aluminum are tough, hard materials with excellent transparency from the ultraviolet to around 10 μm in the infrared. They have not been much used in optical coatings because of the difficulty of thermal evaporation. The process of reactive evaporation of the metal in nitrogen does not work because the nitrogen, unless it is in atomic form, does not readily combine with the metal. The evaporation of aluminum, for example, in a residual atmosphere of nitrogen results in bright aluminum films, whereas evaporation in oxygen gives aluminum oxide. The situation has completely changed with the introduction of the energetic processes and especially ion-assisted deposition. The nitrogen beam from the ion source used in these processes strongly reacts with the metal to form dense, hard, and tough nitride films of good transparency. There is another enormous advantage in these materials. The oxynitrides represent a continuous range of compositions between the pure oxide and the pure nitride. The oxide is of rather lower refractive index, and the refractive index of the oxynitride smoothly ranges with composition from that of the oxide to that of the nitride. The composition of the film is a function of the reacting gas composition, and this can be readily varied to alter the film index in a well-controlled manner. Hwangbo et al. [41] investigated the ion-assisted deposition of aluminum oxynitride. They used aluminum metal as source material. A particularly straightforward way of controlling the index of aluminum oxynitride films from 1.65 to 1.83 at 550 nm was to bombard the growing film with a constant flux of nitrogen from the ion gun and to supply a variable quantity of oxygen to the process simply as a background gas. The reactivity of the oxygen is so great that any small quantity is preferentially taken up by the film. In fact, in the oxynitride process, it is virtually impossible to eliminate oxygen entirely, and so the achievable high index does not quite reach the value that would be associated with the pure nitride. Hwangbo was able to construct simple rugate filters with the sole variable during the process being the background pressure of oxygen, all other quantities, bombardment, evaporation rate, and so on, being held constant. Placido [42] constructed rugate structures of very many accurately controlled cycles from aluminum oxynitride using reactive RF sputtering of aluminum metal in a mixture of oxygen and nitrogen.

Bovard et al. [43] produced silicon nitride films using low-voltage ion plating. Here there was no oxygen in the chamber, and the films were pure nitride giving a refractive index of 2.05 at 550 nm. The range of variation in index from silicon oxynitride films is potentially very great.

Mixtures of materials are receiving attention both in deliberately inhomogeneous films and in homogeneous films where an intermediate index between the two components of the mixture is required. Often such a mixture can replace a rather more difficult single component material.

Jacobsson and Martensson [44] used mixtures of cerium oxide and magnesium fluoride, of zinc sulfide and cryolite, and of germanium and magnesium fluoride, with the relative concentration of the two components varying smoothly throughout the films, to produce inhomogeneous films with a refractive index variation of a prescribed law. Some of the results they obtained for antireflection coatings are mentioned in Chapter 4. To produce the mixture, two separate sources, one for each material, were used; they were simultaneously evaporated but with independent rate controls. Apparently no difficulty in obtaining reasonable films was experienced, the mixing taking place without causing absorption to appear.

Fujiwara [45,46] was interested in the production of homogeneous films for antireflection coatings [47]. The three-layer quarter–half–quarter coating for glass requires a film of intermediate index that is rather difficult to obtain with a simple material, and the solution adopted by Fujiwara was to use a mixture of two materials, one having a refractive index lower than the required value and the other higher. The two combinations that were successfully tried were cerium oxide and cerium fluoride and zinc sulfide and cerium fluoride. These were simply mixed together in powder form in a certain known proportion by weight and then evaporated from a single source. The mixture readily evaporated, giving an index that was sufficiently reproducible for antireflection coating purposes. The range of indices obtainable with the cerium oxide–cerium fluoride mixture was 1.60–2.13, and with the cerium fluoride–zinc sulfide mixture, 1.58–2.40. One interesting feature of the second mixture was that although zinc sulfide on its own is not particularly robust, in the form of a mixture with more than 20% by weight of cerium fluoride, the robustness was greatly increased, the films withstanding boiling in distilled water for 15 minutes without any deterioration. Curves are given for refractive index against mixing ratio in the papers.

Mixtures of zinc sulfide and magnesium fluoride were also studied by Yadava et al. [48]. The refractive index of the mixture varies between the indices of magnesium fluoride and zinc sulfide, depending on the mixing ratio, and the absorption edge varies from that of zinc sulfide to that of magnesium fluoride in a nonlinear fashion. The same authors [48,49] studied the use of assemblies of large numbers of alternate very thin discrete layers of the components instead of mixtures. For a wide range of material combinations, $ZnS–MgF_2$, $ZnS–MgF_2–SiO$, $Ge–ZnS$, and $ZnS–Na_3AlF_6$, for example, the results were similar to those expected from the evaporation of mixtures of the same materials.

Silica is a particularly difficult material to evaporate because of its high melting point and because of its transparency to infrared, which makes it difficult to heat. It was found by Morgan at the Libbey–Owens–Ford Glass Company [50] that silica could be thermally evaporated readily if some pretreatment were carried out. This consisted of combining the silica with a metallic oxide, a vast number of different oxides being suitable. The oxide can be intimately mixed with the silica, coated on the outer surface of silica chunks or, in some cases where the oxide has a rather lower melting temperature than the silica, mixed very crudely. Only a small quantity of the oxide is required, and the evaporation is carried out in the conventional manner from a tungsten source. The oxides mentioned include aluminum, titanium, iron, manganese, cobalt, copper, cerium, and zinc. Along similar lines, it was discovered by workers at Balzers AG [51,52] that cerium oxide mixed with other oxides improves the oxidation and increases the transparency and ease of evaporation. Materials such as titanium dioxide are difficult to evaporate without absorption, and the most successful method is reactive evaporation in oxygen, which produces absorption-free films, although the process is rather time-consuming because the evaporation must proceed slowly. With the addition of a small amount of cerium oxide—the mixture can vary from 1:1 to 8:1 titanium oxide (the monoxide, the dioxide, or even the pure metal) to cerium oxide—hard films free from absorption, even when quickly evaporated at pressures of 1.3×10^{-3} Pa (1.3×10^{-5} mbar or 10^{-5} Torr), are readily obtained. Apparently, this effect is not limited to titanium oxide, and a vast range of different materials that have been successfully tried is given. Other rare earth oxides and mixtures of rare earth oxides can also take the place of the cerium dioxide.

Stetter et al. [20] pointed out the advantage of oxygen-depleted materials as source material for electron beam evaporation, in that composition changes little if at all during evaporation, which leads to more consistent film properties. The extra oxygen is supplied, in the usual way, from the residual atmosphere in the machine. The depleted materials also have higher thermal and electrical conductivities. A mixture of ZrO_2 and $ZrTiO_4$, sintered at high temperature under high vacuum and oxygen-depleted, was developed. This material, designated as Substance no 1, when evaporated from an electron beam system in a residual oxygen pressure of 1.3–2.5×10^{-4} mbar (1.3–2.5×10^{-2} Pa) with substrate temperature of $270°C$, and condensation rate on the order of 10 nm/min, gives homogeneous layers of refractive index 2.15 (at 500 nm). Such a value of index is ideal for the quarter–half–quarter antireflection coating for the visible region. This has prompted further work on mixtures [53], and there are now several similar materials available. H1 is from the zirconia/titania system with index 2.1 at 500 nm and good transparency from 360 nm to 7 μm but with some difficulties in evaporation because of incomplete melting. H2 from the praseodymium/titanium oxide system has a similar index and the advantage of ease of evaporation but suffers from a more restricted range of good transmittance (400 nm to 7 μm) and localized slight absorption in the transparent region. H4 is a lanthanum/titanium oxide combination with again refractive index 2.1 at 500 nm and transmission region from 360 nm to 7 μm, which completely melts and so is normally preferred over the other two materials. M1 is a mixture of praseodymium/aluminum oxide with index on heated substrates of 1.71 at 500 nm and good transparency from 300 nm to longer wavelengths.

Butterfield [54] produced films of a mixture of germanium and selenium. For composition varying from 30 to 50 at.% of germanium, glassy films with refractive index in the range 2.4–3.1, with good transparency from 1.5 to 15 μm, could be produced. The starting material was an alloy of germanium and selenium in the correct proportions, produced by melting the pure substances in an evacuated quartz tube. The evaporation source was a graphite boat. It is likely that much more work will be carried out on mixtures, because of the apparent ease with which the deposition can be performed to give a wide range of refractive indices, many of which are not available by other means. The theory of the optical properties of mixtures is covered in a useful review by Jacobsson [55], who also gives further information on mixtures, and on inhomogeneous layers.

A recent development is a mixture of tantalum pentoxide and titanium dioxide. The mirrors required for the interferometric gravitational wave receivers are required to be of exceptionally low loss, and ion-beam sputtering of silica and tantala has been found to be capable of the required performance. However, optical reflectance is not the sole criterion. The detection of the exceptionally small fringe movements are perturbed by mechanical vibration that must be suppressed over the frequency range of the detection window. An unexpected source of vibration was discovered to be the mirrors themselves. Thermal fluctuations that are always present at any temperature were found to be coupling to mechanical vibrations in the materials of the coatings, principally the tantalum pentoxide layers. Of course, only with an application of this demanding nature would such an effect be detected. It has been found that the inclusion of titania in the tantalum pentoxide results in a significant reduction in the level of induced vibrational noise, and so the latest mirrors employ the mixture rather than the pure material [56–58].

14.2 Measurement of the Optical Properties

Once a suitable method of producing the particular thin film has been determined, the next step is the measurement of the optical properties. Many methods for this exist, and a useful earlier account is given by Heavens [59]. The measurement of the optical constants of thin films is also included in the book by Liddell [60]. A more recent survey is that of Borgogno [61]. Stenzel [62] includes a chapter on optical constant measurement in his book on thin-film materials. Recently, the measurement of the optical properties of thin films has increased in importance to the extent that special purpose instruments are now available. These normally include the extraction software and are essentially push button in operation. As always, however, even when automatic tools are available, some understanding of the nature of the process and its limitations is still necessary. Here we shall be concerned with just a few methods that are frequently used. Many techniques exist. The ones we look at here are mostly routine ones that could exist in any coating operation. We omit those that would be more likely to be found only in the research laboratory.

In all of this, it is important to understand that we never actually measure the optical constants n and k directly. Although thickness d is more susceptible to direct measurement, its value is frequently the product of an indirect process too. The extraction of these properties, and others, involves measurements of thin-film behavior followed by a fitting process in which the parameters of a film model are adjusted so that the calculated behavior of the model matches the measured data. The adjustable parameters of the model are then taken to be the corresponding parameters of the real film. The operation is dependent on a model that closely corresponds to the real film. The appropriateness of the model would be of less importance were we simply trying to recast the measurements in a more convenient form. Even an inadequate model with parameters appropriately adjusted can be expected to reconstitute the original measurements. However, the parameters extracted are rarely used in that role. Rather they are used for predictions of film performance in other situations where film thickness may be quite different and where the film is part of a much more complex structure. This leads to the idea of *stability* of optical constants, a rather different concept from accuracy. The accurate fitting of measured data by using an inappropriate model may reproduce the measurements with immense precision yet yield predictions for other film thicknesses that are seriously in error. Such parameters are lacking in stability. Stable optical constants might reproduce the measured results with only satisfactory precision but would have equal success in a predictive role. A good example might be a case where a film that is really inhomogeneous and free from absorption is modeled by a homogeneous and absorbing film. The extracted film parameters in this case can be completely misleading. It must always be remembered that the film model is of fundamental importance. Some of these problems are discussed in the following section.

As important as the model is the accuracy of the actual measurements. Calibration verification is an indispensable step in the measurement of the performance that will be used for the optical constant extraction. Remember that only two parameters are required to define a straight line, but to verify linearity requires more. Small errors in measurement can have especially serious consequences in the extinction coefficient and/or assessment of inhomogeneity of the film. The samples themselves should be suitable for the quality of measurement. For example, a badly chosen substrate may deflect the beam partially out of the system so that the measurement is deficient, or it may introduce scattering losses that are not characteristic of the film.

The calculation of performance, given the design of an optical coating, is a straightforward matter. Optical constant extraction is quite different. Each film is a separate puzzle. It may be necessary to try different techniques and different models. Repeat films of different thicknesses or on different substrates may be required. Some films may appear to defy rational explanation. A common film defect is a cyclic inhomogeneity that produces measurements that the usual simpler film models are incapable of fitting with sensible results. It is always worthwhile to attempt to recalculate the measurements using the model and extracted parameters to see where deficiencies might lie. Because of all the caveats in this and the previous paragraphs, exact correspondence, however, does not necessarily indicate perfect extraction, but a bad fit is certainly a warning sign.

As we saw in Chapter 2, given the optical constants and thicknesses of any series of thin films on a substrate, the calculation of the optical properties is straightforward. The inverse problem, that of calculating the optical constants and thicknesses of even a single thin film, given the measured optical properties, is much more difficult, and there is no general analytical solution to the problem of inverting the equations. For an ideal thin film, there are at least three parameters involved, n, k, and d, the real and imaginary parts of refractive index and the geometrical thickness, respectively. Both n and k vary with wavelength, which increases the complexity. The traditional methods of measuring optical constants, therefore, mostly rely on special limiting cases that have straightforward solutions.

Nowadays, the model fitting is most often carried out by automatic computer optimization. It was not always so, and there is a long history of analytical methods that are worth examining because they give an insight into the process that is missing from completely automatic computing.

Perhaps the simplest case of all is represented by a quarter wave of material on a substrate, both of which are lossless and dispersionless, that is, k is zero and n is constant with wavelength. The reflectance is given by

$$R = \left(\frac{1 - n_f^2/n_m}{1 + n_f^2/n_m}\right)^2, \tag{14.1}$$

where n_f is the index of the film and n_m that of the substrate and the incident medium is assumed to have an index of unity. Then, n_f is given by

$$n_f = n_m^{1/2}\left(\frac{1 \pm R^{1/2}}{1 \mp R^{1/2}}\right)^{1/2}, \tag{14.2}$$

where the refractive index of the substrate n_m must, of course, be known, and the plus or minus sign will be chosen depending on whether or not the film index is greater or less than that of the substrate. The measurement of reflectance must be reasonably accurate. If, for instance, the refractive index is around 2.3, with a substrate of glass, then the reflectance should be measured to around one third of a percent (absolute ΔR of 0.003) for a refractive index measurement accurate in the second decimal place. It is sometimes claimed that this method gives a more accurate value for refractive index than the original measure of reflectance since the square root of R is used in the calculation. This may be so, but the value obtained for refractive index will be used in the subsequent calculation of the reflectance of a coating, and therefore, the computed figure can be only as good as the original measurement of reflectance.

In the absence of dispersion, the curve of reflectance versus wavelength of the film will be similar to that in Figure 14.2. The extrema correspond to integral numbers of quarter waves, even numbers being half-wave absentees and giving reflectance equal to that of the uncoated substrate, and odd corresponding, to the quarter wave of Equations 14.1 and 14.2. Thus, it is easy to pick out those values of reflectance that correspond to the quarter waves.

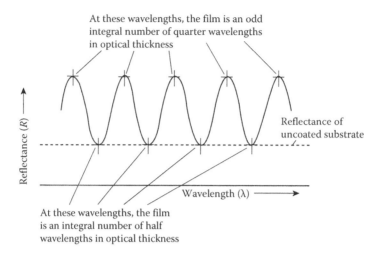

At these wavelengths, the film is an odd integral number of quarter wavelengths in optical thickness

Reflectance of uncoated substrate

At these wavelengths, the film is an integral number of half wavelengths in optical thickness

FIGURE 14.2
Reflectance of a simple thin film.

The technique can be adapted to give results in the presence of slight dispersion. The maxima in Figure 14.2 will now no longer be at the same heights, but provided the index of the substrate is known throughout the range, the heights of the maxima can be used to calculate values for film index at the corresponding wavelengths. Interpolation can then be used to construct a graph of refractive index against wavelength. Results obtained by Hall and Ferguson [63] for MgF$_2$ are shown in Figure 14.1.

This simple method yields results that are usually sufficiently accurate for design purposes. If, however, the dispersion is somewhat greater or if rather more accurate results are required, then the slightly more involved formulas given by Hass et al. [16] must be applied. It is still assumed that the absorption is negligible. If the curve of reflectance or transmittance of a film possessing dispersion is examined, it will be easily seen that the maxima corresponding to the odd quarter-wave thicknesses are displaced in wavelength from the true quarter-wave points, while the half-wave maxima are unchanged. This shift is due to the dispersion, and the measurement of it can yield a more accurate value for the refractive index. In the absence of absorption, the turning values of R, T, $1/R$, and $1/T$ must all coincide. Assuming that the refractive index of the incident medium is unity, that of the substrate n_m and of the film n_f, then their expression for T becomes

$$T = \frac{4}{n_m + 2 + n_m^{-1} + 0.5 n_m^{-1}\left(n_f^2 - 1 - n_m^2 + n_m^2 n_f^{-2}\right)\left[1 - \cos(4\pi n_f d_f / \lambda)\right]}.$$

Since the turning values of T and $1/T$ coincide, the positions of the turning values can be found in terms of d/λ by differentiating the expression for $1/T$ and equating it to zero as follows:

$$\frac{1}{T} = \frac{n_m + 2 + n_m^{-1}}{4} + \frac{1}{8 n_m}\left(n_f^2 - 1 - n_m^2 + n_m^2 n_f^{-2}\right)\left[1 - \cos\left(\frac{4\pi n_f d_f}{\lambda}\right)\right],$$

i.e.,

$$0 = \frac{d(1/T)}{d(d/\lambda)} = 0.25 n'_f\left(n_m^{-1} n_f - n_m n_f^{-3}\right)\left[1 - \cos\left(\frac{4\pi n_f d_f}{\lambda}\right)\right]$$

$$+ 0.5\pi\left(n_m^{-1} n_f^2 - n_m^{-1} - n_m + n_m n_f^{-2}\right)\left(n_f + n'_f \frac{d_f}{\lambda}\right)\sin\left(\frac{4\pi n_f d_f}{\lambda}\right),$$

where $n'_f = dn_f/d(d/\lambda)$. That the equation is satisfied exactly at all half-wave positions can easily be seen since both $\sin(4\pi n_f d_f/\lambda)$ and $(1 - \cos 4\pi n_f d_f/\lambda)$ are zero. At wavelengths corresponding to odd quarter waves, a shift does occur, and this can be determined by manipulating the preceding equation into

$$\tan\left(\frac{2\pi n_f d_f}{\lambda}\right) = -2\pi \frac{n_f^5 - (1 + n_m^2)n_f^3 + n_m^2 n_f}{n_f^4 - n_m^2}\left(\frac{n_f}{n'_f} + \frac{d_f}{\lambda}\right). \tag{14.3}$$

Of course, it is impossible to immediately solve this equation for n_f because there are too many unknowns. Generally, the most useful approach is by successive approximations using the simpler quarter-wave formula (Equation 14.2) to obtain a first approximation for the index and the dispersion. It should be remembered that the reflection of the rear surface of the test glass should be taken into account in the derivation of the reflectance curve. It is also important that the test glass should be free from dispersion to a greater degree than the film; otherwise, it must also be taken into account with consequent complication of the analysis.

If absorption is present, then Equation 14.3 cannot be used. In the case of heavy absorption, it can safely be assumed that there is no interference, and the value of the extinction coefficient can be calculated from the expression

$$\frac{1 - R}{T} = \exp\left(\frac{4\pi k_f d_f}{\lambda}\right)$$

($4\pi k_f d_f/\lambda$ because we are dealing with energies not amplitudes), which gives the following [16] for k_f:

$$k_f = \frac{\lambda}{4\pi d_f \log e}\log\left(\frac{1 - R}{T}\right), \tag{14.4}$$

where the two logarithms are to the same base, usually 10.

The thin-film designer is not too concerned with very accurate values of heavy absorption. Often it is sufficient to merely know that the absorption is high in a given region, and the result given by Equation 14.4 will be more than satisfactory. In regions where the absorption is significant but not great enough to weaken the single-film interference effects, a more accurate method can be used.

We know from Chapter 2 that

$$\frac{T}{1 - R} = \frac{\text{Re}(y_m)}{\text{Re}(BC^*)} = \frac{\text{Re}(n_m)}{\text{Re}(BC^*)}. \tag{14.5}$$

For a single film on a transparent substrate, the values of B and C are given by

$$\begin{bmatrix} B \\ C \end{bmatrix} = \begin{bmatrix} \cos\delta_f & (i\sin\delta_f)/N_f \\ iN_f\sin\delta_f & \cos\delta_f \end{bmatrix}\begin{bmatrix} 1 \\ n_m \end{bmatrix} = \begin{bmatrix} \cos\delta_f + i(n_m/N_f)\sin\delta_f \\ n_m\cos\delta_f + iN_f\sin\delta_f \end{bmatrix}.$$

Now

$$\delta_f = \frac{2\pi N_f d_f}{\lambda} = \varphi - i\psi = \frac{2\pi n_f d_f}{\lambda} - i\frac{2\pi k_f d_f}{\lambda}. \tag{14.6}$$

We shall assume that k_f is small compared with n_f, and this implies that ψ will be small compared with φ. Now for sufficiently small ψ,

$$\cos\delta = \cos\varphi\cosh\psi + i\sin\varphi\sinh\psi \approx \cos\varphi + i\psi\sin\varphi$$

and

$$\sin \delta = \sin \varphi \cosh \psi - i \cos \varphi \sinh \psi = \sin \varphi - i\psi \cos \varphi,$$

yielding the following expressions for B and C:

$$\begin{bmatrix} B \\ C \end{bmatrix} = \begin{bmatrix} [1 - (n_m/n_f)\psi] \cos \varphi - (n_m k_f/n_f^2) \sin \varphi + i[\psi + (n_m/n_f)] \sin \varphi \\ (n_m + n_f\psi) \cos \varphi + k_f \sin \varphi + i(n_f + n_m\psi) \sin \varphi \end{bmatrix}. \tag{14.7}$$

At wavelengths where the optical thickness is an integral number of quarter wavelengths, $\sin \varphi$ or $\cos \varphi$ is zero, and we can neglect terms in $\cos \varphi \sin \varphi$. The value of the real part of (BC^*) is then given by

$$\mathrm{Re}(BC^*) = \cos^2\varphi \left(1 + \frac{n_m}{n_f}\psi\right)(n_m + n_f\psi) + \sin^2\varphi \left(\psi + \frac{n_m}{n_f}\right)(n_f + n_m\psi)$$
$$= \left[n_m + \left(\frac{n_m^2}{n_f} + n_f\right)\psi\right], \tag{14.8}$$

and when substituted in Equation 14.5, it yields

$$\frac{1-R}{T} = 1 + \left(\frac{n_m}{n_f} + \frac{n_f}{n_m}\right)\psi, \tag{14.9}$$

giving for k_f (using the Equation 14.6 in Equation 14.9)

$$k_f = \left(\frac{\lambda}{2\pi d_f[(n_m/n_f) + (n_f/n_m)]}\right)\left(\frac{1-R-T}{T}\right). \tag{14.10}$$

This expression is accurate only close to the turning values of the reflectance or transmittance curves.

In the case of low absorption, the index should also be corrected. Hall and Ferguson [39] give the following expression:

$$n_f = \left[\frac{n_m(1 + \sqrt{R})}{1 - \sqrt{R}}\right]^{1/2} + \frac{\pi k_f d_f}{\lambda}\left[\frac{1 + \sqrt{R}}{1 - \sqrt{R}} - n_m\right], \tag{14.11}$$

where R is the value of reflectance of the film at the reflectance maximum.

In the methods discussed so far, we have been assuming that the thickness of the film is unknown, except inasmuch as it can be deduced from the measurements of reflectance and transmittance, and the extrema have been the principal indicator of film thickness. However, it is possible to measure film thickness in other ways, such as multiple beam interferometry, or electron microscopy, or by using a stylus step-measuring instrument. If there could be an independent accurate measure of physical thickness, then the problem of calculating the optical constants might become much simpler. This was the basis of a technique devised by Hadley (see Heavens [59] for a description). Since two optical constants n_f and k_f are involved at each wavelength, two parameters must be measured, and these can most conveniently be R and T. In the ideal form of the technique, if a value of n_f is now assumed, then by trial and error, one value of k_f can be found, which, together with the known geometrical thickness and the assumed n_f, yields the correct measured value of R, and then a second value of k_f that similarly yields the correct value of T. A different value of n_f will give two further values of k_f and so on. Proceeding thus, we can plot two curves of k_f against n_f, one corresponding to the T values and the other to the R values, and, where they intersect, we have the correct values of n_f and k_f for the film. The angle of intersection of the curves gives an indication of the precision of the result.

Hadley, at a time when such calculations were exceedingly cumbersome, produced a book of curves giving the reflectance and transmittance of films as a function of the ratio of geometrical thickness to wavelength, with n_f and k_f as parameters, which greatly speeded up the process. Nowadays, the method can be readily programmed, and precision, estimates incorporated. This method can be applied to any thickness of film, not just at the extrema, although maximum precision is achieved, as we might expect, near optical thicknesses of odd quarter waves, while, at half-wave optical thicknesses, it is unable to yield any results. As with many other techniques, it suffers from multiple solutions, particularly when the films are thick, and in practice, a range of wavelengths is employed, which adds an element of redundancy and helps eliminate some of the less probable solutions.

Hadley's method involves simple iteration and does not require any very powerful computing facilities. Even in the absence of Hadley's precalculated curves, it can be accommodated on a programmable calculator of modest capacity. It does, however, involve the additional measurement of film thickness, which is of a different character from the measurements of R and T. This is the primary disadvantage. There is a problem with virtually all techniques that make independent measurements of thickness. Unless the thickness is very accurately determined and the model used for the thin film is well chosen, the values of optical constants that are derived may have quite serious errors. The source of the difficulty is that the extrema of the reflectance or transmittance curves are essentially fixed in position by the value of n_f and d_f. There is only a very small influence on the part of k_f. Should the value for d_f be incorrect, then there is no way in which a correct choice of n_f can satisfy both the value and the position of the extremum. What happens, then, is that the extremum position is assured by an apparent dispersion, usually enormous and quite false, and the values of n_f are then seriously in error, sometimes showing abrupt gaps in the curve. The situation is often worse at the half-wave points than at the quarter-wave ones, but even in between the extrema, there are clear errors in level that tend to be alternately too high and then too low in between successive extremum pairs. A technique that has been used to avoid this difficulty is to permit some small variation of d_f around the measured value and to search for a value that removes to the greatest extent the incorrect features of the variation of n_f.

A different approach that was developed by Pelletier et al. [64] in Marseille and requires the use of powerful computing facilities retains the measurement of R and T, but instead of an independent measure of film thickness, adds the measurement of R', the reflectance of the film from the substrate side. Now we have three parameters to calculate at each wavelength and three measurements, and it might appear possible that all three could be calculated by a process of iteration, rather like the Hadley method, but the Marseille group found the possible precision rather poor, and it completely broke down when there was no absorption. To overcome this difficulty, the Marseille method uses the fact that the physical thickness of the film does not vary with wavelength, and therefore, if information over a spectral region is used, there will be sufficient redundancy to permit an accurate estimate of physical thickness. Then once the thickness has been determined, a computer method akin to refinement finds accurate values of the optical constants n_f and k_f over the whole wavelength region. For dielectric layers of use in optical coatings, k_f will usually be small, and often negligible, over at least part of the region, and a preliminary calculation involving an approximate value of n_f is able to yield a value for geometrical thickness, which, in most cases, is sufficiently accurate for the subsequent determination of the optical constants. Given the thickness, R and T, as we have seen, should in fact be sufficient to determine n_f and k_f. However, this would mean discarding the extra information in R', and so the determination of the optical constants uses successive approximations to minimize a figure of merit consisting of a weighted sum of the squares of the differences between measured T, R, and R' and the calculated values of the same quantities using the assumed values of n_f and k_f.

Although seldom necessary, the new values of the optical constants can then be used in an improved estimate of the physical thickness, and the optical constants, recalculated. For an estimate of precision, the changes in n_f and k_f to change the values of T, R, and R' by a prescribed amount, usually 0.3%, are calculated. Invariably, there are regions around the wavelengths for which the film is an integral number of half-waves thick, where the errors are greater than can be accepted and results in these regions are rejected. In practice, the films are deposited over half of a substrate, slightly wedged to eliminate the effects of multiple reflections, and measurements are made of R and R' and T and T', the transmittance measured in the opposite direction (theoretically identical), on both coated and uncoated portions of the substrate. This permits the optical constants of the substrate to be estimated; the redundancy in the measurements of T and T' gives a check on the stability of the apparatus. A very large number of different dielectric thin-film materials have been measured in this way, and a typical result is shown in Figure 14.3.

A particularly useful and straightforward family of techniques are known as envelope methods. The results that they yield are particularly stable. The envelope method was first described in detail by Manifacier et al. [65] and later elaborated by Swanepoel [66].

Let us imagine that we have a homogeneous dielectric film that is completely free from absorption. Let us deposit this film on a transparent substrate and gradually increase the thickness of this film, all the time measuring the reflectance. Let the maximum reflectance be given by R_{max}, and the minimum, by R_{min}. We can plot the locus of the film at one wavelength as the thickness increases, and this will appear as Figure 14.4.

The maximum and minimum reflectances will each be represented by an isoreflectance circle in the admittance plane. Any possible locus for the thin film must then be a circle tangent to both of them. There could be four such loci, but since the incident medium will usually be air, two of the loci must represent characteristic admittances less than unity and can therefore be discarded. There are then two possible remaining loci, both shown in the diagram. Should the addition of the film increase the reflectance above that of the uncoated substrate, then the substrate must be represented by point B in the diagram. There is then only one locus that can represent the film, that is, the locus with extreme points A and B. The reflectances at A and B can then be converted into admittances, and the square root of their product will be the characteristic admittance of the film. Should the film reduce the reflectance below that of the substrate, then the substrate must be represented by point A, and we now have two possible solutions for the admittance of the film. Provided the minimum reflectance is not too low, we should be able to distinguish the correct solution provided we have a sufficiently reasonable idea of the correct value. It is easy to see the danger, however, of a film that is

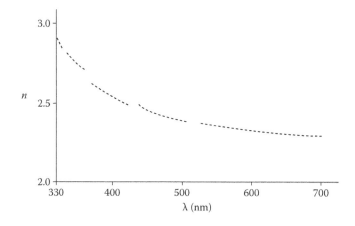

FIGURE 14.3
Refractive index of a film of zinc sulfide. The slight departure from a smooth curve is due to structural imperfections suggesting that even in this case of a very well behaved optical material, there is some very slight residual inhomogeneity. (After E. Pelletier, P. Roche, and B. Vidal, *Nouvelle Revue d'Optique*, 7, 353–362, 1976. Institute of Physics.)

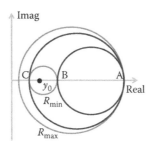

FIGURE 14.4
The circles labeled R_{max} and R_{min} are isoreflectance circles. The two circles tangent to both of them are possible loci for the thin film. Two other possible circular loci, tangent to both circles, are geometrically possible, but would represent admittances less than y_0 and so have been discarded.

acting as a good antireflection coating for the substrate. Then the two possible values will be close together, and it will be very difficult to separate them. In addition, it can be shown that the effect of errors in the measurement of reflectance have a much greater effect on the extracted value of characteristic admittance when the extremum represents a very low value.

So far, to extract the value of the film characteristic admittance, we do not require to calculate the admittance of the substrate separately. Let the film now be slightly inhomogeneous in a simple way where the refractive index uniformly and slowly changes through the film. The locus will no longer be that of a circle but a slowly contracting or expanding spiral, and there will be a gap between the notional half-wave points and the actual substrate admittance. Now, separate knowledge of the substrate admittance will also allow an estimate of the inhomogeneity of the film.

Now let the film be also slightly absorbing. A slightly absorbing film shows very little difference in reflectance when compared with an exactly similar but transparent film. However, this is not the case in transmittance. Transmittance is sensitive to both inhomogeneity and absorptance. The expressions are a little more complicated in transmittance; nevertheless, if we add transmittance measurements T_{max} and T_{min} to our corresponding reflectance measurements, we will be able to distinguish and separately estimate both absorptance and inhomogeneity.

Finally, provided we know the starting and finishing points for the film locus and the number of circles, or the number of exhibited extrema, then we can calculate the optical thickness, and hence, physical thickness of the film.

Unfortunately, we seldom have the necessary information in this form. What we generally have are plots of reflectance and/or transmittance of an already deposited film in terms of wavelength, and similar plots of an uncoated substrate. Figure 14.5 shows typical fringes as a function of wavelength. Two envelope curves that pass through the fringe extrema have been added to these fringes. The basis of the envelope technique is the assumption that these envelope curves in both reflectance and transmittance can be used at any wavelength as substitutes for the R_{max}, R_{min}, T_{max}, and T_{min} that would have been available had the growing film been continuously measured at every wavelength point.

Although, at first sight, this might seem a somewhat inaccurate technique because it depends on envelopes that might be arbitrary, it turns out that films that can be modeled as a straightforward, slightly absorbing, slightly inhomogeneous dielectric film exhibit well-formed fringes that lend themselves to simple envelope curves, such as in Figure 14.5. Films that show fringes that are more variable in their extrema are invariably more complicated in their structure and unable to be represented by a simple model. A great advantage of this technique is that the extracted values of film parameters are exceptionally stable. Predictions of performance that use them do not give results outside what one would expect from the envelope values.

Manifacier et al. [65] considered transmittance curves only and assumed homogeneous and absorbing films, and their analysis was focused on films of index higher than that of the substrate.

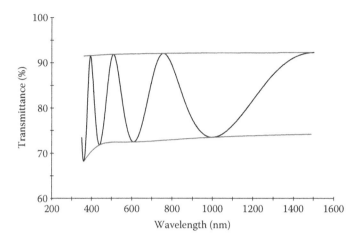

FIGURE 14.5
Fringes measured in transmission of a film of tantalum pentoxide over a substrate of glass. The envelopes of the fringes are shown. With well-behaved films such as this one, the adding of the envelopes is a straightforward process.

Let T_{max} and T_{min} be the envelope values. Then we can write

$$\alpha = \frac{C_1\left[1 - (T_{max}/T_{min})^{1/2}\right]}{C_2\left[1 + (T_{max}/T_{min})^{1/2}\right]}, \tag{14.12}$$

where

$$\alpha = \exp(-4\pi k_f d_f/\lambda), \tag{14.13}$$

$$4\pi n_f d_f/\lambda = m\pi \quad \text{(quarter- or half-wave thickness)},$$

$$\begin{aligned}
C_1 &= (n_f + n_0)/(n_m + n_f),\\
C_2 &= (n_f - n_0)/(n_m - n_f),\\
T_{max} &= 16n_0 n_m n_f^2 \alpha/(C_1 + C_2\alpha)^2,\\
T_{min} &= 16n_0 n_m n_f^2 \alpha/(C_1 - C_2\alpha)^2.
\end{aligned} \tag{14.14}$$

Then, from Equations 14.12 and 14.13, if we define N as

$$N = \frac{n_0^2 + n_m^2}{2} + 2n_0 n_m \frac{T_{max} - T_{min}}{T_{max}T_{min}},$$

n_f is given by

$$n_f = \left[N + \left(N^2 - n_0^2 n_m^2\right)^{1/2}\right]^{1/2}. \tag{14.15}$$

Once n_f has been determined, Equation 14.12 can be used to find a value for α. The thickness d_f can then be found from the wavelengths corresponding to the various extrema, and the extinction coefficient k_f, from the values of d_f and α. The method has the advantage of explicit expressions for the various quantities, which makes it easily implemented on machines as small as programmable calculators. In practice, the method makes use of a mixture of analysis and computer optimization and includes absorption and inhomogeneity, discussed further in the following section.

Computers bring the advantage that we no longer need to devise methods of optical constant measurement with the principal objective of ease of calculation. Instead, methods can be chosen simply based on precision of results, regardless of the complexity of the analytical techniques that are required. This is the approach advocated by Hansen [67], who developed a reflectance attachment making it possible to measure the reflectance of a thin film for virtually any angle of incidence and polarization, the particular measurements carried out being chosen to suit each individual film.

For rapid, straightforward measurement of refractive index, a method due to Abelès [68] is especially useful. It relies on the fact that the reflectance for p-polarization is the same for substrate and film at an angle of incidence that depends only on the indices of film and incident medium, and not at all on either substrate index or film thickness, except, of course, that layers that are a half-wave thick at the appropriate angle of incidence, and the wavelength will give a reflectance equal to the uncoated substrate regardless of index. It is fairly easy to use Snell's law and the expressions for equal p-admittances to give

$$n_f \sin \vartheta_f = n_0 \sin \vartheta_0$$

and

$$n_f / \cos \vartheta_f = n_0 / \cos \vartheta_0,$$

so that

$$n_f = n_0 \tan \vartheta_0. \tag{14.16}$$

The measurement of index reduces to the measurement of the angle ϑ_0, at which the reflectances are equal. Heavens [59] showed that the greatest accuracy of measurement is, once again, obtained when the layer is an odd number of quarter-waves thick at the appropriate angle of incidence. This is because there is then the greatest difference in the reflectances of the coated and uncoated substrates for a given angular misalignment from the ideal. It is possible to achieve an accuracy of around 0.002 in refractive index provided the film and substrate indices are within 0.3 of each other, but not equal. Hacskaylo [69] developed an improved method based on the Abelès' technique. It involves incident light that is linearly polarized with the direction of polarization almost, but not quite, parallel to the plane of incidence. The reflected light is passed through an analyzer, and the analyzer angle, for which the reflected light from the uncoated substrate and from the film-coated substrate are equal, is plotted against the angle of incidence. A very sharp zero at the angle satisfying the Abelès condition is obtained, which permits accuracies of 0.0002–0.0006 in the measurement of indices in the range of 1.2–2.3. It is not necessary for the film index to be close to the substrate index.

Values of R and φ for an opaque surface define, completely and unambiguously, the optical constants of the surface. Absolute reflectance is a difficult measurement, and it is more usual to measure the way in which the unknown surface compares with a known reference—which introduces further difficulties. Phase is even more involved, requiring an interferometric operation as well as a known standard. Phase measurements are therefore quite rare, and routine measurements of reflectance are almost always comparative. A major problem is the calibration and maintenance of suitable standards. There is a way, however, of avoiding such difficulties. At normal incidence, there is only one value of reflectance and one of phase, but at oblique incidence, there are two, one pair for s-polarization and the other for p. In principle, therefore, it should be possible to use one as a reference for the other, and this leads to the method known as ellipsometry

[70–72]. A full description of ellipsometry and its techniques is beyond the scope of this book, but a short discussion is appropriate.

The ellipsometric parameters have already been introduced in the section on retarders in Chapter 9. We recall that they are known as ψ and Δ (psi and delta), where

$$\tan \psi = \left| \frac{\rho_p}{\rho_s} \right| \quad \text{or} \quad \left| \frac{\tau_p}{\tau_s} \right|,$$

$$\Delta = \varphi_p - \varphi_s \quad \text{in transmission,} \tag{14.17}$$

$$\Delta = \varphi_p - \varphi_s \pm \pi \quad \text{in reflection.}$$

Δ is also known as the relative retardation or retardance.

Ellipsometry has many advantages, especially the ability to use a single illuminated spot for both measurements. Then there is the absence of any reference samples that must be maintained, and although high accuracy is required, the measurement is simple, involving straightforward manipulations of polarized light. It should be said, however, that users do usually feel more comfortable with known calibrated samples that increase confidence in the measurements. Disadvantages are that the measurement is at oblique incident, quite far from more normal measurements of performance, making it difficult to exercise instinct in judging the results. A limitation is that there are two parameters only, rather less than the number that must often be established for a complete description of the system. This limitation is frequently addressed by introducing a physical model and by additionally varying the angle of incidence or the wavelength, or both. The combination is known as variable angle spectroscopic ellipsometry, frequently abbreviated to VASE.

Only two parameters, the refractive index and the extinction coefficient, are sufficient to define a simple surface. Since there are two ellipsometric parameters ψ and Δ, then it should be possible to make a determination of the surface parameters from a single ellipsometric measurement. This is indeed the case, and there is a direct analytical connection between the two sets of parameters. Unfortunately, this is not normally the case with a thin film on a substrate. Even with a simple absorbing film on an already characterized substrate, three parameters, n, k, and d, are necessary to define the film. The properties of films that are absorbing may depart only slightly from a surface of bulk material. In such cases, it is often assumed that the extraction techniques used for a simple surface are applicable. The parameters n and k that are extracted in this way are usually referred to as the pseudooptical constants. They exhibit, usually, the gross features of the real optical constants, although they may not be suitable for thin-film calculations and predictions.

In spectroscopic ellipsometry, the wavelength is varied. Since the film physical thickness is not sensitive to wavelength, this introduces an element of redundancy. It is then sufficient to introduce a small amount of additional information. This frequently takes the form of a prescribed spectral variation of optical constants according to some recognized model. Other film parameters may then be also included, and the redundancy in the measurement can be so great that even simple multilayers may be evaluated.

We illustrate the extraction process by considering the simple case of a single wavelength, single angle measurement in reflection of a surface characterized by refractive index n, and extinction coefficient k, where $y = (n - ik)$.

Let the incident medium be of index unity, and let $\varepsilon = \tan \psi \exp i(\Delta \pm \pi)$. Then,

$$\varepsilon = \frac{\rho_p}{\rho_s} = \frac{\left(\eta_{0p} - \eta_p \right)}{\left(\eta_{0p} + \eta_p \right)} \cdot \frac{(\eta_{0s} + \eta_s)}{(\eta_{0s} - \eta_s)}, \tag{14.18}$$

where the symbols may be taken as the modified admittances and the definition of ε is consistent with the usual thin-film sign convention and with the normal sign convention for Δ. Then, replacing the incident medium admittance by unity and recalling that $\eta_p\eta_s = y^2$,

$$\varepsilon = \frac{(1 - y^2) - \left(\eta_p - \eta_s\right)}{(1 - y^2) + \left(\eta_p - \eta_s\right)}. \tag{14.19}$$

Now

$$\eta_s = \frac{\sqrt{y^2 - \sin^2\vartheta_0}}{\cos\vartheta_0} \tag{14.20}$$

and

$$\eta_p = \frac{y^2 \cos\vartheta_0}{\sqrt{y^2 - \sin^2\vartheta_0}}, \tag{14.21}$$

so that after some manipulation, we can write

$$\gamma = \frac{1 - \varepsilon}{1 + \varepsilon} = \frac{\left(\eta_p - \eta_s\right)}{(1 - y^2)}$$

$$= \frac{\dfrac{y^2 \cos\vartheta_0}{\sqrt{y^2 - \sin^2\vartheta_0}} - \dfrac{\sqrt{y^2 - \sin^2\vartheta_0}}{\cos\vartheta_0}}{(1 - y^2)} \tag{14.22}$$

$$= \frac{\sin^2\vartheta_0}{\cos\vartheta_0\sqrt{y^2 - \sin^2\vartheta_0}}.$$

This gives

$$y^2 = \frac{\sin^4\vartheta_0}{\gamma^2\cos^2\vartheta_0} + \sin^2\vartheta_0. \tag{14.23}$$

There will be two solutions, and the fourth quadrant solution will be the correct one.

Much information on the use of spectroscopic ellipsometry in thin-film characterization is found in the book by Tompkins and Hilfiker [72]. Hilfiker et al. [73] also gave a useful account of the use of spectroscopic ellipsometry in the characterization of absorbing coatings.

Ellipsometry is especially useful for the derivation of the optical constants of opaque metal films. Provided they have a suitable thickness, high-performance metal films can also be characterized by a measurement of a surface plasmon resonance, already discussed in Chapter 10. This tool involves a rather simpler optical arrangement than the ellipsometer, but it is more limited in its application. The film in question is deposited over the base of a prism, and the resonance is measured in the normal way. Usually a quite undemanding optical arrangement involving a simple goniometer with laser and collimator and receiver will suffice. The p-polarized resonance has three attributes, the angular position, the resonance width, and the resonance depth. There are three attributes of the metal coating, n, k, and d. n is primarily associated with the resonance width; k, with its position; and d, with its depth, so that the extraction process is a simple process of model fitting.

There is one small problem associated with two possible solutions. The two solutions involve quite distinct values of d, except that when the minimum reflectance is zero, then the two solutions coincide. A simple technique for distinguishing the correct set of values is to ensure that the two thickness values are sufficiently far apart for the correct one to be recognized. This, of course, means that the sample should be prepared so that the minimum reflectance is sufficiently far from zero, yet the resonance is sufficiently well developed for accurate measurement. This is a limitation on the range of usable thicknesses. Alternatively, measurements at more than one wavelength may be performed. The correct solutions will be those with similar values of d. The technique has been used, for example, in studies of the influence of small changes in process parameters on the optical constants of metals [74].

Provided there is some measurable finite transmittance, then metal layers can be characterized by reflectance and transmittance measurements as a function of wavelength. Dobrowolski et al. [75] used a process termed *reverse synthesis* to successfully extract the optical constants of a range of thin-film materials, including metals, from measurements of reflectance and transmittance. Reverse synthesis is essentially a refinement process where parameters of a model are varied to reduce the error between model predictions and actual measurements. The use of a model reduces the parameters to a reasonable number. Dobrowolski's model was a combination of Lorenz and Drude expressions so that both dielectric and metallic layers could be handled.

Unfortunately, the behavior of real thin films is often more complicated than we have been assuming. They are frequently inhomogeneous, that is, their refractive index varies throughout their thickness. They can also be anisotropic, although rather less work has been done on this aspect of their behavior.

Provided that the variation of index throughout the film is either a smooth increase or a smooth decrease, so that there are no extrema within the thickness of the film, the highest and lowest values being at the film boundaries, then we can use a very simple technique to determine the difference in behavior at the quarter-wave and half-wave points, which would be obtained with an inhomogeneous film. We assume that the film is absorption free, and that its properties can be calculated by a multiple-beam approach, which considers the amplitude reflection and trans-mission coefficients at the boundary only. We assume that the index of that part of the film next to the substrate is n_b and that next to the surrounding medium is n_a. The corresponding admittances are y_b and y_a. The only reflections that take place are assumed to be at either of the two interfaces. There is one further complication, also indicated in Figure 14.6, before we can sum the multiple beams to arrive at transmittance and reflectance. A beam propagating from the outer surface of the film to the inner is assumed to suffer no loss by reflection, and therefore, the irradiance is

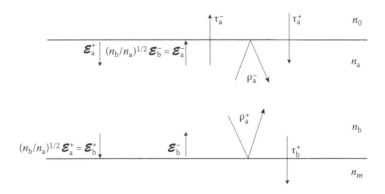

FIGURE 14.6
Inhomogeneous film quantities used in the development in the text of the matrix expression for an inhomogeneous layer.

unaltered. Since irradiance is proportional to the square of the electric amplitude times admittance, a beam that is of amplitude $\mathcal{E}_a^+$, just inside interface a, will have amplitude $(y_a/y_b)\mathcal{E}_a^+ = \mathcal{E}_b^+$ at interface b. The correction will be reversed in travelling from b back to a. This is in addition to any phase changes. The inverse correction applies to magnetic amplitudes. Since the correction cancels out for each double pass, it does not affect the result for resultant reflectance, but it must be taken into account when the multiple beams are being summed for the calculation of transmittance. The derivation of the necessary expressions proceeds as in Chapter 2. Here, for simplicity, we restrict ourselves to normal incidence. Oblique incidence is a simple extension. We start with the total tangential electric and magnetic field amplitudes at interface b:

$$E_b = \mathcal{E}_b^+ + \mathcal{E}_b^-,$$
$$\mathcal{H}_b = y_b\mathcal{E}_b^+ - y_b\mathcal{E}_b^-,$$

giving

$$\mathcal{E}_b^+ = 0.5[(H_b/y_b) + E_b],$$
$$\mathcal{E}_b^- = 0.5[-(H_b/y_b) + E_b].$$

Then, the various rays are transferred to interface a:

$$\mathcal{E}_a^+ = 0.5(y_b/y_a)^{1/2}[(H_b/y_b) + E_b]e^{i\delta},$$
$$\mathcal{E}_a^- = 0.5(y_b/y_a)^{1/2}[-(H_b/y_b) + E_b]e^{-i\delta},$$

giving

$$E_a = \mathcal{E}_a^+ + \mathcal{E}_a^- = \left(\frac{y_b}{y_a}\right)^{1/2}(\cos\delta)E_b + \frac{i\sin\delta}{(y_ay_b)^{1/2}}\mathcal{H}_b,$$

$$\mathcal{H}_a = y_a\mathcal{E}_a^+ - y_a\mathcal{E}_a^- = i(y_ay_b)^{1/2}(\sin\delta)E_b + \left(\frac{y_a}{y_b}\right)^{1/2}(\cos\delta)\mathcal{H}_b.$$

The characteristic matrix for the layer is then given by

$$\begin{bmatrix} \left(\dfrac{y_b}{y_a}\right)^{1/2}\cos\delta & \dfrac{i\sin\delta}{(y_ay_b)^{1/2}} \\[2em] i(y_ay_b)^{1/2}\sin\delta & \left(\dfrac{y_a}{y_b}\right)^{1/2}\cos\delta \end{bmatrix}, \tag{14.24}$$

an expression originally due to Abelès [76,77]. The calculation of inhomogeneous layer properties was considered in detail by Jacobsson [55].

Now we consider cases where the layer is either an odd number of quarter waves or an integral number of half waves. We apply Equation 14.24 in the normal way and find the well-known relations:

$$R = \left[\frac{y_0 - y_ay_b/y_m}{y_0 + y_ay_b/y_m}\right]^2 \quad \text{for a quarter wave} \tag{14.25}$$

and

$$R = \left[\frac{y_0 - y_a y_m / y_b}{y_0 + y_a y_m / y_b}\right]^2 \quad \text{for a half wave.} \tag{14.26}$$

The expression for a quarter-wave layer is indistinguishable from that of a homogeneous layer of admittance $(y_a y_b)^{1/2}$, and so it is impossible to detect the presence of inhomogeneity from the quarter-wave result. The half-wave expression is quite different. Here the layer is no longer an absentee layer and cannot therefore be represented by an equivalent homogeneous layer. The shifting of the reflectance of the half-wave points from the level of the uncoated substrate in absorption-free layers is a sure sign of inhomogeneity and can be used to measure it.

The Hadley method of deriving the optical constants takes no account of inhomogeneity. Any inhomogeneity therefore introduces errors. The Marseille method, however, includes half-wave points and has therefore sufficient information to accommodate inhomogeneity. The matrix expression is a good approximation when the inhomogeneity is not too large and when the admittances y_a and y_b are significantly different from those of substrate and incident medium. To avoid any difficulties due to the model, the Marseille group actually uses a model for the layer consisting of at least 10 homogeneous sublayers with linearly varying values of n but identical values of k and thickness d. The half-wave points still give the principal information on the degree of inhomogeneity. They are also affected by the extinction coefficient k, and this has to be taken into account. One half-wave point within the region of measurement can be used to give a measure of inhomogeneity that is assumed constant over the rest of the region. Several half-wave points can yield values of inhomogeneity that can be fitted to a Cauchy expression, that is, an expression of the following form:

$$\frac{\Delta n}{n} = A + \frac{B}{\lambda^2} + \frac{C}{\lambda^4}. \tag{14.27}$$

Details of the technique are given by Borgogno et al. [78]. Some of their results are shown in Figure 14.7.

The envelope method has also been extended to deal with inhomogeneous films using the inhomogeneous matrix expression for the calculations [79]. The extinction coefficient k, as in the Marseille method, is assumed constant through the film.

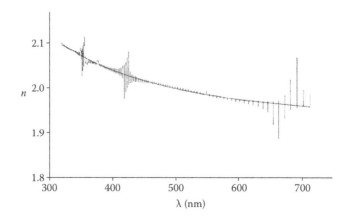

FIGURE 14.7
Values of mean index and the uncertainty n calculated for hafnium oxide using an inhomogeneous film model. The Cauchy coefficients for n are $A = 1.9165$, $B = 2.198 \times 10^4 \, \text{nm}^2$, and $C = -3.276 \times 10^8 \, \text{nm}^4$, and those for $\Delta n / n$ are $A' = -5.39 \times 10^{-2}$, $B' = -1.77 \times 10^3 \, \text{nm}^2$. (After J. P. Borgogno, B. Lazarides, and E. Pelletier, *Applied Optics*, 21, 4020–4029, 1982, With permission of Optical Society of America.)

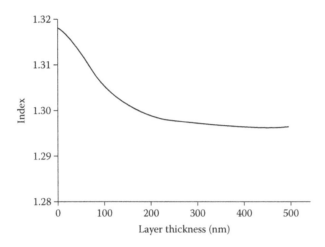

FIGURE 14.8

Graph of the index profile of cryolite layers at $\lambda = 633$ nm, derived from fitting a formula, $n2 = A + [B/(t2+C)]$, where t is the thickness coordinate, to the curves of the variation of reflectance in vacuo of a cryolite film deposited over a zinc sulfide film of varying thickness. $A = 1.6773$, $B = 5.0431 \times 10^2$ nm^2 and $C = 8.2986 \times 10^3$ nm^2. (After R. P. Netterfield, *Applied Optics*, 15:1969–1973, 1976. With permission of Optical Society of America.)

Netterfield [80] measured the variation in reflectance of a film at a single wavelength as it was deposited. If the assumption is made that the part of the film which is already deposited is unaffected by subsequent material, then the values of reflectance associated with extrema can be used to calculate a profile of the refractive index throughout the thickness of the layer. Some results obtained in this way for cryolite are shown in Figure 14.8.

Extinction coefficient is more difficult than refractive index. However, it is not as severe a problem because we do know when it is low, and although the value extracted along with the refractive index may not be completely reliable, it is usual to adjust it according to results obtained with actual coatings.

Some useful techniques are based on optical waveguiding modes. This technique was pioneered by Tien [81]. The principle is simple. Light is coupled, usually by a high-index coupling prism, into a thin-film waveguide constructed from the material under examination that must be of higher index than its substrate and must be thick enough to support at least one mode. Some light is inevitably scattered from the guided mode and can be measured at two separate points along the direction of propagation. This then yields a measure of total loss, and since the distance is relatively large, exceptionally small losses can be measured. A defect of the method is that even in the lowest-order mode, scattering losses are greatly enhanced so that a pessimistic value of extinction coefficient is usual. Nevertheless, extinction coefficients of 0.00001 or lower are readily measured. With some materials, the absorption loss is so low that it can be neglected, and the technique can be used to determine the optimum conditions to minimize scattering losses [82]. The measurement of the coupling angles can also yield accurate values of refractive index. The technique is attractive but operates only at suitable laser wavelengths, and so it is less used as a routine technique but rather as an occasional confirmation, or otherwise, of more conventional measurements.

High-quality films have extinction coefficients that are frequently less than 0.00001. As we have earlier observed, in the normal way, we have difficulty in accurately measuring such low values except with the waveguide techniques just mentioned where the propagation length can be very large compared with film thickness. The more usual measurement of the extinction coefficient involves the difference between reflectance and transmittance, that is, the absorptance of the system. The first requirement is that the substrate should be freer of absorption than the film; otherwise, the accuracy will suffer. Then there is the problem of measurement errors. Reflectance and transmittance must be measured at exactly the same point in the film. Even then, the measurements, even with the best care and attention to detail, are unlikely to be much better than 0.1% absolute in accuracy. Vernhes and Martinu [83] suggest that one of the measurements should be of reflectance and transmittance in p-polarized light at or close to the Brewster condition so that any interference fringes are largely suppressed, making it possible to achieve a more accurate estimation of absorptance. Then with supplementary measures of reflectance and transmittance at closer to normal incidence giving index and thickness, it is possible to achieve much greater accuracy in the extraction of quite low values of extinction coefficient. The method is termed TRACK (transmission, reflection, absorption combination for K evaluation), and it should not be thought that it relieves any requirement for accuracy and care and attention. It is important that the measurements should be accurately p-polarized and exactly at the same point on the film and that the measurements should be as precise as possible.

14.3 Pitfalls in Optical Constant Extraction

As with manufacturing tolerances, computer modeling gives us the opportunity to investigate possible errors and pitfalls in the extraction of optical constants without the discouraging complexity of entirely analytical theoretical approaches. Much of this section heavily relies on an already published paper [84]. Another useful study is by Tikhonravov et al. [85], who clearly showed that systematic rather than random errors in measurement are the culprits. Here we will carry out a few simple computer experiments to illustrate our discussion.

We start with a contrived illustration of accuracy versus stability. We generate transmittance for a dielectric film of index 2.00 with no dispersion and with physical thickness of 510 nm on a BK 7 glass substrate, the rear surface of which is uncoated. We supposedly use a mechanical technique to measure the physical thickness of the film that gives 500 nm, rather than the correct 510 nm. This lack of exact correspondence is not unexpected with mechanical versus optical techniques. We now carry out the extraction process that, since we know the transmittance and the thickness, is quite straightforward. As a test, we compare predictions using the extracted parameters with the input data. The fit is exact, and the accuracy, excellent (Figure 14.9), but when we look at the extracted index (Figure 14.10) we can immediately see that the stability is exceptionally poor. In fact, some similar techniques of extraction involve a deliberate variation of the thickness until a more plausible variation of refractive index is obtained.

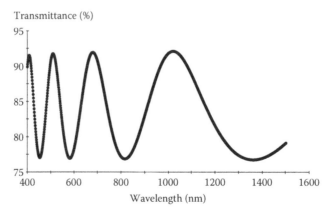

FIGURE 14.9
Extraction process gives the recalculated results shown as black dots superimposed on the input curve shown underneath the dots as a full line, although it can hardly be detected because the dots coincide exactly.

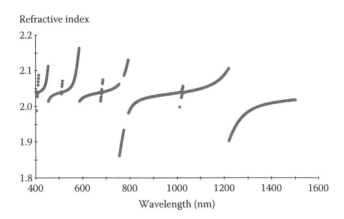

FIGURE 14.10
Extracted values of refractive index that give the results in Figure 14.9 are not encouraging.

Absorption and/or inhomogeneity are two common film conditions that must be handled in the extraction. A simple computer experiment will quickly demonstrate that reflectance measurements are insensitive to absorption, unless it becomes very large, and in fact, reflectance fringes on their own should not be used for the extraction of extinction coefficient. Inhomogeneity, already briefly considered in the previous section, in the absence of absorption affects both transmittance and reflectance equally. That fringes in reflection are insensitive to absorption but sensitive to inhomogeneity, while those in transmission are sensitive to both, permits a method for distinguishing between the two effects.

Let us first consider absorption. We have already mentioned the envelope method, and its advantage of stability. Fringe envelopes are no longer constant with thickness when absorption is high; nevertheless, experience shows that as long as the fringes have reasonably distinct peaks, the method can still be effectively used provided good, distinct fringes are measured. The extraction immediately yields $2\pi kd/\lambda$, and so a measure of physical thickness d has to be extracted before the value of k can be finally determined. The thickness is determined by the fringe peak positions and so is more reliable than a mechanical measurement. We show some calculated heavily absorbing fringes at slightly different thicknesses in Figure 14.11. The envelopes are certainly not constant

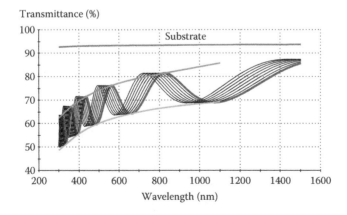

FIGURE 14.11
Fringes in transmission with absorption ($k = 0.02$) and representing differing thicknesses. The thicker curve is that used for n and k extractions with the envelopes shown and results in the very satisfactory optical constant values shown in Figure 14.12.

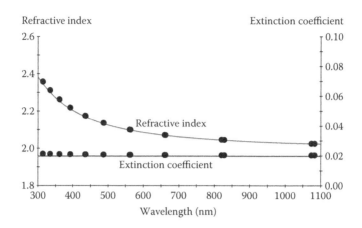

FIGURE 14.12
Full curves represent the optical constants that were used to generate the absorbing data for the extraction while the black dots are the extracted results. The reason for the slight elevation of k compared with its reference value is an extracted value of physical thickness of 397.2 nm, rather than the actual 400.0 nm.

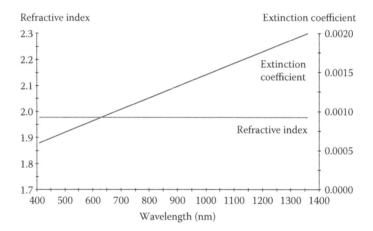

FIGURE 14.13
Here are the parameters resulting from the use of an incorrect model of a homogeneous absorbing film model in extracting the constants of a nonabsorbing inhomogeneous film. The extinction coefficient rising toward longer wavelengths is almost invariably a warning of possible problems.

with thickness, and yet, as Figure 14.12 shows, they still yield good extracted results. Since this is a computer experiment, we can compare the extraction with the values used in the data generation.

A further computer experiment generates transmittance data for a dispersionless inhomogeneous film entirely free from any absorption. The extraction process, however, assumed a homogeneous but absorbing film with the extracted results shown in Figure 14.13. The spurious extinction coefficient rising toward longer wavelengths is a danger signal that almost always indicates a serious problem with the extraction. We emphasize that there was a perfect fit between input data and data calculated using the incorrect extracted parameters. The sole sign of a problem is the unrealistic behavior of the extinction coefficient.

A similar symptom can accompany an inaccuracy in the measurements. Figure 14.14 used a model of a dispersive homogeneous and nonabsorbing film to generated data that were then multiplied by 0.99 to simulate a simple measurement error. Again, the recalculation of the input reveals a perfect fit, but the rising extinction coefficient is a sign of a problem.

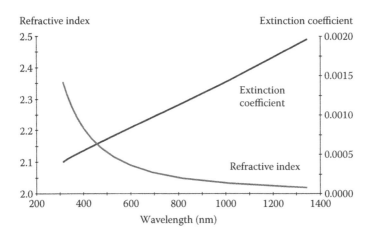

FIGURE 14.14
Optical constants extracted from data generated by a homogeneous nonabsorbing but dispersive film where a scale error was simulated by multiplying the measurements by 0.99. Once again, the rising extinction coefficient with wavelength should be looked upon with deep suspicion.

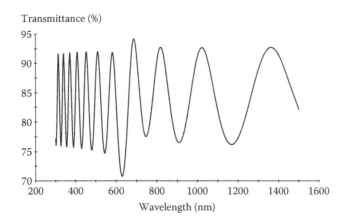

FIGURE 14.15
Fringes were generated by an inhomogeneous film with a cyclic variation in refractive index. Such fringes should not be used in an extraction process.

Finally, we use an example of a thin film with a structure that is a little beyond the capabilities of the simple models normally used in extraction. Figure 14.15 shows fringes that were generated by a homogeneous and absorption-free film exhibiting a slight cyclic variation of index through its thickness. The fringe pattern showing peaks or troughs that are slightly out of line with surrounding fringes is a sure sign that such a film should not be used in an optical constant extraction operation. Its structure is too complex. Nevertheless, for the sake of the experiment, the optical constants were extracted and are shown in Figure 14.16. They are quite useless.

The message in all this is that it is a danger sign that should be heeded if the extracted parameters appear to indicate behavior that is strange and perhaps even unphysical. In all the examples shown, the recalculation of the input data using the extracted parameters was quite acceptable even though they were seriously in error. A good fit is, of course, desirable, and its absence is itself a danger signal, but a good fit is not sufficient evidence that the results should be trusted. Healthy suspicion is always a good accompaniment to any n and k extraction. A wise procedure is to involve more than one sample with different thicknesses in the extraction process.

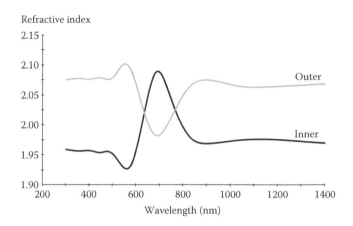

FIGURE 14.16
Reversal in the middle of the range of the sign of the inhomogeneity shown by the outer and inner refractive indices implies that the extracted parameters are useless.

14.4 Measurement of the Mechanical Properties

From the point of view of optical coatings, the importance of the mechanical properties of thin films is primarily in their relation to coating stability, that is, the extent to which coatings will continue to behave as they did when removed from the coating chamber, even when subjected to disturbances of an environmental and/or mechanical nature. There are many factors involved in stability, many of which are neither easy to define nor to measure, and there are still great difficulties to be overcome. The approach used in quality assurance in manufacture is almost entirely empirical. Tests are devised which reproduce, in as controlled a fashion as possible, the disturbances to which the coating will be subjected in practice, and samples are then subjected to these tests and inspected for signs of damage. Sometimes, the tests are deliberately made more severe than those expected in use. Coating performance specifications are usually written in terms of such test levels.

Stress is normally measured by depositing the material on a thin flexible substrate that deforms under the stress applied to it by a deposited film. The deformation is measured, and the value of stress necessary to cause it, calculated. The substrate may be of any suitable material; glass, mica, silica, metal, for example, have all been used. The form of the substrate is often a thin strip, supported so that part of it can deflect. Then, either the deflection is measured in some way or a restoring force is applied to restore the strip to its original position. Usually, the deflection, or the restoring force, is continuously measured during deposition. Optical microscopes, capacitance gauges, piezoelectric devices, and interferometric techniques are some of the successful methods.

A survey of the field of stress measurement in thin films in general is given by Hoffman [86]. A useful paper that deals solely with dielectric films for optical coatings is that by Ennos [87]. Ennos used a thin strip of fused silica as substrate, simply supported at each end on ball bearings so that the center of the strip was free to move. An interferometric technique with a helium–neon laser as the light source was used to measure the movement of the strip. The strip was made of one mirror of a Michelson interferometer of novel design, shown in Figure 14.17. Since the laser light was linearly polarized, the upper surface of the prism was set at the Brewster angle to eliminate losses by reflection of the emergent beam. Apart from the more obvious advantages of large coherence length and high collimation, the laser beam made it possible to line up the interferometer with the bell jar of the machine in the raised position (see Figure 14.17b). No high-quality window in the

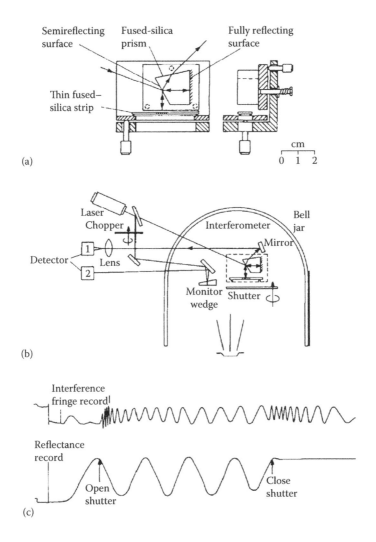

FIGURE 14.17
(a) Film stress interferometer. (b) Experimental arrangement for continuous measurement of film stress during evaporation. (c) Recorder trace of fringe displacement and film reflectance. (After A. E. Ennos, *Applied Optics*, 5, 51–61, 1966. With permission of Optical Society of America.)

machine was necessary, the glass jar of quite poor optical quality proving adequate. To complete the arrangement, the laser light was also directed on a test flat for the optical monitoring of film thickness. A typical record obtained with the apparatus is also shown in Figure 14.17c. The calibration of the fused-silica strip was determined both by calculation and by measurement of deflection under a known applied load.

Curves plotted for a wide range of materials showing the variation of stress in the films during the actual growth as a function both of film thickness and evaporation conditions are included in the paper, some examples being shown in Figure 14.18. It is of particular interest to note the frequent drop in stress when the films are exposed to the atmosphere. This is principally due to the adsorption of water vapor, an effect to be considered further toward the end of this chapter.

The interferometric technique was further improved by Roll [88] and Roll and Hoffman [89]. Then, Ledger and Bastien [90] took the Michelson interferometer of Ennos and replaced it by a cat's-eye interferometer, using circular disks as sensitive elements that are very much less temperature sensitive, and this has enabled the measurement of stress levels in optical films over a wide range of

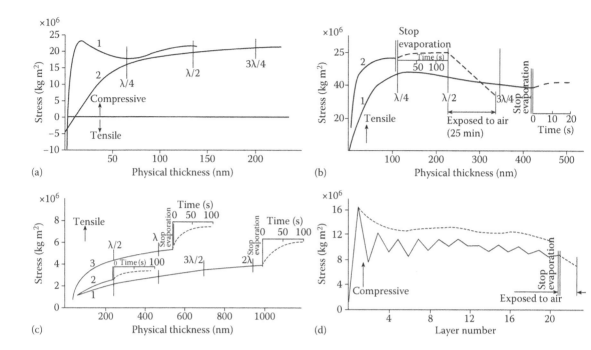

FIGURE 14.18

(a) Film stress in evaporated zinc sulfide on fused silica at ambient temperature. Evaporation rates are 1:0.25 nm/s and 2:2.2 nm/s. (b) Film stress in magnesium fluoride. 1: Direct evaporation from molybdenum; evaporation rate is 4.2 nm/s. 2: Indirect radiative heating; evaporation rate is 1.2 nm/s. (c) Cryolite and chiolite evaporated by indirect radiative heating. 1: Cryolite; evaporation is rate 3.5 nm/s. 2: Chiolite; evaporation rate is 4 nm/s. (d) Zinc sulfide–cryolite multilayer. Twenty-one layers of *(HL)10H*. Resultant average stress after each evaporation plotted. Dashed curve shows upper limit of film stress reached during the warm-up period before the evaporation of a layer commenced. (After A. E. Ennos, *Applied Optics*, 5, 51–61, 1966. With permission of Optical Society of America.)

substrate temperatures. The examination of the differences in thermally induced stress for identical films on different substrate materials, when substrate temperature is varied after deposition, has permitted the measurement of the elastic moduli and thermal expansion coefficients of the thin-film materials. Although the measured value of expansion coefficient for bulk thorium fluoride crystals is small and negative, the values for thorium fluoride thin films were consistently large and positive, varying from 11.1×10^{-6} to $18.1 \times 10^{-6}\,^{\circ}\mathrm{C}^{-1}$. Young's modulus for the same samples varies from 3.9×10^{5} to $6.8 \times 10^{5}\,\mathrm{kg/cm^2}$ (that is, 3.9×10^{10} to $6.8 \times 10^{10}\,\mathrm{Pa}$).

Ledger and Bastien arranged the interferometer so that fringes were counted as they were generated at the center of the interferometer during the deposition of the film and changes in the stress. An asymmetric shape to the fringes permitted the distinction between a fringe appearing and a fringe disappearing. This meant that the stress level would be lost if the fringe count failed at any stage. A group at the Optical Sciences Center [91] modified the interferometer to view a sufficiently large field that included a number of fringes. The fringe pattern was then interpreted in the manner of an interferogram to give the form of the surface of the deformable substrate. This effectively decoupled each measurement from all the others and permitted the stress to be unambiguously determined at any stage even if some intervening measurements were missed or skipped. The interferometer was used in a detailed study of titanium dioxide films deposited by thermal evaporation with or without ion assist.

Thermally evaporated films usually exhibit a tensile stress that is a consequence of the disorder frozen into the film, as freshly arriving material covers what is already existing. An increase in the

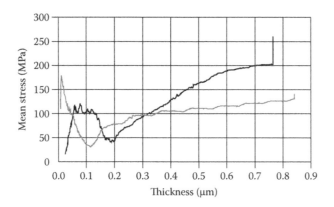

FIGURE 14.19
Mean (tensile) stress as a function of film thickness in titania films deposited at 0.7 nm/s (*gray*) and 0.97 nm/s (*black*). The higher rate of deposition leads to greater tensile stress. The vertical line at the end of each curve is a relaxation thought to be due to the disappearance of the thermal gradient present during deposition. (Courtesy of B. G. Bovard, X. C. d. Lega, S.-H. Hahn et al., Optical Sciences Center, University of Arizona, Tucson, AZ, 1991.)

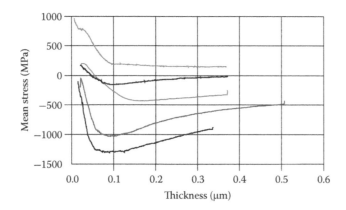

FIGURE 14.20
Mean stress as a function of thickness of a series of titania films deposited by ion-assisted deposition. The background gas was oxygen, and the films were bombarded with 500 eV argon ions at levels from top to bottom of 0.16, 0.32, 0.48, 0.80, and 1.02 mA/cm^2. (Courtesy of B. G. Bovard, X. C. d. Lega, S.-H. Hahn et al., Optical Sciences Center, University of Arizona, Tucson, AZ, 1991.)

rate of deposition gives less time for the material on the surface to reorganize itself and should therefore lead to an increase in tensile stress. This is clearly seen in Figure 14.19.

Under bombardment, the tighter packing of the films leads to an increase in compressive stress because of the transfer of momentum to the growing film (Figure 14.20). In fact, it is possible by careful control of the bombardment to achieve extremely low values. Unfortunately, not all materials exhibit such a simple relationship.

Pulker [92] studied the relationship between stress levels and the microstructure of optical thin films, further developing some ideas of Hoffman. The work was surveyed by Pulker [93]. Good agreement between measured levels of stress and those calculated from the model has been achieved, but perhaps the most spectacular feature has been the demonstration, in accord with the theory, that small amounts of impurity can have a major effect on stress. The impurities congregate at the boundaries of the columnar grains of the films and reduce the forces of attraction between neighboring grains, thus reducing stress. Small amounts of calcium fluoride in magnesium fluoride, around 4 mol%, reduce tensile stress by some 50%. Pellicori [94] showed the beneficial effect of mixtures of fluorides in reducing cracking in low-index films for the infrared.

Windischmann [95] discussed and modeled the stresses in ion beam-sputtered thin films. He identifies momentum transfer as the important parameter. This is in line with conclusions regarding ion-assisted deposition. The results of Figure 14.20 agree with the Windischmann model.

Strauss [96] reviewed mechanical stress in optical coatings.

Abrasion resistance is another mechanical property that is of considerable importance and yet extremely difficult to define in any terms other than empirical. The problem is not any lack of understanding, but rather that abrasion resistance is not a single fundamental property, but a combination of factors such as adhesion, hardness, friction, packing density, brittleness, and so on. Various ways of specifying abrasion resistance exist but all depend on arbitrary empirical standards. The aim is to reproduce under controlled conditions, the essential characteristics of the attack that will be suffered in a practical application. The standard sometimes involves a pad, made from rubber, which may be loaded with a particular grade of emery. The pad is drawn over the surface of the film under a controlled load for a given number of strokes. Signs of visible damage show that the coating has failed the test. Because the pad in early versions of the test was a simple eraser, the test is sometimes known as the eraser test. Similar standard tests may be based on the use of cheesecloth or even of steel wool. Wiper blades and sand slurries have also been used to attempt to reproduce the kind of abrasion that results from wiping in the presence of mud. Most of the tests suffer from the fact that they do not give a measure of the degree of abrasion resistance but are merely of a go/no-go nature. There is a modification of the test, described in Chapter 17, which does permit a measure of abrasion resistance to be derived from the extent of the damage caused by a controlled amount of abrasion. This is still probably the best arrangement yet devised, but even here, the results considerably vary with film thickness and coating design so that it is far from an absolute measure of a fundamental thin-film property. The scratch test, described shortly, is sometimes used to derive an alternative measure of abrasion resistance. Abrasion resistance is therefore primarily a quality-control tool. It will be further considered in Chapter 17.

Adhesion is another important mechanical property that presents difficulties in measurement. What we usually think of as adhesion is the magnitude of the force necessary to detach unit area of the film from the substrate or from a neighboring film in a multilayer. However, accurate measures of this type are impossible. Quality-control testing is, as for many of the other mechanical properties, of a go/no-go nature. A strip of adhesive tape is stuck to the film and removed. The film fails if it delaminates along with the tape. Jacobsson and Kruse [97] have studied the application of a direct-pull technique to optical thin films. In principle, the adhesive forces between film and substrate can be measured by simply applying a pull to a portion of the film until it breaks away, and indeed, this is a technique used for other types of coatings, such as paint films. The test technique is straightforward and consists of cementing the flat end of a small cylinder to the film, and then pulling the cylinder, together with the portion of film under it, off the substrate, in as near normal a direction as possible. The force required to accomplish this is the measure of the force of adhesion. Great attention to detail is required. The end of the cylinder must be true and must be cemented to the film so that the thickness of cement is constant and so that the axis of the cylinder is vertical. The pull applied to the cylinder must have its line of action along the cylinder axis, normal to the film surface.

The precautions to be taken, and the tolerances that must be held, are considered by Jacobsson and Kruse. Their cylindrical blocks were optically polished at the ends, and, in order to more nearly ensure a pull normal to the surface, the film and substrate were cemented between two cylinders, the axes of which were collinear. The mean value of the force of adhesion between 250 nm thick ZnS films and a glass substrate was found to be 2.3×10^7 Pa, which rose to 4.3×10^7 Pa when the glass substrate was subjected to 20 minutes of ion bombardment before coating. Zinc sulfide films evaporated on to a layer of SiO, some 150 nm thick, gave still higher adhesion figures of 5.4×10^7 Pa. The increases in adhesion due to the ion bombardment and the SiO were consistent, and the scatter in successive measures of adhesion was small, some 30% in the worst case.

An alternative method of measuring the force of adhesion is the scratch test, devised by Heavens [98] and improved and studied in detail by Benjamin and Weaver [99,100], who applied it to a range of metal films. Again, in principle, it is a straightforward test that is nevertheless very complex in interpretation. A round-ended stylus is drawn across the film-coated substrate under a series of increasing loads, and the point at which the film under the stylus is removed from the surface is a measure of the adhesion of the film. Benjamin and Weaver were able to show that the plastic deformation of the substrate under the stylus subjected the interface between film and substrate to a shear force, directly related to the load on the stylus by the following expression [99]:

$$F = \left[a / \left(r^2 - a^2 \right)^{1/2} \right] - P, \qquad (14.28)$$

where $a = [W/(\pi P)]^{1/2}$, where P is the indentation hardness of the substrate; r is the radius of the stylus point; a is the radius of the circle of contact; W is the load on the stylus; and F is the shear force.

The shear force is roughly proportional to the root of the load on the stylus. For the film just to be removed by drawing the stylus across it, the shear force had just to be great enough to break the adhesive bonds. Using this apparatus, Benjamin and Weaver were able to confirm, quantitatively, what had been qualitatively observed before, that the adhesion of aluminum deposited at pressures around 1.3×10^{-3} Pa (1.3×10^{-5} mbar or 10^{-5} Torr) on glass was initially poor, of values similar to van der Waals forces, but that after some 200 hours, it improved to reach values consistent with chemical bonding. Aluminum deposited at higher pressures, around 0.13 Pa (1.3×10^{-3} mbar or 10^{-3} Torr), gave consistently high bonding immediately after deposition. This is attributed to the formation of an oxide-bonding layer between aluminum and glass, and a series of experiments demonstrated the importance of such oxide layers in other metal films on glass. On alkali halide crystals, the initial bonding at van der Waals levels showed no subsequent improvement with time. More recently, the scratch test was studied by Laugier [101,102], who included the effects of friction during the scratching action in the analysis. Zinc sulfide was shown to exhibit an unusual aging behavior in that it occurs in two well-defined stages. After a period of some 18–24 hours after deposition, the adhesion increases by as much as a factor of 4 from an initially low figure. After a period of 3 days, the adhesion then begins to increase further and, after a further 7 days, reaches a final maximum that can be some 20 times the initial figure. This is attributed to the formation of zinc oxide at the interface between layer and substrate, first free zinc at the interface combining with oxygen that has diffused through the layer from the outer surface and then later zinc that has diffused to the boundary from within the layer.

Commercial instruments that apply these tests are now available and help standardize the tests as far as is possible.

Unfortunately, none of these adhesion tests is entirely satisfactory. Some of the difficulties are related to consistency of measurement, but the greatest problem is the nature of the adhesion itself. The forces which attach a film to a substrate, or one film to another, are not only all very large (usually greater than 100 ton/in.^{2} or some 10^9 Pa) but also of very short range. In fact, they are principally between one atom and the next. The short range of the forces has two major consequences. First, the forces can be blocked by a single atom or molecule of contaminant, and so adhesion is susceptible to even the slightest contamination. A single monomolecular layer of contaminant is sufficient to completely destroy the adhesion between film and substrate. A small fraction of a monomolecular layer is enough to adversely affect it. Second, although the force of adhesion is large, the work required to detach the coating, the product of the force and its range, can be quite small. Coatings usually fail in adhesion in a progressive manner rather than suddenly and simultaneously over a significant area, and in such peel failures, it is the work, rather than the force, required to detach the coating—the work of adhesion, as it is usually called—that is the important parameter. This work can be considered as the supply of the necessary surface energy associated with the fresh surfaces exposed in the adhesion failure together with any work lost in the plastic deformation of film and/or substrate.

With some metal films, particularly deposited on plastic, there is evidence that an electrostatic double layer gradually forms, which positively contributes to the adhesion. In the tape test, the adhesive forces are comparatively very weak, but their long range, due to the stretching of the adhesive, allows them to be simultaneously applied over a relatively large area. Thus, the film is unlikely to be detached from the substrate unless it is very weakly bonded, and even then, it may not be removed unless there is a stress concentrator that can start the delamination process. Sometimes this is provided by scribing a series of small squares into the coating, and the tape will tend to lift out complete squares.

In the case of the direct-pull technique, it is exceedingly difficult to avoid a progressive failure rather than a simultaneous rupturing of the bonds over the entire area of the pin. Unevenness in the thickness of the adhesive, or a pull that is not completely central, can cause a progressive failure with consequent reduction in the force measured. Even when the greatest care is taken, it is unlikely that the true force of adhesion will be obtained and the test is principally useful as a quality control vehicle. Poor adhesion will tend to give a very much-reduced force.

The scratch test suffers from additional problems. Many of the films used in optical coatings shatter when a sufficiently high load is applied before any delamination from the substrate takes place. Such shattering dissipates additional energy, and thus, film hardness and brittleness enter into the test results. Rarely with dielectric materials does a clean scratch occur. Again, the test becomes useful as a comparison between nominally similar coatings rather than an absolute one. Goldstein and DeLong [103] had some success in the assessment of dielectric films using microhardness testers to scratch the films. Most commercial scratch testers include a microscope, and visual examination of the nature of the failures is an important component of the test. Some also include sensitive acoustical detectors to detect the onset of damage. A stylus skidding over a surface is much quieter than one that is ploughing its way through and shattering the material as it goes.

The chemical resistance of the film is also of some significance, particularly in connection with the effects of atmospheric moisture, to be considered later. In this latter respect, the solubility of the bulk material is a useful guide, although it should always be remembered that in thin-film form, the ratio of surface area to volume can be extremely large, and any tendency toward solubility present in the bulk material, greatly magnified. As in so many other thin-film phenomena, the magnitude of the effect very much depends on the particular thickness of material, on the other materials present in the multilayer, on the particular deposition conditions, as well as on the type of test used. However, a broad classification into moisture-resistant materials (materials such as titanium oxide, silicon oxide, and zirconium oxide), slightly affected (materials such as zinc sulfide) and badly affected (materials such as sodium fluoride), can be made.

14.5 Annealing

Baking has already been mentioned in Chapter 13. Most of the work that has been reported on baking is with regard to narrowband filters, still often constructed from zinc sulfide and cryolite. Meaburn [104] was a particularly early researcher in this area. He found that a process of baking at 90°C for 10 hours enormously improved the stability of narrowband filters of zinc sulfide and cryolite. This was especially so if they were protected afterward by a cemented cover slip.

Title et al. [105] reported a baking process called a *hard bake* with filters similar to those described by Meaburn. In the hard bake, filters were subjected to temperatures around 100°C for a certain time. During the baking process, the peak wavelength moved toward shorter wavelengths. After a critical time, the rate of movement suddenly slowed and the filter became much more stable. Details of the shift and the time were considered proprietary and not included in the published account. This is consistent with a desorption process coupled with a diffusion process to be described shortly.

Richmond [106] and Lee [107] both conducted baking experiments on narrowband filters. They were interested in adsorption and desorption processes in thin films. They found that the baking process did not appear to alter the amount of moisture adsorbed and desorbed by the filters. The stability of the characteristic, in the sense of the total change for a given change in relative humidity, was essentially unaltered. The rate of change, however, was greatly increased so that the characteristic reached equilibrium very much faster. The filters therefore appeared to be much more stable in the laboratory environment.

Müller [108] constructed computer models of the annealing process in thin films. The essential features of the models were thermally activated movements of atoms from a filled site to an available neighboring and vacant site. He found that packing density did not change during this process but that there was a quite definite amalgamation of smaller voids into larger ones. This process appears to be a wandering of the voids through the material of the thin film but is really a process of surface diffusion around the interior of the voids. Once two voids meet, there is an energetic advantage in combining, but once combined, there is no advantage in splitting. Thus, the voids simply increase gradually in size as they reduce in number. The reason for the findings of Richmond and Lee, and probably Title and Meaburn, now become clear. After deposition, the pore-shaped voids in the material are quite irregular in shape, especially at the interfaces between the layers. The annealing or baking process tends to remove the restrictions in the pores so that although their volume is unchanged, their regular shape implies a much faster filling by capillary condensation when exposed to humidity. This means that equilibrium is reached much more rapidly, and the filter appears much more stable when the environmental conditions are stable. In the case of already cemented filters, the effective environment is quite stable, although the filter stability may be disturbed by changes in temperature. However, when the temperature stabilizes, equilibrium is rapidly established once again.

The improved stability of the integral laser mirror is probably also derived at least partly from this decrease in the time constant for it to reach equilibrium. Any drift of the mirrors after alignment in the laser would immediately cause fluctuations, almost invariably reductions, in laser output. If the mirror can reach equilibrium before the final alignment, then, since the environment within the laser is reasonably stable from the point of view of moisture and consequent adsorption, the laser will be stable.

Müller [108] also explained why it is that baking never seems to improve poor adhesion but invariably makes it worse. Here if the bonds that bind atoms together across an interface are weaker than those that bind similar atoms together in either material, then there is an energetic advantage for a void that reaches an interface to remain there. Voids therefore collect at such an interface and gradually weaken the adhesion further.

Amorphous materials tend to exhibit an amorphous to crystalline transition when annealed at elevated temperatures. This transition can be inhibited by mixing with another material of different molecular size. Titanium dioxide can be stabilized in this way by, for example, the addition of silica. Chao et al. [109,110], using ion-beam sputtering, demonstrated that the addition of 17% silica to titania increased the recrystallization temperature from approximately 200°C to over 400°C. Annealing a high-reflectance coating with pure titania high-index layers and silica low-index layers gave rapidly increasing losses at annealing temperatures just above 200°C, but with the 17% silica-doped titania as high index, annealing at 400°C showed a reduction in loss of almost 90%.

There are a few more studies of baking in connection with telecom-quality filters, primarily using energetically deposited materials. Prins et al. [111] found a curious effect that they termed *creep*, although, as they pointed out, it is not creep in the normal sense of the word. Baking narrowband filters that had been energetically deposited, the particular materials were not identified, but it is likely that they were SiO_2 and Ta_2O_5 or, possibly, the chemically similar Nb_2O_5, because at the time, they were the preferred material in that application. These materials become amorphous when energetically deposited. Exposure to a high temperature (1 minute at 340°C, for example) of filters on high expansion coefficient substrates caused an expected immediate shift to shorter wavelengths. However, on cooling back down to room temperature, the original wavelength was not immediately restored. Instead, there was a gradual recovery that occupied around 5 minutes as the wavelength slowly returned toward the original value, although very slight shifts could continue for a period of days. A small permanent shift due to the baking could also be observed. It seems that the relaxation of the films is much slower at lower temperatures. The authors referred to the property of the films as a viscosity that reduced as the temperature increased, and the behavior can certainly be interpreted in that way.

Baking reduces the strain in the films. A useful and informative study of the changes in filters due to baking is due to Brown [112]. Here the filters were definitely constructed from SiO_2 and Ta_2O_5, and the process was ion-beam sputtering, frequently used for telecom-quality filters. Annealing at temperatures of around 500°C induced a shift in the filters toward longer wavelengths. Energetic processes such as ion-beam sputtering induce high levels of compressive strain in the films. Annealing permits redistribution of the material in the film so that it becomes thicker as the strain reduces. The reduction in strain also induces a drop in refractive index since strain birefringence is reduced. The net effect is an increase in optical thickness since the thickness increase dominates. Brown also suggested a packing density effect in which void volume plays a part. It is difficult to say whether these are true voids in the sense of actual empty spaces in the film, or just an expression of the spacing of the elements of the films, but the paper repays close study because of the accurate quantitative nature of the results.

Considerable insight into mechanical properties and annealing behavior in tantala and niobia is afforded by some studies by Çetinörgü-Goldenberg and colleagues [113,114]. Figure 14.21 shows some temperature cycling and annealing results. Both materials were deposited by dual ion-beam

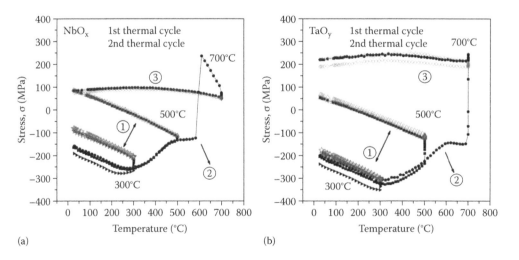

FIGURE 14.21
Plots of stress versus temperature of (a) niobia and (b) tantala thin films deposited by dual ion-beam sputtering. (From E. Çetinörgü-Goldenberg, J.-E. Klemberg-Sapieha, and L. Martinu, *Applied Optics*, 51, 6498–6507, 2012. With permission of Optical Society of America.)

sputtering, and both exhibited compressive stress after deposition, and both exhibited similar behavior. As we would expect, the as-deposited films were amorphous. A gradual increase in the temperature up to 300°C caused an increase in the stress due to differential expansion as might also be expected. A small reduction in the stress level at 300°C then took place. Subsequent temperature cycling between ambient and 300°C showed completely reversible thermal stress with no further annealing-driven shift. A further series of experiments where the 300°C temperature was increased to 500°C, this time in a nitrogen atmosphere, showed further stress reduction at 500°C followed by, once again, reversible linear behavior on cycling between ambient and 500°C with the same slope as in the previous case suggesting no change in the coefficient of thermal expansion. In this case, there was sufficient stress reduction for the level to become tensile for both niobia and tantala at ambient temperature. The films remained amorphous. The situation considerably changed when the upper annealing temperature was raised to 700°C, again in nitrogen. Now the films began to crystallize and the stress levels dramatically changed to a quite high tensile value. After this recrystallization, the films remained in tension, and although the behavior was still essentially reversible on temperature cycling, the variation in stress had a quite different slope, virtually zero, showing that the mechanical properties had significantly changed. The optical properties after recrystallization showed increased losses probably from grain-boundary scattering.

A similar study [115] has been performed on the titania-doped tantala used in the gravitational wave receiver mentioned earlier. This confirms the constancy of the coefficient of thermal expansion up to temperatures of 400°C but finds that the coefficient is reduced by the addition of titania. Young's modulus is, however, affected both by annealing and by the addition of titania.

14.6 Toxicity

In thin-film work, as indeed in any other field where much use is made of a variety of chemicals, the possibility that a material may be toxic should always be borne in mind. Fortunately, most of the materials in common use in thin-film work are reasonably innocuous, but there are occasions where distinctly hazardous materials must be used. The thin-film worker would be wise to check this point before using a new material. The technical literature on thin films, being primarily concerned with physical and chemical properties, seldom mentions the toxic nature of the materials. For example, thorium fluoride, oxyfluoride, and oxide are materials that are extensively covered in the literature, but for a long time, there was little or no mention of the radioactivity of thorium (nowadays well recognized). Recently, there has been a growing realization of the dangers associated with them, and they are gradually being phased out, although there are still some high-power infrared applications where they continue to be necessary. Some of the thallium salts are useful materials for the far infrared, but these are particularly toxic.

Fortunately, manufacturers' literature is becoming a more useful source of information on toxicity, and in any cases of doubt, the manufacturer should always be consulted. As long as toxic material is confined to a bottle, there is little danger, but as soon as the bottle is opened, material can escape. A major objective, in the use of toxic materials, is to confine them in a well-defined space, in which suitable precautions may be taken. If material is allowed to escape from this space, so that dangerous concentrations can exist outside, then it may be impossible to prevent an accident. It may be necessary to include the whole laboratory in the danger zone and to take special precautions in cleaning up on leaving. Special clothing, extending to respirators, may even be required while in the laboratory. On the other hand, machines may be isolated from the remainder of the production area by special dust-containing cabinets complete with air circulation and filtration units.

Most of the material evaporated in a process ends up as a coating on the inside of the machine and on the jigs and fixtures, where it usually forms a powdery deposit. The greatest danger is in the subsequent cleaning. Also, some of the solvents and cleaning fluids that can be used in the process give off harmful vapors. A good rule when dealing with potentially hazardous chemicals is to limit the total quantity on the premises to a minimum and especially the amount that is out of safe storage at any time. This not only puts an upper bound on the magnitude of any major disaster but also, even if no other precautions are taken, minimizes any leakage. It is also good from the psychological point of view. It should also be remembered that many poisons are cumulative in action, and while a slight dose received in the course of a short experiment may not be particularly harmful, the same dose, repeated many times in the course of several years, may do irreparable damage. Thus, the research worker may get away with a particular process that is operated only enough times to prove it, but the production worker will be expected to operate this process day in and day out, possibly for years. The safety standards in the production shop must therefore be of the highest standard, and workers should be aware of them without being dismayed by them. It should be remembered, too, that in an emergency, the laboratory may be rapidly vacated. It is then important, particularly for any emergency workers, that the hazardous materials should be well contained and their situation known. Good housekeeping is indispensable. The thin-film worker in industry should make certain that the medical officer of the works is fully aware of the materials currently in use, so that any necessary precautions can be taken before any trouble occurs.

There are, of course, legal requirements. However, legal requirements may not represent sufficiently prudent precautions. In general, unless positively dangerous materials are involved, the same precautions should be taken as in any chemical laboratory.

14.7 Microstructure and Thin-Film Behavior

One of the most significant features of optical thin films is the way in which their properties and behavior differ from those of identical materials in bulk form. This is, of course, also true for thin films in areas other than optics. Almost always, the performance of the film is poorer than that of the corresponding bulk material. Refractive index is usually lower, although, very occasionally, for some semiconductor materials, it can be slightly higher; losses, greater; durability, less; and stability, inferior. There is also sensitivity to deposition conditions, especially substrate temperature.

Heitmann [116] studied the influence of parameters, such as the residual gas pressure within the chamber and the rate of deposition, on the refractive indices of cryolite and thorium fluoride. Raising the residual gas (nitrogen) pressure from 5.3×10^{-4} Pa (5.3×10^{-6} mbar or 4×10^{-6} Torr) in one case, and 2.6×10^{-4} Pa (2.6×10^{-6} mbar or 2×10^{-6} Torr) in another, to 2.6×10^{-3} Pa (2.6×10^{-5} mbar or 2×10^{-5} Torr) had no measurable effect within the accuracy of the experiment ($\pm 0.1\%$ for thorium fluoride and $\pm 0.3\%$ for cryolite) while a further increase in residual pressure to 2.6×10^{-2} Pa (2.6×10^{-4} mbar or 2×10^{-4} Torr) gave a drop in index of 1.5% for cryolite and 1.4% for thorium fluoride. At this higher pressure, the mean free path of the nitrogen molecules was less than the distance between source and substrate, and the decrease in refractive index was probably caused by increased porosity of the layers. This tends to confirm that the mean free path of the residual gas molecules should be kept longer than the source–substrate distance, but that any further increases in mean free path beyond this have little effect. Heitmann concluded that the mean free path of the molecules is the important parameter, not the ratio of the numbers of evaporant molecules to residual gas molecules impinging on the substrate in unit time, which appeared to have no effect on refractive index. He also found that changes in the rate of deposition, from a quarter wave in 0.5 minute (measured at 632.8 nm) to a quarter wave in 1.5 minutes, caused a decrease in refractive index of 0.6% in both cases, but that a further decrease to a quarter wave in 5 minutes produced only slight variations.

Heitmann's results are probably best interpreted in terms of slight changes in film microstructure, induced by the variations in deposition conditions. Layer microstructure is, in fact, the most significant factor in determining the properties of optical thin films and the way in which they differ from the same material in bulk form. During the past several decades, there has been an increasing interest in the microstructure of, and microstructural effects in, optical thin films.

A useful technique for the study of thin-film structure, which immediately yielded important results, is electron microscopy. Its use in the examination of thin-film coatings has involved the development of techniques for fracturing multilayers and for replicating the exposed sections. Pearson [117], Lissberger and Pearson [118], Pulker and Jung [119], and Pulker and Guenther [120] all made substantial contributions in this area, and their results show that the layers in optical coatings have, almost invariably, a pronounced columnar structure, with the columns running across the films normal to the interfaces. To their investigations, we can add those of Movchan and Demchishin [121] and then of Thornton [122,123], who investigated the effects of substrate temperature and, in Thornton's case, residual gas pressure, on the microstructure of evaporated and sputtered films. This showed that a critical parameter in the vacuum deposition of thin films is the ratio of the temperature of the substrate T_{sub} to the melting temperature T_{melt} of the evaporant. For values of this ratio lower than around 0.5, the structure of the layers is intensely columnar, the columns running along the direction of growth. Increased gas pressure forces the growth into a more pronounced columnar mode even for slightly higher values of substrate temperature.

Because the most useful materials in optical thin films are all of high melting point, substrate temperatures can rarely be higher than a small fraction of the evaporant melting temperature, and so the structure of thin films is almost invariably a columnar one, with the columns running along the direction of growth, normal to the film interfaces. The columns are several tens of nanometers across and roughly cylindrical in shape. They are packed in an approximately hexagonal fashion with gaps in between the columns, which take the form of pores running completely across the film, and there are large areas of column surface that define the pores and are in this way exposed to the surrounding atmosphere. The columnar structure of a film of zinc sulfide is shown in Figure 14.22 [124].

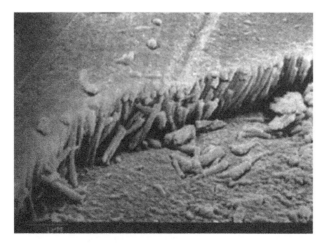

FIGURE 14.22
Columnar structure of a zinc sulfide film. Part of the film has been mechanically removed leaving the columnar structure visible in the cross section. (After I. M. Reid, H. A. Macleod, E. Henderson et al., *Proceedings of the International Conference on Ion Plating and Allied Techniques (IPAT 79)*, pp. 114–118, CEP Consultants, Edinburgh, 1979.)

Packing density p defined as follows:

$$p = \frac{\text{Volume of solid part of film (i.e., columns)}}{\text{Total volume of film (i.e., pores plus columns)}}.$$

It is a very important parameter. It is usually in the range 0.75–1.0 for optical thin films. For thermally evaporated thin films, it is most often 0.8–0.95, and seldom as great as unity. A packing density that is less than unity reduces the refractive index below that of the solid material of the columns. A useful expression that is reasonably accurate for many films, particularly of low index [125,126], connects the index of the film n_f, that of the solid part of the film n_s, and of the voids n_v, with the packing density p:

$$n_f = pn_s + (1 - p)n_v. \tag{14.29}$$

The behavior of films of higher index (2.0 and above) can be rather more complicated, but in many cases, a linear law as in Equation 14.29 is still sufficiently accurate and is therefore often employed. Unfortunately, there are not many actual measurements of index variation with packing density or actual density. Two sets are shown in Figures 14.23 and 14.24 corresponding to

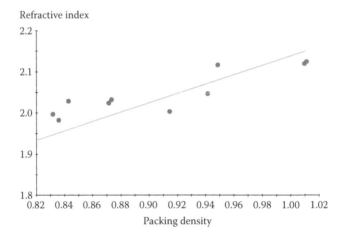

FIGURE 14.23
Comparison between the linear model in Equation 14.29 and measured results from Thielsch et al. [127] for hafnia at 550 nm.

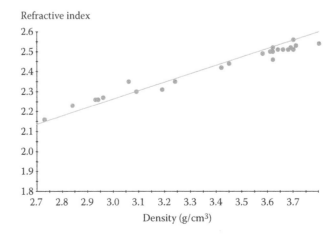

FIGURE 14.24
Comparison between the linear model in Equation 14.29 and measured results from Laube et al. [128] for titania at 550 nm.

hafnia [127] and to titania [128]. These are compared with the linear model of Equation 14.29 with air as the void material.

An alternative approach that is more complicated but can give a better fit in more complicated cases where there is a nonlinear relationship, the model discussed by Harris et al. [126], is relevant. However, if the packing density has been derived using Equation 14.29, then it is reasonable that it should be subsequently used in Equation 14.29.

Packing density is a function of substrate temperature, usually, but not always, increasing with substrate temperature, and of residual gas pressure, decreasing with rising pressure. Film refractive index is therefore also affected by substrate temperature and residual gas pressure. The columns can vary in a cross-sectional area as they grow outwards from the substrate surface, which is one cause of film inhomogeneity. Substrate temperature is a difficult parameter to measure and to control so that consistency in technique—heating for the same period each batch, identical rates of deposition, pumping for the same period before commencing deposition, and so on—is of major importance in assuring a stable and reproducible process. Changing the substrate dimensions, especially substrate thickness, from one run to the next can cause changes in film properties. Such changes are even more marked in the case of reactive processes where the residual gas pressure is raised and where a reaction between evaporant and residual atmosphere takes place at the growing surface of the film. Thus, it should not be surprising that a high proportion of test runs are required in any manufacturing sequence.

Various modeling studies [129–132] have confirmed that the columnar growth results from the limited mobility of the material on the surface of the growing film. It diffuses over the surface under thermal excitation until it is buried by arriving material. Diffusion through the bulk of the material is not significant. Thus, lower substrate temperature and higher rates of deposition lead to more pronounced columns and reduced packing density. The energetic processes involve an element of bombardment of the growing films. The transfer of momentum drives the material deeper into the film and, although the columnar structure may persist to some extent, squeezes out the voids. The packing density is normally close to or equal to unity. The results of the higher packing density are almost all favorable. The consequences, described in this chapter, of the columnar microstructure are all less serious in the energetically deposited films. (See Figure 14.25 [42].)

FIGURE 14.25
Compact microstructure of an aluminum oxynitride rugate structure deposited by RF reactive sputtering of aluminum. The packing density is very high, but some columnar features remain. The fractures at the outer surface tend to be in the nitrogen-rich parts of the rugate cycle leading to the stepped appearance. (Courtesy of Professor Frank Placido.)

A second level of microstructure in thin films is their crystalline state. Although this is less well understood, considerable progress has been made. Optical thin films are deposited from vapor that has been derived from sources at comparatively very high temperature. The substrates on which the films grow are at relatively very low temperature. There is therefore a considerable lack of equilibrium between growing film and arriving vapor. The film material is rapidly cooled or quenched, and this not only influences the formation of the columnar microstructure but also affects the crystalline order. The material that is condensing will attempt to reach the equilibrium form appropriate to the temperature of the substrate, but the correct rearrangement of the molecules will take a certain time, and the film will tend to pass through the higher-temperature forms during this rearrangement. If the rate of cooling is greater than the rate of crystallization, then a higher-temperature form will be frozen into the layer. All this explains some of the curious behavior of thin films. Frequently, there is an inversion in the crystalline structure in that at low substrate temperatures, a predominance of high-temperature crystalline forms are found, whereas at high substrate temperatures, more low-temperature material appears to form. The low substrate temperature leads to a higher quench rate and the rest follows [18]. Amorphous forms, corresponding to a quite high temperature, can often be frozen by very rapid cooling and are enhanced by a higher temperature of the arriving species. For example, sputtering, where additional kinetic energy is possessed by the arriving molecules, often gives amorphous films. The low-voltage ion-plating technique, again with high incident energy, appears virtually invariably to give amorphous films. The high-temperature forms are often only metastable and may change their structure at quite low temperatures leading to problems of various kinds. Some films deposited in amorphous form by sputtering may sometimes be induced to recrystallize, in a manner described as explosive, by a slight mechanical disturbance, such as a scratch, or by laser irradiation [133].

Samarium fluoride has two principal crystalline forms, a hexagonal high-temperature form and an orthorhombic low-temperature form. Table 14.1 shows the results of thermal evaporation and ion-assisted deposition, which both lead to this apparently inverted structure [18]. Zirconia has three principal structures, monoclinic, tetragonal, and cubic in ascending temperature. Klinger and Carniglia [134] found that very thin zirconia shows a cubic structure, but becomes monoclinic when thicker than a quarter wave at 600 nm. This behavior can be explained by a lower rate of quenching when the film is thicker and less thermally conducting. Alumina, normally amorphous in thin-film form, can recrystallize in the electron microscope when subjected to the electron bombardment necessary for viewing [135]. Amorphous zirconia, which can occur when films are very thin, has been shown to exhibit similar behavior [136]. Tang et al. [137] found that ion-assisted titania films switched from an amorphous structure to a crystalline one when the substrate temperature was raised, again lowering the rate of quenching.

Thin films are therefore complicated mixtures of different crystalline phases, some being high-temperature metastable states. Such behavior is clearly very material and process dependent, and each specific system requires individual study. What is a good structure for one application may

TABLE 14.1

Samarium Fluoride (SmF$_3$)

Normal high-temperature form		Hexagonal
Normal low-temperature form		Orthorhombic
Thermal evaporation	Substrate temperature of 100°C	Hexagonal (111)
	Substrate temperature ≥ 200°C	Orthorhombic (111) with some hexagonal
Ion-assisted deposition	Substrate temperature of 100°C	Hexagonal (110) with some (111)
	Higher bombardment at substrate temperature 100°C	Hexagonal (110) with appearance of new peak SmF$_2$ (111)?

Source: L. J. Lingg, Lanthanide trifluoride thin films: Structure, composition and optical properties, PhD Dissertation, University of Arizona, Tucson, AZ, 1990.

not be so for another. The low scattering of the amorphous phases make them attractive for certain applications, but their high-temperature or high-flux behavior may not be as satisfactory. Much more needs to be done to improve our understanding.

The columnar structure and the crystalline structure can be considered as essentially regular intrinsic features of film microstructure. Then, in addition, there are defects that can be thought of as local disturbances of the intrinsic features. A principal and very important class of defect is the nodule. Nodules are inverted conical growths that propagate through the film or multilayer. They can occur in virtually all processes. They start at a seed that is usually a very small defect or irregularity, and it appears that virtually, any irregularity, even minute ones, may act as a seed. Scratches on the substrate, pits, dust, contamination, material particles ejected from the source, and loose accumulations of material in the vapor phase, perhaps even local electric charges, can all cause nodules to start growing. Once the nodule starts, it continues to grow until it forms a domed protrusion at the outer surface of the multilayer. The nodule itself is very much larger than the defect that causes it. It is not, in itself, a contaminant. It is made up of exactly the material of the remainder of the coating. It is simply growing in a different way. The outer surface of the nodule is a quite sharp boundary between it and the remainder of the coating. This sharp boundary is a region of weakness, and there is frequently a fissure around the nodule, either partially or completely, and the nodule may sometimes be detached from the coating, leaving a hole behind. Nodules are present in almost all coatings. The only way of suppressing them appears to be a move toward perfection in the substrate, its surface, and its preparation and in the coating deposition. The incidence of nodules over superpolished substrates, for example, is much reduced compared with conventional substrates. A typical nodule is shown in Figure 14.26 and the hole left by a detached nodule in Figure 14.27. Micrographs of narrowband filters for the visible and infrared are shown in Figures 14.28 and 14.29, respectively.

The variation in refractive index is not the only feature of film behavior associated with the columnar structure. The pores between the columns permit the penetration of atmospheric moisture into the film, where, at low relative humidity, it forms an adsorbed layer over the surfaces of the columns and, at medium relative humidity, actually fills the pores with liquid water

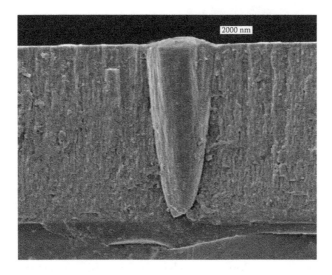

FIGURE 14.26
Nodule. The film is a rugate structure of aluminum oxynitride deposited by RF reactive sputtering of aluminum. The film has been broken across its width to show a cross section that includes a complete nodule. The sharpness of the boundary is clear, and the weakness is shown by the fact that the crack in the film circles around the nodule rather than pass through it. The shape and the domed protrusion at the outer surface (*upper*) of the film system are typical. (Courtesy of Professor Frank Placido.)

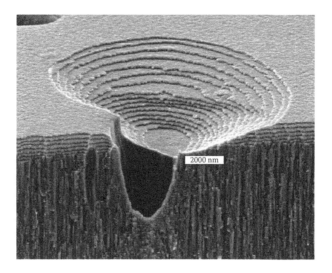

FIGURE 14.27
Hole left by the detachment of a nodule. Part of the outer part of the structure has been removed along with the nodule. The stepped appearance is once again caused by preferential cracking in the nitrogen-rich part of the aluminum oxynitride rugate structure. (Courtesy of Professor Frank Placido.)

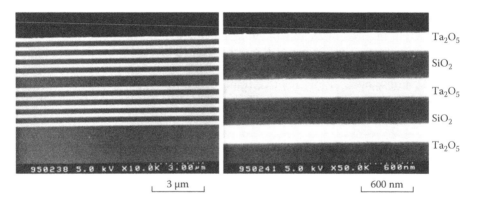

FIGURE 14.28
Micrograph showing the compact amorphous structure of a narrowband filter of silica and tantala produced by ion-assisted deposition using an RF ion gun. (Courtesy of Shincron Co. Ltd., Tokyo.)

due to capillary condensation. Moisture adsorption has been the subject of detailed investigation by Ogura [138] and Oruga and Macleod [139], who used the variation in adsorption with relative humidity to derive information on the pore structure of the films. The moisture, since it has a different refractive index (around 1.33) from the 1.0 of the air that it displaces from the voids, causes an increase in the refractive index of the films. Since the geometrical thickness of the film does not change, the increase of film index during adsorption is accompanied by a corresponding increase in optical thickness. The exposure of a film to the atmosphere, therefore, usually results in a shift of the film characteristic to a longer wavelength. Such shifts in narrowband filters have been the subject of considerable study. Schildt et al. [140] found that for freshly prepared filters of zinc sulfide and magnesium fluoride, constructed for the region of 400–500 nm, the variation in peak wavelength could be expressed as

$$\Delta\lambda = q\log_{10}P,$$

FIGURE 14.29
Structure of a multiple-cavity filter for the far infrared constructed from lead telluride and zinc sulfide. This particular filter was one of a set for the region of 6–18 μm required to have a size of 1.2 mm × 0.45 mm for use in the high-resolution dynamic limb sounder, and the high quality of the diamond-sawn edge of the component is clear from the micrograph. The scale of the micrograph can be assessed from the 4 μm physical thickness of the cavity layers. (Courtesy of Roger Hunneman, University of Reading, Reading, UK.)

where q is a constant varying from around 1.4 for filters that had aged to around 8.3 for freshly prepared filters, and P is the partial pressure of water vapor measured in torrs (P should be replaced by $0.0075 \times P$ if P is measured in pascals or by $0.75 \times P$ if P is measured in millibars) and $\Delta\lambda$ is measured in nanometers. $\Delta\lambda$ was arbitrarily chosen as zero when the pressure was 133 Pa (1.3 mbar or 1 Torr). This relationship was found to hold good for the pressure range of 133–2660 Pa (1.3–26 mbar or 1 to approximately 20 Torr). The filters settled down to the new values of peak wavelength some 10–20 minutes after exposure to a new level of humidity began. They found that the shifted values of peak wavelength could be stabilized by cementing cover slips over the layers using an epoxy resin. Koch [141,142] showed that the characteristics of narrowband filters became quite unstable during adsorption until the filters reached an equilibrium state. Richmond [106], Lee [107], and Macleod and Richmond [143] made detailed studies of the effects of adsorption on the characteristics of narrowband filters. The results are applicable to all types of multilayer coating. The shifts in the characteristics are due, as we have seen, to the filling of the pores of the film with liquid water. In multilayers, the pores of one film are not always directly connected with the pores of the next, and the penetration of atmospheric moisture is frequently a slow and complex process, in which a limited number of penetration pores take part, from which the moisture spreads across the coating in increasing circular patches. The primary entry points for the moisture are thought to be nodules where capillary condensation can take place in the fissures that often surround them. Depending on the materials, the coating may take several weeks to reach equilibrium and, afterward, will exhibit some instability should the environmental conditions change. The patches, which can sometimes be seen with the naked eye as a flecked or mottled appearance, can be made more visible if the coating is viewed in monochromatic light, at or near a wavelength for which there is a rapid variation of transmittance (Figure 14.30). The edge of an edge filter, or the passband of a narrowband filter, is especially suitable. Wet patches show a shift in wavelength that changes them from high to low transmittance, or vice versa, and they can be readily photographed as was done in Figures 14.31 and 14.32.

The drift of the filters toward longer wavelengths, which occurs on exposure to the atmosphere, considerably varies in magnitude with both the materials and the spectral region, and there is frequently considerable hysteresis on desorption. In the infrared, the layers are thick, and many of the semiconductor materials that are used as high-index layers have high packing density. This

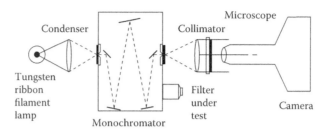

FIGURE 14.30
Sketch of the apparatus for observing moisture-penetration patterns in a multilayer of zinc sulfide and cryolite. Short slits that are virtually pinholes are used in the monochromator. (Reprinted from *Thin Solid Film*, 37, H. A. Macleod and D. Richmond, T Moisture penetration patterns in thin films, 163–169, Copyright (1976), with permission from Elsevier.)

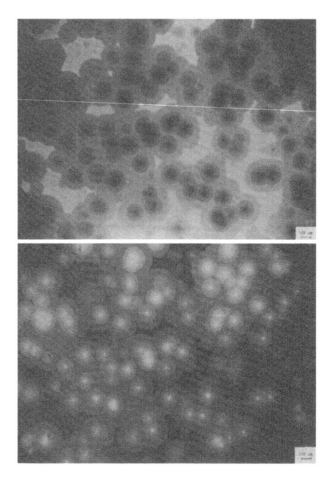

FIGURE 14.31
Photograph of moisture-penetration patterns in a zinc sulfide and cryolite filter some 2 weeks after coating. The relative humidity was approximately 50% during this time. The upper photograph was taken at a wavelength of 488.5 nm and the lower at 512.8 nm. The dark patches of the upper photograph correspond to the light patches of the lower showing that a wavelength shift rather than absorption is responsible for the patterns. (After C. C. Lee, Moisture adsorption and optical instability in thin film coatings, PhD Dissertation, University of Arizona, Tucson, AZ.)

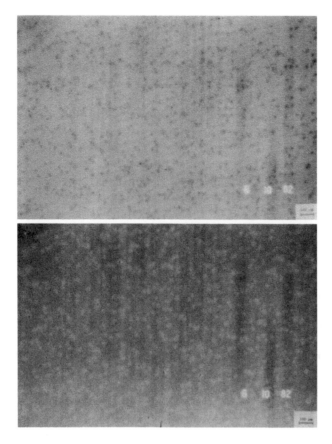

FIGURE 14.32
Moisture-penetration patterns in a multilayer of zirconium dioxide and silicon dioxide. The photographs were taken immediately after removal from the coating chamber. The wavelength for the upper photograph was 543 nm, and that for the lower was 553 nm. (After C. C. Lee, Moisture adsorption and optical instability in thin film coatings, PhD Dissertation, University of Arizona, Tucson, AZ.)

means that moisture-induced drift is less of a general problem than it is in the visible and ultraviolet regions of the spectrum, although it is important in some applications and even quite small amounts of water can cause perceptible absorption in the infrared water bands. In the visible region, drifts can be as high as 10 nm, and sometimes greater, toward longer wavelengths. The gradual stabilization of the coating as it reaches equilibrium is frequently referred to as aging or settling. The energetic processes can usually completely suppress the moisture-induced drifts and have been almost universally adopted for suitable coatings. It should be noted, however, that not all materials respond well to the brutal bombardment that is characteristic of the energetic processes. Metals suffer from the inevitable implantation of the bombarding species. Their optical properties are degraded by the scattering of conduction electrons that results. Fluorides lose fluorine, and so the bombardment must be strictly limited; otherwise, the concentration of vacancy defects becomes too great. Oxygen tends to fill the vacancies and form oxyfluorides that are neither as rugged as the original fluorides nor as useful in the ultraviolet.

It is not simply in generating optical shifts that moisture is a problem for coatings. It has major mechanical and sometimes chemical effects as well. The stress in the coating is transmitted across the gaps between the columns, again by short-range forces. These forces can be very easily blocked by water molecules. An alternative explanation of the phenomenon is that the moisture, which coats the surfaces of the columns, reduces the surface energy to something approaching that of liquid water. Since the surface energy is an important factor in the stress/strain balance in the film, the result of the moisture adsorption is a change in the strain and stress levels. The stress is usually tensile and the moisture reduces it, frequently significantly. We have already mentioned Pulker's [92] work on impurities in thin films and their reduction of stress levels in a similar way.

Adhesion, too, is affected by moisture. The materials used for thin films have usually very high surface energies, and then the work of adhesion is correspondingly high. The presence of liquid water in a film can cause a reduction in the surface energy of the exposed surfaces of at least an order of magnitude. If water is present at the site of an adhesion failure and can take part in a process of bond transfer, rather than bond rupture followed by adsorption, then it will reduce the work of adhesion, and it is more likely that the failure will propagate. There is frequently enough strain energy in a film to supply the required work. The penetration sites for the moisture patches are probably associated with defects that may act as stress concentrators where adhesion failures driven by the internal strain energy in the films may originate. All the ingredients for a moisture-assisted adhesion failure are present, and it is frequently at such sites that delamination is first observed. Blistering is a similar form of adhesion failure frequently associated with moisture penetration sites and a compressively strained film. Uniform strain in a film is translated into a shear stress across its interface that is zero in the center and a maximum at the edge. Thus, the edges of a coating are particularly vulnerable. Defects at the edge act as stress concentrators and if the forces are sufficiently high or the defect is sufficiently large, delamination can begin and gradually propagate from the edge across the film. The presence of moisture encourages such failures. It is important, therefore, that defects at the edge of a coating should be kept to an absolute minimum. Great care should be taken with the fixtures that hold the substrates in place during the coating operation. It is very important that they should be designed to avoid any small scratches or other damage to the edges of the substrates.

We have already mentioned in Chapter 7 that changes in temperature cause changes in the spectral characteristics of coatings, narrowband filters having characteristics that are probably most sensitive to such alterations. We must divide the coatings into those that have been simply thermally evaporated and those that have been produced by an energetic process.

Most of the work that has been reported so far in this chapter has been in respect of conventionally thermally evaporated coatings. For small temperature changes, the principal effect is a simple shift toward longer wavelengths with increasing temperature. For the materials commonly used in the visible region of the spectrum, the shift is on the order of 0.003%/°C, while for infrared filters, it can be greater, and a useful figure is 0.005%/°C, although it can be as high as 0.0125%/°C. It must be emphasized that these figures strongly depend on the particular materials used. Filters of lead telluride and zinc sulfide can actually have negative coefficients greater than 0.01%/°C, and using these materials, it is even possible to design a filter that has zero temperature coefficient [35]. With greater positive changes of, say, 60°C or more, it is usual for any moisture in the filter to partially desorb, causing an abrupt shift toward shorter wavelengths (see Figure 14.33). This shift is not immediately recovered on cooling to room temperature, and so considerable hysteresis is apparent in the behavior [144]. Subsequent temperature cycling, before readsorption of any moisture, will then exhibit no hysteresis. Eventually, if maintained at room temperature, the filter will readsorb moisture and gradually drift back to its initial wavelength. Exposure to higher temperatures, over 100°C, can still cause permanent changes that appear to be related to minute alterations in the structure of the layers, altering the adsorption behavior so that some materials become less ready to adsorb moisture while others show more rapid adsorption [106,107,143]. A frequently applied empirical treatment, already mentioned in Chapter 11, involves baking of filters at elevated temperatures, usually several hundred degrees Celsius, for some hours. The baking process reduces residual absorption, particularly in reactively deposited oxide films, and improves the subsequent stability of the coatings. Part of the baking process appears to involve the opening up of the pores in the films, by smoothing out restrictions, so that moisture adsorption

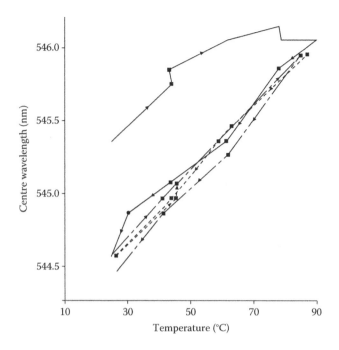

FIGURE 14.33

Record of the variation of peak wavelength with temperature for a filter of design Air | $(HL)^4$ $6H$ $(LH)^4$ | Glass, with L = cryolite and H = zinc sulfide. (After P. Roche, L. Bertrand, and E. Pelletier, *Optica Acta*, 23, 433–444, 1976. With permission of Optical Society of America.)

processes are more rapid and the films reach equilibrium in normal atmospheres much more quickly.

Films that have been deposited by the energetic processes usually exhibit lower temperature coefficients than thermally evaporated, even when the effects of moisture desorption and adsorption in the conventional films are eliminated. This is, at first sight, a quite surprising result. However, the explanation appears to lie in the microstructure. The lateral thermal expansion of the loosely packed columns in the thermally evaporated films enhances the drifts due to temperature changes. In the energetically deposited films, the material is virtually bulk-like in that there are no voids in between any residual columns, and so the material exhibits bulk-like properties. The change in characteristics with a change in temperature now corresponds to what would be expected from bulk materials. Indeed, Takahashi [145] showed that for multiple-cavity narrowband filters, once the design and materials are chosen, the expansion coefficient of the substrate dominates the behavior and can even change the sense of the induced spectral shift. The stress induced in the coating by the differential lateral expansion and contraction of substrate and coating is translated by Poisson's ratio into a swelling or reduction normal to the film surfaces. As a result of this modeling and improved understanding, temperature coefficients of peak wavelength shift at 1550 nm of 3 pm/°C (pm is picometer, i.e., 0.001 nm so that 3 pm/°C at 1550 nm represents 0.0002%/°C) have been routinely achieved in energetically deposited tantala/silica filters for communication purposes, and shifts even lower than 1 pm/°C are possible. The Takahashi model has been further elaborated by Kim and Hwangbo [146]. More information on this model is in Chapter 16.

Coatings that are subjected to very low temperatures usually shift toward shorter wavelengths, consistent with their behavior at elevated temperatures. The actual coatings are not usually mechanically affected. The substrates tend to be more vulnerable. Laminated components, particularly, run the risk of breaking because of differential contraction and/or expansion.

There are losses associated with all layers, which can be divided into scattering and absorption. In absorption, the energy, which is lost from the primary beam, is dissipated within the coating and usually appears as heat. In scattering, the flux lost is deflected and either reemerges from the coating in a different direction or is trapped beyond the critical angle within the coating or substrate. Absorption is a material property that may be intrinsic or due to impurities. A deficiency of oxygen, for example, can cause absorption in most of the refractory oxide materials. Scattering is usually due to defects in the coating that can be classified into volume or surface defects. Surface defects are simply a departure from the smooth flat surfaces of the ideal film. Such departures can be due to roughness of the substrate surface that tends to be reproduced at each interface in a multilayer or to the columnar structure of the layers that results in a nodular appearance of the film boundaries. Volume defects are local variations of optical constants and are usually dust particles, pinholes, or fissures in the coating. Losses in thin films are of particular importance in the laser field where they determine the limiting performance of multilayers. A major problem in the production of high-quality laser coatings is dust that emanates from the sources and from the powdery deposit that forms on the cold walls of the chamber. If this dust can be eliminated, only possible if the strictest attention is paid to detail and the most involved precautions are taken, then the remaining source of scattering loss is the roughness of the interfaces between the layers and between multilayer and substrate. If great care is exercised, then, in the visible and near infrared regions, the total losses, that is, absorption and scattering, can be reduced below 0.001% (for some very special applications, losses toward one tenth of this figure have been achieved), and the power handling capability of the coatings can be on the order of 5 J/cm^2 for pulses of 1 ns or less at 1.06 µm. Useful surveys of scattering in thin-film systems were written by Duparré and Kassam [147], Duparré [148], Duparré and Kaiser [149], and Amra [150,151]. There is a little more information on these topics in Chapter 16.

Many of the topics discussed in this chapter are dealt with in more detail in Pulker [93] and Stenzel [62].

References

1. J. T. Cox and G. Hass. 1958. Antireflection coatings for germanium and silicon in the infrared. *Journal of the Optical Society of America* 48:677–680.
2. H. Bangert and H. Pfefferkorn. 1980. Condensation and stability of ZnS thin films on glass substrates. *Applied Optics* 19:3878–3879.
3. G. Hass, J. B. Ramsay, and R. Thun. 1958. Optical properties and structure of cerium dioxide films. *Journal of the Optical Society of America* 48:324–327.
4. G. Hass. 1952. Preparation, properties and optical applications of thin films of titanium dioxide. *Vacuum* 2:331–345.
5. M. Auwärter. 1960. *Process for the manufacture of thin films*. US Patent, 2,920,002.
6. A. Vogt. 1957. Improvements in or relating to the manufacture of thin light-transmitting layers. UK Patent, 775,002.
7. D. S. Brinsmaid, W. J. Keenan, G. J. Koch et al. 1957. *Method of producing titanium dioxide coatings*. US Patent, 2,784,115.
8. S.-C. Chiao, B. G. Bovard, and H. A. Macleod. 1998. Repeatability of the composition of titanium oxide films produced by evaporation of Ti$_2$O$_3$. *Applied Optics* 37:5284–5290.
9. H. K. Pulker, G. Paesold, and E. Ritter. 1976. Refractive indices of TiO$_2$ films produced by reactive evaporation of various titanium-oxide phases. *Applied Optics* 15:2986–2991.
10. J. H. Apfel. 1980. The preparation of optical coatings for fusion lasers. *Thin Solid Films*. 73:167–168.

11. E. Ritter. 1962. Zür Kentnis des SiO und Si_2O_3-phase in dünnen Schichten. *Optica Acta* 9:197–202.
12. A. P. Bradford, G. Hass, M. McFarland et al. 1965. Effect of ultraviolet irradiation on the optical properties of silicon oxide films. *Applied Optics* 4:971–976.
13. R. A. Mickelsen. 1968. Effects of ultraviolet irradiation on the properties of evaporated silicon oxide films. *Journal of Applied Physics* 39:4594–4600.
14. W. Heitmann. 1971. Reactive evaporation in ionized gases. *Applied Optics* 10:2414–2418.
15. J. Ebert. 1982. Activated reactive evaporation. *Proceedings of SPIE* 325:29–38.
16. G. Hass, J. B. Ramsay, and R. Thun. 1959. Optical properties of various evaporated rare earth oxides and fluorides. *Journal of the Optical Society of America* 49:116–120.
17. D. Smith and P. W. Baumeister. 1979. Refractive index of some oxide and fluoride coating materials. *Applied Optics* 18:111–115.
18. L. J. Lingg. 1990. *Lanthanide trifluoride thin films: Structure, composition and optical properties*. PhD Dissertation. Tucson, AZ: University of Arizona.
19. L. J. Lingg, J. D. Targove, J. P. Lehan et al. 1987. Ion-assisted deposition of lanthanide trifluorides for VUV applications. *Proceedings of SPIE* 818:86–92.
20. F. Stetter, R. Esselborn, N. Harder et al. 1976. New materials for optical thin films. *Applied Optics* 15:2315–2317.
21. P. W. Baumeister and O. Arnon. 1977. Use of hafnium dioxide in mutilayer dielectric reflectors for the near UV. *Applied Optics* 16:439–444.
22. J. T. Cox and G. Hass. 1978. Protected Al mirrors with high reflectance in the 8–12-µm region from normal to high angles of incidence. *Applied Optics* 17:2125–2126.
23. I. Lubezky, E. Ceren, and Z. Klein. 1980. Silver mirrors protected with Yttria for the 0.5 to 14µm region. *Applied Optics* 19:1895.
24. G. A. Al-Jumaily and S. M. Edlou. 1992. Optical properties of tantalum pentoxide coatings deposited using ion beam processes. *Thin Solid Films* 209:223–229.
25. S. M. Edlou, A. Smajkiewicz, and G. Al-Jumaily. 1993. Optical properties and environmental stability of oxide coatings deposited by reactive sputtering. *Applied Optics* 32:5601–5606.
26. Y. Song, T. Sakurai, K. Maruta et al. 2000. Optical and structural properties of dense SiO_2, Ta_2O_5 and Nb_2O_5 thin-films deposited by indirectly reactive sputtering technique. *Vacuum* 59:755–763.
27. H. Szymanowski, O. Zabeida, J. E. Klemberg-Sapieha et al. 2005. Optical properties and microstructure of plasma deposited Ta_2O_5 and Nb_2O_5 films. *Journal of Vacuum Science and Technology A* 23:241–247.
28. R. G. Greenler. 1955. Interferometry in the infrared. *Journal of the Optical Society of America* 45:788–791.
29. T. S. Moss. 1952. Optical properties of tellurium in the infra-red. *Proceedings of the Physical Society B* 65:62–66.
30. C. S. Evans, R. Hunneman, and J. S. Seeley. 1976. Increments at the interface between layers during infra-red filter manufacture. *Optica Acta* 23:297–303.
31. C. S. Evans, R. Hunneman, and J. S. Seeley. 1976. Optical thickness changes in freshly deposited layers of lead telluride. *Journal of Physics D* 9:321–328.
32. C. S. Evans, R. Hunneman, J. S. Seeley et al. 1976. Filters for v2 band of CO_2: Monitoring and control of layer deposition. *Applied Optics* 15:2736–2745.
33. C. S. Evans and J. S. Seeley. 1968. Properties of thick evaporated layers of PbTe. *Journal de Physique Colloques C4* 29:37–42.
34. F. S. Ritchie. 1970. *Multilayer filters for the infrared region 10–100 microns*. PhD Thesis. Reading: University of Reading.
35. J. S. Seeley, R. Hunneman, and A. Whatley. 1981. Far infrared filters for the Galileo-Jupiter and other missions. *Applied Optics* 20:31–39.
36. S. D. Smith and J. S. Seeley. 1968. *Multilayer Filters for the Region 0.8 to 100 Microns*. Bedford, MA: Air Force Cambridge Research Laboratories.
37. Y.-H. Yen, L.-X. Zhu, W.-D. Zhang et al. 1984. Study of PbTe optical coatings. *Applied Optics* 23: 3597–3601.
38. K. G. Zhang, J. S. Seeley, R. Huneman et al. 1989. Optical and semiconductor properties of lead telluride coatings. *Proceedings of SPIE* 1112:393–402.
39. J. F. Hall and W. F. C. Ferguson. 1955. Optical properties of cadmium sulphide and zinc sulphide from 0.6 micron to 14 micron. *Journal of the Optical Society of America* 45:714–718.
40. G. Hass and C. D. Salzberg. 1954. Optical properties of silicon monoxide in the wavelength region from 0.24 to 14.0 microns. *Journal of the Optical Society of America* 44:181–187.

41. C. K. Hwangbo, L. J. Lingg, J. P. Lehan et al. 1989. Reactive ion-assisted deposition of aluminum oxynitride thin films. *Applied Optics* 28:2779–2784.
42. F. Placido. 1997. *RF sputtering of aluminium oxynitride rugates: Micrographs of rugate structures.* Private communication (University of the West of Scotland, Paisley, UK).
43. B. B. Bovard, J. Ramm, R. Hora et al. 1989. Silicon nitride thin films by low voltage reactive ion plating: Optical properties and composition. *Applied Optics* 28:4436–4441.
44. R. Jacobsson and J. O. Martensson. 1966. Evaporated inhomogeneous thin films. *Applied Optics* 5:29–34.
45. S. Fujiwara. 1963. Refractive indices of evaporated cerium dioxide–cerium fluoride films. *Journal of the Optical Society of America* 53:880.
46. S. Fujiwara. 1963. Refractive indices of evaporated cerium fluoride–zinc sulphide films. *Journal of the Optical Society of America* 53:1317–1318.
47. N. Kogaku. 1965. *Surface-coated optical elements.* UK Patent, 1,010,038.
48. V. N. Yadava, S. K. Sharma, and K. L. Chopra. 1974. Optical dispersion of homogeneously mixed ZnS-MgF$_2$ films. *Thin Solid Films* 22:57–66.
49. V. N. Yadava, S. K. Sharma, and K. L. Chopra. 1973. Variable refractive index optical coatings. *Thin Solid Films* 17:243–252.
50. W. L. Morgan. 1949. *Method of coating with quartz by thermal evaporation.* US Patent, 2,463,791.
51. G. Anstalt. 1962. *Improvements in and relating to the oxidation and/or transparency of thin partly oxidic layers.* UK Patent, 895,879.
52. T. Kraus and P. Rheinberger. 1962. *Use of a rare earth metal in vaporizing metals and metal oxides.* US Patent, 3,034,924.
53. M. Friz, F. Koenig, and S. Feiman. 1992. New materials for production of optical coatings. In *35th Annual Technical Conference Proceedings,* pp. 143–148, Society of Vacuum Coaters, Baltimore, MD.
54. A. W. Butterfield. 1974. The optical properties of Ge$_x$Se$_{1-x}$ thin films. *Thin Solid Films* 23:191–194.
55. R. Jacobsson. 1975. Inhomogeneous and coevaporated homogeneous films for optical applications. *Physics of Thin Films* 8:51–98.
56. G. M. Harry, H. Armandula, E. Black et al. 2006. Thermal noise from optical coatings in gravitational wave detectors. *Applied Optics* 45:1569–1574.
57. G. M. Harry, M. R. Abernathy, A. E. Becerra-Toledo et al. 2007. Titania-doped tantala/silica coatings for gravitational-wave detection. *Classical and Quantum Gravity* 24:405–415.
58. G. Harry, T. P. Bodiya, and R. Desalvo, eds. 2012. *Optical Coatings and Thermal Noise in Precision Measurement.* Cambridge, UK: Cambridge University Press.
59. O. S. Heavens. 1964. Measurement of optical constants of thin films. In *Physics of Thin Films,* G. Hass and R. E. Thun (eds), pp. 193–238. New York: Academic Press.
60. H. M. Liddell. 1981. *Computer-Aided Techniques for the Design of Multilayer Filters.* Bristol: Adam Hilger.
61. J.-P. Borgogno. 1995. Spectrophotometric methods for refractive index determination. In *Thin Films for Optical Systems,* F. R. Flory (ed.), pp. 269–328. New York: Marcel Dekker.
62. O. Stenzel. 2014. *Optical Coatings: Material Aspects in Theory and Practice.* Berlin: Springer-Verlag.
63. J. F. Hall, Jr. and W. F. C. Ferguson. 1955. Dispersion of zinc sulfide and magnesium fluoride films in the visible spectrum. *Journal of the Optical Society of America* 45:74–75.
64. E. Pelletier, P. Roche, and B. Vidal. 1976. Détermination automatique des constantes optiques et de l'épaisseur de couches minces: Application aux couches diélectriques. *Nouvelle Revue d'Optique* 7:353–362.
65. J. C. Manifacier, J. Gasiot, and J. P. Fillard. 1976. A simple method for the determination of the optical constants n, k and the thickness of a weakly absorbing thin film. *Journal of Physics E* 9:1002–1004.
66. R. Swanepoel. 1983. Determination of the thickness and optical constants of amorphous silicon. *Journal of Physics E* 16:1214–1222.
67. W. Hansen. 1973. Optical characterization of thin films: Theory. *Journal of the Optical Society of America* 63:793–802.
68. F. Abelès. 1950. La détermination de l'indice et de l'épaisseur des couches minces transparentes. *Journal de Physique et le Radium* 11:310–314.
69. M. Hacskaylo. 1964. Determination of the refractive index of thin dielectric films. *Journal of the Optical Society of America* 54:198–203.
70. R. M. A. Azzam. 1995. Ellipsometry. In *Handbook of Optics,* M. Bass (ed.), pp. 27.1–27.27. New York: McGraw-Hill.

71. J. Rivory. 1995. Ellipsometric measurements. In *Thin Films for Optical Systems*, F. R. Flory (ed.), pp. 299–328. New York: Marcel Dekker.
72. H. G. Tompkins and J. N. Hilfiker. 2016. *Spectroscopic Ellipsometry. Practical Application to Thin Film Characterization.* New York: Momentum Press.
73. J. N. Hilfiker, R. A. Synowicki, and H. G. Tompkins. 2008. Spectroscopic ellipsometry methods for thin absorbing coatings. In *51st Annual Technical Conference Proceedings*, pp. 511–516, Society of Vacuum Coaters, Chicago, IL.
74. C. K. Hwangbo, L. J. Lingg, J. P. Lehan et al. 1989. Ion-assisted deposition of thermally evaporated Ag and Al films. *Applied Optics* 28:2769–2778.
75. J. A. Dobrowolski, F. C. Ho, and A. Waldorf. 1983. Determination of optical constants of thin film coating materials based on inverse synthesis. *Applied Optics* 22:3191–3200.
76. F. Abelès. 1950. Recherches sur la propagation des ondes électromagnétiques sinusoïdales dans les milieus stratifies: Applications aux couches minces: I. *Annales de Physique, 12ième Serie* 5:596–640.
77. F. Abelès. 1950. Recherches sur la propagation des ondes électromagnétiques sinusoïdales dans les milieus stratifiés: Applications aux couches minces: II. *Annales de Physique, 12ième Serie* 5:706–784.
78. J. P. Borgogno, B. Lazarides, and E. Pelletier. 1982. Automatic determination of the optical constants of inhomogeneous thin films. *Applied Optics* 21:4020–4029.
79. D. P. Arndt, R. M. A. Azzam, J. M. Bennett et al. 1984. Multiple determination of the optical constants of thin-film coating materials. *Applied Optics* 23:3571–3596.
80. R. P. Netterfield. 1976. Refractive indices of zinc sulphide and cryolite in multilayer stacks. *Applied Optics* 15:1969–1973.
81. P. K. Tien. 1971. Light waves in thin films and integrated optics. *Applied Optics* 10:2395–2413.
82. A. A. J. Al-Douri and O. S. Heavens. 1983. The measurement of scattering losses in thin dielectric films. *Proceedings of the Royal Society of London A* 388:103–116.
83. R. Vernhes and L. Martinu. 2015. TRACK—A new method for the evaluation of low-level extinction coefficient in optical films. *Optics Express* 23:28501–28521.
84. A. Macleod and C. Clark. 2011. Pitfalls in the characterization of optical thin films. In *54th Annual Technical Conference*, pp. 243–247, Society of Vacuum Coaters, Chicago, IL.
85. A. V. Tikhonravov, M. K. Trubetskov, M. A. Kokarev et al. 2002. Effect of systematic errors in spectral photometric data on the accuracy of determination of optical parameters of dielectric thin films. *Applied Optics* 41:2555–2560.
86. R. W. Hoffman. 1976. Stresses in thin films: The relevance of grain boundaries and impurities. *Thin Solid Films* 34:185–190.
87. A. E. Ennos. 1966. Stresses developed in optical film coatings. *Applied Optics* 5:51–61.
88. K. Roll. 1976. Analysis of stress and strain distribution in thin films and substrates. *Journal of Applied Physics* 47:3224–3229.
89. K. Roll and H. Hoffman. 1976. Michelson interferometer for deformation measurements in an UHV system at elevated temperatures. *Review of Scientific Instruments* 47:1183–1185.
90. A. M. Ledger and R. C. Bastien. 1977. *Intrinsic and Thermal Stress Modeling for Thin-Film Multilayers.* Perkin Elmer, Norwalk, CT.
91. B. G. Bovard, X. C. d. Lega, S.-H. Hahn et al. 1991. *Intrinsic stress in titanium dioxide thin films produced by ion-assisted deposition.* Private communication (Optical Sciences Center, University of Arizona, Tucson, AZ).
92. H. K. Pulker. 1982. Stress, adherence, hardness and density of optical thin films. *Proceedings of SPIE* 325:84–92.
93. H. K. Pulker. 1999. *Coatings on Glass.* nd ed. Amsterdam: Elsevier.
94. S. F. Pellicori. 1984. Stress modification in cerous fluoride films through admixture with other fluoride compounds. *Thin Solid Films* 113:287–295.
95. H. Windischmann. 1987. An intrinsic stress scaling law for polycrystalline thin films prepared by ion beam sputtering. *Journal of Applied Physics* 62:1800–1807.
96. G. N. Strauss. 2003. Mechanical stress in optical coatings. In *Optical Interference Coatings*, N. Kaiser and H. K. Pulker (eds), pp. 207–229. Berlin: Springer-Verlag.
97. R. Jacobsson and B. Kruse. 1973. Measurement of adhesion of thin evaporated films on glass substrates by means of the direct pull method. *Thin Solid Films* 15:71–77.
98. O. S. Heavens. 1950. Some features influencing the adhesion of films produced by vacuum evaporation. *Journal de Physique et le Radium* 11:355–360.

99. P. Benjamin and C. Weaver. 1960. Measurement of adhesion of thin films. *Proceedings of the Royal Society of London, A* 254:163–176.

100. P. Benjamin and C. Weaver. 1960. Adhesion of metal films to glass. *Proceedings of the Royal Society of London, A* 254:177–183.

101. M. Laugier. 1981. Unusual adhesion-aging behaviour in ZnS thin films. *Thin Solid Films* 75:L19–L20.

102. M. Laugier. 1981. The development of the scratch test technique for the determination of the adhesion of coatings. *Thin Solid Films* 76:289–294.

103. I. S. Goldstein and R. DeLong. 1982. Evaluation of microhardness and scratch testing for optical coatings. *Journal of Vacuum Science and Technology* 20:327–330.

104. J. Meaburn. 1967. A search for nebulosity in the high galactic latitude radio spurs. *Zeitschrift für Astrophysik* 65:93–104.

105. A. M. Title, T. P. Pope, and J. P. Andelin. 1974. Drift in interference filters: Part 1. *Applied Optics* 13:2675–2679.

106. D. Richmond. 1976. *Thin film narrow band optical filters*. PhD Thesis. Newcastle upon Tyne: Northumbria University.

107. C. C. Lee. 1983. *Moisture adsorption and optical instability in thin film coatings*. PhD Dissertation. Tucson, AZ: University of Arizona.

108. K.-H. Müller. 1985. A computer model for postdeposition annealing of porous thin films. *Journal of Vacuum Science and Technology, A* 3:2089–2092.

109. S. Chao, W.-H. Wang, M.-Y. Hsu et al. 1999. Characteristics of ion-beam-sputtered high refractive-index TiO_2-SiO_2 mixed films. *Journal of the Optical Society of America A* 16:1477–1483.

110. S. Chao, W.-H. Wang, and C.-C. Lee. 2001. Low-loss dielectric mirror with ion-beam-sputtered TiO_2-SiO_2 mixed films. *Applied Optics* 40:2177–2182.

111. S. L. Prins, A. C. Barron, W. C. Herrmann et al. 2004. Effect of stress on performance of dense wavelength division multiplexing filters: Thermal properties. *Applied Optics* 43:633–637.

112. J. T. Brown. 2004. Center wavelength shift dependence on substrate coefficient of thermal expansion for optical thin-film interference filters deposited by ion-beam sputtering. *Applied Optics* 43:4506–4511.

113. E. Çetinörgü, B. Baloukas, O. Zabeida et al. 2009. Mechanical and thermoelastic characteristics of optical thin films deposited by dual ion beam sputtering. *Applied Optics* 48:4536–4544.

114. E. Çetinörgü-Goldenberg, J.-E. Klemberg-Sapieha, and L. Martinu. 2012. Effect of postdeposition annealing on the structure, composition, and the mechanical and optical characteristics of niobium and tantalum oxide films. *Applied Optics* 51:6498–6507.

115. M. R. Abernathy, J. Hough, I. W. Martin et al. 2014. Investigation of the Young's modulus and thermal expansion of amorphous titania-doped tantala films. *Applied Optics* 53:3196–3202.

116. W. Heitmann. 1968. The influence of various parameters on the refractive index of evaporated dielectric thin films. *Applied Optics* 7:1541–1543.

117. J. M. Pearson. 1970. Electron microscopy of multilayer thin films. *Thin Solid Films* 6:349–358.

118. P. H. Lissberger and J. M. Pearson. 1976. The performance and structural properties of multilayer optical filters. *Thin Solid Films* 34:349–355.

119. H. K. Pulker and E. Jung. 1971. Correlation between film structure and sorption behaviour of vapour deposited ZnS, cryolite and MgF_2 films. *Thin Solid Films* 9:57–66.

120. H. K. Pulker and K. H. Guenther. 1972. Electron optical investigation of cross-sectional structure of vacuum-deposited multilayer systems. *Vakuum-Technik* 21:201–207.

121. B. A. Movchan and A. V. Demchishin. 1969. Study of the structure and properties of thick vacuum condensates of nickel, titanium, tungsten, aluminium oxide and zirconium dioxide. *Fiz Metal Metalloved* 28:653–660.

122. J. A. Thornton. 1974. Influence of apparatus geometry and deposition conditions on the structure and topography of thick sputtered coatings. *Journal of Vacuum Science and Technology* 11:666–670.

123. J. A. Thornton. 1986. The microstructure of sputter-deposited coatings. *Journal of Vacuum Science and Technology, A* 4:3059–3065.

124. I. M. Reid, H. A. Macleod, E. Henderson et al. 1979. The ion plating of optical thin films for the infrared. In *Proceedings of the International Conference on Ion Plating and Allied Techniques (IPAT 79)*, pp. 114–118. Edinburgh: CEP Consultants.

125. K. Kinosita and M. Nishibori. 1969. Porosity of MgF_2 films—Evaluation based on changes in refractive index due to adsorption of vapors. *Journal of Vacuum Science and Technology* 6:730–733.

126. M. Harris, H. A. Macleod, S. Ogura et al. 1979. The relationship between optical inhomogeneity and film structure. *Thin Solid Films* 57:173–178.

127. R. Thielsch, A. Gatto, J. Heber et al. 2002. A comparative study of the UV optical and structural properties of SiO_2, Al_2O_3, and HfO_2 single layers deposited by reactive evaporation, ion-assisted deposition and plasma ion-assisted deposition. *Thin Solid Films* 410:86–93.

128. M. Laube, F. Rauch, C. Ottermann et al. 1996. Density of thin TiO_2 films. *Nuclear Instruments and Methods in Physics Research Section B: Beam Interactions with Materials and Atoms* 113:288–292.

129. K.-H. Müller. 1986. Model for ion-assisted thin-film densification. *Journal of Applied Physics* 59:2803–2807.

130. K.-H. Müller. 1988. Models for microstructure evolution during optical thin film growth. *Proceedings of SPIE* 821:36–44.

131. R. B. Sargent. 1990. Effects of surface diffusion on thin-film morphology: A computer study. *Proceedings of SPIE* 1324:13–31.

132. R. B. Sargent. 1989. *Surface diffusion: A computer study of its effects on thin film morphology*. PhD Dissertation. Tucson, AZ: University of Arizona.

133. R. Messier, T. Takamori, and R. Roy. 1975. Observations on the "explosive" crystallization of non-crystalline Ge. *Solid State Communications* 16:311–314.

134. R. E. Klinger and C. K. Carniglia. 1985. Optical and crystalline inhomogeneity in evaporated zirconia films. *Applied Optics* 24:3184–3187.

135. J. D. Targove. 1987. *The ion-assisted deposition of optical thin films*. PhD Dissertation. Tucson, AZ: University of Arizona.

136. C. Boulesteix and M. Lottiaux. 1987. *Behavior of zirconia film in electron microscope*. Private Communication (University of Aix-Marseille III, Marseille).

137. Q. Tang, K. Kikuchi, S. Ogura et al. 1999. Mechanism of columnar microstructure growth in titanium oxide thin films deposited by ion-beam assisted deposition. *Journal of Vacuum Science and Technology, A* 17:3379–3384.

138. S. Ogura. 1975. *Some features of the behaviour of optical thin films*. PhD Thesis. Newcastle upon Tyne: Northumbria University.

139. S. Ogura and H. A. Macleod. 1976. Water sorption phenomena in optical thin films. *Thin Solid Films* 34:371–375.

140. J. Schildt, A. Steudel, and H. Walther. 1967. The variation of the transmission wavelength of interference filters by the influence of water vapour. *Journal de Physique* 28:C2/276–C2/279.

141. H. Koch. 1965. Optische Untersuchungen zur Wasserdampfsorption in Aufdampfschichten (inbesondere in MgF_2 Schichten). *Physica Status Solidi* 12:533–543.

142. H. Koch. 1967. Über Sorptionsvorgänge beim Belüften von MgF_2 Schichten. In *Proceedings of Colloquium on Thin Films*, pp. 199–203. Budapest: Verlag Kultura.

143. H. A. Macleod and D. Richmond. 1976. Moisture penetration patterns in thin films. *Thin Solid Films* 37:163–169.

144. P. Roche, L. Bertrand, and E. Pelletier. 1976. Influence of temperature on the optical properties of narrowband optical filters. *Optica Acta* 23:433–444.

145. H. Takashashi. 1995. Temperature stability of thin-film narrow-bandpass filters produced by ion-assisted deposition. *Applied Optics* 34:667–675.

146. S.-H. Kim and C. K. Hwangbo. 2004. Temperature dependence of transmission center wavelength of narrow bandpass filters prepared by plasma ion-assisted deposition. *Journal of the Korean Physical Society* 45:93–98.

147. A. Duparré and S. Kassam. 1993. Relation between light scattering and microstructure of optical thin films. *Applied Optics* 32:5475–5480.

148. A. Duparré. 1995. Light scattering of thin dielectric films. In *Handbook of Optical Properties*, vol. 1: *Thin Films for Optical Coatings*, R. E. Hummel and K. H. Guenther (ed.), pp. 273–303. Boca Raton, FL: CRC.

149. A. Duparré and N. Kaiser. 1998. AFM helps engineer low-scatter films. In *Laser Focus World* 34(4): 147–152.

150. C. Amra. 1993. From light scattering to the microstructure of thin-film multilayers. *Applied Optics* 32:5481–5491.

151. C. Amra. 1995. Introduction to light scattering in multilayer optics. In *Thin Films for Optical Systems*, F. R. Flory (ed.), pp. 367–391. New York: Marcel Dekker.

15

Composite, Birefringent, and Metamaterials

So far we have been concentrating on what we might call traditional thin-film materials. Their properties may have been slightly modified by their microstructure, but, largely, they could be described by a single set of optical constants not far removed from those we would expect of the material in a pure state. In this chapter, we take a brief look beyond that model into composite materials, birefringent materials, and metamaterials.

Although there seems to be an almost unlimited number of materials that can be used in thin-film optical coatings, they tend to have quite similar properties, and, in fact, because properties other than optical are also important, the number of materials actually used is still more limited. There is therefore an interest in extending the range of optical properties. Composite materials are essentially combinations of two or more materials to yield properties different from those of the components. Birefringent materials have optical properties that depend on direction. This may be a natural property of the material, or it may depend on anisotropic strain, but the birefringence we are considering here is usually the result of deliberate structuring of the material. Metamaterial is a relatively new term that can encompass all the others but is especially used to denote deliberately engineered materials that exhibit properties quite remote from those displayed by natural materials. A particular property that has attracted considerable attention is negative refraction.

At the time of writing, there has been little use of these materials in mainstream optical coatings. However, there is some potential for use, and so we include some brief discussion in this chapter. We begin with composite materials. There is no single theory to describe the properties of composite materials. The nature of their microstructure can play as important a part as the properties of the individual materials. Often the particular model is chosen only after the composite properties have been measured. We will look at a few of the more important theories.

15.1 Packing Density

We have already discussed packing density p in Chapter 14 in the context of a film of columnar microstructure including voids. It is defined as follows:

$$p = \frac{\text{Volume of solid part of film (i.e., columns)}}{\text{Total volume of film (i.e., pores plus columns)}}.$$

The same parameter can be used in respect to composite materials that consist of one component dispersed in another. However, in a material of lower than unity packing density, we tend to think of the important fraction as being the matrix in which there is a dispersion of voids. The matrix is what we require and the voids represent a perturbation, and packing density represents this situation well. In a composite material, we are more interested in a desired modification of properties brought about by an inclusion, and the concentration of the inclusion is a more useful quantity. We define f the volume fraction, or sometimes inclusion fraction, as

$$f = 1 - p. \tag{15.1}$$

We already examined in Chapter 14 the simple linear law that is frequently used for modeling a reduced packing density. Now we must investigate some more involved models.

15.2 Composite Material Models

A simple model of a dielectric optical material is as an array of dipoles in a matrix of free space where the valence electrons are moving under the influence of the electric field with respect to the solid nuclei and core electrons, the frequency being too high for magnetic interactions. Electromagnetic radiation induces vibration of the dipoles without changing the frequency, and the dipoles first take energy to sustain their vibration from the incident fields but then radiate it back to combine with them. Due to interference of the radiation from the dipoles that are separated by a small fraction of a wavelength, the resultant returned wave mimics the incident primary wave except that it lags in phase. This distributed phase lag is the reason for a slowing down of the resultant combined wave expressed as an increase in the refractive index. Thus, the nature of the distributed dipoles determines the optical properties of the material. It is not difficult to imagine a material where, instead of an array of atoms or molecules, we have particles of one material dispersed in another. The particles are polarizable and so still act as simple dipoles, but their properties differ from those of individual atoms or molecules so that the material with this structure exhibits new optical properties. Such materials are known as composites, but there are other terms that may be applied to them. Cermet, for example, indicates metallic particles dispersed in a dielectric while nanocomposite emphasizes that the particles are very small compared with a wavelength.

We start with a basic optical material consisting of an array of atoms or molecules that act as dipoles. This simple model presents a problem. Not only is each vibrating dipole subjected to the primary field but also to the combined fields of all the other vibrating dipoles. The problem was independently solved by a number of nineteenth-century researchers, the credit usually being assigned to four principal ones, Ottaviano Fabrizio Mossotti (1791–1863), Rudolf Julius Emmanuel Clausius (1822–1888), Ludvig Valentin Lorenz (1829–1891), and Hendrik Antoon Lorentz (1853–1928). Integration over all the individual dipoles of the material is an impossible task, and the contribution of these pioneers was a solution involving treating each individual particle as though it were surrounded by a continuous medium rather than localized particles, the continuous medium having properties derived from the assembly of polarizable particles. This artificial medium is usually called the effective medium. We use our modern ideas and notation in what follows. The key is to think of the particle as surrounded by a sphere, outside of which we have the uniform effective medium, and then the local field at the individual dipole becomes

$$E_{local} = E + \frac{P}{3\varepsilon_0},\tag{15.2}$$

where E is the primary field, P is the polarization of the medium, ε_0 is the permittivity of free space, and the 3 in the denominator is a consequence of the spherical model.

Let us assume isotropic materials with N particles of polarizability α per unit volume. Then, P becomes

$$P = E_{local} \cdot N\alpha\tag{15.3}$$

or

$$N\alpha = \frac{P}{E_{local}}.\tag{15.4}$$

The susceptibility χ is defined as the ratio $P/(\varepsilon_0 E)$. Then, using Equation 15.2, we have

$$P = N\alpha \cdot \left(E + \frac{P}{3\varepsilon_0} \right). \tag{15.5}$$

Dividing by E and substituting χ yields

$$\varepsilon_0 \chi = N\alpha \cdot \left(1 + \frac{\chi}{3} \right), \tag{15.6}$$

so that

$$\frac{\chi}{3 + \chi} = \frac{N\alpha}{3\varepsilon_0}. \tag{15.7}$$

Since the relative permittivity ε_r and the susceptibility χ are related by

$$\chi = \varepsilon_r - 1, \tag{15.8}$$

then

$$\frac{\varepsilon_r - 1}{\varepsilon_r + 2} = \frac{N\alpha}{3\varepsilon_0}. \tag{15.9}$$

Equation 15.7, in various forms, is usually known as the Clausius–Mossotti equation, and Equation 15.2, as Lorenz–Lorentz, but the names are variable and are applied to Equation 15.9 that is the most important form of the relationship. It allows the properties of a material to be derived from the properties of its individual molecules. It also permits the variation of dielectric and optical properties with parameters such as strain to be calculated and is the starting point for the calculation of the properties of composite materials.

Two classic papers at the start of the twentieth century, written by James Clerk Maxwell Garnett [1,2], launched the study of composite materials. He was principally thinking of spherical metal particles dispersed in a dielectric, but the theory is more general. We can think of the material as consisting of a host of relative permittivity ε_m, in which a concentration of particles is dispersed, all the same ellipsoidal shape and characterized by a relative permittivity of ε_i.

The macroscopic internal field inside a uniformly polarized ellipsoid is given by

$$E_i = E - \frac{(\varepsilon_i - \varepsilon_m)}{\varepsilon_m} FE_i, \tag{15.10}$$

where F is a depolarizing factor caused by the accumulation of surface charge and, for a sphere, is $1/3$. The polarization of this ellipsoid is

$$\frac{P}{\varepsilon_0} = (\varepsilon_i - 1)E_i, \tag{15.11}$$

and we can imagine that it replaces an equivalent ellipsoid of host material of polarization:

$$\frac{P}{\varepsilon_0} = (\varepsilon_m - 1)E_i, \tag{15.12}$$

so that the change in polarization is

$$\frac{\Delta P}{\varepsilon_0} = (\varepsilon_i - 1)E_i - (\varepsilon_m - 1)E_i = (\varepsilon_i - \varepsilon_m)E_i. \tag{15.13}$$

Using Equation 15.10 and some manipulation, we obtain

$$\frac{\Delta P}{\varepsilon_0 \varepsilon_m E} = \frac{(\varepsilon_i - \varepsilon_m)}{\varepsilon_m + (\varepsilon_i - \varepsilon_m)F}. \tag{15.14}$$

We can consider E a constant input. The polarization of the entire material will depend on the volume fraction f of the inclusion. If we now change our thinking so that ΔP is the change in polarization over the entire material, then the right-hand side of Equation 15.14 represents the change with a volume fraction of unity. A volume fraction of less than unity then yields a polarization change of

$$\frac{\Delta P}{\varepsilon_0 \varepsilon_m E} = f \frac{(\varepsilon_i - \varepsilon_m)}{\varepsilon_m + (\varepsilon_i - \varepsilon_m)F}. \tag{15.15}$$

We assume that the behavior of the system can be represented as a uniform effective medium, with relative permittivity ε. We can imagine that we start with the surrounding medium of the host material and completely replace it by the effective medium, that is, with volume fraction unity. But the effective medium is actually the host material containing a volume fraction f of inclusion material. This leads from Equation 15.15 to the following:

$$\frac{(\varepsilon - \varepsilon_m)}{\varepsilon_m + (\varepsilon - \varepsilon_m)F} = f \frac{(\varepsilon_i - \varepsilon_m)}{\varepsilon_m + (\varepsilon_i - \varepsilon_m)F}, \tag{15.16}$$

which is of fundamental importance in effective medium theory.

The Maxwell Garnett model assumes spherical inclusions where F is $1/3$, yielding

$$\frac{(\varepsilon - \varepsilon_m)}{(\varepsilon + 2\varepsilon_m)} = f \cdot \frac{(\varepsilon_i - \varepsilon_m)}{(\varepsilon_i + 2\varepsilon_m)}. \tag{15.17}$$

Bragg and Pippard [3] developed their theory in a different way but arrived at what is essentially Equation 15.16, permitting them to introduce a wide range of different shapes for the inclusions, particularly useful being an array of cylinders that can represent the refractive index of a thin film with cylindrical columns. With the axis of the cylinders normal to the electric field, the depolarizing factor is $1/2$.

It is usually fairly easy to see which material should be treated as host and which as inclusion, but there are exceptional cases where the phases, while still separate, are so structured that such identification becomes impossible. This was the problem studied by Bruggeman [4,5]. Bruggeman's model treats each component of a composite material as an inclusion in a matrix of the effective medium. If we start with the effective medium, then the sum of the perturbing contribution of each component should be zero so that the properties remain exactly those of the effective medium. This gives

$$f \cdot \frac{(\varepsilon_i - \varepsilon)}{\varepsilon + (\varepsilon_i - \varepsilon)F} + (1 - f) \frac{(\varepsilon_m - \varepsilon)}{\varepsilon + (\varepsilon_m - \varepsilon)F} = 0. \tag{15.18}$$

Composite optical materials have been in existence ever since the beginning of glass technology. They were well known to the ancient Egyptians, but they still represent an important and developing technological area. For a more extensive treatment, see Berthier and Lafait [6].

Much of the more recent literature uses the alternative sign convention of $(n + ik)$ for the optical constants, but in the use of these expressions, we will normally know whether the imaginary part of the square roots involved represents absorption or gain, and so there is seldom any confusion.

15.3 Birefringent Materials

Birefringent materials exhibit optical properties that are direction sensitive. The subject is vast and a detailed treatment completely beyond the scope of this book. Useful and more detailed treatments are found in Yeh [7] and Hodgkinson and Wu [8]. Here we limit ourselves to a simple discussion of some features of birefringence exhibiting rectangular symmetry.

We can think of a material where the response of the electrons to the oscillating electric field of a propagating light ray varies with direction so that the movement of the electrons is not necessarily exactly in the direction of the exciting electric field. In the most complicated case, there are always three principal and orthogonal directions in which the movement is in the direction of the exciting field. Because the dipole response is different in each of these three directions, the optical constants are different. To make our discussion as simple as possible, we will label these three directions x, y, and z, and they will serve as our reference axes. Polarization is obviously very important, and the optical constants appropriate to any direction vary with polarization. In the most general case of a dielectric material, there are always two directions in the plane containing the maximum and minimum refractive indices where the velocity of the ray does not depend on polarization. These directions are known as the optic axes, and a material with two such axes is known as biaxial. When two of the principal directions have equal properties, there is only one axis, normal to their plane and parallel to the corresponding principal axis, the material then being known as uniaxial.

In the general case where the electric field is not aligned with any of the reference directions, the displacement vector and the electric vector are no longer parallel. E and H are still mutually perpendicular, and power flow direction is still normal to both, but the phase propagation is normal to the magnetic and displacement vector D, also mutually perpendicular, and so is in a different direction. This is a major complication made even worse by the fact that the general case does not always readily collapse to a sensible special case that must therefore be treated separately. In this short discussion, we avoid the general case and treat propagation eigenmodes in a principal plane only.

Let us consider a biaxially birefringent thin-film material with principal axes x, y, and z and surfaces in the x–y plane, where $N_x = n_x - ik_x$, $N_y = n_y - ik_y$, and $N_z = n_z - ik_z$. Let the incident medium be isotropic and free from absorption with index n_0. The direction of propagation of the incident light is in the x–z plane at angle of incidence ϑ_0 with respect to the z-axis. Clearly in this special case, the eigenmodes of polarization are, as usual, s- and p-polarizations.

In s-polarization, the electric field is normal to the x–z plane and, hence, parallel to the y-axis whatever the angle ϑ_0. N_y is therefore the appropriate index and admittance, and so the behavior is exactly as would be expected from an isotropic thin film of index N_y. A ray with this orientation is known as *ordinary* because its behavior is completely conventional.

p-Polarization is different. For any value of ϑ_0 other than zero or 90°, the displacement D is not parallel to the electric field E, although it remains within the x–z plane and, hence, avoids any coupling with s-polarization. The behavior is more complicated than for s-polarization, and so the ray is known by the term *extraordinary*.

In order to carry out calculations for p-polarization, we need an expression for δ and η_p, the tilted p-admittance. For this, we need the phase shift for the component of propagation along the z-axis, and we need to express H_y in terms of E_x to arrive at the tilted admittance.

The phase factor of the wave can be written as follows:

$$\exp\left[i\left\{\omega t - \frac{2\pi N}{\lambda}\left(\alpha x + \gamma z\right)\right\}\right], \tag{15.19}$$

$(\alpha, 0, \gamma)$ being the appropriate direction cosines, and $N\alpha$ is $n_0 \sin \vartheta_0$. Substituting a harmonic wave in Maxwell's equations (and as usual including any conductivity in the permittivity or dielectric function), we have the condition

$$\nabla \times (\nabla \times E) + \varepsilon \mu \frac{\partial^2 E}{\partial t^2} = 0, \tag{15.20}$$

where we will use N_x^2, N_y^2, and N_z^2 as the components of the relative permittivity. Equation 15.20 can be manipulated into the form

$$\begin{bmatrix} \{N_x^2 - N^2\gamma^2\} & N^2\alpha\gamma \\ N^2\alpha\gamma & \{N_z^2 - N^2\alpha^2\} \end{bmatrix} \begin{bmatrix} E_x \\ E_z \end{bmatrix} = 0, \tag{15.21}$$

where the determinant of the leading matrix must be zero for a nonzero solution for E. This leads to

$$N^2\gamma^2 = N_x^2 - \frac{N_x^2}{N_z^2} n_0^2 \sin^2 \vartheta_0. \tag{15.22}$$

We have

$$\delta_p = \frac{2\pi(N\gamma)d}{\lambda}, \tag{15.23}$$

and we have the relationship for H

$$\nabla \times H = \frac{\partial D}{\partial t}, \tag{15.24}$$

giving the other relationship we need,

$$\frac{H_y}{E_x} = \frac{N_x^2}{N\gamma} = \eta_p. \tag{15.25}$$

For more general cases, see Yeh [7] and Hodgkinson and Wu [8].

15.4 Metallic Grid Polarizers

Wire-grid polarizers date back to the nineteenth century when they were actually made with wire. Today they are manufactured by photolithography and etching processes. The accurate modeling of wire-grid polarizers is a difficult and involved process. However, a simplified approach was proposed by Yeh [9,10]. Yeh's model treats the polarizer as a thin anisotropic film and works well, provided the grid spacing is rather smaller than a wavelength.

We represent the structure as a regular stack of plates of one material all of the same properties and thickness and separated by similar plates of a second material again of identical thicknesses but not necessarily equal to that of the first set of plates. We take the x- and z-directions as in the plane of the plates and the y-direction normal to the plates. Boundary conditions are that the electric field parallel to a boundary is continuous across it while the displacement vector normal to the boundary is continuous across it. We also normalize the variables so that the displacement is given by N^2 times the electric vector, N being the complex refractive index. We will represent the different sets of plates by the suffices 1 and 2.

First, we look at E in the x- or z-direction. Since E is parallel to the surfaces, it can be considered as constant throughout the system. The displacement is given by $N_1^2 E$ in material 1 and $N_2^2 E$ in

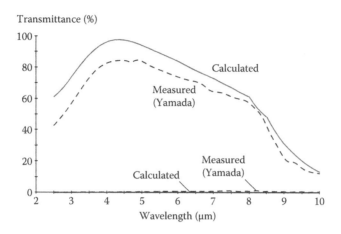

FIGURE 15.1
Comparison of calculated results using Equations 15.26 and 15.27 with measured performance extracted from Yamada et al. (2008. *Optics Letters* 33:258–260.)

material 2. The average displacement divided by the electric field is then the square of the net refractive index, that is,

$$N_x^2 = N_z^2 = \frac{d_1 N_1^2 + d_2 N_2^2}{d_1 + d_2}.$$ (15.26)

With E in the y-direction, it is now the displacement that is continuous, and so the square of refractive index in this case is

$$N_y^2 = \frac{D(d_1 + d_2)}{d_1 \dfrac{D}{N_1^2} + d_2 \dfrac{D}{N_2^2}} = \frac{N_1^2 N_2^2 (d_1 + d_2)}{d_1 N_2^2 + d_2 N_1^2}.$$ (15.27)

The material is uniaxially birefringent with the optic axis in the y-direction.

As a simple example, let us take aluminum at 700 nm with optical constants $(1.83 - i8.31)$ [11] separated by equal thicknesses of air. Then, we have N_x and N_z as $(1.303 - i5.835)$ and N_y as $(1.423 - i0.004)$, clearly metallic in the x-direction (and z-direction) and dielectric with quite low loss in the y-direction. Such a film with 100 nm physical thickness on glass of index 1.52 will give at 700 nm a transmittance with E field in the y-direction of 97.02% and in the x-direction of 0.0015%.

A comparison with actual measured figures is difficult because few publications include the necessary details. We have detailed information concerning tungsten silicide grid polarizers in the infrared in a paper by Yamada et al. [12]. The grids are deposited over silicon that has been antireflected by the addition of a silicon monoxide layer on both sides. Using the optical constants of silicon and silicon monoxide both from Palik [11] and the grid dimensions and the optical constants of tungsten silicide from the paper, we can compare predicted and measured performance in Figure 15.1 with encouraging results.

15.5 Metamaterials

Metamaterials is a word used to indicate engineered materials presenting properties not found in natural materials, and here we are particularly interested in their role in optics. Although all the materials already considered in the chapter could be termed *metamaterials*, the word is most often applied to deliberately structured materials exhibiting strange behavior of permittivity and permeability and supporting unusual propagation of electromagnetic waves. Frequently, the objective is behavior that can be interpreted as negative refraction and especially negative index of refraction.

The idea of negative refractive index dates back to a paper by Veselago [13] published in Russian in 1964 and republished in English in 1968. The paper considers what might happen if a material were to exhibit simultaneously negative permeability and permittivity and showed that the consequence would be a negative refractive index, although the characteristic admittance would remain positive. An interesting application of a parallel-sided slab of such a material would be the focusing of a divergent cone in the manner of a positive lens, although a collimated beam would not be focused. The paper suddenly became much more interesting when, in 2000, a demonstration was made of such a metamaterial for the microwave region [14]. Since then, there has been a large volume of contributions seeking to extend this effect to the optical region. Application as a new kind of lens has been a major objective.

The primary problem is that direct magnetic effects in the optical region are virtually nonexistent. The relative permeability μ_r is unity. The microwave response was achieved with a distributed structure of wires and resonators, inapplicable at optical frequencies. Optical attempts have therefore largely concentrated on the equivalent properties of structured multilayer materials, many consisting of an assembly of dielectric and metal layers, sometimes with etched lateral structures in the metals, others using the properties of birefringent materials.

There is an enormous attraction in a material exhibiting negative refractive index. In a thin-film coating, the characteristic admittance y would remain positive while the phase thickness δ would become negative. The benefits would be enormous. Resonance effects that are no longer localized in wavelength [15], reflectors of unprecedented width [16], complete cancellation of previous errors are just some of them. Unfortunately, the negative index remains elusive, although negative refraction has been convincingly demonstrated. Negative refraction, however, does not require negative index. For a fuller explanation, see Macleod [16,17].

The subject area is huge and much of it concerned with effects somewhat remote from optical coatings and beyond the scope of this book. The ideal negative index material certainly does not exist so far, except in some theoretical studies. However, negative refraction could have some possibly useful applications, and so we concentrate on that aspect here.

We imagine an experiment where a narrow slit on the top surface of a coating is illuminated with collimated light at a given angle of incidence. This produces a ray that propagates through the coating and emerges at the rear surface at a position that can be measured. The relative positions of entry and exit of the ray together with the thickness of the coating and the angle of incidence allow an effective refractive index to be calculated. In some cases, this effective index can be negative. In fact, this behavior is similar to that producing the well-known Goos–Hänchen shift in reflection [18]. The illuminated spot produces an angular spectrum of plane waves that propagate through the coating and suffer phase shifts that depend on their angle of incidence. The ray appears to emerge from the coating where the phases of the various angular spectrum elements coincide.

Let the physical thickness of the sample be d, and the angle of incidence of the illuminating beam in the plane of incidence that is normal both to the surface and the axis of the slit be ϑ_0. If e defines the lateral displacement of the emergent beam with respect to its entrance then, from Snell's law, the effective index of refraction of the material is given by

$$n = \frac{n_0 \sin \vartheta_0 \sqrt{e^2 + d^2}}{e}. \tag{15.28}$$

The phase shift on transmission through the coating φ_ϑ is habitually calculated by definition at a value of e of zero. The phase as a function of e is then given by

$$\varphi = \varphi_\vartheta - \frac{2\pi n_0 \sin \vartheta}{\lambda} e. \tag{15.29}$$

For simplicity, we take the phase coincidence point as zero derivative of phase change with respect to angle, the angle corresponding to that in the incident medium. From Equation 15.29 with a little work, we find

$$e = \frac{\lambda}{2\pi n_0 \cos \vartheta_0} \cdot \frac{d\varphi_{\vartheta_0}}{d\vartheta_0}.$$ (15.30)

Equation 15.30 is completely consistent with a similar displacement expression arrived at by Klinger [19].

A slab of material with no interference simply shows an effective index equal to its actual index. Interference can alter this as shown in Figure 15.2. The behavior shown is typical of dielectrics. Massive shifts can be obtained with dielectric structures such as multiple-cavity filters [19]. Wavelength division multiplexing based on similar effects has also been proposed [20]. It becomes more interesting when metal layers are involved.

The passage of light through a boundary between a metal and a dielectric involves a change in phase, unlike an all-dielectric interface. The slope of the phase change with angle of incidence has opposite sign for s- and p-polarizations with the result is that p-polarization tends to show negative refraction while s-polarization shows positive. We take silver as an example in Figure 15.3, where, with air as both incident and emergent media, and 50.0 nm of silver, at a wavelength of 500 nm, we have a refractive index of almost exactly −1.0. Unfortunately, the transmittance of this film for air termination and p-polarization is only just above 3%. Nevertheless, this behavior of silver, however, has attracted much attention.

One way of improving the transmittance is to use less silver, but that then reduces the thickness of the negative refractor and makes it less attractive as the basis for a lens. Many of the proposed structures therefore involve multiple thin silver layers separated by dielectrics. See Xu et al. [21], for example. Typical of these structures is the following design:

$$\text{Air}|\text{Ag 10 nm}|2.15\ 15\ \text{nm}|\text{Ag 20 nm}|2.15\ 15\ \text{nm}|\text{Ag 10 nm}|\text{Air,}$$ (15.31)

with Ag optical constants at 500 nm of $(0.13 - i2.918)$. The 2.15 index represents Ta_2O_5. The transmittance of this design at 500 nm is around 15%. The effective index is shown in Figure 15.4.

The properties of metal layers are sometimes tuned by creating structures in them by lithography followed by etching. For small-scale structures, this is effectively modifying the metallic optical constants. The principles of their operation remain much the same. Larger structures bring

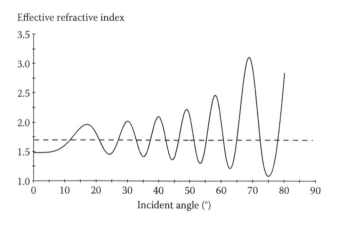

Effective refractive index

Incident angle (°)

FIGURE 15.2

Effective index of 10 μm of material of refractive index 1.70. The dashed line shows the 1.70 index obtained for s-polarization with air incident and perfect matching at the emergent surface. Changing to a terminating medium of air causes interference modifying the index for s-polarization as shown by the full line.

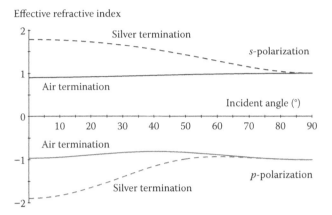

FIGURE 15.3
Effective refractive indices exhibited at 500 nm by a 50 nm thick film of silver terminated by silver to eliminate interference and terminated by air to include it. For p-polarization and air termination, the effective index is close to −1.0.

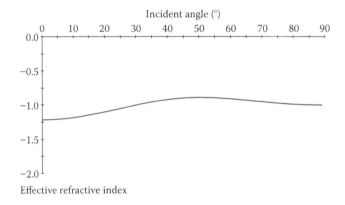

FIGURE 15.4
p-Polarized effective refractive index of the five-layer MDMDM structure at 500 nm. M indicates a metal layer and D a dielectric layer.

an element of diffraction that modifies the properties still further and is beyond the scope of this section. Then metals such as silver do exhibit interference effects at small thicknesses that, with the transmission phase shift at the interfaces, cause a change in transmitted phase with thickness that is opposite to the normal behavior of a dielectric. This behavior occurs at normal incidence as well as oblique and so has something in common with a true negative index material. Unfortunately, their admittance locus is still described clockwise with increasing thickness whereas a true negative index material would have a counterclockwise locus. Then beyond the critical angle, the p-polarized admittance locus of a dielectric material is described counterclockwise but with an imaginary admittance. Thus, although various aspects of the behavior we look for are presented, the complete set in one material is, unfortunately, still missing. Metal–dielectric structures are already used and well understood in optical coatings, and so we cannot expect much in the way of innovations from the examples we have examined so far.

There is perhaps slightly more promise for optical coatings in a different kind of metamaterial. In this case, we have a dielectric matrix containing an array of metal wires arranged so that their length is normal to the film surfaces. This is a uniaxial birefringent material with optic axis along the wires and normal to the film plane. Anodically prepared aluminum oxide possesses a reasonably regular array of deep pores that can be filled with a metal such as silver by an electrolytic process. A detailed description of such a process is given by Sauer et al. [22]. Such films exhibit p-polarized negative refraction [17,23], a primary interest of the metamaterials community. However, these are single films exhibiting optical properties that do not depend on thickness and so are much more attractive as candidates for optical coatings.

The Bragg and Pippard model has been used (Figure 15.5) to calculate the optical properties of such a film with inclusion fraction f of 0.234 [16,24]. The film appears dielectric while the electric vector is in the film plane (x–y plane) but metallic when the electric vector is normal to the plane (z-direction). s-Polarization therefore always sees the dielectric optical constants when the film behaves as a normal dielectric. p-Polarization, on the other hand, is the extraordinary ray, and the corresponding optical constants gradually vary from dielectric to metallic as the angle of incidence increases. This reverses the usual angular shift that is now toward longer wavelengths rather than shorter. Using the metamaterial as the cavity in a metal–dielectric narrowband filter, we find a p-performance that moves toward longer wavelengths with tilt [17]. In a multiple-cavity all-dielectric filter, the truly dielectric layers move in the usual shortwave direction and oppose the behavior of the metamaterial cavities so that it is actually possible to create a design (Figure 15.6)

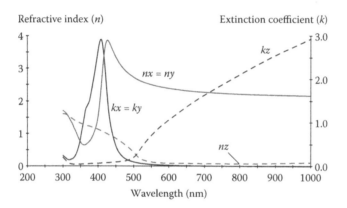

FIGURE 15.5
Optical constants of a composite film consist of silver wires along the z-axis (normal) of an Al_2O_3 matrix. The volume fraction of the silver inclusion f is 0.234, and the Bragg and Pippard composite material model has been used to calculate the optical properties. (From A. Macleod and C. Clark. 2014. Some thoughts on optical thin-film metamaterials. In 57th Annual Technical Conference, pp. 195–199, Society of Vacuum Coaters, Chicago.)

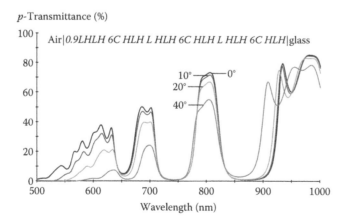

FIGURE 15.6
Three-cavity filter consisting of SiO_2 as L, Ta_2O_5 as H, and the composite as C. The design has been arranged so that the p-polarized passband remains fixed in wavelength as the angle of incidence varies.

that shows zero angular shift for p-polarization [24]. The s-polarization performance shows the conventional behavior.

We recall that p-polarization (and s-polarization) is always defined with respect to the local plane of incidence. Thus, in a normally incident cone of illumination, p-polarization becomes radial polarization, and s-polarization, azimuthal. Whether or not there could be some useful application of this effect remains to be seen.

References

1. J. C. M. Garnett. 1904. Colours in metal glasses and in metallic films. *Philosophical Transactions of the Royal Society* 203:385–420.
2. J. C. M. Garnett. 1906. Colours in metal glasses, in metallic films and in metallic solutions. *Philosophical Transactions of the Royal Society* 205:237–288.
3. W. L. Bragg and A. B. Pippard. 1953. The form birefringence of macromolecules. *Acta Crystallographica* 6:865–867.
4. D. A. G. Bruggeman. 1935. Berechnung verschiedener physikalischer Konstanten von heterogenen Substanzen: 1. Dielektrizitätskonstanten und Leitfahigkeiten der Mischkörper aus isotropen Substanzen. *Annalen der Physik, 5th Series* 24:636–664.
5. D. A. G. Bruggeman. 1935. Berechnung verschiedener physikalischer Konstanten von heterogenen Substanzen: 1. Dielektrizitätskonstanten und Leitfahigkeiten der Mischkörper aus isotropen Substanzen (Schluss). *Annalen der Physik, 5th Series* 24:665–679.
6. S. Berthier and J. Lafait. 1995. Electromagnetic properties of nanocermet thin films. In *Handbook of Optical Properties*, vol. 1: *Thin Films for Optical Coatings*, R. E. Hummel and K. H. Guenther (ed.), pp. 305–352. Boca Raton, FL: CRC Press.
7. P. Yeh. 1988. *Optical Waves in Layered Media*. New York: John Wiley & Sons.
8. I. J. Hodgkinson and Q. H. Wu. 1997. *Birefringent Thin Films and Polarizing Elements*. First ed. Singapore: World Scientific Publishing.
9. P. Yeh. 1978. A new optical model for wire grid polarizers. *Optics Communications* 26:289–292.
10. P. Yeh. 1981. Generalized model for wire grid polarizers. *Proceedings of SPIE* 307:13–21.
11. E. D. Palik, ed. 1985. *Handbook of Optical Constants of Solids I*. Orlando, FL: Academic Press.
12. I. Yamada, K. Kintaka, J. Nishii et al. 2008. Mid-infrared wire-grid polarizer with silicides. *Optics Letters* 33:258–260.
13. V. G. Veselago. 1968. The electrodynamics of substances with simultaneously negative values of ε and μ. *Soviet Physics Uspekhi* 10:509–514.
14. D. R. Smith, W. J. Padilla, D. C. Vier et al. 2000. Composite medium with simultaneously negative permeability and permittivity. *Physical Review Letters* 18:4184–4187.
15. M. Lequime, B. Gralak, S. Guenneau et al. 2014. Negative-index materials: A key to "white" multilayer Fabry–Perot. *Optics Letters* 39:1729–1732.
16. A. Macleod. 2015. Optical coatings and metamaterials. *Proceedings of SPIE* 9558:955802–1 to 955802–9.
17. A. Macleod. 2014. Optical thin-film metamaterials. *Society of Vacuum Coaters Bulletin* 14(Spring):24–31.
18. F. Goos and H. Hänchen. 1947. Ein neuer und fundamentaler Versuch zur Totalreflexion. *Annalen der Physik* 436:333–346.
19. R. E. Klinger, C. A. Hulse, C. K. Carniglia et al. 2006. Beam displacement and distortion effects in narrowband optical thin-film filters. *Applied Optics* 45:3237–3242.
20. B. E. Nelson, M. Gerken, D. A. B. Miller et al. 2000. Use of a dielectric stack as a one-dimensional photonic crystal for wavelength demultiplexing by beam shifting. *Optics Letters* 25:1502–1504.
21. T. Xu, A. Agrawal, M. Abashin et al. 2013. All-angle negative refraction and active flat lensing of ultraviolet light. *Nature* 497:470–474.

22. G. Sauer, G. Brehm, S. Schneider et al. 2002. Highly ordered monocrystalline silver nanowire arrays. *Journal of Applied Physics* 91:3243–3247.
23. J. Yao, Z. Liu, Y. Liu et al. 2008. Optical negative refraction in bulk metamaterials of nanowires. *Science* 321:930.
24. A. Macleod and C. Clark. 2014. Some thoughts on optical thin-film metamaterials. In *57th Annual Technical Conference*, pp. 195–199, Society of Vacuum Coaters, Chicago.

16

Some Coating Properties Important in Systems

When we design and calculate the performance of optical coatings, we normally assume ideal conditions. Light is collimated and monochromatic, optical thicknesses are exact, and the optical constants of the layers are precisely known. In the real world, none of these conditions is fulfilled exactly. Coatings depart from perfection in terms of thickness, optical constants, coating uniformity, scattering, and so on. Measurements use light with finite aperture and bandwidth. Illumination bandwidth, once the filter is installed in an actual system, can be still less well controlled, as well as the environment. Although the effects may sometimes be surprising, they are nevertheless understood and predictable. Here we list some of the more important effects. We simply reference a few because they are dealt with in greater detail elsewhere in this book.

16.1 Measurements and Calculations

One of the first things we do with a thin-film component once it has been constructed is to measure it. The measurement system will be characterized by a spectral bandwidth and a cone of illumination. What influence do these have on the measurement? In this, the concept of coherence length comes to our rescue. The concept of coherence was introduced in Chapter 2 where coherence length was defined as $\lambda^2/\Delta\lambda$ for a spectral bandwidth of $\Delta\lambda$ or $7 \times 10^3 n_e^2\lambda/(n_0^2\vartheta^2)$ for a cone of semiangle ϑ in degrees. These, we emphasize, are rough values. Coherence length is determined by the vanishing of interference effects that disappear gradually rather than abruptly and so cannot have a precise value. Let us assume that the feature we are scanning with our measuring system is a sine- or cosine-shaped fringe. We can arbitrarily take the precision with which we would like to determine the fringe amplitude as 1%. In other words, if the fringe amplitude is 1.0, then we would like our measurement, at the limit, to record an amplitude of 0.99. We construct our fringe from a function involving a cosine and with an amplitude of unity and a total value of 2.0 at g unity. (Note that 2.0 results from the fact that the level cannot become negative.) We scan with a bandwidth of Δg, and at g unity, the value is given by Equation 16.1 that should exhibit a value of 1.99. For this fringe of width unity in g, we must have Δg of not greater than 0.15:

$$\frac{\displaystyle\int_{1-\Delta g/2}^{1+\Delta g/2} [1 - \cos(\pi g)]\, dg}{\Delta g} = 1 + \frac{\sin(\pi\Delta g/2)}{\pi\Delta g/2}. \tag{16.1}$$

Thus, with this criterion, we can say that when we are dealing with scanning in terms of wavelength, the bandwidth should be not greater than 0.15 times the narrowest feature we need to measure. Since $g = \lambda_0/\lambda$, the corresponding coherence length is then about $\lambda_0/\Delta g$ or 6.67 that we can round to $7\lambda_0$. Since the optical thickness corresponding to our fringe is $\lambda_0/4$, we deduce that

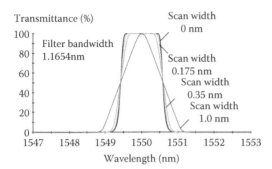 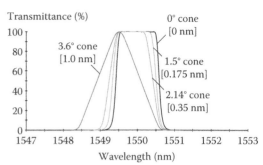

FIGURE 16.1

A narrowband filter of typical of telecommunication applications is scanned with varying spectral bandwidths on the left and varying illuminating cone angles on the right. The effective bandwidths associated with the cones are also shown on the right. The results are similar except for the usual shift to shorter wavelengths always associated with an illuminating cone. Note that 0.175 nm scanning bandwidth is 0.15 times the filter halfwidth.

because of the double traversal of the film, the limiting coherence length should be ≥14 × (total optical path).

We can also use these values to determine the limiting semiangle of an illuminating cone. The coherence length associated with a normally incident cone of illumination can also be expressed as an effective spectral bandwidth.

Let us take a multiple-cavity filter typical of those for telecommunication applications as our test piece. This has been scanned (theoretically) with different bandwidths and with different cone angles in Figure 16.1. The effective index for the cone calculations is roughly taken as 1.7, and the cone is assumed to have uniform power density over its aperture. The results for bandwidth and cone are similar but because the rays in a normally incident cone are all obliquely incident except for the cone axis, the cone results show an additional shift to shorter wavelengths. A slightly more relaxed requirement if the 1% precision is less important is a scanning bandwidth that is one third of the narrowest features. This is illustrated in the figure by the 0.35 nm scanning width.

Normally we use full coherence calculations for our thin films and incoherent calculations for our substrates. How reasonable is this? Let us take an effective bandwidth of 1 nm at 500 nm as our scanning parameters. The coherence length is $500^2/1$ nm = 250 μm. Substrates of thickness greater than 0.125 mm would show no fringes while coating fringes of 7 nm width and greater would show virtually no diminution. That this, our normal approach, is reasonable is demonstrated by Figure 16.2.

16.2 Oblique Incidence and Polarization

Optical coatings can be thought of as a uniaxial birefringent structures with the optic axis normal to the surfaces but with properties considerably complicated by the presence of interference. As soon as oblique incidence is involved, we have polarization eigenmodes that consist of one ray labeled as *s*-polarization and another as *p*-polarization. These two designations depend on the local plane of incidence, *p*-polarization having its electric field parallel to that plane and *s*-polarization normal. In our polarization calculations, we habitually express the polarization state of the light as a combination of these two modes. Their separate behavior is rather different, *p*-polarization usually presenting more complicated, and therefore more interesting, properties.

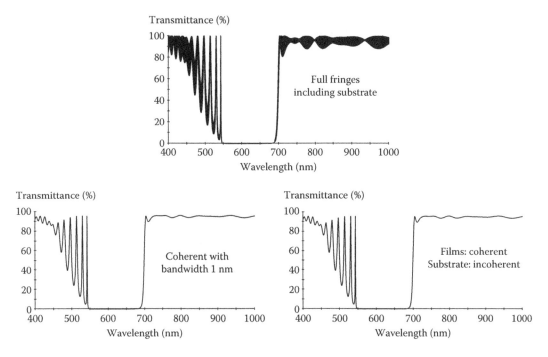

FIGURE 16.2
Calculation of a longwave pass filter under different conditions. Top is the full coherent interference calculation with zero bandwidth showing the substrate fringes as the black bands. Lower left shows the calculation with a bandwidth of 1 nm. Lower right shows the calculation with full coherence for the films and incoherence for the substrate. The two lower curves show no differences.

Before we can calculate or discuss behavior at oblique incidence, we need to be very clear about our sign conventions. They are listed in Chapter 2, but a word or two about them may be useful. In thin-film optics, we frequently deal with illumination that comprises a mixture of normal and slightly oblique incidence. A sign convention that changed as soon as incidence became even slightly oblique would cause immense confusion. At normal incidence, there is no orientational sensitivity in linear polarization, and our normal incidence convention therefore shows no such sensitivity. Our oblique incidence convention is then designed to collapse to our normal incidence one. Oblique incidence calculations can then proceed exactly as at normal incidence, the only changes being the adoption of tilted phase thicknesses and admittances.

Of considerable importance in the manipulation of polarized light is the relative phase shift between its two components. This is usually known as the relative retardation, or retardance, or sometimes as delta, and written as Δ. Conventionally, when we are dealing with surfaces and coatings we use p- and s-polarizations as the two components, and the reference directions for the light become p, s, and propagation directions in that order. It is important that the reference directions should be right-handed, and when we examine our reflection convention, we see that our p, s, and propagation directions are left-handed in reflection. This is a consequence of the parity shift that occurs in a single reflection. This handedness reversal is how we accommodate the parity in the thin-film field. But relative retardation needs a consistently right-handed reference. The solution is simple. In reflection, therefore, and including normal incidence that still involves the parity shift, we define relative retardation as

$$\Delta = \varphi_p - \varphi_s \pm 180°. \tag{16.2}$$

This is completely consistent with the definition used in ellipsometry.

Oblique incidence involves three major effects: a shift toward shorter wavelengths, polarization splitting, and distortion, usually in that order of severity and depending on the specific design of the component. Oblique incidence is dealt with in detail in Chapters 9 and 10, and we will not repeat what was said there. Here we look at several effects of particular important in systems, the maintenance of polarization, roof prism problems, and the effect of illuminating cones.

16.2.1 Polarization Maintenance

In some systems, the polarization of the throughput, whatever form it takes, must not be disturbed, although a reduction in the total power may be acceptable. Circular polarization should remain circular, elliptical should remain elliptical, and the orientation of a linearly polarized beam should remain constant. It is easy to see that this demands an equality of the p- and s-responses that can be expressed in ellipsometric terms as a ψ of 45° and a relative retardation of zero. But a perfect reflector still exhibits a parity shift implying a delta value of 180°, and so we modify our condition so that in reflection, the relative retardation should be 180°. To avoid the use of two parameters rather than one, we visualize a measurement procedure. We illuminate the component at the designated angle of incidence with light linearly polarized at 45° to the plane of incidence. We examine the throughput with an analyzer aligned normal to the expected orientation of the throughput and parallel to it. The ratio of the power in the former (unexpected) direction to that in the latter (expected) plus the leakage, that is, the total power, is a measure of leakage that can be expressed as a percentage. Subtracting this figure from 100% then gives us a measure of polarization maintenance.

If T_p and T_s are the power transmittances for p- and s-polarizations, respectively, and Δ is the relative retardation, then the polarization maintenance parameter in percent is given by

$$PM = 100.0 \cdot \frac{T_p + T_s + 2\sqrt{T_p}\sqrt{T_s}\cos\Delta}{2\left(T_p + T_s\right)},\tag{16.3}$$

and this can be manipulated into

$$PM = 50.0[1 + \sin 2\psi \cos\Delta].\tag{16.4}$$

This assumes no parity shift and so is always valid in transmission. With an odd number of reflections, this becomes

$$PM = 50.0[1 - \sin 2\psi \cos\Delta],\tag{16.5}$$

both results being in percent.

At the center of their reflecting zone, quarter-wave stack reflectors have reasonably good performance in respect of polarization maintenance (Figure 16.3). Extended zone reflectors tend to be rather less favorable.

Metal reflectors show a polarization sensitivity that has been already detailed in Chapter 10. Even uncoated metals, when appreciably tilted, show reflection polarization maintenance of less than 100%, and when overcoated with dielectric protection layers, polarization maintenance can be still poorer. For an interesting discussion of how polarization sensitivity in reflecting optical coatings can affect the image quality of a highly corrected optical instrument (in this case, a reflecting telescope with an element at 45°), see Breckinridge [1].

16.2.2 Roof Prism Problems

Prism systems are frequently used in visual optical instruments to manipulate the image in some way. This might be a translator in a periscope or an erector in a telescope. Internal reflection plays

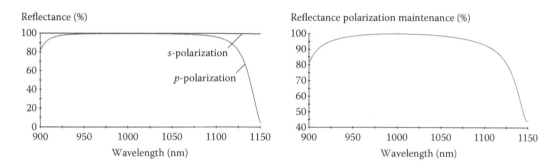

FIGURE 16.3
The left-hand side shows the 45° performance of a quarter-wave stack of 21 layers of 1.45 and 2.15 materials tuned for 45° in air. The right-hand side shows the corresponding polarization maintenance parameter in reflection.

an important part in such systems. The single totally reflecting surface reverses the parity of the image, that is, it changes the handedness of the image. The reversal of parity is usually in itself undesirable in a visual image, but further, it is normally accompanied by a sensitivity of the system to rotation. A reflector that retains parity, and solves most of these problems, is the roof. This consists of two internally reflecting surfaces operating beyond the critical angle and accurately arranged at 90° to each other so that the appearance is exactly that of the roof of a house. Many erecting prism systems have an odd number of reflecting surfaces arranged with coplanar normals. Their reversal of parity and their sensitivity to rotation make them useless in instruments such as binoculars. The replacement of one of the surfaces by a roof restores the parity and removes the rotational sensitivity. The Schmidt–Pechan prism, often used instead of a double Porro prism in compact binoculars, is a good example, as is the Abbe–Koenig (Figure 16.4). However, the roof, in solving the parity and rotation sensitivity problems, introduces another different problem. Half the light contributing to the image meets the two surfaces in reverse order to the other half. At each reflection, there is an effect on the polarization state of the light, even if perfectly collimated. The combined effect of the two surfaces in series depends on the order in

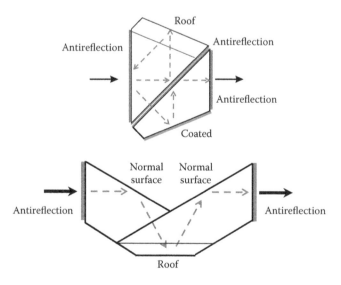

FIGURE 16.4
Schmidt–Pechan prism (*upper*) showing various coatings. Incidence on the roofline is at 22.5°, giving 49.2° on each part of the roof, but the 22.5° incidence on the lower simple surface implies the need for a high-reflectance coating. The Abbe–Koenig prism (*lower*), made up of two cemented prisms, is a little simpler with 30° on the roofline and 52.2° incidence on each flat surface.

which they are encountered. If the resulting polarization states are sufficiently different for the two halves, then the resulting image will consist of a point spread function for each half rather than a single, narrower function for the complete aperture. When the perturbation of polarization essentially consists of two equal parts with identical orientation of polarization but with a half-wave difference in phase, the central maximum of the point spread function disappears, and there is a serious doubling of the image. This is an old problem that was explained in detail by Mahan [2,3]. See Rabinovitch and Toker [4] for some more recent information on this and other similar problems.

To avoid this problem, the relative retardation Δ at each of the reflecting surfaces of the roof should be 180°. Let us take as an example an incident ray direction at 45° to the roof with the plane of incidence bisecting the angle between the two surfaces when the angle of incidence of the ray on each surface is 60°. Let us assume that the light is linearly polarized. Δ of 180° implies that the p-component (note that this is the ellipsometric convention already discussed) of the polarization is flipped. This is equivalent to a rotation of the direction of polarization, around the ray, through twice the angle between it and the s-direction. The two s-directions lie in the planes of the roof, and the resulting double rotation turns each plane of polarization through 180°. One order of surface encounter rotates the polarization through +180°, and the other, through −180°, and so the two alternative passages through the roof yield identical emerging polarizations. This is not the case for a Δ of zero or of some intermediate value. The orientation depends not only on the properties of the surfaces but also on the angle of incidence. The question of allowable tolerances in visual systems was studied by Ito and Noguchi [5], who, using a slightly different theoretical approach, arrived at an identical conclusion. Experimental evidence led Ito to suggest that for best images, the error in Δ should preferably be less than 20° and, under no circumstances, greater than 90°.

At 60° internal incidence, uncoated glass surfaces, of index 1.52 used in a roof, rotate the polarization in opposite directions and can introduce large ellipticity into a linearly polarized input beam. The relative rotation angle is around 66°. This is clearly unsatisfactory. Simple silver coatings, sometimes used to reduce the polarization perturbations of simple plane surfaces at incidence beyond critical, give retardations a little greater than 90° (Figure 16.5) better than uncoated glass but not very satisfactory compared with 180° ± 20°.

A very simple coating consisting of a thin layer of a material such as tantalum pentoxide (around 80 nm in physical thickness) overcoated with opaque silver can achieve a value of Δ of around 180°, at the expense of a small reduction in reflectance, particularly at the blue end of the visible. Figure 16.6

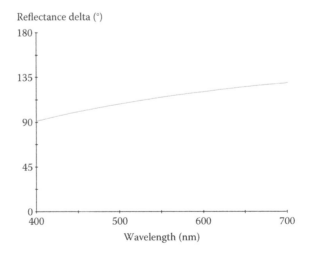

FIGURE 16.5
Relative retardation of a silver coating at 60°, the angle of incidence in a roof used at 45°. Although the difference between the retardation and ideal 180° is less that 90°, it is well outside the optimal 20°.

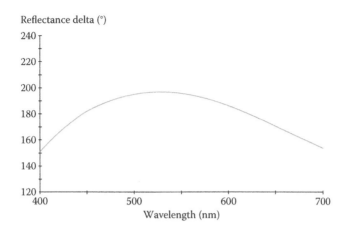

FIGURE 16.6
Relative retardation at 60° internal incidence of a simple two-layer coating of 80 nm of Ta_2O_5 next to the glass covered by a thick layer of silver.

shows such a performance. This is inside the ±20° tolerance suggested by Ito and Noguchi [5]. For improved performance, a dielectric multilayer is required, and with five or so layers, accuracies within 5° can readily be achieved. High-end binoculars and similar instruments will normally use a high-performance dielectric coating.

16.2.3 Cone Response at Oblique Incidence

In practice, collimated monochromatic light does not exist. Illumination may come close to that ideal, but there will be a bundle of incidence angles and a band of wavelengths in the output of any real illuminating system. Most illuminating systems will eventually have a condensing lens that directs the illumination onto a patch on the sample. Each point of the illuminated patch sees uniform power density of illumination emanating from the condenser. This type of illumination is known by the term *Lambertian*. We can consider it as a regular cone of illumination with given semiangle and circular cross section within which the illumination level is constant, and outside of which, it is zero. The principal plane of incidence contains the cone axis and the normal to the surface, and the angle of incidence is that between the cone axis and the surface normal.

The shift in performance due to oblique incidence is a cosine effect that, at small angles, goes as the square of the angle. A 1° cone at normal incidence can be assigned a value of unity while a 1° cone at 3° incidence ranges from $(3 + 1)^2$ to $(3 - 1)^2$, that is, 3.5^2. So roughly, the 1° cone at 3° has a degrading effect similar to that of a 3.5° cone at normal incidence, confirmed by comparing Figure 16.7 with Figure 16.1.

16.2.4 Cone Response of Thin-Film Polarizers

The cone of Figure 16.7, even in its 3° incidence, has vanishingly small polarization sensitivity. Now we turn to cones with much larger incidence where polarization is significant. We examined thin-film polarizers in Chapter 9 and alluded to the problems of illumination cones, and we return to that problem here. The field of the thin-film polarizer is limited because of polarization leakage.

First we must define our polarization. We imagine a linear polarizer in the form of a flat plate that is inserted into the cone normal to the cone axis. The polarization directions in the plate are parallel everywhere. We can visualize planes defined by the lines in the polarizer together with the point at the apex of the cone. The electric vectors are contained everywhere within these planes. We define *p*-polarization in the cone as those polarization directions parallel to the principal plane

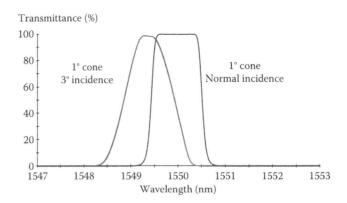

FIGURE 16.7
Same filter as in Figure 16.1 scanned by a 1° cone at normal and at 3° incidence. The performance degradation at 3° incidence is clear.

of incidence and s-polarization as normal to the principal plane of incidence. It is clear that, except for rays in the principal plane of incidence, s- and p-polarizations as defined for the cone will not coincide with the s- and p-directions of the eigenmodes for each ray, which are defined by its local plane of incidence. To perform the calculation, we must resolve the cone p- and s-directions along the local eigenmode directions to form the input and then reverse the process for the output. The calculation is tedious but unambiguous [6] and can be expressed as a limiting value of extinction ratio:

$$R_e = \frac{\Omega^2}{4\tan^2\varphi}, \tag{16.6}$$

where R_e is the limiting extinction ratio, Ω is the cone semiangle in radians, and φ is the angle of incidence of the cone, both angles being measured in the medium of incidence for the coating (that is, the medium of immersion in the case of an immersed polarizer). This expression is reasonably accurate up to a cone semiangle of around 8°. It is a geometrical effect and so is not responsive to design.

We can take our polarizer from Chapter 9 as an example:

$$\text{N-LAK8}|(1.208H\ 1.802L)^{10}\ 1.208H|\text{N-LAK8}. \tag{16.7}$$

Figure 16.8 shows the performance in collimated light and in an illuminating cone of 2° semiangle. The extinction ratio in collimated light is, from Chapter 9, 2.3×10^{-7}, but in the 2° cone at 510 nm, the p-transmittance is 93.23% and the s-transmittance is 0.029% so that the extinction ratio T_s/T_p is now 0.00031. Equation 16.6 gives 0.00030 for a 2° cone of illumination. We can conclude that this polarizer is overdesigned if it is going to be used in a cone of 2° or greater semiangle.

16.2.5 Small Spot Illumination

A related problem involves the illumination of an optical coating over a limited area only. In the normal way, this question would not arise. Illuminated areas are usually enormous compared with the wavelength of light, and there is no perceptible effect. However, a few applications exist, such as optical fiber communication, where illumination areas are much smaller than usual, and so we should briefly look at the problem and any consequences. We retain our x-, y-, and z-axes as

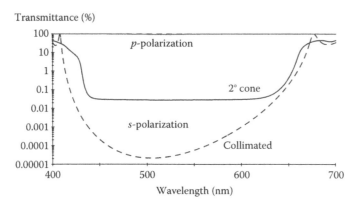

FIGURE 16.8
Transmittance performance of the cube polarizer at 45° in collimated light (*broken line*) and in a 2° illuminating cone (*full line*). The extinction ratio T_s/T_p is 0.00031at 510 nm when illuminated by a 2° cone at 45° incidence.

reference with the z-axis normal to the surface and with positive direction into it and the x–z plane as the plane of incidence. To keep the analysis simple, let us work in two rather than three dimensions so that our patch of illumination is actually a strip of infinite length in the y-direction and bounded in the x-direction. In the simplest case, the strip is evenly illuminated with sharp boundaries. Let the strip width be a with its midpoint at the origin. This is still a linear phenomenon, and the illumination can be represented by a set of components that are all of identical frequency but with wavenumbers such that they interfere to exactly reproduce the illuminated patch. This is once again the angular spectrum that we already discussed in Chapter 15.

We will consider relative phase as contained in the complex amplitude, and we will assume that any polarization is the same for all rays. Let $\mathcal{E}(x)$ be the complex amplitude across the strip, which will be the result of the addition, including phase, of the various components of the angular spectrum. These will all be rays with direction in the plane of incidence. The summation at z = 0 yields

$$\mathcal{E}(x) = \int_{-\infty}^{+\infty} \mathcal{E}(\kappa_x)e^{-i\kappa_x x}\, d\kappa_x, \tag{16.8}$$

which can be recognized as having the form of a Fourier transform. Provided the functions satisfy certain simple conditions, Fourier's integral theorem gives

$$\mathcal{E}(\kappa_x) = \frac{1}{2\pi} \int_{-\infty}^{+\infty} \mathcal{E}(x)e^{i\kappa_x x}\, dx, \tag{16.9}$$

where $\mathcal{E}(\kappa_x)$, the inverse Fourier transform of $\mathcal{E}(x)$, is the x-component of the electric field amplitude of the angular spectrum component. There will be similar expressions for the magnetic field.

The strip has limits $x = \pm a/2$, and we assume that the amplitude of the illumination across the strip is uniform and of zero phase $\mathcal{E}$ that is real. Then,

$$\mathcal{E}(\kappa_x) = \frac{1}{2\pi}\left[\frac{\mathcal{E}e^{i\kappa_x x}}{i\kappa_x}\right]_{-\frac{a}{2}}^{+\frac{a}{2}} = \mathcal{E}_0 \text{sinc}\frac{\kappa_x a}{2}, \tag{16.10}$$

$\mathcal{E}_0$ being the amplitude of the normally incident component. The irradiance, or power per unit area, as a function of angle then becomes

$$\text{Irradiance} = I_0 \text{sinc}^2(\kappa a/2) = I_0 \text{sinc}^2(\pi n_0 \sin \vartheta_0 a/\lambda), \tag{16.11}$$

which is equivalent to Fraunhofer diffraction.

A more realistic distribution of illumination when lasers are involved in the system has a Gaussian form. The Fourier transform of a Gaussian function is another Gaussian function so that the two functions

$$\exp\left(-\frac{\mu^2 \kappa_x^2}{2}\right) \quad \text{and} \quad \exp\left(-\frac{x^2}{2\mu^2}\right) \tag{16.12}$$

with appropriate multiplying constants, are transform pairs. Of course, a Gaussian function has no sharp cutoff, but we can still use a as a measure of the width of the slice, but now it can indicate that width at which the amplitude falls to $1/e$, and the irradiance to $1/e^2$, of its maximum value:

$$\mathcal{E}(x) \propto \exp\left(-\frac{4x^2}{a^2}\right). \tag{16.13}$$

From Equation 16.12, the components of the angular spectrum are of the form

$$\mathcal{E}_0 \exp\left(-\frac{a^2 \kappa_x^2}{16}\right) \tag{16.14}$$

or, in terms of irradiance,

$$I_0 \exp\left(-\frac{a^2 \kappa_x^2}{8}\right). \tag{16.15}$$

The angle at which the amplitude falls to $1/e$, and the irradiance to $1/e^2$, of its maximum is given by

$$\kappa_x = \frac{2\pi n_0 \sin \vartheta_0}{\lambda} = \frac{4}{a}, \tag{16.16}$$

$$n_0 \sin \vartheta_0 = \frac{2\lambda}{\pi a}. \tag{16.17}$$

This is a simple two-dimensional representation of a Gaussian beam. The two distributions are plotted together in Figure 16.9 as a function of $\kappa a/2$. Over the central part of the plot, there is little difference. At larger incidence, the Gaussian has a more regular shape. Thus, small spot illumination is equivalent to a cone of illumination.

We work with the Gaussian. We have already seen that a useful technique for simplifying the analysis of tilt-induced spectral displacement uses the concept of effective index n_e, the index of that single layer exhibiting the same degree of spectral shift. The effective index is within the range of the indices of the coating materials. A reasonable value for approximate results is the geometric mean of the extreme values.

The shift in the wavelength of any feature will be such that the phase thickness of our effective layer remains constant:

$$\frac{2\pi n_e d \cos \vartheta_e}{\lambda - \Delta\lambda} = \frac{2\pi n_e d}{\lambda},$$

i.e.,

$$\frac{\lambda - \Delta\lambda}{\lambda} = \cos \vartheta_e \approx 1 - \frac{\vartheta_e^2}{2} \approx 1 - \frac{n_0^2 \vartheta_0^2}{2 n_e^2}, \tag{16.18}$$

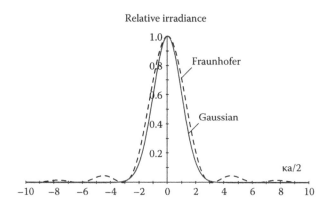

FIGURE 16.9
Comparison of Fraunhofer (*broken line*) and Gaussian (*solid line*) angular spectra.

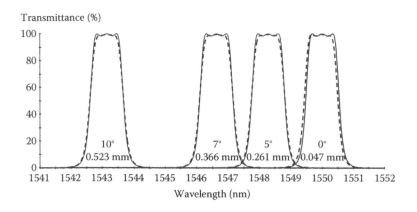

FIGURE 16.10
Performance of a simple three-cavity narrowband filter with varying incidence and varying spot widths assuming a Gaussian cone of illumination. The spot widths are calculated on the basis of one third of the halfwidth.

where we have used approximations of the second order. The quantity $\Delta\lambda/\lambda$ can be set to be the allowable shift before the characteristic is unacceptably degraded, and then we can derive a value for the allowable spot size. At normal incidence, this is

$$\frac{a}{\lambda} = \frac{\sqrt{2}}{\pi n_e} \left(\frac{\lambda}{\Delta\lambda}\right)^{\frac{1}{2}}. \tag{16.19}$$

Oblique incidence is more complicated. Let us suppose that the cone is tilted at an angle greater than the cone semiangle. Then, we have

$$\frac{\Delta\lambda}{\lambda} = \frac{\vartheta_{e\,max}^2 - \vartheta_{e\,min}^2}{2} = \frac{n_0^2}{2n_e^2}\left(\vartheta_{0\,max}^2 - \vartheta_{0\,min}^2\right). \tag{16.20}$$

The factor in parentheses on the right-hand side is twice the beam divergence time twice the mean tilt angle. This gives

$$\frac{a}{\lambda} = \frac{4n_0}{n_e^2} \cdot \frac{\vartheta_{tilt}}{\pi} \cdot \frac{\lambda}{\Delta\lambda}, \tag{16.21}$$

where ϑ_{tilt} is expressed in radians.

Figure 16.10 demonstrates the application of the expressions. The halfwidth of the filter is 1.06 nm, and the criterion for establishing the limiting spot widths shown was one third of the halfwidth. Note the similarity between these results and those of Figure 16.1. The normal incidence results show the typical offset with a cone at normal incidence. This offset disappears once the tilt becomes greater than the semiangle of the cone.

16.3 Surface Figure and Uniformity

For a normal, simple, and untreated optical surface, the geometrical figure and the apparent figure are identical. This is still the case when the surface is coated with a uniform, opaque metallic layer. When the surface is coated with a coating that exhibits interference effects, however, any error in uniformity will likely cause the apparent surface figure to differ from the geometric, and the difference will usually vary with wavelength. The effect is discussed in some detail in Chapter 6, and it will not be repeated here. We mention the effect here for the sake of completeness and particularly note the implication for testing. There is no guarantee that the apparent surface figure measured at one wavelength will be identical to that measured at another. We also note that the quarter-wave stack reflector is relatively benign in this respect because the light is essentially reflected at the front interface. Extended zone systems where light penetrates into the coating tend to show a considerably enhanced effect.

16.4 Contamination Sensitivity

Optical coatings are rarely used in an ideal environment. They are subjected to all kinds of environmental disturbances ranging from abrasion to high temperature and humidity. These cause performance degradation that mostly originates in an actual irreversible and usually visible destruction of the layers. However, performance may be degraded in a rather less spectacular way by the simple acquisition of a contaminant that may have no aggressive effect on the layers other than a reduction of the level of performance of the coating as a whole. The action of water vapor that is adsorbed by a process of capillary condensation and causes a spectral shift of the coating is well known. Here we are concerned with much smaller amounts of absorbing material, such as carbon, in the form of submolecular thicknesses either at some point during the construction of the coating or, more usually, over the surface after deposition.

Although there are many tests for the assessment of the resistance of a coating to most environmental disturbances, there is no standard test for the measurement of susceptibility to contamination. Yet it can be shown that the response of coatings can enormously vary, depending on many factors including design, wavelength, and even on errors committed during deposition. The reason may be that, often, careful cleaning will restore the performance, but this does not avoid the degradation in between cleanings, and cleanings that are more frequent are required for coatings that are more susceptible.

Fortunately, it is possible to make some predictions of coating response to low levels of contamination and, especially, to make assessments of comparative sensitivity [7,8]. Electric field distribution and potential absorption are the keys to understanding the phenomenon.

If the contaminating layer is on the front surface, then it receives the full irradiance that enters the multilayer, and the admittance at the contamination determines the reflectance as well as the potential absorptance. The key expressions involving absorptance A and potential absorptance $\mathcal{A}$ have already been derived in Chapter 2:

$$\mathcal{A} = \left(\frac{2\pi nkd}{\lambda}\right)\left(\frac{2}{\mathrm{Re}(Y)}\right) \tag{16.22}$$

and

$$A = (1 - R)\mathcal{A}. \tag{16.23}$$

Then, we can write

$$
\begin{aligned}
A &= (1 - R)\mathcal{A} \\
&= \left(\frac{4\pi nkd}{\lambda}\right)\left(\frac{1}{\mathrm{Re}(Y)}\right)\left\{1 - \frac{[y_0 - \mathrm{Re}(Y)]^2 + [\mathrm{Im}(Y)]^2}{[y_0 + \mathrm{Re}(Y)]^2 + [\mathrm{Im}(Y)]^2}\right\} \\
&= \left(\frac{4\pi nkd}{\lambda}\right)\left\{\frac{4y_0}{[y_0 + \mathrm{Re}(Y)]^2 + [\mathrm{Im}(Y)]^2}\right\},
\end{aligned} \tag{16.24}
$$

and Equation 16.24 permits us to put on the admittance diagram contours of absorption due to contamination on the outer surface. Before we draw actual lines, we need to define some of the quantities. It is simplest to use numbers that allow us to easily scale the diagram. We therefore simplify the expression by defining H by

$$H = \frac{4\pi nkd}{\lambda}, \tag{16.25}$$

and, replacing Y by $x - iz$, we find

$$\frac{A}{H} = \frac{4y_0}{(y_0 + x)^2 + z^2}. \tag{16.26}$$

H is a measure of the absorption capacity of the film, while A is the actual absorptance. We can think of A/H as a measure of the sensitivity to absorptance of an optical coating [9]. This sensitivity is purely a function of the optical admittance of the complete coating. Further, from Equation 16.26, contours of constant sensitivity are circles in the admittance plane that are centered on point $-y_0$ and exhibit decreasing sensitivity with increasing radius.

To simplify matters still further, we take the value of y_0 as 1.00. The contour lines for this case are then as shown in Figure 16.11.

As an example of the magnitude of H, we can take the values of amorphous carbon given by Palik [10–12], that is, optical constants of $2.26 - i1.025$ at 1000 nm, and assume a thickness of 0.1 nm. A plot of H is shown in Figure 16.12, and over most of the wavelength region shown, it is between 0.003 and 0.006. It should be noted, however, that although carbon is commonly found over virtually any surface, it may be at least partially oxidized or bound to hydrogen, reducing its capacity for absorption.

Antireflection coatings all attempt to terminate their loci at point $(y_0, 0)$. This implies a value of A/H of $1/(y_0)$, that is, 1.00 for y_0 of unity, and from (Figure 16.12), this gives, for a perfect antireflection coating, a range of absorptance across the visible region from around 0.3% to 0.6% with a film of carbon that is 0.1 nm thick. A slightly less than perfect coating will exhibit figures a little greater or less than these. It all depends on the admittance at termination. Typical results for a four-layer antireflection coating over the visible region are shown in Figure 16.13. The design of

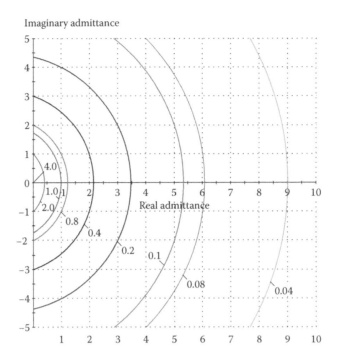

FIGURE 16.11
Circles of constant contamination sensitivity A/H in the admittance plane calculated for an incident admittance of 1.0. Greatest sensitivity corresponds to the origin where the value is 4.0. (From A. Macleod, *Society of Vacuum Coaters Bulletin*, 24–25, 28, 2006.)

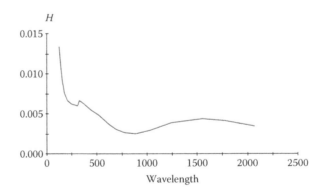

FIGURE 16.12
Plot of H against wavelength for 0.1 nm thickness of carbon film.

the coating has little influence on this result, and all coatings that have precisely zero reflectance will have exactly the same level of sensitivity.

Reflectors exhibit much greater variation. A dielectric reflector that is made up of quarter-wave layers and terminates with a final high-admittance layer will end its locus to the far right of the diagram, and the sensitivity to contamination will be much reduced. This, however, is not so for extended-zone high-reflectance coatings. In such coatings, at least part of the high-reflectance zone involves the inner part of the coating, and the outer part exhibits an admittance that circles around from far to the right to very near the imaginary axis. The value of A/H can then be almost as large

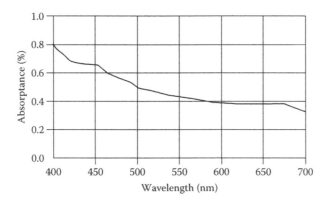

FIGURE 16.13
Absorptance produced by a layer of carbon of thickness 0.1 nm in front of a four-layer antireflection coating for the visible region.

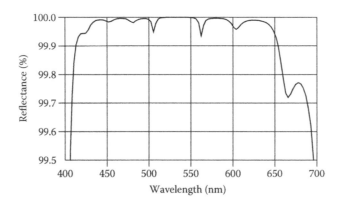

FIGURE 16.14
Reflectance of an extended-zone high-reflectance coating for the visible region. The coating consists of two mutually displaced quarter-wave stacks making up a total of 39 layers.

as 4.0, so that over parts of the visible region, the absorptance due to the 0.1 nm thickness of carbon can rise to between 1.0% and 2.0%. This is illustrated by a 39-layer extended zone reflector with performance as in Figure 16.14 and absorptance behavior as in Figure 16.15.

Aluminum reflectors are normally protected by a thin layer of low index, most often a half wave in thickness, although a quarter wave may also be used. The quarter-wave thickness gives a greater fall in reflectance at the reference wavelength and a higher electric field. The sensitivity to contamination of the two coatings is quite different and shown in Figure 16.16.

The simple quarter-wave stack is of enormous importance as the most common high-performance reflector. We have seen how poor the extended-zone high reflector is. What can we deduce about the quarter-wave stack? We can take the contamination figures as at 1000 nm. At the center wavelength, where all layers are quarter waves, the admittance presented by a quarter-wave stack Y is real. The absorptance of the layer, using the 1000 nm figures and assuming air as incident medium, is therefore given from Equations 16.25 and 26.26, by

$$A = \frac{0.0116}{(1 + Y)^2} . \tag{16.27}$$

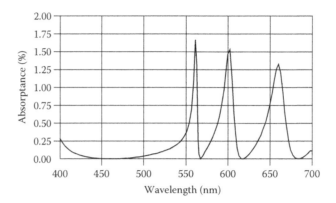

FIGURE 16.15
Absorptance produced by 0.1 nm of carbon deposited over the outer surface of the reflector of Figure 16.14.

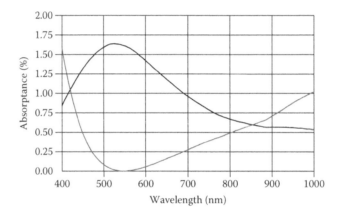

FIGURE 16.16
Effect of contamination by 0.1 nm thick film of carbon on aluminum reflector with quarter wave of silica protecting layer (*upper curve*) and half wave of silica (*lower curve*).

We take a quarter-wave stack of silica and titania and calculate the absorptance as a function of the (odd) number of layers assuming titania outermost. The result is shown as the dashed line in Figure 16.17. The results were also calculated using the full matrix theory. Agreement is excellent up to around 15 layers, and then the full calculation shows a leveling off. The effect is due to the failure of the thin-layer approximation. The admittance locus of the very thin contamination layer is shifted to the extreme right, and now, even though it is exceedingly thin, it swings round toward the imaginary axis. The potential absorptance rises, and it is when multiplied by the decreasing $(1-R)$ factor, a constant is obtained. This constant level is very small, less than 10 parts per billion. Equation 16.27 shows that for a quarter-wave stack terminated by a low-admittance layer, where Y would be very small, the limiting absorptance would be 0.0116 or 1.16%. Accurate calculation confirms this.

As the wavelength changes, however, the admittance locus for the quarter-wave stack begins to unwind. The major effect is that the value of $Re(Y)$ decreases. This is also accompanied by a slight decrease in reflectance. The result is a considerable increase in the level of absorption associated with the contamination layer. Figure 16.18 shows the rapid increase in absorptance up to 500 parts per million from the less than 10 parts per billion at the center wavelength.

Thermally evaporated coatings are known to be affected by moisture. The moisture enters in localized spots and spreads out in the form of circular patches of increasing diameter. This changes

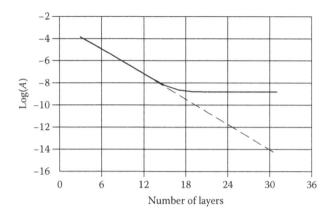

FIGURE 16.17
Predicted absorptance, plotted as log(A), of a quarter-wave stack as a function of the odd number of layers. The dashed line is the simple theory. The full line is calculated using the full matrix theory. The wavelength was assumed to be 1000 nm, and the carbon contaminant, 0.1 nm thick.

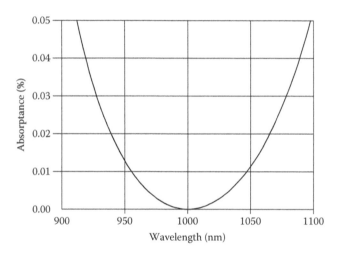

FIGURE 16.18
Absorptance of the quarter-wave stack with contamination layer as a function of wavelength.

the field distribution in a coating and therefore alters the absorptance associated with a contamination layer (Figure 16.19). Monitoring errors that have no perceptible effect on the reflectance of a quarter-wave stack can have major effects on the sensitivity to contamination.

Some additional information on contamination sensitivity at interfaces within the coating are included in the article by Macleod and Clark [7].

16.5 Scattering

Scattering is a difficult topic [13–17], and any detailed discussion is beyond the scope of this book. In scattering, light is lost from the specular directions. Surfaces and bulk material exhibit imperfections of one kind or another that disturb their uniform nature. The defects that matter most are typically spaced much wider apart than the atoms or molecules of the intrinsic material and

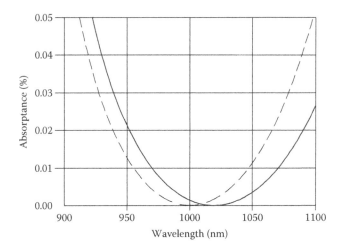

FIGURE 16.19
The bold line shows absorptance of a contamination layer over a wet patch in a quarter-wave stack. The dashed line shows the absorptance when deposited over a dry area.

act as dipoles that absorb and reradiate the light. Unlike the regular units, the atoms and molecules, there is no regular phase relationship in their outputs, and so the scattering is incoherent. A principal problem is the impossibility of an exact specification of the form either of the solid parts or of the surfaces of the films, and we must therefore take refuge in statistical parameters that depend on the scale with which we measure the imperfections. Scattering theory is therefore approximate. It cannot match the incredible accuracy of the specular theory. From the point of view of the specular properties, scattering represents a loss, much like absorption, and it goes with the square of the electric field amplitude. Unlike absorption, the scattered light does go somewhere else. Some may be trapped in waveguide modes, but scattered light can cause problems. An array filter laid over an array receiver can, by its scattering, cause an unwanted coupling among the elements of the array, for example. Surface roughness is the major contributor to scattering in optical coatings, and we emphasize this in what follows.

We know how to define our specular properties such as reflectance and transmittance. What of scattering? We can imagine that the exciting light will illuminate a spot of quite limited area on the coating, and then the scattered light will form a distribution of power that can be expressed in the form of power per unit solid angle. Also, rather like reflectance and transmittance, we can compare the scattered power with that incident on the illuminated spot. This leads to various definitions. In these, we write the total power incident on the spot as P_{inc} and the power that the receiver measures as P_s. Then the area of the receiver subtends a solid angle of $\Delta\Omega$ in the direction given by angle ϑ with respect to the normal.

The bidirectional scattering distribution function (*BSDF*) is given by

$$BSDF = \frac{P_s}{(P_{inc}\cos\vartheta)\Delta\Omega} = \frac{(P_s/\Delta\Omega)}{P_{inc}\cos\vartheta}, \tag{16.28}$$

the second definition being the common one. The denominator of the first version recalls the $A\Omega$ product, important in the throughput of a well-designed optical system.

Angle-resolved scattering (*ARS*) simply omits the cosine in Equation 16.28 to give

$$ARS = \frac{(P_s/\Delta\Omega)}{P_{inc}} \tag{16.29}$$

and is often preferred because it is a little closer to what is actually measured. It is sometimes called the cosine-corrected *BSDF*.

Total integrated scatter (*TIS*) is a measure of the total scattering into the appropriate hemisphere, invariably the reflected one, and is particularly popular because of its direct connection to surface roughness. The incident beam is usually normally directed onto the surface, and the scattered light is collected, taking care to exclude the specularly reflected beam. The preferred form normalizes the integrated scatter by dividing by the specularly reflected light for a completely smooth surface, rather than the incident light, and then the total integrated scatter purely becomes a surface property related to the surface roughness.

$$TIS = \frac{\int \frac{P_s}{\Delta \Omega} \, d\Omega}{P_{\text{reflected}}} \simeq \left(\frac{4\pi\sigma}{\lambda}\right)^2, \tag{16.30}$$

where σ is the root-mean-square (rms) surface roughness. Of course, this particularly applies to a coated or uncoated surface that can reasonably be described by this single parameter.

The major scattering contribution in optical coatings is from the roughness at the various interfaces that can be classified in two primary ways. The roughness may be largely due to the state of the substrate surface so that it is reproduced at each interface. We use the term *correlated* to describe this roughness. On the other hand, the roughness at the various interfaces may be wholly due to the immediately underlying material so that roughness at one interface does not reproduce in any degree that at another, and we describe this as uncorrelated. Sometimes we may have a mixture of these two. The scattering properties of these two classifications show some significant differences.

Scattering distribution is much more complex than the specular properties, and the theory is involved and difficult. To illustrate the effects, we use the model published by Elson [16] and a multiple-cavity narrowband filter as the subject. Elson models the roughness of a surface as a combination of both a long-range roughness and a short range. In the correlation function, the long-range roughness is represented by an exponential term, and the short-range roughness, by a Gaussian.

$$G(\tau) = \delta_L^2 \exp(-|\tau|/\sigma_L) + \delta_S^2 \exp\left[-(\tau/\sigma_S)^2\right], \tag{16.31}$$

where τ is the lag length, δ is the appropriate rms roughness, and σ is the correlation length with L and S denoting long-range and short-range roughness, respectively.

We begin with the uncorrelated case where the intrinsic roughness of the thin films themselves is dominant. For the most significant scattering, the exciting light must enter the coating, and so we shall see larger scattering effects associated with those wavelengths where the light can readily penetrate into the interior of the coating. Then the light is scattered in all possible directions, and the light that we measure must emerge from the coating and will be greatest in those directions where emergence is easier. Thus, in the scattering distribution, we shall tend to see an emphasis on the easy entrance and exit wavelengths and angles of incidence. The plot will look roughly like the product of these two factors. The variation of scattering with direction can be large, and in distributions, it is normal to plot the logarithm of the scattering magnitude.

Figure 16.20 shows the transmittance in decibels of our narrowband filter as a function of both wavelength and angle of incidence (measured in glass). Figure 16.21 shows a scattering calculation where we have 0.5 nm rms roughness with correlation length of 2000 nm for long range and 200 nm for short range. All the examples of this section use these values. The light is incident normally, and the scattered *s*-polarized light is measured as a function of emergent angle. Easy input corresponds to zero angle in Figure 16.20 and can be seen echoed in Figure 16.21 as a set of vertical lines. The angular variation in Figure 16.20 can also be clearly seen.

It is a little easier to see the difference between scatter due to uncorrelated and correlated roughness when we look at the calculations of reflected ARS from a high reflectance coating. The coating is a 21-layer quarter-wave stack of SiO_2 and Ta_2O_5 centered on 1000 nm. In the case of

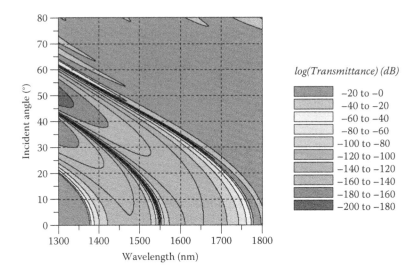

FIGURE 16.20
s-Transmittance in decibles of the three-cavity narrowband filter showing the variation with wavelength and angle. The incident angle is referred to a medium of glass.

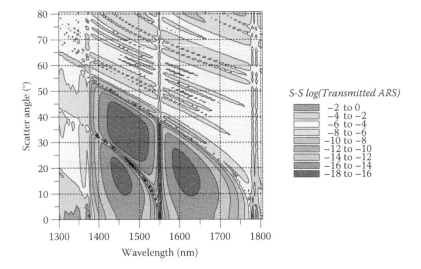

FIGURE 16.21
Distribution of s-polarized ARS from the filter of Figure 16.20 with uncorrelated surface roughness. The illumination is incident normally. Note not only the reproduction of the main features of Figure 16.20, but also, superimposed over it, the vertical features that mark the wavelengths of easy entry.

uncorrelated surface roughness (Figure 16.22), the scattering tends to be higher where the reflectance is lower so that the light can penetrate into the coating. In the case of correlated roughness, the opposite tends to be true (Figure 16.23). There the scattering is higher where the reflectance is higher.

Often we are less interested in the scattering distribution than in the effect of the loss on the specular characteristics. Here we can recognize two primary effects. Again we concentrate on surface roughness. We can characterize the roughness as being small scale with small correlation length or long scale with large length, small and large being with respect to the wavelength [18,19]. Small-scale roughness largely acts as an inhomogeneous transition layer between the layers on

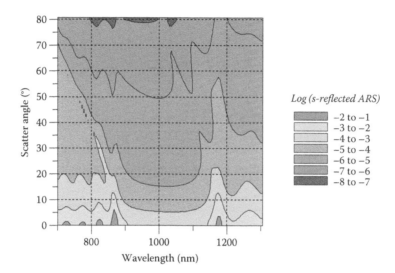

FIGURE 16.22
Reflected *s*-polarized ARS from a quarter-wave stack reflector designed for 1000 nm and with uncorrelated surface roughness. The scattering is low over the central high-reflectance zone but high in the reflectance dips outside the high-reflectance zone.

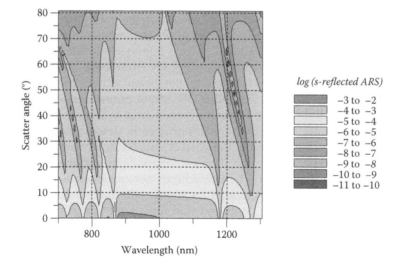

FIGURE 16.23
Reflected *s*-polarized ARS from the quarter-wave stack reflector of Figure 16.22 but with correlated surface roughness. Unlike Figure 16.22, the scattering is high over the central high-reflectance zone and low in the reflectance dips outside the high-reflectance zone. The scattering tends to follow the regions of higher reflectance. It also increases toward shorter wavelengths.

either side of the interface and with thickness 2δ, where δ, as before, is the rms roughness. No additional loss is involved in this case, although reflectance and transmittance will be affected if δ is at all significant because of the antireflecting effect even if only slight. The case of large correlation length is similar except that now, we have significant incoherent scattering that does represent a loss. Since the scattering goes with the square of the electric field as does absorption, as long as we are interested only in the specular properties, we can represent the loss as an extinction coefficient in the transition layer.

Carniglia and Jensen [18] gives expressions for the transition layer. Its thickness is still 2δ, but the index and extinction coefficient are given by

$$n = \left(\frac{n_p^2 + n_q^2}{2}\right)^{1/2},$$

$$k = \frac{\pi\left(n_p - n_q\right)^2\left(n_p + n_q\right)}{2n} \cdot \frac{\delta}{\lambda},$$

(16.32)

where n_p and n_q are the materials on either side of the interface. Frequently, these expressions are all that are required.

16.6 Temperature Shifts

As already discussed in Chapter 14, filters are frequently treated at elevated temperatures to stabilize them before they are incorporated in a system. Then, in systems, they are sometimes subjected to elevated temperatures. Our understanding of the effects is rather poor and largely empirical. We have better understanding of the reversible shifts with temperature changes in the optical characteristics of coatings. The breakthrough in understanding was a paper by Takahashi. [20]. Note the author's name is misprinted as Takashashi in the publication. The Takahashi model deals with coatings on a rigid substrate. It does not apply to plastic substrates that dissipate strain energy. It is assumed in the model that the substrate dominates in that the film system in any change in temperature takes its lateral dimensions totally from those of the substrate.

The Takahashi model (Figure 16.24) adds to the linear expansion of film thickness and temperature-induced change in refractive index, the strain birefringence caused by differential expansion of substrate and film and the Poisson's ratio change in layer thickness. This shows that the properties of the substrate are of primary importance in determining temperature-induced changes in optical coatings. By choosing the correct substrate, we can achieve virtually zero temperature coefficients of spectral change. This is of extreme importance in telecommunication applications. The original Takahashi model treated narrowband filter designs as a succession of layer pairs. Kim and Hwangbo [21] elaborated the model and made it more general.

It turns out that dense energetically deposited thin films tend to expand rather more than do their normal substrates, causing a net shift in their characteristics toward longer wavelengths especially when deposited on low-expansion materials such as fused silica. This was rather surprising because fused silica had been thought to be an ideal substrate material. Experiments

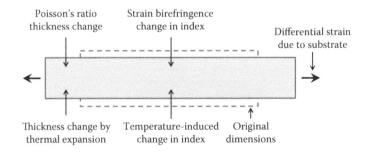

FIGURE 16.24
The Takahashi model [20] of temperature shift adds strain birefringence and the effect of Poisson's ratio to the simple model of thermal expansion and thermally induced refractive index change.

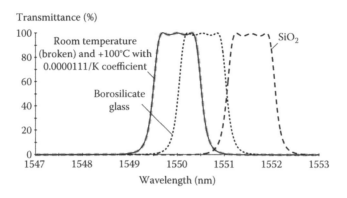

FIGURE 16.25

Transmittance of a simple three-cavity filter for telecommunication applications consisting of tantala and silica layers on different substrates and raised through 100°C. The broken line on the left is the room temperature performance over which, virtually exactly superimposed, is the raised temperature characteristic (*solid line*) of the filter on a high-expansion coefficient glass. The coefficient shown is that of the special glass substrate, not the filter characteristic.

TABLE 16.1

Film Constants Used in Figure 16.25

Material	Expansion Coefficient (ppm/K)	dn/dT (ppm/K)	Poisson's Ratio
Ta_2O_5	1.1	20	0
SiO_2	0.55	9	0.05

Note: ppm, parts per million.

showed that for good temperature stability, substrates of rather high expansion coefficients were required. The glass manufacturers responded with new and tougher high-expansion coefficient materials that are nowadays the preferred substrate materials for the narrowband filters used in the telecommunications industry as multiplexers and demultiplexers.

Figure 16.25 shows calculations of induced changes in a narrowband filter on different substrates using an elaboration of the Takahashi model. The high-coefficient glass is typical of the WMS series produced by Ohara or the S series of Schott. Note that fused silica gives maximum shift. The values for the various thermooptical constants of the thin films vary with preparation conditions. Those used in the calculations (Table 16.1) were derived from Takahashi [20] and are those adopted by Kim and Hwangbo [21

16.7 Stray Light

Rejected light in an optical filter is most usually reflected. The rejected light is therefore not suppressed but is simply redirected. Care should always be taken to ensure that it is trapped so that it cannot return, especially at a different angle of incidence. Figure 16.26 illustrates the worst case of what might happen when light with a wavelength shorter than the cutoff is returned to a longwave pass filter at oblique incidence so that it is now transmitted rather than rejected. Baffles are important traps for stray light in optical instruments. The filter is the blocked version from Figure 7.11.

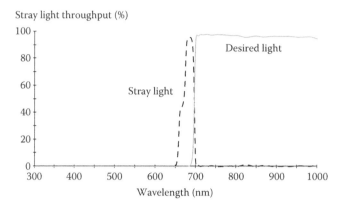

FIGURE 16.26
The calculation assumes all the light reflected normally from the longwave pass filter returns at 30° incidence. The result is the broken curve showing the amount transmitted. Shorter wavelength light is suppressed by the absorption glass-blocking filter.

Stray light is also a consequence of multiple reflections between optical surfaces in an instrument. For example, light that is reflected back and forth between the surfaces of a compound lens may emerge along with the imaging light to cause glare or ghosts. It is relatively easy to calculate the possible levels of stray light and, in turn, to determine the specification necessary for the antireflection coatings.

A more serious effect, also completely predictable, occurs when we have two high-rejection coatings in series where the rejection is by reflection with no absorption loss. The calculation is simplest if the two coatings have the same degree of rejection. If all beams are collected, then the combined transmittance is just one half of the individual transmittance rather than the desired product of the transmittances.

We emphasize once more that the these effects are completely predictable.

16.8 Laser Damage

Laser damage is still a very active specialized research topic. This section is a very short summary of some of the important points. The best current detailed account of the various aspects of laser damage is that of Ristau [22].

We can instinctively understand what is meant by the term *laser damage threshold*. Clearly, it is the level of illumination at which the component first exhibits damage. But it turns out that even this apparently simple concept is far from simple. First of all, what is damage? We tend to think of damage as serious impairment, but in laser damage, it is usually interpreted as a permanent detectable change in a component. Then power level, duration, pulse shape, wavelength, whether or not the irradiation is on top of a spot already irradiated are just some of the factors of influence. The field of laser-induced damage still has a large empirical content, but this should not be taken to mean a lack of understanding. Understanding of the phenomenon is actually very advanced, but, rather like abrasion resistance, although we understand it in considerable detail, there are so many contributing factors that it is virtually impossible to accurately quantify it with a single parameter. Nevertheless, despite the complications, we still like to assign a value of laser damage threshold to a coating. In most of the topics we have so far discussed in this book, we are able to make very good quantitative estimates of coating performance. Our model of the optical behavior of optical coatings is incredibly accurate and reliable. The situation completely changes when we consider the laser damage threshold. So far, the only way of accurately determining the laser damage threshold of a given coating is to actually damage it, under as far as possible the conditions that would be applied in use.

Soileau [23] gives a useful rule of thumb that is probably good to within roughly an order of magnitude. It assumes good practice in the construction of the component

$$E_d = (10 \text{ J/cm}^2)(t_p/1 \text{ ns})^{1/2}, \tag{16.33}$$

where E_d is the fluence at damage threshold in joules per square centimeter and t_p is the pulse duration in nanoseconds. Soileau strongly advises that this figure should not be used for the prediction of laser damage threshold in a component, but rather for the identification of any laser-containing system where laser damage might become an issue.

In terms of materials, hafnium oxide is considered the material of choice for high laser resistance with silicon dioxide as its accompanying low index.

The best bulk crystals can exhibit intrinsic damage thresholds that are ultimately connected with multiphoton events causing the raising of electrons into the conduction band so that there is dielectric breakdown. This can also be thought of as a maximum electric field problem. Damage in thin-film systems, on the other hand, is most frequently caused by the defects in the films so that the intrinsic level is not reached.

The intrinsic damage level is ultimately a problem of the electric field in the material, and this can cause dielectric breakdown or the generation of sufficient local heat to disrupt the material, and in these cases, it is the square of the electric field that is important, and, of course, irradiance is proportional to the square of the field. Doubling the electric field increases the danger by a factor of 4. High-index materials, because of their longer wavelength absorption onset, are generally more vulnerable than low-index materials, and some improvements in laser damage threshold can be achieved by altering the design so as to move the highest fields from the high into the low index. Note that because the field distribution varies with wavelength, this technique is suitable for a quite narrow range of wavelengths, preferably one single wavelength. It involves reducing the high-index thicknesses in the outermost layers and increasing the low, so that the basic period remains a half wave. Such designs are especially employed in the very large lasers for fusion power research.

In continuous wave applications, particularly in the infrared, thermal effects associated with absorption, either local or general, appear to be the principal source of damage, small defects appearing less important. In most other cases, local defects are the problem. One can imagine a defect situated at or near an interface. The laser can so heat this defect that a plasma is produced that lifts the coating and produces a blister that is usually detached from the coating to leave a crater often with a flat base [24]. The particular nature of the defects may considerably vary, from inclusions to cracks or fissures, but considerable attention in recent years has been paid to the nodules that tend to grow through the films from any substrate imperfections. These nodules are poorly connected thermally to the film, and this is suspected to be an important factor in the initiation of damage. Then they can act as a microlens to concentrate the power [24]. In those spectral regions where water strongly absorbs, considerable importance is attached to the presence of liquid water within the films. In other parts of the spectrum, its role is less clear, but it may well play a part. Laser conditioning [24] can help reduce the effect of damage-prone defects including nodules. It consists of a gradual exposure of the component to increasing fluence that appears to cause a kind of annealing accompanied by sometimes actual ejection of nodules. Once a nodule is ejected, its power concentration disappears and the damage threshold is improved.

The addition of a half-wave silica layer to the outside of a reflector can have beneficial effects. There are two reasons advanced for this. An outermost layer of high index can exhibit low resistance to any plasma that might be generated at its surface. Low-index materials such as SiO_2 have appreciably greater resistance, and since the half wave is an absentee layer, it has no effect on the interference properties at the design wavelength. It also appears to improve the lateral thermal conductance. Stolz points out that on occasions, there can be a disadvantage in that its surface roughness can add additional scattering that can modify the beam profile and cause problems in the system downstream [24].

It also appears that substrate cleaning is, unsurprisingly, of great importance as is the use of metallic targets in both evaporation and sputtering processes. Source material and targets of dielectric material that can be readily used in ion-beam sputtering appear to produce more particles in the coatings with consequent deterioration of their laser damage threshold.

Ristau [22] is the latest and most complete text, but there also is useful information in Koslowski [25] and Stolz and Génin [26].

References

1. J. B. Breckinridge, W. S. T. Lam, and R. A. Chipman. 2015. Polarization aberrations in astronomical telescopes: The point spread function. *Publications of the Astronomical Society of the Pacific* 127:445–468.
2. A. I. Mahan. 1945. Focal plane anomalies in roof prisms. *Journal of the Optical Society of America* 35:623–645.
3. A. I. Mahan. 1947. Focal plane anomalies in telescopic systems. *Journal of the Optical Society of America* 37:852–867.
4. K. Rabinovitch and G. Toker. 1994. Polarization effects in optical thin films. *Proceedings of SPIE* 2253:89–102.
5. T. Ito and M. Noguchi. 2001. *Viewing optical instrument having roof prism and a roof prism*. US Patent, 6,304,395.
6. A. Macleod. 2009. Thin film polarizers and polarizing beam splitters. *Society of Vacuum Coaters Bulletin* (Summer):24–27.
7. A. Macleod and C. Clark. 1997. How sensitive are coatings to contamination? In *Eleventh International Conference on Vacuum Web Coatings*, pp. 176–186, Bakish Materials Corporation, Englewood, NJ.
8. H. A. Macleod and C. Clark. 1997. Electric field distribution as a tool in optical coating design. In *40th Annual Technical Conference Proceedings*, pp. 221–26, Society of Vacuum Coaters, New Orleans, LA.
9. A. Macleod. 2006. Potential transmittance. *Society of Vacuum Coaters Bulletin* (Fall):24–25, 28.
10. E. D. Palik, ed. 1985. *Handbook of Optical Constants of Solids I*. Orlando, FL: Academic Press.
11. E. D. Palik. 1991. *Handbook of Optical Constants of Solids II*. San Diego, CA: Academic Press.
12. E. D. Palik. 1998. *Handbook of Optical Constants of Solids III*. San Diego, CA: Academic Press.
13. C. Amra. 1995. Introduction to light scattering in multilayer optics. In *Thin Films for Optical Systems*, F. R. Flory (ed.), pp. 367–391. New York: Marcel Dekker.
14. A. Duparré. 1995. Light scattering of thin dielectric films. In *Handbook of Optical Properties*, vol. 1: *Thin Films for Optical Coatings*, R. E. Hummel and K. H. Guenther (ed.), pp. 273–303. Boca Raton, FL: CRC Press.
15. J. M. Elson, J. P. Rahn, and J. M. Bennett. 1980. Light scattering from multilayer optics: Comparison of theory and experiment. *Applied Optics* 19:669–679.
16. J. M. Elson. 1995. Multilayer-coated optics: Guided-wave coupling and scattering by means of interface random roughness. *Journal of the Optical Society of America A* 12:729–742.
17. S. Schröder, S. Gliech, and A. Duparré. 2005. Measurement system to determine the total and angle-resolved light scattering of optical components in the deep-ultraviolet and vacuum-ultraviolet spectral regions. *Applied Optics* 44:6093–6107.
18. C. K. Carniglia and D. G. Jensen. 2002. Single-layer model for surface roughness. *Applied Optics* 41:3167–3171.
19. A. V. Tikhonravov, M. K. Trubetskov, A. A. Tikhonravov et al. 2003. Effects of interface roughness on the spectral properties of thin films and multilayers. *Applied Optics* 42:5140–5148.
20. H. Takashashi. 1995. Temperature stability of thin-film narrow-bandpass filters produced by ion-assisted deposition. *Applied Optics* 34:667–675.

21. S.-H. Kim and C. K. Hwangbo. 2004. Temperature dependence of transmission center wavelength of narrow bandpass filters prepared by plasma ion-assisted deposition. *Journal of the Korean Physical Society* 45:93–98.
22. D. Ristau, ed. 2015. *Laser-Induced Damage in Optical Materials*. Boca Raton, FL: CRC Press.
23. M. J. Soileau. 2015. Laser-induced damage phenomena in optics: A historical overview. In *Laser-Induced Damage in Optical Materials*, D. Ristau (ed.), pp. 3–7. Boca Raton, FL: CRC Press.
24. C. J. Stolz. 2015. High-power coatings for NIR lasers. In *Laser-Induced Damage in Optical Materials*, D. Ristau (ed.), pp. 385–409. Boca Raton, FL: CRC Press.
25. M. Koslowski. 1995. Damage-resistant laser coatings. In *Thin Films for Optical Systems*, F. R. Flory (ed.), pp. 521–549. New York: Marcel Dekker.
26. C. J. Stolz and F. Y. Génin. 2003. Laser resistant coatings. In *Optical Interference Coatings*, N. Kaiser and H. K. Pulker (eds.), pp. 310–333. Berlin: Springer-Verlag.

17

Specification of Filters and Coatings

Ideally, if a filter or coating is to be manufactured for a customer for a given application, then the performance required by the customer, and the design, manufacturing and test methods, should all be defined, even if only implicitly. These details form different aspects of the specification of the filter.

There is no standard method for setting up the specification of an optical filter or coating, the problem being much the same as for any other device. There are three main aspects to be considered: performance, manufacture, and testing. How detailed these are will depend on many different circumstances. Here we imagine a project sufficiently demanding for considerable details to be required.

The performance specification will list the details of the performance required from the filter and will often be the customer's specification. The manufacturing specification will define the design and detail the steps involved in the manufacture of the filter. The test specification will lay down the tests to be carried out on the filter to ensure that it meets the performance requirements. These two latter aspects will mainly be the concern of the manufacturer. In the following notes, a few of the more important points are mentioned, but they do not form a complete guide to the writing of specifications, which is a complete subject in its own right.

There can sometimes be vagueness in the statement of requirements, and this is particularly true when dealing with matters connected with polarization and phase. Doubts about requirements including the meanings of terms are best resolved at the start rather than toward the end of a project.

Optical filter specifications can conveniently be divided into two sections, one concerned with optical properties, and the other, with physical or environmental properties. We shall first of all consider the optical properties.

17.1 Optical Properties

17.1.1 Performance Specification

The performance specification of a filter is essentially a statement of the capabilities of the filter in a language that can be readily interpreted by system designer, customer, and filter manufacturer alike. It can sometimes be prepared by a filter manufacturer from knowledge of the known achievable performance. This can be either for a customer or, possibly, without having a particular application in mind, as in the case of a standard product in a catalog. We shall say little about the latter here. Probably more often, the performance specification will be written by the system designer and will state a level of performance required from a filter in order to achieve a desired level of performance from a system. In writing such a specification, an answer must first of all be given to the question, what is the filter for? The purpose of the filter should be set down as clearly and concisely as possible, and this will form the basis for the work on the performance specification. Unless completely secret, this can usefully form a preamble to the specification. There is really no systematic method for specifying the details of performance. Sometimes it happens that the performance of the system in which the filter is to be used must be of a certain definite level; otherwise, there will be no point in proceeding further. The filter performance requirements can then be quite readily set down. Often, however, it will not be quite so simple. No absolute requirement for performance may exist, only that the performance should be as high as possible within allowable limits of complexity or perhaps price. In such a case, the performance of the system with different levels of filter performance must be balanced against cost and system complexity and a decision made as to what is reasonable. The final specification will be a compromise between what is desirable and what is achievable. This will often need the input of much design and manufacturing information and close contact between customer and manufacturer. It should always be remembered in this that specifications that cannot be met in practice can be of only academic interest.

By way of an example, let us briefly consider the case where a spectral line must be picked out against a continuum. Clearly, a narrowband filter will be required, but what will be the required bandwidth and type of filter? The energy from the line to be transmitted by the filter will depend on the peak transmittance (assuming that the peak of the filter can always be tuned to the line in question), while the energy from the continuum will depend on the total area under the transmission curve, including the rejection region at wavelengths far removed from the peak. The narrower the passband, the higher the contrast between the line and the continuum, especially as narrowing the passband generally also improves the rejection. However, the narrower the passband, because of the increased difficulty of manufacture, the higher the price, and further, because of the increased sensitivity to the lack of collimation, the larger the tolerable focal ratio. This latter point implies that for the same field of view, a filter with a narrower bandwidth must be made larger to permit the use of the larger focal ratio, which, in turn, will still further increase the difficulties of manufacture and, possibly, the complexity of the entire system. Another way of improving the performance of the filter is by increasing the steepness of edge of the passband while still retaining the same bandwidth. A rectangular passband shape gives higher contrast than a simple Fabry–Perot of identical halfwidth and usually possesses the additional advantage that the rejection remote from the peak of the filter is also rather greater. This edge steepness can be specified by quoting the necessary tenth peak bandwidth or even the hundredth peak bandwidth. Again, inevitably, the steeper the edges, the more difficult the manufacture and the higher the price.

Because filters, as with any manufactured product, cannot be made exactly to a specification in absolute terms, some tolerances must always be stated. For a narrowband filter, the principal parameters that should be given tolerances are peak wavelength, peak transmittance, and bandwidth. Since in almost all applications the higher the peak transmittance, the better, it is usually sufficient to state a lower limit for it. There are two aspects of peak wavelength tolerance. The first is uniformity of peak wavelength over the surface of the filter. There will always be some grading of the films, although perhaps small, and a limit must be put on this. The effect is similar to that of an incident cone of illumination (discussed in Chapters 8 and 16), and it is usually best to limit the uniformity errors in the specification to not more than one third of the halfwidth. The second aspect is error in the mean peak wavelength measured over the whole area of the filter. The tolerance for this is usually made positive so that the filter can always be tuned to the correct wavelength by tilting. For a given bandwidth, the amount of tilt that can be tolerated in any application will be determined to a great extent by the aperture and field of the system, since the total range of angles of incidence that can be accepted by a filter falls as the tilt angle is increased.

The bandwidth of the filter should also be specified, and a tolerance put on it, but because of the difficulty of controlling bandwidth very accurately, it is not usually desirable to tie it up too tightly, and the tolerance should be kept as wide as possible, not normally less than 0.2 times the nominal figure unless there is a very good reason for it.

One other important parameter involved in the optical performance specification is rejection in the stopping zones, which may be defined in a number of different ways. Either the average transmittance over a range, or absolute transmittance at any wavelength in the range, can be given an upper limit. The first would usually apply where the interfering source is a continuum, and the second, where it is a line source, in which case the wavelengths involved should be stated, if known.

Yet another entirely different method of specifying filter performance is by drawing maximum and minimum envelopes of transmittance against wavelength. The performance of the filter must not fall outside the region laid down by the envelopes. It is important that the acceptance angle of the filter also be stated. This type of specification is rather more definite than the first type mentioned earlier. A disadvantage, however, is that it may be rather too severe since everything is stated in absolute terms when average values may be just as good. A further point is that it is strictly impossible to devise a test to determine whether or not a filter meets an absolute specification of this type. Finite bandwidth of the measuring apparatus will ultimately be involved. It is advisable, therefore, if specifying a filter in this way, to include a note to the effect that the performance specified at each wavelength is the average over a certain definite interval.

There is little else that can be said in general terms about the optical performance specification. In any one application, these factors will assume different relative importance, and each case must to a very great extent be considered on its own merits. Clearly, this is an area where it is of prime importance that the system designer work very closely with the filter designer.

If there is one particular area where problems occur most often, it is that of polarization and related phase shifts. First of all what is the sign convention? Does a reflectance phase difference of 90° between p- and s-polarizations really mean 90° for $\varphi_p - \varphi_s$, or could it be the relative retardation, or would ±90° be satisfactory?

17.1.2 Manufacturing Specification

We shall now briefly consider the manufacturing specification containing the filter design together with details of the manufacturing method. In most cases, this will be intended for the use of the machine operator.

First, the filter design, including the materials, will be given. Most filters contain no more than three different thin-film materials having relatively low, medium, and high refractive index. Designs are usually written in terms of quarter-wave optical thicknesses at a reference wavelength λ_0 using the symbols L, M, and H. Typical designs may be written as follows:

$$L|Ge|LHLHHLH, \quad L = ZnS, H = Ge,$$

$$M|Si|MHLHHLH, \quad L = CaF_2, M = ZnS, H = Ge,$$

the substrates being indicated by the symbols $|Ge|$ and $|Si|$. Next, the constructional details should be written down. These consist of the monitoring method to be used, including the wavelengths, and the form of the signals together with other important details such as substrate temperature and special types of evaporation sources. It will be found useful to arrange the whole manufacturing specification in the form of a table that can be issued to the machine operators for use as a checklist. Operators should always be encouraged to critically observe the operation of the machine so that faults or anomalies can be spotted at an early stage, and it is a help in this if they are expected to list comments in appropriate places on the form. It will also be found convenient to give each filter production batch a different reference number. Once the filters are produced, the completed specification form can then be filed by the machine operator to form the machine logbook. Additional information such as pumping performance can also be recorded on the sheets, useful from the maintenance point of view.

For calculation purposes, there is no consensus on whether the incident medium should be at the top or at the foot of a table of design. For manufacture, however, the first layer to be deposited is necessarily next to the substrate, and it is usual to list the layers in tables of manufacturing instructions from innermost, next to the substrate, to outermost.

Software products can assist in setting up the manufacturing specification, especially the sequence of monitoring signals. In some cases, these can be automatically fed into the deposition controller so that the printed copy can be simply for reference and record keeping.

17.1.3 Test Specification

Probably the most important specification of all is the test specification. This lays down the set of tests that will be carried out on the filters to measure the performance. It should always be remembered that although the filter will have been designed to meet a particular performance specification, it is only the performance laid down in the test specification that can actually be guaranteed, and although it may seem obvious, the test specification must be written with the requirements of the performance specification always in mind. In fact, it is possible to simply specify the performance of a filter as that which will pass the appropriate test specification. It will sometimes be found that the test specification, if it exists at all, is a rather loose document or that sometimes the customer's performance specification will serve both roles. If so, then someone somewhere along the line will be interpreting the performance specification in order to decide on the tests to be applied, and it is always better to have the tests and the method of interpretation in writing.

The first essential in any test specification is a definite statement of the performance or the make and type of the test equipment to be used. This ensures that results can be repeated if necessary, even if remote from the original testing site. Next, the various tests together with the appropriate acceptance levels can be set down.

It is in the measurement of such factors as uniformity where the tests and the method of interpretation are particularly important. Absolute uniformity is impossible to measure in the ordinary way. The performance would have to be measured at every point on the coating with an infinitesimally small measuring beam. A simpler and usually satisfactory method is to check the performance, for example, the peak wavelength, at the center of the filter and at four approximately equally spaced areas around the circumference, using a specified area of measuring beam. The spread over the filter is taken to be the spread in the values of peak wavelength, or other performance attributes, over the five separate measurements. The spectrometer used for the measurement will also have a finite bandwidth, and features of the filter that are rather less than this will, in general, not be picked up. This particularly applies to the measurement of rejection. Rejection must usually be measured over a very wide region, and for the test to be completed in a reasonable time, a fast scanning speed must be used, which, in turn, requires a broad bandwidth. This averages the measurement over a finite region and is therefore unsatisfactory if the energy that is to be rejected has a line rather than a continuous spectrum. In such a case, the lines should be defined, and the tests include more careful measurements at the defined lines. A technique for measuring the rejection of films using a Fourier transform spectrometer was suggested by Bousquet and Richier [1]. While this is difficult to apply in the visible region, the availability of commercial Fourier transform spectrometers for the infrared makes it a feasible technique for infrared filters.

Of course, inevitably, the more extensive the testing that must be carried out on each individual filter, the more expensive that filter is going to be. Performance testing of low-price standard filters is, in the main, carried out on a batch basis, with, at the most, only a few details being checked on each individual filter. This is a point that should be borne in mind by a prospective customer buying a standard filter from a catalog that a superlative level of performance cannot be absolutely guaranteed from a single given filter, which, by its price, cannot have had more than the basic testing carried out on it.

So far, we have dealt with the directly measurable optical performance of the filter, but there are additional properties of a subjective nature and rather more difficult to measure. These are connected with the quality and finish of the films and substrates. Substrates are specified as for any optically worked component; details such as flatness or curvature of surface, degree of polish and allowable blemishes, sleeks, and the like can all be stated. We shall not consider substrates further here. There is a specification, used particularly in the United States, MIL-E13830 A, which gives a useful set of standards for optical components including substrates.

The quality of the coating can be measured by the presence or absence of defects such as pinholes, stains, spatter marks, and uncoated areas. Pinholes are important for two reasons. First, they are actually small uncoated, or partially uncoated, areas and, as such, will allow extra light to be transmitted in the rejection regions, reducing the overall performance of the filter. Second, and this is especially so for filters for the visible region, they are unsightly and detract from the appearance. In fact, they usually look worse to the eye than the effect they actually have on performance. Apart from the purely subjective appearance, the permissible level of pinholes can be defined based on a given maximum number of a certain size per unit area, calculated to reduce the rejection in the stop bands by not more than a given amount. To calculate this figure, a minimum area of filter that will be used at any one time must be assumed. This will depend on the application, but in the absence of any definite information on this, a suitable figure is 5 mm × 5 mm. Obviously, the smaller this area, the lower the size of the largest pinhole. Of course, the actual counting of pinholes in any filter would involve a prohibitive amount of labor, and in practice, with visible filters, the measurement is often carried out visually, comparing the filter with limit samples. A simple fixture consisting of a light box with sets of filters laid out on it, some just inside, some on, and some just outside the limit, can be readily constructed. For infrared filters on transparent substrates, this method can also be applied, but for filters on opaque substrates, it is easier to measure actual rejection performance.

Spatter marks are caused by fragments of material ejected from the sources. In themselves, unless gigantic, they have little effect on the optical performance. A major danger is that the fragments may be removed later, leaving pinholes. Sometimes, however, spatter causes nodular growth with their associated problems, or if many spatter fragments are present, the scattering losses may rise. The incidence can be tied down just as with pinholes, but as the specular optical performance is little affected, unless the number of marks is enormous, the basis for deciding what is permissible is usually subjective. If the spatter is causing pinholes, they will be dealt with separately. Often specifications will state that there must be no spatter marks visible to the naked eye, but this is vague, particularly when dealing with inspectors with no optical experience. Disagreements can arise between manufacturer and customer especially when, as can happen, the customer's inspectors use an eyeglass to assist the naked eye. The best course is probably to relate the test to agreed limit samples when it can be carried out in exactly the same way as for pinholes.

Stains can be caused in a number of ways. A not uncommon reason is a faulty substrate. One type of mark that is often seen, especially when antireflection coatings are involved, is due to a defect in the optical working. The polishing process partly consists of a smoothing out of irregularities in the surface by removal of material. If the grinding, which always precedes the polishing, has been too coarse, then the deeper pits during the polishing can be filled in with material that is only loosely bonded to the surface, although the polish will usually appear completely satisfactory to the eye. In the heating and then coating of the surface, this poorly bonded material breaks away, leaving a patch of surface that is etched in appearance and often possesses well-defined boundaries. The only remedy for this type of blemish is improved polishing techniques. Other stains that may appear can be caused by faulty substrate cleaning. If water, or even alcohol, is allowed to dry on a surface without wiping, watermarks appear. Droplets should always be removed from the surface by a final vapor cleaning stage, by blowing with clean air (great care must be taken to make sure the air is clean and does not carry oil with it), or by wiping with a clean tissue or cloth during the cleaning process. Water should never be allowed to dry on the surface by itself. Stains, unless particularly bad, do not usually affect the optical performance to anything like the extent their appearance would suggest (except in the case of very high-performance components such as Fabry–Perot interferometer plates or laser mirrors), and the basis for judging them is again subjective.

Finally, the filter must be held in a jig during coating so that at least some uncoated areas must exist. These usually take the form of a ring around the periphery of the filter, perhaps around 0.5 mm wide. There will be a slight taper in the coating at the very edge, which must also be allowed for, the combined taper and uncoated area forming a strip perhaps 1.0 mm in width. The uncoated area actually serves a useful purpose because mechanical mounts can grip the component at this point without damaging the coating. Damage near the edge is dangerous because it is there that delamination is frequently initiated. Jigs that allow the substrates to chatter as they rotate can cause such defects. Uncoated areas should not occur within the boundary of the filter proper; when they do, it is usually a sign of adhesion failures that may recur. They may be due to substrate contamination or to moisture penetration with weakening of adhesion, as described in Chapter 14, but they are always a cause for the rejection of the component. Blisters, too, which are a slightly different version of the same fault, are a cause for immediate rejection.

17.2 Physical Properties

As far as the physical properties of the filter are concerned, there are two primary aspects. First, the dimensions of the filter must meet the requirements laid down. This is purely a matter of mechanical tolerances that we need not go into any further here. Second, the filter must be capable of withstanding, as far as possible, the handling it will receive in service and of resisting any attack from the environment. The assessment of the ruggedness, or robustness, of the coating will now be considered in greater detail.

The approach almost invariably used in defining and testing the ruggedness of a coating is to combine the performance and test specifications. A series of controlled tests reproducing typical conditions likely to be met in practice is set up, and then performance is defined as being a measure of the ability to pass the particular tests. This avoids the difficulty in setting up a more general performance specification.

There is one basic difference between the tests of optical performance and those we are about to discuss. Optical tests, except, perhaps, for laser damage threshold, are usually nondestructive in nature while tests of ruggedness are, in the main, destructive. The filters are deliberately tested to cause damage, and the extent of the damage, or the point at which damage can be detected, if it can be measured, is used as a measure of the ruggedness of the filter. Thus, it is not possible to carry out the whole series of tests on the actual filter that is to be supplied to the customer, and it is normal to use a system of batch testing. A number of filters are made in a batch, and either one or perhaps two are chosen at random for testing. Provided these test filters are found acceptable, then the complete batch is assumed satisfactory. This arrangement is, of course, not peculiar to thin-film devices. Another aspect of this batch testing is involved in what is known as a type test. Often if a large number of filters, all the same type and characteristic, are involved, a series of very extensive and severe tests will be carried out on a sample of filters from a number of production batches. The test results will then be assumed to apply to the entire production of this type of filter. Once the filters have passed this type test, normal production testing is carried out on a reduced scale. It is imperative that once the type test has been successful, there are no subsequent changes, even of a minor nature, to the production process; otherwise, the type test becomes invalidated.

17.2.1 Abrasion Resistance

Coatings on exposed surfaces, such as the antireflection coating on a lens, will probably require cleaning from time to time. Cleaning usually consists of some sort of rubbing action with a cloth or perhaps lens tissue. Often there may be dust or grit on the surface of the lens, which may not be removed before rubbing. The result of such treatment is abrasion, and it is important to have the abrasion resistance of exposed coatings as high as possible. An absolute measure of abrasion resistance is not at all easy to establish because of the difficulty of defining it in absolute terms, and the approach is to reproduce, under controlled conditions, abrasion similar to that likely to be met in practice, only rather more severe. The degree to which the coating withstands the treatment is then a guide to its performance in actual use. In the United Kingdom, a great deal of work was carried out on standardizing this test by the Sira Institute (formerly, the British Scientific Instrument Research Association). Their method involved a standard pad made from rubber loaded with emery powder, which, with a precise load, is drawn across the surface a given number of times under test—typically 20 times with a loading of 5 lb/in.^{2} (34.47 kPa). Their work was mainly directed toward the assessment of the performance of magnesium fluoride single-layer antireflection coatings for the visible. It has been established that sufficiently rugged coatings of this type do not show signs of damage under the normal test conditions given earlier. Abrasion resistance, however, has been found to be not just a function of the film material but also of the thickness. Multilayer coatings are generally much more prone to damage than either of the component materials in single-layer form. It is therefore necessary to establish fresh standards for each and every type of coating.

The US military specification MIL-E-12397B that dates from 1954 and is still active specifies the composition of an eraser containing ground pumice abrasive material for testing coated optical components. This eraser is used in many of the US military specifications that include abrasion testing. An important related specification is MIL-C-675C. This specification strictly applies to single-layer magnesium fluoride antireflection coatings but, nevertheless, is quoted as the standard for a wide range of optical coatings including multilayers, and although it is strictly a military specification, it is frequently applied to optical coatings in general.

There are difficulties in achieving exactly the same abrading performance from different batches of abrading pad. Similar tests using pads that may or may not include abrading particles are widely used. It is not uncommon to find similar tests using rough cloth and even steel wool.

Unfortunately, such tests do not normally produce an actual measure of the abrasion resistance, but merely decide whether or not a given coating is acceptable. Because of this, some investigations into a better arrangement were carried out by Holland and van Dam [2]. Their test is based on the principle that a measurement of abrasion resistance must involve actual damage to the films. The measure of the damage can then be taken as a measure of the abrasion resistance. Their method was to subject the films to abrasive action that varied in intensity over the surface and that was, at its most intense point, sufficiently severe completely to remove the coating. The point at which the coating just stopped being completely removed was then found. Of course, the method is still relative in that a different standard must be set up for every thin-film combination, but it does permit comparison of the abrasion resistance of similar coatings, impossible with the previous method. The apparatus is shown in Figure 17.1. It consists of a reciprocating arm carrying the abrasive pad of the Sira type (0.25 in. [6.35 mm] diameter) and loaded with 5.5 lb (2.495 kg). The table carrying the sample under test rotates approximately once for every three strokes of the pad. The pad traces out a series of spirals on the surface of the sample, and the geometry is arranged so that the diameter of the abraded area is approximately 1.25 in. (31.75 mm). The abrasion takes the form of a gradual fall off in intensity toward the outside of the circle, and the test is arranged to carry on for such a time that the central area of the coating is completely removed, while the outside, not at all. Holland and van Dam found that some 200 strokes were sufficient to do this with single layers of magnesium fluoride. They then defined the abrasion resistance measure of the coating by

$$w = \left(d^2/D^2\right) \times 100\%, \tag{17.1}$$

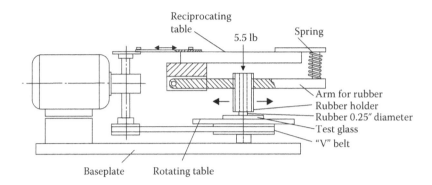

FIGURE 17.1
Schematic arrangement of an abrasion machine. The reciprocating table is supported by two horizontal bars not shown in the diagram. (After L. Holland and E. W. van Dam, *Journal of the Optical Society of America*, 46, 773–777, 1956. With permission of Optical Society of America.)

where d is the diameter of the circle where the coating has been completely removed and D is the diameter of the area that has been subjected to abrasion. Holland and van Dam particularly studied the case, as had Sira, of the single-layer magnesium fluoride antireflection coating for the visible region, and they quote a wide range of most interesting results.

They investigated many different conditions of evaporation including angle of incidence and substrate temperature. A common value for the abrasion resistance of a typical magnesium fluoride layer of thickness to give antireflection in the green is between two and five, depending on the exact conditions of deposition. Best results were obtained when the substrate temperature during evaporation was 300°C, and the glow discharge cleaning before coating lasted for 10 minutes. There was a significant reduction in abrasion resistance if either the temperature were allowed to drop to 260°C or there were only 5 minutes of glow discharge cleaning. They also found that the abrasion resistance of the film is considerably increased by burnishing with a Selvyt cloth (a popular cloth for cleaning optics) or by baking further at 400°C in air after deposition. Another significant result obtained concerns the occurrence of a critical angle of vapor incidence during film deposition, beyond which the abrasion resistance falls off extremely rapidly. This critical angle slightly varies with film thickness but is approximately 40° for thicknesses in excess of 300 nm and rises as the thickness decreases.

The test appears never to have received general recognition in specifications. It should be extremely useful as a quality-control test in manufacture, especially as a reduction in quality can be detected long before it drops below the level of the normal abrasion test, and remedial action can be taken before any coatings are even rejected.

17.2.2 Adhesion

Adhesion has already been discussed in Chapter 14. In the simplest type of adhesion test, a piece of adhesive tape is stuck down on the surface of the coating and pulled off. Whether or not this removes the film is taken as an indication of whether the adhesion of the film to the substrate is less than or greater than that of the tape to the film. The test is again of the go–no-go, or binary, type.

It is important if consistent results are to be obtained that some precautions are taken in carrying out the test. The first is that the tape should have a consistent peel adhesion rating, which should be stated in the specification. Peel adhesion is measured by sticking a freshly cut piece of tape on a clean surface, usually metal, and then steadily pulling it off, normal to the surface. The tension per unit tape width, often expressed in newtons per 10 mm or sometimes pounds per inch or grams per inch, is the measure of the peel adhesion rating of the tape. The rating obtained in this way is usually virtually the same as the rating obtained when the tape is removed from a thin-film coating. Some precautions in applying the test are necessary. Fresh tape should always be used. The tape should be firmly stuck to the coating, exerting a little pressure and smoothing it down. It should be steadily removed, pulling it at right angles to the surface, and never snatched off, which would put an uncontrolled impulsive load on the film and would certainly lead to inconsistent results. The same thickness of tape should be used for all testing. With thicker tape of the same peel adhesion rating, the test would be slightly less severe. The width of the tape, however, does not matter. A rating around 1200 gm/in. width (0.77 N/cm) is often used. If necessary, the adhesion rating of any tape can be checked using a spring balance. For obvious reasons, the test is often called the *Scotch tape test*.

The major problem with this test is the great difference in the nature of the adhesion between tape and film and film and substrate. The tape adhesive force is relatively low, but the range of the force, because the adhesive stretches, is large. The forces holding the film to the substrate are large but the range is small. The test therefore detects those areas where the adhesion is virtually nonexistent rather than areas where it is lower than normal.

Attempts have been made to devise quantitative techniques for adhesion measurement, and a number of these have been discussed in greater detail in Chapter 14. The simplest and most straightforward is the direct-pull test, involving the attachment of the flat end of a cylindrical pin to the coating, followed by measurement of the force necessary to pull it off. Provided the coating is detached with the pin, the force required divided by the area of the pin is then the measure of adhesion. Because the thin-film adhesive forces have such a short range, it is almost impossible to carry out the test in such a way as to completely avoid any progressive fracturing of the film bonds, and so the results will usually show an adhesive force that is lower than is actually the case. Of course, in a test of film quality, this lowering of the test result is a much better feature than one that artificially inflates it.

An alternative test that has some advantages as well as disadvantages is the scratch test, in which a loaded stylus is drawn across the coating with gradually increasing load. At each stroke, the coating is examined under a microscope for signs of damage. The load at which the coating is completely removed is taken as the measure of adhesion. The Goldstein and DeLong [3] technique involving the use of a microhardness tester as a scratch tester has also been mentioned in Chapter 14.

17.2.3 Environmental Resistance

One further aspect of thin-film performance is also of very great importance. This is the resistance that the film assembly offers to environmental attack. Probably the universally important aspect of the environmental performance of a coating is its resistance to the effects of humidity, but depending on the application, its resistance to other agents, such as temperature, vibration, shock, and corrosive fluids such as salt water, may also be important.

There are two possible approaches. Either the filter may be expected to satisfactorily operate while actually undergoing the test or it may only be expected to withstand the test conditions without suffering any permanent damage, although the performance need not be adequate during the actual application of the test. The latter is usual as far as interference filters are concerned, and in such a case, the specification is known as a *derangement specification* because it is sufficient that the performance is not permanently damaged by the application of the test conditions. Derangement specifications are easier to apply than the other type because the normal performance measuring equipment can be used remote from the environmental test chamber. However, the user of the coating needs to be aware of the nature of the specification: whether it is of the operational or derangement class.

Of all the agents likely to cause damage, atmospheric moisture is probably the most dangerous. For most applications, particularly where severe environments are excluded, it will be found sufficient for the filter to be tested by exposing it for 24 hours to an atmosphere of relative humidity of 98% ± 2% at a temperature of 50°C ± 2°C. It is often found that although the coatings are not removed by this test, they are softened, and it is normal to carry out this test before the adhesion or abrasion-resistance tests, which can follow on immediately after.

A great deal of work has been carried out by government bodies on the environmental testing of equipment and components for the services. This has resulted in specifications that are equivalent to the most severe conditions ever likely to be met in both tropical and polar climates. These specifications include in the UK DEF133 and DTD1085 for aircraft equipment. Relevant specifications in the United States include MIL-C-675, MIL-C-14806, MIL-C-48497, and MIL-M-13508. The tests vary from one specification to another but can include exposure to the effects of high humidity and temperature cycling over periods of 28 days, exposure conditions equivalent to dust storms, exposure to fungus attack, vibration and shock, exposure to salt, fog and rain, and immersion in salt water. It is not always possible for coatings to meet all tests in these specifications, and concessions are often given if the coatings are to be enclosed within an instrument. Humidity and exposure to salt fog and water are particularly severe tests. Fungus does not normally represent as severe a problem to the coatings as it does to the substrates. Certain types of glass can be damaged by fungus, and in such cases, coatings, even if they themselves are not attacked, will suffer along with the substrates. Most instruments likely to be exposed to sand or dust are adequately sealed since their performance is likely to suffer if dust or sand is permitted to enter. Thus, dust storms are usually a danger only to those elements with surfaces on the outside of an instrument.

References

1. P. Bousquet and R. Richier. 1972. Etude du flux parasite transmis par un filtre optique à partir de la détermination de sa fonction de transfert. *Optics Communications* 5:27–30.
2. L. Holland and E. W. van Dam. 1956. Wear resistance of magnesium fluoride films on glass. *Journal of the Optical Society of America* 46:773–777.
3. I. S. Goldstein and R. DeLong. 1982. Evaluation of microhardness and scratch testing for optical coatings. *Journal of Vacuum Science and Technology* 20:327–330.

18

Characteristics of Thin-Film Dielectric Materials

This list gives some details of the more common thin-film dielectric materials. It is not a definitive list but is intended to show the wide range of available materials. The metals exhibit enormous dispersion, and so an abbreviated table of values is of little use. For extended tables of the optical constants of metals, consult Hass and Hadley [1] and Palik [2–4]. Surveys of many thin-film materials are given by Ritter [5,6] and by Palik [2–4]. For a fuller account of the fluorides of the rare earths, consult Lingg [7].

In most cases, the materials in the table can be deposited by many different processes. Where thermal evaporation is possible, it is the main process listed. Many of the materials with the principal exception of the fluorides can be sputtered in their dielectric form by either RF sputtering or neutral ion-beam sputtering. A few materials, the nitrides especially, are not capable of evaporation or reactive evaporation and require an energetic process such as ion-assisted deposition.

The optical properties of thin films are very dependent on deposition conditions and other factors. The values quoted in Table 18.1 should be simply interpreted as values that were reported at some time and not as necessarily intrinsic and repeatable properties of the materials.

TABLE 18.1

Material Optical Properties

Material	Deposition Technique	Refractive Index	Region of Transparency	Remarks	References
Aluminum oxide (Al$_2$O$_3$)	E-beam	1.62 at 0.6 μm; 1.59 at 1.6 μm T_s = 300°C 1.62 at 0.6 μm; 1.59 at 1.6 μm T_s = 40°C		Can also be produced by anodic oxidation of Al in ammonium tartrate solution [8]	Cox et al. [9]
Aluminum oxynitride (AlO$_x$N$_y$)	E-beam evaporation of Al with nitrogen ion assist and oxygen background	1.71–1.93 at 350 nm; 1.65–1.83 at 550 nm	<300 nm–6.5 μm	Index continuously varies as function of composition	Hwangbo et al. [10]; Targove et al. [11]
Antimony trioxide (Sb$_2$O$_3$)	Molybdenum boat	2.20 at 366 nm; 2.04 at 546 nm	300 nm–1 μm	Important to avoid overheating otherwise decomposes	Jenkins [12]
Antimony sulfide (Sb$_2$S$_3$)		3.0 at 589 nm	500 nm–10 μm	Brief note in Heavens et al. [13] (p. 189)	Heavens [14]; Billings and Hyman [15]
Beryllium oxide (BeO)	Tantalum boat; reactive evaporation of Be metal in activated oxygen	1.82 at 193 nm; 1.72 at 550 nm	190 nm–infrared (IR)	Highly toxic	Ebert [16]
Bismuth oxide (Bi$_2$O$_3$)	E-beam [17]; also reactive sputtering of bismuth in oxygen [18]	2.7 at 600 nm; 2.2 at 9 μm (E-beam) 2.45 at 550 nm (Sputter)	<550 nm–12 μm	Good IR material but less abrasion resistant than other oxides [17]	Kruschwitz and Pawlewicz [17]; Holland and Siddall [18]
Bismuth trifluoride (BiF$_3$)	Graphite Knudsen cell	1.74 at 1 μm; 1.65 at 10 μm	260 nm–20 μm		Moravec et al. [19]
Cadmium sulfide (CdS)	Quartz crucible with spiral filament in contact with charge	2.6 at 600 nm; 2.27 at 7 μm	600 nm–7 μm	Avoid overheating; filament temperature must be ≤1025°C	Heavens [14]; Hall and Ferguson [20]
Cadmium telluride (CdTe)	Molybdenum boat	3.05 in near IR			Ennos [21] (brief)
Calcium fluoride (CaF$_2$)	Molybdenum or tantalum boat; E-beam [17]	1.23–1.26 at 546 nm (porous) 1.40 at 600 nm; 1.32 at 9 μm (E-beam)	150 nm–12 μm		Heavens [14]; Kruschwitz and Pawlewicz [17]; Ennos [21]; Heavens and Smith [22]

(Continued)

TABLE 18.1 (CONTINUED)

Material Optical Properties

Material	Deposition Technique	Refractive Index	Region of Transparency	Remarks	References
Ceric oxide (CeO$_2$)	Tungsten boat	2.2 at 550 nm; 2.18 at 550 nm; T_s = 50°C; 2.42 at 550 nm; T_s = 350°C; 2.2 in near IR	400 nm–16 μm	Tends to form inhomogeneous layers; suffers from moisture adsorption	Ritter [23]; Smith and Baumeister [24]; Hass et al. [25]; Cox and Hass [26]
Cerous fluoride (CeF$_3$)	Tungsten boat; E-beam [17]	1.63 at 550 nm; 1.59 at 2 μm; 1.57 at 9 μm (E-beam)	300 nm–12 μm	Hot substrate; crazes on cold substrate [23]; high tensile stress	Kruschwitz and Pawlewicz [17]; Ennos [21]; Ritter [23]; Smith and Baumeister [24]; Hass et al. [27]
Chiolite (5NaF-3AlF$_3$)	Howitzer or tantalum boat			Similar to cryolite	Ennos [21]
Chromium oxide (Cr$_2$O$_3$)	E-beam	2.242 at 700 nm; 2.1 at 8 μm	<600 nm–8 μm		Kruschwitz and Pawlewicz [17]
Cryolite (Na$_3$AlF$_6$)	Howitzer or tantalum boat	1.35 at 550 nm	<200 nm–14 μm	Slightly hygroscopic; soft, easily damaged	Heavens [14]; Ennos [21]; Heavens and Smith [22]; Ritter [23]; Pelletier et al. [28]; Netterfield [29]
Gadolinium fluoride (GdF$_3$)	E-beam	1.55 at 400 nm	140 nm to >12 μm		Lingg [7]
Germanium (Ge)	E-beam or graphite boat	4.25 in IR (usually slightly higher than bulk value)	1.7–100 μm	Absorption band centered at approximately 25 μm	Ennos [21]; Ritter [23]
Hafnium dioxide (HfO$_2$)	E-beam	2.088 at 350 nm; 2.00 at 500 nm; 1.88 at 8 μm	220 nm–12 μm		Kruschwitz and Pawlewicz [17]; Smith and Baumeister [24]; Borgogno et al. [30]; Baumeister and Arnon [31]
Hafnium fluoride (HfF$_4$)	E-beam	1.57 at 600 nm; 1.46 at 10 μm	<600 nm–12 μm		Kruschwitz and Pawlewicz [17]

(Continued)

TABLE 18.1 (CONTINUED)

Material Optical Properties

Material	Deposition Technique	Refractive Index	Region of Transparency	Remarks	References
Lanthanum fluoride (LaF₃)	Tungsten boat; E-beam [17]	1.59 at 550 nm; 1.57 at 2 μm; 1.52 at 9 μm (E-beam)	200 nm–12 μm	Heated substrate	Kruschwitz and Pawlewicz [17]; Ritter [23]; Smith and Baumeister [24]; Hass et al. [27]; Bourg et al. [32]; Targove et al. [33]
Lanthanum oxide (La₂O₃)	Tungsten boat	1.95 at 550 nm; 1.86 at 2 μm	350 nm to >2 μm	Hot substrate (~300°C)	Ritter [23]; Smith and Baumeister [24]; Hass et al. [27]
Lead chloride (PbCl₂)	Platinum or molybdenum boat	2.3 at 550 nm; 2.0 at 10 μm	300 nm to >14 μm		Ennos [21]; Penselin and Steudel [34]
Lead fluoride (PbF₂)	Platinum boat; E-beam [17]	1.75 at 550 nm; 1.70 at 1 μm; 1.3 at 10 μm (E-beam)	240 nm to >20 μm		Kruschwitz and Pawlewicz [17]; Ennos [21]; Ritter [23]; Stiftung [35]; Les et al. [36]
Lead telluride (PbTe)	Tantalum boat	5.5 in IR	3.4 μm to >30 μm	Avoid overheating; hot substrate (see text)	Smith and Seeley [37]; Len et al. [38]; Ritchie [39]
Lithium fluoride (LiF)	Tantalum boat	1.36–1.37 at 546 nm	110 nm–7 μm		Heavens [14]; Schulz [40]
Lutetium fluoride (LuF₃)	E-beam	1.51 at 400 nm	140 nm–12 μm		Lingg [7]
Magnesium fluoride (MgF₂)	Tantalum boat	1.38 at 550 nm; 1.35 at 2 μm	210 nm–10 μm	Films on heated substrates much more rugged; high tensile stress	Heavens [14]; Ennos [21]; Heavens and Smith [22]; Smith and Baumeister [24]; Borgogno et al. [30]; Hall et al. [41]; Wood et al. [42]; Hall [43]
Magnesium oxide (MgO)	E-beam	1.7 at 550 nm; T_s = 50°C; 1.74 at 550 nm; T_s = 300°C	210 nm–8 μm		Pulker [44]
Neodymium fluoride (NdF₃)	Tungsten boat; E-beam [17]	1.60 at 550 nm; 1.58 at 2 μm; 1.60 at 9 μm (E-beam)	220 nm–12 μm	Hot substrate: 300°C	Kruschwitz and Pawlewicz [17]; Ritter [23]; Smith and Baumeister [24]; Hass et al. [27]

(Continued)

TABLE 18.1 (CONTINUED)

Material Optical Properties

Material	Deposition Technique	Refractive Index	Region of Transparency	Remarks	References
Neodymium oxide (Nd_2O_3)	Tungsten boat	2.0 at 550 nm; 1.95 at 2 μm	400 to >2 μm	Hot substrate: 300°C; decomposes at high boat temperature	Ritter [23]; Hass et al. [27]
Niobium oxide (Nb_2O_5)	Sputtering usually ion-beam	2.316 at 632.8 nm	380–	Used for high-quality precise coatings	Lee et al. [45]; Filmetrics [46]
Praseodymium oxide (Pr_6O_{11})	Tungsten boat	1.92 at 500 nm; 1.83 at 2 μm	400 to >2 μm	Hot substrate: 300°C	Hass et al. [27]
Samarium fluoride (SmF_3)	E-beam	1.56 at 400 nm	160 nm to >12 μm		Lingg [7]
Scandium oxide (Sc_2O_3)	E-beam	1.86 at 550 nm	350 nm–13 μm		Arndt et al. [47]
Silicon (Si)	E-beam with water-cooled hearth; sputtering	3.5 in IR	1.1–14 μm		Ritter [23]
Silicon monoxide (SiO)	Tantalum boat or howitzer	2.0 at 550 nm; 1.7 at 6 μm	500 nm–8 μm	Fast evaporation at low pressure	Ennos [21] (brief); Cox et al. [9]; Heavens [14]; Ritter [23]; Borgogno et al. [30]; Hass and Salzberg [48]
Disilicon trioxide (Si_2O_3)	Tantalum boat or howitzer	1.52–1.55 at 550 nm	300 nm–8 μm		Cox et al. [9]; Ritter [23]; UK Patent, 775,002 [49]; Ritter [50]; Okamoto and Hishinuma [51]; Bradford et al. [52]; Bradford and Hass [53]; Auwärter [54]
Silicon dioxide (Si_2)	E-beam; mixture in tungsten boat	1.46 at 500 nm; 1.445 at 1.6 μm	<200 nm–8 μm (in thin films)		Cox et al. [9]; Ritter [23]; Reichelt [55]; UK Patent, 632,442 [56]
Silicon nitride (Si_3N_4)	Low-voltage reactive ion plating	2.06 at 500 nm	320 nm–7 μm		Bovard et al. [57]
Sodium fluoride (NaF)	Tantalum boat	1.34 in visible	<250 nm–14 μm		Heavens [14] (brief)

(Continued)

TABLE 18.1 (CONTINUED)

Material Optical Properties

Material	Deposition Technique	Refractive Index	Region of Transparency	Remarks	References
Strontium fluoride (SrF$_2$)	E-beam	1.46 at 600 nm; 1.3 at 10 μm	<600 nm to >12 μm		Kruschwitz and Pawlewicz [17]
Tantalum pentoxide (Ta$_2$O$_5$)	E-beam and sputtering	2.16 at 550 nm; 1.95 at 8 μm	300 nm–10 μm		Kruschwitz and Pawlewicz [17]; Smith and Baumeister [24]
Tellurium (Te)	Tantalum boat	4.9 at 6 μm	3.4 μm–20 μm		Ennos [21]; Ritter [23]; Moss [58]; Greenler [59]
Titanium dioxide (TiO$_2$)	Reactive evaporation of TiO, Ti$_2$O$_3$, or Ti$_3$O$_5$ in O$_2$; E-beam reactive evaporation	2.2–2.7 at 550 nm depending on structure	350 nm–12 μm	Can also be produced by subsequent oxidation of Ti film	Heavens [14]; Ritter [23]; UK Patent, 775,002 [49]; Auwärter [54]; Reichelt [55]; Hass [60]; Brinsmaid et al. [61]; Anstalt [62]; Pulker et al. [63]; Heitmann [64]; Chiao et al. [65]
Thallous chloride (TlCl)	Tantalum boat	2.6 at 12 μm	Visible to >20 μm		Ennos [21]; UK Patent, 970,071 [66]
Thorium oxide (ThO$_2$)	E-beam	1.8 at 550 nm; 1.75 at 2 μm	250 nm–15 μm	Radioactive	Ennos [21]; Ritter [23]; Heitmann and Ritter [67]; Heitmann [68]; Behrndt and Doughty [69]
Thorium fluoride (ThF$_4$)	Tantalum boat	1.52 at 400 nm; 1.51 at 750 nm	200 nm to >15 μm	Radioactive (Note: Thorium oxyfluoride (ThOF$_2$) actually forms ThF$_4$ when evaporated.)	Ennos [21]; Ritter [23]; Heitmann and Ritter [67]; Heitmann [68]; Behrndt and Doughty [69]; Ledger and Bastien [70]
Ytterbium fluoride (YbF$_3$)	E-beam	1.52 at 600 nm; 1.48 at 10 μm	<600 nm–12 μm		Kruschwitz and Pawlewicz [17]
Yttrium oxide (Y$_2$O$_3$)	E-beam	1.82 at 550 nm; 1.69 at 9 μm	250 nm–12 μm		Kruschwitz and Pawlewicz [17]; Smith and Baumeister [24]; Borgogno et al. [30]; Lubezky et al. [71]

(Continued)

TABLE 18.1 (CONTINUED)

Material Optical Properties

Material	Deposition Technique	Refractive Index	Region of Transparency	Remarks	References
Zinc selenide (ZnSe)	Platinum or tantalum boat	2.58 at 633 nm	600 nm to >15 μm		Heitmann [68]
Zinc sulfide (ZnS)	Tantalum boat or howitzer	2.35 at 550 nm; 2.2 at 2.0 μm	380 nm–25 μm		Heavens [14]; Ennos [21]; Ritter [23]; Cox and Hass [26]; Netterfield [29]; Ritchie[39]; Hall et al. [41]; Behrndt and Doughty [69]
Zirconium dioxide (ZrO$_2$)	E-beam	2.1 at 550 nm; 2.05 at 9.0 μm	340 nm–12 μm		Kruschwitz and Pawlewicz [17]; Smith and Baumeister [24]; Hass and Salzberg [48]
Substance H1[a] (zirconia/titania)	Tungsten boat or E-beam	2.1 at 550 nm	360 nm–7 μm	Does not melt completely [72]	Fritz et al. [72]; Stetter et al. [73]
Substance H2[a] (mixed praseodymium and titanium oxides)	E-beam	2.1 at 550 nm	400–7 μm	Some weak absorption bands in visible [72]	Fritz et al. [72]
Substance H4[a] (lanthanum and titanium oxide)	E-beam with molybdenum liner	2.1 at 500 nm; T_s = 300°C	360 nm–7 μm		Fritz et al. [72]
Substance M1[a] (mixed praseodymium and aluminum oxides)	E-beam	1.71 at 500 nm; T_s = 300°C	300 nm–9 μm		Fritz et al. [72]

[a] Substance H1, substance H2, substance H4, and substance M1 are members of the Patinal® series of optical coating materials manufactured by E. Merck, Darmstadt, Germany.

References

1. G. Hass and L. Hadley. 1972. Optical constants of metals. In *American Institute of Physics Handbook*, D. E. Gray (ed.), pp. 6.124–6.156. New York: McGraw-Hill.
2. E. D. Palik, ed. 1985. *Handbook of Optical Constants of Solids I*. Orlando, FL: Academic Press.
3. E. D. Palik. 1991. *Handbook of Optical Constants of Solids II*. San Diego, CA: Academic Press.
4. E. D. Palik. 1998. *Handbook of Optical Constants of Solids III*. San Diego, CA: Academic Press.
5. E. Ritter. 1975. Dielectric film materials for optical applications. In *Physics of Thin Films*, G. Hass, M. H. Francombe, and R. W. Hoffman (eds), pp. 1–49. New York: Academic Press.
6. E. Ritter. 1976. Optical film materials and their applications. *Applied Optics* 15:2318–2327.
7. L. J. Lingg. 1990. *Lanthanide trifluoride thin films: Structure, composition and optical properties*. PhD Dissertation. Tucson, AZ: University of Arizona.
8. G. Hass. 1949. On the preparation of hard oxide films with precisely controlled thickness on evaporated aluminum mirrors. *Journal of the Optical Society of America* 39:532–539.
9. J. T. Cox, G. Hass, and J. B. Ramsay. 1964. Improved dielectric films for multilayer coatings and mirror protection. *Journal de Physique* 25:250–254.
10. C. K. Hwangbo, L. J. Lingg, J. P. Lehan et al. 1989. Reactive ion-assisted deposition of aluminum oxynitride thin films. *Applied Optics* 28:2779–2784.
11. J. D. Targove, L. J. Lingg, J. P. Lehan et al. 1987. Preparation of aluminum nitride and oxynitride thin films by ion-assisted deposition. In *Materials Modification and Growth Using Ion Beams Symposium*, pp. 311–316, Materials Research Society, Pittsburgh, PA.
12. F. A. Jenkins. 1958. Extension du domaine spectral de pouvoir réflecteur élevé des couches multiples diélectriques. *Journal de Physique et le Radium* 19:301–306.
13. O. S. Heavens, J. Ring, and S. D. Smith. 1957. Interference filters for the infra-red. *Spectrochimica Acta* 10:179–194.
14. O. S. Heavens. 1960. Optical properties of thin films. *Reports on Progress in Physics* 23:1–65.
15. B. H. Billings and M. Hyman Jr. 1947. The infra-red refractive index and dispersion of evaporated stibnite thin films. *Journal of the Optical Society of America* 37:119–121.
16. J. Ebert. 1982. Activated reactive evaporation. *Proceedings of SPIE* 325:29–38.
17. J. D. T. Kruschwitz and W. T. Pawlewicz. 1997. Optical and durability properties of infrared transmitting thin films. *Applied Optics* 36:2157–2159.
18. L. Holland and G. Siddall. 1958. Heat-reflecting windows using gold and bismuth oxide films. *British Journal of Applied Physics* 9:359–361.
19. T. J. Moravec, R. A. Skogman, and E. Bernal G. 1979. Optical properties of bismuth trifluoride thin films. *Applied Optics* 18:105–110.
20. J. F. Hall and W. F. C. Ferguson. 1955. Optical properties of cadmium sulphide and zinc sulphide from 0.6 micron to 14 micron. *Journal of the Optical Society of America* 45:714–718.
21. A. E. Ennos. 1966. Stresses developed in optical film coatings. *Applied Optics* 5:51–61.
22. O. S. Heavens and S. D. Smith. 1957. Dielectric thin films. *Journal of the Optical Society of America* 47:469–472.
23. E. Ritter. 1961. Gesichtspunkte bei der Stoffauswahl für dünne Schichten in der Optik. *Zeitschrift für Angewandte Mathematische Physik* 12:275–276.
24. D. Smith and P. W. Baumeister. 1979. Refractive index of some oxide and fluoride coating materials. *Applied Optics* 18:111–115.
25. G. Hass, J. B. Ramsay, and R. Thun. 1958. Optical properties and structure of cerium dioxide films. *Journal of the Optical Society of America* 48:324–327.
26. J. T. Cox and G. Hass. 1958. Antireflection coatings for germanium and silicon in the infrared. *Journal of the Optical Society of America* 48:677–680.
27. G. Hass, J. B. Ramsay, and R. Thun. 1959. Optical properties of various evaporated rare earth oxides and fluorides. *Journal of the Optical Society of America* 49:116–120.
28. E. Pelletier, P. Roche, and B. Vidal. 1976. Détermination automatique des constantes optiques et de l'épaisseur de couches minces: Application aux couches diélectriques. *Nouvelle Revue d'Optique* 7:353–362.
29. R. P. Netterfield. 1976. Refractive indices of zinc sulphide and cryolite in multilayer stacks. *Applied Optics* 15:1969–1973.

30. J. P. Borgogno, B. Lazarides, and E. Pelletier. 1982. Automatic determination of the optical constants of inhomogeneous thin films. *Applied Optics* 21:4020–4029.
31. P. W. Baumeister and O. Arnon. 1977. Use of hafnium dioxide in mutilayer dielectric reflectors for the near uv. *Applied Optics* 16:439–444.
32. A. Bourg, N. Barbaroux, and M. Bourg. 1965. Propriétés optiques et structure de couches minces de fluorure de lanthane. *Optica Acta*:151–160.
33. J. D. Targove, J. P. Lehan, L. J. Lingg et al. 1987. Ion-assisted deposition of lanthanum fluoride thin films. *Applied Optics* 26:3733–3737.
34. S. Penselin and A. Steudel. 1955. Fabry-Perot-Interferometerverspiegelungen aus dielektrischen Vielfachschichten. *Zeitschrift für Physik* 142:21–41.
35. Carl Zeiss Stiftung. 1965. *Interference filters*. UK Patent, 994,638.
36. Z. Lès, F. Lès, and L. Gabla. 1963. Semitransarent metallic-dielectric mirrors with low absorption coefficient in the ultra-violet region of the spectrum (3200–2400A). *Acta Physica Polonica* 23:211–214.
37. S. D. Smith and J. S. Seeley. 1968. *Multilayer Filters for the Region 0.8 to 100 Microns*. Bedford, MA: Air Force Cambridge Research Laboratories.
38. Y.-H. Yen, L.-X. Zhu, W.-D. Zhang et al. 1984. Study of PbTe optical coatings. *Applied Optics* 23:3597–3601.
39. F. S. Ritchie. 1970. *Multilayer filters for the infrared region 10–100 microns*. PhD Thesis. Reading: University of Reading.
40. L. G. Schulz. 1949. The structure and growth of evaporation LiF and NaCl films on amorphous substrates. *Journal of Chemical Physics* 17:1153–1162.
41. J. F. Hall, Jr. and W. F. C. Ferguson. 1955. Dispersion of zinc sulfide and magnesium fluoride films in the visible spectrum. *Journal of the Optical Society of America* 45:74–75.
42. O. R. Wood, II, H. G. Craighead, J. E. Sweeney et al. 1984. Vacuum ultraviolet loss in magnesium fluoride films. *Applied Optics* 23:3644–3649.
43. J. F. Hall. 1957. Optical properties of magnesium fluoride films in the ultraviolet. *Journal of the Optical Society of America* 47:662–665.
44. H. K. Fulker. 1979. Characterization of optical thin films. *Applied Optics* 18:1969–1977.
45. C.-C. Lee, C.-L. Tien, and J.-C. Hsu. 2002. Internal stress and optical properties of Nb_2O_5 thin films deposited by ion-beam sputtering. *Applied Optics* 41:2043–2047.
46. Filmetrics. 2017. *Refractive index of Nb_2O_5, Niobium pentoxide*. Available from http://www.filmetrics.com /refractive-index-database/Nb2O5/Niobium-pentoxide.
47. D. P. Arndt, R. M. A. Azzam, J. M. Bennett et al. 1984. Multiple determination of the optical constants of thin-film coating materials. *Applied Optics* 23:3571–3596.
48. G. Hass and C. D. Salzberg. 1954. Optical properties of silicon monoxide in the wavelength region from 0.24 to 14.0 microns. *Journal of the Optical Society of America* 44:181–187.
49. A. Vogt. 1957. *Improvements in or relating to the manufacture of thin light transmitting layers*. UK Patent, 775,002.
50. E. Ritter. 1962. Zür Kentnis des SiO und Si_2O_3-phase in dünnen Schichten. *Optica Acta* 9:197–202.
51. E. Okamoto and Y. Hishinuma. 1965. Properties of evaporated thin films of Si_2O_3. *Transactions of the Third International Vacuum Congress* 2:49–56.
52. A. P. Bradford, G. Hass, M. McFarland et al. 1965. Effect of ultraviolet irradiation on the optical properties of silicon oxide films. *Applied Optics* 4:971–976.
53. A. P. Bradford and G. Hass. 1963. Increasing the far-ultra-violet reflectance of silicon oxide protected aluminium mirrors by ultraviolet irradiation. *Journal of the Optical Society of America* 53:1096–1100.
54. M. Auwärter. 1960. *Process for the manufacture of thin films*. US Patent, 2,920,002.
55. W. Reichelt. 1965. Fortschritte in der Herstellung von Oxydschichten für optische und elektrische Zwecke. *Transactions of the Third International Vacuum Congress* 2:25–29.
56. Libbey-Owens-Ford. 1947. *Method of coating with quartz by thermal evaporation*. UK Patent, 632,442.
57. B. B. Bovard, J. Ramm, R. Hora et al. 1989. Silicon nitride thin films by low voltage reactive ion plating: Optical properties and composition. *Applied Optics* 28:4436–4441.
58. T. S. Moss. 1952. Optical properties of tellurium in the infra-red. *Proceedings of the Physical Society B* 65:62–66.
59. R. G. Greenler. 1955. Interferometry in the infrared. *Journal of the Optical Society of America* 45:788–791.
60. G. Hass. 1952. Preparation, properties and optical applications of thin films of titanium dioxide. *Vacuum* 2:331–345.

61. D. S. Brinsmaid, W. J. Keenan, G. J. Koch et al. 1957. *Method of producing titanium dioxide coatings*. US Patent, 2,784,115.
62. Geraetebau Anstalt. 1962. *Improvements in and relating to the oxidation and/or transparency of thin partly oxidic layers*. UK Patent, 895,879.
63. H. K. Pulker, G. Paesold, and E. Ritter. 1976. Refractive indices of TiO$_2$ films produced by reactive evaporation of various titanium-oxide phases. *Applied Optics* 15:2986–2991.
64. W. Heitmann. 1971. Reactive evaporation in ionized gases. *Applied Optics* 10:2414–2418.
65. S.-C. Chiao, B. G. Bovard, and H. A. Macleod. 1998. Repeatability of the composition of titanium oxide films produced by evaporation of Ti$_2$O$_3$. *Applied Optics* 37:5284–5290.
66. Perkin Elmer. 1961. *Infrared filters*. UK Patent, 970,071.
67. W. Heitmann and E. Ritter. 1968. Production and properties of vacuum evaporated films of thorium fluoride. *Applied Optics* 7:307–309.
68. W. Heitmann. 1966. Extrem hochreflektierende dielektrische Spiegelschichten mit Zincselenid. *Zeitschrift für Angewandte Physik* 21:503–508.
69. K. H. Behrndt and D. W. Doughty. 1966. Fabrication of multilayer dielectric films. *Journal of Vacuum Science and Technology* 3:264–272.
70. A. M. Ledger and R. C. Bastien. 1977. *Intrinsic and Thermal Stress Modeling for Thin-Film Multilayers*. Norwalk, CT: Perkin Elmer.
71. I. Lubezky, E. Ceren, and Z. Klein. 1980. Silver mirrors protected with Yttria for the 0.5 to 14µm region. *Applied Optics* 19:1895.
72. M. Fritz, F. Koenig, E. Merck et al. 1992. New materials for production of optical coatings. In *35th Annual Technical Conference Proceedings*, pp. 143–147, Society of Vacuum Coaters, Albuquerque, NM.
73. F. Stetter, R. Esselborn, N. Harder et al. 1976. New materials for optical thin films. *Applied Optics* 15:2315–2317.

Index

Printed and bound by CPI Group (UK) Ltd, Croydon, CR0 4YY

01/11/2024

01782600-0015